Information Security and Cryptography
Texts and Monographs

Phong Q. Nguyen • Brigitte Vallée
Editors

The LLL Algorithm

Survey and Applications

Editors

Dr. Phong Q. Nguyen
INRIA Research Director
École Normale Supérieure
Département d'Informatique
Paris, France
phong.nguyen@ens.fr

Dr. Brigitte Vallée
CNRS Research Director
and Research Director
Département d'Informatique
Université de Caen, France
brigitte.vallee@info.unicaen.fr

ISSN 1619-7100
ISBN 978-3-642-26164-0 e-ISBN 978-3-642-02295-1
DOI 10.1007/978-3-642-02295-1
Springer Heidelberg Dordrecht London New York

ACM Computing Classification (1998): F.2, F.1, E.3, G.1

Cover design: KuenkelLopka GmbH

Printed on acid-free paper

Springer is a part of Springer Science+Business Media (www.springer.com)

Preface

Computational aspects of geometry of numbers have been revolutionized by the Lenstra–Lenstra–Lovász lattice reduction algorithm (LLL), which has led to breakthroughs in fields as diverse as computer algebra, cryptology, and algorithmic number theory. After its publication in 1982, LLL was immediately recognized as one of the most important algorithmic achievements of the twentieth century, because of its broad applicability and apparent simplicity. Its popularity has kept growing since, as testified by the hundreds of citations of the original article, and the ever more frequent use of LLL as a synonym to lattice reduction.

As an unfortunate consequence of the pervasiveness of the LLL algorithm, researchers studying and applying it belong to diverse scientific communities, and seldom meet. While discussing that particular issue with Damien Stehlé at the 7th Algorithmic Number Theory Symposium (ANTS VII) held in Berlin in July 2006, John Cremona accurately remarked that 2007 would be the 25th anniversary of LLL and this deserved a meeting to celebrate that event. The year 2007 was also involved in another arithmetical story. In 2003 and 2005, Ali Akhavi, Fabien Laguillaumie, and Brigitte Vallée with other colleagues organized two workshops on cryptology and algorithms with a strong emphasis on lattice reduction: CAEN '03 and CAEN '05, CAEN denoting both the location and the content (*Cryptologie et Algorithmique En Normandie*). Very quickly after the ANTS conference, Ali Akhavi, Fabien Laguillaumie, and Brigitte Vallée were thus readily contacted and reacted very enthusiastically about organizing the LLL birthday conference. The organization committee was formed.

Within a couple of months, the three L's, Arjen and Hendrik Lenstra, and László Lovász, kindly accepted to participate, which provided confidence to the organizing team. At the same time, a program committee was created. Its members – Karen Aardal, Shafi Goldwasser, Phong Nguyen, Claus Schnorr, Denis Simon, and Brigitte Vallée – come from diverse fields, so as to represent as many LLL-practitioners as possible. They invited speakers to give overview talks at the conference.

The anniversary conference eventually took place between 29th June and 1st July 2007, at the University of Caen. During these three days, 14 invited talks were given on topics closely related to the LLL algorithm. A poster session gathered 12 presentations on ongoing research projects. Overall, 120 researchers from 16 countries and very diverse scientific backgrounds attended the event. And naturally, a birthday party was set and the three L's blew out the candles of their algorithm's birthday cake!

Unlike many other domains, the community misses a reference book dealing with almost all aspects of lattice reduction. One important goal of the conference was to provide such material, which may be used by both junior and senior researchers, and hopefully even useful for undergraduate students. The contributors were selected to make such a collective book possible. This book is a brief (and inevitably incomplete) snapshot of the research, which was sparked by the publication of the LLL algorithm in 1982. The survey articles were written to be accessible by a large audience, with detailed motivations, explanations, and examples. We hope they will help pursuing further research on this very rich topic. Each article of the present book can be read independently and provides an introductory overview of the results obtained in each particular area in the past 25 years.

The first contribution of this book, by Ionica Smeets and in collaboration with Arjen Lenstra, Hendrik Lenstra, László Lovász, and Peter van Emde Boas, describes the genesis of the LLL algorithm. The rest of the book may be informally divided into five chapters, each one essentially matching a session of the anniversary conference.

The first chapter deals with algorithmic aspects of lattice reduction, independently of applications. The first article of that chapter, by Phong Nguyen, introduces lattices, and surveys the main provable algorithms for finding the shortest vector in a lattice, either exactly or approximately. It emphasizes a somewhat overlooked connection between lattice algorithms and Hermite's constant, that is, between computational and mathematical aspects of the geometry of numbers. For instance, LLL is presented as an (efficient) algorithmic version of Hermite's inequality on Hermite's constant. The second article, by Brigitte Vallée and Antonio Vera, surveys the probabilistic analysis of several lattice reduction algorithms, in particular LLL and Gauss' algorithm. Different random models for the input bases are considered and the result introduces sophisticated analytic tools as complex and functional analysis. The third article, by Claus Schnorr, surveys provable and heuristic algorithmic variations around LLL, to make the algorithm more efficient or with better outputs. For example, the fruitful notion of blockwise reduction is a natural generalization of LLL. The fourth article, by Damien Stehlé, surveys all aspects of floating-point lattice reduction. The different analyses exhibit the parameters that play an important role when relating the execution time of the floating-point versions of LLL to the quality of the output. Both provable and heuristic versions of the algorithm are considered.

The second chapter is concerned with the applications of lattice reduction in the vast field of algorithmic number theory. Guillaume Hanrot's article describes several efficient algorithms to solve diverse Diophantine approximation problems. For example, these algorithms relying on lattice reduction tackle the problems of approximating real numbers by rational and algebraic numbers, of disclosing linear relations and of solving several Diophantine equations. Denis Simon's paper contains a collection of examples of problems in number theory that are solved efficiently via lattice reduction. Among others, it introduces a generalization of the

LLL algorithm to reduce indefinite quadratic forms. Finally, the article by Jürgen Klüners surveys the original application of the LLL, namely factoring polynomials with rational coefficients. It compares the original LLL factoring method and the recent one developed by Mark von Hoeij, which relies on the knapsack problem.

The third chapter contains a single article, by Karen Aardal and Friedrich Eisenbrand. It surveys the application of the LLL algorithm to integer programming, recalling Hendrik Lenstra's method – an ancestor of the LLL algorithm, and describing recent advances.

The fourth chapter is devoted to an important area where lattices have been applied with much success, both in theory and practice: cryptology. Historically, LLL and lattices were first used in cryptology for "destructive" purposes: one of the very first applications of LLL was a practical attack on the Merkle–Hellman knapsack public-key cryptosystem. The success of reduction algorithms at breaking various cryptographic schemes since the discovery of LLL have arguably established lattice reduction techniques as the most popular tool in public-key cryptanalysis. Alexander May's article surveys one of the major applications of lattices to cryptanalysis: lattice attacks on the RSA cryptosystem, which started in the late eighties with Håstad's work, and has attracted much interest since the mid-nineties with Coppersmith's method to find small roots of polynomials. The other two articles of the chapter deal instead with "positive" applications of lattices to cryptography. The NTRU paper by Jeff Hoffstein, Nick Howgrave-Graham, Jill Pipher, and William Whyte gives an excellent example of an efficient cryptosystem whose security relies on the concrete hardness of lattice problems. The paper by Craig Gentry surveys security proofs of non-lattice cryptographic schemes in which lattices make a surprising appearance. It is perhaps worth noting that lattices are used both to attack RSA in certain settings, and to prove the security of industrial uses of RSA.

The final chapter of the book focuses on the complexity of lattice problems. This area has attracted much interest since 1996, when Miklós Ajtai discovered a fascinating connection between the worst-case and average-case complexity of certain lattice problems. The contribution of Daniele Micciancio deals with (lattice-based) cryptography from worst-case complexity assumptions. It presents recent cryptographic primitives whose security can be proven under worst-case assumptions: any instance of some well-known hard problem can be solved efficiently with access to an oracle breaking random instances of the cryptosystem. Daniele Micciancio's article contains an insightful discussion on the concrete security of lattice-based cryptography. The last two articles of the book, by respectively Subhash Khot and Oded Regev, are complementary. The article by Subhash Khot surveys inapproximability results for lattice problems. And the article by Oded Regev surveys the so-called limits to inapproximability results for lattice problems, such as the proofs that some approximation lattice problems belong to the complexity class coNP. It also shows how one can deduce zero-knowledge proof systems from the previous proofs.

Acknowledgements We, the editors, express our deep gratitude to the organizing committee comprised of Ali Akhavi, Fabien Laguillaumie, and Damien Stehlé. We also acknowledge with gratitude the various forms of support received from our sponsors; namely, CNRS, INRIA, Université de Caen, Mairie de Caen, Pôle TES, as well as several laboratories and research groups (LIP, GREYC, LIAFA, Laboratoire Elie Cartan, LIENS, GDR IM, ECRYPT, Orange Labs). Together with all participants, we were naturally extremely happy to benefit from the presence of the three **L**'s and our thanks are extended to Peter van Emde Boas for providing invaluable historical material. We also wish to thank all the speakers and participants of the conference LLL+25. Finally, we are indebted to Loick Lhote for his extensive help in the material preparation of this book.

Paris, August 2009

Phong Nguyen and Brigitte Vallée
Caen

Foreword

I have been asked by my two co-L's to write a few words by way of introduction, and consented on the condition of being allowed to offer a personal perspective.

On 1 September 2006, the three of us received an e-mail from Brigitte Vallée. John Cremona, she wrote, had suggested the idea of celebrating the 25th anniversary of the publication of "the LLL paper," and together with Ali Akhavi, Fabien Laguillaumie, and Damien Stehlé, she had decided to follow up on his suggestion. As it was "not possible to celebrate this anniversary without (...) the three L's of LLL," she was consulting us about suitable dates. I was one of the two L's who were sufficiently flattered to respond immediately, and the dates chosen turned out to be convenient for number three as well.

In her very first e-mail, Brigitte had announced the intention of including a historical session in the meeting, so that we would have something to do other than cutting cakes and posing for photographers. Hints that some of my own current work relates to lattices were first politely disregarded, and next, when I showed some insistence, I was referred to the Program Committee, consisting of Karen Aardal, Shafi Goldwasser, Phong Nguyen, Claus Schnorr, Denis Simon, and Brigitte herself. This made me realize which role I was expected to play, and I resolved to wait another 25 years with the new material.

As the meeting came nearer, it transpired that historical expertise was not represented on the Program Committee, and with a quick maneuver I seized unrestricted responsibility for organizing the historical session. I did have the wisdom of first securing the full cooperation of LLL's court archivist Peter van Emde Boas. How successful the historical session was, reported on by Ionica Smeets in the present volume, is not for me to say. I did myself learn a few things I was not aware of, and do not feel ashamed of the way I played my role.

All three L's extended their stay beyond the historical session. Because of the exemplary way in which the Program Committee had acquitted themselves in this job, we can now continue to regard ourselves as universal experts on all aspects of lattice basis reduction and its applications.

John Cremona, apparently mortified at the way his practical joke had run out of hand, did not show up, and he was wrong. John, it is my pleasure to thank you most cordially on behalf of all three L's. Likewise, our thanks are extended not only to everybody mentioned above, but also to all others who contributed to the success of the meeting, as speakers, as participants, as sponsors, or invisibly behind the scenes.

Leiden,
August 2008 *Hendrik Lenstra*

Contents

List of Contributors

Karen Aardal Delft Institute of Applied Mathematics, TU Delft, Mekelweg 4, 2628 CD Delft, The Netherlands and CWI, Science Park 123, 1098 XG Amsterdam, The Netherlands, k.i.aardal@tudelft.nl

Friedrich Eisenbrand EPFL, MA C1 573 (Bâtiment MA), Station 8, CH-1015 Lausanne, Switzerland, friedrich.eisenbrand@epfl.ch

Peter van Emde Boas ILLC, Depts. of Mathematics and Computer Science, Faculty of Sciences, University of Amsterdam, The Netherlands, peter@bronstee.com

Craig Gentry Stanford University, USA, cgentry@cs.stanford.edu

Guillaume Hanrot INRIA/LORIA, Projet CACAO - Bâtiment A, 615 rue du jardin botanique, F-54602 Villers-lès-Nancy Cedex, France, hanrot@loria.fr

Jeff Hoffstein NTRU Cryptosystems, 35 Nagog Park, Acton, MA 01720, USA, jhoffstein@ntru.com

Nick Howgrave-Graham NTRU Cryptosystems, 35 Nagog Park, Acton, MA 01720, USA, nhowgravegraham@ntru.com

Subhash Khot New York University, New York, NY-10012, USA, khot@cs.nyu.edu

Jürgen Klüners Mathematisches Institut, Universität Paderborn, Warburger Str. 100, 30098 Paderborn, Germany. klueners@math.uni-paderborn.de

Arjen K. Lenstra EPFL IC LACAL, Station 14, Lausanne, Switzerland, arjen.lenstra@epfl.ch

Hendrik W. Lenstra Mathematisch Instituut, Universiteit Leiden, Postbus 9512, 2300 RA Leiden, The Netherlands, hwl@math.leidenuniv.nl

László Lovász Eötvös Loránd Tudományegyetem, Számitógéptudományi Tanszék, Pázmány Péter sétány 1/C, H-1117 Budapest, Hungary, lovasz@cs.elte.hu

Alexander May Horst Görtz Institute for IT-Security, Faculty of Mathematics, Ruhr-University Bochum, Germany, alex.may@ruhr-uni-bochum.de

Daniele Micciancio Department of Computer Science and Engineering, University of California at San Diego, La Jolla CA 92093, USA, daniele@cs.ucsd.edu

Phong Nguyen Department of Computer Science, Ecole Normale Supérieure de Paris, 45 rue d'Ulm, 75230 Paris Cedex 05, France, Phong.Nguyen@ens.fr

Jill Pipher NTRU Cryptosystems, 35 Nagog Park, Acton, MA 01720, USA, jpipher@ntru.com

Oded Regev School of Computer Science, Tel-Aviv University, Tel-Aviv 69978, Israel, odedr@post.tau.ac.il

Claus Peter Schnorr Fachbereich Informatik und Mathematik, Universität Frankfurt, PSF 111932, D-60054 Frankfurt am Main, Germany, schnorr@cs.uni-frankfurt.de

Ionica Smeets Mathematisch Institut, Universiteit Leiden, Niels Bohrweg 1, 2333 CA Leiden, Netherlands, smeets@math.leidenuniv.nl

Denis Simon Université de Caen, LMNO, Bd Maréchal Juin BP 5186 – 14032 Caen Cedex, France, simon@math.unicaen.fr

Damien Stehlé CNRS/Universities of Macquarie, Sydney and Lyon/INRIA/ÉNS Lyon, Dept of Mathematics and Statistics, University of Sydney, NSW 2008, Australia, damien.stehle@gmail.com

Brigitte Vallée Laboratoire GREYC, CNRS UMR 6072, Université de Caen and ENSICAEN, F-14032 Caen, France, brigitte.vallee@info.unicaen.fr

Antonio Vera Laboratoire GREYC, CNRS UMR 6072, Université de Caen and ENSICAEN, F-14032 Caen, France, antonio.vera@info.unicaen.fr

William Whyte NTRU Cryptosystems, 35 Nagog Park, Acton, MA 01720, USA, wwhyte@ntru.com

Chapter 1
The History of the LLL-Algorithm

Ionica Smeets
In collaboration with Arjen Lenstra, Hendrik Lenstra, László Lovász, and Peter van Emde Boas

Abstract The 25th birthday of the LLL-algorithm was celebrated in Caen from 29th June to 1st July 2007. The three day conference kicked off with a historical session of four talks about the origins of the algorithm. The speakers were the three L's and close bystander Peter van Emde Boas. These were the titles of their talks.

- *A tale of two papers* – Peter van Emde Boas.
- *The early history of LLL* – Hendrik Lenstra.
- *The ellipsoid method and basis reduction* – László Lovász.
- *Polynomial factorization and lattices in the very early 1980s* – Arjen Lenstra.

This chapter is based on those talks, conversations with these four *historic* characters, the notes that Peter van Emde Boas and Arjen Lenstra wrote for the preproceedings, and many artifacts from the phenomenal archive of Van Emde Boas.

Fig. 1.1 On both pictures you see from left to right Peter van Emde Boas, László Lovász, Hendrik Lenstra, and Arjen Lenstra. Alexander Schrijver took the first picture in Bonn on 27th February 1982. For the poster of the conference, Van Emde Boas was digitally removed from this picture. The second picture was taken by Ghica van Emde Boas at Le moulin de Bully on 29th June 2007

I. Smeets (✉)
Mathematisch Instituut, Universiteit Leiden, Niels Bohrweg 1, 2333 CA Leiden, the Netherlands, e-mail: ionica.smeets@gmail.com

P.Q. Nguyen and B. Vallée (eds.), *The LLL Algorithm*, Information Security and Cryptography, DOI 10.1007/978-3-642-02295-1_1,

Skinny Triangles and Lattice Reduction

One possible starting point for the LLL-algorithm is May 1980. At that time, Peter van Emde Boas was visiting Rome. While he was there he discussed the following problem with Alberto Marchetti-Spaccamela.

Question 1 *Given three points with rational coordinates in the plane, is it possible to decide in polynomial time whether there exists a point with integral coefficients lying within the triangle defined by these points?*

This question seemed easy to answer: for big triangles the answer will be "yes" and for small triangles there should be only a small number of integer points close to it that need checking. But for extremely long and incredibly thin triangles this does not work; see Fig. 1.2.

It is easy to transform such a skinny triangle into a "rounder" one, but this transformation changes the lattice too; see Fig. 1.3. Van Emde Boas and Marchetti-Spaccamela did not know how to handle these skewed lattices. Back in Amsterdam, Van Emde Boas went to Hendrik Lenstra with their question. Lenstra immediately replied that this problem could be solved with lattice reduction as developed by Gauss almost two hundred years ago. The method is briefly explained below.

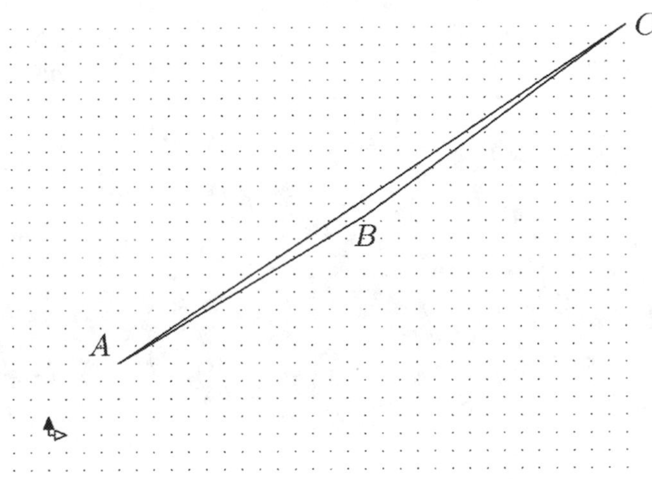

Fig. 1.2 The problematic triangles almost look like a line: they are incredibly thin and very, very long. This picture should give you an idea; in truly interesting cases the triangle is much thinner and longer. In the lower left corner you see the standard basis for the integer lattice

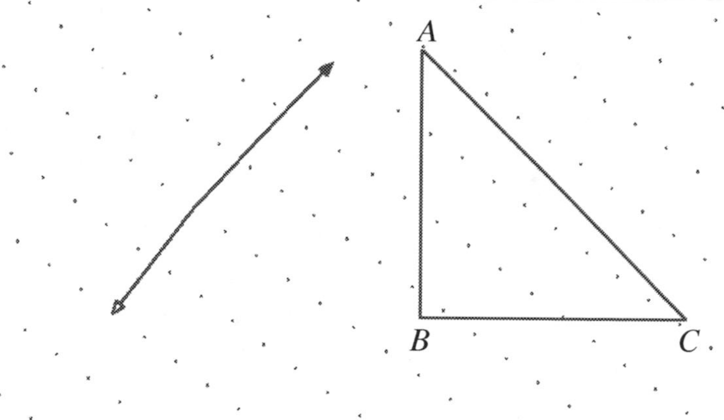

Fig. 1.3 The triangle from Fig. 1.2 transformed into a right-angled isosceles triangle, the skewed lattice and the transformed standard basis. Now the transformed basis looks thin and long

Method for Answering Question 1.
First apply a linear transformation that changes the triangle into a right-angled isosceles triangle. This transforms the integer lattice into a lattice with some given basis of two rational vectors.
Find a reduced basis (b_1, b_2) for this new lattice: b_1 is a shortest nonzero vector in the lattice and b_2 is a shortest vector in the lattice that is linearly independent of b_1. Compute $b_2^* = b_2 - \frac{\langle b_1, b_2 \rangle}{\langle b_1, b_1 \rangle} b_1$.
If the triangle is sufficiently large compared to $||b_2^*||$, then there is a lattice point in the triangle.
Otherwise, check if lines parallel to b_1 (with successive distances $||b_2^*||$) contain points in the triangle. Remember that in this case the size of the triangle is small compared to $||b_2^*||$, so the number of lines to be checked is small.

Van Emde Boas wrote to Marchetti in the summer of 1980: "Solution: the answer is yes." In his letter he explained how the method worked. When Marchetti-Spaccamela was visiting Amsterdam in October of the same year, he paid Hendrik Lenstra a visit to talk about the solution. Together with Van Emde Boas, he went to Lenstra's office. Hendrik Lenstra vividly remembers his initial feelings about this visit: "I felt a bit like a dentist. I had dealt with this problem before, so why were they asking the same question again? I told them the solution and they apparently understood it, but then they refused to walk out of my office. I had work to do and I felt that they were imposing upon my time. I was too naive to realize that this was my real work."

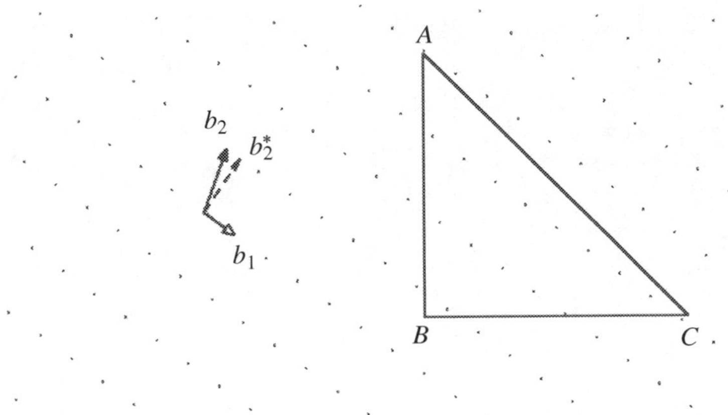

Fig. 1.4 The skewed lattice, its reduced basis (b_1, b_2) and the orthogonal projection b_2^*.

Lenstra opened his mouth about to say "Go away," but he phrased this in a slightly more polite manner as: "Why is this question about the triangle interesting in the first place?" His visitors answered that it was just a special case of integer programming with a fixed number of variables. "And then I stared at it and asked, can you not do that in the same way?" Van Emde Boas recalls: "At this point I had to leave the meeting to teach a class. When I came back three quarters of an hour later, Hendrik had given the answer that it really works for any dimension." This resulted in Lenstra's integer linear programming algorithm.

Linear programming, sometimes known as linear optimization, is the problem of maximizing or minimizing a linear function over a convex polyhedron specified by linear nonnegativity constraints. Integer linear programming is a special case of linear programming in which all variables are required to take on integer values only.

Integer Programming

In the early eighties, Hendrik Lenstra was not doing integer programming at all. He was among other things working on primality testing and Euclidean number fields. "I would probably not have found the integer linear programming algorithm if I had not been asked this question about a triangle in a integer lattice." The generalized question can be stated as follows.

Question 2 *Let n and m be positive integers, A an $m \times n$-matrix with integral entries, and $b \in \mathbb{Z}^m$. Is there a vector $x \in \mathbb{Z}^n$ satisfying the system of m inequalities $Ax \leq b$? So if $K = \{x \in \mathbb{R}^n : Ax \leq b\}$, then the question is whether $\mathbb{Z}^n \cap K$ is nonempty.*

The integer linear programming algorithm essentially consists of three stages.

Integer Linear Programming.
We may assume the problem is reduced to the case $0 < \text{vol } K < \infty$, thus K is bounded and has positive volume.

1. Find a linear transformation τ such that τK is *round*. If we put
$$B(p,z) = \{x \in \mathbb{R}^n : |x - p| \leq z\} \quad \text{for } p \in \mathbb{R}^n, \quad z \in \mathbb{R}_{>0},$$
then the formal definition of *round* is that there are spheres $B(p,r)$ and $B(p,R)$ with $B(p,r) \subset \tau K \subset B(p,R)$ and $\frac{R}{r} \leq c_1$, where c_1 is a constant depending only on n.
2. Find a reduced basis for $\tau\mathbb{Z}^n$.
3. *Either* find a point in $\tau\mathbb{Z}^n \cap \tau K$ *or* reduce the problem to a bounded number of problems in $n-1$ dimensions.

There are three versions of this algorithm: the first preprint appeared in April 1981 [3], to be followed by an improved version in November of the same year [4]. The final version was published in 1983 in *Mathematics of Operations Research* [5], the year after the LLL-algorithm appeared [8]. Lenstra:

> The reason that there are so many versions is that Lovász kept improving parts of the algorithm. He started with the first step. I had a very naive and straightforward way of finding

Fig. 1.5 Hendrik Lenstra using his hands to explain the algorithm to Alberto Marchetti-Spaccamela, Amsterdam on 21st October 1980

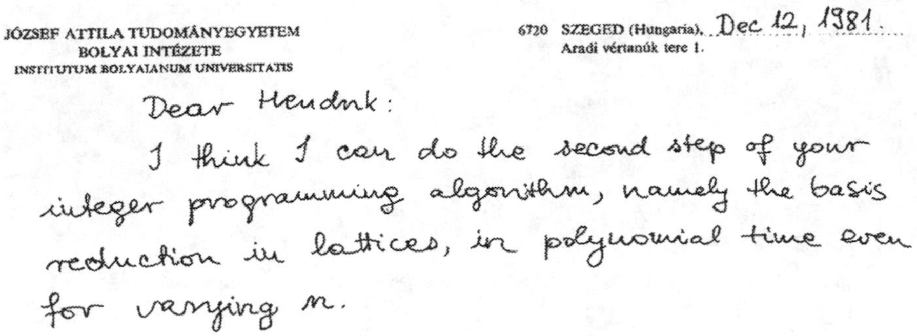

JÓZSEF ATTILA TUDOMÁNYEGYETEM
BOLYAI INTÉZETE
INSTITUTUM BOLYAIANUM UNIVERSITATIS

6720 SZEGED (Hungaria), Dec 12, 1981.
Aradi vértanúk tere 1.

Dear Hendrik:

I think I can do the second step of your integer programming algorithm, namely the basis reduction in lattices, in polynomial time even for varying n.

Fig. 1.6 The beginning of the letter from Lovász in which he explains the basis reduction algorithm

> the needed transformation, and this method was polynomial only for fixed n. Lovász found an algorithm to do this in polynomial time even for varying n.

Lovász later improved the second step, the basis reduction algorithm. In the introduction to the first preprint of his paper, Lenstra expressed some dissatisfaction with his complexity analysis of this step.

> It is not easy to bound the running time of this algorithm in a satisfactory way. We give an argument which shows that it is polynomially bounded, for fixed n. But the degree of this polynomial is an exponential function of n, and we feel that there is still room for improvement.

At the time Lenstra believed this problem was caused by his analysis, not by the algorithm. But Lovász improved the algorithm instead of the analysis.

In a letter dated 12th December 1981, Lovász explains the basis reduction algorithm. He defines two concepts that are at the core of the LLL-algorithm. Let $(b_1, \ldots, b_n)$ be an ordered basis for $\mathbb{R}^n$. We say that it is *straightened* if for every $1 \leq i < k \leq n$ and

$$b_k = \sum_{j=1}^{i} \lambda^i_{jk} b_j + b_k^{(i)}, \text{ where } b_j^T b_k^{(i)} = 0 \text{ for } j = 1, \ldots, i,$$

one has

$$\left|\lambda^i_{ik}\right| \leq \frac{1}{2} \quad \text{(only the last coefficient!)}.$$

We say that $(b_1, \ldots, b_n)$ is *weakly greedy* if

$$(b_1 \wedge b_2 \wedge \cdots \wedge b_i \wedge b_{i+2})^2 \geq \frac{3}{4}(b_1 \wedge b_2 \wedge \cdots \wedge b_i \wedge b_{i+1})^2 \qquad (1.1)$$

holds for every $0 \leq i \leq n-2$, where

$$(b_1 \wedge \cdots \wedge b_k)^2 = \det\left((b_i^T b_j)_{i,j=1}^k\right).$$

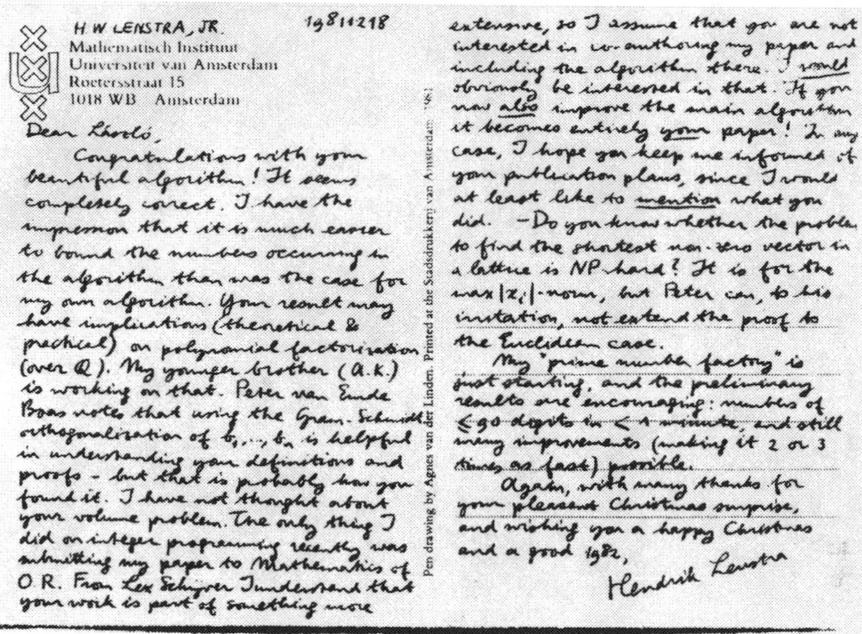

H. W. LENSTRA, JR. 19811218
Mathematisch Instituut
Universiteit van Amsterdam
Roetersstraat 15
1018 WB Amsterdam

Dear László,

Congratulations with your beautiful algorithm! It seems completely correct. I have the impression that it is much easier to bound the numbers occurring in the algorithm than was the case for my own algorithm. Your result may have implications (theoretical & practical) on polynomial factorization (over $\mathbb{Q}$). My younger brother (A.K.) is working on that. Peter van Emde Boas notes that using the Gram-Schmidt orthogonalization of $b_1, \ldots, b_n$ is helpful in understanding your definitions and proofs – but that is probably how you found it. I have not thought about your volume problem. The only thing I did on integer programming recently was submitting my paper to Mathematics of O.R. From Lex Schrijver I understand that your work is part of something more extensive, so I assume that you are not interested in co-authoring my paper and including the algorithm there. I would obviously be interested in that. If you now also improve the main algorithm it becomes entirely your paper! In any case, I hope you keep me informed of your publication plans, since I would at least like to mention what you did. – Do you know whether the problem to find the shortest non-zero vector in a lattice is NP-hard? It is for the $\max|x_i|$-norm, but Peter can, to his irritation, not extend the proof to the Euclidean case.

My "prime number factory" is just starting, and the preliminary results are encouraging: numbers of ≤ 90 digits in < 1 minute, and still many improvements (making it 2 or 3 times as fast) possible.

Again, with many thanks for your pleasant Christmas surprise, and wishing you a happy Christmas and a good 1982,

Hendrik Lenstra

Fig. 1.7 The postcard that Hendrik Lenstra sent to László Lovász on 18th December 1981. Notice that the shortest vector problem he mentions is still open after 25 years

Lovász wrote:

> Thus the basis algorithm is the following: start with a basis of the lattice. Look for an i, $0 \leq i \leq n-2$ violating (1.1). If such an i exists, interchange b_{i+1} and b_{i+2}, and look for another i. If no such i exists, straighten out the basis and start all over again. Stop if no exchange has been made after straightening.

A few days after Lenstra got the letter from Lovász, he sent an excited postcard to Hungary, see Fig. 1.7: "Dear László, Congratulations with your beautiful algorithm! [...] Your result may have implications (theoretical & practical) on polynomial factorization (over $\mathbb{Q}$). My younger brother (A.K.) is working on that." More on this polynomial factorization is in section "Polynomial Factorization". First Lovász explains why he was working on lattice basis reduction.

The Ellipsoid Method

László Lovász started his talk in Caen by declaring that he was not really interested in trying to improve Lenstra's algorithm. In fact, he was interested in a tiny little detail in the ellipsoid method. It all started around 1978 with the paper

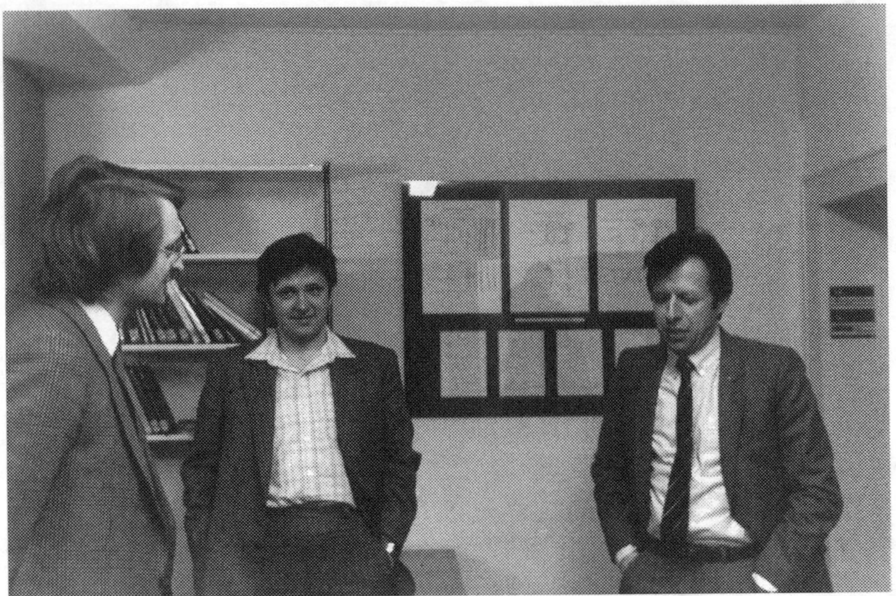

Fig. 1.8 Hendrik Lenstra, László Lovász, and their host Bernhard Korte in Bonn (February 1982)

A polynomial algorithm in linear programming from Leonid Khachiyan (sometimes spelled Hačijan) [2]. Lovász: "The ellipsoid method was developed by Soviet scientists in the second half of the seventies. Khachiyan noticed that this algorithm can be applied to solve linear programming in polynomial time, which was a big unsolved problem. All of a sudden there was a big interest in these things."

Peter van Emde Boas remembers how the ellipsoid method first arrived in the west as a rumor and how "Khachiyan conquered the world and everyone became crazy." In those days, there was no email or internet and the iron curtain made things even more difficult.

Lovász:

> I was living in Hungary, but I had the possibility to travel every now and then. In 1978–1979, I spent a year in Canada and in the summer I was in Stanford. There I met Peter Gács and someone sent us Khachiyan's paper. We read and understood it. On the way back to Hungary, I took the opportunity to visit Amsterdam and Bonn. You tried to minimize the number of times you passed the iron curtain, because that was somehow limited. In Amsterdam I met Lex Schrijver and in Bonn Martin Grötschel. I told them both about the ellipsoid method and they became very enthusiastic about it and we started working on it.

Lovász and Gács wrote a report [7] about Khachiyan's paper that explained the ideas and convinced the operations research and computer science communities that the algorithm was correct.

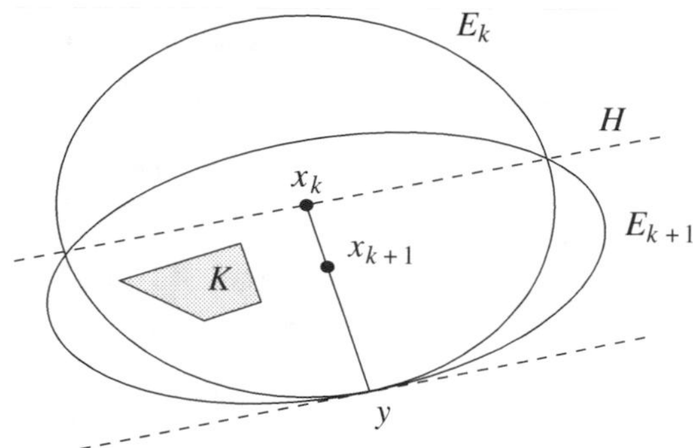

Fig. 1.9 An illustration of one step of the ellipsoid method

The Ellipsoid Method (As Described in [1]).
There is a simple geometric idea behind the ellipsoid method. We start with a convex body K in R^n, included in a big ellipsoid E_0, and a linear objective function $c^T x$. In the kth step, there is an ellipsoid E_k, which includes the set K_k of those points x of K for which $c^T x$ is at least as large as the best found so far. We look at the center x_k of E_k.
If x_k is not an element of K, then we take a hyperplane through x_k which avoids K. This hyperplane H cuts E_k into two halves; we pick that one which includes K_k and include it in a new ellipsoid E_{k+1}, which is essentially the ellipsoid of least volume containing this half of E_k, except for an allowance for rounding errors. The ellipsoid E_{k+1} can be geometrically described as follows. Let $F = E_k \cap H$, and let y be the point where a hyperplane parallel to H touches our half of E_k. Then the center of this smallest ellipsoid divides the segment $x_k y$ in ratio $1{:}n$, the ellipsoid intersects H in F, and touches E_k in y. The ellipsoid E_{k+1} then arises by blowing up and rounding; see Fig. 1.9.
If $x_k \in K$, then we cut with the hyperplane $c^T x = c^T x_k$ similarly.
The volumes of the ellipsoids E_k will tend to 0 exponentially fast and this guarantees that those centers x_k which are in K will tend to an optimum solution exponentially fast.

Consider the following problems for K, a nonempty convex compact set in $\mathbb{R}^n$.

1. *Strong optimization problem*: given a vector $c \in \mathbb{R}^n$, find a vector x in K which maximizes $c^T x$ on K.
2. *Strong separation problem*: given a vector $y \in \mathbb{R}^n$, decide if $y \in K$, and if not, find a hyperplane that separates y from K; more exactly find a vector $c \in \mathbb{R}^n$ such that $c^T y > \max\{c^T x | x \in K\}$.

In 1980, Grötschel, Lovász, and Schrijver proved the following theorem [1].

Theorem 1 *Let $\mathcal{K}$ be a class of convex bodies. There is a polynomial algorithm to solve the separation problem for the members of $\mathcal{K}$ if and only if there is a polynomial algorithm to solve the optimization problem for the members of $\mathcal{K}$.*

The proof uses the ellipsoid method. Lovász:

> Other people also noticed that the main interest of the ellipsoid method is not in practical applications for linear programming, but in theoretical applications for combinatorial optimization problems. We decided to write a book about this. For this book we wanted to make everything as nice as possible, but there was one annoying little gap.

In combinatorial applications, K is typically given by a system of linear inequalities, with rational coefficients, such that each defining inequality can be written down using a polynomial number of digits. We want to know whether the ellipsoid method terminates. If the solution set K is full-dimensional, then $\text{vol}(K) > 0$ and one can prove that $\log(1/\text{vol}(K))$ is bounded by a polynomial in the dimension n and the length of the rest of input for K. So the ellipsoid method terminates after a polynomial number of steps in this case. If K is not full-dimensional (so $\text{vol}(K) = 0$), the ellipsoid method may go on forever. In many interesting applications, it is impossible to tell from the input of K whether $\text{vol}(K) = 0$, but luckily we can determine that this must be the case if the ellipsoids become smaller than the computable lower bound for $\text{vol}(K)$. In this case we can use diophantine rounding as follows.

If $\text{vol}(K) = 0$, then K lies in a hyperplane, and one would like to do the ellipsoid algorithm in dimension $n - 1$. For this, one needs to find a hyperplane containing K. If we do the ellipsoid algorithm in dimension n, we get smaller and smaller ellipsoids that may never have their center in K. After some steps, we do find a hyperplane that approximates K, see Fig. 1.10. All vertices of K are close to this hyperplane given by the equality

$$\alpha_1 x_1 + \cdots + \alpha_n x_n = \alpha_0.$$

We want to round this to a hyperplane containing all the vertices of K

$$\frac{p_1}{q} x_1 + \cdots + \frac{p_n}{q} x_n = \frac{p_0}{q}.$$

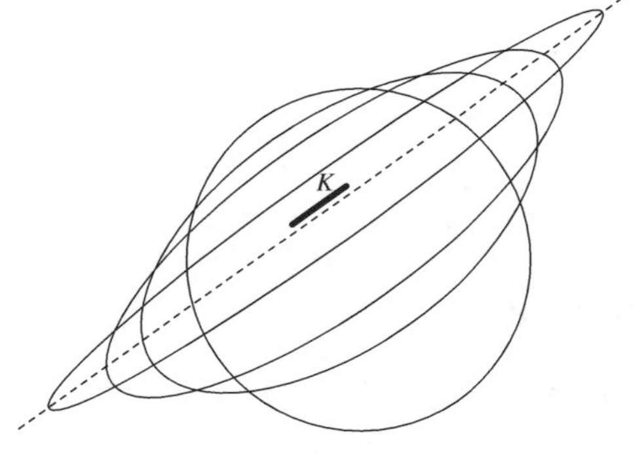

Fig. 1.10 Some of the (nonconsecutive) ellipsoids found in the ellipsoid method. The hyperplane returned by the ellipsoid method approximates K

To make this rounding work, we need the following condition

$$\left|\alpha_i - \frac{p_i}{q}\right| \leq \frac{\varepsilon}{q}$$

for some ε that can be computed from the problem. This is classic simultaneous Diophantine approximation. The question for Lovász was how to do this algorithmically.

> I started to play around with 1, $\sqrt{2}$, and $\sqrt{3}$ on my TI59 calculator. It was very easy to come up with ideas, it was clear that you wanted to subtract integer multiples of these numbers from each other. Whatever rule I chose, things started nicely, but after some point the process slowed down. I played around for a fairly long time until I found a way that did not slow down and seemed to make steady progress. This was of course just experimentation.
>
> I recalled that Diophantine approximation is discussed in the context of lattices and I realized that the real issue is trying to find short vectors in lattices. I remembered that when I was in Bonn six months earlier, Hendrik Lenstra gave a lecture about integer programming in which he also talked about finding short vectors in lattices. So this was really the way to go.

It took Lovász quite some time to generalize his rule for 1, $\sqrt{2}$, and $\sqrt{3}$ to higher dimensions."It seemed that the less greedy you were, the better it worked. So I swapped only neighboring vectors and swapped only when you really made progress by a constant factor. And then I sent my letter to Hendrik."

Hendrik Lenstra emphasized in his talk why these rules make LLL fast:

> Consider the sublattices L_j spanned by the first j basisvectors, $L_j = \mathbb{Z}b_1 + \cdots + \mathbb{Z}b_j$. It is really through these sublattices that you see the progress that you are making in your algorithm. In the LLL-algorithm, you only swap neighboring vectors b_i and b_{i+1}, so only L_i changes and all L_j with $j \neq i$ remain the same. Throughout the entire process, none of the determinants $d(L_j)$ ever gets larger.
>
> In my original algorithm I was too greedy. If there was at some stage a very short vector at the end of the basis, I would immediately swap it up front. This makes L_1 better, but all the intermediate L_j with $1 < j < n$ may become worse and you lose all control.

Polynomial Factorization

Arjen Lenstra's connection with the LLL-algorithm began while he still was a student. He opened his talk in Caen with: "My perspective is more or less that of a surprised bystander while all this violence was going on around me." It started with a report from Hendrik Lenstra on Euclidean number fields of large degree [6]. This report from 1976 contained a large number of tables of irreducible monic polynomials over the integers. The algebraic number fields generated by the roots of these polynomials were not isomorphic. The question was if other polynomials generated the same number fields as the polynomials in the table. In those days, to answer such a question, you had to factor polynomials over number fields. For the course *Programming Methods*, Arjen Lenstra and fellow students Henk Bos and Rudolf Mak

$$\begin{array}{llr}
\delta & \delta+1,\ \ 1,\ \ 1 & 5 \\
0 & \left\{\begin{array}{lllllllll} 1, & 0, & 0, & 0, & 1, & -3, & 3, & -2, & 1 \\ 1 & -2 & 2 & 1 & -4 & 3 & 1 & -2 & 1 \end{array}\right. & 5 \\
\varepsilon & 1,\ \ \varepsilon,\ \ 1 & 5
\end{array} \tag{1.2}$$

Fig. 1.11 In the tables of Arjen Lenstra's copy of the report on Euclidean number fields [6] there were polynomials pencilled in. The question was if these pencilled-in polynomials generated the same number fields as the ones above them

set out to study and implement methods to factor univariate polynomials over algebraic number fields. This was usually done using the Berlekamp-Hensel approach suggested by Zassenhaus [10].

Polynomial Factorization for $f \in \mathbb{Q}[X]$ with Berlekamp-Hensel.
We may assume that f is square-free, as it is easy to remove repeated factors of f.

1. Take a prime p such that f modulo p remains square-free and use Berlekamp's algorithm to factor f modulo p.
2. Apply the Hensel-lemma to lift these factors to factors modulo p^k for a sufficiently large k.
3. Try products of the modular factors to find the "true" factorization.

The big problem with this Berlekamp-Hensel approach was that the last step could be exponential in the degree of f, as there are irreducible polynomials that split into many factors modulo any prime. Arjen Lenstra: "No one tried to do anything about this exponential step, all people tried to do was convince themselves that it was indeed very, very much exponential. They were busy generating polynomials that were extremely bad cases for this Berlekamp-Hensel approach."

Generalizing this approach from the rationals to algebraic number fields was according to Arjen Lenstra: "sticking your head in the sand and hoping that it would work."

Polynomial Factorization for $f \in \mathbb{Q}(\alpha)[X]$ with the Zassenhaus Approach as Described in [9].
Let g be a monic irreducible polynomial of degree d over the integers, let $g(\alpha) = 0$, and let f be a square-free polynomial to be factored over $\mathbb{Q}(\alpha)$.

1. If there is a prime p such that g modulo p is irreducible and f modulo p is square-free
   (a) Factor f over the finite field $(\mathbb{Z}/p\mathbb{Z})[X]/(g(X))$; the resulting factorization modulo g and p corresponds to the factorization of f over $\mathbb{Q}(\alpha)$.
   (b) Follow the usual Berlekamp-Hensel method.

2. If there is no prime p such that g modulo p is irreducible, then take a prime p with $\gcd(p, \Delta(f)) = 1$ and $\gcd(p, \Delta(g)) = 1$.
   (a) Factor f over several finite fields, one for each irreducible factor of g modulo p.
   (b) Lift the factors of g modulo p to factors of g modulo p^k for a sufficiently large k.
   (c) Working modulo p^k, lift the factors of f from step 2a to factors modulo the lifted factors of g from step 2b.
   (d) Use Chinese remaindering to combine the resulting modular factorizations of f to factors of f modulo a high power of p.
   (e) Try combinations of these factors.

Notice that this algorithm is exponential in the product of the degrees of g and f. The students got around halfway implementing this approach in their project. Peter van Emde Boas was one of the supervisors of this project, and when he later became Arjen's master thesis advisor, he decided that completing this project would be the perfect task for Arjen.

There were many problems. One of them was that they had to use the fairly newly developed programming language ALGOL68. Lenstra: "It was developed in

Fig. 1.12 Arjen Lenstra defending his master thesis on 10th December 1980 – 1 year and 2 days before Lóvász posted the letter with his basis reduction algorithm. The committee was formed by Th. J. Dekker, Peter van Emde Boas, and Hendrik Lenstra (behind the desk). Also visible is Pia Pfluger from the Numerical Mathematics group

Amsterdam and if we did not use it, no-one would use it. It worked great, but it had a rather impossible two-level grammar and a seven-pass compiler. The ALGOL68 punch-card jobs had very long turnaround times, rebooting took hours, and we were only allowed limited time on the computers." Another obvious problem was that the algorithms could be exponential, but in practice they often worked. Arjen Lenstra: "I managed to answer the isomorphism questions and thus to complete my master thesis, but it was a rather unsatisfactory method.

When I discussed this with Hendrik, he asked why we used this silly Chinese remaindering and why we combined all those primes in the number field. He suggested that it might be possible to replace the Chinese remaindering by a lattice step." To explain how this lattice step works we assume without loss of generality that the minimum polynomial g has a monic linear factor h_k modulo some power p^k of p. Furthermore, let c in $\mathbb{Z}[\alpha]$ be a coefficient of a factor of f over $\mathbb{Q}(\alpha)$. There is an integer ℓ and a polynomial t of degree at most $d-1$ in $\mathbb{Z}[\alpha]$ such that

$$c = c_k + \ell \cdot p^k + t \cdot h_k, \tag{1.3}$$

where c_k is an integer value that will appear as one of the coefficients of the (combinations of) factors of f modulo h_k and p^k.

From (1.3) it follows that we should consider the d-dimensional lattice spanned by the vectors

$$\begin{array}{l}
(p^k, \ 0, \ \ 0, \ \ldots 0, \ \ 0), \\
(h_{k0}, 1, \ \ 0, \ \ldots 0, \ \ 0), \\
(0, \ \ h_{k0}, 1, \ 0, \ \ldots \ 0), \\
\qquad\qquad \vdots \\
(0, \ \ 0, \ \ \ldots 0, \ h_{k0}, 1),
\end{array}$$

where $h_k = \alpha + h_{k0}$. One may expect that for large enough k, the coefficient c will be the unique shortest vector that is congruent to c_k modulo the lattice as generated above. If we reduce the lattice basis, we find a fundamental domain and when k tends to infinity this domain should spread in all directions to make sure c is contained in it.

Arjen: "I used the lattice basis reduction from Hendrik's paper on integer linear programming [5]. This reduction algorithm did not run in polynomial time, but who cares about such petty issues when dealing with an algorithm that runs in exponential time anyhow? So, the lattice approach was implemented, and it turned out to work beautifully."

The next goal was to prove that the lattice approach always works as expected, including an estimate what value of k one should use to be able to derive valid irreducibility results. Arjen Lenstra:

> I started to think about this and I was not very good at these things. My lack of understanding of the situation reached its zenith when, in my confusion, I added an extra vector and used a $d+1$–dimensional lattice instead of the normal d–dimensional one. I was trying to prove that every vector in my lattice was very long, but this $d+1$–dimensional lattice always contained a short vector: g itself. This observation baffled me for a while, but then quickly led to the desired result: apparently the property I needed was coprimality with g over the

$$\left.\begin{array}{l} b_i^* := b_i; \\ \left.\begin{array}{l}\mu_{ij} := (b_i, b_j^*)/B_j; \\ b_i^* := b_i^* - \mu_{ij} b_j^*\end{array}\right\} \text{ for } j = 1, 2, \ldots, i-1; \\ B_i := (b_i^*, b_i^*)\end{array}\right\} \text{ for } i = 1, 2, \ldots, n;$$

$k := 2$

(1) perform $(*)$ for $l = k-1$;
if $B_k < \left(\frac{3}{4} - \mu_{kk-1}^2\right) B_{k-1}$, go to (2);
perform $(*)$ for $l = k-2, k-3, \ldots, 1$;
if $k = n$, terminate;
$k := k+1$;
go to (1);

(2) $\mu := \mu_{kk-1}$; $B := B_k + \mu^2 B_{k-1}$; $\mu_{kk-1} : -\mu B_{k-1}/B$;
$B_k := B_{k-1} B_k / B$; $B_{k-1} := B$;

$$\begin{pmatrix} b_{k-1} \\ b_k \end{pmatrix} := \begin{pmatrix} b_k \\ b_{k-1} \end{pmatrix};$$

$$\begin{pmatrix} \mu_{k-1j} \\ \mu_{kj} \end{pmatrix} := \begin{pmatrix} \mu_{kj} \\ \mu_{k-1j} \end{pmatrix} \quad \text{for } j = 1, 2, \ldots, k-2;$$

$$\begin{pmatrix} \mu_{ik-1} \\ \mu_{ik} \end{pmatrix} := \begin{pmatrix} 1 & \mu_{kk-1} \\ 0 & 1 \end{pmatrix} \begin{pmatrix} 0 & 1 \\ 1 & -\mu \end{pmatrix} \begin{pmatrix} \mu_{ik-1} \\ \mu_{ik} \end{pmatrix} \quad \text{for } i = k+1, k+2, \ldots, n;$$

if $k > 2$, then $k := k-1$;
go to (1).

$(*)$ If $|\mu_{kl}| > \frac{1}{2}$, then:

$$\begin{cases} r := \text{integer nearest to } \mu_{kl}; b_k := b_k - r b_l; \\ \mu_{kj} := \mu_{kj} - r\mu_{lj} \quad \text{for } j = 1, 2, \ldots, l-1; \\ \mu_{kl} := \mu_{kl} - r. \end{cases}$$

Fig. 1.13 The basis reduction algorithm as published in the LLL-article. The figure containing pseudocode for the algorithm was added after a suggestion by the referee

integers, yet a factor h_k in common with g modulo p^k. This property I could then indeed use to derive the lower bound proof – a very inelegant proof that is now happily lost in oblivion. In any case, I now knew for sure that we could factor polynomials over algebraic number fields faster than before. How much faster precisely no one seemed to care, as the overall algorithm was still exponential in the degree of f.

The initially disturbing observation had an interesting side-result, namely that if we do the entire method for a polynomial g that is *not* irreducible and use the d-dimensional lattice, we find a factor of g. This implied that if one lifts far enough, the combinatorial search in Berlekamp-Hensel can be avoided at the cost of shortest vector computations in various lattices. Furthermore, by pushing k even further, the shortest vector computations can be replaced by lattice basis reductions. Cute, but useless, as neither the shortest vector nor lattice basis reduction methods ran in polynomial time.

When Lovász sent his letter that lattice basis reduction could be done in polynomial time, Hendrik Lenstra started to look for an error in the proof that the factorization algorithm ran in polynomial time. A few days after he mailed his postcard to Lovász (see Fig. 1.7), Hendrik Lenstra sent a much longer letter, starting: "Ever since I got your letter I have been in a state of surprise, since it seems that your basis reduction algorithm implies that there is a polynomial algorithm for factorization in $\mathbb{Q}[X]$. For several days I have been looking for an error, and not having found one I am writing for your opinion." At that time, factoring polynomials over the rationals was so firmly established as something that could not be done in polynomial time, that something else must be spoiling their factorization algorithm. For a moment Hendrik Lenstra believed he found the wrongdoer in the prime p you needed to maintain square-freeness. However, he proved that this p can be bounded in such a way that Berlekamp runs in polynomial time, deterministically. And so, as Arjen Lenstra put it: "We were suddenly looking at this weird result that polynomials could be factored in polynomial time."

The LLL-Article

On 12th May 1982, after five months of polishing the algorithm, refining the analysis and many letters to-and-fro, Hendrik Lenstra wrote to Lovász: "Perhaps we should start thinking about where to send our paper. I am personally inclined to send it to a pure mathematics journal rather than a computer science journal. This maximizes the probability of getting sensible comments from the referee. [...] What do you think of Mathematische Annalen?" Lenstra admitted in Caen that there was another reason he wanted to send the article to a pure mathematics journal: "In those days pure mathematicians were not used to doing complexity analyses of algorithms, it was considered the domain of computer scientists. I felt this was a beautiful area that – in this case – gave rise to fantastical problems in number theory and that mathematicians should be more aware of this field. This seemed a good opportunity, as we had a pretty amazing result that nobody had expected."

The unexpected result of polynomial factorization became the title of the paper. Peter van Emde Boas asked the audience in Caen what they thought of when they heard *LLL-algorithm*: was it "basis reduction" or "factoring polynomials"? All hands rose for "basis reduction." So in hindsight maybe the title should have been something like "A new basis reduction algorithm."

On 2nd July 1982, Hendrik Lenstra submitted the article to Mathematische Annalen. The article went rather swiftly through the refereeing process and appeared later that year [8]. The algorithm has made a great impact. In September 2007, the article has 486 citations on ISI Web of Knowledge. As you can see in the rest of this book, research on the LLL-algorithm and its applications are very much alive.

References

1. M. Grötschel, L. Lovász, and A. Schrijver. The ellipsoid method and its consequences in combinatorial optimization. *Combinatorica*, 1(2): 169–197, 1981.
2. L. G. Hačijan. A polynomial algorithm in linear programming. *Dokl. Akad. Nauk SSSR*, 244(5): 1093–1096, 1979.

3. H. W. Lenstra, Jr. Integer programming with a fixed number of variables. *Report 81-03 (First version)*, University of Amsterdam, April 1981.
4. H. W. Lenstra, Jr. Integer programming with a fixed number of variables. *Report 81-03 (Second version)*, University of Amsterdam, November 1981.
5. H. W. Lenstra, Jr. Integer programming with a fixed number of variables. *Math. Oper. Res.*, 8(4): 538–548, 1983.
6. H. W. Lenstra, Jr. Euclidean number fields of large degree. *Report 76-09*, University of Amsterdam, May 1976.
7. P. Gács, and L. Lovász. Khachiyan's algorithm for linear programming. *Math. Programming Stud.*, 14: 61–68, 1981.
8. A. K. Lenstra, H. W. Lenstra, Jr., and L. Lovász. Factoring polynomials with rational coefficients. *Math. Ann.*, 261(4): 515–534, 1982.
9. P. J. Weinberger and L. P. Rothschild. Factoring polynomials over algebraic number fields. *ACM Trans. Math. Software*, 2(4): 335–350, 1976.
10. H. Zassenhaus. On Hensel factorization. I. *J. Number Theory*, 1: 291–311, 1969.

Chapter 2
Hermite's Constant and Lattice Algorithms

Phong Q. Nguyen

Abstract We introduce lattices and survey the main provable algorithms for solving the shortest vector problem, either exactly or approximately. In doing so, we emphasize a surprising connection between lattice algorithms and the historical problem of bounding a well-known constant introduced by Hermite in 1850, which is related to sphere packings. For instance, we present the Lenstra–Lenstra–Lovász algorithm (LLL) as an (efficient) algorithmic version of Hermite's inequality on Hermite's constant. Similarly, we present blockwise generalizations of LLL as (more or less tight) algorithmic versions of Mordell's inequality.

Introduction

Informally, a *lattice* is an infinite arrangement of points in $\mathbb{R}^n$ spaced with sufficient regularity that one can shift any point onto any other point by some symmetry of the arrangement. The simplest example of a lattice is the hypercubic lattice $\mathbb{Z}^n$ formed by all points with integral coordinates. Geometry of numbers [1–4] is the branch of number theory dealing with lattices (and especially their connection with convex sets), and its origins go back to two historical problems:

1. *Higher-dimensional generalizations of Euclid's algorithm.* The elegance and simplicity of Euclid's greatest common divisor algorithm motivate the search for generalizations enjoying similar properties. By trying to generalize previous work of Fermat and Euler, Lagrange [5] studied numbers that can be represented by quadratic forms at the end of the eighteenth century: given a triplet $(a, b, c) \in \mathbb{Z}^3$, identify which integers are of the form $ax^2 + bxy + cy^2$, where $(x, y) \in \mathbb{Z}^2$. Fermat had for instance characterized numbers that are sums of two squares: $x^2 + y^2$, where $(x, y) \in \mathbb{Z}^2$. To answer such questions, Lagrange invented a generalization [5, pages 698–700] of Euclid's algorithm to binary quadratic forms. This algorithm is often attributed (incorrectly) to Gauss [6], and was generalized in

P.Q. Nguyen
INRIA, Ecole normale supérieure, Département d'informatique, 45 rue d'Ulm, 75005 Paris, France
e-mail: http://www.di.ens.fr/~pnguyen/

P.Q. Nguyen and B. Vallée (eds.), *The LLL Algorithm*, Information Security and Cryptography, DOI 10.1007/978-3-642-02295-1_2,

the nineteenth century by Hermite [7] to positive definite quadratic forms of arbitrary dimension. Let $q(x_1,\ldots,x_n) = \sum_{1\leq i,j\leq n} q_{i,j}x_ix_j$ be a positive definite quadratic form over $\mathbb{R}^n$, and denote by $\Delta(q) = \det_{1\leq i,j\leq n} q_{i,j} \in \mathbb{R}^+$ its discriminant. Hermite [7] used his algorithm to prove that there exist $x_1,\ldots,x_n \in \mathbb{Z}$ such that

$$0 < q(x_1,\ldots,x_n) \leq (4/3)^{(n-1)/2}\Delta(q)^{1/n}. \tag{2.1}$$

If we denote by $\|q\|$ the minimum of $q(x_1,\ldots,x_n)$ over $\mathbb{Z}^n \setminus \{0\}$, (2.1) shows that $\|q\|/\Delta(q)^{1/n}$ can be upper bounded independently of q. This proves the existence of Hermite's constant γ_n defined as the supremum of this ratio over all positive definite quadratic forms:

$$\gamma_n = \max_{q \text{ positive definite over } \mathbb{R}^n} \frac{\|q\|}{\Delta(q)^{1/n}}, \tag{2.2}$$

because it turns out that the supremum is actually reached. The inequality (2.1) is equivalent to Hermite's inequality on Hermite's constant:

$$\gamma_n \leq (4/3)^{(n-1)/2},\ n \geq 1, \tag{2.3}$$

which can be rewritten as

$$\gamma_n \leq \gamma_2^{n-1},\ n \geq 1, \tag{2.4}$$

because Lagrange [5] showed that $\gamma_2 = \sqrt{4/3}$. Though Hermite's constant was historically defined in terms of positive definite quadratic forms, it can be defined equivalently using lattices, due to the classical connection between lattices and positive definite quadratic forms, which we will recall precisely in section "Quadratic Forms."

2. *Sphere packings.* This famous problem [8] asks what fraction of $\mathbb{R}^n$ can be covered by equal balls that do not intersect except along their boundaries. The problem is open as soon as $n \geq 4$ (see Fig. 2.1 for the densest packing for $n = 2$), which suggests to study simpler problems.

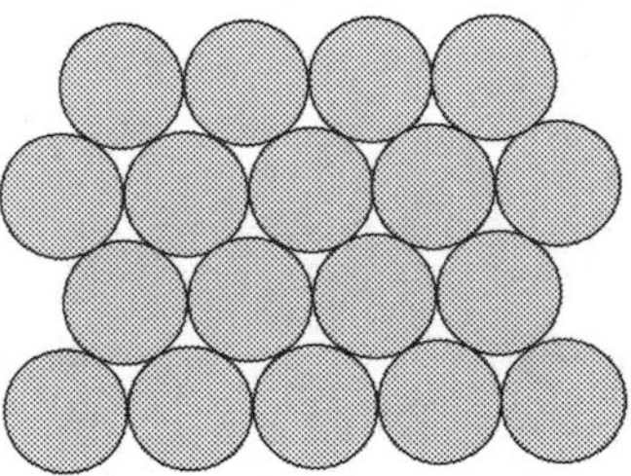

Fig. 2.1 The densest packing in dimension two: the hexagonal lattice packing

Of particular interest is the lattice packing problem, which asks what is the densest packing derived from lattices (such as the packing of Fig. 2.1): any full-rank lattice L induces a packing of $\mathbb{R}^n$ whose centers are the lattice points, and the diameter of the balls is the minimal distance $\lambda_1(L)$ between two lattice points. The density $\delta(L)$ of the lattice packing is equal to the ratio between the volume of the n-dimensional ball of diameter $\lambda_1(L)$ and the volume of any fundamental domain of L (i.e., the volume of the compact set $\mathbb{R}^n/L$). There is the following simple relationship between Hermite's constant γ_n and the supremum $\delta_n = \max_L \delta(L)$ over all full-rank lattices L of $\mathbb{R}^n$, due to the alternative lattice-based definition of γ_n previously mentioned:

$$\gamma_n = 4\left(\frac{\delta_n}{v_n}\right)^{2/n}, \tag{2.5}$$

where v_n denotes the volume of the n-dimensional unit ball. Thus, the problem of finding the maximal density of lattice packings is equivalent to finding the exact value of Hermite's constant γ_n, which is currently open for $n \geq 9$, $n \neq 24$.

Lattice algorithms deal with integral lattices, which are usually represented by a matrix with integer coefficients. This means that the lattice L is formed by all integral linear combinations of the row vectors of a given integral matrix B:

$$L = \{a_1\mathbf{b}_1 + \cdots + a_n\mathbf{b}_n, a_i \in \mathbb{Z}\},$$

where $\mathbf{b}_1, \mathbf{b}_2, \ldots, \mathbf{b}_n \in \mathbb{Z}^m$ denote the row vectors of B. The most famous lattice problem is the so-called *shortest vector problem* (SVP), which asks to find a shortest nonzero vector in L, that is, a nonzero vector of the form $a_1\mathbf{b}_1 + \cdots + a_n\mathbf{b}_n$ (where $a_i \in \mathbb{Z}$) and of minimal Euclidean norm $\lambda_1(L)$. SVP can be viewed as a geometric generalization of gcd computations: Euclid's algorithm actually computes the smallest (in absolute value) nonzero linear combination of two integers, as $\gcd(a,b)\mathbb{Z} = a\mathbb{Z} + b\mathbb{Z}$, which means that we are replacing the integers a and b by an arbitrary number of vectors $\mathbf{b}_1, \ldots, \mathbf{b}_n$ with integer coordinates.

When the vectors $\mathbf{b}_i$'s span a low-dimensional space, one can solve SVP as efficiently as Euclid's algorithm. But when the dimension increases, NP-hardness looms (see [9]), which gives rise to two types of algorithms:

(a) *Exact algorithms.* These algorithms provably find a shortest vector, but they are expensive, with a running time at least exponential in the dimension. Intuitively, these algorithms perform an exhaustive search of all extremely short lattice vectors, whose number is exponential in the dimension (in the worst case): in fact, there are lattices for which the number of shortest lattice vectors is already exponential. The best deterministic algorithm is Kannan's enumeration [10, 11], with super-exponential worst-case complexity, namely $n^{n/(2e)+o(n)}$ polynomial-time operations (see [12, 13]), where n denotes the lattice dimension. The best randomized algorithm is the sieve of Ajtai, Kumar, and Sivakumar (AKS) [14, 15], with exponential worst-case complexity of $2^{O(n)}$ polynomial-time operations (where $O()$ can be taken to be 5.9 [15]): this algorithm also requires exponential space, whereas enumeration requires only negligible space.

(b) *Approximation algorithms*. The Lenstra–Lenstra–Lovász algorithm (LLL) and other efficient lattice reduction algorithms known provide only an approximation of SVP, in the sense that the norm of the nonzero output vector can be upper bounded using some function of the dimension, either absolutely or relatively to the minimal norm $\lambda_1(L)$. We will see that all polynomial-time approximation algorithms known [16–19] can be viewed as (more or less tight) algorithmic versions of upper bounds on Hermite's constant. For instance, LLL can be viewed as an algorithmic version of Hermite's inequality (2.3): it can be used to find efficiently $x_1, \ldots, x_n \in \mathbb{Z}$ satisfying essentially (2.1), which corresponds to short lattice vectors within Hermite's inequality. Similarly, the recent blockwise algorithm of Gama and Nguyen [19] can be viewed as an algorithmic version of Mordell's inequality, which itself is a generalization of Hermite's inequality (2.3).

In high dimension (say, higher than 150), only approximation algorithms are practical, but both categories are in fact complementary: all exact algorithms known first apply an approximation algorithm (typically at least LLL) as a preprocessing, while all approximation algorithms known call many times an exact algorithm in low dimension as a subroutine.

In this article, we will survey the main provable algorithms for solving the shortest vector problem, either exactly or approximately. This is related to Hermite's constant as follows:

- The analysis of exact algorithms involves counting the number of lattice points inside balls, for which good estimates are related to Hermite's constant.
- All approximation algorithms known are rather designed to find short nonzero lattice vectors in an absolute sense: the fact that the norm of the output is also relatively close to the first minimum can be viewed as a by-product. This means that any proof of correctness of the algorithm will have to include a proof that the output lattice vector is short in an absolute sense, which gives rise to an upper bound on Hermite's constant. In fact, it turns out that all approximation algorithms known are related (in a more or less tight manner) to a classical upper bound on Hermite's constant.

The rest of the article is organized as follows. Section "Background and Lattices" introduces lattices and their mathematical background. Section "Lattice Reduction" introduces lattice reduction and the main computational problems. Subsequent sections present the main lattice algorithms. Section "Two-Dimensional Case" deals with the two-dimensional case: Lagrange's algorithm. Section "Hermite's Inequality and the Lenstra–Lenstra–Lovász Algorithm" deals with the first efficient approximation algorithm in high dimension: the LLL algorithm. Section "Solving Exact SVP" deals with exact algorithms for SVP, which all use the LLL algorithm. Finally, section "Mordell's Inequality and Blockwise Algorithms" deals with polynomial-time generalizations of LLL that have a better approximation factor.

Background on Lattices

Notation

We consider $\mathbb{R}^n$ with its usual topology of an Euclidean vector space. We use bold letters to denote vectors, usually in row notation. The Euclidean inner product of two vectors $\mathbf{x} = (x_i)_{i=1}^n$ and $\mathbf{y} = (y_i)_{i=1}^n$ is denoted by

$$\langle \mathbf{x}, \mathbf{y} \rangle = \sum_{i=1}^{n} x_i y_i .$$

The corresponding Euclidean norm is denoted by

$$\|\mathbf{x}\| = \sqrt{x_1^2 + \cdots + x_n^2}.$$

Denote by $\mathcal{B}(\mathbf{x}, r)$ the open ball of radius r centered at $\mathbf{x}$:

$$\mathcal{B}(\mathbf{x}, r) = \{\mathbf{y} \in \mathbb{R}^n : \|\mathbf{x} - \mathbf{y}\| < r\}.$$

Definition 1. A subset D of $\mathbb{R}^n$ is called *discrete* when it has no limit point, that is, for all $x \in D$, there exists $\rho > 0$ such that $\mathcal{B}(x, \rho) \cap D = \{x\}$.

As an example, $\mathbb{Z}^n$ is discrete (because $\rho = 1/2$ clearly works), while $\mathbb{Q}^n$ and $\mathbb{R}^n$ are not. The set $\{1/n : n \in \mathbb{N}^*\}$ is discrete, but the set $\{0\} \cup \{1/n : n \in \mathbb{N}^*\}$ is not. Any subset of a discrete set is discrete.

For any ring R, we denote by $\mathcal{M}_{n,m}(R)$ (resp. $\mathcal{M}_n(R)$) the set of $n \times m$ (resp. $n \times n$) matrices with coefficients in R. $GL_n(R)$ denotes the group of invertible matrices in the ring $\mathcal{M}_n(R)$. For any subset S of $\mathbb{R}^n$, we define the linear span of S, denoted by $\text{span}(S)$, as the minimal vector subspace (of $\mathbb{R}^n$) containing S.

Definition 2. Let $\mathbf{b}_1, \ldots, \mathbf{b}_m$ be in $\mathbb{R}^n$. The vectors $\mathbf{b}_i$'s are said to be *linearly dependent* if there exist $x_1, \ldots, x_m \in \mathbb{R}$, which are not all zero and such that

$$\sum_{i=1}^{m} x_i \mathbf{b}_i = 0.$$

Otherwise, they are said to be *linearly independent.*

Definition 3. The *Gram determinant* of $\mathbf{b}_1, \ldots, \mathbf{b}_m \in \mathbb{R}^n$, denoted by $\Delta(\mathbf{b}_1, \ldots, \mathbf{b}_m)$, is the determinant of the $m \times m$ Gram matrix $\left(\langle \mathbf{b}_i, \mathbf{b}_j \rangle\right)_{1 \le i,j \le m}$.

We list basic properties of the Gram determinant:

- The Gram determinant $\Delta(\mathbf{b}_1, \ldots, \mathbf{b}_m)$ is always ≥ 0. It is equal to zero if and only if the $\mathbf{b}_i$'s are linearly dependent.

- The Gram determinant is invariant by any permutation of the m vectors, and by any integral linear transformation of determinant ± 1, such as adding to one of the vectors a linear combination of the others.
- The Gram determinant has a very important geometric interpretation: when the $\mathbf{b}_i$'s are linearly independent, $\sqrt{\Delta(\mathbf{b}_1, \ldots, \mathbf{b}_m)}$ is the m-dimensional volume $\text{vol}(\mathbf{b}_1, \ldots, \mathbf{b}_m)$ of the parallelepiped $\{\sum_{i=1}^{m} x_i \mathbf{b}_i : 0 \le x_i \le 1, 1 \le i \le m\}$ spanned by the $\mathbf{b}_i$'s.

Denote by v_n the volume of the n-dimensional unit ball $\mathcal{B}(0,1)$. Then

$$v_n = \frac{\pi^{n/2}}{\Gamma(1+n/2)} \sim \left(\frac{2e\pi}{n}\right)^{n/2} \frac{1}{\sqrt{\pi n}}, \tag{2.6}$$

where $\Gamma(x) = \int_0^\infty t^{x-1} e^{-t}\, dt$.

Lattices

Definition 4. A *lattice* of $\mathbb{R}^n$ is a discrete subgroup of $(\mathbb{R}^n, +)$; that is any subgroup of $(\mathbb{R}^n, +)$ which has the discreteness property.

Notice that an additive group is discrete if and only if 0 is not a limit point, which implies that a lattice is any nonempty set $L \subseteq \mathbb{R}^n$ stable by subtraction (in other words: for all $\mathbf{x}$ and $\mathbf{y}$ in L, $\mathbf{x}-\mathbf{y}$ belongs to L), and such that $L \cap \mathcal{B}(0, \rho) = \{0\}$ for some $\rho > 0$.

With this definition, the first examples of lattices that come to mind are the zero lattice $\{0\}$ and the *lattice of integers* $\mathbb{Z}^n$. Our definition implies that any subgroup of a lattice is a lattice, and therefore, any subgroup of $(\mathbb{Z}^n, +)$ is a lattice. Such lattices are called *integral lattices*. As an example, consider two integers a and $b \in \mathbb{Z}$: the set $a\mathbb{Z} + b\mathbb{Z}$ of all integral linear combinations of a and b is a subgroup of $\mathbb{Z}$, and therefore a lattice; it is actually the set $\gcd(a,b)\mathbb{Z}$ of all multiples of the gcd of a and b. For another example, consider n integers $a_1, \ldots, a_n$, together with a modulus M. Then the set of all $(x_1, \ldots, x_n) \in \mathbb{Z}^n$ such that $\sum_{i=1}^{n} a_i x_i \equiv 0 \pmod{M}$ is a lattice in $\mathbb{Z}^n$ because it is clearly a subgroup of $\mathbb{Z}^n$.

We give a few basic properties of lattices:

Lemma 1. *Let L be a lattice in $\mathbb{R}^n$.*

1. *There exists $\rho > 0$ such that for all $\mathbf{x} \in L$:*

$$L \cap \mathcal{B}(\mathbf{x}, \rho) = \{\mathbf{x}\}.$$

2. *L is closed.*
3. *For all bounded subsets S of $\mathbb{R}^n$, $L \cap S$ is finite.*
4. *L is countable.*

Notice that a set that satisfies either property 1 or 3 is necessarily discrete, but an arbitrary discrete subset of $\mathbb{R}^n$ does not necessarily satisfy property 1 nor 3. It is the group structure of lattices that allows such additional properties.

Bases

Let $\mathbf{b}_1, \ldots, \mathbf{b}_m$ be arbitrary vectors in $\mathbb{R}^n$. Denote by $\mathcal{L}(\mathbf{b}_1, \ldots, \mathbf{b}_m)$ the set of all integral linear combinations of the $\mathbf{b}_i$'s:

$$\mathcal{L}(\mathbf{b}_1, \ldots, \mathbf{b}_m) = \left\{ \sum_{i=1}^{m} n_i \mathbf{b}_i : n_1, \ldots, n_m \in \mathbb{Z} \right\} . \tag{2.7}$$

This set is a subgroup of $\mathbb{R}^n$, but it is not necessarily discrete. For instance, one can show that $\mathcal{L}((1), (\sqrt{2}))$ is not discrete because $\sqrt{2} \notin \mathbb{Q}$. However, the following elementary result gives sufficient conditions for this set to be discrete:

Theorem 1. *The subgroup $\mathcal{L}(\mathbf{b}_1, \ldots, \mathbf{b}_m)$ is a lattice in either of the following two cases:*

1. $\mathbf{b}_1, \ldots, \mathbf{b}_m \in \mathbb{Q}^n$.
2. $\mathbf{b}_1, \ldots, \mathbf{b}_m \in \mathbb{R}^n$ *are linearly independent.*

Proof. Case 1 is trivial. Now consider Case 2, and let $L = \mathcal{L}(\mathbf{b}_1, \ldots, \mathbf{b}_m)$. It suffices to show that 0 is not a limit point of L. Consider the parallelepiped P defined by

$$P = \left\{ \sum_{i=1}^{m} x_i \mathbf{b}_i : |x_i| < 1 \right\} .$$

As the $\mathbf{b}_i$'s are linearly independent, $L \cap P = \{0\}$. Besides, there exists $\rho > 0$ such that $B(0, \rho) \subseteq P$, which shows that 0 cannot be a limit point of L. □

Definition 5. When $L = \mathcal{L}(\mathbf{b}_1, \ldots, \mathbf{b}_m)$ is a lattice, we say that L is spanned by the $\mathbf{b}_i$'s, and that the $\mathbf{b}_i$'s are *generators*. When the $\mathbf{b}_i$'s are further linearly independent, we say that $(\mathbf{b}_1, \ldots, \mathbf{b}_m)$ is a *basis* of the lattice L, in which case each lattice vector decomposes itself uniquely as an integral linear combination of the $\mathbf{b}_i$'s:

$$\forall \mathbf{v} \in L, \ \exists! v_1, \ldots, v_m \in \mathbb{Z} \text{ s.t. } \mathbf{v} = \sum_{i=1}^{m} v_i \mathbf{b}_i .$$

Bases and sets of generators are useful to represent lattices and to perform computations. One will typically represent a lattice on a computer by some lattice basis, which can itself be represented by a matrix with real coefficients. In practice, one will usually restrict to integral lattices, so that the underlying matrices are integral matrices.

Definition 6. We define the *dimension* or *rank* of a lattice L in $\mathbb{R}^n$, denoted by $\dim(L)$, as the dimension d of its linear span denoted by $\mathrm{span}(L)$. The lattice is said to be *full-rank* when $d = n$: in the remaining, we usually denote the dimension by n when the lattice is full-rank, and by d otherwise.

The dimension is the maximal number of linearly independent lattice vectors. Any lattice basis of L must have exactly d elements. There always exist d linearly independent lattice vectors; however, such vectors do not necessarily form a basis, as opposed to the case of vector spaces. But the following theorem shows that one can always derive a lattice basis from such vectors:

Theorem 2. *Let L be a d-dimensional lattice of $\mathbb{R}^n$. Let $\mathbf{c}_1, \ldots, \mathbf{c}_d \in L$ be linearly independent vectors. There exists a lower triangular matrix $(u_{i,j}) \in \mathcal{M}_d(\mathbb{R})$ such that the vectors $\mathbf{b}_1, \ldots, \mathbf{b}_d$ defined as $\mathbf{b}_i = \sum_{j=1}^{i} u_{i,j}\mathbf{c}_j$ are linearly independent and such that $L = \mathcal{L}(\mathbf{b}_1, \ldots, \mathbf{b}_d)$.*

This proves the unconditional existence of lattice bases:

Corollary 1. *Any lattice of $\mathbb{R}^n$ has at least one basis.*

Thus, even if sets of the form $\mathcal{L}(\mathbf{b}_1, \ldots, \mathbf{b}_m)$ may or may not be lattices, all lattices can be written as $\mathcal{L}(\mathbf{b}_1, \ldots, \mathbf{b}_m)$ for some linearly independent $\mathbf{b}_i$'s. Corollary 1 together with Theorem 1 give an alternative definition of a lattice: a nonempty subset L of $\mathbb{R}^n$ is a lattice if only if there exist linearly independent vectors $\mathbf{b}_1, \mathbf{b}_2, \ldots, \mathbf{b}_d$ in $\mathbb{R}^n$ such that

$$L = \mathcal{L}(\mathbf{b}_1, \ldots, \mathbf{b}_d).$$

This characterization suggests that lattices are discrete analogues of vector spaces.

The following elementary result shows the relationship between two bases:

Theorem 3. *Let $(\mathbf{b}_1, \ldots, \mathbf{b}_d)$ be a basis of a lattice L in $\mathbb{R}^n$. Let $\mathbf{c}_1, \ldots, \mathbf{c}_d$ be vectors of L. Then there exists a unique $d \times d$ integral matrix $U = (u_{i,j})_{1 \le i,j \le d} \in \mathcal{M}_d(\mathbb{Z})$ such that $\mathbf{c}_i = \sum_{j=1}^{d} u_{i,j}\mathbf{b}_j$ for all $1 \le i \le d$. And $(\mathbf{c}_1, \ldots, \mathbf{c}_d)$ is a basis of L if and only if the matrix U has determinant ± 1.*

As a result, as soon as the lattice dimension is ≥ 2, there are infinitely many lattice bases.

Quadratic Forms

Historically, lattices were first studied in the language of positive definite quadratic forms. Let $(\mathbf{b}_1, \ldots, \mathbf{b}_d)$ be a basis of a lattice L in $\mathbb{R}^n$. Then the function

$$q(x_1, \ldots, x_d) = \| \sum_{i=1}^{d} x_i \mathbf{b}_i \|^2, \tag{2.8}$$

defines a positive definite quadratic form over $\mathbb{R}^d$.

Reciprocally, let q be a positive definite quadratic form over $\mathbb{R}^d$. Then Cholesky factorization shows the existence of linearly independent vectors $\mathbf{b}_1, \ldots, \mathbf{b}_d$ of $\mathbb{R}^d$ such that (2.8) holds for all $(x_1, \ldots, x_d) \in \mathbb{R}^d$.

Volume and the Gaussian Heuristic

Let $(\mathbf{b}_1, \ldots, \mathbf{b}_d)$ and $(\mathbf{c}_1, \ldots, \mathbf{c}_d)$ be two bases of a lattice L in $\mathbb{R}^n$. By Theorem 3, there exists a $d \times d$ integral matrix $U = (u_{i,j})_{1 \le i,j \le d} \in \mathcal{M}_d(\mathbb{Z})$ of determinant ± 1 such that $\mathbf{c_i} = \sum_{j=1}^{d} u_{i,j} \mathbf{b}_j$ for all $1 \le i \le d$. It follows that the Gram determinant of those two bases are equal:

$$\Delta(\mathbf{b}_1, \ldots, \mathbf{b}_d) = \Delta(\mathbf{c}_1, \ldots, \mathbf{c}_d) > 0,$$

which gives rise to the following definition:

Definition 7. The *volume* (or *determinant*) of the lattice L is defined as

$$\mathrm{vol}(L) = \Delta(\mathbf{b}_1, \ldots, \mathbf{b}_d)^{1/2},$$

which is independent of the choice of the basis $(\mathbf{b}_1, \ldots, \mathbf{b}_d)$ of the lattice L.

We prefer the name *volume* to the name *determinant* because of its geometric interpretation: it corresponds to the d-dimensional volume of the parallelepiped spanned by any basis. In the mathematical literature, the lattice volume we have just defined is sometimes alternatively called co-volume, because it is also the volume of the torus $\mathrm{span}(L)/L$. For full-rank lattices, the volume has the following elementary properties:

Lemma 2. *Let L be a full-rank lattice in $\mathbb{R}^n$. Then:*

1. *For any basis $(\mathbf{b}_1, \ldots, \mathbf{b}_n)$ of L, $\mathrm{vol}(L) = |\det(\mathbf{b}_1, \ldots, \mathbf{b}_n)|$.*
2. *For any $r > 0$, denote by $s_L(r)$ the number of $\mathbf{x} \in L$ such that $\|\mathbf{x}\| \le r$. Then*

$$\lim_{r \to \infty} \frac{s_L(r)}{r^n v_n} = 1/\mathrm{vol}(L).$$

The second statement of Lemma 2 says that, as the radius r grows to infinity, the number of lattice vectors inside the ball (centered at zero) of radius r is asymptotically equivalent to the ratio between the volume $r^n v_n$ of the n-dimensional ball of radius r and the volume of the lattice. This suggests the following heuristic, known as the Gaussian Heuristic:

Definition 8. Let L be a full-rank lattice in $\mathbb{R}^n$, and C be a measurable subset of $\mathbb{R}^n$. The *Gaussian Heuristic* "predicts" that the number of points of $L \cap C$ is roughly $\mathrm{vol}(C)/\mathrm{vol}(L)$.

We stress that this is only a heuristic: there are cases where the heuristic is proved to hold, but there are also cases where the heuristic is proved to be incorrect.

Given a lattice L, how does one compute the volume of L? If an explicit basis of L is known, this amounts to computing a determinant: for instance, the volume of the hypercubic lattice $\mathbb{Z}^n$ is clearly equal to one. But if no explicit basis is known, one can sometimes use full-rank sublattices, as we will see in the next subsection.

Sublattices

Definition 9. Let L be a lattice in $\mathbb{R}^n$. A *sublattice* of L is a lattice M included in L: clearly, the sublattices of L are the subgroups of L. If the rank of M is equal to the rank of L, we say that M is a *full-rank* sublattice of L.

Lemma 3. *Let L be a lattice in $\mathbb{R}^n$. A sublattice M of L is full-rank if and only if the group index $[L : M]$ is finite, in which case we have*

$$\mathrm{vol}(M) = \mathrm{vol}(L) \times [L : M].$$

As an illustration, consider n integers $a_1, \ldots, a_n$, together with a modulus M. We have seen in section "Lattices" that the set L of all $(x_1, \ldots, x_n) \in \mathbb{Z}^n$ such that $\sum_{i=1}^n a_i x_i \equiv 0 \pmod{M}$ is a lattice in $\mathbb{Z}^n$ because it is a subgroup of $\mathbb{Z}^n$. But there seems to be no trivial basis of L. However, note that $L \subseteq \mathbb{Z}^n$ and that the dimension of L is n because L contains all the vectors of the canonical basis of $\mathbb{R}^n$ multiplied by M. It follows that

$$\mathrm{vol}(L) = [\mathbb{Z}^n : L].$$

Furthermore, the definition of L clearly implies that

$$[\mathbb{Z}^n : L] = M / \gcd(M, a_1, a_2, \ldots, a_n).$$

Hence,

$$\mathrm{vol}(L) = \frac{M}{\gcd(M, a_1, a_2, \ldots, a_n)}.$$

Definition 10. A sublattice M of L is said to be *primitive* if there exists a subspace E of $\mathbb{R}^n$ such that $M = L \cap E$.

It follows from Theorem 2 that:

Lemma 4. *A sublattice M of L is primitive if and only if every basis of M can be completed to a basis of L, that is, for any basis $(\mathbf{b}_1, \ldots, \mathbf{b}_r)$ of M, there exist $\mathbf{b}_{r+1}, \ldots, \mathbf{b}_d \in L$ such that $(\mathbf{b}_1, \ldots, \mathbf{b}_d)$ is a basis of L.*

Definition 11. Let $\mathbf{b}_1, \ldots, \mathbf{b}_k \in L$. They are *primitive vectors* of L if and only if $\mathcal{L}(\mathbf{b}_1, \ldots, \mathbf{b}_k)$ is a primitive sublattice of L.

In particular, any nonzero shortest vector of L is primitive.

Projected Lattices

Let L be a lattice in $\mathbb{R}^n$. The (orthogonal) projection of L over a subspace of $\mathbb{R}^n$ is a subgroup of $\mathbb{R}^n$, but it is not necessarily discrete. However, with suitable choices of the subspace, one can ensure discreteness, in which case the projection is a lattice:

Lemma 5. *Let L be a d-rank lattice in $\mathbb{R}^n$, and M be a r-rank primitive sublattice of L: $1 \leq r \leq d$. Let π_M denote the orthogonal projection over the orthogonal supplement of the linear span of M. Then $\pi_M(L)$ is a lattice of $\mathbb{R}^n$, of rank $d - r$, and of volume* $\mathrm{vol}(L)/\mathrm{vol}(M)$.

Proof. Let $(\mathbf{b}_1, \ldots, \mathbf{b}_r)$ be a basis of M. As M is primitive sublattice of L, this basis can be extended to a basis of L: there exist $\mathbf{b}_{r+1}, \ldots, \mathbf{b}_d \in L$ such that $(\mathbf{b}_1, \ldots, \mathbf{b}_d)$ is a basis of L. Clearly, the set $\pi_M(L)$ is equal to $\mathcal{L}(\pi_M(\mathbf{b}_{r+1}), \ldots, \pi_M(\mathbf{b}_d))$. As $\pi_M(\mathbf{b}_{r+1}), \ldots, \pi_M(\mathbf{b}_{r+1})$ are linearly independent, the subgroup $\mathcal{L}(\pi_M(\mathbf{b}_{r+1}), \ldots, \pi_M(\mathbf{b}_d))$ is a lattice, and so is $\pi_M(L)$. □

The following corollary will be used many times in lattice reduction:

Corollary 2. *Let $(\mathbf{b}_1, \ldots, \mathbf{b}_d)$ be a basis of a lattice L in $\mathbb{R}^n$. For $1 \leq i \leq d$, let π_i denote the orthogonal projection over the orthogonal supplement of the linear span of $\mathbf{b}_1, \ldots, \mathbf{b}_{i-1}$; in particular, π_1 is the identity. Then $\pi_i(L)$ is a lattice of $\mathbb{R}^n$, of rank $d - i + 1$, and of volume* $\mathrm{vol}(L)/\mathrm{vol}(\mathcal{L}(\mathbf{b}_1, \ldots, \mathbf{b}_{i-1}))$.

We will often use the notation π_i.

It is classical to prove statements by induction on the lattice rank using projected lattices, such as in the classical proof of Hermite's inequality: see Theorem 8 of section "Hermite's Inequality and the Lenstra–Lenstra–Lovász Algorithm." More precisely, for any basis $(\mathbf{b}_1, \ldots, \mathbf{b}_d)$ of L, we have $\dim(\pi_2(L)) = \dim(L) - 1$, and any nonzero vector $\mathbf{v} \in \pi_2(L)$ can be lifted into a nonzero vector $\mathbf{u} \in L$ such that $\mathbf{v} = \pi_2(\mathbf{u})$ and $\|\mathbf{u}\|^2 \leq \|\mathbf{v}\|^2 + \|\mathbf{b}_1\|^2/4$. This means that if one can find a short vector in $\pi_2(L)$, then one can also find a reasonably short vector in L.

Duality

Let L be a lattice in $\mathbb{R}^n$. The *dual lattice* of L is defined as

$$L^{\times} = \{\mathbf{y} \in \mathrm{span}(L) \text{ such that} \langle \mathbf{x}, \mathbf{y} \rangle \in \mathbb{Z} \text{ for all } \mathbf{x} \in L\}.$$

Lemma 6. *If L is a d-rank lattice of $\mathbb{R}^n$, then $L^\times$ is a d-rank lattice of $\mathbb{R}^n$ such that*

$$\mathrm{vol}(L) \times \mathrm{vol}(L^\times) = 1.$$

Duality also allows to consider sublattices of lower dimension, which can be used in proofs by induction, such as the classical proof of Mordell's inequality (see section "Classical Proofs of Mordell's Inequality"). For instance, if L is a d-rank lattice and $\mathbf{v}$ is a nonzero vector of $L^\times$, then $L \cap H$ is a $(d-1)$-rank sublattice of L, where $H = \mathbf{v}^\perp$ denotes the hyperplane orthogonal to $\mathbf{v}$.

Gram–Schmidt and Triangularization

Definition 12. Let $\mathbf{b}_1, \ldots, \mathbf{b}_d$ be linearly independent vectors in $\mathbb{R}^n$. Their *Gram–Schmidt orthogonalization* (GSO) is the orthogonal family $(\mathbf{b}_1^\star, \ldots, \mathbf{b}_d^\star)$ defined as follows: $\mathbf{b}_1^\star = \mathbf{b}_1$ and more generally $\mathbf{b}_i^\star = \pi_i(\mathbf{b}_i)$ for $1 \le i \le d$, where π_i denotes (as in Corollary 2) the orthogonal projection over the orthogonal supplement of the linear span of $\mathbf{b}_1, \ldots, \mathbf{b}_{i-1}$.

We have the recursive formula

$$\mathbf{b}_i^\star = \mathbf{b}_i - \sum_{j=1}^{i-1} \mu_{i,j} \mathbf{b}_j^\star, \quad \text{where } \mu_{i,j} = \frac{\langle \mathbf{b}_i, \mathbf{b}_j^\star \rangle}{\left\| \mathbf{b}_j^\star \right\|^2} \text{ for all } 1 \le j < i \le d \tag{2.9}$$

The main reason why the Gram–Schmidt orthogonalization is widely used in lattice reduction is because it allows to triangularize the basis. More precisely, the family $(\mathbf{b}_1^\star/\|\mathbf{b}_1^\star\|, \ldots, \mathbf{b}_d^\star/\|\mathbf{b}_d^\star\|)$ is an orthonormal basis of $\mathbb{R}^n$. And if we express the vectors $\mathbf{b}_1, \ldots, \mathbf{b}_d$ with respect to the orthonormal basis $(\mathbf{b}_1^\star/\|\mathbf{b}_1^\star\|, \ldots, \mathbf{b}_d^\star/\|\mathbf{b}_d^\star\|)$ (rather than the canonical basis), we obtain the following lower-triangular matrix, with diagonal coefficients $\|\mathbf{b}_1^\star\|, \ldots, \|\mathbf{b}_d^\star\|$:

$$\begin{pmatrix} \|\mathbf{b}_1^\star\| & 0 & \ldots & 0 \\ \mu_{2,1}\|\mathbf{b}_1^\star\| & \|\mathbf{b}_2^\star\| & \ddots & \\ & \ddots & \ddots & \vdots \\ \vdots & \ddots & \ddots & 0 \\ \mu_{d,1}\|\mathbf{b}_1^\star\| & \ldots & \mu_{d,d-1}\|\mathbf{b}_{d-1}^\star\| & \|\mathbf{b}_d^\star\| \end{pmatrix} \tag{2.10}$$

This can be summarized by the matrix equality $B = \mu B^\star$, where B is the $d \times n$ matrix whose rows are $\mathbf{b}_1, \ldots, \mathbf{b}_d$, $B^\star$ is the $d \times n$ matrix whose rows are $\mathbf{b}_1^\star, \ldots, \mathbf{b}_d^\star$, and μ is the $d \times d$ lower-triangular matrix whose diagonal coefficients are all equal to 1, and whose off-diagonal coefficients are the $\mu_{i,j}$'s. It follows that the lattice L spanned by the $\mathbf{b}_i$'s satisfies

$$\mathrm{vol}(L) = \prod_{i=1}^{d} \|\mathbf{b}_i^\star\| \tag{2.11}$$

Notice that the GSO family depends on the order of the vectors. If $\mathbf{b}_i \in \mathbb{Q}^n$, then $\mathbf{b}_i^\star \in \mathbb{Q}^n$ and $\mu_{i,j} \in \mathbb{Q}$. The GSO of $(\mathbf{b}_1, \ldots, \mathbf{b}_d)$ is $(\mu_{i,j})_{1\le j<i\le d}$ together with $(\|\mathbf{b}_i^\star\|)_{1\le i\le d}$. Geometrically, $\|\mathbf{b}_i^\star\|$ is the distance of $\mathbf{b}_i$ to span$\mathbf{b}_1, \ldots, \mathbf{b}_{i-1}$.

The basis triangularization could have been obtained with other factorizations. For instance, if we had used Iwasa's decomposition of the row matrix B corresponding to $(\mathbf{b}_1, \ldots, \mathbf{b}_d)$, we would have obtained $B = UDO$, where U is a lower-triangular matrix with unit diagonal, D is diagonal, and O is an orthogonal matrix. In other words, U would be the matrix defined by the $\mu_{i,j}$'s (lower-triangular with unit diagonal, where the remaining coefficients are the $\mu_{i,j}$'s), D would be the diagonal matrix defined by the $\|\mathbf{b}_i^\star\|$'s, and O would be the row representation of $(\mathbf{b}_1^\star/\|\mathbf{b}_1^\star\|, \ldots, \mathbf{b}_d^\star/\|\mathbf{b}_d^\star\|)$.

Finally, it is worth noting that Gram–Schmidt orthogonalization is related to duality as follows. For any $i \in \{2, \ldots, d\}$, the vector $\mathbf{b}_i^\star/\|\mathbf{b}_i^\star\|^2$ is orthogonal to $\mathbf{b}_1, \ldots, \mathbf{b}_{i-1}$ and we have $\langle \mathbf{b}_i^\star/\|\mathbf{b}_i^\star\|^2, \mathbf{b}_i\rangle = 1$, which implies that

$$\mathbf{b}_i^\star/\|\mathbf{b}_i^\star\|^2 \in \pi_j(\mathcal{L}(\mathbf{b}_1, \ldots, \mathbf{b}_i))^\times, \forall j \in \{1, \ldots, i\}.$$

Lattice Reduction

A fundamental result of linear algebra states that any finite-dimensional vector space has a basis. We earlier established the analogue result for lattices: any lattice has a basis. In the same vein, a fundamental result of bilinear algebra states that any finite-dimensional Euclidean space has an orthonormal basis, that is, a basis consisting of unit vectors that are pairwise orthogonal. A natural question is to ask whether lattices also have orthonormal bases, or at least, orthogonal bases. Unfortunately, it is not difficult to see that even in dimension two, a lattice may not have an orthogonal basis, and this is in fact a typical situation. Informally, the goal of lattice reduction is to circumvent this problem: more precisely, the theory of lattice reduction shows that in any lattice, there is always a basis, which is not that far from being orthogonal. Defining precisely what is meant exactly by not being far from being orthogonal is tricky, so for now, let us just say that such a basis should consist of reasonably short lattice vectors, which implies that geometrically such vectors are not far from being orthogonal to each other.

Minkowski's Minima

To explain what is a reduced basis, we need to define what is meant by short lattice vectors. Let L be a lattice of dimension ≥ 1 in $\mathbb{R}^n$. There exists a nonzero vector $\mathbf{u} \in L$. Consider the closed hyperball $\mathcal{B}$ of radius $\|\mathbf{u}\|$ centered at zero. By Lemma

1, $L \cap \mathcal{B}$ is finite and contains $\mathbf{u}$, so it must have a shortest nonzero vector. The Euclidean norm of that shortest nonzero vector is called the *first minimum* of L, and is denoted by $\lambda_1(L) > 0$ or $\|L\|$. By definition, any nonzero vector $\mathbf{v}$ of L satisfies $\|\mathbf{v}\| \geq \lambda_1(L)$, and $\lambda_1(L)$ is the minimal distance between two distinct lattice vectors. And there exists $\mathbf{w} \in L$ such that $\|\mathbf{w}\| = \lambda_1(L)$: any such $\mathbf{w}$ is called a shortest vector of L, and it is not unique as $-\mathbf{w}$ would also be a shortest vector. The *kissing number* of L is the number of shortest vectors in L: it is upper bounded by some exponential function of the lattice dimension (see [8]).

We noticed that if $\mathbf{w}$ is a shortest vector of L, then so is $-\mathbf{w}$. Thus, one must be careful when defining the *second-to-shortest* vector of a lattice. To circumvent this problem, Minkowski [1] defined the other minima as follows.

Definition 13. Let L be a lattice of $\mathbb{R}^n$. For all $1 \leq i \leq \dim(L)$, the ith *minimum* $\lambda_i(L)$ is defined as the minimum of $\max_{1 \leq j \leq i} \|\mathbf{v}_j\|$ over all i linearly independent lattice vectors $\mathbf{v}_1, \ldots, \mathbf{v}_i \in L$.

Clearly, the minima are increasing: $\lambda_1(L) \leq \lambda_2(L) \leq \cdots \leq \lambda_d(L)$. And the Gram–Schmidt triangularization implies:

Lemma 7. *If $(\mathbf{b}_1, \ldots, \mathbf{b}_d)$ is a basis of a lattice L, then its GSO satisfies for all $1 \leq i \leq d$*

$$\lambda_i(L) \geq \min_{i \leq j \leq d} \|\mathbf{b}_j^\star\|.$$

It is not difficult to see that there always exist linearly independent lattice vectors $\mathbf{v}_1, \ldots, \mathbf{v}_d$ reaching simultaneously the minima, that is, $\|\mathbf{v}_i\| = \lambda_i(L)$ for all i. However, surprisingly, as soon as $\dim(L) \geq 4$, such vectors do not necessarily form a lattice basis. The canonical example is the four-dimensional lattice L defined as the set of all $(x_1, x_2, x_3, x_4) \in \mathbb{Z}^4$ such that $\sum_{i=1}^4 x_i$ is even. It is not difficult to see that $\dim(L) = 4$ and that all the minima of L are equal to $\sqrt{2}$. Furthermore, it can be checked that the following row vectors form a basis of L:

$$\begin{pmatrix} 1 & -1 & 0 & 0 \\ 1 & 1 & 0 & 0 \\ 1 & 0 & 1 & 0 \\ 1 & 0 & 0 & 1 \end{pmatrix}.$$

The basis proves in particular that $\mathrm{vol}(L) = 2$. However, the following row vectors are linearly independent lattice vectors, which also reach all the minima:

$$\begin{pmatrix} 1 & -1 & 0 & 0 \\ 1 & 1 & 0 & 0 \\ 0 & 0 & 1 & 1 \\ 0 & 0 & 1 & -1 \end{pmatrix}.$$

But they do not form a basis, as their determinant is equal to 4: another reason is that for all such vectors, the sum of the first two coordinates is even, and that property also holds for any integral linear combination of those vectors, but clearly not for all vectors of the lattice L. More precisely, the sublattice spanned by those four row vectors has index two in the lattice L.

Nevertheless, in the lattice L, there still exists at least one basis that reaches all the minima simultaneously, and we already gave one such basis. This also holds for any lattice of dimension ≤ 4, but it is no longer true in dimension ≥ 5, as was first noticed by Korkine and Zolotarev in the nineteenth century, in the language of quadratic forms. More precisely, it can easily be checked that the lattice spanned by the rows of the following matrix

$$\begin{pmatrix} 2 & 0 & 0 & 0 & 0 \\ 0 & 2 & 0 & 0 & 0 \\ 0 & 0 & 2 & 0 & 0 \\ 0 & 0 & 0 & 2 & 0 \\ 1 & 1 & 1 & 1 & 1 \end{pmatrix}$$

has no basis reaching all the minima (which are all equal to two).

Hermite's Constant and Minkowski's Theorems

Now that successive minima have been defined, it is natural to ask how large those minima can be. Hermite [7] was the first to prove that the quantity $\lambda_1(L)/\text{vol}(L)^{1/d}$ could be upper bounded over all d-rank lattices L.

Definition 14. The supremum of $\lambda_1(L)^2/\text{vol}(L)^{2/d}$ over all d-rank lattices L is denoted by γ_d, and called *Hermite's constant* of dimension d.

The use of quadratic forms in [7] explains why Hermite's constant refers to $\max_L \lambda_1(L)^2/\text{vol}(L)^{2/d}$ and not to $\max_L \lambda_1(L)/\text{vol}(L)^{1/d}$. It can be noted that γ_d could also be equivalently defined as the supremum of $\lambda_1(L)^2$ over all d-rank lattices L of unit volume.

It is known that γ_d is reached, that is, for all $d \geq 1$, there is a d-rank lattice L such that $\gamma_d = \lambda_1(L)^2/\text{vol}(L)^{2/d}$, and any such lattice is called *critical*. But finding the exact value of γ_d is a very difficult problem, which has been central in Minkowski's geometry of numbers. The exact value of γ_d is known only for $1 \leq d \leq 8$ (see the book [20] for proofs) and very recently also for $d = 24$ (see [21]): the values are summarized in the following table.

d	2	3	4	5	6	7	8	24
γ_d	$2/\sqrt{3}$	$2^{1/3}$	$\sqrt{2}$	$8^{1/5}$	$(64/3)^{1/6}$	$64^{1/7}$	2	4
Approximation	1.1547	1.2599	1.4142	1.5157	1.6654	1.8114	2	4

Furthermore, the list of all critical lattices (up to scaling and isometry) is known for each of those dimensions.

However, rather tight asymptotical bounds are known for Hermite's constant. More precisely, we have

$$\frac{d}{2\pi e} + \frac{\log(\pi d)}{2\pi e} + o(1) \le \gamma_d \le \frac{1.744d}{2\pi e}(1 + o(1)).$$

For more information on the proof of those bounds: see [22, Chap. 2] for the lower bound (which comes from the Minkowski–Hlawka theorem), and [8, Chap. 9] for the upper bound. Thus, γ_d is essentially linear in d. It is known that $\gamma_d^d \in \mathbb{Q}$ (because there is always an integral critical lattice), but it is unknown if $(\gamma_d)_{d \ge 1}$ is an increasing sequence.

Hermite's historical upper bound [7] on his constant was exponential in the dimension:

$$\gamma_d \le (4/3)^{(d-1)/2}.$$

The first linear upper bound on Hermite's constant is due to Minkowski, who viewed it as a consequence of his Convex Body Theorem:

Theorem 4 (Minkowski's Convex Body Theorem). *Let L be a full-rank lattice of $\mathbb{R}^n$. Let C be a measurable subset of $\mathbb{R}^n$, convex, symmetric with respect to 0, and of measure $> 2^n \mathrm{vol}(L)$. Then C contains at least a nonzero point of L.*

This theorem is a direct application of the following elementary lemma (see [2]), which can be viewed as a generalization of the pigeon-hole principle:

Lemma 8 (Blichfeldt). *Let L be a full-rank lattice in $\mathbb{R}^n$, and F be a measurable subset of $\mathbb{R}^n$ with measure $> \mathrm{vol}(L)$. Then F contains at least two distinct vectors whose difference belongs to L.*

Indeed, we may consider $F = \frac{1}{2}C$, and the assumption in Theorem 4 implies that the measure of F is $> \mathrm{vol}(L)$. From Blichfeldt's lemma, it follows that there exist $\mathbf{x}$ and $\mathbf{y}$ in F such that $\mathbf{x} - \mathbf{y} \in L \setminus \{0\}$. But

$$\mathbf{x} - \mathbf{y} = \frac{1}{2}(2\mathbf{x} - 2\mathbf{y}),$$

which belongs to C by convexity and symmetry with respect to 0. Hence, $\mathbf{x} - \mathbf{y} \in C \cap (L \setminus \{0\})$, which completes the proof of Theorem 4.

One notices that the bound on the volumes in Theorem 4 is the best possible, by considering

$$C = \left\{ \sum_{i=1}^{n} x_i \mathbf{b}_i \ : |x_i| < 1 \right\},$$

where the $\mathbf{b}_i$'s form an arbitrary basis of the lattice. Indeed, the measure of this C is exactly $2^n \mathrm{vol}(L)$, but by definition of C, no nonzero vector of L belongs to C.

In Theorem 4, the condition on the measure of C is a strict inequality, but it is not difficult to show that the strict inequality can be relaxed to an inequality $\ge 2^n \mathrm{vol}(L)$ if C is further assumed to be compact. By choosing for C a closed hyperball of sufficiently large radius (so that the volume inequality is satisfied), one obtains the following corollary:

Corollary 3. *Any d-dimensional lattice L of $\mathbb{R}^n$ contains a nonzero $\mathbf{x}$ such that*

$$\|\mathbf{x}\| \leq 2\left(\frac{\mathrm{vol}(L)}{v_d}\right)^{\frac{1}{d}},$$

where v_d denotes the volume of the closed unitary hyperball of $\mathbb{R}^d$. In other words,

$$\gamma_d \leq \left(\frac{4}{v_d}\right)^{2/d}, \quad d \geq 1.$$

Note that if the Gaussian heuristic (see Definition 8 of section "Volume and the Gaussian Heuristic") held for all hyperballs, we would expect $\lambda_1(L)$ to be close to $(\mathrm{vol}(L)/v_d)^{1/d} \approx \sqrt{d/(2\pi e)}\mathrm{vol}(L)^{1/d}$ by (2.6). This means that the proved upper bound is only twice as large as the heuristic estimate from the Gaussian heuristic.

Using well-known formulas for v_d, one can derive a linear bound on Hermite's constant, for instance

$$\forall d,\ \gamma_d \leq 1 + \frac{d}{4}.$$

Notice that this bound is reached by $L = \mathbb{Z}^d$.

Now that we know how to bound the first minimum, it is natural to ask if a similar bound can be obtained for the other minima. Unfortunately, one cannot hope to upper bound separately the other minima, because the successive minima could be unbalanced. For instance, consider the rectangular two-rank lattice L spanned by the following row matrix:

$$\begin{pmatrix} \varepsilon & 0 \\ 0 & 1/\varepsilon \end{pmatrix},$$

where $\varepsilon > 0$ is small. The volume of L is one, and by definition of L, it is clear that $\lambda_1(L) = \varepsilon$ and $\lambda_2(L) = 1/\varepsilon$ if $\varepsilon \leq 1$. Here, $\lambda_2(L)$ can be arbitrarily large compared to the lattice volume, while $\lambda_1(L)$ can be arbitrarily small compared to the upper bound given by Hermite's constant.

However, it is always possible to upper bound the geometric mean of the first consecutive minima, as summarized by the following theorem (for an elementary proof, see [2?]):

Theorem 5 (Minkowski's Second Theorem). *Let L be a d-rank lattice of $\mathbb{R}^n$. Then for any integer r such that $1 \leq r \leq d$,*

$$\left(\prod_{i=1}^{r} \lambda_i(L)\right)^{1/r} \leq \sqrt{\gamma_d}\,\mathrm{vol}(L)^{1/d}.$$

Rankin's Constant

In 1953, Rankin [24] introduced the following generalization of Hermite's constant. For any n-rank lattice L and $1 \leq m \leq n$, the Rankin invariant $\gamma_{n,m}(L)$ is defined as

$$\begin{aligned}\gamma_{n,m}(L) &= \min_{\substack{\mathbf{x}_1,\dots,\mathbf{x}_m \in L \\ \mathrm{vol}(\mathbf{x}_1,\dots,\mathbf{x}_m) \neq 0}} \left(\frac{\mathrm{vol}(\mathbf{x}_1,\dots,\mathbf{x}_m)}{\mathrm{vol}(L)^{m/n}}\right)^2 \\ &= \min_{\substack{S \text{ sublattice of } L \\ \dim S = m}} \left(\frac{\mathrm{vol}(M)}{\mathrm{vol}(L)^{m/n}}\right)^2 \end{aligned} \tag{2.12}$$

Using a family of linearly independent lattice vectors simultaneously reaching all the minima and Theorem 5, one obtains

$$\gamma_{n,m}(L) \leq \left(\frac{\prod_{i=1}^{m} \lambda_i(L)}{\mathrm{vol}(L)^{m/n}}\right)^2 \leq \gamma_n^m.$$

It follows that Rankin's constant $\gamma_{n,m} = \max \gamma_{n,m}(L)$ over all n-rank lattices L is well-defined, and we have $\gamma_{n,m} \leq \gamma_n^m$. This upper bound is not tight: using HKZ reduction (which we will define later) as in [17, 18], it can be shown that for $1 \leq m \leq n/2$,

$$\gamma_{n,m} \leq O(n)^{(n-m)\times(1/(n-1)+1/(n-2)+\cdots+1/(n-m))} \tag{2.13}$$

Rankin's constants satisfy the following three relations, which are proved in [20, 24]:

$$\forall n \in \mathbb{N},\ \gamma_{n,n} = 1,\ \gamma_{n,1} = \gamma_n \tag{2.14}$$

$$\forall n, m \text{ with } m < n\ \gamma_{n,m} = \gamma_{n,n-m} \tag{2.15}$$

$$\forall r \in [m+1, n-1],\ \gamma_{n,m} \leq \gamma_{r,m}(\gamma_{n,r})^{m/r} \tag{2.16}$$

The only known values of Rankin's constants are $\gamma_{4,2} = \frac{3}{2}$, which is reached for the $\mathbb{D}_4$ lattice, and those corresponding to the nine Hermite constants known. In the definition of $\gamma_{n,m}(L)$, the minimum is taken over sets of m linearly independent vectors of L, but we may restrict the definition to primitive sets of L or pure sublattices of L, as for any sublattice S of L, there exists a pure sublattice S_1 of L with $\mathrm{span}(S) = \mathrm{span}(S_1)$ and $\mathrm{vol}(S)/\mathrm{vol}(S_1) = [S : S_1]$. If $\mathrm{vol}(S)$ is minimal, then $[S : S_1] = 1$ so $S = S_1$ is pure.

Thunder [25] and Bogulavsky [26] proved the following lower bound on Rankin's constant, as a generalization of Minkowski–Hlawka's theorem:

$$\gamma_{n,m} \geq \left(n \frac{\prod_{j=n-m+1}^{n} Z(j)}{\prod_{j=2}^{m} Z(j)} \right)^{\frac{2}{n}} \tag{2.17}$$

where $Z(j) = \zeta(j)\Gamma(\frac{j}{2})/\pi^{\frac{j}{2}}$ and ζ is Riemann's zeta function: $\zeta(j) = \sum_{p=1}^{\infty} p^{-j}$. This shows that for $1 \leq m \leq n/2$,

$$\gamma_{n,m} \geq \Omega(n)^{m(n-m+1)/n} \tag{2.18}$$

Hermite–Korkine–Zolotarev (HKZ) Reduction

Hermite [7] introduced the following weak reduction notion in the language of quadratic forms:

Definition 15. A basis $(\mathbf{b}_1, \ldots, \mathbf{b}_d)$ of a lattice is *size-reduced* if its Gram–Schmidt orthogonalization satisfies, for all $1 \leq j < i \leq d$,

$$|\mu_{i,j}| \leq \frac{1}{2}. \tag{2.19}$$

Geometrically, this means that the projection $\mathbf{b}_i - \mathbf{b}_i^\star$ of $\mathbf{b}_i$ over the linear span of $\mathbf{b}_1, \ldots, \mathbf{b}_{i-1}$ is inside the parallelepiped $\mathcal{P} = \{\sum_{j=1}^{i-1} x_i \mathbf{b}_i,\ |x_j| \leq 1/2\}$ spanned by $\mathbf{b}_1^\star, \ldots, \mathbf{b}_{i-1}^\star$ with coefficients $\leq 1/2$ in absolute value, one tries to reduce the component of $\mathbf{b}_i$ over the linear span of $\mathbf{b}_1, \ldots, \mathbf{b}_{i-1}$. Then (2.19) implies for all $1 \leq i \leq d$:

$$\|\mathbf{b}_i^\star\|^2 \leq \|\mathbf{b}_i\|^2 \leq \|\mathbf{b}_i^\star\|^2 + \frac{1}{4}\sum_{j=1}^{i-1} \|\mathbf{b}_j^\star\|^2. \tag{2.20}$$

Korkine and Zolotarev [27, 28] strengthened Hermite's size-reduction as follows:

Definition 16. A basis $(\mathbf{b}_1, \ldots, \mathbf{b}_d)$ of a lattice is *Hermite–Korkine–Zolotarev-reduced (HKZ-reduced)* if it is size-reduced and such that for all $1 \leq i \leq d$, $\|\mathbf{b}_i^\star\| = \lambda_1(\pi_i(L))$.

Note that $\mathbf{b}_i^\star \in \pi_i(L)$ and $\mathbf{b}_i^\star \neq 0$, so it is natural to ask that $\|\mathbf{b}_i^\star\| = \lambda_1(\pi_i(L))$. Note also that the condition $\|\mathbf{b}_d^\star\| = \lambda_1(\pi_d(L))$ is necessarily satisfied.

HKZ-reduced bases have two interesting properties. The first is that an HKZ-reduced basis provides a very good approximation to the successive minima:

Theorem 6. *Let $(\mathbf{b}_1, \ldots, \mathbf{b}_d)$ be an HKZ-reduced basis of a lattice L, then for all index i such that $1 \leq i \leq d$,*

$$\frac{4}{i+3} \leq \left(\frac{\|\mathbf{b}_i\|}{\lambda_i(L)}\right)^2 \leq \frac{i+3}{4}$$

The upper bound is easy to prove and can be attributed to Mahler [29]: it suffices to notice that $\|\mathbf{b}_i^\star\| = \lambda_1(\pi_i(L)) \leq \lambda_i(L)$ (where the right-hand inequality can be proved by considering a set of linearly independent vectors reaching all the minima simultaneously), and to use the right-hand inequality of (2.20). The lower bound is proved in [30]: first, notice that HKZ-reduction implies that for all $1 \leq j \leq i$, $\|\mathbf{b}_j^\star\| \leq \|\mathbf{b}_i\|$, therefore $\|\mathbf{b}_j\|^2/\|\mathbf{b}_i\|^2 \leq (j+3)/4$ by (2.20). It should be noted that it is not necessarily true that $\|\mathbf{b}_i\| \geq \lambda_i(L)$ because it does not necessarily hold that $\|\mathbf{b}_2\| \leq \|\mathbf{b}_3\| \leq \cdots \leq \|\mathbf{b}_d\|$. Thus, the gap between an HKZ-reduced basis and the successive minima of a lattice is at most polynomial, namely less than $\sqrt{(i+3)/4}$. The article [30] shows that the bounds of Theorem 6 are not far from being tight in the worst case.

The second interesting property of HKZ-reduced bases is that they have local properties. Indeed, if $(\mathbf{b}_1, \ldots, \mathbf{b}_d)$ is HKZ-reduced, then $(\pi_i(\mathbf{b}_i), \pi_i(\mathbf{b}_{i+1}), \ldots, \pi_i(\mathbf{b}_j))$ is HKZ-reduced for all $1 \leq i \leq j \leq d$. Thus, by studying low-dimensional HKZ-reduced bases, one can deduce properties holding for any dimension. For instance, any two-dimensional HKZ-reduced basis $(\mathbf{c}_1, \mathbf{c}_2)$ satisfies $\|\mathbf{c}_1\|/\|\mathbf{c}_2^\star\| \leq \sqrt{4/3}$, which implies that ay HKZ-reduced basis $(\mathbf{b}_1, \ldots, \mathbf{b}_d)$ satisfies $\|\mathbf{b}_i^\star\|/\|\mathbf{b}_{i+1}^\star\| \leq \sqrt{4/3}$ for all $1 \leq i \leq d$. It is by using such reasonings that Korkine and Zolotarev found better upper bounds on Hermite's constant than Hermite's inequality.

Algorithmic Lattice Problems

In the previous section, we presented lattice reduction from a mathematical point of view. In this section, we introduce the main algorithmic problems for lattices.

Representation

In practice, one deals only with *rational lattices*, that is, lattices included in $\mathbb{Q}^n$. In this case, by a suitable multiplication, one needs only to be able to deal with integral lattices, those which are included in $\mathbb{Z}^n$. Such lattices are usually represented by a basis, that is, a matrix with integral coefficients. When we explicitly give such a matrix, we will adopt a row representation: the row vectors of the matrix will be the basis vectors. The size of the lattice is measured by the dimensions of the matrix (the number d of rows, which correspond to the lattice dimension, and the number n of columns), and the maximal bit-length $\log B$ of the matrix coefficients; thus, the whole matrix can be stored using $dn \log B$ bits.

Lattice problems are often relative to norms: here, we will only be concerned with the Euclidean norm. Before describing hard problems, let us recall two easy problems that can be solved in deterministic polynomial time:

- Given a generating set of an integral lattice L, find a basis of the lattice L.
- Given a basis of an integral lattice $L \subseteq \mathbb{Z}^n$ and a target vector $\mathbf{v} \in \mathbb{Z}^n$, decide if $\mathbf{v} \in L$, and if so, find the decomposition of $\mathbf{v}$ with respect to the basis.

The Shortest Vector Problem (SVP)

The most famous lattice problem is the following:

Problem 1 (Shortest Vector Problem (SVP)). Given a basis of a d-rank integral lattice L, find $\mathbf{u} \in L$ such that $\|\mathbf{u}\| = \lambda_1(L)$.

In its exact form, this problem is known to be NP-hard under randomized reductions (see the survey [9]), which suggests to relax the problem. There are two approximation versions of SVP: approx-SVP (ASVP) and Hermite-SVP (HSVP), which are defined below.

Problem 2 (Approximate Shortest Vector Problem (ASVP)). Given a basis of a d-rank integral lattice L and an approximation factor $f \geq 1$, find a nonzero $\mathbf{u} \in L$ such that $\|\mathbf{u}\| \leq f\lambda_1(L)$.

Problem 3 (Hermite Shortest Vector Problem (HSVP) [31]). Given a basis of a d-rank integral lattice L and an approximation factor $f > 0$, find a nonzero $\mathbf{u} \in L$ such that $\|\mathbf{u}\| \leq f\mathrm{vol}(L)^{1/d}$.

When $f = 1$, ASVP is exactly SVP. As opposed to SVP and ASVP, it is possible to easily check a solution to HSVP: indeed, given $\mathbf{u}$, L and f, one can check in polynomial time whether or not $\mathbf{u} \in L$ and $\|\mathbf{u}\| \leq f\mathrm{vol}(L)^{1/d}$. By definition of Hermite's constant, if one can solve ASVP with an approximation factor f, then one can solve HSVP with a factor $f\sqrt{\gamma_d}$. Reciprocally, it was shown in [32] that if one has access to an oracle solving HSVP with a factor f, then one can solve ASVP with a factor f^2 in polynomial time using a number of oracle queries linear in the dimension d. Hence, solving ASVP with an approximation factor polynomial in the dimension is equivalent to solving HSP with an approximation factor polynomial in the dimension.

Hardness results for SVP are surveyed in [9, 33], so let us just briefly summarize. SVP was conjectured NP-hard as early as 1981 [34] (see also [32]). Ajtai showed NP-hardness under randomized reductions in 1998 [35], but the historical conjecture with deterministic reductions remains open. The best result so far [12] suggests that it is unlikely that one can efficiently approximate SVP to within quasi-polynomial factors. But NP-hardness results have limits: essentially, approximating SVP within a factor $\sqrt{d/\log d}$ is unlikely to be NP-hard. More precisely, Aharonov and Regev [37] showed that there exists a constant c such that approximating SVP with a factor $c\sqrt{d}$ is in the l′ intersection NP∩coNP, while Goldreich and Goldwasser [38] showed that each constant c approximating SVP with a factor $c\sqrt{d/\log d}$ is in NP∩coAM.

We will present the main algorithms for solving SVP, either exactly or approximately, but we can already summarize the situation. The LLL algorithm [16] (section "The LLL Algorithm") solves ASVP with factor $(4/3 + \varepsilon)^{(d-1)/2}$, and HSVP with factor $(4/3 + \varepsilon)^{(d-1)/4}$, in time polynomial in $1/\varepsilon$ and the size of the lattice basis. This algorithm is used in the best exact-SVP algorithms:

- Kannan's deterministic algorithm [10] has super-exponential complexity $2^{O(d \log d)}$ polynomial-time operations (see [12] for a tight analysis of the constant).
- The randomized algorithm of Ajtai et al. [14] has exponential complexity $2^{O(d)}$ polynomial-time operations.

The best polynomial-time algorithms known to approximate SVP (better than LLL) are blockwise algorithms that use such exact-SVP algorithms in low dimension: indeed, in dimension d, one can use a subroutine an exact-SVP algorithm in dimension $k = f(d)$, if the function $f(d)$ is sufficiently small that the cost of the subroutine remains polynomial in d. For instance, the super-exponential running-time $2^{O(k \log k)}$ of Kannan's algorithm [10] remains polynomial in d if we select $k = \log d / \log \log d$.

With a number of calls to the SVP-oracle in dimension $\leq k$, Schnorr [17] showed one could approximate SVP with a factor $(2k)^{2d/k}$ and HSVP with a factor $(2k)^{d/k}$. Gama et al. [18] proved that Schnorr's analysis [17] was not optimal: one can raise to the power $\ln 2 \approx 0.69 < 1$ both approximation factors. Gama et al. [18] also presented a slightly better variant: it can approximate SVP with a factor $O(k)^{d/k}$ and HSVP with a factor $O(k)^{d/(2k)}$, still with a polynomial number of calls to the SVP-oracle in dimension $\leq k$. The best blockwise algorithm known is Gama–Nguyen's slide algorithm [19], which approximates SVP with a factor $((1+\varepsilon)\gamma_d)^{(d-k)/(k-1)}$ and HSVP with a factor $\sqrt{(1+\varepsilon)\gamma_d}^{(d-1)/(k-1)}$, with a polynomial (in $1/\varepsilon$ and the size of the lattice basis) number of calls to a SVP-oracle in dimension $\leq k$. When k is fixed, the approximation factors of all these blockwise algorithms remain exponential in d, like for LLL. But if one takes $k = \log d$ and use the AKS algorithm [14] as a SVP-subroutine, one obtains a randomized polynomial-time algorithm approximating SVP and HSP with slightly sub-exponential factors: $2^{O(d \log \log d / \log d)}$.

The Closest Vector Problem

The closest vector problem can be viewed as a homogeneous problem: one is looking for the radius of the smallest hyperball (centered at zero) intersecting the lattice nontrivially. One obtains a nonhomogeneous version by considering hyperballs centered at any point of the space, rather than zero. For any point $\mathbf{x}$ of $\mathbb{R}^n$, and a lattice L of $\mathbb{R}^n$, we will thus denote by $\text{dist}(\mathbf{x}, L)$ the minimal distance between $\mathbf{x}$ and a lattice vector of L. The corresponding computational problem is the following:

Problem 4 (Closest Vector Problem (CVP)). Given a basis of a d-rank integer lattice $L \subseteq \mathbb{Z}^n$, and a point $\mathbf{x} \in \mathbb{Z}^n$, find $\mathbf{y} \in L$ such that $\|\mathbf{x} - \mathbf{y}\| = \text{dist}(\mathbf{x}, L)$.

Similarly to SVP/ASVP, one can define the following approximate version:

Problem 5 (Approximate Closest Vector Problem (ACVP)). Given a basis of a d-rank integer lattice $L \subseteq \mathbb{Z}^n$, a point $\mathbf{x} \in \mathbb{Z}^n$, and an approximation factor $f \geq 1$, find $\mathbf{y} \in L$ such that $\|\mathbf{x} - \mathbf{y}\| \leq f \times \text{dist}(\mathbf{x}, L)$.

In this article, we will not further discuss CVP: we only survey SVP algorithms.

The Two-Dimensional Case

Lagrange's Reduction and Hermite's Constant in Dimension Two

Lagrange [5] formalized for the first time a reduction notion for rank-two lattices, in the language of quadratic forms. This reduction notion is so natural that all other reduction notions usually match in dimension two.

Definition 17. Let L be a two-rank lattice of $\mathbb{R}^n$. A basis $(\mathbf{b}_1, \mathbf{b}_2)$ of L is said to be *Lagrange-reduced* (or simply *L-reduced*) if and only if $\|\mathbf{b}_1\| \leq \|\mathbf{b}_2\|$ and $|\langle \mathbf{b}_1, \mathbf{b}_2 \rangle| \leq \|\mathbf{b}_1\|^2/2$.

Geometrically, this means that $\mathbf{b}_2$ is inside the disc of radius $\|\mathbf{b}_1\|$ centered at the origin, and that the angle $(\mathbf{b}_1, \mathbf{b}_2)$ modulo π is between $\pi/3$ and $2\pi/3$. Note that the second condition $|\langle \mathbf{b}_1, \mathbf{b}_2 \rangle| \leq \|\mathbf{b}_1\|^2/2$ is equivalent to size-reduction.

The definition implies that it is trivial to check whether a given basis is L-reduced or not. The following result shows that this reduction notion is optimal in a natural sense:

Theorem 7. *Let $(\mathbf{b}_1, \mathbf{b}_2)$ be a basis of a two-rank lattice L of $\mathbb{R}^n$. The basis $(\mathbf{b}_1, \mathbf{b}_2)$ is Lagrange-reduced if and only if $\|\mathbf{b}_1\| = \lambda_1(L)$ and $\|\mathbf{b}_2\| = \lambda_2(L)$.*

Assuming this result, it is clear that there always exist L-reduced bases. And by definition, the first vector of any such basis satisfies

$$\|\mathbf{b}_1\| \leq (4/3)^{1/4} \text{vol}(L)^{1/2}.$$

In particular, one can deduce the inequality $\gamma_2 \leq \sqrt{4/3}$. But one also knows that $\gamma_2 \geq \sqrt{4/3}$, by considering the hexagonal lattice spanned by $(\mathbf{b}_1, \mathbf{b}_2)$ such that $\|\mathbf{b}_1\| = \|\mathbf{b}_2\|$ and $\langle \mathbf{b}_1, \mathbf{b}_2 \rangle = \|\mathbf{b}_1\|^2/2$, which is the equality case of Lagrange's reduction.

In other words, one can arguably summarize Lagrange's reduction by a single equality

$$\gamma_2 = \sqrt{4/3}.$$

Lagrange's Algorithm

Lagrange's algorithm [5] solves the two-rank lattice reduction problem: it finds a basis achieving the first two minima, in a running time similar to Euclid's algorithm. It is often incorrectly attributed to Gauss [6]. Lagrange's algorithm can be viewed as a two-dimensional generalization of the centered variant of Euclid's algorithm (Algorithm 1).

Input: $(n, m) \in \mathbb{Z}^2$.
Output: $\gcd(n, m)$.
1:
2: **if** $|n| \leq |m|$ **then**
3:     swap n and m.
4: **end if**
5:
6: **while** $m \neq 0$ **do**
7:     $r \longleftarrow n - qm$ where $q = \lfloor \frac{n}{m} \rceil$.
8:     $n \longleftarrow m$
9:     $m \longleftarrow r$
10: **end while**
11: Output $|n|$.

Algorithm 1: Euclid's centered algorithm

This algorithm corresponds to a reduction in dimension one. Indeed, the gcd is simply the first minimum of the lattice $n\mathbb{Z} + m\mathbb{Z}$ spanned by n and m. The only difference with the classical Euclidean algorithm is in Step 7, where one takes for q the closest integer to $\frac{n}{m}$, rather than its integral part. This amounts to selecting the integer q to minimize $|n - qm|$, which guarantees $|n - qm| \leq \frac{|m|}{2}$. It is easy to show that Euclid's centered algorithm has quadratic complexity without fast integer arithmetic.

Lagrange's algorithm (Algorithm 2) is a natural generalization in dimension two.

Input: a basis $(\mathbf{u}, \mathbf{v})$ of a two-rank lattice L.
Output: an L-reduced basis of L, reaching $\lambda_1(L)$ and $\lambda_2(L)$.
if $\|\mathbf{u}\| < \|\mathbf{v}\|$ **then**
    swap $\mathbf{u}$ and $\mathbf{v}$
end if
repeat
    $\mathbf{r} \longleftarrow \mathbf{u} - q\mathbf{v}$ where $q = \left\lfloor \frac{\langle \mathbf{u}, \mathbf{v} \rangle}{\|\mathbf{v}\|^2} \right\rceil$.
    $\mathbf{u} \longleftarrow \mathbf{v}$
    $\mathbf{v} \longleftarrow \mathbf{r}$
until $\|\mathbf{u}\| \leq \|\mathbf{v}\|$
Output $(\mathbf{u}, \mathbf{v})$.

Algorithm 2: Lagrange's reduction algorithm

The analogy is clear: Step 5 selects the integer q such that $\mathbf{r} = \mathbf{u} - q\mathbf{v}$ is as short as possible. This is precisely the case when the orthogonal projection of $\mathbf{r}$ over $\mathbf{v}$ is as short as possible, and this projection can have length less than $\leq \|\mathbf{v}\|/2$. This can be viewed geometrically, and an elementary computation shows that $q = \left\lfloor \frac{\langle \mathbf{u}, \mathbf{v} \rangle}{\|\mathbf{v}\|^2} \right\rceil$ works.

One can show that Lagrange's algorithm has quadratic complexity (in the maximal bit-length of the coefficients of the input basis) without fast integer arithmetic, see [39]. For further generalizations of Lagrange's algorithm, see [39, 40].

Gram–Schmidt Orthogonalization and Size-Reduction

If $\mathbf{b}_1, \ldots, \mathbf{b}_d \in \mathbb{Z}^n$ have norms bounded by B, the computation of all Gram–Schmidt coefficients (i.e., of the rational numbers $\mu_{i,j}$ and $\|\mathbf{b}_i^\star\|^2$) can be done in time $O(d^5 \log^2 B)$ without fast arithmetic.

From the triangular representation of the basis, it is very easy to see how to size-reduce a basis (See Algorithm 3): the vectors $\mathbf{b}_i$'s are modified, but not their projections $\mathbf{b}_i^\star$.

Input: A basis $(\mathbf{b}_1, \ldots, \mathbf{b}_d)$ of a lattice L.
Output: A size-reduced basis $(\mathbf{b}_1, \ldots, \mathbf{b}_d)$.
Compute all the Gram–Schmidt coefficients $\mu_{i,j}$.
for $i = 2$ to d **do**
  for $j = i - 1$ downto 1 **do**
    $\mathbf{b}_i \longleftarrow \mathbf{b}_i - \lfloor \mu_{i,j} \rceil \mathbf{b}_j$
    for $k = 1$ to j **do**
      $\mu_{i,k} \longleftarrow \mu_{i,k} - \lfloor \mu_{i,j} \rceil \mu_{j,k}$
    end for
  end for
end for

Algorithm 3: A size-reduction algorithm

Hermite's Inequality and the Lenstra–Lenstra–Lovász Algorithm

All the algorithms of this section can be viewed as algorithmic versions of the following elementary result:

Theorem 8 (Hermite's inequality [7]). *For all integer $d \geq 2$,*

$$\gamma_d \leq \gamma_2^{d-1}. \tag{2.21}$$

Proof. We give a proof by induction, slightly different from the historical proof of Hermite. As the inequality is trivial for $d = 2$, assume that it holds for $d-1$. Consider a shortest nonzero vector $\mathbf{b}_1$ of a d-rank lattice L. Denote by $L' = \pi_2(L)$ the $(d-1)$-rank lattice obtained by projecting L over $\mathbf{b}_1^\perp$. Its volume is $\mathrm{vol}(L') = \mathrm{vol}(L)/\|\mathbf{b}_1\|$. Let $\mathbf{b}_2'$ be a shortest nonzero vector of L'. The induction assumption ensures that

$$\|\mathbf{b}_2'\| \leq (4/3)^{(d-2)/4}\mathrm{vol}(L')^{1/(d-1)}.$$

We can lift $\mathbf{b}_2'$ (by size-reduction) into a nonzero vector $\mathbf{b}_2 \in L$ such that $\|\mathbf{b}_2\|^2 \leq \|\mathbf{b}_2'\|^2 + \|\mathbf{b}_1\|^2/4$. As $\mathbf{b}_1$ cannot be longer than $\mathbf{b}_2$, we deduce

$$\|\mathbf{b}_1\| \leq \sqrt{4/3}\|\mathbf{b}_2'\| \leq (4/3)^{d/4}\mathrm{vol}(L')^{1/(d-1)},$$

which can be rewritten as

$$\|\mathbf{b}_1\| \leq (4/3)^{(d-1)/4}\mathrm{vol}(L)^{1/d},$$

which completes the proof. In retrospect, one notices that with the inequality $\|\mathbf{b}_1\| \leq \sqrt{4/3}\|\mathbf{b}_2^\star\|$, one has in fact proved the inequality

$$\gamma_d \leq (4\gamma_{d-1}/3)^{(d-1)/d}.$$

By composing all these inequalities, one indeed obtains Hermite's inequality

$$\gamma_d \leq (4/3)^{(d-1)/d+(d-2)/d+\cdots+1/d} = (4/3)^{(d-1)/2}.$$

The historical proof given by Hermite in his first letter [7] to Jacobi also proceeded by induction, but in a slightly different way. Hermite considered an arbitrary primitive vector $\mathbf{b}_1$ of the lattice L. If $\mathbf{b}_1$ satisfies Hermite's inequality, that is, if $\|\mathbf{b}_1\| \leq (4/3)^{(d-1)/4}\mathrm{vol}(L)^{1/d}$, there is nothing to prove. Otherwise, one applies the induction assumption to the projected lattice $L' = \pi_2(L)$: one knows that there exists a primitive vector $\mathbf{b}_2^\star \in L'$ satisfying Hermite's inequality: $\|\mathbf{b}_2^\star\| \leq (4/3)^{(d-2)/4}\mathrm{vol}(L')^{1/(d-1)}$. One can lift this vector $\mathbf{b}_2^\star \in L'$ into a primitive vector $\mathbf{b}_2 \in L$ such that $\|\mathbf{b}_2\|^2 \leq \|\mathbf{b}_2^\star\|^2 + \|\mathbf{b}_1\|^2/4$. As $\mathbf{b}_1$ does not satisfy Hermite's inequality, one notices that $\|\mathbf{b}_2\| < \|\mathbf{b}_1\|$: one can therefore replace $\mathbf{b}_1$ by $\mathbf{b}_2$ and start again. But this process cannot go on indefinitely: indeed, there are only finitely many vectors of L that have norm $\leq \|\mathbf{b}_1\|$. Hence, there must exist a nonzero vector $\mathbf{b}_1 \in L$ satisfying Hermite's inequality. □

The inequality (2.21) suggests to use two-dimensional reduction to find in any d-rank lattice a nonzero vector of norm less than

$$\sqrt{\gamma_2^{d-1}}\mathrm{vol}(L)^{1/d} = (4/3)^{(d-1)/4}\mathrm{vol}(L)^{1/d}.$$

This is somewhat the underlying idea behind all the algorithms of this section: Hermite's algorithms and the LLL algorithm. In fact, the proof of (2.21) that we gave provides such an algorithm, implicitly. This algorithm makes sure that the basis is size-reduced and that all the local bases $(\pi_i(\mathbf{b}_i), \pi_i(\mathbf{b}_{i+1})) = (\mathbf{b}_i^\star, \mathbf{b}_{i+1}^\star + \mu_{i+1,i}\mathbf{b}_i^\star)$ are L-reduced: these local bases correspond to the 2×2 matrices on the diagonal, when we represent the basis in triangular form. In other words, the reduced bases obtained are size-reduced and such that for all $1 \leq i \leq d$:

$$\|\mathbf{b}_{i+1}^\star\|^2 \geq \frac{3}{4}\|\mathbf{b}_i^\star\|^2, \tag{2.22}$$

that is, the decrease of the norms of the Gram–Schmidt vectors (which are the diagonal coefficients in the triangular representation) is at most geometric, which is sometimes called Siegel's condition [2]. It is then easy to see that the first vector of such a basis satisfies

$$\|\mathbf{b}_1\| \leq (4/3)^{(d-1)/4}\text{vol}(L)^{1/d},$$

as announced. But it is unknown if this algorithm and those of Hermite are polynomial time: the LLL algorithm guarantees a polynomial running-time by relaxing inequalities (2.22).

Hermite's Algorithms

We now describe the first reduction algorithms in arbitrary dimension, described by Hermite in his famous letters [7] to Jacobi, in the language of quadratic forms. They are very close to the algorithm underlying the proof of (2.21), but they do not explicitly rely on Lagrange's algorithm, although they try to generalize it. They were historically presented in a recursive way, but they can easily be made iterative, just like LLL.

Input: A basis $(\mathbf{b}_1, \ldots, \mathbf{b}_d)$ of a d-rank lattice L.
Output:

if $d = 1$ **then**
  output $\mathbf{b}_1$
end if
Apply recursively the algorithm to the basis $(\pi_2(\mathbf{b}_2), \ldots, \pi_2(\mathbf{b}_d))$ of the projected lattice $\pi_2(L)$.
Lift the vectors $(\pi_2(\mathbf{b}_2), \ldots, \pi_2(\mathbf{b}_d))$ into $\mathbf{b}_2, \ldots, \mathbf{b}_d \in L$ in such a way that they are size-reduced with respect to $\mathbf{b}_1$.
if $\mathbf{b}_1$ satisfies Hermite's inequality, that is $\|\mathbf{b}_1\| \leq (4/3)^{(d-1)/4}\text{vol}(L)^{1/d}$ **then**
  Output $(\mathbf{b}_1, \ldots, \mathbf{b}_d)$
end if
Swap $\mathbf{b}_1$ and $\mathbf{b}_2$ since $\|\mathbf{b}_2\| < \|\mathbf{b}_1\|$, and restart from the beginning.

Algorithm 4: A simplified version of Hermite's first reduction algorithm, described in the first letter to Jacobi [7]

Hermite's first algorithm was described in the first letter [7] to Jacobi: Algorithm 4 is a simplified version of this algorithm; Hermite's historical algorithm actually uses duality, which we ignore for simplicity. It is easy to see that Algorithm 4 terminates, and that the output basis $(\mathbf{b}_1, \ldots, \mathbf{b}_d)$ satisfies the following reduction notion (which we call H1):

- The basis is size-reduced.
- For all i, $\mathbf{b}_i^\star$ verifies Hermite's inequality in the projected lattice $\pi_i(L)$:

$$\|\mathbf{b}_i^\star\| \leq (4/3)^{(d-i)/4} \mathrm{vol}(\pi_i(L))^{1/(d-i+1)}$$

Notice that this reduction notion is rather weak: for instance, the orthogonality defect of a H1-reduced basis can be arbitrarily large, as soon as the dimension is greater than 3, as shown by the following triangular basis (where $\varepsilon > 0$ tends to 0):

$$\begin{pmatrix} 1 & 0 & 0 \\ 1/2 & \varepsilon & 0 \\ 1/2 & \varepsilon/2 & 1/\varepsilon \end{pmatrix}.$$

By the way, Hermite notices himself that his first algorithm does not match with Lagrange's algorithm in dimension two. It seems to be one of the reasons why he presents a second algorithm (Algorithm 5) in his second letter [7] to Jacobi.

Input: a basis $(\mathbf{b}_1, \ldots, \mathbf{b}_d)$ of a lattice L.
Output: a size-reduced basis $(\mathbf{b}_1, \ldots, \mathbf{b}_d)$ such that for all i, $\|\mathbf{b}_i^\star\|/\|\mathbf{b}_{i+1}^\star\| \leq \gamma_2 = \sqrt{4/3}$. In particular, each $\mathbf{b}_i^\star$ satisfies Hermite's inequality in the projected lattice $\pi_i(L)$.
if $d = 1$ **then**
    output $\mathbf{b}_1$
end if
By making swaps if necessary, ensure that $\|\mathbf{b}_1\| \leq \|\mathbf{b}_i\|$ for all $i \geq 2$.
Apply recursively the algorithm to the basis $(\pi_2(\mathbf{b}_2), \ldots, \pi_2(\mathbf{b}_d))$ of the projected lattice $\pi_2(L)$.
Lift the vectors $(\pi_2(\mathbf{b}_2), \ldots, \pi_2(\mathbf{b}_d))$ to $\mathbf{b}_2, \ldots, \mathbf{b}_d \in L$ in such a way that they are size-reduced with respect to $\mathbf{b}_1$.
if $\|\mathbf{b}_1\| \leq \|\mathbf{b}_i\|$ for all $i \geq 2$ **then**
    output $(\mathbf{b}_1, \ldots, \mathbf{b}_d)$
else
    restart from the beginning.
end if

Algorithm 5: Hermite's second reduction algorithm, described in his second letter to Jacobi [7]

It is easy to see that this algorithm terminates and that the output basis $(\mathbf{b}_1, \ldots, \mathbf{b}_d)$ satisfies the following reduction notion (which we call H2):

- The basis is size-reduced.
- For all i, $\mathbf{b}_i^\star$ has minimal norm among all the vectors of the basis $(\pi_i(\mathbf{b}_i), \pi_i(\mathbf{b}_{i+1}) \ldots, \pi_i(\mathbf{b}_d))$ of the projected lattice $\pi_i(L)$, that is $\|\mathbf{b}_i^\star\| \leq \|\pi_i(\mathbf{b}_j)\|$ for all $1 \leq i \leq j \leq d$.

Notice that an H2-reduced basis necessarily satisfies (2.22), that is, for all i

$$\|\mathbf{b}_i^\star\|/\|\mathbf{b}_{i+1}^\star\| \leq \gamma_2 = \sqrt{4/3}.$$

This implies that its orthogonality defect is bounded:

$$\prod_{i=1}^{d} \|\mathbf{b}_i^\star\| \leq (4/3)^{d(d-1)/4}\mathrm{vol}(\mathcal{L}(\mathbf{b}_1,\ldots,\mathbf{b}_d)).$$

And this also shows that an H2-reduced basis is necessarily H1-reduced.

Hermite's second algorithm is very close to the so-called deep insertion variant of LLL by Schnorr and Euchner [41]: both algorithms want to achieve the same reduction notion.

The LLL Algorithm

Surprisingly, it is unknown if Hermite's algorithms are polynomial time for varying dimension. It is also the case for Lenstra's algorithm [42], which is a relaxed variant of Hermite's second algorithm, where the inequalities $\|\mathbf{b}_i^\star\| \leq \|\pi_i(\mathbf{b}_j)\|$ are replaced by $c\|\mathbf{b}_i^\star\| \leq \|\pi_i(\mathbf{b}_j)\|$, where c is a constant such that $1/4 < c < 1$. However, Lenstra proved that his algorithm was polynomial time for any fixed dimension, which was sufficient for his celebrated result on integer programming [42].

It is Lenstra et al. [16] who invented in 1982 the first polynomial-time reduction algorithm outputting basis nearly as reduced as Hermite's. This algorithm, known as LLL or L^3, is essentially a relaxed variant of Hermite's second algorithm: László Lovász discovered that a crucial modification guaranteed a polynomial running-time; more precisely, compared to the H2 reduction notion, one replaces for each i all the inequalities $\|\mathbf{b}_i^\star\| \leq \|\pi_i(\mathbf{b}_j)\|$ by a single inequality $c\|\mathbf{b}_i^\star\| \leq \|\pi_i(\mathbf{b}_{i+1})\|$, where c is a constant such that $1/4 < c < 1$. The final algorithm was published in [16].

Let δ be a real in $[\frac{1}{4}, 1]$. A numbered basis $(\mathbf{b}_1,\ldots,\mathbf{b}_d)$ of L is said to be *LLL-reduced* with factor δ if it is size-reduced, and if it satisfies *Lovász' condition*: for all $1 < i \leq d$,

$$\left\|\mathbf{b}_{i+1}^\star + \mu_{i+1,i}\mathbf{b}_i^\star\right\|^2 \geq \delta\|\mathbf{b}_i^\star\|^2.$$

Let us explain this mysterious condition. As Gram–Schmidt orthogonalization depends on the order of the vectors, its vectors change if $\mathbf{b}_i$ and $\mathbf{b}_{i+1}$ are swapped; in fact, only $\mathbf{b}_i^\star$ and $\mathbf{b}_{i+1}^\star$ can possibly change. And the new $\mathbf{b}_i^\star$ is simply $\mathbf{b}_{i+1}^\star + \mu_{i+1,i}\mathbf{b}_i^\star$; therefore, Lovász' condition means that by swapping $\mathbf{b}_i$ and $\mathbf{b}_{i+1}$, the norm of $\mathbf{b}_i^\star$ does not decrease too much, where the loss is quantified by δ: one cannot gain much on $\|\mathbf{b}_i^\star\|$ by swap. In other words,

$$\delta\|\mathbf{b}_i^\star\|^2 \leq \|\pi_i(\mathbf{b}_{i+1})\|^2,$$

which illustrates the link with the H2 reduction notion. The most natural value for the constant δ is therefore $\delta = 1$ (in dimension 2, this matches with Lagrange's reduction), but then, it is unknown if such a reduced basis can be computed in polynomial time. The LLL-reduction was initially[1] presented in [16] with the factor $\delta = \frac{3}{4}$, so that in the literature, LLL-reduction usually means LLL-reduction with the factor $\delta = \frac{3}{4}$.

Lovász' condition can also be rewritten equivalently: for all i,

$$\|\mathbf{b}_{i+1}^\star\|^2 \geq \left(\delta - \mu_{i+1,i}^2\right)\|\mathbf{b}_i^\star\|^2,$$

which is a relaxation of (2.22). Thus, LLL reduction guarantees that each $\mathbf{b}_{i+1}^\star$ cannot be much shorter than $\mathbf{b}_i^\star$: the decrease is at most geometric. This proves the following result:

Theorem 9. *Assume that* $\frac{1}{4} < \delta \leq 1$, *and let* $\alpha = 1/(\delta - \frac{1}{4})$. *Let* $(\mathbf{b}_1, \ldots, \mathbf{b}_d)$ *be an LLL-reduced basis with factor* δ *of a lattice* L *in* $\mathbb{R}^n$. *Then*

1. $\|\mathbf{b}_1\| \leq \alpha^{(d-1)/4}(\mathrm{vol}L)^{1/d}$.
2. *For all* $i \in \{1, \ldots, d\}$, $\|\mathbf{b}_i\| \leq \alpha^{(d-1)/2}\lambda_i(L)$.
3. $\|\mathbf{b}_1\| \times \cdots \times \|\mathbf{b}_d\| \leq \alpha^{d(d-1)/4} \det L$.

Thus, an LLL-reduced basis provides an approximation of the lattice reduction problem. By taking δ very close to 1, one falls back on Hermite's inequality in an approximate way, where the constant $4/3$ is replaced by $4/3 + \varepsilon$.

The other interest of this reduction notion is that there exists a simple algorithm to compute such reduced bases, and which is rather close to Hermite's second algorithm (Algorithm 5). In its simplest form, the LLL algorithm corresponds to Algorithm 6.

Input: a basis $(\mathbf{b}_1, \ldots, \mathbf{b}_d)$ of a lattice L.
Output: the basis $(\mathbf{b}_1, \ldots, \mathbf{b}_d)$ is LLL-reduced with factor δ.
1: Size-reduce $(\mathbf{b}_1, \ldots, \mathbf{b}_d)$ (using Algorithm 3).
2: **if** there exists an index j which does not satisfy Lovász' condition **then**
3: swap $\mathbf{b}_j$ and $\mathbf{b}_{j+1}$, then return to Step 1.
4: **end if**

Algorithm 6: The basic LLL algorithm

Compared to this simple version, the so-called iterative versions of the LLL algorithm consider instead the smallest index j not satisfying Lovász' condition: in contrast, Hermite's second algorithm considered the greatest index j refuting H2.

[1] This simplifies the exposition.

Theorem 10. *Assume that* $\frac{1}{4} < \delta < 1$. *If each* $\mathbf{b}_i \in \mathbb{Q}^n$, *Algorithm 6 computes an LLL-reduced basis in time polynomial in the maximal bit-length of the coefficients of the* $\mathbf{b}_i$*'s, the lattice rank* d, *and the space dimension* n.

Let us sketch a proof of this fundamental result, assuming to simplify that $\mathbf{b}_i \in \mathbb{Z}^n$. First of all, it is clear that if the algorithm terminates, then the output basis is LLL-reduced with factor δ. To see why the algorithm terminates, let us analyze each swap (Step 3). When $\mathbf{b}_j$ and $\mathbf{b}_{j+1}$ are swapped, only $\mathbf{b}_j^\star$ and $\mathbf{b}_{j+1}^\star$ can be modified among all the Gram–Schmidt vectors. Let us therefore denote by $\mathbf{c}_j^\star$ and $\mathbf{c}_{j+1}^\star$ the new Gram–Schmidt vectors after swapping. As the product of all the Gram–Schmidt vector norms is equal to $\mathrm{vol}(L)$, we have

$$\|\mathbf{c}_j^\star\| \times \|\mathbf{c}_{j+1}^\star\| = \|\mathbf{b}_j^\star\| \times \|\mathbf{b}_{j+1}^\star\|.$$

As Lovász' condition is not satisfied, $\|\mathbf{c}_j^\star\|^2 < \delta\|\mathbf{b}_j^\star\|^2$. Hence,

$$\|\mathbf{c}_j^\star\|^{2(d-j+1)}\|\mathbf{c}_{j+1}^\star\|^{2(d-j)} < \delta\|\mathbf{b}_j^\star\|^{2(d-j+1)}\|\mathbf{b}_{j+1}^\star\|^{2(d-j)}.$$

This suggests to consider the following quantity:

$$D = \|\mathbf{b}_1^\star\|^{2d}\|\mathbf{b}_2^\star\|^{2(d-1)} \times \cdots \times \|\mathbf{b}_d^\star\|^2.$$

At each swap, D decreases by a factor $\delta < 1$. Notice that D can be decomposed as a product of d Gram determinants $D_i = \Delta(\mathbf{b}_1, \ldots, \mathbf{b}_i)$ for i going through 1 to d. Therefore, D is in fact an integer, as $\mathbf{b}_i \in \mathbb{Z}^n$. It follows that the number of swaps is at most logarithmic in the initial value of D, which can be upper bounded by B^{2d}, where B is the maximum of the initial norms $\|\mathbf{b}_i\|$. To bound the complexity of the algorithm, one also needs to upper bound the size of the rational coefficients $\mu_{i,j}$ and $\|\mathbf{b}_i^\star\|^2$ during the reduction. A careful analysis based on the D_i's shows that all the $\mu_{i,j}$'s always have polynomial size (see [16, 32, 43, 44]).

By coupling Theorem 9 with Theorem 10, we can summarize the LLL result as follows:

Corollary 4. *There exists an algorithm which, given as input a basis of a* d*-dimensional integer lattice* $L \subseteq \mathbb{Z}^n$ *and a reduction factor* $\varepsilon > 0$, *outputs a basis* $(\mathbf{b}_1, \ldots, \mathbf{b}_d)$ *of* L, *in time polynomial in* $1/\varepsilon$ *and the size of the basis, such that*

$$\|\mathbf{b}_1\|/\mathrm{vol}(L)^{1/d} \leq \left((1+\varepsilon)\sqrt{4/3}\right)^{(d-1)/2},$$

$$\|\mathbf{b}_i\|/\lambda_i(L) \leq \left((1+\varepsilon)\sqrt{4/3}\right)^{d-1}, \quad 1 \leq i \leq d,$$

$$\left(\prod_{i=1}^{d}\|\mathbf{b}_i\|\right)/\mathrm{vol}(L) \leq \left((1+\varepsilon)\sqrt{4/3}\right)^{d(d-1)/2}.$$

Solving Exact SVP

In this section, we survey the two main algorithms for finding the shortest vector in a lattice: enumeration [10,45,46] and sieving [14], which both use the LLL algorithm in their first stage. In section "Mordell's Inequality and Blockwise Algorithms", we use such algorithms in low dimension as subroutines to obtain polynomial-time algorithms with better approximation factors than LLL.

Enumeration Algorithms

The simplest method consists in enumerating the coordinates of a shortest lattice vector, and this idea goes back to the early 1980s with Pohst [45], Kannan [10], and Fincke-Pohst [46]. More precisely, by using LLL-reduced bases or other reduced bases not far from being orthogonal, it is possible to exhaustive search the projections of any shortest vector in the projected lattices $\pi_i(L)$.

Consider a basis $(\mathbf{b}_1, \ldots, \mathbf{b}_d)$ of a lattice L. Let $\mathbf{x} \in L$ be a (nonzero) shortest vector of L: $\mathbf{x} = x_1\mathbf{b}_1 + \cdots + x_d\mathbf{b}_d$, where the x_i's are integers. We have

$$\mathbf{x} = \sum_{i=1}^{d} x_i \mathbf{b}_i = \sum_{i=1}^{d} x_i \left(\mathbf{b}_i^\star + \sum_{j=1}^{i-1} \mu_{i,j} \mathbf{b}_j^\star \right) = \sum_{j=1}^{d} \left(x_j + \sum_{i=j+1}^{d} \mu_{i,j} x_i \right) \mathbf{b}_j^\star.$$

It follows that the projections of $\mathbf{x}$, together with their norms, are given by

$$\pi_k(\mathbf{x}) = \sum_{j=k}^{d} \left(x_j + \sum_{i=j+1}^{d} \mu_{i,j} x_i \right) \mathbf{b}_j^\star, \quad 1 \le k \le d, \tag{2.23}$$

$$\|\pi_k(\mathbf{x})\|^2 = \sum_{j=k}^{d} \left(x_j + \sum_{i=j+1}^{d} \mu_{i,j} x_i \right)^2 \|\mathbf{b}_j^\star\|^2, \quad 1 \le k \le d. \tag{2.24}$$

Now, let B be an upper bound on $\lambda_1(L) = \|\mathbf{x}\|$, we take $B = \sqrt{\gamma_d}\,\mathrm{vol}(L)^{1/d}$, but we could also have taken $B = \|\mathbf{b}_1\|$; if ever one knows a better upper bound B, which might be the case for special lattices, then this will decrease the running time of enumeration. Using (2.24), the d inequalities $\|\pi_k(\mathbf{x})\| \le B$ enable us to exhaustive search of the coordinates $x_d, x_{d-1}, \ldots, x_1$ of $\mathbf{x}$:

$$\sum_{j=k}^{d} \left(x_j + \sum_{i=j+1}^{d} \mu_{i,j} x_i \right)^2 \|\mathbf{b}_j^\star\|^2 \le B^2, \quad 1 \le k \le d,$$

which can be rewritten as

$$\left| x_k + \sum_{i=k+1}^{d} \mu_{i,j} x_i \right| \leq \frac{\sqrt{B^2 - \sum_{j=k+1}^{d} \left(x_j + \sum_{i=j+1}^{d} \mu_{i,j} x_i \right)^2 \|\mathbf{b}_j^\star\|^2}}{\|\mathbf{b}_k^\star\|},$$
$$1 \leq k \leq d. \tag{2.25}$$

We start with (2.25), with $k = d$, that is, $|x_d| \leq B/\|\mathbf{b}_d^\star\|$. This allows to exhaustive search of the integer x_d. Now assume that the projection $\pi_{k+1}(\mathbf{x})$ has been guessed for some k: the integers $x_{k+1}, \dots, x_d$ are known. Then (2.25) enables to compute an interval I_k such that $x_k \in I_k$, and therefore to exhaustive search x_k. For a full description of an exact algorithm implementing this exhaustive search, we refer to [41].

Rigorous Upper Bounds

We start with an elementary result:

Lemma 9. *Let* $(\mathbf{b}_1, \dots, \mathbf{b}_d)$ *be an LLL-reduced basis and* $B = \|\mathbf{b}_1\|$. *Then for each* $(x_{k+1}, \dots, x_d) \in \mathbb{Z}^{d-k}$, *the number of* $x_k \in \mathbb{Z}$ *satisfying* (2.25) *is at most*

$$\lfloor 2\|\mathbf{b}_1\|/\|\mathbf{b}_k^\star\| \rfloor + 1 = 2^{O(k)}.$$

This implies that if $(\mathbf{b}_1, \dots, \mathbf{b}_d)$ is an LLL-reduced basis and $B = \|\mathbf{b}_1\|$, then the cost of enumeration is, up to a polynomial-time multiplicative factor,

$$\prod_{k=1}^{d} 2^{O(k)} = 2^{O(d^2)}.$$

Kannan [10–12] showed how to decrease $2^{O(d^2)}$ to $2^{O(d \log d)}$ using a stronger reduction notion than LLL, close to HKZ-reduction. More precisely, Kannan used quasi-HKZ-reduction, which means that $(\pi_2(\mathbf{b}_2), \dots, \pi_2(\mathbf{b}_d))$ is HKZ-reduced, and that $\|\mathbf{b}_1\|$ is not much longer than $\|\mathbf{b}_2^\star\|$. And Kannan [10] noticed that by applying recursively the enumeration algorithm, one could transform an LLL-reduced basis into a quasi-HKZ-reduced basis in $2^{O(d \log d)}$ polynomial-time operations. Kannan [10]'s recursive enumeration algorithm has therefore a total complexity of $2^{O(d \log d)}$ polynomial-time operations. Recently, Hanrot and Stehlé [12, 13] showed that the worst-case complexity of Kannan's algorithm is $d^{d/(2e)+o(d)}$ polynomial-time operations.

Unfortunately, the practical interest of Kannan's algorithm is unclear. More precisely, Nguyen and Vidick [15] provides experimental evidence that for dimensions of practical interest, the $2^{O(d \log d)}$ polynomial-time operations of Kannan [10] are much slower than the $2^{O(d^2)}$ polynomial-time operations of basic enumeration from an LLL-reduced basis. This can be explained as follows: in both cases, the polynomial-time operations and the $O()$ constants are not the same.

Heuristic Estimates

The previous analysis gave only upper bounds. To provide an intuition on the exact cost of enumeration, we now give a heuristic analysis. The cost of enumeration is $\sum_{k=1}^{d} N_k$ up to a multiplicative polynomial-time factor, where N_k is the number of $(x_k, \ldots, x_d) \in \mathbb{Z}^{d-k+1}$ satisfying (2.25). Thus, N_k is exactly the number of vectors in $\pi_k(L)$ of norm $\leq B$. By the Gaussian heuristic (see Definition 8 of section "Volume and the Gaussian Heuristic"), we hope that $N_k \approx H_k$ defined by

$$H_k = \frac{B^{d-k+1} v_{d-k+1}}{\mathrm{vol}(\pi_k(L))} = \frac{B^{d-k+1} v_{d-k+1} \mathrm{vol}(\mathbf{b}_1, \ldots, \mathbf{b}_{k-1})}{\mathrm{vol}(L)} \tag{2.26}$$

Let us try to estimate (2.26) for typical reduced bases. It has been reported (see [31,47]) that for most practical reduction algorithms in high dimension, except when the lattice has a very special structure, applying the reduction algorithm to a sufficiently randomized input basis gives rise to a reduced basis such that $\|\mathbf{b}_i^\star\|/\|\mathbf{b}_{i+1}^\star\| \approx q$, where q depends on the algorithm:

- for LLL, $q \approx 1.02^2 \approx 1.04$ in high dimension.
- for BKZ-20 [41], $q \approx 1.025$.

It follows that $\|\mathbf{b}_1\| \approx q^{(d-1)/2} \mathrm{vol}(L)^{1/d}$ and

$$\frac{\mathrm{vol}(\mathbf{b}_1, \ldots, \mathbf{b}_{k-1})}{\mathrm{vol}(L)} \approx \frac{\|\mathbf{b}_1\|^{k-1}}{q^{1+2+\cdots+k-2} \mathrm{vol}(L)} = \frac{\|\mathbf{b}_1\|^{k-1}}{q^{(k-2)(k-1)/2} \mathrm{vol}(L)}.$$

Then (2.26) becomes

$$H_k \approx \frac{B^{d-k+1} v_{d-k+1} \|\mathbf{b}_1\|^{k-1}}{q^{(k-2)(k-1)/2} \mathrm{vol}(L)}. \tag{2.27}$$

The complexity will depend on the choice of the upper bound B:

- If one takes $B = \|\mathbf{b}_1\|$, then (2.27) becomes

$$H_k \approx \frac{\|\mathbf{b}_1\|^d v_{d-k+1}}{q^{(k-2)(k-1)/2} \mathrm{vol}(L)} = \frac{q^{d(d-1)/2} v_{d-k+1}}{q^{(k-2)(k-1)/2}}$$

$$= q^{[d(d-1)-(k-2)(k-1)]/2} v_{d-k+1}$$

Thus,

$$H_k \lessapprox q^{d^2/2+o(d^2)}.$$

- If one takes $B = \sqrt{\gamma_d}\mathrm{vol}(L)^{1/d}$, then $\sqrt{\gamma_d} = \Theta(\sqrt{d})$ implies that (2.27) becomes

$$H_k \approx \frac{\|\mathbf{b}_1\|^{k-1} 2^{O(d)}}{q^{(k-2)(k-1)/2}\mathrm{vol}(L)^{(k-1)/d}} = \frac{q^{(k-1)(d-1)/2} 2^{O(d)}}{q^{(k-2)(k-1)/2}}$$

$$= q^{(k-1)(d-k+1)/2} 2^{O(d)},$$

where the right-hand term is always less than $q^{d^2/8-1/2} 2^{O(d)}$, because $(k-1)(d-k+1)$ is maximized for $k = d/2$. Hence,

$$H_k \lessapprox q^{d^2/8} 2^{O(d)}.$$

In both cases, $\max_k H_k$ is super-exponential in d, but the exponentiation base ($q^{1/2}$ or $q^{1/8}$) is very close to 1.

A Heuristic Lower Bound

One might wonder if Kannan's worst-case complexity of $d^{d/(2e)+o(d)}$ polynomial-time operations can be improved using a different reduction notion. By definition of Rankin's constant, we have:

$$H_k \geq \frac{B^{d-k+1} v_{d-k+1} \sqrt{\gamma_{d,k-1}(L)}\mathrm{vol}(L)^{(k-1)/d}}{\mathrm{vol}(L)} = \frac{B^{d-k+1} v_{d-k+1} \sqrt{\gamma_{d,k-1}(L)}}{\mathrm{vol}(L)^{d-k+1}}.$$

If we take $B = \sqrt{\gamma_d}\mathrm{vol}(L)^{1/d}$, we obtain

$$H_k \geq \sqrt{\gamma_d}^{d-k+1} v_{d-k+1} \sqrt{\gamma_{d,k-1}(L)}.$$

Now recall that $\sqrt{\gamma_d} = \Theta(\sqrt{d})$, which implies that

$$H_k \geq v_{d-k+1} \Theta(\sqrt{d})^{d-k+1} \sqrt{\gamma_{d,k-1}(L)}.$$

An elementary (but tedious) computation shows that as d grows to infinity, for all $1 \leq k \leq d$,

$$v_{d-k+1} \Theta(\sqrt{d})^{d-k+1} = 2^{\Theta(d)}.$$

Hence:

$$H_k \geq 2^{\Theta(d)} \sqrt{\gamma_{d,k-1}(L)}.$$

But using (2.18) with $m = \lfloor n/2 \rceil$, we know that

$$\max_{k=2}^{d} \gamma_{d,k-1} \geq \Omega(d)^{d/4+o(d)}.$$

Therefore,

$$H_{\lfloor d/2 \rceil} \geq 2^{\Theta(d)} d^{d/8+o(d)}.$$

This suggests that, independently of the quality of the reduced basis, the complexity of enumeration will be at least $d^{d/8}$ polynomial-time operations for many lattices.

Sieve Algorithms

In 2001, Ajtai et al. [14] discovered a randomized algorithm, which is asymptotically much better than Kannan's deterministic super-exponential algorithm [10]. Indeed, the AKS algorithm outputs with overwhelming probability a shortest vector of a lattice L in $2^{O(d)}$ polynomial-time operations. Running time apart, the algorithm is interesting because it is based on totally different principle: sieving.

We just give the main idea, making significant simplifications: for more details, see [14] or [15], which presents the most practical variant known of AKS. This heuristic variant [15] has complexity $(4/3)^d$ polynomial-time operations, but the output is not guaranteed to be a shortest vector.

Consider a ball S centered at the origin and of radius r such that $\lambda_1(L) \leq r \leq O(\lambda_1(L))$. Then $|L \cap S| = 2^{O(d)}$. If we could exhaustive search $L \cap S$, we could output the shortest vector within $2^{O(d)}$ polynomial-time operations. Enumeration algorithms do perform an exhaustive search of $L \cap S$, but to do so, they also require to go through all the points of $\cup_{1 \leq k \leq d} \pi_k(L) \cap S$. Because $\sum_{k=1}^{d} |\pi_k(L) \cap S| = 2^{O(d \log d)}$ in the worst case for HKZ-reduced bases, and the worst-case complexity of Kannan's algorithm is $2^{O(d \log d)}$, rather than $2^{O(d)}$, up to some polynomial-time factor .

The main idea of sieve algorithms is to do a randomized sampling of $L \cap S$, without going through the much larger set $\cup_{1 \leq k \leq d} \pi_k(L) \cap S$. If sampling was such that each point of $L \cap S$ was output with probability roughly $|L \cap S|^{-1}$, and if $N \gg |L \cap S|$, then one of N samples would be a shortest vector with probability close to 1. Unfortunately, it is unclear if this property is satisfied by the AKS sampling. However, it can be shown that there exists $\mathbf{w} \in L \cap S$ such that both $\mathbf{w}$ and $\mathbf{w} + \mathbf{s}$, where $\mathbf{s}$ is a shortest vector, can be output with nonzero probability. Thus, by computing the shortest difference between the N sampled vectors in $L \cap S$, where $N \gg |L \cap S|$, one obtains a shortest vector of L with probability close to 1.

However, sampling directly in a ball centered at 0 and of radius r such that $\lambda_1(L) \leq r \leq O(\lambda_1(L))$ is difficult. But, starting with an LLL-reduced basis, it is easy to sample with a radius $2^{O(d)}\lambda_1(L)$. To decrease the factor $2^{O(d)}$ to $O(1)$, one uses a sieve, which is the most expensive stage of the algorithm.

Sieving iteratively shortens the vectors of S by a geometric factor of at least γ (such that $0 < \gamma < 1$) at each iteration; thus, a linear number of sieve iterations suffices to decrease the multiplicative factor $2^{O(d)}$ to $O(1)$. At each iteration, each vector output by the sieve is a subtraction of two input vectors. In other words, the sieve will select a subset C of the initial set S, and the output set will be obtained by subtracting a vector of C to each vector of $S \setminus C$. By volume arguments, one can choose a set C, which is never too large, so that the number of samples does not decrease too much. Intuitively, one uses the fact that for any $0 < \gamma < 1$, a ball of radius R can be recovered by at most an exponential number of balls of radius γR.

We just described the principles of the AKS algorithm [14], but the proved algorithm is a bit more complex, and its analysis is nontrivial.

HKZ Reduction

It is easy to see that any exact SVP algorithm allows to find an HKZ-reduced basis, within the same asymptotic running time, by calling the algorithm a linear number of times. For instance, one can do as follows:

- Call the SVP algorithm on L to obtain a shortest vector $\mathbf{b}_1$ of the lattice L.
- Extend $\mathbf{b}_1$ into a basis $(\mathbf{b}_1, \mathbf{c}_2, \ldots, \mathbf{c}_d)$ of L and compute a basis of the projected lattice $\pi_2(L)$.
- Call the SVP algorithm on $\pi_2(L)$ to obtain a shortest vector $\mathbf{b}'_2$ of the projected lattice $\pi_2(L)$.
- Lift $\mathbf{b}'_2$ into a vector $\mathbf{b}_2$ of L by adding an appropriate multiple of $\mathbf{b}_1$ so that $(\mathbf{b}_1, \mathbf{b}_2)$ is size-reduced.
- Extend $(\mathbf{b}_1, \mathbf{b}_2)$ into a basis $(\mathbf{b}_1, , \mathbf{b}_2, \mathbf{c}_3, \ldots, \mathbf{c}_d)$ of L and use this basis to compute a basis of the projected lattice $\pi_3(L)$. And so on.

Mordell's Inequality and Blockwise Algorithms

We saw in section "Hermite's Inequality and the Lenstra–Lenstra–Lovász Algorithm" the LLL algorithm [16] (see Corollary 4): given a basis of an d-dimensional integer lattice $L \subseteq \mathbb{Z}^n$ and a reduction factor $\varepsilon > 0$, LLL outputs (in time polynomial in $1/\varepsilon$ and the size of the basis) a reduced basis $(\mathbf{b}_1, \ldots, \mathbf{b}_d)$ whose first vector is provably short, namely,

$$\|\mathbf{b}_1\|/\mathrm{vol}(L)^{1/d} \leq \left((1+\varepsilon)\sqrt{4/3}\right)^{(d-1)/2}, \tag{2.28}$$

$$\|\mathbf{b}_1\|/\lambda_1(L) \leq \left((1+\varepsilon)\sqrt{4/3}\right)^{d-1}. \tag{2.29}$$

We noted that the first inequality (2.28) was reminiscent of Hermite's inequality [7] on γ_d:

$$\gamma_d \leq \left(\sqrt{4/3}\right)^{d-1} = \gamma_2^{d-1}, \quad \text{(Hermite's inequality)} \tag{2.30}$$

which means that L has a nonzero vector of norm $\leq (\sqrt{4/3})^{(d-1)/2}\mathrm{vol}(L)^{1/d}$. Thus, we viewed LLL as an algorithmic version of Hermite's inequality (2.21), and this connection was strengthened by the fact that LLL is a variant of an algorithm introduced by Hermite [7] to prove (2.21), based on Lagrange's two-dimensional algorithm [5].

The second inequality (2.29) means that LLL approximates the shortest vector problem (SVP) within an exponential factor. On the other hand, we saw in section "Solving Exact SVP" the best algorithms for exact-SVP, which are exponential: Kannan's deterministic algorithm [10] requires $2^{O(d\log d)}$ polynomial-time operations, and the AKS probabilistic algorithm [14] requires $2^{O(d)}$ polynomial-time operations.

A natural question is whether the upper bounds of (2.28) or (2.29) can be decreased in polynomial time. The only polynomial-time algorithms achieving better inequalities than (2.28) or (2.29) are blockwise generalizations of LLL: Schnorr's algorithm [17], the transference algorithm by Gama et al. [18], and Gama–Nguyen's slide algorithm [19], the latter one offering better theoretical guarantees than the first two. Blockwise algorithms rely on a SVP-subroutine [10, 14] (see section "Solving Exact SVP") computing shortest vectors in smaller lattices of dimension $\leq k$, where k is an additional input parameter referred to as the blocksize. Note that the exponential cost of the SVP-subroutine can be kept polynomial in the size of the basis if the blocksize k is sufficiently small: namely, $k = O(\log d)$ (resp. $k = O(\log d/\log\log d)$) suffices with AKS [14] (respectively [10]) as the SVP subroutine. As the cost of the SVP-subroutine is exponential in the blocksize, it is important to use the SVP-subroutine as efficiently as possible for a given output quality.

In this section, we will describe Gama–Nguyen's slide algorithm [19], which improves [17, 18], and is simpler in several respects. For instance, it might be argued that the inequalities achieved by [17, 18] are not very natural: more precisely, in Schnorr's algorithm [17], k must be even, d must be a multiple of $k/2$, and the upper bound of (2.29) is replaced by $\sqrt{2\gamma_{k/2}\alpha_{k/2}}((1+\varepsilon)\beta_{k/2})^{d/k-1}$, where $\alpha_{k/2}$ and $\beta_{k/2}$ are technical constants bounded in [13, 17, 18]; and in the GHKN algorithm [18], the upper bound of (2.28) is replaced by $\gamma_{k-1}^{(d+k-1)/(4(k-1))}((1+\varepsilon)\gamma_k)^{k(d-k+1)/(4(k-1)^2)}$, while the upper bound of (2.29) is replaced by the square of the previous expression. The new algorithm [19] is a blockwise algorithm achieving better and more "natural" upper bounds, corresponding to the following classical generalization of Hermite's inequality (2.21), known as Mordell's inequality [20, 48]:

Theorem 11 (Mordell's Inequality [48]). *For all integers d and k such that $2 \leq k \leq d$:*

$$\gamma_d \leq \gamma_k^{(d-1)/(k-1)} \tag{2.31}$$

This implies that any d-rank lattice L has a nonzero vector of norm:

$$\leq \sqrt{\gamma_k}^{(d-1)/(k-1)} \mathrm{vol}(L)^{1/d}.$$

By analogy with the LLL case, Mordell's inequality (2.31) suggests that there might exist a blockwise reduction algorithm calling polynomially many times a SVP-subroutine in dimension $\leq k$, and which outputs a basis whose first vector $\mathbf{b}_1 \in L$ would satisfy

$$\|\mathbf{b}_1\|/\mathrm{vol}(L)^{1/d} \leq \sqrt{(1+\varepsilon)\gamma_k}^{(d-1)/(k-1)} \tag{2.32}$$

Such an algorithm would be a polynomial-time version of Mordell's inequality, just as LLL is a polynomial-time version of Hermite's inequality. And an old result of Lovász [32] shows that by calling d times such an algorithm, we would also obtain a nonzero lattice vector $\mathbf{b}_1 \in L$ satisfying

$$\|\mathbf{b}_1\|/\lambda_1(L) \leq ((1+\varepsilon)\gamma_k)^{(d-1)/(k-1)} \tag{2.33}$$

Note that (2.28) and (2.29) are exactly the $k = 2$ case of (2.32) and (2.33). Unfortunately, the classical proof [20] of Mordell's inequality (2.31) does not give such an algorithm. And the blockwise algorithms [17, 18] turn out to be loose algorithmic versions of Mordell's inequality: for any k, the best upper bounds known on $\|\mathbf{b}_1\|$ for [17, 18] are worse than (2.32) and (2.33). For instance, the best upper bound known on $\|\mathbf{b}_1\|/\lambda_1(L)$ for Schnorr's algorithm is essentially $\left((1+\varepsilon)(k/2)^{2\ln 2}\right)^{d/k-1}$.

Slide reduction [19] is an algorithmic version of Mordell's inequality in the following sense: given a basis of an d-dimensional integer lattice $L \subseteq \mathbb{Z}^n$, a blocksize k dividing d, a reduction factor $\varepsilon > 0$, and a SVP-subroutine computing shortest vectors in any lattice of dimension $\leq k$, slide reduction outputs (in time polynomial in the size of the basis and $1/\varepsilon$) a basis whose first vector $\mathbf{b}_1$ satisfies (2.32) and the following inequality:

$$\|\mathbf{b}_1\|/\lambda_1(L) \leq ((1+\varepsilon)\gamma_k)^{(d-k)/(k-1)}, \tag{2.34}$$

and the number of calls to the SVP-subroutine is polynomial in the size of the basis and $1/\varepsilon$. Surprisingly, (2.34) is slightly better than the speculated inequality (2.33), by a multiplicative factor close to γ_k. Hence, slide reduction is theoretically better than Schnorr's algorithm [17] and Gama *et al.*'s transference algorithm [18] for any fixed k, but does not improve the asymptotical sub-exponential approximation factor when $k = O(\log d)$.

Like all known proofs of Mordell's inequality, slide reduction is based on duality. Furthermore, it was proved in [19] that in the worst case, (2.32) and (2.34) are essentially tight: namely, there exist slide reduced bases such that these upper bounds become lower bounds if we replace γ_k by a slightly smaller linear function of k, namely $\gamma_k/2$ or even $(1-\varepsilon')k/(2\pi e)$ for all $\varepsilon' > 0$. Ajtai proved [49] an analogue result for Schnorr's algorithm [17], without effective constants.

Classical Proofs of Mordell's Inequality

We give here the classical argument showing Mordell's inequality (2.31), such as the one given in [20, Theorem 2.3.1]: this argument can actually be found earlier than Mordell's article [48], for instance when Korkine and Zolotarev [28] determined the value of γ_4 by showing first that $\gamma_4 \leq \gamma_3^{3/2}$, and also somewhat implicitly in Hermite's first letter [7].

We first notice that it suffices to show the inequality for $k = d - 1$: indeed, if (2.31) holds for $k = d - 1$, then by applying recursively the inequality, we obtain (2.31) for all k. In fact, Mordell's inequality is equivalent to showing that the sequence $(\gamma_d^{1/(d-1)})_{d \geq 2}$ decreases.

Let L be a d-rank lattice. Let $\mathbf{x}$ be a shortest nonzero vector of the dual lattice $L^\times$ and let H be the hyperplane $\mathbf{x}^\perp$. Denote by M the $(d-1)$-rank lattice $L \cap H$. Then $\mathrm{vol}(M) = \mathrm{vol}(L)\|\mathbf{x}\|$ and $\|\mathbf{x}\| \leq \sqrt{\gamma_d}\mathrm{vol}(L^\times)^{1/d} = \sqrt{\gamma_d}\mathrm{vol}(L)^{-1/d}$; therefore,

$$\mathrm{vol}(M) \leq \sqrt{\gamma_d}\mathrm{vol}(L)^{1-1/d}.$$

In particular,

$$\lambda_1(M) \leq \sqrt{\gamma_{d-1}}\left(\sqrt{\gamma_d}\mathrm{vol}(L)^{1-1/d}\right)^{1/(d-1)} = \sqrt{\gamma_{d-1}}\sqrt{\gamma_d}^{1/(d-1)}\mathrm{vol}(L)^{1/d}.$$

Furthermore, we have $\lambda_1(L) \leq \lambda_1(M)$. Hence, by definition of γ_d,

$$\sqrt{\gamma_d} \leq \sqrt{\gamma_{d-1}}\sqrt{\gamma_d}^{1/(d-1)}.$$

The proof of (2.31) is now over, since we can rewrite the previous inequality as

$$\gamma_d \leq \gamma_{d-1}^{(d-1)/(d-2)}.$$

This classical proof of Mordell's inequality cannot be directly translated into a recursive algorithm: indeed, it considers shortest vectors in the $(d-1)$-rank lattice M, and also in the d-rank lattice $L^\times$. In the next subsection, we slightly modify the argument so that only $(d-1)$-rank lattices are considered, which naturally gives rise to algorithms.

Mordell's Inequality by Reduction

We introduce the following reduction notion, which we dub Mordell's reduction because it is inspired by Mordell's inequality or rather its proof:

Definition 18. Let $d \geq 2$. A basis $(\mathbf{b}_1, \ldots, \mathbf{b}_d)$ of a lattice L is *Mordell-reduced* with factor $\varepsilon \geq 0$ if and only if the following two conditions hold:

$$\|\mathbf{b}_1\| = \lambda_1(\mathcal{L}(\mathbf{b}_1, \ldots, \mathbf{b}_{d-1})) \tag{2.35}$$

and

$$1/\|\mathbf{b}_d^\star\| \leq (1+\varepsilon)\lambda_1(\pi_2(L)^\times), \tag{2.36}$$

where $\pi_2(L)$ denotes the orthogonal projection of L over the hyperplane $\mathbf{b}_1^\perp$, and $\mathbf{b}_d^\star$ denotes as usual the component of $\mathbf{b}_d$ which is orthogonal to the hyperplane spanned by $\mathbf{b}_1, \ldots, \mathbf{b}_{d-1}$.

The inequality (2.36) is motivated by the fact that $\mathbf{b}_d^\star / \|\mathbf{b}_d^\star\|^2 \in \pi_2(L)^\times$ (which we previously mentioned at the end of section "Gram–Schmidt and Triangularization" giving a link between duality and Gram–Schmidt orthogonalization), because the vector is orthogonal with $\mathbf{b}_1, \ldots, \mathbf{b}_{d-1}$, and its dot product with $\mathbf{b}_d$ is equal to 1.

Note that there always exist Mordell-reduced bases for all $\varepsilon \geq 0$. Indeed, consider an HKZ-reduced basis $(\mathbf{b}_1, \ldots, \mathbf{b}_d)$ of L. Then (2.35) holds. Next, consider a shortest vector $\mathbf{c}$ in $\pi_2(L)^\times$ and modify $\mathbf{b}_2, \ldots, \mathbf{b}_d$ in such a way that $\mathbf{b}_d^\star / \|\mathbf{b}_d^\star\| = \mathbf{c}$ and $(\mathbf{b}_1, \ldots, \mathbf{b}_d)$ remains a basis of L: then both (2.36) and (2.35) hold.

Mordell's reduction has the following properties:

Lemma 10. *Let* $(\mathbf{b}_1, \ldots, \mathbf{b}_d)$ *be a Mordell-reduced basis of* L *with factor* $\varepsilon \geq 0$ *and* $d \geq 3$. *Then*

1. *Primal inequality:*

$$\|\mathbf{b}_1\| \leq \sqrt{\gamma_{d-1}}^{(d-1)/(d-2)} \left(\prod_{i=2}^{d-1} \|\mathbf{b}_i^\star\| \right)^{1/(d-2)}. \tag{2.37}$$

2. *Dual inequality:*

$$\left(\prod_{i=2}^{d-1} \|\mathbf{b}_i^\star\| \right)^{1/(d-2)} \leq \left((1+\varepsilon)\sqrt{\gamma_{d-1}}\right)^{(d-1)/(d-2)} \|\mathbf{b}_d^\star\|. \tag{2.38}$$

3. *Primal–dual inequality:*

$$\|\mathbf{b}_1^\star\| / \|\mathbf{b}_d^\star\| \leq ((1+\varepsilon)\gamma_{d-1})^{(d-1)/(d-2)}. \tag{2.39}$$

4. *Relaxed Mordell's inequality:*

$$\|\mathbf{b}_1\| \leq \left((1+\varepsilon)^{1/d}\sqrt{\gamma_{d-1}}\right)^{(d-1)/(d-2)} \mathrm{vol}(L)^{1/d}. \tag{2.40}$$

Proof. Equation (2.37) follows from $\|\mathbf{b}_1\| = \lambda_1(\mathcal{L}(\mathbf{b}_1, \ldots, \mathbf{b}_d))$ and the definition of γ_d. Indeed, we have

$$\|\mathbf{b}_1\| \leq \sqrt{\gamma_{d-1}} \left(\prod_{i=1}^{d-1} \|\mathbf{b}_i^\star\| \right)^{1/(d-1)} .$$

Therefore,

$$\|\mathbf{b}_1\|^{d-1} \leq \sqrt{\gamma_{d-1}}^{d-1} \prod_{i=1}^{d-1} \|\mathbf{b}_i^\star\|,$$

which can be rewritten as (2.37). Similarly, $1/\|\mathbf{b}_d^\star\| \leq (1+\varepsilon)\lambda_1(\pi_2(L)^\times)$ implies that

$$1/\|\mathbf{b}_d^\star\| \leq (1+\varepsilon)\sqrt{\gamma_{d-1}} \left(\prod_{i=2}^{d} 1/\|\mathbf{b}_i^\star\| \right)^{1/(d-1)} ;$$

therefore,

$$\prod_{i=2}^{d} \|\mathbf{b}_i^\star\| \leq \left((1+\varepsilon)\sqrt{\gamma_{d-1}}/\|\mathbf{b}_d^\star\|\right)^{d-1} ,$$

which implies (2.38). And (2.39) follows from multiplying (2.37) and (2.38). Furthermore, we have

$$\begin{aligned}
\mathrm{vol}(L) &= \prod_{i=1}^{d} \|\mathbf{b}_i^\star\| \\
&= \|\mathbf{b}_d^\star\| \times \|\mathbf{b}_1^\star\| \times \prod_{i=2}^{d-1} \|\mathbf{b}_i^\star\| \\
&\geq \frac{\left(\prod_{i=2}^{d-1} \|\mathbf{b}_i^\star\|\right)^{1/(d-2)}}{((1+\varepsilon)\sqrt{\gamma_{d-1}})^{(d-1)/(d-2)}} \times \|\mathbf{b}_1^\star\| \times \prod_{i=2}^{d-1} \|\mathbf{b}_i^\star\| \quad \text{by (2.38)} \\
&= \frac{\|\mathbf{b}_1^\star\|}{((1+\varepsilon)\sqrt{\gamma_{d-1}})^{(d-1)/(d-2)}} \times \left(\prod_{i=2}^{d-1} \|\mathbf{b}_i^\star\| \right)^{1+1/(d-2)} \\
&\geq \frac{\|\mathbf{b}_1^\star\|}{((1+\varepsilon)\sqrt{\gamma_{d-1}})^{(d-1)/(d-2)}} \times \left(\frac{\|\mathbf{b}_1^\star\|}{\sqrt{\gamma_{d-1}}^{(d-1)/(d-2)}} \right)^{(d-2)+1} \quad \text{by (2.37)} \\
&= \frac{\|\mathbf{b}_1^\star\|^d}{(1+\varepsilon)^{(d-1)/(d-2)} \sqrt{\gamma_{d-1}}^{(1+(d-2)+1)(d-1)/(d-2)}} \\
&= \frac{\|\mathbf{b}_1^\star\|^d}{(1+\varepsilon)^{(d-1)/(d-2)} \sqrt{\gamma_{d-1}}^{d(d-1)/(d-2)}},
\end{aligned}$$

which proves (2.40). □

Theorem 12. *Let $k \geq 2$. Let $(\mathbf{b}_1, \ldots, \mathbf{b}_{2k})$ be a basis of a lattice L such that $(\mathbf{b}_1, \ldots, \mathbf{b}_{k+1})$ is Mordell-reduced and $\mathbf{b}_{k+1}^\star$ is a shortest vector in the projected* *lattice* $\pi_{k+1}(L)$. *Then*

$$\frac{\prod_{i=1}^{k} \|\mathbf{b}_i^\star\|}{\prod_{i=k+1}^{2k} \|\mathbf{b}_i^\star\|} \leq ((1+\varepsilon)\gamma_k)^{k^2/(k-1)}. \tag{2.41}$$

Proof. As $\mathbf{b}_{k+1}^\star$ is a shortest vector of the projected lattice $\pi_{k+1}(L)$, we can apply (2.37) to obtain

$$\|\mathbf{b}_{k+1}^\star\| \leq \sqrt{\gamma_k}^{k/(k-1)} \left(\prod_{i=k+2}^{2k} \|\mathbf{b}_i^\star\| \right)^{1/(k-1)};$$

therefore, we can lower bound the denominator of (2.41) as

$$\prod_{i=k+1}^{2k} \|\mathbf{b}_i^\star\| \geq \|\mathbf{b}_{k+1}^\star\| \times \left(\frac{\|\mathbf{b}_{k+1}^\star\|}{\sqrt{\gamma_k}^{k/(k-1)}} \right)^{k-1} = \|\mathbf{b}_{k+1}^\star\|^k / \sqrt{\gamma_k}^k. \tag{2.42}$$

On the other hand, $(\mathbf{b}_1, \ldots, \mathbf{b}_{k+1})$ is Mordell-reduced, so (2.38) implies that

$$\prod_{i=2}^{k} \|\mathbf{b}_i^\star\| \leq \left((1+\varepsilon)\sqrt{\gamma_k}\right)^k \|\mathbf{b}_{k+1}^\star\|^{k-1},$$

and (2.39) implies that

$$\|\mathbf{b}_1^\star\| \leq ((1+\varepsilon)\gamma_k)^{k/(k-1)} \times \|\mathbf{b}_{k+1}^\star\|.$$

By multiplying the previous two inequalities, we can upper bound the numerator of (2.41) as

$$\prod_{i=1}^{k} \|\mathbf{b}_i^\star\| \leq \|\mathbf{b}_{k+1}^\star\|^k \times ((1+\varepsilon)\gamma_k)^{k/(k-1)} \times \left((1+\varepsilon)\sqrt{\gamma_k}\right)^k. \tag{2.43}$$

Hence, (2.43) and (2.42) imply that

$$\begin{aligned} \frac{\prod_{i=1}^{k} \|\mathbf{b}_i^\star\|}{\prod_{i=k+1}^{2k} \|\mathbf{b}_i^\star\|} &\leq ((1+\varepsilon)\gamma_k)^{k/(k-1)} \times \left((1+\varepsilon)\sqrt{\gamma_k}\right)^k \times \sqrt{\gamma_k}^k \\ &= ((1+\varepsilon)\gamma_k)^{k+k/(k-1)} \\ &= ((1+\varepsilon)\gamma_k)^{k^2/(k-1)}, \end{aligned}$$

which proves (2.41). □

We later show (and it is not difficult to see) that there exist bases satisfying the assumptions of Theorem 12 for any $\varepsilon \geq 0$: by taking $\varepsilon = 0$, this proves that for all $k \geq 2$

$$\gamma_{2k,k} \leq \gamma_k^{k^2/(k-1)}.$$

Blockwise Reduction

For any basis $B = [\mathbf{b}_1, \ldots, \mathbf{b}_d]$, we use the notation $B_{[i,j]}$ for the projected block $[\pi_i(\mathbf{b}_i), \ldots, \pi_i(\mathbf{b}_j)]$, where π_i is the orthogonal projection over $\mathrm{span}(\mathbf{b}_1, \ldots, \mathbf{b}_{i-1})^\perp$. When looking at the lower-triangular representation of B, $B_{[i,j]}$ corresponds to the (lower-triangular) submatrix of the lower-triangular matrix within row i to row j. Note that $B_{[i,j]}$ always represents a linearly independent family of $j - i + 1$ vectors, whose first vector is $\mathbf{b}_i^\star$. For example, $B_{[i,i]} = [\mathbf{b}_i^\star]$ and $B_{[1,i]} = [\mathbf{b}_1, \ldots, \mathbf{b}_i]$ for all $i \in [1, d]$. If B has integer coefficients, then $B_{[i,j]}$ has rational coefficients if $i > 1$ and integer coefficients if $i = 1$. As an important particular case, if T is a lower triangular matrix (such as the μ matrix of the Gram–Schmidt orthogonalization), then $T_{[i,j]}$ is simply the inner triangular matrix within the indices $[i, j]$.

In the LLL algorithm, vectors are considered two by two. At each loop iteration, the two-dimensional lattice $L_i = [\pi_i(\mathbf{b}_i), \pi_i(\mathbf{b}_{i+1})]$ is partially reduced (through a swap) to decrease $\|\mathbf{b}_i^\star\|$ by at least some geometric factor. When all such lattices are almost reduced, every ratio $\|\mathbf{b}_i^\star\|/\|\mathbf{b}_{i+1}^\star\|$ is roughly less than $\gamma_2 = \sqrt{\frac{4}{3}}$.

In blockwise generalizations of LLL, we select an integer $k \geq 2$ dividing d, called the blocksize. Then, the vectors $\mathbf{b}_i^\star$ are "replaced" by k-dimensional blocks $S_i = B_{[ik-k+1,ik]}$, where $1 \leq i \leq \frac{d}{k}$. The analogue of the two-dimensional L_i in LLL are the $2k$-dimensional large blocks $L_i = B_{[ik-k+1,ik+k]}$, where $1 \leq i \leq \frac{d}{k} - 1$. The link between the small blocks $S_1, \ldots, S_{d/k}$ and the large blocks $L_1, \ldots, L_{d/k-1}$ is that S_i consists of the first k vectors of L_i, while S_{i+1} is the projection of the last k vectors of L_i over $\mathrm{span}(S_i)^\perp$. As a result, $\mathrm{vol}(L_i) = \mathrm{vol}(S_i) \times \mathrm{vol}(S_{i+1})$. By analogy with LLL, the blockwise algorithm will perform operations on each large block L_i so that $\mathrm{vol}(S_i)/\mathrm{vol}(S_{i+1})$ can be upper bounded.

Gama and Nguyen [19] introduced the following blockwise version of Mordell's reduction (in fact, the reduction in [19] is a bit stronger, but the difference is minor and not relevant):

Definition 19. Let $d \geq 2$ and $k \geq 2$ dividing d. A basis $(\mathbf{b}_1, \ldots, \mathbf{b}_d)$ of a lattice L is *block-Mordell-reduced* with factor $\varepsilon \geq 0$ and blocksize k if and only if it is size-reduced and the following two conditions hold:

- For each $i \in \{1, \ldots, d/k - 1\}$, the block $B_{[ik-k+1,ik+1]}$ is Mordell-reduced.
- We have, $\|\mathbf{b}_{d-k+1}^\star\| = \lambda_1(\mathcal{L}(B_{[d-k+1,d]}))$.

This is equivalent to asking that the basis is size-reduced and the following two conditions hold:

1. *Primal conditions*: for each $j \in \{1,\ldots,d\}$ such that $j \equiv 1 \pmod{k}$,

$$\|\mathbf{b}_j^\star\| = \lambda_1(\mathcal{L}(B_{[j,j+k-1]})). \tag{2.44}$$

   Note that $B_{[j,j+k-1]}$ is one of the small blocks S_i, namely $S_{1+(j-1)/k}$.
2. *Dual conditions*: for each $j \in \{1,\ldots,d-k\}$ such that $j \equiv 1 \pmod{k}$,

$$1/\|\mathbf{b}_{j+k}^\star\| \le (1+\varepsilon)\lambda_1(\mathcal{L}(B_{[j+1,j+k]})^\times). \tag{2.45}$$

   Note that $B_{[j+1,j+k]}$ is not one of the small blocks S_i, because there is a shift of index: the block starts at index $j+1$ rather than j.

Let us explain the intuition behind block-Mordell reduction. Conditions (2.44) and (2.45) imply that each vector $\mathbf{b}_j^\star$ such that $j \in \{k,\ldots,d\}$ and $j \equiv 1 \pmod{k}$ is neither too large, nor too short:

- Not too large because $\|\mathbf{b}_j^\star\| = \lambda_1(\mathcal{L}(B_{[j,j+k-1]}))$;
- Not too short because $1/\|\mathbf{b}_j^\star\| \le (1+\varepsilon)\lambda_1(\mathcal{L}(B_{[j-k+1,j]})^\times)$.

These conditions are inspired by the fact that $\mathbf{b}_j^\star$ is connected to two natural k-rank lattices:

- $\mathbf{b}_j^\star$ belongs to the projected lattice $\mathcal{L}(B_{[j,j+k-1]})$: it is in fact the first vector of $B_{[j,j+k-1]}$.
- $\mathbf{b}_j^\star/\|\mathbf{b}_j^\star\|^2$ belongs to the dual-projected lattice $\mathcal{L}(B_{[j-k+1,j]})^\times$: see the end of section "Gram–Schmidt and Triangularization" for links between duality and Gram–Schmidt orthogonalization.

We now give elementary properties of block-Mordell-reduced bases, which follow from Mordell reduction:

Lemma 11. *Let $(\mathbf{b}_1,\ldots,\mathbf{b}_d)$ be a block-Mordell-reduced basis of a lattice L with factor $\varepsilon \ge 0$ and blocksize $k \ge 2$ dividing d. Then,*

1. *Primal inequality: for each $j \in \{1,\ldots,d\}$ such that $j \equiv 1 \pmod{k}$,*

$$\|\mathbf{b}_j^\star\| \le \sqrt{\gamma_k}^{k/(k-1)} \left(\prod_{i=j+1}^{j+k-1} \|\mathbf{b}_i^\star\|\right)^{1/(k-1)}. \tag{2.46}$$

2. *Dual inequality: for each $j \in \{1,\ldots,d-k\}$ such that $j \equiv 1 \pmod{k}$,*

$$\left(\prod_{i=j+1}^{j+k-1} \|\mathbf{b}_i^\star\|\right)^{1/(k-1)} \le \left((1+\varepsilon)\sqrt{\gamma_k}\right)^{k/(k-1)} \|\mathbf{b}_{j+k}^\star\|. \tag{2.47}$$

3. *Primal-dual inequality: for each $j \in \{1,\ldots,d-k\}$ such that $j \equiv 1 \pmod{k}$,*

$$\|\mathbf{b}_j^\star\|/\|\mathbf{b}_{j+k}^\star\| \le ((1+\varepsilon)\gamma_k)^{k/(k-1)}. \tag{2.48}$$

4. *Half-volume inequality: for each* $j \in \{1, \ldots, d-k\}$ *such that* $j \equiv 1 \pmod{k}$,

$$\frac{\prod_{i=j}^{j+k-1} \|\mathbf{b}_i^\star\|}{\prod_{i=j+k}^{j+2k-1} \|\mathbf{b}_i^\star\|} \leq ((1+\varepsilon)\gamma_k)^{k^2/(k-1)}. \tag{2.49}$$

Proof. Equation (2.46) follows from (2.37), (2.47) follows from (2.38), (2.48) follows from (2.39), and (2.49) follows from (2.41). □

Theorem 13. *Let* $(\mathbf{b}_1, \ldots, \mathbf{b}_d)$ *be a block-Mordell-reduced basis of a lattice* L *with factor* $\varepsilon \geq 0$ *and blocksize* $k \geq 2$ *dividing* d. *Then,*

$$\|\mathbf{b}_1\|/\mathrm{vol}(L)^{1/d} \leq \sqrt{\gamma_k}^{(d-1)/(k-1)} \times \sqrt{1+\varepsilon}^{(d-k)/(k-1)}. \tag{2.50}$$

$$\|\mathbf{b}_1\|/\lambda_1(L) \leq ((1+\varepsilon)\gamma_k)^{(d-k)/(k-1)}, \tag{2.51}$$

Proof. We have,

$$\mathrm{vol}(L) = \prod_{i=1}^{d/k} \mathrm{vol}(S_i),$$

where, by (2.49), for each $i \in \{1, \ldots, d/k-1\}$:

$$\mathrm{vol}(S_i)/\mathrm{vol}(S_{i+1}) \leq ((1+\varepsilon)\gamma_k)^{k^2/(k-1)}.$$

This implies that, similar to LLL,

$$\mathrm{vol}(S_1) \leq ((1+\varepsilon)\gamma_k)^{k^2/(k-1)\times(d/k-1)/2}\,\mathrm{vol}(L)^{1/(d/k)}.$$

And (2.44) implies that $\|\mathbf{b}_1^\star\| = \lambda_1(\mathcal{L}(B_{[1,k]})) = \lambda_1(S_1)$; therefore,

$$\begin{aligned}
\|\mathbf{b}_1^\star\| &\leq \sqrt{\gamma_k}\,\mathrm{vol}(S_1)^{1/k} \\
&\leq \sqrt{\gamma_k}\,((1+\varepsilon)\gamma_k)^{k/(k-1)\times(d/k-1)/2}\,\mathrm{vol}(L)^{1/d} \\
&= \sqrt{\gamma_k}^{1+(d-k)/(k-1)}(1+\varepsilon)^{(d-k)/(2(k-1))}\mathrm{vol}(L)^{1/d} \\
&= \sqrt{\gamma_k}^{(d-1)/(k-1)}(1+\varepsilon)^{(d-k)/(2(k-1))}\mathrm{vol}(L)^{1/d},
\end{aligned}$$

which implies (2.50). Now, consider a shortest vector $\mathbf{u}$ of L. Then $\|\mathbf{u}\| = \lambda_1(L)$ and $\mathbf{u}$ can be written as $\mathbf{u} = \sum_{i=1}^m \alpha_i \mathbf{b}_i$, where each $\alpha_i \in \mathbb{Z}$ and $\alpha_m \neq 0$. If we let $q = \lfloor (m-1)/k \rfloor$, then $\pi_{qk+1}(\mathbf{u})$ is a nonzero vector of $L(B_{[qk+1,qk+k]})$. But by definition of block-Mordell reduction, $\mathbf{b}_{qk+1}^\star$ is a shortest vector of $L(B_{[qk+1,qk+k]})$; therefore,

$$\|\mathbf{b}_{qk+1}^\star\| \leq \|\pi_{qk+1}(\mathbf{u})\| \leq \|\mathbf{u}\| = \lambda_1(L),$$

which implies that

$$\|\mathbf{b}_1\|/\lambda_1(L) \leq \|\mathbf{b}_1\|/\|\mathbf{b}^\star_{qk+1}\|.$$

However, note that

$$\frac{\|\mathbf{b}_1\|}{\|\mathbf{b}^\star_{qk+1}\|} = \prod_{i=0}^{q-1} \frac{\|\mathbf{b}^\star_{ik+1}\|}{\|\mathbf{b}^\star_{(i+1)k+1}\|},$$

which, by (2.48), is

$$\leq \left(((1+\varepsilon)\gamma_k)^{k/(k-1)}\right)^q = ((1+\varepsilon)\gamma_k)^{qk/(k-1)},$$

where $qk \leq d - k$. Hence,

$$\|\mathbf{b}_1\|/\lambda_1(L) \leq ((1+\varepsilon)\gamma_k)^{(d-k)/(k-1)},$$

which proves (2.51). □

The Slide Algorithm

Gama and Nguyen [19] presented a polynomial-time algorithm to block-Mordell-reduce a basis, using an SVP-oracle in dimension $\leq k$: Algorithm 7 is a simplified version, to make exposition easier. By an SVP-oracle, [19] means any algorithm which, given as input the Gram matrix of a basis $(\mathbf{b}_1, \ldots, \mathbf{b}_k)$ of an integer lattice L, outputs $(u_1, \ldots, u_k) \in \mathbb{Z}^k$ such that $\|\sum_{i=1}^k u_i \mathbf{b}_i\| = \lambda_1(L)$.

Input: a basis $(\mathbf{b}_1, \ldots, \mathbf{b}_d)$ of a lattice L, together with a reduction factor $\varepsilon \geq 0$ and a blocksize $k \geq 2$ dividing d.
Output: the basis $(\mathbf{b}_1, \ldots, \mathbf{b}_d)$ is block-Mordell-reduced with factor ε and blocksize $k \geq 2$.
1: LLL-reduce $(\mathbf{b}_1, \ldots, \mathbf{b}_d)$ using Algorithm 6.
2: **if** there exists $j \in \{1, \ldots, d\}$ such that $j \equiv 1 \pmod{k}$ and j does not satisfy (2.44) **then**
3: Use an SVP-oracle in dimension $\leq k$ to locally HKZ-reduce the block $B_{[j.j+k-1]}))$, which implies that (2.44) holds; then return to Step 1. Basis vectors outside the block $B_{[j.j+k-1]}))$ are not modified.
4: **end if**
5: **if** there exists $j \in \{1, \ldots, d-k\}$ such that $j \equiv 1 \pmod{k}$ and j does not satisfy (2.45) **then**
6: Use an SVP-oracle in dimension $\leq k$ to reduce the block $B_{[j+1.j+k]}$ in such a way that $1/\|\mathbf{b}^\star_{j+k}\| = \lambda_1(\mathcal{L}(B_{[j+1.j+k]})^\times)$, which implies that (2.45) holds; then return to Step 1. Basis vectors outside the block $B_{[j+1.j+k]}$ are not modified.
7: **end if**

Algorithm 7: The basic slide algorithm [19]

Tests in Steps 2 and 5 are performed using an SVP-oracle in dimension k. We will not describe the local reductions performed in Steps 3 and 6: they are natural and are presented in [19]. Their cost is a linear number of calls to an SVP-oracle in dimension $\leq k$, together with polynomial-time operations, like an HKZ-reduction of a k-dimensional basis.

What is clear is that if the slide algorithm of Fig. 7 terminates, then the final basis is block-Mordell-reduced with factor ε and blocksize k. What is less clear is why the algorithm terminates, and what is its complexity. By analogy with the complexity analysis of LLL, one considers the following integral potential:

$$D' = \prod_{i=1}^{d/k} \mathrm{vol}(\mathcal{L}(B_{[1,ik]}))^2 \in \mathbb{Z}^+.$$

Then D can be rewritten as

$$D' = \prod_{i=1}^{d/k} \prod_{j=1}^{i} \mathrm{vol}(S_j)^2 = \prod_{j=1}^{d/k} \mathrm{vol}(S_j)^{2(d/k+1-j)}, \tag{2.52}$$

which is the blockwise analogue of $D = \|\mathbf{b}_1^\star\|^{2d} \|\mathbf{b}_2^\star\|^{2(d-1)} \times \cdots \times \|\mathbf{b}_d^\star\|^2$, which was used for analyzing LLL. Clearly, $\log D'$ is initially polynomial in the size of the basis.

We use D' to show that the number of times that the slide algorithm (Algorithm 7) goes through Step 1 is polynomially bounded, just as D was used to show that number of swaps in LLL was polynomially bounded. Let us look at the operations of Algorithm 7, which could possibly modify the integer D': it turns out that only Steps 1 and 6 can modify D', because Step 3 only modifies one block S_i (for some i), but the volume of this block cannot change, as the volume of the whole lattice remains the same. We discuss Steps 1 and 6 separately:

- Step 1 is an LLL reduction, which performs only size-reductions and swaps. Size-reductions do not modify any of the $\mathbf{b}_i^\star$, and therefore cannot modify D'. And we note that swaps of vectors $\mathbf{b}_{i-1}$ and $\mathbf{b}_i$ can modify D' only if $i \equiv 1 \pmod{k}$. When this is the case, $i = 1 + k\ell$ for some integer $\ell \geq 1$, and we see that the last vector of the block $S_{\ell-1}$ is the projection of $\mathbf{b}_{i-1}$, while the first vector of the block S_ℓ is the projection of $\mathbf{b}_i$. This means that in (2.52) of D', only $\mathrm{vol}(S_{\ell-1})$ and $\mathrm{vol}(S_\ell)$ may change. On the other hand, $\mathrm{vol}(S_{\ell-1}) \times vol(S_\ell)$ remains the same because $\mathrm{vol}(L) = \prod_{j=1}^{d/k} \mathrm{vol}(S_j)$ cannot change. But if LLL swapped $\mathbf{b}_{i-1}$ and $\mathbf{b}_i$, this means that Lovász' condition failed for $(i-1, i)$, which implies that $\|\mathbf{b}_{i-1}^\star\|$ will decrease strictly (in fact, by some multiplicative factor < 1): in this case, $\mathrm{vol}(S_{\ell-1})$ will decrease, and therefore D'. Hence, only two situations can occur:

Case 1: Step 1 never swaps vectors $\mathbf{b}_{i-1}$ and $\mathbf{b}_i$ such that $i \equiv 1 \pmod{k}$, in which case D' does not change. Here, the swaps are always within a block S_ℓ, never between two consecutive blocks $S_{\ell-1}$ and S_ℓ.

Case 2: Step 1 swaps at least once a pair of vectors $\mathbf{b}_{i-1}$ and $\mathbf{b}_i$ such that $i \equiv 1 \pmod{k}$, in which case D' decreases by some multiplicative factor < 1 depending on ε. This means that this situation occurs at most polynomially many times.

- Step 6 modifies the block $B_{[j+1,j+k]}$ so that $1/\|\mathbf{b}^\star_{j+k}\| = \lambda_1(\mathcal{L}(B_{[j+1,j+k]})^\times)$, which implies (2.45). As $j \equiv 1 \pmod{k}$, we may write $j = 1 + k\ell$ for some integer $\ell \geq 0$. We see that in (2.52) of D', only $\mathrm{vol}(S_{\ell+1})$ and $\mathrm{vol}(S_{\ell+2})$ change. On the other hand, $\mathrm{vol}(S_{\ell+1}) \times vol(S_{\ell+2})$ remains the same because $\mathrm{vol}(L) = \prod_{j=1}^{d/k} \mathrm{vol}(S_j)$ cannot change. Before Step 6, (2.45) did not hold, which means that $1/\|\mathbf{b}^\star_{j+k}\| > (1+\varepsilon)\lambda_1(\mathcal{L}(B_{[j+1,j+k]})^\times)$. But after Step 6, we have $1/\|\mathbf{b}^\star_{j+k}\| = \lambda_1(\mathcal{L}(B_{[j+1,j+k]})^\times)$, which implies that $1/\|\mathbf{b}^\star_{j+k}\|$ decreases by a multiplicative factor $\leq 1/(1+\varepsilon) < 1$. As $\mathbf{b}^\star_{j+k}$ is the first vector of $S_{\ell+2}$, this means that $\mathrm{vol}(S_{\ell+2})$ increases by a multiplicative factor $\geq 1+\varepsilon$, and therefore $\mathrm{vol}(S_{\ell+1})$ decreases by a multiplicative factor $\leq 1/(1+\varepsilon) < 1$. Hence, D' also decreases by a multiplicative factor $\leq 1/(1+\varepsilon)^2 < 1$. Thus, the number of times Step 6 is performed is at most polynomial in $1/\varepsilon$ and the size of the basis.

We showed that the steps of the slide algorithm (Algorithm 7) either preserve or decrease the integer D' by a mulplicative factor < 1 depending on ε. As $D' \geq 1$ and $\log D'$ is initially polynomial in the size of the basis, this means that number of steps for which there is a strict decrease is at most polynomial in $1/\varepsilon$ and the size of the basis. On the other hand, it is not difficult to see that the number of consecutive steps for which D' is preserved is also polynomially bounded: for instance, once Steps 6 are all performed, then all the blocks S_i are HKZ-reduced, which implies that during Step 1, Case 1 cannot occur.

We have seen the main argument why the slide algorithm is polynomial: the number of steps is polynomial. Like in LLL, it would remain to check that all the numbers remain polynomially bounded, which is done in [19]. We only have sketched a proof of the following result:

Theorem 14 ([19]). *There exists an algorithm which, given as input a basis of a d-dimensional integer lattice $L \subseteq \mathbb{Z}^n$, a reduction factor $\varepsilon > 0$, a blocksize $k \geq 2$ dividing d, and access to an SVP-oracle in dimension $\leq k$, outputs a block-Mordell-reduced basis of L with factor ε and blocksize k, such that*

1. *The number of calls to the SVP-oracle is polynomial in the size of the input basis and $1/\varepsilon$.*
2. *The size of the coefficients given as input to the SVP-oracle is polynomial in the size of the input basis.*
3. *Apart from the calls to the SVP-oracle, the algorithm only performs arithmetic operations on rational numbers of size polynomial in the size of the input basis, and the number of arithmetic operations is polynomial in $1/\varepsilon$ and the size of the basis.*

Acknowledgements We thank Nicolas Gama and Damien Stehlé for helpful comments.

References

1. H. Minkowski. *Geometrie der Zahlen*. Teubner, Leipzig, 1896
2. C. L. Siegel. *Lectures on the Geometry of Numbers*. Springer, 1989
3. M. Gruber and C. G. Lekkerkerker. *Geometry of Numbers*. North-Holland, 1987
4. J. Cassels. *An Introduction to the Geometry of Numbers*. Springer, 1997
5. L. Lagrange. Recherches d'arithmétique. *Nouv. Mém. Acad.*, 1773
6. C. Gauss. *Disquisitiones Arithmeticæ*. Leipzig, 1801
7. C. Hermite. Extraits de lettres de M. Hermite à M. Jacobi sur différents objets de la théorie des nombres, deuxième lettre. *J. Reine Angew. Math.*, 40:279–290, 1850. Also available in the first volume of Hermite's complete works, published by Gauthier-Villars
8. J. Conway and N. Sloane. *Sphere Packings, Lattices and Groups*. Springer, 1998. Third edition
9. S. Khot. *Inapproximability results for computational problems on lattices*. Springer, 2009. In this book
10. R. Kannan. Improved algorithms for integer programming and related lattice problems. In *Proc. of 15th STOC*, pages 193–206. ACM, 1983
11. R. Kannan. Minkowski's convex body theorem and integer programming. *Math. Oper. Res.*, 12(3):415–440, 1987
12. G. Hanrot and D. Stehlé. Improved analysis of Kannan's shortest lattice vector algorithm. In *Advances in Cryptology – Proc. CRYPTO 2007*, volume 4622 of *Lecture Notes in Computer Science*, pages 170–186. Springer, 2007
13. G. Hanrot and D. Stehlé. Worst-case Hermite-Korkine-Zolotarev reduced lattice bases. *CoRR*, abs/0801.3331, 2008
14. M. Ajtai, R. Kumar, and D. Sivakumar. A sieve algorithm for the shortest lattice vector problem. In *Proc. 33rd STOC*, pages 601–610. ACM, 2001
15. P. Q. Nguyen and T. Vidick. Sieve algorithms for the shortest vector problem are practical. *J. of Mathematical Cryptology*, 2(2):181–207, 2008
16. A. K. Lenstra, H. W. Lenstra, Jr., and L. Lovász. Factoring polynomials with rational coefficients. *Mathematische Ann.*, 261:513–534, 1982
17. C. P. Schnorr. A hierarchy of polynomial lattice basis reduction algorithms. *Theor. Comput. Sci.*, 53:201–224, 1987
18. N. Gama, N. Howgrave-Graham, H. Koy, and P. Q. Nguyen. Rankin's constant and blockwise lattice reduction. In *Proc. of Crypto '06*, volume 4117 of *LNCS*, pages 112–130. Springer, 2006
19. N. Gama and P. Q. Nguyen. Finding short lattice vectors within Mordell's inequality. In *STOC '08 – Proc. 40th ACM Symposium on the Theory of Computing*. ACM, 2008.
20. J. Martinet. *Perfect lattices in Euclidean spaces*, volume 327 of *Grundlehren der Mathematischen Wissenschaften*. Springer, Berlin, 2003
21. H. Cohn and A. Kumar. The densest lattice in twenty-four dimensions. *Electron. Res. Announc. Amer. Math. Soc.*, 10:58–67 (electronic), 2004
22. J. Milnor and D. Husemoller. *Symmetric Bilinear Forms*. Springer, 1973
23. D. Micciancio and S. Goldwasser. *Complexity of Lattice Problems: A Cryptographic Perspective*. Kluwer Academic Publishers, 2002
24. R. A. Rankin. On positive definite quadratic forms. *J. London Math. Soc.*, 28:309–314, 1953
25. J. L. Thunder. Higher-dimensional analogs of Hermite's constant. *Michigan Math. J.*, 45(2):301–314, 1998
26. M. I. Boguslavsky. Radon transforms and packings. *Discrete Appl. Math.*, 111(1–2):3–22, 2001
27. A. Korkine and G. Zolotareff. Sur les formes quadratiques positives ternaires. *Math. Ann.*, 5:581–583, 1872
28. A. Korkine and G. Zolotareff. Sur les formes quadratiques. *Math. Ann.*, 6:336–389, 1873
29. K. Mahler. A theorem on inhomogeneous diophantine inequalities. In *Nederl. Akad. Wetensch., Proc.*, volume 41, pages 634–637, 1938

30. J. C. Lagarias, H. W. Lenstra, Jr., and C. P. Schnorr. Korkin-Zolotarev bases and successive minima of a lattice and its reciprocal lattice. *Combinatorica*, 10:333–348, 1990
31. N. Gama and P. Q. Nguyen. Predicting lattice reduction. In *Advances in Cryptology – Proc. EUROCRYPT '08*, Lecture Notes in Computer Science. Springer, 2008
32. L. Lovász. *An Algorithmic Theory of Numbers, Graphs and Convexity*, volume 50. SIAM Publications, 1986. CBMS-NSF Regional Conference Series in Applied Mathematics
33. O. Regev. *On the Complexity of Lattice Problems with Polynomial Approximation Factors*. Springer, 2009. In this book
34. P. Emde Boas. Another NP-complete problem and the complexity of computing short vectors in a lattice. Technical report, Mathematische Instituut, University of Amsterdam, 1981. Report 81-04. Available at `http://turing.wins.uva.nl/~peter/`
35. M. Ajtai. The shortest vector problem in L_2 is NP-hard for randomized reductions. In *Proc. of 30th STOC*. ACM, 1998. Available at [36] as TR97-047
36. ECCC. `http://www.eccc.uni-trier.de/eccc/`. The Electronic Colloquium on Computational Complexity
37. D. Aharonov and O. Regev. Lattice problems in NP $\cap$ coNP. *J. ACM*, 52(5):749–765 (electronic), 2005
38. O. Goldreich and S. Goldwasser. On the limits of non-approximability of lattice problems. In *Proc. of 30th STOC*. ACM, 1998. Available at [36] as TR97-031
39. I. A. Semaev. A 3-dimensional lattice reduction algorithm. In *Proc. of CALC '01*, volume 2146 of *LNCS*. Springer, 2001
40. P. Q. Nguyen and D. Stehlé. Low-dimensional lattice basis reduction revisited (extended abstract). In *Proc. of the 6th Algorithmic Number Theory Symposium (ANTS VI)*, volume 3076 of *LNCS*, pages 338–357. Springer, 2004. Full version to appear in ACM Transactions on Algorithms, 2009
41. C. P. Schnorr and M. Euchner. Lattice basis reduction: improved practical algorithms and solving subset sum problems. *Math. Programming*, 66:181–199, 1994
42. H. W. Lenstra, Jr. Integer programming with a fixed number of variables. Technical report, Mathematisch Instituut, Universiteit van Amsterdam, April 1981. Report 81-03
43. H. Cohen. *A Course in Computational Algebraic Number Theory*. Springer, 1995. Second edition
44. C. Dwork. *Lattices and Their Application to Cryptography*. Stanford University, 1998. Lecture Notes, Spring Quarter. Several chapters are translations of Claus Schnorr's 1994 lecture notes *Gittertheorie und algorithmische Geometrie, Reduktion von Gitterbasen und Polynomidealen*
45. M. Pohst. On the computation of lattice vectors of minimal length, successive minima and reduced bases with applications. *ACM SIGSAM Bull.*, 15(1):37–44, 1981
46. U. Fincke and M. Pohst. Improved methods for calculating vectors of short length in a lattice, including a complexity analysis. *Math. Comp.*, 44(170):463–471, 1985
47. P. Q. Nguyen and D. Stehlé. LLL on the average. In *Proc. of ANTS-VII*, volume 4076 of *LNCS*. Springer, 2006
48. L. J. Mordell. Observation on the minimum of a positive quadratic form in eight variables. *J. London Math. Soc.*, 19:3–6, 1944
49. M. Ajtai. The worst-case behavior of Schnorr's algorithm approximating the shortest nonzero vector in a lattice. In *Proc. 35th Annual ACM Symposium on Theory of Computing*, pages 396–406 (electronic), ACM, 2003

Chapter 3
Probabilistic Analyses of Lattice Reduction Algorithms

Brigitte Vallée and Antonio Vera

Abstract The general behavior of lattice reduction algorithms is far from being well understood. Indeed, many experimental observations, regarding the execution of the algorithms and the geometry of their outputs, pose challenging questions, which remain unanswered and lead to natural conjectures yet to be settled. This survey describes complementary approaches which can be adopted for analyzing these algorithms, namely, dedicated modeling, probabilistic methods, and a dynamical systems approach. We explain how a mixed methodology has already proved fruitful for small dimensions p, corresponding to the variety of Euclidean algorithms ($p = 1$) and to the Gauss algorithm ($p = 2$). Such small dimensions constitute an important step in the analysis of lattice reduction in any (high) dimension, since the celebrated LLL algorithm, due to Lenstra, Lenstra, and Lovász, precisely involves a sequence of Gauss reduction steps on sublattices of a large lattice.

General Context

The present study surveys the main works aimed at understanding, both from a theoretical and an experimental viewpoint, how the celebrated LLL algorithm designed by Lenstra, Lenstra, and Lovász performs in practice. The goal is to precisely quantify the probabilistic behavior of lattice reduction and attain a justification of many of the experimental facts observed. Beyond its intrinsic theoretical interest, such a justification is important as a fine understanding of the lattice reduction process conditions algorithmic improvements in major application areas, most of them being described in this book: cryptography (see [28, 31]), computational number theory (see [21, 22, 35]), integer programming (see [1]), etc. The results obtained in this perspective may then be applied for developing a *general algorithmic strategy* for lattice reduction.

B. Vallée (✉)
Laboratoire GREYC, CNRS UMR 6072, Université de Caen and ENSICAEN, F-14032 Caen, France,
e-mail: brigitte.vallee@info.unicaen.fr

P.Q. Nguyen and B. Vallée (eds.), *The LLL Algorithm*, Information Security and Cryptography, DOI 10.1007/978-3-642-02295-1_3,

Varied Approaches

We briefly describe now three different points of view: dedicated modeling, probabilistic methods, and dynamical systems approach.

Dedicated Modeling. Probabilistic models are problem-specific in the various applications of lattice reduction. For each particular area, special types of lattice bases are used as input models, which induce rather different quantitative behaviors. An analysis of the lattice reduction algorithms under such probabilistic models aims at characterizing the behavior of the main parameters – principally, the number of iterations, the geometry of reduced bases, and the evolution of densities during an execution.

Probabilistic Methods. The probabilistic line of investigation has already led to tangible results under the (somewhat unrealistic) models where vectors of the input basis are independently chosen according to a distribution that is rotationally invariant. In particular, the following question has been answered: what is the probability for an input basis to be already reduced? A possible extension of this study to realistic models and to the complete algorithm (not just its input distribution) is discussed here.

Dynamical Systems Approach. Thanks to earlier results, the dynamics of Euclid's algorithm is now well-understood – many results describe the probabilistic behavior of that algorithm, based on *dynamical systems theory* as well as related tools, like transfer operators. These techniques are then extended to dimension $p = 2$ (Gauss' algorithm). We examine here the possible extensions of the "dynamical analysis methodology" to higher dimensions. The first step in such an endeavor should describe the dynamical system for the LLL algorithm, which is probably a complex object, for $p > 2$.

Historical and Bibliographic Notes

Over the past 20 years, there have been several parallel studies dedicated to the probabilistic behavior of lattice reduction algorithms, in the two-dimensional case as well as in the general case.

The Two-Dimensional Case. The history of the analysis of lattice reduction algorithms starts before 1982, when Lagarias [23] performs in 1980 a first (worst–case) analysis of the Gauss algorithms in two and three dimensions. In 1990, Vallée [38] exhibits the exact worst–case complexity of the Gauss algorithm. In the same year, Flajolet and Vallée [16] perform the first probabilistic analysis of the Gauss algorithm: they study the mean value of the number of iterations in the uniform model. Then, in 1994, Daudé et al. [14] obtain a complete probabilistic analysis of the Gauss algorithm, with a "dynamical approach," but still under the uniform model. The same year, Laville and Vallée [24] study the main output parameters of the algorithm (the first minimum, Hermite's defect), under the uniform model, still. In 1997,

Vallée [39] introduces the model "with valuation" for the Sign Algorithm: this is an algorithm for comparing rationals, whose behavior is similar to the Gauss algorithm. In 2000, Flajolet and Vallée [17] precisely study all the constants that appear in the analysis of the Sign Algorithm. Finally, in 2007, Vallée and Vera [45, 47] study all the main parameters of the Gauss algorithm (execution parameters and output parameters) in the general model "with valuation."

The Dynamical Analysis Methodology. From 1995, Vallée has built a general method for analyzing a whole class of gcd algorithms. These algorithms are all based on the same principles as the Euclid algorithms (divisions and exchanges), but they differ on the kind of division performed. This method, summarized for instance in [37], views an algorithm as a dynamical system and uses a variety of tools, some of them coming from analysis of algorithms (generating functions, singularity analysis, etc.) and other ones being central in dynamical systems, like transfer operators. The interest of such an analysis becomes apparent in the work about the Gauss Algorithm [14], previously described, which is in fact the first beginning of dynamical analysis. The dynamical systems underlying the Gauss algorithms are just extensions of systems associated to the (centered) Euclid algorithms, which first need a sharp understanding. This is why Vallée returns to the one-dimensional case, first performs average-case analysis for a large variety of Euclidean algorithms and related parameters of interest: number of iterations [41], bit-complexity (with Akhavi) [5], and bit-complexity of the fast variants of the Euclid algorithms (with the CAEN group) [10]. From 2003, Baladi et al. [6, 27] also obtain distributional results on the main parameters of the Euclid algorithms – number of iterations, size of the remainder at a fraction of the execution, and bit-complexity – and show that they all follow asymptotic normal laws.

It is now natural to expect that most of the principles of dynamical analysis can be applied to the Gauss algorithm. The first work in this direction is actually done by Vallée and Vera, quite recently (2007), and completes the first work [14].

The General Case. The first probabilistic analysis of the LLL algorithm is performed by Daudé and Vallée on 1994 [15] under the "random ball model." These authors obtain an upper bound for the mean number of iterations of the algorithm. Then, in 2002, Akhavi [3] studies the probabilistic behavior of a random basis (again, under the random ball model) and he detects two different regimes, according to the dimension of the basis relative to the dimension of the ambient space. In 2006, Akhavi et al. [4] improve on the previous study, while generalizing it to other randomness models (the so-called spherical models): they exhibit a limit model when the ambient dimension becomes large. These studies illustrate the importance of the model "with valuation" for the local bases associated to the input.

In 2003, Ajtai [2] exhibits a randomness model of input bases (which is called the Ajtai model in this paper), under which the probabilistic behavior of the LLL algorithm is close to the worst-case behavior. In 2006, Nguyen et al. [30] study random lattices together with their parameters relevant to lattice reduction algorithms. In 2006, Nguyen and Stehlé [30] conduct many experiments for the LLL algorithms under several randomness models. They exhibit interesting experimental phenomena and provide conjectures that would explain them.

The Two-Dimensional Case as a Main Tool for the General Case. This paper describes a first attempt to apply the dynamical analysis methodology to the LLL algorithm: the LLL algorithm is now viewed as a whole dynamical system that runs in parallel many two-dimensional dynamical systems and "gathers" all the dynamics of these small systems. This (perhaps) makes possible to use the precise results obtained on the Gauss algorithm – probabilistic and dynamic – as a main tool for describing the probabilistic behavior of the LLL algorithm and its whole dynamics.

Plan of the Survey

Section "The Lattice Reduction Algorithm in the Two-Dimensional Case" explains why the two-dimensional case is central, introduces the lattice reduction in this particular case, and presents the Gauss algorithm, which is our main object of study. Section "The LLL Algorithm" is devoted to a precise description of the LLL algorithm in general dimension; it introduces the main parameters of interest: the output parameters, which describe the geometry of the output bases, and the execution parameters, which describe the behavior of the algorithm itself. The results of the main experiments conducted regarding these parameters on "useful" classes of lattices are also reported there. Finally, we introduce variants of the LLL algorithm, where the role of the Gauss algorithm becomes more apparent than in standard versions. Section "What is a Random (Basis of a) Lattice?" describes the main probabilistic models of interest that appear in "real life" applications – some of them are given because of their naturalness, while other ones are related to actual applications of the LLL algorithm. Section "Probabilistic Analyses of the LLL Algorithm in the Spherical Model" is devoted to a particular class of models, the so-called spherical models, which are the most natural models (even though they do not often surface in actual applications). We describe the main results obtained under this model: the distribution of the "local bases," the probability of an initial reduction, and mean value estimates of the number of iterations and of the first minimum.

The first step towards a precise study of other, more "useful," models is a fine understanding of the two-dimensional case, where the mixed methodology is employed. In Section "Returning to the Gauss Algorithm", we describe the dynamical systems that underlie the (two) versions of the Gauss algorithms, together with two (realistic) input probabilistic models of use: the model "with valuation" and the model "with fixed determinant." Sections "Analysis of Lattice Reduction in Two-Dimensions: The Output Parameters" and "Analysis of the Execution Parameters of the Gauss Algorithm" on the precise study of the main parameters of interest – either output parameters or execution parameters – under the model "with valuation." Finally, Section "First Steps in the Probabilistic Analysis of the LLL Algorithm" returns to the LLL algorithm and explains how the results of Sections "Returning to the Gauss Algorithm – Analysis of the Execution Parameters of the Gauss Algorithm" could (should?) be used and/or extended to higher dimensions.

The Lattice Reduction Algorithm in the Two-Dimensional Case

A lattice $\mathcal{L} \subset \mathbb{R}^n$ of dimension p is a discrete additive subgroup of $\mathbb{R}^n$. Such a lattice is generated by integral linear combinations of vectors from a family $B := (b_1, b_2, \ldots b_p)$ of $p \le n$ linearly independent vectors of $\mathbb{R}^n$, which is called a basis of the lattice $\mathcal{L}$. A lattice is generated by infinitely many bases that are related to each other by integer matrices of determinant ± 1. Lattice reduction algorithms consider a Euclidean lattice of dimension p in the ambient space $\mathbb{R}^n$ and aim at finding a "reduced" basis of this lattice, formed with vectors almost orthogonal and short enough. The LLL algorithm designed in [25] uses as a sub-algorithm the lattice reduction algorithm for two dimensions (which is called the Gauss algorithm):[1] it performs a succession of steps of the Gauss algorithm on the "local bases," and it stops when all the local bases are reduced (in the Gauss sense). This is why it is important to precisely describe and study the two-dimensional case. This is the purpose of this section: it describes the particularities of the lattices in two dimensions, provides two versions of the two-dimensional lattice reduction algorithm, namely the Gauss algorithm, and introduces its main parameters of interest.

We also see in this article that the Gauss algorithm solves the reduction problem in an optimal sense: it returns a minimal basis, after a number of iterations, which is at most linear with respect to the input size. This type of algorithms can be generalized in small dimensions. For instance, in the three-dimensional case, Vallée in 1987 [42] or Semaev more recently [33] provide optimal algorithms, which directly find a minimal basis, after a linear number of iterations. However, algorithms of this quality no longer exist in higher dimensions, and the LLL algorithm can be viewed as an approximation algorithm that finds a good basis (not optimal generally speaking) after a polynomial number of iterations (not linear generally speaking).

Lattices in Two-Dimensions

Up to a possible isometry, a two-dimensional lattice may always be considered as a subset of $\mathbb{R}^2$. With a small abuse of language, we use the same notation for denoting a complex number $z \in \mathbb{C}$ and the vector of $\mathbb{R}^2$ whose components are $(\Re z, \Im z)$. For a complex z, we denote by $|z|$ both the modulus of the complex z and the Euclidean norm of the vector z; for two complex numbers u, v, we denote by $(u \cdot v)$ the scalar product between the two vectors u and v. The following relation between two complex numbers u, v will be very useful in the sequel

$$\frac{v}{u} = \frac{(u \cdot v)}{|u|^2} + i\frac{\det(u, v)}{|u|^2}. \tag{3.1}$$

[1] It seems that the Gauss algorithm, as it is described here, is not actually due to Gauss, but due to Lagrange.

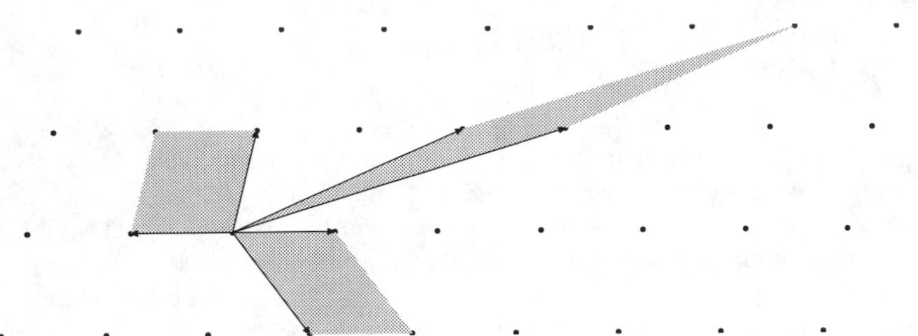

Fig. 3.1 A lattice and three of its bases represented by the parallelogram they span. The basis on the *left* is minimal (reduced), while the two other ones are skew

A *lattice* of two-dimensions in the complex plane $\mathbb{C}$ is the set $\mathcal{L}$ of elements of $\mathbb{C}$ (also called vectors) defined by

$$\mathcal{L} = \mathbb{Z}u \oplus \mathbb{Z}v = \{au + bv; \quad a, b \in \mathbb{Z}\},$$

where (u, v), called a *basis*, is a pair of $\mathbb{R}$-linearly independent elements of $\mathbb{C}$. Remark that in this case, due to (3.1), one has $\Im(v/u) \neq 0$.

Amongst all the bases of a lattice $\mathcal{L}$, some that are called reduced enjoy the property of being formed with "short" vectors. In dimension 2, the best reduced bases are *minimal* bases that satisfy optimality properties: define u to be a first minimum of a lattice $\mathcal{L}$ if it is a nonzero vector of $\mathcal{L}$ that has smallest Euclidean norm; the length of a first minimum of $\mathcal{L}$ is denoted by $\lambda_1(\mathcal{L})$. A second minimum v is any shortest vector amongst the vectors of the lattice that are linearly independent of one of the first minimum u; the Euclidean length of a second minimum is denoted by $\lambda_2(\mathcal{L})$. Then a basis is *minimal* if it comprises a first and a second minimum (See Fig. 3.1). In the sequel, we focus on particular bases that satisfy one of the two following properties:

(P) It has a positive determinant [i.e., $\det(u, v) > 0$ or $\Im(v/u) > 0$]. Such a basis is called *positive*.

(A) It has a positive scalar product [i.e., $(u \cdot v) \geq 0$ or $\Re(v/u) \geq 0$]. Such a basis is called *acute*.

Without loss of generality, we may always suppose that a basis is acute (resp. positive), as one of (u, v) and $(u, -v)$ is.
The following result gives characterizations of minimal bases. Its proof is omitted.

Proposition 1. [Characterizations of minimal bases.]

(P) [Positive bases.] *Let (u, v) be a positive basis. Then the following two conditions (a) and (b) are equivalent:*

(a) *The basis (u, v) is minimal*

(b) *The pair (u, v) satisfies the three simultaneous inequalities:*

$$(P_1): \quad \left|\frac{v}{u}\right| \geq 1, \quad (P_2): \quad \left|\Re\left(\frac{v}{u}\right)\right| \leq \frac{1}{2}, \quad \text{and} \quad (P_3): \ \Im\left(\frac{v}{u}\right) > 0.$$

(A) [Acute bases.] *Let* (u, v) *be an acute basis. Then the following two conditions* (a) *and* (b) *are equivalent:*

> PGAUSS(u, v)
>
> **Input.** A positive basis (u, v) of $\mathbb{C}$ with $|v| \leq |u|$, $|\tau(v, u)| \leq (1/2)$.
>
> **Output.** A positive minimal basis (u, v) of $\mathcal{L}(u, v)$ with $|v| \geq |u|$.
>
> `While` $|v| < |u|$ `do`
>
> $\quad (u, v) := (v, -u)$;
>
> $\quad q := \lfloor \tau(v, u) \rceil$;
>
> $\quad v := v - qu$;

(a) *The basis* (u, v) *is minimal*

(b) *The pair* (u, v) *satisfies the two simultaneous inequalities:*

$$(A_1): \quad \left|\frac{v}{u}\right| \geq 1 \qquad \text{and} \qquad (A_2): \ 0 \leq \Re\left(\frac{v}{u}\right) \leq \frac{1}{2}.$$

The Gaussian Reduction Schemes

There are two reduction processes, according as one focuses on positive bases or acute bases. Accordingly, as we study the behavior of the algorithm itself, or the geometric characteristics of the output, it will be easier to deal with one version than with the other one: for the first case, we will choose the acute framework, and for the second case, the positive framework.

The Positive Gauss Algorithm

The positive lattice reduction algorithm takes as input a positive arbitrary basis and produces as output a positive minimal basis. The positive Gauss algorithm aims at satisfying simultaneously the conditions (P) of Proposition 1. The conditions (P_1) and (P_3) are simply satisfied by an exchange between vectors followed by a sign change $v := -v$. The condition (P_2) is met by an integer translation of the type

$$v := v - qu \qquad \text{with} \quad q := \lfloor \tau(v, u) \rceil, \quad \tau(v, u) := \Re\left(\frac{v}{u}\right) = \frac{(u \cdot v)}{|u|^2}, \tag{3.2}$$

where $\lfloor x \rceil$ represents the integer nearest[2] to the real x. After this translation, the new coefficient $\tau(v, u)$ satisfies $0 \leq |\tau(v, u)| \leq (1/2)$.

---

[2] The function $\lfloor x \rceil$ is extended to the negative numbers with the relation $\lfloor x \rceil = -\lfloor -x \rceil$.

On the input pair $(u, v) = (v_0, v_1)$, the positive Gauss Algorithm computes a sequence of vectors v_i defined by the relations

$$v_{i+1} = -v_{i-1} + q_i\, v_i \qquad \text{with} \quad q_i := \lfloor \tau(v_{i-1}, v_i) \rceil\,. \tag{3.3}$$

Here, each quotient q_i is an integer of $\mathbb{Z}$, the final pair (v_p, v_{p+1}) satisfies the conditions (P) of Proposition 1, and $P(u, v) := p$ denotes the number of iterations. Each step defines a unimodular matrix $\mathcal{M}_i$ with $\det \mathcal{M}_i = 1$,

$$\mathcal{M}_i = \begin{pmatrix} q_i & -1 \\ 1 & 0 \end{pmatrix}, \qquad \text{with} \quad \begin{pmatrix} v_{i+1} \\ v_i \end{pmatrix} = \mathcal{M}_i \begin{pmatrix} v_i \\ v_{i-1} \end{pmatrix},$$

so that the algorithm produces a matrix $\mathcal{M}$ for which

$$\begin{pmatrix} v_{p+1} \\ v_p \end{pmatrix} = \mathcal{M} \begin{pmatrix} v_1 \\ v_0 \end{pmatrix} \qquad \text{with} \quad \mathcal{M} := \mathcal{M}_p \cdot \mathcal{M}_{p-1} \cdot \ldots \cdot \mathcal{M}_1. \tag{3.4}$$

The Acute Gauss Algorithm

The acute reduction algorithm takes as input an arbitrary acute basis and produces as output an acute minimal basis. This AGAUSS algorithm aims at satisfying simultaneously the conditions (A) of Proposition 1. The condition (A_1) is simply satisfied by an exchange, and the condition (A_2) is met by an integer translation of the type

$$v := \varepsilon(v - qu) \qquad \text{with} \quad q := \lfloor \tau(v, u) \rceil\,, \qquad \varepsilon = \operatorname{sign}\left(\tau(v, u) - \lfloor \tau(v, u) \rceil\right),$$

where $\tau(v, u)$ is defined as in (3.2). After this transformation, the new coefficient $\tau(v, u)$ satisfies $0 \leq \tau(v, u) \leq (1/2)$.

```
AGAUSS(u, v)
Input. An acute basis (u, v) of ℂ with |v| ≤ |u|, 0 ≤ τ(v, u) ≤ (1/2).
Output. An acute minimal basis (u, v) of ℒ(u, v) with |v| ≥ |u|.
While |v| < |u| do
        (u, v) := (v, u);
        q := ⌊τ(v, u)⌉ ; ε := sign (τ(v, u) − ⌊τ(v, u)⌉);
        v := ε(v − qu);
```

On the input pair $(u, v) = (w_0, w_1)$, the Gauss Algorithm computes a sequence of vectors w_i defined by the relations $\quad w_{i+1} = \varepsilon_i (w_{i-1} - \widetilde{q}_i\, w_i)$ with

$$\widetilde{q}_i := \lfloor \tau(w_{i-1}, w_i) \rceil\,, \qquad \varepsilon_i = \operatorname{sign}\left(\tau(w_{i-1}, w_i) - \lfloor \tau(w_{i-1}, w_i) \rceil\right). \tag{3.5}$$

Here, each quotient $\widetilde{q}_i$ is a positive integer, $p \equiv P(u, v)$ denotes the number of iterations [this equals the previous one], and the final pair (w_p, w_{p+1}) satisfies the conditions (A) of Proposition 1. Each step defines a unimodular matrix $\mathcal{N}_i$ with $\det \mathcal{N}_i = -\varepsilon_i = \pm 1$,

$$\mathcal{N}_i = \begin{pmatrix} -\varepsilon_i \, \widetilde{q}_i & \varepsilon_i \\ 1 & 0 \end{pmatrix}, \qquad \text{with} \quad \begin{pmatrix} w_{i+1} \\ w_i \end{pmatrix} = \mathcal{N}_i \begin{pmatrix} w_i \\ w_{i-1} \end{pmatrix},$$

so that the algorithm produces a matrix $\mathcal{N}$ for which

$$\begin{pmatrix} w_{p+1} \\ w_p \end{pmatrix} = \mathcal{N} \begin{pmatrix} w_1 \\ w_0 \end{pmatrix} \qquad \text{with} \quad \mathcal{N} := \mathcal{N}_p \cdot \mathcal{N}_{p-1} \cdot \ldots \cdot \mathcal{N}_1.$$

Comparison Between the Two Algorithms

These algorithms are closely related, but different. The AGAUSS Algorithm can be viewed as a folded version of the PGAUSS Algorithm, in the sense defined in [7]. We shall come back to this fact in Section "Relation with the Centered Euclid Algorithm", and the following is true.

Consider two bases: a positive basis (v_0, v_1) *and an acute basis* (w_0, w_1)*, which satisfy* $w_0 = v_0$ *and* $w_1 = \eta_1 \, v_1$ *with* $\eta_1 = \pm 1$*. Then the sequences of vectors* (v_i) *and* (w_i) *computed by the two versions of the Gauss algorithm (defined in (3.3) and (3.5)) satisfy* $w_i = \eta_i \, v_i$ *for some* $\eta_i = \pm 1$ *and the quotient* $\widetilde{q}_i$ *is the absolute value of quotient* q_i.

Then, when studying the two kinds of parameters – execution parameters or output parameters – the two algorithms are essentially the same. As already said, we shall use the PGAUSS Algorithm for studying the output parameters, and the AGAUSS Algorithm for the execution parameters.

Main Parameters of Interest

The size of a pair $(u, v) \in \mathbb{Z}[i] \times \mathbb{Z}[i]$ is

$$\ell(u, v) := \max\{\ell(|u|^2), \ell(|v|^2)\} \approx 2 \max\{\ell(|u|), \ell(|v|)\},$$

where $\ell(x)$ is the binary length of the integer x. The Gram matrix $G(u, v)$ is defined as

$$G(u, v) = \begin{pmatrix} |u|^2 & (u \cdot v) \\ (u \cdot v) & |v|^2 \end{pmatrix}.$$

In the following, we consider subsets Ω_M, which gather all the (valid) inputs of size M relative to each version of the algorithm. They will be endowed with some discrete probability $\mathbb{P}_M$, and the main parameters become random variables defined on these sets.

All the computations of the Gauss algorithm are done on the Gram matrices $G(v_i, v_{i+1})$ of the pair (v_i, v_{i+1}). The *initialization* of the Gauss algorithm *computes* the Gram Matrix of the initial basis: it computes three scalar products, which takes a *quadratic time*[3] with respect to the length of the input $\ell(u,v)$. After this, all the computations of the *central part* of the algorithm *are directly done* on these matrices; more precisely, each step of the process is an Euclidean division between the two coefficients of the first line of the Gram matrix $G(v_i, v_{i-1})$ of the pair (v_i, v_{i-1}) for obtaining the quotient q_i, followed with the computation of the new coefficients of the Gram matrix $G(v_{i+1}, v_i)$, namely

$$|v_{i+1}|^2 := |v_{i-1}|^2 - 2q_i\,(v_i \cdot v_{i-1}) + q_i^2 |v_i|^2, \qquad (v_{i+1} \cdot v_i) := q_i\,|v_i|^2 - (v_{i-1} \cdot v_i).$$

Then the cost of the ith step is proportional to $\ell(|q_i|) \cdot \ell(|v_{i-1}|^2)$, and the bit-complexity of the central part of the Gauss Algorithm is expressed as a function of

$$B(u,v) = \sum_{i=1}^{P(u,v)} \ell(|q_i|) \cdot \ell(|v_{i-1}|^2), \tag{3.6}$$

where $P(u,v)$ is the number of iterations of the Gauss Algorithm. In the sequel, B will be called the bit-complexity.

The bit-complexity $B(u,v)$ is one of our main parameters of interest, and we compare it to other simpler costs. Define three new costs, the quotient bit-cost $Q(u,v)$, the difference cost $\underline{D}(u,v)$, and the approximate difference cost D:

$$Q(u,v) = \sum_{i=1}^{P(u,v)} \ell(|q_i|), \qquad \underline{D}(u,v) = \sum_{i=1}^{P(u,v)} \ell(|q_i|)\left[\,\ell(|v_{i-1}|^2) - \ell(|v_0|^2)\right], \tag{3.7}$$

$$D(u,v) := \sum_{i=1}^{P(u,v)} \ell(|q_i|) \lg \left|\frac{v_{i-1}}{v}\right|^2,$$

which satisfy $D(u,v) - \underline{D}(u,v) = \Theta(Q(u,v))$ and

$$B(u,v) = Q(u,v)\,\ell(|u|^2) + D(u,v) + [\underline{D}(u,v) - D(u,v)]\,. \tag{3.8}$$

We are then led to study two main parameters related to the bit-cost, which may be of independent interest:

(a) The additive costs, which provide a generalization of costs P and Q. They are defined as the sum of elementary costs, which depend only on the quotients q_i. More precisely, from a positive elementary cost c defined on $\mathbb{N}$, we consider the total cost on the input (u,v) defined as

$$C_{(c)}(u,v) = \sum_{i=1}^{P(u,v)} c(|q_i|)\,. \tag{3.9}$$

[3] We consider the naive multiplication between integers of size M, whose bit-complexity is $O(M^2)$.

When the elementary cost c satisfies $c(m) = O(\log m)$, the cost C is said to be of moderate growth.

(b) The sequence of the ith length decreases d_i for $i \in [1..p]$ (with $p := P(u, v)]$) and the total length decrease $d := d_p$, defined as

$$d_i := \left|\frac{v_i}{v_0}\right|^2, \qquad d := \left|\frac{v_p}{v_0}\right|^2. \tag{3.10}$$

Finally, the configuration of the output basis $(\widehat{u}, \widehat{v})$ is described via its Gram–Schmidt orthogonalized basis, that is, the system $(\widehat{u}^\star, \widehat{v}^\star)$, where $\widehat{u}^\star := \widehat{u}$ and $\widehat{v}^\star$ is the orthogonal projection of $\widehat{v}$ onto the orthogonal of $<\widehat{u}>$. There are three main output parameters closely related to the minima of the lattice $\mathcal{L}(u, v)$,

$$\lambda(u, v) := \lambda_1(\mathcal{L}(u, v)) = |\widehat{u}|, \qquad \mu(u, v) := \frac{|\det(u, v)|}{\lambda(u, v)} = |\widehat{v}^\star|, \tag{3.11}$$

$$\gamma(u, v) := \frac{\lambda^2(u, v)}{|\det(u, v)|} = \frac{\lambda(u, v)}{\mu(u, v)} = \frac{|\widehat{u}|}{|\widehat{v}^\star|}. \tag{3.12}$$

We return later to these output parameters and shall explain in Section "A Variation for the LLL Algorithm: The Odd-Even Algorithm" why they are so important in the study of the LLL algorithm. We now return to the general case of lattice reduction.

The LLL Algorithm

We provide a description of the LLL algorithm, introduce the parameters of interest, and explain the bounds obtained in the worst-case analysis. Then, we describe the results of the main experiments conducted for classes of "useful" lattices. Finally, this section presents a variant of the LLL algorithm, where the Gauss algorithm plays a more apparent rôle: it appears to be well-adapted to (further) analyses.

Description of the Algorithm

We recall that the LLL algorithm considers a Euclidean lattice given by a system B formed of p linearly independent vectors in the ambient space $\mathbb{R}^n$. It aims at finding a reduced basis, denoted by $\widehat{B}$ formed with vectors almost orthogonal and short enough. The algorithm (see Figure 3.2) deals with the matrix $\mathcal{P}$, which expresses the system B as a function of the Gram–Schmidt orthogonalized system B^*; the coefficient $m_{i,j}$ of matrix $\mathcal{P}$ is equal to $\tau(b_i, b_j^\star)$, with τ defined in (3.2). The algorithm performs two main types of operations (see Figure 3.2):

$$\mathcal{P} := \begin{array}{c} \\ b_1 \\ b_2 \\ \vdots \\ b_i \\ b_{i+1} \\ \vdots \\ b_p \end{array} \begin{array}{c} \begin{array}{cccccccc} b_1^\star & b_2^\star & \dots & b_i^\star & b_{i+1}^\star & \dots & b_p^\star \end{array} \\ \begin{pmatrix} 1 & 0 & \dots & 0 & 0 & 0 & 0 \\ m_{2,1} & 1 & \dots & 0 & 0 & 0 & 0 \\ \vdots & \vdots & \ddots & \vdots & \vdots & \vdots & \vdots \\ m_{i,1} & m_{i,2} & \dots & 1 & 0 & 0 & 0 \\ m_{i+1,1} & m_{i+1,2} & \dots & m_{i+1,i} & 1 & 0 & 0 \\ \vdots & \vdots & \vdots & \vdots & \vdots & \ddots & \vdots \\ m_{p,1} & m_{p,2} & \dots & m_{p,i} & m_{p,i+1} & \dots & 1 \end{pmatrix} \end{array}$$

$$U_k := \begin{array}{c} \\ u_k \\ v_k \end{array} \begin{array}{c} \begin{array}{cc} b_k^\star & b_{k+1}^\star \end{array} \\ \begin{pmatrix} 1 & 0 \\ m_{k+1,k} & 1 \end{pmatrix} \end{array}$$

LLL (t) $[t > 1]$

Input. A basis B of a lattice L of dimension p.
Output. A reduced basis $\widehat{B}$ of L.

```
Gram computes the basis B* and the matrix P.
i := 1;
While i < p do
        1– Diagonal-Size-Reduction (b_{i+1})
        2– Test if local basis U_i is reduced : Is |v_i| > (1/t)|u_i|?
                if yes : Other-size-reduction (b_{i+1})
                            i := i + 1;
                if not: Exchange b_i and b_{i+1}
                            Recompute (B*, P);
                            If  i ≠ 1 then  i := i − 1;
```

Fig. 3.2 The LLL algorithm: the matrix $\mathcal{P}$, the local bases U_k, and the algorithm itself

1. *Size-reduction of vectors.* The vector b_i is size-reduced if all the coefficients $m_{i,j}$ of the ith row of matrix $\mathcal{P}$ satisfy $|m_{i,j}| \leq (1/2)$ for all $j \in [1..i-1]$. Size-reduction of vector b_i is performed by integer translations of b_i with respect to vectors b_j for all $j \in [1..i-1]$.
   As subdiagonal coefficients play a particular rôle (as we shall see later), the operation `Size-reduction` (b_i) is subdivided into two main operations:
   `Diagonal-size-reduction` (b_i);
   $b_i := b_i - \lfloor m_{i,i-1} \rceil b_{i-1}$;
   followed with
   `Other-size-reduction` (b_i);
   `For` $j := i-2$ `downto` 1 do $b_i := b_i - \lfloor m_{i,j} \rceil b_j$.
2. *Gauss-reduction of the local bases.* The ith local basis U_i is formed with the two vectors u_i, v_i, defined as the orthogonal projections of b_i, b_{i+1} on the orthogonal of the subspace $\langle b_1, b_2, \dots, b_{i-1} \rangle$. The LLL algorithm performs the PGAUSS

algorithm [integer translations and exchanges] on local bases U_i, but there are three differences with the PGAUSS algorithm previously described:

(*a*) The output test is *weaker* and depends on a parameter $t > 1$: the classical Gauss output test $|v_i| > |u_i|$ is replaced by the output test $|v_i| > (1/t)|u_i|$.
(*b*) The operations that are performed during the PGAUSS algorithm on the local basis U_i are then *reflected* on the system (b_i, b_{i+1}): if $\mathcal{M}$ is the matrix built by the PGAUSS algorithm on (u_i, v_i), then it is applied to the system (b_i, b_{i+1}) in order to find the new system (b_i, b_{i+1}).
(*c*) The PGAUSS algorithm is performed on the local basis U_i *step by step*. The index i of the local basis visited begins at $i = 1$, ends at $i = p$, and is incremented (when the test in Step 2 is positive) or decremented (when the test in Step 2 is negative and the index i does not equal 1) at each step. This defines a random walk. The length K of the random walk is the number of iterations, and the number of steps K^- where the test in step 2 is negative satisfies

$$K \leq (p-1) + 2K^-. \tag{3.13}$$

The LLL algorithm considers the sequence ℓ_i formed with the lengths of the vectors of the Gram orthogonalized basis $B^\star$ and deals with the Siegel ratios r_i's between successive Gram orthogonalized vectors, namely

$$r_i := \frac{\ell_{i+1}}{\ell_i}, \qquad \text{with} \quad \ell_i := |b_i^\star|. \tag{3.14}$$

The steps of Gauss reduction aim at obtaining lower bounds on these ratios. In this way, the interval $[a, A]$ with

$$a := \min\{\ell_i;\quad 1 \leq i \leq p\}, \qquad A := \max\{\ell_i;\quad 1 \leq i \leq p\}, \tag{3.15}$$

tends to be narrowed as, all along the algorithm, the minimum a is increasing and the maximum A is decreasing. This interval $[a, A]$ plays an important rôle because it provides an approximation for the first minimum $\lambda(\mathcal{L})$ of the lattice (i.e., the length of a shortest nonzero vector of the lattice), namely

$$\lambda(\mathcal{L}) \leq A\sqrt{p}, \qquad \lambda(\mathcal{L}) \geq a. \tag{3.16}$$

At the end of the algorithm, the basis $\widehat{B}$ satisfies the following:[4] each local bases is reduced in the t-Gauss meaning. It satisfies conditions that involve the subdiagonal matrix coefficients $\widehat{m}_{i+1,i}$ together with the sequence $\widehat{\ell}_i$, namely the t-Lovász conditions, for any i, $1 \leq i \leq p-1$,

[4] All the parameters relative to the output basis $\widehat{B}$ are denoted with a hat.

$$|\widehat{m}_{i+1,i}| \leq \frac{1}{2}, \qquad t^2\,(\widehat{m}^2_{i+1,i}\,\widehat{\ell}^2_i + \widehat{\ell}^2_{i+1}) \geq \widehat{\ell}^2_i, \tag{3.17}$$

which *imply* the s-Siegel conditions, for any $i, 1 \leq i \leq p-1$,

$$|\widehat{m}_{i+1,i}| \leq \frac{1}{2}, \quad \widehat{r}_i := \frac{\widehat{\ell}_{i+1}}{\widehat{\ell}_i} \geq \frac{1}{s}, \quad \text{with } s^2 = \frac{4t^2}{4-t^2} \quad \text{and} \quad s = \frac{2}{\sqrt{3}} \quad \text{for } t = 1. \tag{3.18}$$

A basis fulfilling conditions (3.18) is called s-Siegel reduced .

Main Parameters of Interest

There are two kinds of parameters of interest for describing the behavior of the algorithm: the output parameters and the execution parameters.

Output Parameters

The geometry of the output basis is described with three main parameters – the Hermite defect $\gamma(B)$, the length defect $\theta(B)$, or the orthogonality defect $\rho(B)$. They satisfy the following (worst-case) bounds that are functions of parameter s, namely

$$\gamma(B) := \frac{|\widehat{b}_1|^2}{(\det \mathcal{L})^{2/p}} \leq s^{p-1}, \qquad \theta(B) := \frac{|\widehat{b}_1|}{\lambda(\mathcal{L})} \leq s^{p-1}, \tag{3.19}$$

$$\rho(B) := \frac{\prod_{i=1}^{d} |\widehat{b}_i|}{\det \mathcal{L}} \leq s^{p(p-1)/2}.$$

This proves that the output satisfies good Euclidean properties. In particular, the length of the first vector of $\widehat{B}$ is an approximation of the first minimum $\lambda(\mathcal{L})$ – up to a factor that exponentially depends on dimension p.

Execution Parameters

The execution parameters are related to the execution of the algorithm itself : the length of the random walk (equal to the number of iterations K), the size of the integer translations, the size of the rationals $m_{i,j}$ along the execution.

The product D of the determinants D_j of beginning lattices $\mathcal{L}_j := \langle b_1, b_2, \ldots, b_j \rangle$, defined as

$$D_j := \prod_{i=1}^{j} \ell_i, \qquad D = \prod_{j=1}^{p-1} D_j = \prod_{j=1}^{p-1}\prod_{i=1}^{j} \ell_i,$$

is never increasing all along the algorithm and is strictly decreasing, with a factor of $(1/t)$, for each step of the algorithm when the test in 2 is negative. In this case, the exchange modifies the length of ℓ_i and ℓ_{i+1} – without modifying their product, equal to the determinant of the basis U_i. The new ℓ_i, denoted by $\check{\ell}_i$, is the old $|v_i|$, which is at most $(1/t)|u_i| = (1/t)\ell_i$. Then the ratio between the new determinant $\check{D}_i$ and the old one satisfies $\check{D}_i/D_i \leq (1/t)$, while the other D_j are not modified.

Then, the ratio between the final $\widehat{D}$ and the initial D satisfies $(\widehat{D}/D) \leq (1/t)^{K^-}$, where K^- denotes the number of indices of the random walk when the test in 2 is negative (see Section "Description of the Algorithm"). With the following bounds on the initial D and the final $\widehat{D}$, as a function of variables a, A, defined in (3.15),

$$D \leq A^{p(p-1)/2}, \qquad \widehat{D} \geq a^{p(p-1)/2},$$

together with the expression of K as a function of K^- given in (3.13), the following bound on K is derived,

$$K \leq (p-1) + p(p-1)\log_t \frac{A}{a}. \tag{3.20}$$

In the same vein, another kind of bound involves $N := \max |b_i|^2$ and the first minimum $\lambda(\mathcal{L})$, (see [15]),

$$K \leq \frac{p^2}{2} \log_t \frac{N\sqrt{p}}{\lambda(\mathcal{L})}.$$

In the case when the lattice is integer (namely $\mathcal{L} \subset \mathbb{Z}^n$), this bound is slightly better and becomes

$$K \leq (p-1) + p(p-1)\frac{M}{\lg t}.$$

It involves $\lg t := \log_2 t$ and the binary size M of B, defined as $M := \max \ell(|b_i|^2)$, where $\ell(x)$ is the binary size of integer x.

All the previous bounds are *proven upper bounds* on the main parameters. It is interesting to compare these bounds to *experimental mean values* obtained on a variety of lattice bases that actually occur in applications of lattice reduction.

Experiments for the LLL Algorithm

In [30], Nguyen and Stehlé have made a great use of their efficient version of the LLL algorithm [29] and conducted for the first time extensive experiments on the two major types of useful lattice bases: the Ajtai bases, and the knapsack-shape bases, which will be defined in the next section. Figures 3.3 and 3.4 show some of the main experimental results. These experimental results are also described in the survey written by D. Stehlé in these proceedings [36].

Main parameters.	$\widehat{r_i}$	γ	θ	ρ	K
Worst-case (Proven upper bounds)	$1/s$	s^{p-1}	s^{p-1}	$s^{p(p-1)/2}$	$\Theta(Mp^2)$
Random Ajtai bases (Experimental mean values)	$1/\alpha$	α^{p-1}	$\alpha^{(p-1)/2}$	$\alpha^{p(p-1)/2}$	$\Theta(Mp^2)$
Random knapsack–shape bases (Experimental mean values)	$1/\alpha$	α^{p-1}	$\alpha^{(p-1)/2}$	$\alpha^{p(p-1)/2}$	$\Theta(Mp)$

Fig. 3.3 Comparison between proven upper bounds and experimental mean values for the main parameters of interest. Here p is the dimension of the input (integer) basis and M is the binary size of the input (integer) basis: $M := \Theta(\log N)$, where $N := \max |b_i|^2$

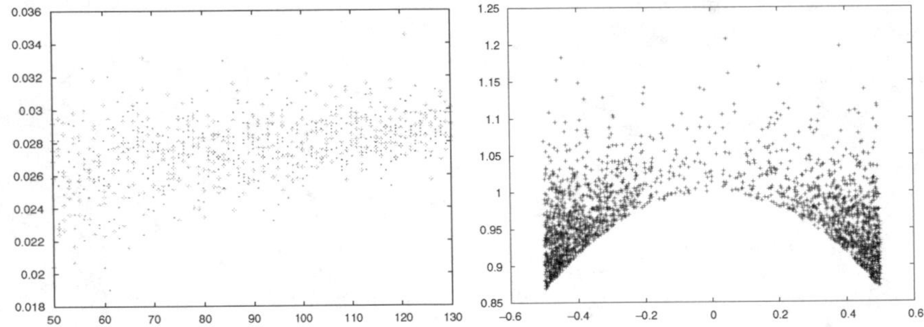

Fig. 3.4 *Left*: experimental results for $\log_2 \gamma$. The experimental value of parameter $[1/(2p)]\,\mathbb{E}[\log_2 \gamma]$ is close to 0.03, so that α is close to 1.04. *Right*: the output distribution of "local bases"

Output geometry. The geometry of the output local basis $\widehat{U_k}$ seems to depend neither on the class of lattice bases nor on index k of the local basis (along the diagonal of $\mathcal{P}$), except for very extreme values of k. We consider the complex number $\widehat{z_k}$ that is related to the output local basis $\widehat{U_k} := (\widehat{u_k}, \widehat{v_k})$ via the equality $\widehat{z_k} := \widehat{m}_{k,k+1} + i\widehat{r_k}$. Because of the t-Lovász conditions on $\widehat{U_k}$, described in (3.17), the complex number $\widehat{z_k}$ belongs to the domain

$$\mathcal{F}_t := \{z \in \mathbb{C}; \quad |z| \geq 1/t,\ |\Re(z)| \leq 1/2\},$$

and the geometry of the output local basis $\widehat{U_k}$ is characterized by a distribution, which much "weights" the "corners" of $\mathcal{F}_t$ defined by $\mathcal{F}_t \cap \{z; \Im z \leq 1/t\}$ [see Fig. 3.4 (right)]. The (experimental) mean values of the output Siegel ratios $\widehat{r_k} := \Im(\widehat{z_k})$ appear to be of the same form as the (proven) upper bounds, with a ratio α (close to 1.04), which replaces the ratio s_0 close to 1.15 when t_0 is close to 1. As a consequence, the (experimental) mean values of parameters $\gamma(B)$ and $\rho(B)$ appear to be of the same form as the (proven) upper bounds, with a ratio α (close to 1.04) that replaces the ratio s_0 close to 1.15.

For parameter $\theta(B)$, the situation is slightly different. Remark that the estimates on parameter θ are not only a consequence of the estimates on the Siegel ratios, but they also depend on estimates that relate the first minimum and the determinant. Most of the lattices are (probably) *regular*: this means that the average value of the ratio between the first minimum $\lambda(\mathcal{L})$ and $\det(\mathcal{L})^{1/p}$ is of polynomial order with respect to dimension p. This regularity property should imply that the experimental mean value of parameter θ is of the same form as the (proven) upper bound, but now with a ratio $\alpha^{1/2}$ (close to 1.02), which replaces the ratio s_0 close to 1.15.

Open Question. Does this constant α admit a mathematical definition, related for instance to the underlying dynamical system [see Sections "Returning to the Gauss Algorithm and First Steps in the Probabilistic Analysis of the LLL Algorithm"?

Execution parameters. Regarding the number of iterations, the situation differs according to the types of bases considered. For the Ajtai bases, the number of iterations K exhibits experimentally a mean value of the same order as the proven upper bound, whereas, in the case of the knapsack-shape bases, the number of iterations K has an experimental mean value of smaller order than the proven upper bound.

Open question. Is it true for the "actual" knapsack bases that come from cryptographic applications? [See Section "Probabilistic Models: Continuous or Discrete"]

All the remainder of this survey is devoted to presenting a variety of methods that could (should?) lead to explaining these experiments. One of our main ideas is to use the Gauss algorithm as a central tool for this purpose. This is why we now present a variant of the LLL algorithm, where the Gauss algorithm plays a more apparent rôle.

A Variation for the LLL Algorithm: The Odd-Even Algorithm

The original LLL algorithm performs the Gauss Algorithm *step by step*, but does not perform the *whole* Gauss algorithm on local bases. This is due to the definition of the random walk of the indices on the local bases (See Section "Description of the Algorithm"). However, this is not the only strategy for reducing all the local bases. There exists for instance a variant of the LLL algorithm, introduced by Villard [48], which performs a succession of phases of two types, the odd ones and the even ones. We adapt this variant and choose to perform the AGAUSS algorithm, because we shall explain in Section "Returning to the Gauss Algorithm" that it has a better "dynamical" structure.

During one even (respectively, odd) phase (see Figure 3.5), the *whole* AGAUSS algorithm is performed on all local bases U_i with even (respectively, odd) indices. Since local bases with odd (respectively, even) indices are "disjoint," it is possible to perform these Gauss algorithms *in parallel*. This is why Villard has introduced this algorithm. Here, we will use this algorithm in Section "First Steps in the Probabilistic Analysis of the LLL Algorithm", when we shall explain the main principles for a dynamical study of the LLL algorithm.

```
Odd–Even LLL (t)        [t > 1]

Input. A basis B of a lattice L of dimension p.
Output. A reduced basis B̂ of L.
Gram computes the basis B* and the matrix P.
While B is not reduced do
        Odd Phase (B):
                For i = 1 to ⌊n/2⌋ do
                        Diagonal-size-reduction (b_{2i});
                        M_i := t–AGAUSS (U_{2i−1});
                        (b_{2i−1}, b_{2i}) := (b_{2i−1}, b_{2i})^t M_i;
                For i = 1 to n do Other-size-reduction (b_i);
                Recompute B*, P;
        Even Phase (B):
                For i = 1 to ⌊(n − 1)/2⌋ do
                        Diagonal-size-reduction (b_{2i+1});
                        M_i := t–AGAUSS (U_{2i});
                        (b_{2i}, b_{2i+1}) := (b_{2i}, b_{2i+1})^t M_i;
                For i = 1 to n do Other-size-reduction (b_i);
                Recompute B*, P;
```

Fig. 3.5 Description of the Odd–Even variant of the LLL algorithm, with its two phases, the Odd Phase and the Even Phase

Consider, for an odd index k, two successive bases $U_k := (u_k, v_k)$ and $U_{k+2} := (u_{k+2}, v_{k+2})$. Then, the Odd Phase of the Odd–Even LLL algorithm (completely) reduces these two local bases (in the t-Gauss meaning) and computes two reduced local bases denoted by $(\widehat{u}_k, \widehat{v}_k)$ and $(\widehat{u}_{k+2}, \widehat{v}_{k+2})$, which satisfy in particular

$$|\widehat{v}_k^{\star}| = \mu(u_k, v_k), \qquad |\widehat{u}_{k+2}| = \lambda(u_{k+2}, v_{k+2}),$$

where parameters λ, μ are defined in (3.11). During the Even phase, the LLL algorithm considers (in parallel) all the local bases with an even index. Now, at the beginning of the following Even Phase, the (input) basis U_{k+1} is formed (up to a similarity) from the two previous output bases, as $u_{k+1} = \widehat{v}_k^{\star}$, $v_{k+1} = \nu\widehat{v}_k^{\star} + \widehat{u}_{k+2}$, where ν is a real number of the interval $[-1/2, +1/2]$. Then, the initial Siegel ratio r_{k+1} of the Even Phase can be expressed with the output lengths of the Odd Phase, as

$$r_{k+1} = \frac{\lambda(u_{k+2}, v_{k+2})}{\mu(u_k, v_k)}.$$

This explains the important rôle that is played by these parameters λ, μ. We study these parameters in Section "Analysis of Lattice Reduction in Two-Dimensions: The Output Parameters".

What is a Random (Basis of a) Lattice?

We now describe the main probabilistic models, addressing the various applications of lattice reduction. For each particular area, there are special types of input lattice bases that are used and this leads to different probabilistic models dependent upon the specific application area considered. Cryptology is a main application area, and it is crucial to describe the major "cryptographic" lattices, but there also exist other important applications.

There are various types of "interesting" lattice bases. Some of them are also described in the survey of Stehlé in this book [36].

Spherical Models

The most natural way is to choose independently p vectors in the n-dimensional unit ball, under a distribution that is invariant by rotation. This is the spherical model introduced for the first time in [15], then studied in [3,4] (See Section "Probabilistic Analyses of the LLL Algorithm in the Spherical Model"). This model does not seem to have surfaced in practical applications (except perhaps in integer linear programming), but it constitutes a reference model, to which it is interesting to compare the realistic models of use.

We consider distributions $\nu_{(n)}$ on $\mathbb{R}^n$ that are invariant by rotation, and satisfy $\nu_{(n)}(0) = 0$, which we call "simple spherical distributions." For a simple spherical distribution, the angular part $\theta_{(n)} := b_{(n)}/|b_{(n)}|$ is uniformly distributed on the unit sphere $\S_{(n)} := \{x \in \mathbb{R}^n : \|x\| = 1\}$. Moreover, the radial part $|b_{(n)}|^2$ and the angular part are independent. Then, a spherical distribution is completely determined by the distribution of its radial part, denoted by $\rho_{(n)}$.

Here, the beta and gamma distribution play an important rôle. Let us recall that, for strictly positive real numbers $a, b \in \mathbb{R}^{+\star}$, the beta distribution of parameters (a, b) denoted by $\beta(a, b)$ and the gamma distribution of parameter a denoted by $\gamma(a)$ admit densities of the form

$$\beta_{a,b}(x) = \frac{\Gamma(a+b)}{\Gamma(a)\Gamma(b)} x^{a-1}(1-x)^{b-1}\, \mathbf{1}_{(0,1)}(x), \qquad \gamma_a(x) = \frac{e^{-x}x^{a-1}}{\Gamma(a)}\, \mathbf{1}_{[0,\infty)}(x). \tag{3.21}$$

We now describe three natural instances of simple spherical distributions.

1. The first instance of a simple spherical distribution is the uniform distribution in the unit ball $\mathcal{B}_{(n)} := \{x \in \mathbb{R}^n : \|x\| \le 1\}$. In this case, the radial distribution $\rho_{(n)}$ equals the beta distribution $\beta(n/2, 1)$.
2. A second instance is the uniform distribution on the unit sphere $\mathbb{S}_{(n)}$, where the radial distribution $\rho_{(n)}$ is the Dirac measure at $x = 1$.
3. A third instance occurs when all the n coordinates of the vector $b_{(n)}$ are independent and distributed with the standard normal law $\mathcal{N}(0, 1)$. In this case, the radial distribution $\rho_{(n)}$ has a density equal to $2\gamma_{n/2}(2t)$.

When the system $B_{p,(n)}$ is formed with p vectors (with $p \leq n$), which are picked up randomly from $\mathbb{R}^n$, independently, and with the same simple spherical distribution $\nu_{(n)}$, we say that the system $B_{p,(n)}$ is distributed under a "spherical model." Under this model, the system $B_{p,(n)}$ (for $p \leq n$) is almost surely linearly independent.

Ajtai Bases

Consider an integer sequence $a_{i,p}$ defined for $1 \leq i \leq p$, which satisfies the conditions

$$\text{For any } i, \quad \frac{a_{i+1,p}}{a_{i,p}} \to 0 \qquad \text{when } p \to \infty.$$

A sequence of Ajtai bases $B := (B_p)$ relative to the sequence $a = (a_{i,p})$ is defined as follows: the basis B_p is of dimension p and is formed by vectors $b_{i,p} \in \mathbb{Z}^p$ of the form

$$b_{i,p} = a_{i,p}\, e_i + \sum_{j=1}^{i-1} a_{i,j,p}\, e_j, \quad \text{with } a_{i,j,p} = \text{rand}\left(-\frac{a_{j,p}}{2}, \frac{a_{j,p}}{2}\right) \quad \text{for } j < i.$$

[Here, (e_j) (with $1 \leq j \leq p$) is the canonical basis of $\mathbb{R}^p$]. Remark that these bases are already size-reduced, as the coefficient $m_{i,j}$ equals $a_{i,j,p}/a_{j,p}$. However, all the input Siegel ratios r_i, defined in (3.14) and here equal to $a_{i+1,p}/a_{i,p}$, tend to 0 when p tends to ∞. Then, such bases are not reduced "at all," and this explains why similar bases have been used by Ajtai in [2] to show the tightness of worst-case bounds of [32].

Variations Around Knapsack Bases and Their Transposes

This last type gathers various shapes of bases, which are all formed by "bordered identity matrices"; see Fig. 3.6.

1. The knapsack bases themselves are the rows of the $p \times (p+1)$ matrices of the form of Fig. 3.6a, where I_p is the identity matrix of order p and the components $(a_1, a_2, \ldots a_p)$ of vector A are sampled independently and uniformly in $[-N, N]$ for some given bound N. Such bases often occur in cryptanalyses of knapsack-based cryptosystems or in number theory (reconstructions of minimal polynomials and detections of integer relations between real numbers).
2. The bases relative to the transposes of matrices described in Fig. 3.6b arise in searching for simultaneous Diophantine approximations (with $q \in \mathbb{Z}$) or in discrete geometry (with $q = 1$).

$$(A|I_p) \quad \left(\begin{array}{c|c} y & 0 \\ \hline x & qI_p \end{array}\right) \quad \left(\begin{array}{c|c} I_p & H_p \\ \hline 0_p & qI_p \end{array}\right) \quad \left(\begin{array}{c|c} q & 0 \\ \hline x & I_{n-1} \end{array}\right)$$

(a) (b) (c) (d)

Fig. 3.6 Different kinds of lattice bases useful in applications. Type (**a**) Knapsack bases; Type (**b**) bases used for factoring polynomials, for solving Diophantine equations; Type (**c**) Bases for NTRU; Type (**d**) bases related to random lattices

3. The NTRU cryptosystem was first described in terms of polynomials over finite fields, but the public-key can be seen [12] as the lattice basis given by the rows of the matrix $(2p \times 2p)$ described in Fig. 3.6c, where q is a small power of 2 and H_p is a circulant matrix whose line coefficients are integers of the interval $]-q/2, q/2]$.

Random Lattices

There is a natural notion of random lattice, introduced by Siegel [34] in 1945. The space of (full-rank) lattices in $\mathbb{R}^n$ modulo scale can be identified with the quotient $\mathbb{X}_n = SL_n(\mathbb{R})/SL_n(\mathbb{Z})$. The group $G_n = SL_n(\mathbb{R})$ possesses a unique (up to scale) bi-invariant Haar measure, which projects to a finite measure on the space $\mathbb{X}_n$. This measure ν_n (which can be normalized to have total volume 1) is by definition the unique probability on $\mathbb{X}_n$, which is invariant under the action of G_n: if $A \subseteq \mathbb{X}_n$ is measurable and $g \in G_n$, then $\nu_n(A) = \nu_n(gA)$. This gives rise to a natural notion of random lattices. We come back to this notion in the two-dimensional case in Section "Relation with Eisenstein Series".

Probabilistic Models: Continuous or Discrete

Except two models – the spherical model or the model of random lattices – that are *continuous* models, all the other ones (the Ajtai model or the various knapsack-shape models) are discrete models. In these cases, it is natural to build probabilistic models that preserve the "shape" of matrices and replace discrete coefficients by continuous ones. This allows to use in the probabilistic studies all the continuous tools of (real and complex) analysis.

1. A first instance is the Ajtai model relative to sequence $a := (a_{i,p})$, for which the continuous version of dimension p is as follows:

$$b_{i,p} = a_{i,p}\, e_i + \sum_{j=1}^{i-1} x_{i,j,p}\, a_{j,p}\, e_j, \quad \text{with } x_{i,j,p} = \text{rand}\,(-1/2, 1/2)$$

for all $j < i \le p$.

2. We may also replace the discrete model associated to knapsack bases of Fig. 3.6a by the continuous model, where A is replaced by a real vector x uniformly chosen in the ball $\|x\|_\infty \leq 1$ and I_p is replaced by ρI_p, with a small positive constant $0 < \rho < 1$. Generally speaking, choosing continuous random matrices independently and uniformly in their "shape" class leads to a class of "knapsack-shape" lattices.

Remark 1. It is very unlikely that such knapsack-shape lattices share all the same properties as the knapsack lattices that come from the actual applications – for instance, the existence of an unusually short vector (significantly shorter than expected from Minkowski's theorem).

Conversely, we can associate to any continuous model a discrete one: consider a domain $\mathcal{X} \subset \mathbb{R}^n$ with a "smooth" frontier. For any integer N, we can "replace" a (continuous) distribution in the domain $\mathcal{X}$ relative to some density f of class $\mathcal{C}^1$ by the distribution in the discrete domain

$$\mathcal{X}_N := \mathcal{X} \cap \frac{\mathbb{Z}^n}{N},$$

defined by the restriction f_N of f to $\mathcal{X}_N$. When $N \to \infty$, the distribution relative to density f_N tends to the distribution relative to f, due to the Gauss principle, which relates the volume of a domain $\mathcal{A} \subset \mathcal{X}$ (with a smooth frontier $\partial\mathcal{A}$) and the number of points in the domain $\mathcal{A}_N := \mathcal{A} \cap \mathcal{X}_N$,

$$\frac{1}{N^n}\operatorname{card}(\mathcal{A}_N) = \operatorname{Vol}(\mathcal{A}) + O\left(\frac{1}{N}\right) \operatorname{Area}(\partial\mathcal{A}).$$

We can apply this framework to any (simple) spherical model and also to the models that are introduced for the two-dimensional case.

In the same vein, we can consider a discrete version of the notion of a random lattice: consider the set $\mathcal{L}(n, N)$ of the n-dimensional integer lattices of determinant N. Any lattice of $\mathcal{L}(n, N)$ can be transformed into a lattice of $\mathbb{X}_n$ (defined in 4.4) by the homothecy Ψ_N of ratio $N^{-1/n}$. Goldstein and Mayer [20] show that for large N, the following is true: given any measurable subset $A_n \subseteq \mathbb{X}_n$ whose boundary has zero measure with respect to ν_n, the proportion of lattices of $\mathcal{L}(n, N)$ whose image by Ψ_N lies in A_n tends to $\nu_n(A)$ as N tends to infinity. In other words, the image by Ψ_N of the uniform probability on $\mathcal{L}(n, N)$ tends to the measure ν_n.

Thus, to generate lattices that are random in a natural sense, it suffices to generate uniformly at random a lattice in $\mathcal{L}(n, N)$ for large N. This is particularly easy when $N = q$ is prime. Indeed, when q is a large prime, the vast majority of lattices in $\mathcal{L}(n, q)$ are lattices spanned by rows of the matrices described in Fig. 3.6d, where the components x_i (with $i \in [1..n-1]$) of the vector x are chosen independently and uniformly in $\{0, \ldots, q-1\}$.

Probabilistic Analyses of the LLL Algorithm in the Spherical Model

In this section, the dimension of the ambient space is denoted by n, and the dimension of the lattice is denoted by p, and a basis of dimension p in $\mathbb{R}^n$ is denoted by $B_{p,(n)}$. The codimension g, equal by definition to $n - p$, plays a fundamental rôle here. We consider the case where n tends to ∞ while $g := g(n)$ is a fixed function of n (with $g(n) \leq n$). We are interested in the following questions:

1. Consider a real $s > 1$. What is the probability $\pi_{p,(n),s}$ that a random basis $B_{p,(n)}$ was already s-reduced in the Siegel sense [i.e., satisfy the relations (3.18)]?
2. Consider a real $t > 1$. What is the average number of iterations of the LLL(t) algorithm on a random basis $B_{p,(n)}$?
3. What is the mean value of the first minimum of the lattice generated by a random basis $B_{p,(n)}$?

This section answers these questions in the case when $B_{p,(n)}$ is randomly chosen under a spherical model, and shows that there are two main cases according to the codimension $g := n - p$.

Main Parameters of Interest

Let $B_{p,(n)}$ be a linearly independent system of vectors of $\mathbb{R}^n$ whose codimension is $g = n - p$. Let $B^\star_{p,(n)}$ be the associated Gram–Schmidt orthogonalized system. We are interested by comparing the lengths of two successive vectors of the orthogonalized system, and we introduce several parameters related to the Siegel reduction of the system $B_{p,(n)}$.

Definition 1. *To a system $B_{p,(n)}$ of p vectors in $\mathbb{R}^n$, we associate the Gram–Schmidt orthogonalized system $B^\star_{p,(n)}$ and the sequence $\underline{r}_{j,(n)}$ of Siegel ratios, defined as*

$$\underline{r}_{j,(n)} := \frac{\ell_{n-j+1,(n)}}{\ell_{n-j,(n)}}, \quad \text{for } g+1 \leq j \leq n-1,$$

together with two other parameters

$$\mathcal{M}_{g,(n)} := \min\{\underline{r}^2_{j,(n)};\ g+1 \leq j \leq n-1\} \quad \mathcal{I}_{g,(n)} := \min\left\{j : \underline{r}^2_{j,(n)} = \mathcal{M}_{g,(n)}\right\}.$$

The parameter $\mathcal{M}_{g,(n)}$ is the reduction level, *and the parameter $\mathcal{I}_{g,(n)}$ is the* index of worst local reduction.

Remark 2. The ratio $\underline{r}_{j,(n)}$ is closely related to the ratio r_i defined in Section "Description of the Algorithm" [see (3.14)]. There are two differences: the rôle of the ambient dimension n is made apparent, and the indices i and j are related via

$\underline{r}_j := r_{n-j}$. The rôle of this "time inversion" will be explained later. The variable $\mathcal{M}_{g,(n)}$ is the supremum of the set of those $1/s^2$ for which the basis $B_{n-g,(n)}$ is s-reduced in the Siegel sense. In other words, $1/\mathcal{M}_{g,(n)}$ denotes the infimum of values of s^2 for which the basis $B_{n-g,(n)}$ is s-reduced in the Siegel sense. This variable is related to our initial problem due to the equality

$$\pi_{n-g,(n),s} := \mathbb{P}[B_{n-g,(n)} \text{is } s\text{–reduced}] = \mathbb{P}\left[\mathcal{M}_{g,(n)} \geq \frac{1}{s^2}\right],$$

and we wish to evaluate the limit distribution (if it exists) of $\mathcal{M}_{g,(n)}$ when $n \to \infty$. The second variable $\mathcal{I}_{g,(n)}$ denotes the smallest index j for which the Siegel condition relative to the index $n-j$ is the weakest. Then $n - \mathcal{I}_{g,(n)}$ denotes the largest index i for which the Siegel condition relative to index i is the weakest. This index indicates where the limitation of the reduction comes from.

When the system $B_{p,(n)}$ is chosen at random, the Siegel ratios, the reduction level, and the index of worst local reduction are random variables, well-defined whenever $B_{p,(n)}$ is a linearly independent system. We wish to study the asymptotic behavior of these random variables (with respect to the dimension n of the ambient space) when the system $B_{p,(n)}$ is distributed under a so-called (concentrated) spherical model, where the radial distribution $\rho_{(n)}$ fulfills the following *Concentration Property* $\mathcal{C}$.

Concentration Property $\mathcal{C}$. *There exist a sequence* $(a_n)_n$ *and constants* d_1, d_2, $\alpha>0$, $\theta_0 \in (0,1)$ *such that, for every* n *and* $\theta \in (0, \theta_0)$, *the distribution function* $\rho_{(n)}$ *satisfies*

$$\rho_{(n)}\left(a_n(1+\theta)\right) - \rho_{(n)}\left(a_n(1-\theta)\right) \geq 1 - d_1\,\mathrm{e}^{-nd_2\theta^{\alpha}}. \tag{3.22}$$

In this case, it is possible to transfer results concerning the uniform distribution on $\mathbb{S}_{(n)}$ [where the radial distribution is Dirac] to more general spherical distributions, provided that the radial distribution be concentrated enough. This *Concentration Property* $\mathcal{C}$ holds in the three main instances previously described of simple spherical distributions.

We first recall some definitions of probability theory, and define some notations:

A sequence (X_n) *of real random variables converges in distribution towards the real random variable* X *iff the distribution function* F_n *of* X_n *is pointwise convergent to the distribution function* F *of* X *on the set of continuity points of* F. *A sequence* (X_n) *of real random variables converges in probability to a constant* a *if, for any* $\varepsilon > 0$, *the sequence* $\mathbb{P}[|X_n - a| > \varepsilon]$ *tends to 0. The two situations are respectively denoted as*

$$X_n \xrightarrow[n]{(d)} X, \qquad X_n \xrightarrow[n]{proba.} a.$$

We now state the main results of this section, and provide some hints for the proof.

Theorem 1. (Akhavi et al. [4] 2005)
Let $B_{p,(n)}$ be a random basis with codimension $g := n - p$ under a concentrated spherical model. Let $s > 1$ be a real parameter, and suppose that the dimension n of the ambient space tends to ∞.

i. If $g := n - p$ tends to infinity, then the probability $\pi_{p,(n),s}$ that $B_{p,(n)}$ is already s–reduced tends to 1.
ii. If $g := n - p$ is constant, then the probability $\pi_{p,(n),s}$ that $B_{p,(n)}$ is already s-reduced converges to a constant in $(0, 1)$ (depending on s and g). Furthermore, the index of worst local reduction $\mathcal{I}_{g,(n)}$ converges in distribution.

The Irruption of β and γ Laws

When dealing with the Gram–Schmidt orthogonalization process, beta and gamma distributions are encountered in an extensive way. We begin to study the variables $Y_{j,(n)}$ defined as

$$Y_{j,(n)} := \frac{\ell^2_{j,(n)}}{|b_{j,(n)}|^2} \qquad \text{for } j \in [2..n],$$

and we show that they admit beta distributions.

Proposition 2. (Akhavi et al. [4] 2005)

1. *Under any spherical model, the variables $\ell^2_{j,(n)}$ are independent.*
   Moreover, the variable $Y_{j,(n)}$ follows the beta distribution $\beta((n - j + 1)/2, (j - 1)/2)$ for $j \in [2..n]$, and the set $\{Y_{j,(n)}, |b_{k,(n)}|^2;\ (j, k) \in [2..n] \times [1..n]\}$ is formed with independent variables.
2. *Under the random ball model $\mathbb{U}_n$, the variable $\ell^2_{j,(n)}$ follows the beta distribution $\beta((n - j + 1)/2, (j + 1)/2)$.*

Proposition 2 is now used for showing that, under a concentrated spherical model, the beta and gamma distributions will play a central rôle in the analysis of the main parameters of interest introduced in Definition 1.

Denote by $(\eta_i)_{i\geq 1}$ a sequence of independent random variables where η_i follows a Gamma distribution $\gamma(i/2)$ and consider, for $k \geq 1$, the following random variables

$$\mathcal{R}_k = \eta_k/\eta_{k+1},\ \mathcal{M}_k = \min\{\mathcal{R}_j;\ j \geq k+1\},\ \mathcal{I}_k = \min\{j \geq k+1;\ \mathcal{R}_j = \mathcal{M}_k\}.$$

We will show in the sequel that they intervene as the limits of variables (of the same name) defined in Definition 1. There are different arguments in the proof of this fact.

(a) Remark first that, for the indices of the form $n - i$ with i fixed, the variable $\underline{r}^2_{n-i,(n)}$ tends to 1 when $n \to \infty$. It is then convenient to extend the tuple $(\underline{r}_{j,(n)})$ (only defined for $j \leq n - 1$) into an infinite sequence by setting $\underline{r}_{k,(n)} := 1$ for any $k \geq n$.

(b) Second, the convergence

$$\mathcal{R}_j \xrightarrow[j]{a.s.} 1, \qquad \sqrt{k}(\mathcal{R}_k - 1) \xrightarrow[k]{(d)} \mathcal{N}(0,4),$$

leads to consider the sequence $(\mathcal{R}_k - 1)_{k\geq 1}$ as an element of the space $\mathcal{L}_q$, for $q > 2$. We recall that

$$\mathcal{L}_q := \{x, \|x\|_q < +\infty\}, \text{ with } \|x\|_q := \left(\sum_{i\geq 1} |x_i|^q\right)^{1/q}, \text{ for } x = (x_i)_{i\geq 1}.$$

(c) Finally, classical results about independent gamma and beta distributed random variables, together with the weak law of large numbers and previous Proposition 2, prove that

$$\text{For each } j \geq 1, \ \underline{r}_{j,(n)}^2 \xrightarrow[n]{(d)} \mathcal{R}_j. \tag{3.23}$$

This suggests that the minimum $\mathcal{M}_{g,(n)}$ is reached by the $\underline{r}_{j,(n)}^2$ corresponding to smallest indices j and motivates the "time inversion" done in Definition 1.

The Limit Process

It is then possible to prove that the processes $R_{(n)} := (\underline{r}_{k,(n)} - 1)_{k\geq 1}$ converge (in distribution) to the process $R := (\mathcal{R}_k - 1)_{k\geq 1}$ inside the space $\mathcal{L}_q$ when the dimension n of the ambient space tends to ∞. As $\mathcal{M}_{g,(n)}$ and $\mathcal{I}_{g,(n)}$ are continuous functionals of the process $R_{(n)}$, they also converge in distribution, respectively, to $\mathcal{M}_g$ and $\mathcal{I}_g$.

Theorem 2. (Akhavi et al. [4] 2005) *For any concentrated spherical distribution, the following holds:*

1. *The convergence* $(\underline{r}_{k,(n)}^2 - 1)_{k\geq 1} \xrightarrow[n]{(d)} (\mathcal{R}_k - 1)_{k\geq 1}$ *holds in any space* $\mathcal{L}_q$, *with* $q > 2$.
2. *For any fixed* k, *one has* $\mathcal{M}_{k,(n)} \xrightarrow[n]{(d)} \mathcal{M}_k, \mathcal{I}_{k,(n)} \xrightarrow[n]{(d)} \mathcal{I}_k$.
3. *For any sequence* $n \mapsto g(n)$ *with* $g(n) \leq n$ *and* $g(n) \to \infty$, *the convergence* $\mathcal{M}_{g(n),(n)} \xrightarrow[n]{proba.} 1$ *holds.*

This result solves our problem and proves Theorem 1. We now give some precisions on the limit processes $\sqrt{\mathcal{R}_k}$, $\sqrt{\mathcal{M}_k}$, and describe some properties of the distribution function F_k of $\sqrt{\mathcal{M}_k}$, which is of particular interest due to the equality $\lim_{n\to\infty} \pi_{n-k,(n),s} = 1 - F_k(1/s)$.

Proposition 3. (Akhavi et al. [4] 2005) *The limit processes $\sqrt{\mathcal{R}_k}, \sqrt{\mathcal{M}_k}$ admit densities that satisfy the following:*

1. *For each k, the density φ_k of $\sqrt{\mathcal{R}_k}$ is*

$$\varphi_k(x) = 2B\left(\frac{k}{2}, \frac{k+1}{2}\right) \frac{x^{k-1}\mathbf{1}_{[0,\infty[}(x)}{(1+x^2)^{k+(1/2)}}, \quad \textit{with } B(a,b) := \frac{\Gamma(a)\Gamma(b)}{\Gamma(a+b)}. \tag{3.24}$$

2. *For each k, the random variables $\sqrt{\mathcal{M}_k}, \mathcal{M}_k$ have densities, which are positive on $(0,1)$ and zero outside. The distribution functions F_k, G_k satisfy for x near 0, and for each k,*

$$\Gamma\left(\frac{k+2}{2}\right) F_k(x) \sim x^{k+1}, \qquad G_k(x) = F_k(\sqrt{x}).$$

There exists τ such that, for each k and for $x \in [0,1]$ satisfying $|x^2 - 1| \le (1/\sqrt{k})$,

$$0 \le 1 - F_k(x) \le \exp\left[-\left(\frac{\tau}{1-x^2}\right)^2\right].$$

3. *For each k, the cardinality of the set $\{j \ge k+1; \quad \mathcal{R}_j = \mathcal{M}_k\}$ is almost surely equal to 1.*

In particular, for a full-dimensional lattice,

$$\lim_{n\to\infty} \pi_{n,(n),s} \sim_{s\to\infty} 1 - \frac{1}{s}, \quad \lim_{n\to\infty} \pi_{n,(n),s} \le \exp\left[-\left(\frac{\tau s^2}{s^2-1}\right)^2\right] \quad \text{when } s \to 1.$$

Figure 3.7 shows some experiments in the case of a full-dimensional lattice ($g = 0$). In this case, the density g_0 of $\mathcal{M}_0$ is proven to be $\Theta(1/\sqrt{x})$ when $x \to 0$ and tends rapidly to 0 when $x \to 1$. Moreover, the same figure shows that the worst reduction level for a full-dimensional lattice is almost always very small: that means that the first index i where the test in step 2 of the LLL algorithm (see Section "Description of the Algorithm") is negative is very close to n.

These (probabilistic) methods do not provide any information about the speed of convergence of $\pi_{n-g,(n)}$ towards 1 when n and g tend to ∞. In the case of the random ball model, Akhavi directly deals with the beta law of the variables ℓ_i and observes that

$$\begin{aligned} 1 - \pi_{p,(n),s} &\le \sum_{i=1}^{p-1} \mathbb{P}\left[\ell_{i+1} \le \frac{1}{s}\ell_i\right] \le \sum_{i=1}^{p-1} \mathbb{P}\left[\ell_{i+1} \le \frac{1}{s}\right] \\ &\le \sum_{i=1}^{p-1} \exp\left[\frac{n}{2} H\left(\frac{i}{n}\right)\right]\left(\frac{1}{s}\right)^{n-i}, \end{aligned}$$

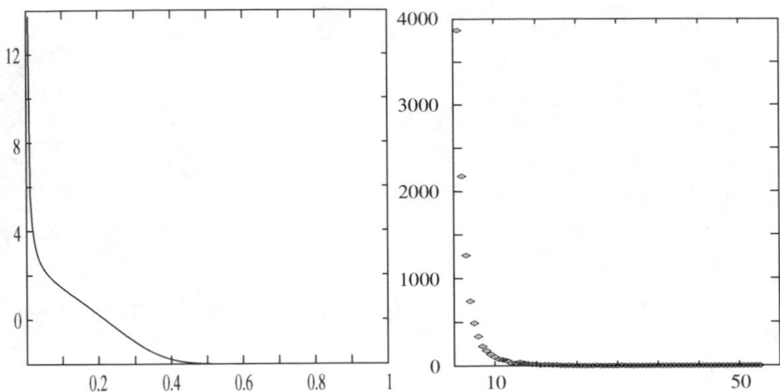

Fig. 3.7 *Left*: simulation of the density of $\mathcal{M}_0$ with 10^8 experiments. *Right*: the histogram of $\mathcal{I}_0$ provided by 10^4 simulations. For any g, the sequence $k \mapsto \P[\mathcal{I}_g = k]$ seems to be rapidly decreasing

where H is the entropy function defined as $H(x) = -x \log x - (1-x)\log(1-x)$, for $x \in [0,1]$, which satisfies $0 \le H(x) \le \log 2$. This proves :

Proposition 4. (Akhavi [3] 2000) *Under the random ball model, the probability that a basis $B_{p,(n)}$ be reduced satisfies, for any n, for any $p \le n$, for any $s > 1$,*

$$1 - \pi_{p,(n),s} \le \frac{1}{s-1} (\sqrt{2})^n \left(\frac{1}{s}\right)^{n-p}.$$

In particular, for any $s > \sqrt{2}$, the probability that $B_{cn,(n)}$ be s-reduced tends exponentially to 1, provided $1-c$ is larger than $1/(2\lg s)$.

A First Probabilistic Analysis of the LLL Algorithm

In the case of the random ball model, Daudé and Vallée directly deal with the beta law of the variables ℓ_i and obtain estimates for the average number of iterations K and the first minimum $\lambda(\mathcal{L})$. They consider the case of the full-dimensional lattices, namely the case when $p = n$. However, their proof can be extended to the case of a basis $B_{p,(n)}$ in the random ball model with $p \le n$.

Using properties of the beta function, they first obtain a simple estimate for the distribution for the parameter ℓ_i,

$$\mathbb{P}[\ell_i \le u] \le (u\sqrt{n})^{n-i+1}$$

and deduce that the random variable $a := \min \ell_i$ satisfies

$$\mathbb{P}[a \le u] \le \sum_{i=1}^{p} \mathbb{P}[\ell_i \le u] \le (2\sqrt{n})u^{n-p+1}, \quad \mathbb{E}\left[\log\left(\frac{1}{a}\right)\right] \le \frac{1}{n-p+1}\left[\frac{1}{2}\log n + 2\right].$$

The result then follows from (3.16) and (3.20). It shows that, as previously, there are two regimes according to the dimension p of the basis relative to the dimension n of the ambient space.

Theorem 3. (Daudé and Vallée [15] 1994) *Under the random ball model, the number of iterations K of the LLL algorithm on $B_{p,(n)}$ has a mean value satisfying*

$$\mathbb{E}_{p,(n)}[K] \le p - 1 + \frac{p(p-1)}{n-p+1}\left(\frac{1}{\log t}\right)\left[\frac{1}{2}\log n + 2\right].$$

Furthermore, the first minimum of the lattice generated by $B_{p,(n)}$ satisfies

$$\mathbb{E}_{p,(n)}[\lambda(\mathcal{L})] \ge \frac{n-p+1}{n-p+2}\left(\frac{1}{2\sqrt{n}}\right)^{1/(n-p+1)}.$$

In the case when $p = cn$, with $c < 1$,

$$\mathbb{E}_{cn,(n)}[K] \le \frac{cn}{1-c}\left(\frac{1}{\log t}\right)\left[\frac{1}{2}\log n + 2\right],$$

$$\mathbb{E}_{cn,(n)}[\lambda(\mathcal{L})] \ge \exp\left[\frac{1}{2(1-c)n}\log\frac{1}{4n}\right].$$

Conclusion of the Probabilistic Study in the Spherical Model

In the spherical model, and when the ambient dimension n tends to ∞, all the local bases (except perhaps the "last" ones) are s-Siegel reduced. For the last ones, at indices $i := n - k$, for fixed k, the distribution of the ratio r_i admits a density φ_k, which is given by Proposition 5.5. Both when $x \to 0$ and when $x \to \infty$, the density φ_k has a behavior of power type $\varphi_k(x) = \Theta(x^{k-1})$ for $x \to 0$, and $\varphi_k(x) = \Theta(x^{-k-2})$ for $x \to \infty$. It is clear that the potential degree of reduction of the local basis of index k is decreasing when k is decreasing. It will be interesting in the sequel to consider local bases with an initial density of this power type. However, the exponent of the density and the index of the local basis may be chosen independent, and the exponent is no longer integer. This type of choice provides a class of input local bases with different potential degree of reduction and leads to the so-called model "with valuation," which will be introduced in the two-dimensional case in Section "Probabilistic Models for Two-Dimensions" and studied in Sections "Analysis of Lattice Reduction in Two-Dimensions: The Output Parameters" and "Analysis of the Execution Parameters of the Gauss Algorithm".

Returning to the Gauss Algorithm

We return to the two-dimensional case, and describe a complex version for each of the two versions of the Gauss algorithm. This leads to consider each algorithm as a dynamical system, which can be seen as a (complex) extension of a (real) dynamical system relative to a centered Euclidean algorithm. We provide a precise description of the linear fractional transformations (LFTs) used by each algorithm. We finally describe the (two) classes of probabilistic models of interest.

The complex Framework

Many structural characteristics of lattices and bases are invariant under linear transformations – similarity transformations in geometric terms – of the form $S_\lambda : u \mapsto \lambda u$ with $\lambda \in \mathbb{C} \setminus \{0\}$.

(a) A first instance is the execution of the Gauss algorithm itself: it should be observed that translations performed by the Gauss algorithms depend only on the quantity $\tau(v,u)$ defined in (3.2), which equals $\Re(v/u)$. Furthermore, exchanges depend on $|v/u|$. Then, if v_i (or w_i) is the sequence computed by the algorithm on the input (u,v), defined in (3.3) and (3.5), the sequence of vectors computed on an input pair $S_\lambda(u,v)$ coincides with the sequence $S_\lambda(v_i)$ (or $S_\lambda(w_i)$). This makes it possible to give a formulation of the Gauss algorithm entirely in terms of complex numbers.

(b) A second instance is the characterization of minimal bases given in Proposition 2.1 that only depends on the ratio $z = v/u$.

(c) A third instance are the main parameters of interest: the execution parameters D, C, d defined in (3.7), (3.9), and (3.10) and the output parameters λ, μ, γ defined in (3.11) and (3.12). All these parameters admit also complex versions: for $X \in \{\lambda, \mu, \gamma, D, C, d\}$, we denote by $X(z)$ the value of X on basis $(1,z)$. Then, there are close relations between $X(u,v)$ and $X(z)$ for $z = v/u$:

$$X(z) = \frac{X(u,v)}{|u|}, \text{ for } X \in \{\lambda, \mu\}, \quad X(z) = X(u,v), \text{ for } X \in \{D, C, d, \gamma\}.$$

It is thus natural to consider lattice bases taken up to equivalence under similarity, and it is sufficient to restrict attention to lattice bases of the form $(1,z)$. We denote by $L(z)$ the lattice $\mathcal{L}(1,z)$. In the complex framework, the geometric transformation effected by each step of the algorithm consists of an inversion-symmetry $S : z$

$\mapsto 1/z$, followed by a translation $z \mapsto T^{-q}z$ with $T(z) = z + 1$, and a possible sign change $J : z \mapsto -z$.

The upper half plane $\mathbb{H} := \{z \in \mathbb{C};\ \Im(z) > 0\}$ plays a central rôle for the PGAUSS Algorithm, while the right half plane $\{z \in \mathbb{C};\ \Re(z) \geq 0, \Im(z) \neq 0\}$ plays a central rôle in the AGAUSS algorithm. Remark just that the right half plane is the union $\mathbb{H}_+ \cup J\mathbb{H}_-$, where $J : z \mapsto -z$ is the sign change and

$$\mathbb{H}_+ := \{z \in \mathbb{C};\ \Im(z) > 0,\ \Re(z) \geq 0\},\ \mathbb{H}_- := \{z \in \mathbb{C};\ \Im(z) > 0,\ \Re(z) \leq 0\}.$$

The Complex Versions for the GAUSS *Algorithms*

In this complex context, the PGAUSS algorithm brings z into the vertical strip $\mathcal{B} = \mathcal{B}_+ \cup \mathcal{B}_-$, with

$$\mathcal{B} = \left\{ z \in \mathbb{H};\quad |\Re(z)| \leq \frac{1}{2} \right\}, \qquad \mathcal{B}_+ := \mathcal{B} \cap \mathbb{H}_+, \quad \mathcal{B}_- := \mathcal{B} \cap \mathbb{H}_-,$$

reduces to the iteration of the mapping

$$U(z) = -\frac{1}{z} + \left\lfloor \Re\left(\frac{1}{z}\right) \right\rceil = -\left(\frac{1}{z} - \left\lfloor \Re\left(\frac{1}{z}\right) \right\rceil\right). \tag{3.25}$$

and stops as soon as z belongs to the domain $\mathcal{F} = \mathcal{F}_+ \cup \mathcal{F}_-$, with

$$\mathcal{F} = \left\{ z \in \mathbb{H};\quad |z| \geq 1, |\Re z| \leq \frac{1}{2} \right\}, \quad \mathcal{F}_+ := \mathcal{F} \cap \mathbb{H}_+,\ \mathcal{F}_- := \mathcal{F} \cap \mathbb{H}_-. \tag{3.26}$$

Such a domain, represented in Fig. 3.8, is closely related to the classical fundamental domain $\widehat{\mathcal{F}}$ of the upper half plane $\mathbb{H}$ under the action of the group

$$PSL_2(\mathbb{Z}) := \{h : z \mapsto h(z);\quad h(z) = \frac{az + b}{cz + d},\quad a, b, c, d \in \mathbb{Z},\quad ad - bc = 1\}.$$

More precisely, the difference $\mathcal{F} \setminus \widehat{\mathcal{F}}$ is contained in the frontier of $\mathcal{F}$.

Consider the pair $(\mathcal{B}, U)$, where the map $U : \mathcal{B} \to \mathcal{B}$ is defined in (3.25) for $z \in \mathcal{B} \setminus \mathcal{F}$ and extended to $\mathcal{F}$ with $U(z) = z$ for $z \in \mathcal{F}$. This pair $(\mathcal{B}, U)$ defines a dynamical system,[5] and $\mathcal{F}$ can be seen as a "hole": as the PGAUSS algorithm terminates, there exists an index $p \geq 0$, which is the first index for which $U^p(z)$ belongs to $\mathcal{F}$. Then, any complex number of $\mathcal{B}$ gives rise to a trajectory $z, U(z), U^2(z), \ldots, U^p(z)$, which "falls" in the hole $\mathcal{F}$, and stays inside $\mathcal{F}$ as soon as it attains $\mathcal{F}$. Moreover, as $\mathcal{F}$ is, up to its frontier, a fundamental domain of the

[5] We will see a formal definition of a dynamical system in Section "Analysis of the Execution Parameters of the Gauss Algorithm".

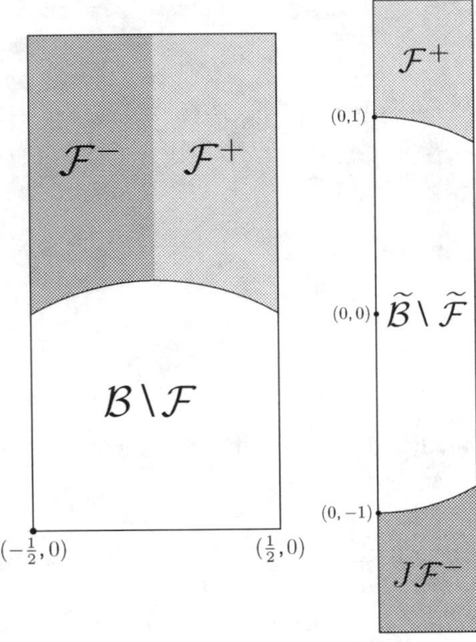

Fig. 3.8 The fundamental domains $\mathcal{F}, \widetilde{\mathcal{F}}$ and the strips $\mathcal{B}, \widetilde{\mathcal{B}}$ defined in Section "The Complex Versions for the GAUSS Algorithms"

upper half plane $\mathbb{H}$ under the action of $PSL_2(\mathbb{Z})$, there exists a topological tessellation of $\mathbb{H}$ with transforms of $\mathcal{F}$ of the form $h(\mathcal{F})$ with $h \in PSL_2(\mathbb{Z})$. We will see later in Section "The LFTs Used by the AGAUSS Algorithm. The COREGAUSS Algorithm" that the geometry of $\mathcal{B} \setminus \mathcal{F}$ is compatible with this tessellation.

In the same vein (see Figure 3.8), the AGAUSS algorithm brings z into the vertical strip

$$\mathcal{B} := \left\{ z \in \mathbb{C}; \quad \Im(z) \neq 0, \quad 0 \leq \Re(z) \leq \frac{1}{2} \right\} = \mathcal{B}_+ \cup J\mathcal{B}_-,$$

reduces to the iteration of the mapping

$$\widetilde{U}(z) = \varepsilon\left(\frac{1}{z}\right)\left(\frac{1}{z} - \left\lfloor \Re\left(\frac{1}{z}\right)\right\rceil\right), \qquad \text{with} \quad \varepsilon(z) := \text{sign}(\Re(z) - \lfloor \Re(z) \rceil), \tag{3.27}$$

and stops as soon as z belongs to the domain $\widetilde{\mathcal{F}}$

$$\widetilde{\mathcal{F}} = \left\{ z \in \mathbb{C}; \quad |z| \geq 1, \quad 0 \leq \Re(z) \leq \frac{1}{2} \right\} = \mathcal{F}_+ \cup J\mathcal{F}_-. \tag{3.28}$$

Consider the pair $(\widetilde{\mathcal{B}}, \widetilde{U})$, where the map $\widetilde{U} : \widetilde{\mathcal{B}} \to \widetilde{\mathcal{B}}$ is defined in (3.27) for $z \in \widetilde{\mathcal{B}} \setminus \widetilde{\mathcal{F}}$ and extended to $\widetilde{\mathcal{F}}$ with $\widetilde{U}(z) = z$ for $z \in \widetilde{\mathcal{F}}$. This pair $(\widetilde{\mathcal{B}}, \widetilde{U})$ also defines a dynamical system, and $\widetilde{\mathcal{F}}$ can also be seen as a "hole."

Relation with the Centered Euclid Algorithm

It is clear (at least in an informal way) that each version of Gauss algorithm is an extension of the (centered) Euclid algorithm:

- For the PGAUSS algorithm, it is related to a Euclidean division of the form $v = qu + r$ with $|r| \in [0, +u/2]$
- For the AGAUSS algorithm, it is based on a Euclidean division of the form $v = qu + \varepsilon r$ with $\varepsilon := \pm 1, r \in [0, +u/2]$

If, instead of pairs, that are the old pair (u, v) and the new pair (r, u), one considers rationals, namely the old rational $x = u/v$ or the new rational $y = r/u$, each Euclidean division can be written with a map that expresses the new rational y as a function of the old rational x, as $y = V(x)$ (in the first case) or $y = \widetilde{V}(x)$ (in the second case). With $\mathcal{I} := [-1/2, +1/2]$ and $\widetilde{\mathcal{I}} := [0, 1/2]$, the maps $V : \mathcal{I} \to \mathcal{I}$ or $\widetilde{V} : \widetilde{\mathcal{I}} \to \widetilde{\mathcal{I}}$ are defined as follows

$$V(x) := \frac{1}{x} - \left\lfloor \frac{1}{x} \right\rceil, \quad \text{for } x \neq 0, \qquad V(0) = 0, \tag{3.29}$$

$$\widetilde{V}(x) = \varepsilon\left(\frac{1}{x}\right)\left(\frac{1}{x} - \left\lfloor \frac{1}{x} \right\rceil\right), \quad \text{for } x \neq 0, \qquad \widetilde{V}(0) = 0. \tag{3.30}$$

[Here, $\varepsilon(x) := \text{sign}(x - \lfloor x \rceil)$].

This leads to two (real) dynamical systems $(\mathcal{I}, V)$ and $(\widetilde{\mathcal{I}}, \widetilde{V})$ whose graphs are represented in Fig. 3.9. Remark that the tilded system is obtained by a folding of the untilded one (or unfolded one), first along the x axis, then along the y axis, as it is explained in [7]. The first system is called the F-EUCLID system (or algorithm), while the second one is called the U-EUCLID system (or algorithm).

Of course, there are close connections between U and $-V$, on the one hand, and $\widetilde{U}$ and $\widetilde{V}$, on the other hand: even if the complex systems $(\mathcal{B}, U)$ and $(\widetilde{\mathcal{B}}, \widetilde{U})$ are

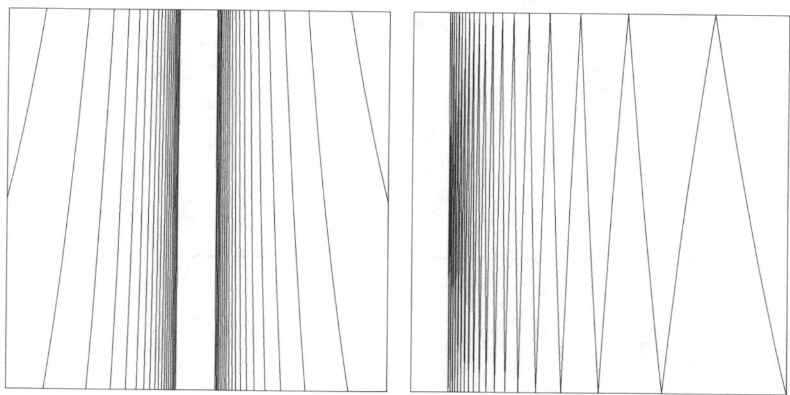

Fig. 3.9 The two dynamical systems underlying the centered Euclidean algorithms

defined on strips formed with complex numbers z that are not real (i.e., $\Im z \neq 0$), they can be extended to real inputs "by continuity": This defines two new dynamical systems $(\underline{\mathcal{B}}, \underline{U})$ and $(\underline{\widetilde{\mathcal{B}}}, \underline{\widetilde{U}})$, and the real systems $(\mathcal{I}, -V)$ and $(\widetilde{\mathcal{I}}, \widetilde{V})$ are just the restriction of the extended complex systems to real inputs. Remark now that the fundamental domains $\mathcal{F}, \widetilde{\mathcal{F}}$ are no longer "holes" as any real irrational input stays inside the real interval and never "falls" in them. On the contrary, the trajectories of rational numbers end at 0, and finally each rational is mapped to $i\infty$.

The LFTs Used by the PGAUSS *Algorithm*

The complex numbers that intervene in the PGAUSS algorithm on the input $z_0 = v_1/v_0$ are related to the vectors (v_i) defined in (3.3) via the relation $z_i = v_{i+1}/v_i$. They are directly computed by the relation $z_{i+1} := U(z_i)$, so that the old z_{i-1} is expressed with the new one z_i as

$$z_{i-1} = h_{[m_i]}(z_i), \qquad \text{with} \quad h_{[m]}(z) := \frac{1}{m - z}.$$

This creates a continued fraction expansion for the initial complex z_0, of the form

$$z_0 = \cfrac{1}{m_1 - \cfrac{1}{m_2 - \cfrac{1}{\ddots \over m_p - z_p}}} = h(z_p), \qquad \text{with} \quad h := h_{[m_1]} \circ h_{[m_2]} \circ \ldots h_{[m_p]},$$

which expresses the input $z = z_0$ as a function of the output $\widehat{z} = z_p$. More generally, the ith complex number z_i satisfies

$$z_0 = h_i(z_i), \qquad \text{with} \quad h_i := h_{[m_1]} \circ h_{[m_2]} \circ \ldots h_{[m_i]}.$$

Proposition 5. (Folklore) *The set $\mathcal{G}$ of LFTs $h : z \mapsto (az + b)/(cz + d)$ defined with the relation $z = h(\widehat{z})$, which sends the output domain $\mathcal{F}$ into the input domain $\mathcal{B} \setminus \mathcal{F}$, is characterized by the set $\mathcal{Q}$ of possible quadruples (a, b, c, d). A quadruple $(a, b, c, d) \in \mathbb{Z}^4$ with $ad - bc = 1$ belongs to $\mathcal{Q}$ if and only if one of the three conditions is fulfilled*

1. *$(c = 1$ or $c \geq 3)$ and $(|a| \leq c/2)$*
2. *$c = 2$, $a = 1$, $b \geq 0$, $d \geq 0$*
3. *$c = 2$, $a = -1$, $b \geq 0$, $d < 0$*

There exists a bijection between $\mathcal{Q}$ and the set $\mathcal{P} = \{(c, d) \in \mathbb{Z}^2;\ c \geq 1, \gcd(c, d) = 1\}$. On the other hand, for each pair (a, c) in the set

$$\mathcal{C} := \{(a, c);\ \ \frac{a}{c} \in [-1/2, +1/2],\ c \geq 1; \gcd(a, c) = 1\}, \tag{3.31}$$

Fig. 3.10 *Left*: the "central" festoon $\mathcal{F}_{(0,1)}$. *Right*: three festoons of the strip $\mathcal{B}$, relative to $(0,1)$, $(1,3)$, $(-1,3)$ and the two half-festoons at $(-1,2)$ and $(1,2)$

each LFT of $\mathcal{G}$, which admits (a,c) as coefficients can be written as $h = h_{(a,c)} \circ T^q$, with $q \in \mathbb{Z}$ and $h_{(a,c)}(z) = (az + b_0)/(cz + d_0)$, with $|b_0| \leq |a/2|, |d_0| \leq |c/2|$.

Definition 2. [Festoons] *If $\mathcal{G}_{(a,c)}$ denotes the set of LFTs of $\mathcal{G}$, which admit (a,c) as coefficients, the domain*

$$\mathcal{F}_{(a,c)} = \bigcup_{h \in \mathcal{G}_{(a,c)}} h(\mathcal{F}) = h_{(a,c)}\left(\bigcup_{q \in \mathbb{Z}} T^q \mathcal{F}\right) \tag{3.32}$$

gathers all the transforms of $h(\mathcal{F})$ which belong to $\mathcal{B} \setminus \mathcal{F}$ for which $h(i\infty) = a/c$. It is called the festoon of a/c.

Remark that, in the case when $c = 2$, there are two half-festoons at $1/2$ and $-1/2$ (See Fig. 3.10).

The LFTs Used by the AGAUSS *Algorithm. The* COREGAUSS *Algorithm*

In the same vein, the complex numbers that intervene in the AGAUSS algorithm on the input $z_0 = w_1/w_0$ are related to the vectors (w_i) defined in (3.5) via the relation $z_i = w_{i+1}/w_i$. They are computed by the relation $z_{i+1} := \widetilde{U}(z_i)$, so that the old z_{i-1} is expressed with the new one z_i as

$$z_{i-1} = h_{\langle m_i, \varepsilon_i \rangle}(z_i), \qquad \text{with} \quad h_{\langle m, \varepsilon \rangle}(z) := \frac{1}{m + \varepsilon z}.$$

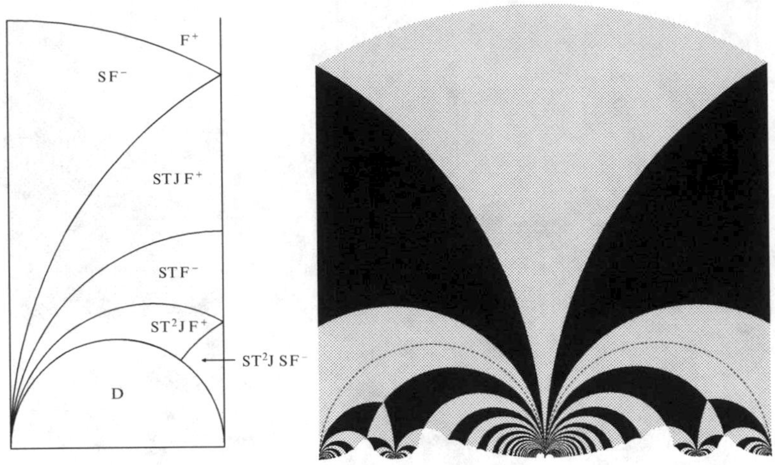

Fig. 3.11 *Left*: the six domains which constitute the domain $\mathcal{B}_+ \setminus \mathcal{D}_+$. *Right*: the disk $\mathcal{D}$ is not compatible with the geometry of transforms of the fundamental domains $\mathcal{F}$

This creates a continued fraction expansion for the initial complex z_0, of the form

$$z_0 = \cfrac{1}{m_1 + \cfrac{\epsilon_1}{m_2 + \cfrac{\epsilon_2}{\ddots \; \overline{m_p + \epsilon_p z_p}}}} = \widetilde{h}(z_p) \quad \text{with} \quad \widetilde{h} := h_{\langle m_1, \epsilon_1 \rangle} \circ h_{\langle m_2, \epsilon_2 \rangle} \circ \ldots h_{\langle m_p, \epsilon_p \rangle}.$$

More generally, the i th complex number z_i satisfies

$$z_0 = \widetilde{h}_i(z_i) \qquad \text{with} \quad \widetilde{h}_i := h_{\langle m_1, \varepsilon_1 \rangle} \circ h_{\langle m_2, \varepsilon_2 \rangle} \circ \ldots h_{\langle m_i, \varepsilon_i \rangle}. \tag{3.33}$$

We now explain the particular rôle that is played by the disk $\mathcal{D}$ of diameter $\widetilde{\mathcal{I}} = [0, 1/2]$. Figure 3.11 shows that the domain $\widetilde{\mathcal{B}} \setminus \mathcal{D}$ decomposes as the union of six transforms of the fundamental domain $\widetilde{\mathcal{F}}$, namely

$$\widetilde{\mathcal{B}} \setminus \mathcal{D} = \bigcup_{h \in \mathcal{K}} h(\widetilde{\mathcal{F}}), \qquad \text{with} \quad \mathcal{K} := \{I, S, STJ, ST, ST^2J, ST^2JS\}. \tag{3.34}$$

This shows that the disk $\mathcal{D}$ itself is also a union of transforms of the fundamental domain $\widetilde{\mathcal{F}}$. Remark that the situation is different for the PGAUSS algorithm, as the frontier of $\mathcal{D}$ lies "in the middle" of transforms of the fundamental domain $\mathcal{F}$ (see Fig. 3.11).

As Fig. 3.12 shows it, there are two main parts in the execution of the AGAUSS Algorithm, according to the position of the current complex z_i with respect to the disk $\mathcal{D}$ of diameter $[0, 1/2]$ whose alternative equation is

COREGAUSS(z)
Input. A complex number in $\mathcal{D}$.
Output. A complex number in $\widetilde{\mathcal{B}} \setminus \mathcal{D}$.
`While` $z \in \mathcal{D}$ do $z := \widetilde{U}(z)$;

FINALGAUSS(z)
Input. A complex number in $\widetilde{\mathcal{B}} \setminus \mathcal{D}$.
Output. A complex number in $\widetilde{\mathcal{F}}$.
`While` $z \notin \widetilde{\mathcal{F}}$ do $z := \widetilde{U}(z)$;

AGAUSS(z)
Input. A complex number in $\widetilde{\mathcal{B}} \setminus \widetilde{\mathcal{F}}$.
Output. A complex number in $\widetilde{\mathcal{F}}$.
COREGAUSS (z);
FINALGAUSS (z);

Fig. 3.12 The decomposition of the AGAUSS Algorithm into two parts: its core part (the COREGAUSS Algorithm) and its final part (the FINALGAUSS Algorithm)

$$\mathcal{D} := \left\{ z;\ \Re\left(\frac{1}{z}\right) \geq 2 \right\}.$$

While z_i belongs to $\mathcal{D}$, the quotient (m_i, ε_i) satisfies $(m_i, \varepsilon_i) \geq (2, +1)$ (wrt the lexicographic order), and the algorithm uses at each step the set

$$\mathcal{H} := \{h_{\langle m,\varepsilon \rangle};\quad (m, \varepsilon) \geq (2, +1)\}$$

so that $\mathcal{D}$ can be written as

$$\mathcal{D} = \bigcup_{h \in \mathcal{H}^+} h(\widetilde{\mathcal{B}} \setminus \mathcal{D}) \qquad \text{with} \quad \mathcal{H}^+ := \sum_{k \geq 1} \mathcal{H}^k. \tag{3.35}$$

The part of the AGAUSS algorithm performed when z_i belongs to $\mathcal{D}$ is called the COREGAUSS algorithm. The total set of LFTs used by the COREGAUSS algorithm is then the set $\mathcal{H}^+ = \cup_{k \geq 1} \mathcal{H}^k$. As soon as z_i does not any longer belong to $\mathcal{D}$, there are two cases. If z_i belongs to $\widetilde{\mathcal{F}}$, then the algorithm ends. If z_i belongs to $\widetilde{\mathcal{B}} \setminus (\widetilde{\mathcal{F}} \cup \mathcal{D})$, there remains at most two iterations (due to (3.34) and Fig. 3.11), that constitutes the FINALGAUSS algorithm, which uses the set $\mathcal{K}$ of LFTs, called the final set of LFTs and described in (3.34). Finally, we have proven the decomposition of the AGAUSS Algorithm, as is described in Fig. 3.12, and summarized in the following proposition:

Proposition 6. (Daudé et al. [14] (1994), Flajolet and Vallée [16, 17] (1990–1999)) *The set $\widetilde{\mathcal{G}}$ formed by the LFTs that map the fundamental domain $\widetilde{\mathcal{F}}$ into the set $\widetilde{\mathcal{B}} \setminus \widetilde{\mathcal{F}}$ decomposes as $\widetilde{\mathcal{G}} = (\mathcal{H}^\star \cdot \mathcal{K}) \setminus \{I\}$, where*

$$\mathcal{H}^\star := \sum_{k \geq 0} \mathcal{H}^k, \qquad \mathcal{H} := \{h_{\langle m,\varepsilon \rangle};\quad (m, \varepsilon) \geq (2, +1)\},$$
$$\mathcal{K} := \{I, S, STJ, ST, ST^2J, ST^2JS\}.$$

Here, if $\mathcal{D}$ denotes the disk of diameter $[0, 1/2]$, then $\mathcal{H}^+$ is the set formed by the LFTs that map $\widetilde{\mathcal{B}} \setminus \mathcal{D}$ into $\mathcal{D}$ and $\mathcal{K}$ is the final set formed by the LFTs that map $\widetilde{\mathcal{F}}$ into $\widetilde{\mathcal{B}} \setminus \mathcal{D}$. Furthermore, there is a characterization of $\mathcal{H}^+$ due to Hurwitz, which involves the golden ratio $\phi = (1 + \sqrt{5})/2$:

$$\mathcal{H}^+ := \left\{ h(z) = \frac{az+b}{cz+d}; \quad (a,b,c,d) \in \mathbb{Z}^4, b,d \geq 1, ac \geq 0, \right.$$
$$\left. |ad - bc| = 1, |a| \leq \frac{|c|}{2}, b \leq \frac{d}{2}, -\frac{1}{\phi^2} \leq \frac{c}{d} \leq \frac{1}{\phi} \right\}.$$

Comparing the CoreGauss *Algorithm and the* F-Euclid *Algorithm*

The CoreGauss algorithm has a nice structure as it uses ateach step the same set $\mathcal{H}$. This set is exactly the set of LFTs that are used by the F-Euclid Algorithm, closely related to the dynamical system defined in (3.30). Then, the CoreGauss algorithm is just a lifting of this F-Euclid Algorithm, while the final steps of the AGauss algorithm use different LFT's, and are not similar to a lifting of a Euclidean Algorithm. This is why the CoreGauss algorithm is interesting to study: we will see in Section "Analysis of the Execution Parameters of the Gauss Algorithm" why it can be seen as an exact generalization of the F-Euclid algorithm.

For instance, if R denotes the number of iterations of the CoreGauss algorithm, the domain $[R \geq k+1]$ gathers the complex numbers z for which $\widetilde{U}^k(z)$ are in $\mathcal{D}$. Such a domain admits a nice characterization, as a union of disjoint disks, namely

$$[R \geq k+1] = \bigcup_{h \in \mathcal{H}^k} h(\mathcal{D}), \tag{3.36}$$

which is represented in Figure 3.13. The disk $h(\mathcal{D})$ for $h \in \mathcal{H}^+$ is the disk whose diameter is the interval $[h(0), h(1/2)] = h(\widetilde{\mathcal{I}})$. Inside the F-Euclid dynamical system, the interval $h(\widetilde{\mathcal{I}})$ (relative to a LFT $h \in \mathcal{H}^k$) is called a fundamental interval (or a cylinder) of depth k: it gathers all the real numbers of the interval $\widetilde{\mathcal{I}}$ that have the same continued fraction expansion of depth k. This is why the disk $h(\mathcal{D})$ is called a fundamental disk.

This figure shows in a striking way the efficiency of the algorithm, and asks natural questions: Is it possible to estimate the probability of the event $[R \geq k+1]$? Is it true that it is geometrically decreasing? With which ratio? We return to these questions in Section "Analysis of the Execution Parameters of the Gauss Algorithm".

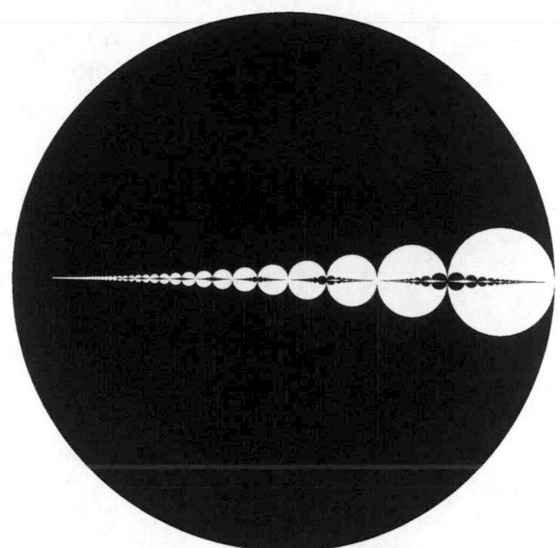

Fig. 3.13 The domains $[R = k]$ alternatively in *black* and *white*. The figure suggests that reduction of almost-collinear bases is likely to require a large number of iterations

Worst-Case Analysis of the Gauss Algorithm

Before beginning our probabilistic studies, we recall the worst-case behavior of execution parameters and give a proof in the complex framework.

Theorem 4. (Vallée [38] 1991) *Consider the* AGAUSS *Algorithm, with an input* (u, v) *of length* $\max(|u|, |v|)$ *at most equal to* N*. Then, the maximum number of iterations* P_N*, and the maximum value* C_N *of any additive cost* C *of moderate growth*[6] *are* $\Theta(\log N)$*, while the maximal value* B_N *of the bit-complexity* B *is* $\Theta(\log^2 N)$*. More precisely, the maximal value* P_N *of the number of iterations* P *satisfies*

$$P_N \sim_{N\to\infty} \frac{1}{\log(1+\sqrt{2})} \log N.$$

Proof. We here use the complex framework of the AGAUSS algorithm, and the study of the maximum number of iterations is the complex version of Vallée's result, initially performed in the vectorial framework [38].

Number of iterations. It is sufficient to study the number R of iterations of the COREGAUSS Algorithm as it is related to the total number of iterations P via the inequality $P \leq R + 2$. The inclusion

$$[R \geq k+1] \subset \left\{ z; \quad |\Im(z)| \leq \frac{1}{2}\left(\frac{1}{1+\sqrt{2}}\right)^{2k-1} \right\} \tag{3.37}$$

[6] This means that the elementary cost c satisfies $c(q) = O(\log q)$ (see Section "Main Parameters of Interest").

will lead to the result: as any nonreal complex $z = v/u$ relative to an integer pair (u, v) has an imaginary part at least equal to $1/|u|^2$, then z belongs to the domain $[R \leq k]$ as soon as $|u|^2 \leq 2(1+\sqrt{2})^{2k-1}$.

We now prove Relation (3.37): Indeed, we know from (3.36) that the domain $[R \geq k+1]$ is the union of transforms $h(\mathcal{D})$ for $h \in \mathcal{H}^k$, where $\mathcal{D}$ and $\mathcal{H}$ are defined in Proposition 6. The largest such disk $h(\mathcal{D})$ is obtained when all the quotients (m, ε) are the smallest ones, that is, when all $(m, \varepsilon) = (2, +1)$. In this case, the coefficients (c, d) of h are the terms A_k, A_{k+1} of the sequence defined by

$$A_0 = 0, \quad A_1 = 1, \qquad \text{and} \quad A_{k+1} = 2A_k + A_{k-1} \quad \text{for } k \geq 1,$$

which satisfy $A_k \geq (1+\sqrt{2})^{k-2}$. Then, the largest such disk has a radius at most equal to $(1/2)(1+\sqrt{2})^{1-2k}$.

Additive costs. As we restrict ourselves to costs c of moderate growth, it is sufficient to study the cost C relative to the step cost $c(q) := \log q$.

Consider the sequence of vectors $w_0 = u, w_1 = v, \ldots, w_{k+1}$ computed by the AGAUSS algorithm on the input (u, v) with $M := \ell(|u|^2)$. We consider the last step as a special case, and we use for it the (trivial) upper bound $|m_{k+1}| \leq |u|^2$; for the other steps, we consider the associated complex numbers z_i defined by $z_{i-1} = h_i(z_i)$ [where the LFT h_i has a digit q_i at least equal to 2] and the complex $\check{z} := z_k$ before the last iteration that belongs to $\widetilde{\mathcal{B}} \setminus \widetilde{\mathcal{F}}$. Then the expression $z = z_0 = h(\check{z})$ involves the LFT $h := h_1 \circ h_2 \ldots \circ h_k$, which corresponds to the algorithm except its last step. As any complex $z = v/u$ relative to an integer pair (u, v) has an imaginary part at least equal to $1/|u|^2$, one has

$$\frac{1}{|u|^2} \leq |\Im h(\check{z})| = |\Im(\check{z})| \cdot |h'(\check{z})| \leq \prod_{i=1}^{k} |h_i'(z_i)| \leq \prod_{i=1}^{k} \frac{1}{|q_i - (1/2)|^2} \leq 2^k \prod_{i=1}^{k} \frac{1}{q_i^2}.$$

This proves that the cost $C(u, v)$ relative to $c(q) = \log q$ satisfies $C(u, v) = O(M)$.
Bit-complexity. The result is obtained, thanks to (3.8). ■

Probabilistic Models for Two-Dimensions

We now return to our initial motivation, and begin our probabilistic studies. As we focus on the invariance of algorithm executions under similarity transformations, we assume that the two random variables $|u|$ and $z = v/u$ are independent and consider densities F on pairs of vectors (u, v), which only depend on the ratio $z = v/u$, of the form $F(u, v) = f(v/u)$. Moreover, it is sufficient to consider pairs (u, v) with a first vector u of the form $u = (N, 0)$. Finally, we define in a generic way the discrete model Ω_N as

$$\Omega_N := \left\{ z = \frac{v}{u}; \quad u = (N, 0), v = (a, b), \ (a, b, N) \in \mathbb{Z}^3, \ z \in \mathcal{X} \right\},$$

and there are three main cases, according to the algorithm of interest, namely $\mathcal{X} = \mathcal{B} \setminus \mathcal{F}$ for PGAUSS, $\mathcal{X} = \widetilde{\mathcal{B}} \setminus \widetilde{\mathcal{F}}$ for AGAUSS, or $\mathcal{X} = \mathcal{D}$ for COREGAUSS.

In each case, the complex $z = v/u$ belongs to $\mathbb{Q}[i] \cap \mathcal{X}$ and is of the form $(a/N) + i(b/N)$. Our discrete probabilistic models are defined as the restrictions to Ω_N of a continuous model defined on $\mathcal{X}$. More precisely, we choose a density f on $\mathcal{X}$, and consider its restriction on Ω_N. Normalized by the cardinality $|\Omega_N|$, this gives rise to a density f_N on Ω_N, which we extend on $\mathcal{X}$ as follows: $f_N(x) := f_N(\omega)$ as soon as x belongs to the square of center $\omega \in \Omega_N$ and edge $1/N$. We obtain, in such a way, a family of functions f_N defined on $\mathcal{X}$. When the integer N tends to ∞, this discrete model "tends" to the continuous model relative to the density f (as we already explained in Section "Probabilistic Models: Continuous or Discrete").

It is sometimes more convenient to view these densities as functions defined on $\mathbb{R}^2$, and we will denote by the same symbol the function f viewed as a function of two real variables x, y. It is clear that the rôles of two variables x, y are not of the same importance. In our asymptotic framework, where the size M becomes large, the variable $y = \Im(z)$ plays the crucial rôle, while the variable $x = \Re(z)$ plays an auxiliary rôle. This is why the two main models that are now presented involve densities $f(x, y)$, which depend only on y.

The Model with "Valuation"

In Section "Probabilistic Analyses of the LLL Algorithm in the Spherical Model", it is shown that each input local basis U_{n-k} in the spherical model with ambient dimension n admits (for $n \to \infty$) a distribution with a density φ_k defined in (3.24). We are then led to consider the two-dimensional bases (u, v), which follow the so-called model of valuation r (with $r > -1$), for which

$$\mathbb{P}\left[(u, v);\quad \frac{|\det(u, v)|}{\max(|u|, |v|)^2} \le y\right] = \Theta(y^{r+1}), \qquad \text{when} \quad y \to 0.$$

We note that, when the valuation r tends to -1, this model tends to the "one-dimensional model," where u and v are collinear. In this case, the Gauss Algorithm "tends" to the Euclidean Algorithm, and it is important to precisely describe the transition. This model "with valuation" was already presented in [39] in a slightly different context, but not actually studied there.

The model with valuation defines a scale of densities for which the weight of skew bases may vary. When r tends to -1, almost all the input bases are formed of vectors which form a very small angle, and with a high probability, they represent hard instances for reducing the lattice.

In the complex framework, a density f on the set $\mathcal{S} \subset \mathbb{C} \setminus \mathbb{R}$ is of valuation r (with $r > -1$) if it is of the form

$$f(z) = |\Im(z)|^r \cdot g(z), \qquad \text{where} \quad g(z) \neq 0 \quad \text{for } \Im(z) = 0. \tag{3.38}$$

Such a density is called of type (r, g). We often deal with the standard density of valuation r, denoted by f_r,

$$f_r(z) = \frac{1}{A(r)} |\Im(z)|^r, \qquad \text{with} \quad A(r) = \iint_{\mathcal{B}\setminus\mathcal{F}} y^r \, dx dy. \tag{3.39}$$

Of course, when $r = 0$, we recover the uniform distribution on $\mathcal{B} \setminus \mathcal{F}$ with $A(0) = (1/12)(2\pi + 3\sqrt{3})$. When $r \to -1$, then $A(r)$ is $\Theta[(r+1)^{-1}]$. More precisely

$$A(r) \sim \frac{1}{r+1}, \qquad r \to -1.$$

The (continuous) model relative to a density f is denoted with an index of the form $\langle f \rangle$, and when the valuation is the standard density of valuation r, the model is denoted with an index of the form (r). The discrete models are denoted by two indices, the integer size M and the index that describes the function f, as previously.

The Ajtai Model in Two-Dimensions

This model (described in the general case in Section "Ajtai Bases") corresponds to bases (u, v) for which the determinant $\det(u, v)$ satisfies

$$\frac{|\det(u, v)|}{\max(|u|, |v|)^2} = y_0 \qquad \text{for some } y_0 \in]0, 1].$$

In the complex framework, this leads to densities $f(z)$ on $\mathcal{B} \setminus \mathcal{F}$ (or on the tilde corresponding domain) of the form $f(z) = \mathtt{Dirac}(y_0)$ for some $y_0 \in]0, 1]$. When y_0 tends to 0, then the model also tends to the "one-dimensional model" (where u and v are collinear) and the Gauss Algorithm also "tends" to the Euclidean Algorithm. As in the model "with valuation," it is important to precisely describe this transition and compare to the result of Goldstein and Mayer [20].

Analysis of Lattice Reduction in Two-Dimensions: The Output Parameters

This section describes the probabilistic behavior of output parameters: we first analyze the output densities, then we focus on the geometry of our three main parameters defined in (3.11) and (3.12). We shall use the PGAUSS Algorithm for studying the output parameters.

Output Densities

For studying the evolution of distributions (on complex numbers), we are led to study the LFTs h used in the Gauss algorithm [Section "Returning to the Gauss Algorithm"], whose set is $\mathcal{G}$ for the PGAUSS Algorithm [Section "The LFTs Used by the PGAUSS Algorithm"]. We consider the two-variables function $\underline{h}$ that corresponds to the complex mapping $z \mapsto h(z)$. More precisely, we consider the function $\underline{h}$, which is conjugated to $(h,h) : (u,v) \mapsto (h(u),h(v))$ with respect to map Φ, namely $\underline{h} = \Phi^{-1} \circ (h,h) \circ \Phi$, where mappings Φ, Φ^{-1} are linear mappings $\mathbb{C}^2 \to \mathbb{C}^2$ defined as

$$\Phi(x,y) = (z = x+iy, \bar{z} = x-iy), \qquad \Phi^{-1}(z,\bar{z}) = \left(\frac{z+\bar{z}}{2}, \frac{z+\bar{z}}{2i}\right).$$

As Φ and Φ^{-1} are linear mappings, the Jacobian $J\underline{h}$ of the mapping $\underline{h}$ satisfies

$$J\underline{h}(x,y) = |h'(z) \cdot h'(\bar{z})| = |h'(z)|^2, \tag{3.40}$$

as h has real coefficients. Let us consider any measurable set $\mathcal{A} \subset \mathcal{F}$, and study the final density $\widehat{f}$ on $\mathcal{A}$. It is brought by all the antecedents $h(\mathcal{A})$ for $h \in \mathcal{G}$, which form disjoints subsets of $\mathcal{B} \setminus \mathcal{F}$. Then,

$$\iint_{\mathcal{A}} \widehat{f}(\widehat{x},\widehat{y})\, \mathrm{d}\widehat{x}\mathrm{d}\widehat{y} = \sum_{h \in \mathcal{G}} \iint_{\underline{h}(\mathcal{A})} f(x,y)\, \mathrm{d}x\mathrm{d}y.$$

Using the expression of the Jacobian (3.40), and interchanging integral and sum lead to the equality

$$\iint_{\mathcal{A}} \widehat{f}(\widehat{x},\widehat{y})\mathrm{d}\widehat{x}\,\mathrm{d}\widehat{y} = \iint_{\mathcal{A}} \left(\sum_{h \in \mathcal{G}} |h'(\widehat{z})|^2 f \circ \underline{h}(\widehat{x},\widehat{y}) \right) \mathrm{d}\widehat{x}\mathrm{d}\widehat{y}.$$

Finally, we have proven:

Theorem 5. (Vallée and Vera [45, 47] 2007) *The output density $\widehat{f}$ of each of the three algorithms satisfies the following:*

i. The output density $\widehat{f}$ of the PGAUSS *Algorithm on the fundamental domain $\mathcal{F}$ is expressed as a function of the input density f on $\mathcal{B} \setminus \mathcal{F}$ as*

$$\widehat{f}(z) = \sum_{h \in \mathcal{G}} |h'(z)|^2 f \circ h(z),$$

where $\mathcal{G}$ is the set of LFTs used by the PGAUSS *algorithm described in Proposition 5.*

ii. *The output density* $\widehat{f}$ *of the* AGAUSS *Algorithm on the fundamental domain* $\widetilde{\mathcal{F}}$ *is expressed as a function of the input density* f *on* $\widetilde{\mathcal{B}} \setminus \widetilde{\mathcal{F}}$ *as*

$$\widehat{f}(z) = \sum_{h \in \widetilde{\mathcal{G}}} |h'(z)|^2 \, f \circ h(z),$$

where $\widetilde{\mathcal{G}}$ *is the set of LFTs used by the* AGAUSS *algorithm defined in Proposition 6.*

iii. *The output density* $\widehat{f}$ *of the* COREGAUSS *Algorithm on the domain* $\widetilde{\mathcal{B}} \setminus \mathcal{D}$ *can be expressed as a function of the input density* f *on* $\mathcal{D}$ *as*

$$\widehat{f}(z) = \sum_{h \in \mathcal{H}^+} |h'(z)|^2 \, f \circ h(z),$$

where $\mathcal{H}$ *is the set of LFTs used by each step of the* COREGAUSS *algorithm defined in Proposition 6. and* $\mathcal{H}^+ := \cup_{k \geq 1} \mathcal{H}^k$.

Relation with Eisenstein Series

We now analyze an important particular case, where the initial density is the standard density of valuation r defined in (3.39). As each element of $\mathcal{G}$ gives rise to a unique pair (c, d) with $c \geq 1, \gcd(c, d) = 1$ [see Section "The LFTs Used by the PGAUSS Algorithm"] for which

$$|h'(\widehat{z})| = \frac{1}{|c\widehat{z} + d|^4}, \qquad f_r \circ \underline{h}(\widehat{x}, \widehat{y}) = \frac{1}{A(r)} \frac{\widehat{y}^r}{|c\widehat{z} + d|^{2r}}, \tag{3.41}$$

the output density on $\mathcal{F}$ is
$$\widehat{f}_r(\widehat{x}, \widehat{y}) = \frac{1}{A(r)} \sum_{\substack{(c,d)=1 \\ c \geq 1}} \frac{\widehat{y}^r}{|c\widehat{z} + d|^{4+2r}}. \tag{3.42}$$

It is natural to compare this density with the density relative to the measure relative to "random lattices" defined in Section "Random Lattices". In the particular case of two-dimensions, the fundamental domain for the action of $PSL_2(\mathbb{Z})$ on $\mathbb{H}$ equals $\mathcal{F}$ up to its frontier. Moreover, the measure of density $f(z) = \Im(z)^{-2}$ is invariant under the action of $PSL_2(\mathbb{Z})$: indeed, for any LFT h with $\det h = \pm 1$, one has $|\Im(h(z))| = |\Im(z)| \cdot |h'(z)|$, so that

$$\iint_{h(\mathcal{A})} \frac{1}{y^2} \, \mathrm{d}x\mathrm{d}y = \iint_{\mathcal{A}} |h'(z)|^2 \frac{1}{\Im(h(z))^2} \, \mathrm{d}x\mathrm{d}y = \iint_{\mathcal{A}} \frac{1}{y^2} \, \mathrm{d}x\mathrm{d}y.$$

Then, the probability ν_2 defined in Section "Random Lattices" is exactly the measure on $\mathcal{F}$ of density

$$\eta(x,y) := \frac{3}{\pi}\frac{1}{y^2} \qquad \text{as} \qquad \iint_{\mathcal{F}} \frac{1}{y^2}\,\mathrm{d}x\mathrm{d}y = \frac{\pi}{3}. \tag{3.43}$$

If we make apparent this density η inside the expression of $\widehat{f_r}$ provided in (3.42), we obtain:

Theorem 6. (Vallée and Vera [45, 47] 2007) *When the initial density on $\mathcal{B} \setminus \mathcal{F}$ is the standard density of valuation r, denoted by f_r and defined in (3.39), the output density of the* PGAUSS *algorithm on $\mathcal{F}$ involves the Eisenstein series E_s of weight $s = 2 + r$: With respect to the Haar measure ν_2 on $\mathcal{F}$, whose density η is defined in (3.43), the output density $\widehat{f_r}$ is expressed as*

$$\widehat{f_r}(x,y)\,\mathrm{d}x\mathrm{d}y = \frac{\pi}{3A(r)} F_{2+r}(x,y)\,\eta(x,y)\,\mathrm{d}x\mathrm{d}y,$$

$$\textit{where} \quad F_s(x,y) = \sum_{\substack{(c,d)=1\\ c\geq 1}} \frac{y^s}{|cz+d|^{2s}}$$

is closely related to the classical Eisenstein series E_s of weight s, defined as

$$E_s(x,y) := \frac{1}{2} \sum_{\substack{(c,d)\in\mathbb{Z}^2\\ (c,d)\neq(0,0)}} \frac{y^s}{|cz+d|^{2s}} = \zeta(2s)\cdot\left[F_s(x,y) + y^s\right].$$

When $r \to -1$, classical results about Eisenstein series prove that

$$E_s(x,y) \sim_{s\to 1} \frac{\pi}{2(s-1)} \qquad \textit{so that} \qquad \lim_{r\to -1} \frac{\pi}{3A(r)} F_{2+r}(x,y) = 1.$$

Then, when r tends to -1, the output distribution relative to an input distribution, which is standard and of valuation r, tends to the distribution ν_2 relative to random lattices.

The series E_s are Mass forms (see for instance the book [8]): they play an important rôle in the theory of modular forms, because E_s is an eigenfunction for the Laplacian, relative to the eigenvalue $s(1-s)$. The irruption of Eisenstein series in the lattice reduction framework is unexpected, and at the moment, it is not clear how to use the (other) classical well-known properties of the Eisenstein series E_s for studying the output densities.

Geometry of the Output Parameters

The main output parameters are defined in (3.11,3.12). For $X \in \{\lambda, \mu, \gamma\}$, we denote by $X(z)$ the value of X on basis $(1, z)$, and there are close relations between

$X(u,v)$ and $X(z)$ for $z = v/u$:

$$\lambda(u,v) = |u| \cdot \lambda(z), \qquad \mu(u,v) = |u| \cdot \mu(z), \qquad \gamma(u,v) = \gamma(z).$$

Moreover, the complex versions of parameters λ, μ, γ can be expressed with the input–output pair $(z, \widehat{z})$.

Proposition 7. *If $z = x + iy$ is an initial complex number of $\mathcal{B} \setminus \mathcal{F}$ leading to a final complex $\widehat{z} = \widehat{x} + i\widehat{y}$ of $\mathcal{F}$, then the three main output parameters defined in (3.11) and (3.12) admit the following expressions:*

$$\det L(z) = y, \qquad \lambda^2(z) = \frac{y}{\widehat{y}}, \qquad \mu^2(z) = y\widehat{y}, \qquad \gamma(z) = \frac{1}{\widehat{y}}.$$

The following inclusions hold:

$$[\lambda(z) = t] \subset \left[\Im(z) \geq \frac{\sqrt{3}}{2}t^2\right], \qquad [\mu(z) = u] \subset \left[\Im(z) \leq \frac{2}{\sqrt{3}}u^2\right]. \tag{3.44}$$

If z leads to $\widehat{z}$ by using the LFT $h \in \mathcal{G}$ with $z = h(\widehat{z}) = (a\widehat{z} + b)/(c\widehat{z} + d)$, then

$$\lambda(z) = |cz - a|, \qquad \gamma(z) = \frac{|cz - a|^2}{y}, \qquad \mu(z) = \frac{y}{|cz - a|}.$$

Proof. If the initial pair (v_1, v_0) is written as in (3.4) as

$$\begin{pmatrix} v_1 \\ v_0 \end{pmatrix} = \mathcal{M}^{-1} \begin{pmatrix} v_{p+1} \\ v_p \end{pmatrix}, \quad \text{with } \mathcal{M}^{-1} := \begin{pmatrix} a & b \\ c & d \end{pmatrix} \quad \text{and } z = h(\widehat{z}) = \frac{a\widehat{z} + b}{c\widehat{z} + d},$$

then the total length decrease satisfies

$$\frac{|v_p|^2}{|v_0|^2} = \frac{|v_p|^2}{|cv_{p+1} + dv_p|^2} = \frac{1}{|c\widehat{z} + d|^2} = |h'(\widehat{z})|, \tag{3.45}$$

[we have used the fact that $\det \mathcal{M} = 1$.] This proves that $\lambda^2(z)$ equals $|h'(\widehat{z})|$ as soon as $z = h(\widehat{z})$. Now, for $z = h(\widehat{z})$, the relations

$$y = \frac{\widehat{y}}{|c\widehat{z} + d|^2}, \qquad \widehat{y} = \frac{y}{|cz - a|^2}$$

easily lead to the result. ■

Domains Relative to the Output Parameters

We now consider the following well-known domains defined in Fig. 3.14. The Ford disk $\mathtt{Fo}(a,c,\rho)$ is a disk of center $(a/c, \rho/(2c^2))$ and radius $\rho/(2c^2)$: it is tangent to $y = 0$ at point $(a/c, 0)$. The Farey disk $\mathtt{Fa}(a,c,t)$ is a disk of center $(a/c, 0)$ and radius t/c. Finally, the angular sector $\mathtt{Se}(a,c,u)$ is delimited by two lines that intersect at a/c, and form with the line $y = 0$ angles equal to $\pm\arcsin(cu)$. These domains intervene for defining the three main domains of interest.

Theorem 7. (Laville and Vallée [24] (1990), Vallée and Vera [45] (2007)) *The domains relative to the main output parameters, defined as*

$$\Gamma(\rho) := \{z \in \mathcal{B} \setminus \mathcal{F};\quad \gamma(z) \le \rho\}, \qquad \Lambda(t) := \{z \in \mathcal{B} \setminus \mathcal{F};\quad \lambda(z) \le t\},$$

$$M(u) := \{z \in \mathcal{B} \setminus \mathcal{F};\quad \mu(z) \le u\},$$

are described with Ford disks $\mathtt{Fo}(a,c,\rho)$, Farey disks $\mathtt{Fa}(a,c,t)$, and angular sectors $\mathtt{Se}(a,c,u)$. More precisely, if $\mathcal{F}_{(a,c)}$ denotes the Festoon relative to pair (a,c) defined in (3.32) and if the set $\mathcal{C}$ is defined as in (3.31), one has:

$$\Gamma(\rho) = \bigcup_{(a,c)\in\mathcal{C}} \mathtt{Fo}(a,c,\rho) \cap \mathcal{F}_{(a,c)}, \qquad \Lambda(t) = \bigcup_{(a,c)\in\mathcal{C}} \mathtt{Fa}(a,c,t) \cap \mathcal{F}_{(a,c)},$$

$$M(u) = \bigcup_{(a,c)\in\mathcal{C}} \mathtt{Se}(a,c,u) \cap \mathcal{F}_{(a,c)}.$$

$$\mathtt{Fo}(a,c,\rho) := \left\{(x,y);\quad y>0,\quad \left(x-\frac{a}{c}\right)^2 + \left(y - \frac{\rho}{2c^2}\right)^2 \le \left(\frac{\rho}{2c^2}\right)^2\right\}$$

$$\mathtt{Fa}(a,c,t) := \left\{(x,y);\quad y>0,\quad \left(x-\frac{a}{c}\right)^2 + y^2 \le \left(\frac{t}{c}\right)^2\right\}$$

$$\mathtt{Se}(a,c,u) := \left\{(x,y);\quad y>0,\quad y \le \frac{|c|u}{\sqrt{1-c^2u^2}}\left|x-\frac{a}{c}\right|\right\} \qquad \text{for } |c|u<1$$

$$\mathtt{Se}(a,c,u) := \{(x,y);\quad y>0,\} \qquad \text{for } |c|u \ge 1$$

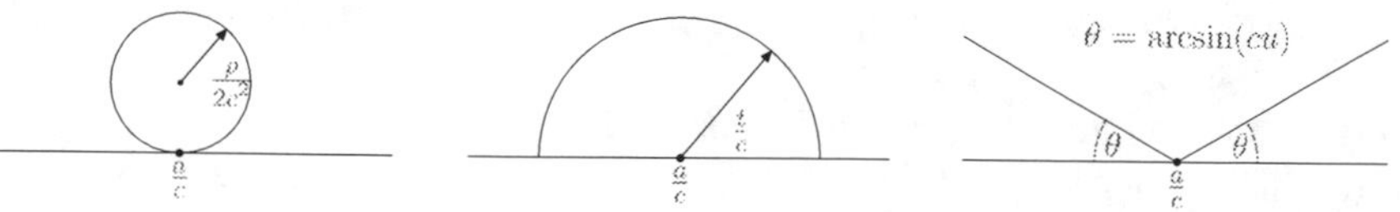

Fig. 3.14 The three main domains of interest: the Ford disks $\mathtt{Fo}(a,c,\rho)$, the Farey disks $\mathtt{Fa}(a,c,t)$, and the angular sectors $\mathtt{Se}(a,c,u)$

Each of these descriptions of Λ, Γ, M can be transformed in a description that no more involves the festoons. It involves, for instance, a subfamily of Farey disks (for Λ), or a subfamily of angular sectors (for M) [see Fig. 3.15].

Consider the set $\mathcal{P} := \{(c,d);\ \ c,d \geq 1, (c,d) = 1\}$, already used in Section "The LFTs Used by the PGAUSS Algorithm", and its subset $\mathcal{P}(t)$ defined as

$$\mathcal{P}(t) := \{(c,d);\quad c,d \geq 1, ct \leq 1, dt \leq 1, (c+d)t > 1, (c,d) = 1\}.$$

Consider a pair $(c,d) \in \mathcal{P}(t)$. There exists a unique pair (a,b) for which the rationals a/c and b/d belong to $[-1/2, +1/2]$ and satisfy $ad - bc = 1$. We then associate to the pair (c,d) the intersection of the vertical strip $\{(x,y);\ \ (a/c) \leq x \leq (b/d)\}$ with $\mathcal{B} \setminus \mathcal{F}$, and we denote it by $\mathcal{S}(c,d)$. Remark that the definition of $\mathcal{P}(t)$ implies that the only rationals of the strip $\mathcal{S}(c,d)$ with a denominator at most $(1/t)$ are a/c and b/d.

Domain $\Lambda(t)$. For any $t > 0$ and any pair $(c,d) \in \mathcal{P}(t)$, there exists a characterization of the intersection of the domain $\Lambda(t)$ with the vertical strip $\mathcal{S}(c,d)$, provided in [24], which does not depend any longer on the festoons, namely

$$\Lambda(t) \cap \mathcal{S}(c,d) = \underline{\mathtt{Fa}}(a,c,t) \cup \underline{\mathtt{Fa}}(b,d,t) \cup \mathtt{Fa}(a+b,c+d,t). \tag{3.46}$$

Here, the pair (a,b) is the pair associated to (c,d), the domains $\underline{\mathtt{Fa}}(a,c,t)$, $\underline{\mathtt{Fa}}(b,d,t)$ are the intersections of Farey disks $\mathtt{Fa}(a,c,t)$, $\mathtt{Fa}(b,d,t)$ with the strip $\mathcal{S}(c,d)$. The domain in (3.46) is exactly the union of the two disks $\underline{\mathtt{Fa}}(a,c,t)$ and $\underline{\mathtt{Fa}}(b,d,t)$ if and only if the condition $(c^2+d^2+cd)t^2 \geq 1$ holds, but the Farey disk relative to the median $(a+b)/(c+d)$ plays a rôle otherwise. The proportion of pairs $(c,d) \in \mathcal{P}(t)$ for which the condition $(c^2+d^2+cd)t^2 \geq 1$ holds tends to $2 - (2\pi)/(3\sqrt{3}) \approx 0.7908$ when $t \to 0$.

Then, the following inclusions hold (where the "left" union is a disjoint union)

$$\bigcup_{\substack{(a,c)\in\mathcal{C} \\ c\leq 1/(2t)}} \mathtt{Fa}(a,c,t) \subset \Lambda(t) \subset \bigcup_{\substack{(a,c)\in\mathcal{C} \\ c\leq 2/(\sqrt{3}t)}} \mathtt{Fa}(a,c,t). \tag{3.47}$$

Domain $M(u)$. For any $u > 0$ and any pair $(c,d) \in \mathcal{P}(u)$, there exists a characterization of the intersection of the domain $M(u)$ with the vertical strip $\mathcal{S}(c,d)$, provided in [47], which does not depend any longer on the festoons, namely

$$M(u) \cap \mathcal{S}(c,d) = \underline{\mathtt{Se}}(a,c,u) \cap \underline{\mathtt{Se}}(b,d,u) \cap \mathtt{Se}(b-a,d-c,u). \tag{3.48}$$

Here, the pair (a,b) is the pair associated to (c,d), the domains $\underline{\mathtt{Se}}(a,c,u)$, $\underline{\mathtt{Se}}(b,d,u)$ are the intersections of $\mathtt{Se}(a,c,u)$, $\mathtt{Se}(b,d,u)$ with the strip $\mathcal{S}(c,d)$. The domain in (3.48) is exactly the triangle $\underline{\mathtt{Se}}(a,c,u) \cap \underline{\mathtt{Se}}(b,d,u)$ if and only if one of the two conditions $(c^2+d^2-cd)u^2 \leq (3/4)$ or $cd\,u^2 \leq (1/2)$ holds, but this is a "true" quadrilateral otherwise. The proportion of pairs $(c,d) \in \mathcal{P}(u)$ for which the condition $[(c^2 + d^2 - cd)u^2 \leq (3/4)$ or $cd\, u^2 \leq (1/2)]$ holds tends to $(1/2) + (\pi\sqrt{3}/12) \approx 0.9534$ when $u \to 0$.

Distribution Functions of Output Parameters: Case of Densities with Valuations

Computing the measure of disks and angular sectors with respect to a standard density of valuation r leads to the estimates of the main output distributions. We first present the main constants that will intervene in our results.

Constants of the Analysis

The measure of a disk of radius ρ centered on the real axis equals $2A_2(r)\,\rho^{r+2}$. The measure of a disk of radius ρ tangent to the real axis equals $A_1(r)\,(2\rho)^{r+2}$. Such measures involve constants $A_1(r),\ A_2(r)$, which are expressed with the β law, already defined in (3.21) as

$$A_1(r) := \frac{\sqrt{\pi}}{A(r)} \frac{\Gamma(r+3/2)}{\Gamma(r+3)}, \qquad A_2(r) := \frac{\sqrt{\pi}}{2A(r)} \frac{\Gamma((r+1)/2)}{\Gamma(r/2+2)}. \tag{3.49}$$

For a triangle with basis a on the real axis and height h, this measure equals $A_3(r)\,a\,h^{r+1}$, and involves the constant

$$A_3(r) := \frac{1}{A(r)} \frac{1}{(r+2)(r+1)}. \tag{3.50}$$

For (α, β) that belongs to the triangle $\mathcal{T} := \{(\alpha, \beta); 0 < \alpha, \beta < 1, \alpha + \beta > 1\}$, we consider the continuous analogs of the configurations previously described:

Disks. We consider the figure obtained with three disks $D_\alpha, D_\beta, D_{\alpha+\beta}$ when these disks satisfy the following: For any $\delta, \eta \in \{\alpha, \beta, \alpha+\beta\}$, the center x_δ is on the real axis, the distance between x_δ and x_η equals $1/(\delta\eta)$ and the radius of D_δ equals $1/\delta$. We can suppose $x_\alpha < x_{\alpha+\beta} < x_\beta$. Then, the configuration $D(\alpha, \beta)$ is defined by the intersection of the union $\cup_\delta D_\delta$ with the vertical strip $\langle x_\alpha, x_\beta \rangle$. The constant $A_4(r)$ is defined as the integral

$$A_4(r) = \frac{1}{A(r)} \iint_{\mathcal{T}} \mathrm{d}\alpha \mathrm{d}\beta \left(\iint_{D(\alpha,\beta)} y^r \,\mathrm{d}x\mathrm{d}y \right). \tag{3.51}$$

Sectors. In the same vein, we consider the figure obtained with three sectors $S_\alpha, S_\beta, S_{\beta-\alpha}$ when these sectors satisfy the following:[7] for any $\delta \in \{\alpha, \beta, \beta - \alpha\}$, the sector S_δ is delimited by two half lines, the real axis (with a positive orientation) and another half-line, that intersect at the point x_δ of the real axis. For any $\delta, \eta \in \{\alpha, \beta, \beta - \alpha\}$, the distance between x_δ and x_η equals $1/(\delta\eta)$. We can suppose $x_{\beta-\alpha} < x_\alpha < x_\beta$; in this case, the angle of the sector S_δ equals $\arcsin \delta$ for $\delta \in \{\beta - \alpha, \alpha\}$ and equals $\pi - \arcsin \delta$ for $\delta = \beta$. The configuration $S(\alpha, \beta)$ is defined by the intersection of the intersection $\cap_\delta S_\delta$ with the vertical strip $\langle x_\alpha, x_\beta \rangle$. The constant $A_5(r)$ is defined as the integral

$$A_5(r) = \frac{1}{A(r)} \iint_{\mathcal{T}} \mathrm{d}\alpha \mathrm{d}\beta \left(\iint_{S(\alpha,\beta)} y^r \,\mathrm{d}x\mathrm{d}y \right). \tag{3.52}$$

Theorem 8. (Vallée and Vera [45,47] 2007) *When the initial density on $\mathcal{B} \setminus \mathcal{F}$ is the standard density of valuation r, the distribution of the three main output parameters involves the constants $A_i(r)$ defined in (3.49), (3.50), (3.51), and (3.52) and satisfies the following:*

1. *For parameter γ, there is an exact formula for any valuation r and any $\rho \leq 1$,*

$$\mathbb{P}_{(r)}[\gamma(z) \leq \rho] = A_1(r) \cdot \frac{\zeta(2r+3)}{\zeta(2r+4)} \cdot \rho^{r+2} \quad \textit{for} \quad \rho \leq 1$$

2. *For parameter λ, there are precise estimates for any fixed valuation $r > -1$, when $t \to 0$,*

$$\mathbb{P}_{(r)}[\lambda(z) \leq t] \sim_{t \to 0} \frac{\zeta(r+1)}{\zeta(r+2)} A_2(r) \cdot t^{r+2} \; \textit{for} \quad r > 0,$$

$$\mathbb{P}_{(r)}[\lambda(z) \leq t] \sim_{t \to 0} \frac{1}{\zeta(2)} A_2(0) \cdot t^2 |\log t| \; \textit{for} \quad r = 0,$$

$$\mathbb{P}_{(r)}[\lambda(z) \leq t] \sim_{t \to 0} \frac{1}{\zeta(2)} A_4(r) \cdot t^{2r+2} \quad \textit{for} \quad r < 0.$$

Moreover, for any fixed valuation $r > -1$ and any $t > 0$, the following inequality holds

$$\mathbb{P}_{(r)}[\lambda(z) \leq t] \geq \frac{1}{A(r)} \frac{1}{r+1} \left(\frac{\sqrt{3}}{2} \right)^{r+1} t^{2r+2}. \tag{3.53}$$

3. *For parameter μ, there is a precise estimate for any fixed valuation $r > -1$, when $u \to 0$,*

[7] The description is given in the case when $\beta > \alpha$.

$$\mathbb{P}_{(r)}[\mu(z) \le u] \sim_{u \to 0} \frac{1}{\zeta(2)} A_5(r) \cdot u^{2r+2}.$$

Moreover, for any fixed valuation $r > -1$ and any $u > 0$, the following inequalities hold:

$$A_3(r) \left(\frac{\sqrt{3}}{2}\right)^{r+1} \cdot u^{2r+2} \le \mathbb{P}_{(r)}[\mu(z) \le u] \le A_3(r) \cdot u^{2r+2}. \qquad (3.54)$$

Proof. [Sketch] If φ denotes the Euler quotient function, there are exactly $\varphi(c)$ coprime pairs (a, c) with $a/c \in]-1/2, +1/2]$. Then, the identity

$$\sum_{c \ge 1} \frac{\varphi(c)}{c^s} = \frac{\zeta(s-1)}{\zeta(s)}, \qquad \text{for} \quad \Re s > 2,$$

explains the occurrence of the function $\zeta(s-1)/\zeta(s)$ in our estimates. Consider two examples:

(a) For $\rho \le 1$, the domain $\Gamma(\rho)$ is made with disjoint Ford disks of radius $\rho/(2c^2)$. An easy application of previous principles leads to the result.

(b) For $\Lambda(t)$, these same principles together with relation (3.47) entail the following inequalities

$$t^{r+2}\left(\sum_{c \le 1/(2t)} \frac{\varphi(c)}{c^{r+2}}\right) \le \frac{1}{A_2(r)} \mathbb{P}_{(r)}[\lambda(z) \le t] \le t^{r+2}\left(\sum_{c \le 2/(\sqrt{3}t)} \frac{\varphi(c)}{c^{r+2}}\right),$$

and there are several cases when $t \to 0$ according to the sign of r. For $r > 0$, the Dirichlet series involved are convergent. For $r \le 0$, we consider the series

$$\sum_{c \ge 1)} \frac{\varphi(c)}{c^{r+2+s}} = \frac{\zeta(s+r+1)}{\zeta(s+r+2)},$$

(which has a pôle at $s = -r$), and classical estimates entail an estimate for

$$\sum_{c \le N} \frac{\varphi(c)}{c^{r+2}} \sim_{N \to \infty} \frac{1}{\zeta(2)} \frac{N^{-r}}{|r|}, \text{ (for } r < 0\text{), and} \sum_{c \le N} \frac{\varphi(c)}{c^2} \sim_{N \to \infty} \frac{1}{\zeta(2)} \log N.$$

For domain $M(u)$, the study of quadrilaterals can be performed in a similar way. Furthermore, the height of each quadrilateral of $M(u)$ is $\Theta(u^2)$, and the sum of the bases a equal 1. Then $\mathbb{P}_{(r)}[\mu(z) \le u] = \Theta(u^{2r+2})$. Furthermore, using the inclusions of (3.44) leads to the inequality. ■

Interpretation of the Results

We provide a first interpretation of the main results described in Theorem 8.

1. For any $y_0 \geq 1$, the probability of the event $[\widehat{y} \geq y_0]$ is
$$\mathbb{P}_{(r)}[\widehat{y} \geq y_0] = \mathbb{P}_{(r)}\left[\gamma(z) \leq \frac{1}{y_0}\right] = A_1(r)\frac{\zeta(2r+3)}{\zeta(2r+4)}\frac{1}{y_0^{r+2}}.$$
This defines a function of the variable $y_0 \mapsto \psi_r(y_0)$, whose derivative is a power function of variable y_0, of the form $\Theta(y_0^{-r-3})$. This derivative is closely related to the output density $\widehat{f_r}$ of Theorem 6 via the equality
$$\psi_r'(y_0) := \int_{-1/2}^{+1/2} \widehat{f_r}(x, y_0)\, \mathrm{d}x.$$
Now, when $r \to -1$, the function $\psi_r'(y)$ has a limit, which is exactly the density η, defined in (3.43), which is associated to the Haar measure ν_2 defined in Sections "Random Lattices and Relation with Eisenstein Series".
2. The regime of the distribution function of parameter λ changes when the sign of valuation r changes. There are two parts in the domain $\Lambda(t)$: the lower part, which is the horizontal strip $[0 \leq \Im(z) \leq (2/\sqrt{3})t^2]$, and the upper part defined as the intersection of $\Lambda(t)$ with the horizontal strip $[(2/\sqrt{3})t^2 \leq \Im(z) \leq t]$. For negative values of r, the measure of the lower part is dominant, while for positive values of r, it is the upper part that has a dominant measure. For $r = 0$, there is a phase transition between the two regimes: this occurs in particular in the usual case of a uniform density.
3. In contrast, the distribution function of parameter μ has always the same regime. In particular, for negative values of valuation r, the distribution functions of the two parameters, λ and μ, are of the same form.
4. The bounds (3.53, 3.54) prove that for any $u, t \in [0, 1]$, the probabilities $\mathbb{P}[\lambda(z) \leq t]$, $\mathbb{P}[\mu(z) \leq u]$ tend to 1, when the valuation r tends to -1. This shows that the limit distributions of λ and μ are associated to the Dirac measure at 0.
5. It is also possible to conduct these studies in the discrete model defined in Section "Probabilistic Models for Two-Dimensions". It is not done here, but this type of analysis will be performed in the following section.

Open question. Is it possible to describe the distribution function of parameter γ for $\rho > 1$? Figure 3.15 [top] shows that its regime changes at $\rho = 1$. This will be important for obtaining a precise estimate of the mean value $\mathbb{E}_{(r)}[\gamma]$ as a function of r and comparing this value to experiments reported in Section "A Variation for the LLL Algorithm: The Odd-Even Algorithm".

The corners of the fundamental domain. With Theorem 8, it is possible to compute the probability that an output basis lies in the corners of the fundamental domain, and to observe its evolution as a function of valuation r. This is a first step for a sharp understanding of Fig. 3.4 [right].

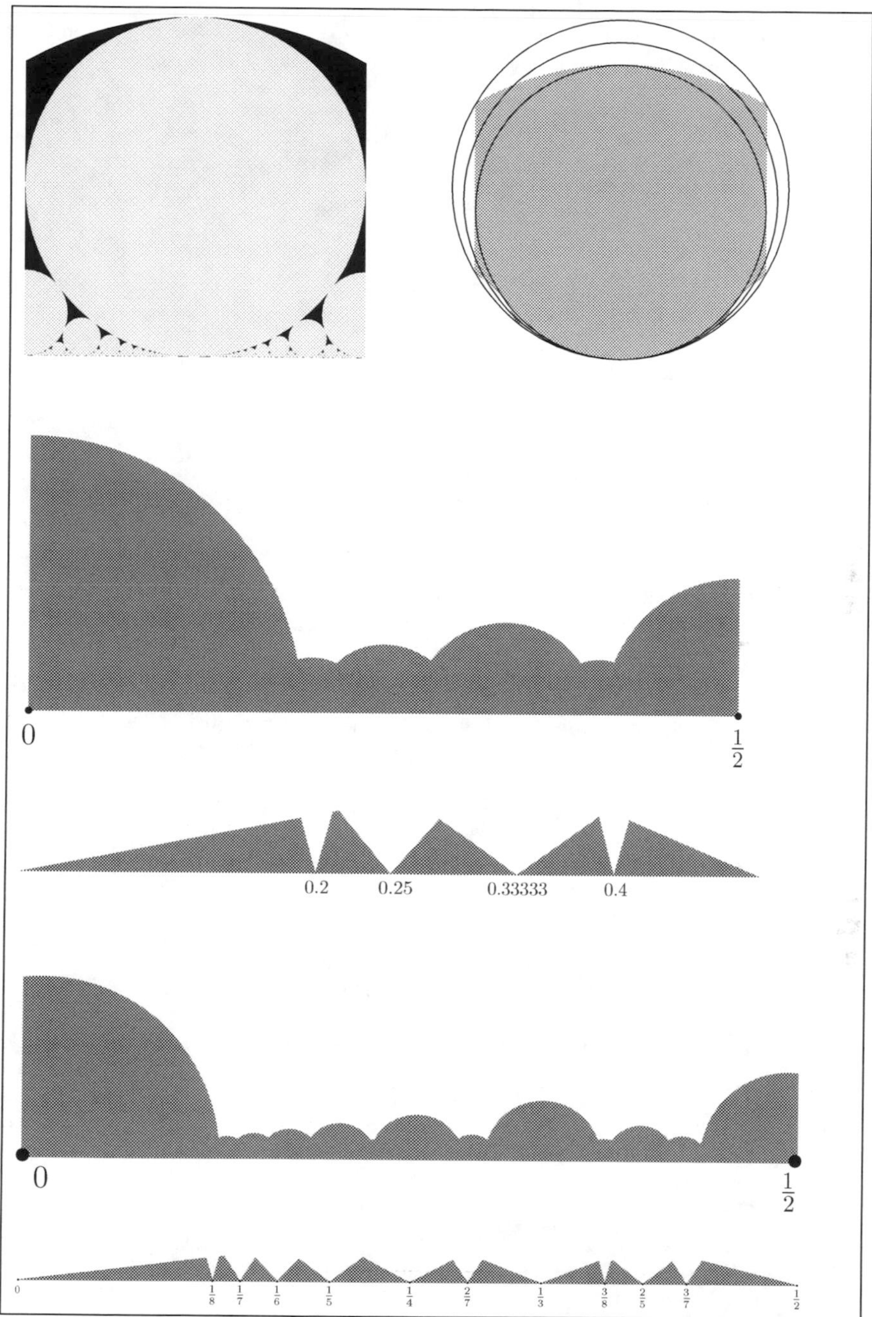

Fig. 3.15 Above: the domain $\Gamma(\rho) := \{z;\ \ \gamma(z) \leq \rho\}$. On the *left*, $\rho = 1$ (in *white*). On the *right*, the domain $\mathcal{F}_{(0,1)} \cap \text{Fo}(0, 1, \rho)$ for $\rho = 1, \rho_0 = 2/\sqrt{3}, \rho_1 = (1 + \rho_0)/2$. – In the *middle*: the domain $\Lambda(t) \cap \mathcal{B}_+$, with $\Lambda(t) := \{z;\ \ \lambda(z) \leq t\}$ and the domain $M(u) \cap \mathcal{B}_+$ with $M(u) := \{z;\ \mu(z) \leq u\}$ for $u = t = 0.193$.– Below: the same domains for $u = t = 0.12$

Proposition 8. *When the initial density on $\mathcal{B}\backslash\mathcal{F}$ is the standard density of valuation r, the probability for an output basis to lie on the corners of the fundamental domain is equal to*

$$C(r) := 1 - A_1(r) \cdot \frac{\zeta(2r+3)}{\zeta(2r+4)},$$

where $A_1(r)$ is defined in Section "Distribution Functions of Output Parameters: Case of Densities with Valuations". There are three main cases of interest for $1 - C(r)$, namely

$$[r \to -1] : \frac{3}{\pi}, \; [r = 0] : \frac{3\pi}{2\pi + 3\sqrt{3}} \frac{\zeta(3)}{\zeta(4)}, \; [r \to \infty] : \sqrt{\frac{\pi}{r}} e^{-3/2}.$$

Distribution Functions of Output Parameters: Case of Fixed Determinant

Computing the measure of disks and angular sectors with respect to the measure concentrated on the line $y = y_0$ leads to the estimates of the main output distributions. We here focus on the parameter γ.

The intersection of the disk $\mathtt{Fo}(a, c, \rho)$ with the line $y = y_0$ is nonempty as soon as y_0 is less than ρ/c^2. The intersection $\Gamma(\rho) \cap [y = y_0]$ is just "brought" by the Ford disks for which the integer c is less than $x_0 = \sqrt{\rho/y_0}$. Then, for $\rho < 1$, the Ford disks $\mathtt{Fo}(a, c, \rho)$ are disjoint and

$$\mathbb{P}_{[y_0]}[\gamma(z) \le \rho] = 2\rho\, S_g(x_0) \qquad \text{with} \quad S_g(x_0) = \frac{1}{x_0} \sum_{c \le x_0} \frac{\varphi(c)}{c}\, g\left(\frac{c}{x_0}\right),$$

and $g(t) = \sqrt{1 - t^2}$. For any function g smooth enough, one has

$$\lim_{x \to \infty} S_g(x) = \frac{1}{\zeta(2)} \int_0^1 g(t)\, \mathrm{d}t.$$

This proves that when y_0 tends to 0, the probability $\mathbb{P}_{[y_0]}[\gamma(z) \le \rho]$ tends to $(3/\pi)\rho$. We recover the result of [20] in the two-dimensional case.

A Related Result which also Deals with Farey Disks

For analyzing integer factoring algorithms, Vallée was led in 1988 to study the set of "small quadratic residues" defined as

$$\mathcal{B} = \mathcal{B}(N, h, h') := \{x \in [1..N]; \quad x^2 \bmod N \in [h, h']\}, \qquad \text{for } h' - h = 8N^{2/3},$$

and its distribution in $[1..N]$. She described in [43,44] a polynomial–time algorithm, called the Two-Thirds Algorithm, which draws elements from $\mathcal{B}$ in a quasi-uniform way.[8] This was (for her) a main tool for obtaining a *provable* complexity bound for integer factoring algorithms based on congruences of squares. Fifteen years later, Coron in [13], then Gentry in [19], discovered that such an algorithm also plays a central rôle in cryptography, more precisely in security proofs (see the survey of Gentry [18] in these proceedings). Furthermore, Gentry in [19] modified Vallée's algorithm and obtained an algorithm that draws elements from $\mathcal{B}$ in an exact uniform way. This constitutes a main step in the security proof of Rabin partial-domain-hash signatures.

The main idea of Vallée, which has been later adapted and made more precise by Gentry, is to perform a local study of the set $\mathcal{B}$. In this way, she refines ideas of the work done in [46]. This last work was one of the first works that relates general small modular equations to lattices, and was further generalized ten years later by Coppersmith [11]. Consider an integer x_0, for which the rational $2x_0/N$ is close to a rational a/c with a small denominator c. Then, the set of elements of $\mathcal{B}$ near x_0 can be easily described with the help of the lattice $\underline{L}(x_0)$ generated by the pair of vectors $(2x_0, 1), (N, 0)$. More precisely, the following two conditions are equivalent:

1. $x = x_0 + u$ belongs to $\mathcal{B}$
2. There exists w such that the point (w, u) belongs to $\underline{L}(x_0)$ and lies between two parabolas with respective equations

$$w + u^2 + x_0^2 = h, \qquad w + u^2 + x_0^2 = h'.$$

This equivalence is easy to obtain (just expand x^2 as $(x_0+u)^2 = x_0^2+2x_0u+u^2$) and gives rise to an efficient drawing algorithm of $\mathcal{B}$ near x_0, *provided that* the lattice $\underline{L}(x_0)$ has a sufficiently short vector in comparison to the gap $h'-h$ between the two parabolas. Vallée proved that this happens when the complex $z_0 = 2x_0/N + i/N$ relative to the input basis of $\underline{L}(x_0)$ belongs to a Farey disk $\mathtt{Fa}(a, c, t)$, with $t = (h'-h)/N = 4N^{-1/3}$. In 1988, the rôle played by Farey disks (or Farey intervals) was surprising, but now, from previous studies performed in Section "Domains Relative to the Output Parameters", we know that these objects are central in such a result.

Analysis of the Execution Parameters of the Gauss Algorithm

We finally focus on parameters that describe the execution of the algorithm: we are mainly interested in the bit-complexity, but we also study additive costs that may be of independent interest. We here use an approach based on tools that come both from dynamical system theory and analysis of algorithms. We shall deal here with the

[8] We use the term quasi-uniform to mean that the probability that $x \in \mathcal{B}$ is drawn in between $\ell_1/|\mathcal{B}|$ and $\ell_2/|\mathcal{B}|$, for constants independent on x and N.

CoreGauss algorithm, using the decomposition provided in Section "Returning to the Gauss Algorithm" Proposition 6.

Dynamical Systems

A dynamical system is a pair formed by a set X and a mapping $W : X \to X$ for which there exists a (finite or denumerable) set $\mathcal{Q}$ (whose elements are called digits) and a topological partition $\{X_q\}_{q\in\mathcal{Q}}$ of the set X in subsets X_q such that the restriction of W to each element X_q of the partition is of class $\mathcal{C}^2$ and invertible. Here, we deal with the so-called complete dynamical systems, where the restriction of $W|_{X_q} : X_q \to X$ is surjective. A special rôle is played by the set $\mathcal{H}$ of branches of the inverse function W^{-1} of W that are also naturally numbered by the index set $\mathcal{Q}$: we denote by $h_{\langle q\rangle}$ the inverse of the restriction $W|_{X_q}$, so that X_q is exactly the image $h_{\langle q\rangle}(X)$. The set $\mathcal{H}^k$ is the set of the inverse branches of the iterate W^k; its elements are of the form $h_{\langle q_1\rangle} \circ h_{\langle q_2\rangle} \circ \cdots \circ h_{\langle q_k\rangle}$ and are called the inverse branches of depth k. The set $\mathcal{H}^\star := \cup_{k\geq 0}\mathcal{H}^k$ is the semi-group generated by $\mathcal{H}$.

Given an initial point x in X, the sequence $\mathcal{W}(x) := (x, Wx, W^2x, \ldots)$ of iterates of x under the action of W forms the trajectory of the initial point x. We say that the system has a hole Y if any point of X eventually falls in Y: for any x, there exists $p \in \mathbb{N}$ such that $W^p(x) \in Y$.

We will study here two dynamical systems, respectively, related to the F-Euclid algorithm and to the CoreGauss algorithm, previously defined (in an informal way) in Section "Returning to the Gauss Algorithm".

Case of the F-Euclid **Algorithm.** Here, X is the interval $\widetilde{\mathcal{I}} = [0, 1/2]$. The map W is the map $\widetilde{V}$ defined in Section "Relation with the Centered Euclid Algorithm". The set $\mathcal{Q}$ of digits is the set of pairs $q = (m, \varepsilon)$ with the condition $(m, \varepsilon) \geq (2, +1)$ (with respect to the lexicographic order). The inverse branch $h_{\langle m,\varepsilon\rangle}$ is a LFT, defined as $h_{\langle m,\varepsilon\rangle}(z) = 1/(m + \varepsilon z)$. The topological partition is defined by $X_{(m,\varepsilon)} = h_{\langle m,\varepsilon\rangle}(\widetilde{\mathcal{I}})$.

Case of the CoreGauss **Algorithm.** Here, X is the vertical strip $\widetilde{\mathcal{B}}$. The map W is equal to the identity on $\widetilde{\mathcal{B}} \setminus \mathcal{D}$ and coincides with the map $\widetilde{U}$ defined in Section "The Complex Versions for the Gauss Algorithms" otherwise. The set $\mathcal{Q}$ of digits is the set of pairs $q = (m, \varepsilon)$ with the condition $(m, \varepsilon) \geq (2, +1)$ (with respect to the lexicographic order). The inverse branch $h_{\langle m,\varepsilon\rangle}$ is a LFT defined as $h_{\langle m,\varepsilon\rangle}(z) = 1/(m + \varepsilon z)$. The topological partition is defined by $X_{(m,\varepsilon)} = h_{\langle m,\varepsilon\rangle}(\widetilde{\mathcal{B}})$ and drawn in Fig. 16. The system has a hole, namely $\widetilde{\mathcal{B}} \setminus \mathcal{D}$.

Transfer Operators

The main study in dynamical systems concerns itself with the interplay between properties of the transformation W and properties of trajectories under iteration of the transformation. The behavior of typical trajectories of dynamical systems

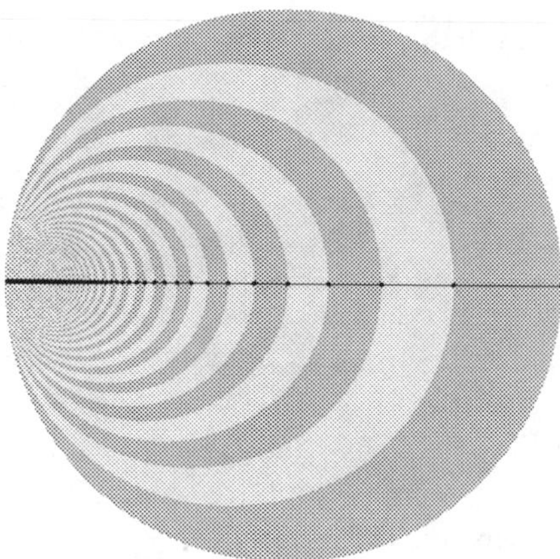

Fig. 3.16 The topological partitions of the COREGAUSS dynamical system. The intersection of this partition with the real axis gives rise to the topological partition of the F-EUCLID dynamical system

is more easily explained by examining the flow of densities. The time evolution governed by the map W modifies the density, and the successive densities $f_0, f_1, f_2, \ldots, f_n, \ldots$ describe the global evolution of the system at discrete times $t = 0, t = 1, t = 2, \ldots$.

Consider the (elementary) operator $\mathbf{X}_{s,[h]}$, relative to an inverse branch $h \in \mathcal{H}$, which acts on functions $f : X \to \mathbb{R}$, depends on some parameter s, and is formally defined as

$$\mathbf{X}_{s,[h]}[f](x) = J(h)(x)^s \cdot f \circ h(x), \quad \text{where } J(h) \text{ is the Jacobian of branch } h. \tag{3.55}$$

The operator $\mathbf{X}_{1,[h]}$ expresses the part of the new density f_1, which is brought when the algorithm uses the branch h, and the operator that takes into account all the inverse branches of the set $\mathcal{H}$, defined as

$$\mathbf{H}_s := \sum_{h \in \mathcal{H}} \mathbf{X}_{s,[h]}, \tag{3.56}$$

is called the transfer operator. For $s = 1$, the operator $\mathbf{H}_1 = \mathbf{H}$ is the density transformer, (or the Perron–Frobenius operator) which expresses the new density f_1 as a function of the old density f_0 via the relation $f_1 = \mathbf{H}[f_0]$. The operators defined in (3.56) are called transfer operators. For $s = 1$, they coincide with density transformers, and for other values of s, they can be viewed as extensions of density transformers. They play a central rôle in studies of dynamical systems.

We will explain how transfer operators are a convenient tool for studying the evolution of the densities, in the two systems of interest.

Case of the F-EUCLID **system.** This system is defined on an interval, and the Jacobian $J(h)(x)$ is just equal to $|h'(x)|$. Moreover, because of the precise expression of the set $\mathcal{H}$, one has, for any $x \in \widetilde{\mathcal{I}} = [0, 1/2]$,

$$\mathbf{H}_s[f](x) = \sum_{(m,\varepsilon)\geq(2,1)} \left(\frac{1}{m+\varepsilon x}\right)^{2s} \cdot f\left(\frac{1}{m+\varepsilon x}\right). \tag{3.57}$$

The main properties of the F-EUCLID algorithm are closely related to spectral properties of the transfer operator $\mathbf{H}_s$ when it acts on a convenient functional space. We return to this fact in Section "Functional Analysis".

Case of the COREGAUSS **algorithm.** We have seen in Section "Output Densities" that the Jacobian of the transformation $(x, y) \mapsto \underline{h}(x, y) = (\Re h(x+iy), \Im h(x+iy))$ equals $|h'(x+iy)|^2$. It would be natural to consider an (elementary) transfer operator $\mathbf{Y}_{s,[h]}$, of the form

$$\mathbf{Y}_{s,[h]}[f](z) = |h'(z)|^s \cdot f \circ h(z).$$

In this case, the sum of such operators, taken over all the LFTs that intervene in one step of the COREGAUSS algorithm, and viewed at $s = 2$, describes the new density that is brought at each point $z \in \widetilde{\mathcal{B}} \setminus \mathcal{D}$ during this step, when the density on $\mathcal{D}$ is f.

However, such an operator does not possess "good" properties, because the map $z \mapsto |h'(z)|$ is not analytic. It is more convenient to introduce another elementary operator $\underline{\mathbf{X}}_{s,[h]}$, which acts on functions F of two variables, and is defined as

$$\underline{\mathbf{X}}_{2s,[h]}[F](z, u) = \check{h}(z)^s \cdot \check{h}(u)^s \cdot F(h(z), h(u)),$$

where $\check{h}$ is the analytic extension of $|h'|$ to a complex neighborhood of $\widetilde{\mathcal{I}} := [0, 1/2]$. Such an operator acts on analytic functions, and the equalities, which relate $F(z, u)$ and its diagonal f defined by $f(z) := F(z, \bar{z})$,

$$\underline{\mathbf{X}}_{s,[h]}[F](z, \bar{z}) = \mathbf{Y}_{s,[h]}[f](z), \quad \underline{\mathbf{X}}_{s,[h]}[F](x, x) = \mathbf{X}_{s,[h]}[f](x) \tag{3.58}$$

prove that the elementary operators $\underline{\mathbf{X}}_{s,[h]}$ are extensions of the operators $\mathbf{X}_{s,[h]}$ that are well-adapted to our purpose. Furthermore, they are also well-adapted to deal with densities with valuation. Indeed, when applied to a density f of valuation r, of the form $f(z) = F(z, \bar{z})$, where $F(z, u) = |z - u|^r L(z, u)$ involves an analytic function L, which is nonzero on the diagonal $z = u$, one has

$$\underline{\mathbf{X}}_{2s,[h]}[F](z, \bar{z}) = |y|^r \, \underline{\mathbf{X}}_{2s+r,[h]}[L](z, \bar{z}).$$

Finally, for the COREGAUSS Algorithm, we shall deal with the operator $\underline{\mathbf{H}}_s$ defined as $\underline{\mathbf{H}}_s = \sum_{h\in\mathcal{H}} \underline{\mathbf{X}}_{s,[h]}$, which, in this case, admits a nice expression

$$\underline{\mathbf{H}}_s[F](z,u) = \sum_{(m,\varepsilon)\geq(2,1)} \left(\frac{1}{m+\varepsilon z}\right)^s \left(\frac{1}{m+\varepsilon u}\right)^s \cdot F\left(\frac{1}{m+\varepsilon z}, \frac{1}{m+\varepsilon u}\right). \tag{3.59}$$

Because of (3.58), this is an extension of the operator $\mathbf{H}_s$, defined in (3.57), which satisfies the equality

$$\underline{\mathbf{H}}_s[F](x,x) = \mathbf{H}_s[f](x), \qquad \text{when } f \text{ is the diagonal map of } F.$$

The operators $\mathbf{X}_{s,[h]}$, underlined or not, satisfy a crucial relation of composition due to multiplicative properties of the derivative of $g \circ h$. We easily remark that

$$\mathbf{X}_{s,[h]} \circ \mathbf{X}_{s,[g]} = \mathbf{X}_{s,[g\circ h]}, \qquad \underline{\mathbf{X}}_{s,[h]} \circ \underline{\mathbf{X}}_{s,[g]} = \underline{\mathbf{X}}_{s,[g\circ h]}.$$

We recall that the set $\mathcal{H}^+ = \cup_{k>0}\mathcal{H}^k$ is the set of the transformations describing the whole executions of our two algorithms of interest. Then, the transfer operator relative to $\mathcal{H}^+$, denoted by $\mathbf{G}_s$ (for the EUCLID Algorithm) or $\underline{\mathbf{G}}_s$ (for the COREGAUSS Algorithm), satisfies

$$\mathbf{G}_s = \mathbf{H}_s \circ (I - \mathbf{H}_s)^{-1} \qquad \text{or} \qquad \underline{\mathbf{G}}_s = \underline{\mathbf{H}}_s \circ (I - \underline{\mathbf{H}}_s)^{-1}, \tag{3.60}$$

and the assertion (3) of Theorem 6 can be re-written as

Theorem 9. [Dynamical version of Theorem 6]. *Consider the* COREGAUSS *algorithm, with its input density f on $\mathcal{D}$ and its output density $\widehat{f}$ on $\widetilde{\mathcal{B}} \setminus \mathcal{D}$, viewed as functions of two complex variables $z, \bar{z}$, namely $f(x,y) = F(z,\bar{z})$, $\widehat{f}(x,y) = \widehat{F}(z,\bar{z})$.*
Then, one has $\widehat{F} = \underline{\mathbf{G}}_2[F]$, where the operator $\underline{\mathbf{G}}_2$ is the "total" density transformer of the COREGAUSS *algorithm, which is related to the density transformer $\underline{\mathbf{H}}_2$ via the equality $\underline{\mathbf{G}}_2 = \underline{\mathbf{H}}_2 \circ (I - \underline{\mathbf{H}}_2)^{-1}$. When the input density F is of type (r, L), then the equality $\widehat{F}(z,\bar{z}) = y^r \underline{\mathbf{G}}_{2+r}[L]$ holds.*

Consider the COREGAUSS algorithm with an initial density, standard of valuation r. Such a density is defined on the input disk $\mathcal{D}$ and involves constant $A_0(r)$ [related with constant $A_2(r)$ defined in (3.49)] under the form

$$\frac{y^r}{A_0(r)} \qquad \text{with} \quad A_0(r) = \frac{1}{4^{r+2}} A_2(r) = \frac{\sqrt{\pi}}{4^{r+2}} \frac{\Gamma((r+1)/2)}{\Gamma(r/2+2)}. \tag{3.61}$$

Remark that $A_0(r) \sim 1/(r+1)$ when $r \to -1$. Then, the Hurwitz characterization provided in Proposition 6 gives rise to a nice expression for the output density $\widehat{F}$ in the case of a standard input density of valuation r, namely

$$\widehat{F}_r(z,\bar{z}) = \frac{1}{A_0(r)} \frac{1}{\zeta(2r+4)} \sum_{\substack{c,d\geq 1 \\ d\phi<c<d\phi^2}} \frac{y^r}{|cz+d|^{2r+4}}.$$

Execution Parameters in the Complex Framework

We are now interested in the study of the following costs:

1. Any additive cost $C_{(c)}$, defined in (3.9), relative to a cost c of moderate growth. There are two particular cases of interest: the number of iterations P, relative to $c = 1$, and the length Q of the continued fraction, relative to the case when c is the binary length ℓ,
$$Q(u,v) = \sum_{i=1}^{P(u,v)} \ell(|q_i|).$$
2. The bit-complexity B defined in Section "Main Parameters of Interest". It is explained (see 3.8) that the cost B decomposes as
$$B(u,v) = Q(u,v)\,\ell(|u|^2) + D(u,v) + \Theta\,(Q(u,v))\,, \tag{3.62}$$
where Q is the length of the continued fraction, already studied in (1), and cost D is defined by
$$D(u,v) = 2\sum_{i=1}^{P(u,v)} \ell(|q_i|)\lg\left|\frac{v_{i-1}}{v}\right|.$$
It is then sufficient to study costs Q and D.

All these costs are invariant by similarity, that is, $X(\lambda u, \lambda v) = X(u,v)$ for $X \in \{Q, D, P\}$ and $\lambda \in \mathbb{C}^{\star}$. If, with a small abuse of notation, we let $X(z) := X(1,z)$, we are led to study the main costs of interest in the complex framework. We first provide precise expressions for all these costs in the complex framework.

An additive cost $C_{(c)}$, defined more precisely in (3.9), is related to an elementary cost c defined on quotients q. Such a cost can be defined on $\mathcal{H}$ via the equality $c(h) = c(q)$ for $h = h_{\langle q\rangle}$, and is extended to the total set of LFTs in a linear way: for $h = h_1 \circ h_2 \circ \ldots \circ h_p$, we define $c(h)$ as $c(h) := c(h_1) + c(h_2) + \ldots + c(h_p)$. This gives rise to another definition for the complex version of cost defined by $C(z) := C(1,z)$. If an input $z \in \mathcal{D}$ leads to an output $\widehat{z} \in \widetilde{\mathcal{B}} \setminus \mathcal{D}$ by using the LFT $h \in \mathcal{G}$ with $z = h(\widehat{z})$, then $C(z)$ equals $c(h)$.

In the same vein as in (3.45), the ith length decrease can be expressed with the derivative of the LFT $g_i := h_i^{-1}$ (with h_i defined in (3.33)) as

$$\frac{|v_i|^2}{|v_0|^2} = \frac{1}{|g_i'(z)|} = |c_i z - a_i|^2 \text{ so that } 2\lg\left(\frac{|v_i|}{|v_0|}\right) = -\lg|g_i'(z)| = -\lg|c_i z - a_i|^2,$$

where a_i, c_i are coefficients of the LFT h_i. Finally, the complex versions of cost D is

$$D(z) = \sum_{i=1}^{P(z)} \ell(|q_i|)\lg|h_{i-1}'(z_{i-1})| = -2\sum_{i=1}^{P(z)} \ell(|q_i|)\,\lg|c_{i-1}z - a_{i-1}|. \tag{3.63}$$

The main idea of the dynamical analysis methodology is to use the transfer operators (introduced for studying dynamical systems) in the analysis of algorithms; for this aim, we modify the operators $\underline{\mathbf{X}}_{s,[h]}$ defined in Section "Transfer Operators" in such a way that they become "generating operators" that play the same role as generating functions in analytic combinatorics. In fact, these operators generate themselves... generating functions of the main costs of interest.

Generating Operators for Additive Costs C and Cost D

We now explain how to modify transfer operator in the two main cases: additive cost C and cost D.

Case of additive costs. It is natural to add a new parameter w inside the transfer operator $\underline{\mathbf{X}}_{s,[h]}$ for "marking" the cost: we consider the two-parameters operator $\underline{\mathbf{X}}_{s,w,(c),[h]}$ defined as

$$\underline{\mathbf{X}}_{2s,w,(c),[h]}[F](z,u) = \exp[wc(h)] \cdot \check{h}(z)^s \cdot \check{h}(u)^s \cdot F(h(z),h(u)).$$

Of course, when $w = 0$ or $c = 0$, we recover the operator $\underline{\mathbf{X}}_{2s,[h]}$. When the cost c is additive, that is, $c(g \circ h) = c(g) + c(h)$, the composition relation

$$\underline{\mathbf{X}}_{s,w,(c),[h]} \circ \underline{\mathbf{X}}_{s,w,(c),[g]} = \underline{\mathbf{X}}_{s,w,(c),[g \circ h]}$$

entails, an extension of (3.60) as

$$\underline{\mathbf{G}}_{s,w,(c)} = \underline{\mathbf{H}}_{s,w,(c)} \circ (I - \underline{\mathbf{H}}_{s,w,(c)})^{-1}, \tag{3.64}$$

where the operators $\underline{\mathbf{G}}_{s,w,(c)}, \underline{\mathbf{H}}_{s,w,(c)}$ are defined in the same vein as in (3.56). In particular,

$$\underline{\mathbf{H}}_{s,w,(c)}[F](z,u) = \sum_{(m,\varepsilon)\geq(2,1)} \exp[wc(m,\varepsilon)] \left(\frac{1}{m+\varepsilon z}\right)^s \left(\frac{1}{m+\varepsilon u}\right)^s \tag{3.65}$$

$$\cdot F\left(\frac{1}{m+\varepsilon z}, \frac{1}{m+\varepsilon u}\right). \tag{3.66}$$

The operator $\underline{\mathbf{G}}_{s,w,(c)}$ generates the moment generating function of the cost $C_{(c)}$, as we will see now. The moment generating function $\mathbb{E}_{\langle f \rangle}(\exp[wC_{(c)}])$ is defined as

$$\mathbb{E}_{\langle f \rangle}(\exp[wC_{(c)}]) := \sum_{h \in \mathcal{H}+} \exp[wc(h)] \cdot \mathbb{P}_{\langle f \rangle}[C = c(h)]$$

$$= \sum_{h \in \mathcal{H}+} \exp[wc(h)] \iint_{h(\widetilde{\mathcal{B}} \setminus \mathcal{D})} \underline{f}(x,y)\, \mathrm{d}x\mathrm{d}y.$$

Using a change of variables and the expression of the Jacobian leads to

$$\mathbb{E}_{\langle f\rangle}(\exp[wC_{(c)}]) = \sum_{h\in\mathcal{H}^+} \exp[wc(h)] \iint_{\widetilde{\mathcal{B}}\setminus\mathcal{D}} |h'(z)|^2 f(h(z), h(\bar{z}))\,\mathrm{d}x\mathrm{d}y$$

$$= \iint_{\widetilde{\mathcal{B}}\setminus\mathcal{D}} \underline{\mathbf{G}}_{2,w,(c)}[f](z,\bar{z})\,\mathrm{d}x\mathrm{d}y.$$

Now, when the density F is of type (r, L), using relation (3.41) leads to

$$\mathbb{E}_{\langle f\rangle}(\exp[wC_{(c)}]) = \iint_{\widetilde{\mathcal{B}}\setminus\mathcal{D}} y^r\, \underline{\mathbf{G}}_{2+r,w,(c)}[L](z,\bar{z})\,\mathrm{d}x\mathrm{d}y. \tag{3.67}$$

The expectation $\mathbb{E}_{\langle f\rangle}[C_{(c)}]$ is just obtained by taking the derivative with respect to w (at $w = 0$). This is why we introduce the functional $W_{(c)}$, which takes the derivative with respect to w, at $w = 0$. It then "weights" the operator $\underline{\mathbf{X}}_{s,[h]}$ with the cost $c(h)$, as

$$W_{(c)}\mathbf{X}_{s,[h]} := \frac{\partial}{\partial w}\underline{\mathbf{X}}_{s,w,(c),[h]}|_{w=0} = c(h)\mathbf{X}_{s,[h]}.$$

When extended via linearity, it defines the generating operator of cost C as

$$\underline{\mathbf{G}}_{s,C} := W_{(c)}[\underline{\mathbf{G}}_s] = W_{(c)}\left[\underline{\mathbf{H}}_s \circ (I - \underline{\mathbf{H}}_s)^{-1}\right]. \tag{3.68}$$

This provides an alternative expression for the expectation of any additive cost:

$$\mathbb{E}_{\langle f\rangle}[C_{(c)}] = \iint_{\widetilde{\mathcal{B}}\setminus\mathcal{D}} \underline{\mathbf{G}}_{2,C}[F](z,\bar{z})\,\mathrm{d}x\mathrm{d}y = \iint_{\widetilde{\mathcal{B}}\setminus\mathcal{D}} y^r\, \underline{\mathbf{G}}_{2+r,C}[L](z,\bar{z})\,\mathrm{d}x\mathrm{d}y, \tag{3.69}$$

the last equality holding for a density F of type (r, L).

Case of Cost D. Remark that, in (3.63), the quantity $\lg|h_i'(z_i)|\cdot|h_i'(z)|^s$ is just the derivative of $(1/\log 2)|h_i'(z)|^s$ with respect to s. This is why we introduce another functional Δ, in the same vein as previously, where the functional W relative to the cost was introduced. To an operator $\underline{\mathbf{X}}_{s,[h]}$, we associate an operator $\Delta\underline{\mathbf{X}}_{s,[h]}$ defined as

$$\Delta\underline{\mathbf{X}}_{s,[h]} = \frac{1}{\log 2}\frac{\partial}{\partial s}\underline{\mathbf{X}}_{s,[h]}.$$

The functional Δ weights the operator $\underline{\mathbf{X}}_{s,[h]}$ with the weight $-\lg|h'|$.

Now, with the help of these two functionals $W := W_{(\ell)}$ and Δ, we can build the generating operator for D. The decomposition of the set $\mathcal{H}^+$ as $\mathcal{H}^+ := \mathcal{H}^\star\cdot\mathcal{H}\cdot\mathcal{H}^\star$ gives rise to the parallel decomposition of the operators (in the reverse order). If we weight the second factor with the help of $W := W_{(\ell)}$, we obtain the operator

$$(I - \underline{\mathbf{H}}_s)^{-1} \circ W[\underline{\mathbf{H}}_s] \circ (I - \underline{\mathbf{H}}_s)^{-1} = W[(I - \underline{\mathbf{H}}_s)^{-1}],$$

which is the "generating operator" of the cost $Q(z)$. If, in addition of weighting the second factor with the help of W, we take the derivative Δ of the third one, then we obtain the operator

$$\underline{\mathbf{G}}_{s,D} := (I - \underline{\mathbf{H}}_s)^{-1} \circ W\,[\underline{\mathbf{H}}_s] \circ \Delta\left[(I - \underline{\mathbf{H}}_s)^{-1}\right],$$

$$\underline{\mathbf{G}}_{s,D} = (I - \underline{\mathbf{H}}_s)^{-1} \circ W\,[\underline{\mathbf{H}}_s] \circ (I - \underline{\mathbf{H}}_s)^{-1} \circ \Delta\,[\underline{\mathbf{H}}_s] \circ (I - \underline{\mathbf{H}}_s)^{-1}, \qquad (3.70)$$

which is the "generating operator" of the cost $D(z)$, as the equalities hold,

$$\mathbb{E}_{\langle f \rangle}[D] := \iint_{\mathcal{D}} D(z)\, F(z,\bar{z})\, \mathrm{d}x\mathrm{d}y,$$

$$= \iint_{\widetilde{\mathcal{B}}\setminus\mathcal{D}} \underline{\mathbf{G}}_{2,D}[F](z,\bar{z})\, \mathrm{d}x\mathrm{d}y = \iint_{\widetilde{\mathcal{B}}\setminus\mathcal{D}} y^r\, \underline{\mathbf{G}}_{2+r,D}[L](z,\bar{z})\, \mathrm{d}x\mathrm{d}y, \qquad (3.71)$$

the last equality holding for a density F of type (r, L).

Case of costs C, B in the Euclid Algorithm. These functionals W, Δ are also central in the analysis of the bit-complexity of the Euclid Algorithm [5,27]. One deals in this case with the Dirichlet series relative to cost X, for $X \in \{\mathrm{Id}, C_{(c)}, B\}$, defined as

$$F_X(s) := \sum_{\substack{(u,v)\in\mathbb{Z}^2 \\ v/u \in \widetilde{\mathcal{I}},\, \gcd(u,v)=1}} \frac{X(u,v)}{v^{2s}}.$$

These series admit alternative expressions that involve the quasi-inverse $(I - \mathbf{H}_s)^{-1}$ of the plain operator $\mathbf{H}_s$, together with functionals $W_{(c)}$ and Δ. Finally, the following equalities

$$F_{\mathrm{Id}}(s) = \mathbf{G}_s[1](0), \qquad F_C(s) = \mathbf{G}_{s,C}[1](0), \qquad F_B(s) = -\mathbf{G}_{s,D}[1](0). \quad (3.72)$$

hold, and involve the non-underlined[9] versions $\mathbf{G}_s, \mathbf{G}_{s,C}, \mathbf{G}_{s,D}$ of the generating operators $\underline{\mathbf{G}}_s, \underline{\mathbf{G}}_{s,C}, \underline{\mathbf{G}}_{s,D}$ defined in (3.68, 3.70).

Functional Analysis

We need precise information on the quasi-inverse $(I - \underline{\mathbf{H}}_s)^{-1}$, which is omnipresent in the expressions of our probabilistic studies (see 3.67, 3.69, 3), as the quasi-inverse $(I - \underline{\mathbf{H}}_s)^{-1}$ was already omnipresent in the probabilistic analyses of the F-EUCLID Algorithm.

[9] These operators are defined in the same vein as underlined versions, replacing each occurrence of the underlined operator $\underline{\mathbf{H}}_s$ by the plain operator $\mathbf{H}_s$.

It is first needed to find convenient functional spaces where the operators $\mathbf{H}_s, \underline{\mathbf{H}}_s$ and its variants $\underline{\mathbf{H}}_{s,w,(c)}$ will possess good spectral properties. Consider the open disk $\mathcal{V}$ of diameter $[-1/2, 1]$ and the functional spaces $A_\infty(\mathcal{V})$, $B_\infty(\mathcal{V})$ of all functions f (of one variable) or F (of two variables) that are holomorphic and continuous on the frontier: $A_\infty(\mathcal{V})$ is the space of functions f holomorphic in the domain $\mathcal{V}$ and continuous on the closure $\bar{\mathcal{V}}$, while $B_\infty(\mathcal{V})$ is the space of functions F holomorphic in the domain $\mathcal{V}\times\mathcal{V}$ and continuous on the closure $\bar{\mathcal{V}}\times\bar{\mathcal{V}}$. Endowed with the sup-norm, these are Banach spaces; for $\Re(s) > (1/2)$, the transfer operator $\mathbf{H}_s$ acts on $A_\infty(\mathcal{V})$, the transfer operator $\underline{\mathbf{H}}_s$ acts on $B_\infty(\mathcal{V})$, and these are compact operators. Furthermore, when weighted by a cost of moderate growth [i.e., $c(h_{\langle q\rangle}) = O(\log q)$], for w close enough to 0, and $\Re s > (1/2)$, the operator $\underline{\mathbf{H}}_{s,w,(c)}$ also acts on $B_\infty(\mathcal{V})$, and is also compact.

In the case of the F-EUCLID Algorithm, the spectral properties of the transfer operator defined in (3.57) play a central rôle in the analysis of the algorithm. For real s, the transfer operator $\mathbf{H}_s$ has a unique dominant eigenvalue $\lambda(s)$, which is real and separated from the remainder of the spectrum by a spectral gap. For $s = 1$, the dominant eigenvalue of the density transformer $\mathbf{H}$ satisfies $\lambda(1) = 1$, and the dominant eigenfunction $\psi(x)$ (which is then invariant under the action of $\mathbf{H}$) admits a closed form that involves the golden ratio $\phi = (1+\sqrt{5})/2$,

$$\psi(x) = \frac{1}{\log\phi}\left(\frac{1}{\phi + x} + \frac{1}{\phi^2 - x}\right).$$

This is the analog (for the F-EUCLID algorithm) of the celebrated Gauss density associated with the standard Euclid algorithm and equal to $(1/\log 2)1/(1+x)$.

Moreover, the quasi-inverse $(I - \mathbf{H}_s)^{-1}$ has a pôle at $s = 1$, and satisfies

$$(I - \mathbf{H}_s)^{-1}[f](z) \sim_{s\to 1} \frac{1}{s-1}\frac{1}{h(\mathcal{E})}\,\psi(z)\int_{\widetilde{\mathcal{I}}} f(x)\,\mathrm{d}x, \tag{3.73}$$

where the constant $h(\mathcal{E})$ is the entropy of the F-EUCLID dynamical system, and satisfies

$$h(\mathcal{E}) = |\lambda'(1)| = \frac{\pi^2}{6\log\phi} \approx 3.41831. \tag{3.74}$$

The operator $\underline{\mathbf{H}}_{s,w,(c)}$ also possesses nice spectral properties (see [40], [9]): for a complex number s close enough to the real axis, with $\Re s > (1/2)$, it has a unique dominant eigenvalue, denoted by $\lambda_{(c)}(s,w)$, which is separated from the remainder of the spectrum by a spectral gap. This implies the following: for any fixed s close enough to the real axis, the quasi-inverse $w \mapsto (I - \underline{\mathbf{H}}_{s,w,(c)})^{-1}$ has a dominant pôle located at $w = w_{(c)}(s)$ defined by the implicit equation $\lambda_{(c)}(s, w_{(c)}(s)) = 1$. More precisely, when $w = 0$, one recovers the plain operator $\underline{\mathbf{H}}_s$, which has the same dominant eigenvalue $\lambda(s)$ as the operator $\mathbf{H}_s$. For $s = 1$, it has a dominant eigenvalue $\lambda(1) = 1$ with a dominant eigenfunction $\underline{\psi}$, which is an extension of the invariant density ψ of the F-EUCLID Algorithm, and satisfies $\underline{\psi}(x,x) = \psi(x)$. An exact expression for $\underline{\psi}$ is provided in [40],

$$\underline{\psi}(z,u) = \frac{1}{\log\phi}\frac{1}{u-z}\left(\log\frac{\phi+u}{\phi+z} + \log\frac{\phi^2-u}{\phi^2-z}\right) \quad \text{for } z \neq u, \text{ and } \underline{\psi}(z,z) = \psi(z). \tag{3.75}$$

Near $s = 1$, the quasi-inverse satisfies

$$(I - \underline{\mathbf{H}}_s)^{-1}[F](z,u) \sim_{s\to 1} \frac{1}{s-1}\frac{1}{h(\mathcal{E})} I[F]\,\underline{\psi}(z,u), \quad \text{with } I[F] := \int_{\widetilde{\mathcal{I}}} F(x,x)\,dx. \tag{3.76}$$

We consider, in the sequel of this section, the CoreGauss algorithm with an initial density, standard of valuation r. Such a density is defined as $y^r/A_0(r)$, with $A_0(r)$ defined in (3.61). In this case, there are nice expressions for the moment generating functions $\mathbb{E}_{(r)}[\exp(wC)]$, for the expectations $E_{(r)}[C], \mathbb{E}_{(r)}[D]$, described in (3.67, 3.69, 3), where we let $L = 1$.

Probabilistic Analysis of the F-Euclid ***Algorithm***

We wish to compare the behavior of the two algorithms, the CoreGauss Algorithm and the F-Euclid Algorithm, and we first recall here the main facts about the probabilistic behavior of the F-Euclid Algorithm.

Theorem 10. (Akhavi and Vallée [5] (1998), Vallée [37, 41] (2003-2007)) *On the set ω_N formed with input pairs (u,v) for which $u/v \in \widetilde{\mathcal{I}}$ and $|v| \le N$, the mean number of iterations P, the mean value of a cost C of moderate growth, the mean value of the bit-complexity B satisfy, when $M \to \infty$,*

$$\mathbb{E}_N[P] \sim \frac{2\log 2}{h(\mathcal{E})} \lg N, \ \mathbb{E}_M[C_{(c)}] \sim \frac{2\log 2}{h(\mathcal{E})} \mathbb{E}[c] \lg N, \ \mathbb{E}_M[B] \sim \frac{\log 2}{h(\mathcal{E})} \mathbb{E}[\ell] \lg^2 N.$$

Here, $h(\mathcal{E})$ denotes the entropy of the F-Euclid *dynamical system, described in (3.74), and $\mathbb{E}[c]$ denotes the mean value of the step-cost c with respect to the invariant density ψ. This is a constant of Khinchin's type, of the form*

$$\mathbb{E}[c] := \sum_{h\in\mathcal{H}} \int_{h(\widetilde{\mathcal{I}})} \ell(h)\psi(x)\,dx.$$

In particular, when c is the binary length ℓ, there is a nice formula for $\mathbb{E}[\ell]$, namely

$$\mathbb{E}[\ell] = \frac{1}{\log\phi}\log\prod_{k\ge 1}\frac{2^k\phi^2+\phi}{2^k\phi^2-1} \approx 2.02197.$$

Proof (Sketch). One deals with the Dirichlet series $F_X(s)$ relative to cost X, defined in (3.72). Using the spectral relation (3.73) together with Tauberian Theorems leads to the asymptotic study of the coefficients of the series and provides the result. ■

Moreover, there exist also more precise distributional results [6, 27] which show that all these costs $P, C_{(c)}$, together with a regularized version of B, admit asymptotic Gaussian laws for $M \to \infty$.

What can be expected about the probabilistic behavior of the COREGAUSS Algorithm? On the one hand, there is a strong formal similarity between the two algorithms, as the COREGAUSS Algorithm can be viewed as a lifting of the F-EUCLID Algorithm. On the other hand, important differences appear when we consider algorithms: the F-EUCLID algorithm never terminates, except on rational inputs that fall in the hole $\{0\}$, while the COREGAUSS Algorithm always terminates, except for irrational real inputs. However, it is clear that these differences disappear when we restrict to rational inputs, real or complex ones. In this case, both algorithms terminate, and it is quite interesting to determine if there exists a precise transition between these two (discrete) algorithms.

Distribution of Additive Costs

We wish to prove that $k \mapsto \mathbb{P}_{(r)}[C_{(c)} = k]$ has a geometrical decreasing, with a precise estimate for the ratio. For this purpose, we use the moment generating function $\mathbb{E}_{(r)}(\exp[wC_{(c)}])$ of the cost $C_{(c)}$, for which we have provided an alternative expression in (3.67). We first study any additive cost, then we focus on the number of iterations.

General additive cost. The asymptotic behavior of the probability $\mathbb{P}_{(r)}[C_{(c)} = k]$ (for $k \to \infty$) is obtained by extracting the coefficient of $\exp[kw]$ in the moment generating function. Then the asymptotic behavior of $\mathbb{P}_{(r)}[C_{(c)} = k]$ is related to singularities of $\mathbb{E}_{(r)}(\exp[wC_{(c)}])$. This series has a pôle at $e^{w_{(c)}(r+2)}$, where $w = w_{(c)}(s)$ is defined by the spectral equation $\lambda_{(c)}(s, w) = 1$ that involves the dominant eigenvalue $\lambda_{(c)}(s, w)$ of the operator $\underline{\mathbf{H}}_{s,w,(c)}$, which is described in (3.65). Then, with classical methods of analytic combinatorics, we obtain:

Theorem 11. (Daudé et al. [14] (1994), Vallée and Vera [45] (2007)) *Consider the* COREGAUSS *algorithm, when its inputs are distributed inside the disk $\mathcal{D}$ with the continuous standard density of valuation r. Then, any additive cost $C_{(c)}$ defined in (3.9), associated to a step-cost c of moderate growth asymptotically, follows a geometric law.*

The ratio of this law, equal to $\exp[-w_{(c)}(r+2)]$, *is related to the solution $w_{(c)}(s)$ of the spectral relation $\lambda_{(c)}(s, w) = 1$, which involves the dominant eigenvalue of the transfer operator $\underline{\mathbf{H}}_{s,w,(c)}$. It satisfies, for any cost c of moderate growth, $w_{(c)}(r+2) = \Theta(r+1)$ when $r \to -1$. More precisely, one has*

$$\mathbb{P}_{(r)}[C_{(c)} = k] \sim_{k\to\infty} a(r) \exp[-kw_{(c)}(r+2)], \qquad \textit{for } k \to \infty, \qquad (3.77)$$

where $a(r)$ is a strictly positive constant that depends on cost c and valuation r.

Number of iterations. In the particular case of a constant step-cost $c = 1$, the cost $C_{(c)}$ is just the number R of iterations and the operator $\underline{\mathbf{H}}_{s,w,(1)}$ reduces to $e^w \cdot \underline{\mathbf{H}}_s$. In this case, there exists a nice alternative expression for the mean number of iterations of the COREGAUSS algorithm which uses the characterization of Hurwitz (recalled in Proposition 6.2). Furthermore, the probability of the event $[R \geq k+1]$ can be expressed in an easier way using (3.36), as

$$\mathbb{P}_{(r)}[R \geq k+1] = \frac{1}{A_0(r)} \sum_{h \in \mathcal{H}^k} \iint_{h(\mathcal{D})} y^r \mathrm{d}x\mathrm{d}y = \frac{1}{A_0(r)} \iint_{\mathcal{D}} y^r\, \underline{\mathbf{H}}_{2+r}^k\,[1](z)\,\mathrm{d}x\mathrm{d}y,$$

where $A_0(r)$ is defined in (3.61). This leads to the following result:

Theorem 12. (Daudé et al. [14] (1994), Vallée [40] (1996)) *Consider the* CORE-GAUSS *algorithm, when its inputs are distributed inside the disk $\mathcal{D}$ with the continuous standard density of valuation r. Then, the expectation of the number R of iterations admits the following expression:*

$$\mathbb{E}_{(r)}[R] = \frac{2^{2r+4}}{\zeta(2r+4)} \sum_{\substack{c,d \geq 1 \\ d\phi < c < d\phi^2}} \frac{1}{(cd)^{2+r}}.$$

Furthermore, for any fixed valuation $r > -1$, the number R of iterations asymptotically follows a geometric law

$$\mathbb{P}_{(r)}[R \geq k+1] \sim_{k \to \infty} \widetilde{a}(r)\, \lambda(2+r)^k,$$

where $\lambda(s)$ is the dominant eigenvalue of the transfer operator $\underline{\mathbf{H}}_s$ and $\widetilde{a}(r)$ is a strictly positive constant that depends on the valuation r.

It seems that there does not exist any close expression for the dominant eigenvalue $\lambda(s)$. However, this dominant eigenvalue is polynomial-time computable, as it is proven by Lhote [26]. In [17], numerical values are computed in the case of the uniform density, that is, for $\lambda(2)$ and $\mathbb{E}_{(0)}[R]$,

$$\mathbb{E}_{(0)}[R] \approx 1.08922, \qquad \lambda(2) \approx 0.0773853773.$$

For $r \to -1$, the dominant eigenvalue $\lambda(2+r)$ tends to $\lambda(1) = 1$ and $\lambda(2+r) - 1 \sim \lambda'(1)(1+r)$. This explains the evolution of the behavior of the Gauss Algorithm when the data become more and more concentrated near the real axis.

Mean Bit-Complexity

We are now interested in the study of the bit-complexity B,[10] and we focus on a standard density of valuation r. We start with the relation between B, and costs C, D

[10] We study the central part of the bit-complexity, and do not consider the initialization process, where the Gram matrix is computed; see Section "Main Parameters of Interest".

recalled in Section "Execution Parameters in the Complex Framework", together with the expressions of the mean values of parameters C, D obtained in (3.69, 3). We state three main results: the first one describes the evolution in the continuous model when the valuation r tends to -1; the second one describes the evolution of the discrete model when the integer size M tends to ∞, the valuation being fixed; finally, the third one describes the evolution of the discrete model when the valuation r tends to -1 *and* the integer size M tends to ∞.

Theorem 13. (Vallée and Vera [45, 47], 2007) *Consider the* CoreGauss *Algorithm, where its inputs are distributed inside the input disk $\mathcal{D}$ with the standard density of valuation $r > -1$. Then, the mean value $\mathbb{E}(r)[C]$ of any additive cost C of moderate growth, and the mean value $E_{(r)}[D]$ of cost D are well-defined and satisfy when $r \to -1$,*

$$\mathbb{E}_{(r)}[C] \sim \frac{1}{r+1} \frac{\mathbb{E}[c]}{h(\mathcal{E})}, \qquad \mathbb{E}_{(r)}[D] \sim -\frac{1}{(r+1)^2} \frac{1}{\log 2} \frac{\mathbb{E}[\ell]}{h(\mathcal{E})}.$$

When r tends to -1, the output density, associated with an initial density of valuation r, tends to $\dfrac{1}{h(\mathcal{E})} \dfrac{1}{y} \underline{\psi}$, where $\underline{\psi}$ is the invariant density for $\underline{\mathbf{H}}_1$ described in (3.75).

Remark that the constants that appear here are closely related to those which appear in the analysis of the Euclid algorithm (Theorem 10). More precisely, the asymptotics are almost the same when we replace $1/(r+1)$ (in Theorem 13) by $\log N$ (in Theorem 10). Later, Theorem 15 will make precise this observation.

Proof. For any valuation r, the variables C, D are integrable on the disk $\mathcal{D}$: this is due to the fact that, for $X \in \{\mathrm{Id}, C, D\}$, the integrals taken over the horizontal strip $\mathcal{H}_N := \mathcal{D} \cap \{z;\ |\Im z| \le (1/N)\}$ satisfy, with $M = \log N$,

$$\frac{1}{A_0(r)} \iint_{\mathcal{H}_N} y^r\, X(z)\, \mathrm{d}x \mathrm{d}y = \frac{M^{e(X)}}{N^{r+1}}\, O\left(1 + (r+1)M\right),$$

where the exponent $e(X)$ depends on cost X; one has $e(\mathrm{Id}) = 0, e(C) = 1, e(D) = 2$. This proves that cost X is integrable on $\mathcal{D}$. Furthermore, when $r \to -1$, relations (3.76, 3.73) prove the following behaviors:

$$\underline{\mathbf{G}}_{2+r,\mathrm{Id}}[F] \sim \frac{1}{r+1} \frac{1}{h(\mathcal{E})}\, I[F]\, \underline{\psi},$$

$$\underline{\mathbf{G}}_{2+r,C_{(c)}}[F] \sim \frac{1}{(r+1)^2} \frac{\mathbb{E}[c]}{h(\mathcal{E})^2}\, I[F]\, \underline{\psi}, \quad \underline{\mathbf{G}}_{2+r,D}[F] \sim -\frac{1}{(r+1)^3} \frac{\mathbb{E}[\ell]}{h(\mathcal{E})^2}\, I[F]\, \underline{\psi},$$

where the integral $I[F]$ is defined in (3.76) and $\underline{\psi}$ is described in (3.75). The first equality, together with the definition of $A_0(r)$ and the fact that $A_0(r) \sim (r+1)^{-1}$

for $r \to -1$, implies the equality

$$\iint_{\widetilde{\mathcal{B}}\setminus\mathcal{D}} \frac{1}{y}\, \Psi(z,\bar{z})\, \mathrm{d}x\mathrm{d}y = h(\mathcal{E}).$$

Using a nice relation between $I[\Psi]$ and $h(\mathcal{E})$ finally leads to the result. ■

It is now possible to transfer this analysis to the discrete model defined in Section "Probabilistic Models for Two-Dimensions", with the Gauss principle recalled in Section "Probabilistic Models: Continuous or Discrete".

Theorem 14. (Vallée and Vera [45, 47], 2007) *Consider the* CoreGauss *Algorithm, where its integer inputs (u, v) of length $M := \max\{\ell(|u|^2, \ell(|v|^2\}$ are distributed inside the input disk $\mathcal{D}$ with the standard density of valuation $r > -1$. Then, the mean value $\mathbb{E}_{(r,M)}[X]$ of cost X – where X is any additive cost C of moderate growth, or cost D – tends to the mean value $\mathbb{E}_{(r)}[X]$ of cost X, when $M \to \infty$. More precisely,*

$$\mathbb{E}_{(r,M)}[X] = \mathbb{E}_{(r)}[X] + \frac{M^{e(X)}}{N^{r+1}}\, O\left(\max\{1,(r+1)M\}\right),$$

where the exponent $e(X)$ depends on cost X and satisfies $e(C) = 1, e(D) = 2$. The mean value $\mathbb{E}_{(r,M)}[B]$ of the bit-complexity B satisfies, for any fixed $r > -1$, when $M \to \infty$,

$$\mathbb{E}_{(r,M)}[B] \sim \mathbb{E}_{(r)}[Q] \cdot M.$$

In particular, the mean bit-complexity is linear with respect to M.

Finally, the last result describes the transition between the CoreGauss algorithm and the F-Euclid Algorithm, obtained when the valuation r tends to -1, and the integer size M tends to $= \infty$:

Theorem 15. (Vallée and Vera [45, 47], 2007) *Consider the* CoreGauss *Algorithm, where its integer inputs (u, v) of length $M := \max\{\ell(|u|^2, \ell(|v|^2\}$ are distributed inside the input disk $\mathcal{D}$ with the standard density of valuation $r > -1$. When the integer size M tends to ∞ and the valuation r tends to -1, with $(r+1)M = \Omega(1)$, the mean value $\mathbb{E}_{(r,M)}[X]$ of cost X, where X can be any additive cost C of moderate growth, or cost D, satisfies*

$$\mathbb{E}_{(r,M)}[X] = \mathbb{E}_{(r)}[X]\left[1 + O\left(\frac{(M(r+1))^{e(X)+1}}{N^{r+1}}\right)\right]\left[\frac{1}{1-N^{-(r+1)}}\right],$$

where the exponent $e(X)$ depends on cost X and satisfies $e(C) = 1, e(D) = 2$.

Then, if we let $(r+1)M =: M^\alpha \to \infty$ *(with* $0 < \alpha < 1$*), then the mean values satisfy*

$$\mathbb{E}_{(r,M)}[C] \sim \frac{\mathbb{E}[c]}{h(\mathcal{E})} M^{1-\alpha}, \qquad \mathbb{E}_{(r,M)}[D] \sim -\frac{\mathbb{E}[\ell]}{h(\mathcal{E})} \frac{1}{\log 2} M^{2-2\alpha}$$

$$\mathbb{E}_{(r,M)}[B] \sim \frac{\mathbb{E}[\ell]}{h(\mathcal{E})} M^{2-\alpha}.$$

If now $(r+1)M$ *is* $\Theta(1)$*, then*

$$\mathbb{E}_{(r,M)}[C] = \Theta(M), \qquad \mathbb{E}_{(r,M)}[D] = \Theta(M^2), \qquad \mathbb{E}_{(r,M)}[B] = \Theta(M^2).$$

Open question. Provide a precise description of the phase transition for the behavior of the bit-complexity between the Gauss algorithm for a valuation $r \to -1$ and the Euclid algorithm: determine the constant hidden in the Θ term as a function of $(r+1)M$.

First Steps in the Probabilistic Analysis of the LLL Algorithm

We return now to the LLL algorithm and explain how the previous approaches can be applied for analyzing the algorithm.

Evolution of Densities of the Local Bases

The LLL algorithm aims at reducing all the local bases U_k (defined in Section "Description of the Algorithm") in the Gauss meaning. For obtaining the output density at the end of the algorithm, it is interesting to describe the evolution of the distribution of the local bases along the execution of the algorithm. The variant ODDEVEN described in Section "A Variation for the LLL Algorithm: The Odd-Even Algorithm" is well-adapted to this purpose.

In the first Odd Phase, the LLL algorithm first deals with local bases with odd indices. Consider two successive bases U_k and U_{k+2}, respectively, endowed with some initial densities F_k and F_{k+2}. Denote by z_k and z_{k+2} the complex numbers associated with local bases (u_k, v_k) and (u_{k+2}, v_{k+2}) via relation (3.1). Then, the LLL algorithm reduces these two local bases (in the Gauss meaning) and computes two reduced local bases denoted by $(\widehat{u}_k, \widehat{v}_k)$ and $(\widehat{u}_{k+2}, \widehat{v}_{k+2})$, which satisfy[11] in particular

$$|\widehat{v}_k^\star| = |u_k| \cdot \mu(z_k), \qquad |\widehat{u}_{k+2}| = |u_{k+2}| \cdot \lambda(z_{k+2}).$$

[11] The notation $\star$ refers to the Gram–Schmidt process as in Sections "The Lattice Reduction Algorithm in the Two-Dimensional Case and The LLL Algorithm".

Then, Theorem 8 provides insights on the distribution of $\mu(z_k), \lambda(z_{k+2})$. As, in our model, the random variables $|u_k|$ and z_k (respectively, $|u_{k+2}|$ and z_{k+2}) are independent (see Section "Probabilistic Models for Two-Dimensions"), we obtain a precise information on the distribution of the norms $|\widehat{v}_k^\star|, |\widehat{u}_{k+2}|$.

In the first Even Phase, the LLL algorithm considers the local bases with an even index. Now, the basis U_{k+1} is formed (up to a similarity) from the two previous output bases, as

$$u_{k+1} = |\widehat{v}_k^\star|, \qquad v_{k+1} = \nu|\widehat{v}_k^\star| + i\,|\widehat{u}_{k+2}|,$$

where ν *can be assumed to follow a (quasi-)uniform law* on $[-1/2, +1/2]$. Moreover, at least at the beginning of the algorithm, the two variables $|\widehat{v}_k^\star|, |\widehat{u}_{k+2}|$ are independent. All this allows to obtain precise information on the new input density F_{k+1} of the local basis U_{k+1}. We then hope to "follow" the evolution of densities of local bases along the whole execution of the LLL algorithm.

Open question: Is this approach robust enough to "follow" the evolution of densities of local bases along the whole execution of the LLL algorithm? Of course, in the "middle" of the algorithm, the two variables $\widehat{v}_k^\star, \widehat{u}_{k+2}$ are no longer independent. Are they independent enough, so that we can apply the previous method? Is it true that the variables ν at the *beginning* of the phase are almost uniformly distributed on $[-1/2, +1/2]$? Here, some experiments will be of great use.

The Dynamical System Underlying the ODD–EVEN–LLL *Algorithm*

We consider two dynamical systems, the Odd dynamical system (relative to the Odd phases) and the Even dynamical system (relative to the Even phases). The Odd (respectively, Even) dynamical system performs (in parallel) the same operations as the AGAUSS dynamical system, on each complex number z_i of odd (respectively, even) indices. Between the end of one phase and the beginning of the following phase, computations in the vein of Section "Evolution of Densities of the Local Bases" take place.

The dynamics of each system, Odd or Even, is easily deduced from the dynamics of the AGAUSS system. In particular, there is an Even Hole and an Odd Hole, which can be described as a function of the hole of the AGAUSS system. But the main difficulty for analyzing the ODD–EVEN Algorithm will come from the difference on the geometry of the two holes – the Odd one and the Even one. This is a work in progress!

References

1. K. AARDAL AND F. EISENBRAND. The LLL algorithm and integer programming, *This book*
2. M. AJTAI. Optimal lower bounds for the Korkine-Zolotareff parameters of a lattice and for Schnorr's algorithm for the shortest vector problem. Theory of Computing 4(1): 21–51 (2008)

3. A. Akhavi. Random lattices, threshold phenomena and efficient reduction algorithms, *Theoretical Computer Science*, 287, (2002), 359–385
4. A. Akhavi, J.-F. Marckert , and A. Rouault. On the reduction of a random basis, *Proceedings of SIAM-ALENEX/ANALCO'07.* New-Orleans, January 07, long version to appear in *ESAIM Probability and Statistics*
5. A. Akhavi and B. Vallée. Average bit-complexity of Euclidean algorithms, In *Proceedings of ICALP'2000* – Genève, 14 pages, LNCS, 373–387, (1853)
6. V. Baladi and B. Vallée. Euclidean algorithms are Gaussian, *Journal of Number Theory*, 110(2), (2005), 331–386
7. J. Bourdon, B. Daireaux, and B. Vallée. Dynamical analysis of α-Euclidean algorithms, *Journal of Algorithms*, 44, (2002), 246–285
8. D. Bump. *Automorphic Forms and Representations*, Cambridge University Press, Cambridge, (1996)
9. F. Chazal, V. Maume-Deschamps, and B. Vallée. Erratum to "Dynamical sources in information theory: fundamental intervals and word prefixes", *Algorithmica*, 38, (2004), 591–596
10. E. Cesaratto, J. Clément, B. Daireaux, L. Lhote, V. Maume-Deschamps and B. Vallée. Analysis of fast versions of the Euclid algorithm, Journal of Symbolic Computation, 44 (2009) pp 726-767
11. D. Coppersmith. Small solutions to polynomial equations, and low exponent RSA vulnerabilities, *Journal of Cryptology*, 10(4), (1997), 233–260
12. D. Coppersmith and A. Shamir. Lattice attacks on NTRU, *Proceedings of Eurocrypt 1997*, LNCS, 1233, 52–61, Springer, Berlin, (1997)
13. J.-S. Coron. Security proof for partial-domain hash signature schemes. In *Proceedings of Crypto 2002*, LNCS, 2442, 613–626, Springer, Berlin, (2002)
14. H. Daudé, P. Flajolet, and B. Vallée. An average-case analysis of the Gaussian algorithm for lattice reduction, *Combinatorics, Probability and Computing* 6, (1997), 397–433
15. H. Daudé and B. Vallée. An upper bound on the average number of iterations of the LLL algorithm. *Theoretical Computer Science* 123(1), (1994), 95–115
16. P. Flajolet and B. Vallée. Gauss' reduction algorithm : an average case analysis, *Proceedings of IEEE-FOCS 90*, St-Louis, Missouri, 2, 830–39
17. P. Flajolet and B. Vallée. Continued fractions, comparison algorithms and fine structure constants *Constructive, Experimental and Non-Linear Analysis*, Michel Thera, Editor, Proceedings of Canadian Mathematical Society, 27, 53–82, (2000)
18. C. Gentry. The geometry of provable security: some proofs of security in which lattices make a surprise appearance, *This book*
19. C. Gentry. How to compress rabin ciphertexts and signatures (and more), *Proceedings of Crypto'04*, 179–200, Springer, Berlin, (2004)
20. D. Goldstein and A. Mayer. On the equidistribution of Hecke points, *Forum Mathematicum*, 15, (2003), 165–189
21. G. Hanrot. LLL: a tool for effective diophantine approximation, *This book*
22. J. Klüners. The van Hoeij algorithm for factoring polynomials, *This book*
23. J. C. Lagarias. Worst-case complexity bounds for algorithms in the theory of integral quadratic forms. *Journal of Algorithms* 1(2), (1980), 142–186
24. H. Laville and B. Vallée. Distribution de la constante d'Hermite et du plus court vecteur dans les réseaux de dimension 2, *Journal de Théorie des nombres de Bordeaux* 6, (1994), 135–159
25. A. K. Lenstra, H. W. Lenstra, and L. Lovász. Factoring polynomials with rational coefficients. *Mathematische Annalen* 261, (1982), 513–534
26. L. Lhote. Computation of a class of continued fraction constants. *Proceedings of ALENEX-ANALCO'04*, 199–210
27. L. Lhote and B. Vallée. Gaussian laws for the main parameters of the Euclid algorithm, *Algorithmica* (2008), 497–554
28. A. May. Using LLL reduction for solving RSA and factorization problems, *This book*

29. P. NGUYEN AND D. STEHLÉ. Floating-point LLL revisited, *Proceedings of Eurocrypt 2005* , LNCS, 3494, 215–233, Springer, Berlin, (2005)
30. P. NGUYEN AND D. STEHLÉ. LLL on the average, *Proceedings of the 7th Algorithmic Number Theory Symposium (ANTS VII)*, LNCS, 4076, 238–256, Springer, Berlin, (2006)
31. J. HOFFSTEIN, N. HOWGRAVE-GRAHAM, J. PIPHER, AND W. WHYTE. NTRUEncrypt and NTRUSign, *This book.*
32. C. P. SCHNORR. A hierarchy of polynomial lattice basis reduction algorithms, *Theoretical Computer Science*, 53, (1987), 201–224
33. I. SEMAEV. A 3-dimensional lattice reduction algorithm, *Proceedings of the 2001 Cryptography and Lattices Conference (CALC'01)*, LNCS, 2146, 181–193, Springer, Berlin, (2001)
34. C. L. SIEGEL. A mean value theorem in geometry of numbers, *Annals in Mathematics*, 46(2), (1945), 340–347
35. D. SIMON. Selected applications of LLL in number theory, *This book*
36. D. STEHLÉ. Floating point LLL: theoretical and practical aspects, *This book.*
37. B. VALLÉE. Euclidean Dynamics, *Discrete and Continuous Dynamical Systems*, 15(1), (2006), 281–352
38. B. VALLÉE. Gauss' algorithm revisited. *Journal of Algorithms* 12, (1991), 556–572
39. B. VALLÉE. Algorithms for computing signs of 2×2 determinants: dynamics and average-case analysis, *Proceedings of ESA'97* (5th Annual European Symposium on Algorithms) (Graz, September 97), LNCS, 1284, 486–499
40. B. VALLÉE. Opérateurs de Ruelle-Mayer généralisés et analyse en moyenne des algorithmes de Gauss et d'Euclide, *Acta Arithmetica* 81.2, (1997), 101–144
41. B. VALLÉE. Dynamical analysis of a class of Euclidean algorithms, *Theoretical Computer Science* 297(1–3), (2003), 447–486
42. B. VALLÉE. An affine point of view on minima finding in integer lattices of lower dimensions. *Proceedings of EUROCAL'87*, LNCS, 378, 376–378, Springer, Berlin, (1987)
43. B. VALLÉE. Generation of elements with small modular squares and provably fast integer factoring algorithms, *Mathematics of Computation*, 56(194), (1991), 823–849
44. B. VALLÉE. Provably fast integer factoring algorithm with quasi-uniform quadratic residues, *Proceedings of ACM-STOC-89*, Seattle, 98–106
45. B. VALLÉE AND A. VERA. Lattice reduction in two-dimensions: analyses under realistic probabilistic models, Proceedings of the AofA'07 conference, *Discrete Mathematics and Theoretical Computer Science*, Proc. AH, (2007), 181–216
46. B. VALLÉE, M. GIRAULT, AND P. TOFFIN. How to guess ℓ-th roots modulo n by reducing lattices bases, *Proceedings of AAECC-88*, Rome, LNCS, (357), 427–442
47. A. VERA. Analyses de l'algorithme de Gauss. Applications à l'analyse de l'algorithme LLL, PhD Thesis, University of Caen, (July 2009)
48. G. VILLARD. Parallel lattice basis reduction. *Proceedings of International Symposium on Symbolic and Algebraic Computation*, Berkeley, ACM, (1992)

Chapter 4
Progress on LLL and Lattice Reduction

Claus Peter Schnorr

Abstract We review variants and extensions of the LLL-algorithm of Lenstra, Lenstra Lovász, extensions to quadratic indefinite forms and to faster and stronger reduction algorithms. The LLL-algorithm with Householder orthogonalisation in floating-point arithmetic is very efficient and highly accurate. We review approximations of the shortest lattice vector by feasible lattice reduction, in particular by block reduction, primal–dual reduction and random sampling reduction. Segment reduction performs LLL-reduction in high dimension, mostly working with a few local coordinates.

Introduction

A *lattice basis* of dimension n consists of n linearly independent real vectors $\mathbf{b}_1, \ldots, \mathbf{b}_n \in \mathbb{R}^m$. The basis $B = [\mathbf{b}_1, \ldots, \mathbf{b}_n]$ generates the *lattice* $\mathcal{L}$ consisting of all integer linear combinations of the basis vectors. Lattice reduction transforms a given basis of $\mathcal{L}$ into a basis with short and nearly orthogonal vectors.

Unifying two traditions. The LLL-algorithm of Lenstra, Lenstra Lovász [47] provides a reduced basis of proven quality in polynomial time. Its inventors focused on Gram–Schmidt orthogonalisation (GSO) in exact integer arithmetic, which is only affordable for small dimension n. With floating-point arithmetic (*fpa*) available, it is faster to orthogonalise in *fpa*. In numerical mathematics, the orthogonalisation of a lattice basis B is usually done by QR-factorization $B = QR$ using Householder reflections [76]. The LLL-community has neglected this approach as it requires square roots and does not allow exact integer arithmetic. We are going to unify these two separate traditions.

Practical experience with the GSO in *fpa* led to the LLL-algorithm of [66] implemented by Euchner that tackles the most obvious problems in limiting *fpa*-errors by heuristic methods. It is easy to see that the heuristics works and short lattice vectors are found. The LLL of [66] with some additional accuracy measures performs well

C.P. Schnorr
Fachbereich Informatik und Mathematik, Universität Frankfurt, PSF 111932, D-60054 Frankfurt am Main, Germany,
e-mail: schnorr@cs.uni-frankfurt.de, http: www.mi.informatik.uni-frankfurt.de

P.Q. Nguyen and B. Vallée (eds.), *The LLL Algorithm*, Information Security and Cryptography, DOI 10.1007/978-3-642-02295-1_4,

in double *fpa* up to dimension 250. But how does one proceed in higher dimensions Schnorr [65] gives a proven LLL in approximate rational arithmetic. It uses an iterative method by Schulz to recover accuracy and is not adapted to *fpa*. Its time bound in bit operations is a polynomial of degree 7, respectively, $6 + \varepsilon$ using school-, respectively FFT-multiplication while the degree is 9, respectively, $7 + \varepsilon$ for the original LLL.

In 1996, Rössner implemented in his thesis a continued fraction algorithm applying [31] and Householder orthogonalization (HO) instead of GSO. This algorithm turned out to be very stable in high dimension [63]. Our experience has well confirmed the higher accuracy obtained with HO, e.g., by attacks of *May* and *Koy* in 1999 on NTRU- and GGH-cryptosystems. The accuracy and stability of the QR-factorization of a basis B with HO has been well analysed for backwards accuracy [26, 34]. These results do not directly provide bounds for worst case forward errors. As fully proven *fpa*-error bounds are in practice far to pessimistic, we will use strong heuristics.

Outline. Section "LLL-Reduction and Deep Insertions" presents the LLL in ideal arithmetic and discusses various extensions of the LLL by deep insertions that are more powerful than LLL swaps. Section "LLL-Reduction of Quadratic and Indefinite Forms" extends the LLL to arbitrary quadratic forms, indefinite forms included.

Section "LLL Using Householder Orthogonalization" presents $\mathbf{LLL}_H$, an LLL with HO together with an analysis of forward errors. Our heuristics assumes that small error vectors have a negligible impact on the vector length as correct and error vectors are very unlikely to be near parallel in high dimension. It is important to weaken size-reduction under *fpa* such that the reduction becomes independent of *fpa*-errors. This also prevents infinite cycling of the algorithm. Fortunately, the weakened size-reduction has a negligible impact on the quality of the reduced basis. $\mathbf{LLL}_H$ of [70] and the L^2 of [56] are adapted to *fpa*, they improve the time bounds of the theoretical LLL of [65], and are well analysed. L^2 provides provable correctness; it is quadratic in the bit length of the basis. However, it loses about half of the accuracy compared to $\mathbf{LLL}_H$, and it takes more time.

Sections "Semi Block $2k$-Reduction Revisited" end the next two sections survey the practical improvements of the LLL that strengthen LLL-reduction to find shorter lattice vectors and better approximations of the shortest lattice vector. We revisit block reduction from [64], extend it to Koy's primal–dual reduction, and combine it with random sampling reduction of [69].

The last two sections survey recent results of [70]. They speed up $\mathbf{LLL}_H$ for large dimension n to LLL-type segment reduction that goes back to an idea of Schönhage [71]. This reduces the time bound of $\mathbf{LLL}_H$ to a polynomial of degree 6, respectively, $5 + \varepsilon$, and preserves the quality of the reduced bases. Iterating this method by iterated subsegments performs LLL-reduction in $O(n^{3+\varepsilon})$ arithmetic steps using large integers and *fpa*-numbers. How this can be turned into a practical algorithm is still an open question.

For general background on lattices and lattice reduction see [13, 16, 52, 54, 59]. Here is a small selection of applications in number theory [8, 14, 47, 72] computational theory [11, 27, 31, 45, 49, 53] cryptography [1, 9, 10, 17, 54, 58, 61] and complexity theory [2, 12, 20, 25, 36, 54]. Standard program packages are LIDIA, Magma and NTL.

Notation, GSO and GNF. Let $\mathbb{R}^m$ denote the real vector space of dimension m with *inner product* $\langle \mathbf{x}, \mathbf{y} \rangle = \mathbf{x}^t \mathbf{y}$. A vector $\mathbf{b} \in \mathbb{R}^m$ has *length* $\|\mathbf{b}\| = \langle \mathbf{b}, \mathbf{b} \rangle^{1/2}$. A sequence of linearly independent vectors $\mathbf{b}_1, \ldots, \mathbf{b}_n \in \mathbb{R}^m$ is a *basis*, written as matrix $B = [\mathbf{b}_1, \ldots, \mathbf{b}_n] \in \mathbb{R}^{m \times n}$ with columns $\mathbf{b}_i$. The basis B generates the lattice $\mathcal{L} = \mathcal{L}(B) = \{B\mathbf{x} \,|\, \mathbf{x} \in \mathbb{Z}^n\} = \sum_{i=1}^n \mathbf{b}_i \mathbb{Z} \subset \mathbb{R}^m$. It has *dimension* $\dim \mathcal{L} = n$. Let $\mathbf{q}_i$ denote the orthogonal projection of $\mathbf{b}_i$ in $\text{span}(\mathbf{b}_1, \ldots, \mathbf{b}_{i-1})^\perp$, $\mathbf{q}_1 = \mathbf{b}_1$. The *orthogonal vectors* $\mathbf{q}_1, \ldots, \mathbf{q}_n \in \mathbb{R}^m$ and the *Gram–Schmidt coefficients* $\mu_{j,i}$, $1 \le i, j \le n$ of the basis $\mathbf{b}_1, \ldots, \mathbf{b}_n$ satisfy for $j = 1, \ldots, n$: $\mathbf{b}_j = \sum_{i=1}^j \mu_{j,i} \mathbf{q}_i$, $\mu_{j,j} = 1$, $\mu_{j,i} = 0$ for $i > j$, $\mu_{j,i} = \langle \mathbf{b}_j, \mathbf{q}_i \rangle / \langle \mathbf{q}_i, \mathbf{q}_i \rangle$, $\langle \mathbf{q}_j, \mathbf{q}_i \rangle = 0$ for $j \ne i$. The basis $B \in \mathbb{R}^{m \times n}$ has a unique QR-decomposition $B = QR$, where $Q \in \mathbb{R}^{m \times n}$ is *isometric* (i.e., Q preserves the inner product, $\langle \mathbf{x}, \mathbf{y} \rangle = \langle Q\mathbf{x}, Q\mathbf{y} \rangle$, Q can be extended to an orthogonal matrix $Q' \in \mathbb{R}^{m \times m}$) and $R = [r_{i,j}] \in \mathbb{R}^{n \times n}$ is upper-triangular ($r_{i,j} = 0$ for $i > j$) with positive diagonal entries $r_{1,1}, \ldots, r_{n,n} > 0$. Hence $Q = [\mathbf{q}_1 / \|\mathbf{q}_1\|, \ldots, \mathbf{q}_n / \|\mathbf{q}_n\|]$, $\mu_{j,i} = r_{i,j} / r_{i,i}$, $\|\mathbf{q}_i\| = r_{i,i}$ and $\|\mathbf{b}_i\|^2 = \sum_{j=1}^i r_{j,i}^2$. Two bases $B = QR$, $B' = Q'R'$ are *isometric* iff $R = R'$, or equivalently iff $B^t B = B'^t B'$. We call R the *geometric normal form* (GNF) of the basis, $\text{GNF}(B) := R$. The GNF is preserved under isometric transforms Q, i.e., $\text{GNF}(QB) = \text{GNF}(B)$.

The Successive Minima. The jth successive minimum $\lambda_j(\mathcal{L})$ of a lattice $\mathcal{L}$, $1 \le j \le \dim \mathcal{L}$, is the minimal real number ρ for which there exist j linearly independent lattice vectors of length $\le \rho$; λ_1 is the length of the shortest nonzero lattice vector.

Further Notation. $\text{GL}_n(\mathbb{Z}) = \{T \in \mathbb{Z}^{n \times n} \mid \det T = \pm 1\}$,
$R^{-t} = (R^{-1})^t = (R^t)^{-1}$ is the inverse transpose of $R \in \mathbb{R}^{n \times n}$,
$d_i = \det([\mathbf{b}_1, \ldots, \mathbf{b}_i]^t [\mathbf{b}_1, \ldots, \mathbf{b}_i])$, $d_0 = 1$, $\det \mathcal{L}(B) = \det(B^t B)^{1/2} = d_n^{1/2}$,
$\pi_i : \mathbb{R}^m \to \text{span}(\mathbf{b}_1, \ldots, \mathbf{b}_{i-1})^\perp$ is the orthogonal projection, $\mathbf{q}_i = \pi_i(\mathbf{b}_i)$,
$\gamma_n = \max_{\mathcal{L}} \lambda_1^2(\mathcal{L}) / \det \mathcal{L}^{2/n}$ over all lattices $\mathcal{L}$ of $\dim \mathcal{L} = n$ is the Hermite constant,

$$U_k = \begin{bmatrix} & & 1 \\ & \cdot & \\ 1 & & \end{bmatrix} \in \mathbb{Z}^{k \times k};$$ $B := BU_n / B := U_m B$ reverses the order of columns/rows of $B \in \mathbb{R}^{m \times n}$, we write matrices $A = [a_{i,j}]$ with capital letter A, and small letter entries $a_{i,j}$, $I_n \in \mathbb{Z}^{n \times n}$ denotes the unit matrix, let $\varepsilon \in \mathbb{R}$, $0 \le \varepsilon \approx 0$.

Details of Floating Point Arithmetic. We use the *fpa* model of Wilkinson [76]. An *fpa* number with $t = 2t' + 1$ *precision bits* is of the form $\pm 2^e \sum_{i=-t'}^{t'} b_i 2^i$, where $b_i \in \{0, 1\}$ and $e \in \mathbb{Z}$. It has bit length $t + s + 2$ for $|e| < 2^s$, two signs included. We denote the set of these numbers by $\mathbb{FL}_t$. Standard double length *fpa* has $t = 53$ precision bits, $t + s + 2 = 64$. Let $fl : \mathbb{R} \supset [-2^{2^s}, 2^{2^s}] \ni r \mapsto \mathbb{FL}_t$ approximate real numbers by *fpa* numbers. A step $c := a \circ b$ for $a, b, c \in \mathbb{R}$ and a binary operation $\circ \in \{+, -, \cdot, /\}$ translates under *fpa* into $\bar{a} := fl(a)$, $\bar{b} := fl(b)$, $\bar{c} := fl(\bar{a} \circ \bar{b})$, respectively into $\bar{a} := fl(\circ(\bar{a}))$ for unary operations $\circ \in \{\lceil\ \rfloor, \sqrt{\ }\}$. Each *fpa* operation induces a normalized relative error bounded in magnitude by 2^{-t}: $|fl(\bar{a} \circ \bar{b}) - \bar{a} \circ \bar{b}|/|\bar{a} \circ \bar{b}| \leq 2^{-t}$. If $|\bar{a} \circ \bar{b}| > 2^{2^s}$ or $|\bar{a} \circ \bar{b}| < 2^{-2^s}$, then $fl(\bar{a} \circ \bar{b})$ is undefined due to an *overflow* respectively *underflow*.

Usually, one requires that $2^s \leq t^2$ and thus, $s \leq 2\log_2 t$, for brevity, we identify the bit length of *fpa*-numbers with t, neglecting the minor $(s+2)$-part. We use approximate vectors $\bar{\mathbf{h}}_l$, $\bar{\mathbf{r}}_l \in \mathbb{FL}_t^m$ for HO under *fpa* and exact basis vectors $\mathbf{b}_l \in \mathbb{Z}^m$.

LLL-Reduction and Deep Insertions

We describe reduction of the basis $B = QR$ in terms of the GNF $R = [r_{i,j}] \in \mathbb{R}^{n \times n}$.

Standard reductions. A lattice basis $B = QR \in \mathbb{R}^{m \times n}$ is *size-reduced* (for ε) if

$$|r_{i,j}|/r_{i,i} \leq \frac{1}{2} + \varepsilon \quad \text{for all } j > i. \quad (\textit{occasionally we neglect } \varepsilon)$$

$B = QR$ is LLL-*reduced* (or an LLL-*basis*) for $\delta \in (\eta^2, 1]$, $\eta = \frac{1}{2} + \varepsilon$, if B is size-reduced and

$$\delta r_{i,i}^2 \leq r_{i,i+1}^2 + r_{i+1,i+1}^2 \quad \text{for } i = 1, \ldots, n-1.$$

A basis $B = QR$ is HKZ-*reduced* (or an HKZ-*basis*) if B is size-reduced, and each diagonal coefficient $r_{i,i}$ of the GNF $R = [r_{i,j}] \in \mathbb{R}^{n \times n}$ is minimal under all transforms in $\mathrm{GL}_n(\mathbb{Z})$ that preserve $\mathbf{b}_1, \ldots, \mathbf{b}_{i-1}$.

LLL-bases satisfy $r_{i,i}^2 \leq \alpha\, r_{i+1,i+1}^2$ for $\alpha := 1/(\delta - \eta^2)$. This yields Theorem 1. Lenstra, Lenstra, and Lovász [47] introduced LLL-bases focusing on $\delta = 3/4$, $\epsilon = 0$ and $\alpha = 2$.

HKZ-bases are due to Hermite [33] and Korkine-Zolotareff [38]. LLL/HKZ-bases $B = QR$ are preserved under isometry. LLL/HKZ-reducedness is a property of the GNF R.

Theorem 1. **[47]** *An LLL-basis* $B \in \mathbb{R}^{m \times n}$ *of lattice* $\mathcal{L}$ *satisfies for* $\alpha = 1/(\delta - \eta^2)$

1. $\|\mathbf{b}_1\|^2 \leq \alpha^{\frac{n-1}{2}} (\det \mathcal{L})^{2/n}$, **2.** $\|\mathbf{b}_1\|^2 \leq \alpha^{n-1} \lambda_1^2$

3. $\|\mathbf{b}_i\|^2 \leq \alpha^{i-1} r_{i,i}^2$, **4.** $\alpha^{-i+1} \leq \|\mathbf{b}_i\|^2 \lambda_i^{-2} \leq \alpha^{n-1}$ *for* $i = 1, \ldots, n$.

If an LLL-basis satisfies the stronger inequalities $r_{i,i}^2 \leq \bar{\alpha} r_{i+1,i+1}^2$ for all i and some $1 < \bar{\alpha} < \alpha$, then the inequalities **1–4** hold with α^{n-1} replaced by $\bar{\alpha}^{n-1}$ and with α^{i-1} in **3, 4** replaced by $\bar{\alpha}^{i-1}/(4\bar{\alpha} - 4)$.

The last five sections survey variants of LLL- and HKZ-bases that either allow faster reduction for large dimension n or provide a rather short vector $\mathbf{b}_1$. For these bases we either modify clause 1 or clause 2 of Theorem 1. Either clause is sufficient on account of an observation of Lovász [49] pp.24 ff, that the following problems are polynomial time equivalent for all lattices $\mathcal{L}(B)$ given B:

1. Find $\mathbf{b} \in \mathcal{L}$, $\mathbf{b} \neq \mathbf{0}$ with $\|\mathbf{b}\| \leq n^{O(1)} \lambda_1(\mathcal{L})$.

2. Find $\mathbf{b} \in \mathcal{L}$, $\mathbf{b} \neq \mathbf{0}$ with $\|\mathbf{b}\| \leq n^{O(1)} (\det \mathcal{L})^{1/n}$.

Theorem 2. [48] *An HKZ-basis $B \in \mathbb{R}^{m \times n}$ of lattice $\mathcal{L}$ satisfies*

$$4/(i+3) \leq \|\mathbf{b}_i\|^2 \lambda_i^{-2} \leq (i+3)/4 \quad \textit{for } i = 1, \dots, n.$$

The algorithms for HKZ-reduction [21, 35], exhaustively enumerate, for various l, all lattice vectors $\mathbf{b}$ such that $\|\pi_l(\mathbf{b})\| \leq \|\pi_l(\mathbf{b}_l)\|$ for the current $\mathbf{b}_l$. Their theoretical analysis has been stepwise improved [32, 35] to a proven $mn^{\frac{n}{2e}+o(n)}$ time bound for HKZ-reduction [28]. The enumeration **ENUM** of [66, 67] is particularly fast in practice, see [1] for a survey and [28, 60] for heuristic and experimental results.

Algorithm 1: LLL *in ideal arithmetic*

INPUT $\mathbf{b}_1, \dots, \mathbf{b}_n \in \mathbb{Z}^m$ a basis with $M_0 = \max\{\|\mathbf{b}_1\|, \dots, \|\mathbf{b}_n\|\}$, δ with $\frac{1}{4} < \delta < 1$

1. $l := 1$, *# at stage l $\mathbf{b}_1, \dots, \mathbf{b}_{\max(l-1,1)}$ is an LLL-basis with given GNF.*

2. WHILE $l \leq n$ DO

2.1 *#compute $\mathbf{r}_l = col(l, R)$:*

FOR $i = 1, \dots, l-1$ DO [$r_{i,l} := (\langle \mathbf{b}_i, \mathbf{b}_l \rangle - \sum_{k=1}^{i-1} r_{k,i} r_{k,l}) / r_{i,i}$],

$r_{l,l} := |\, \|\mathbf{b}_l\|^2 - \sum_{k=1}^{l-1} r_{k,l}^2 \,|^{1/2}$, $\mathbf{r}_l := (r_{1,l}, \dots, r_{l,l}, 0, \dots 0)^t \in \mathbb{R}^n$.

2.2 *#size-reduce $\mathbf{b}_l$ and $\mathbf{r}_l$, $\lceil r \rfloor = \lceil r - \frac{1}{2} \rceil$ denotes the nearest integer to $r \in \mathbb{R}$:*

FOR $i = l-1, \dots, 1$ DO $\mathbf{b}_l := \mathbf{b}_l - \lceil r_{i,l}/r_{i,i} \rfloor \mathbf{b}_i$, $\mathbf{r}_l := \mathbf{r}_l - \lceil r_{i,l}/r_{i,i} \rfloor \mathbf{r}_i$.

2.3 IF $l > 1$ and $\delta r_{l-1,l-1}^2 > r_{l-1,l}^2 + r_{l,l}^2$

THEN swap $\mathbf{b}_{l-1}, \mathbf{b}_l$, $l := l - 1$ ELSE $l := l + 1$.

3. OUTPUT LLL-basis $B = [\mathbf{b}_1, \dots, \mathbf{b}_n]$, $R = [\mathbf{r}_1, \dots, \mathbf{r}_n]$ for δ.

Comments. **LLL** performs simultaneous column operations on R and B; it swaps columns $\mathbf{r}_{l-1}, \mathbf{r}_l$ and $\mathbf{b}_{l-1}, \mathbf{b}_l$ if this shortens the length of the first column of the submatrix $\begin{bmatrix} r_{l-1,l-1} & r_{l-1,l} \\ 0 & r_{l,l} \end{bmatrix}$ of R by the factor $\sqrt{\delta}$. To enable a swap, the entry $r_{l-1,l}$ is first reduced to $|r_{l-1,l}| \leq \frac{1}{2} |r_{l-1,l-1}|$ by transforming $\mathbf{r}_l := \mathbf{r}_l - \lceil r_{l-1,l}/r_{l-1,l-1} \rfloor \mathbf{r}_{l-1}$. At stage l, we get $\mathbf{r}_l = \mathrm{col}(l, R)$ of $R = \mathrm{GNF}(B)$, and we have $\mathbf{r}_{l-1}$ from a previous stage. The equation $\mathrm{GNF}([\mathbf{b}_1, \dots \mathbf{b}_l]) = [\mathbf{r}_1, \dots \mathbf{r}_l]$ is preserved during simultaneous size-reduction of $\mathbf{r}_l$ and $\mathbf{b}_l$.

Each swap in step 2.3 decreases $\mathcal{D}^{(1)} := \prod_{i=1}^{n-1} d_i$ by the factor δ. As initially $\mathcal{D}^{(1)} \leq M_0^{n^2}$ and $\mathcal{D}^{(1)}$ remains, integer **LLL** performs $\leq n^2 \log_{1/\delta} M_0$ rounds, denoting $M_0 = \max\{\|\mathbf{b}_1\|, \dots, \|\mathbf{b}_n\|\}$ for the input basis. Each round performs $O(nm)$ arithmetic steps; **LLL** runs in $O(n^3 m \log_{1/\delta} M_0)$ arithmetic steps.

Time bound in exact rational/integer arithmetic. The rationals $r_{l,l}^2$, $\mu_{l,i} = r_{i,l}/r_{i,i}$ can easily be obtained within **LLL** in exact rational/integer arithmetic. Moreover, the integer $\mu_{j,i} d_i$ has bit length $O(n \log_2 M_0)$ throughout **LLL**-computations. This yields

Theorem 3. **LLL** *runs in* $O(n^5 m(\log_{1/\delta} M_0)^3)$, *respectively* $O(n^{4+\varepsilon} m (\log_{1/\delta} M_0)^{2+\varepsilon})$ *bit operations, under school-, respectively FFT-multiplication.*

The degree of this polynomial time bound (in $n, m, \log M_0$) is $9/7 + \varepsilon$ under school-/FFT-multiplication. The degree reduces to $7/6 + \varepsilon$ by computing the GNF, respectively, the GSO in floating-point arithmetic (*fpa*). This is done by $\mathbf{LLL}_H$ [70] of section "What is a Random (Basic of a) Lattice?" and the L^2 of [56]; both use *fpa* numbers of bit length $O(n + \log_2 M_0)$. LLL-type segment reduction **SLLL** of [70, Section 9] further decreases the time bound degree to 6/5 + ε. The factor m in the time bounds can be reduced to m by performing the reduction under random projections [6]. The theoretical, less practical algorithms of [65, 73] approach the time bound of $\mathbf{LLL}_H$ by approximate rational, respectively very long integer arithmetic.

A canonical order of the input basis vectors. While the **LLL** depends heavily on the order of the basis vectors, this order can be made nearly canonical by swapping before step 2.3 $\mathbf{b}_l$ and $\mathbf{b}_j$ for some $j \geq l$ that minimizes the value $r_{l-1,l}^2 + r_{l,l}^2$ resulting from the swap. This facilitates a subsequent swap of $\mathbf{b}_{l-1}, \mathbf{b}_l$ by step 2.3. Moreover, this tends to decrease the number of rounds of the **LLL** and anticipates subsequent deep insertions into the LLL-basis $\mathbf{b}_1, \ldots, \mathbf{b}_{l-1}$ via the following new step 2.3.

New step 2.3, *deep insertion at* (j, l) [SE91], (the old step 2.3 is restricted to *depth* $l - j = 1$):

```
IF ∃j, 0 < j < l such that δr²_{j,j} > Σ_{i=j}^{l} r²_{i,l}
  THEN for the smallest such j do (b_j, ..., b_l) := (b_l, b_j, b_{j+1}, ..., b_{l-1}), l := j
THEN ELSE l := l + 1.
```

LLL *with deep insertions* is efficient in practice, and runs with some restrictions in polynomial time. Deep insertion at (j, l) decreases $r_{j,j}$ and $d_j = r_{j,j}^2 \cdots r_{n,n}^2$ and provides $r_{j+k,j+k}^{new}$ close to $r_{j+k-1,j+k-1}$ for $j + k < l$. This is because $\sum_{i=j}^{l} r_{i,l}^2 < \delta r_{j,j}^2$ and the $\langle \pi_j(\mathbf{b}_l), \mathbf{b}_{j+k-1} \rangle$ are quite small.

Deep insertion at (j, l) can be strengthened by decreasing $\|\pi_j(\mathbf{b}_l)\|$ through additional, more elaborate size-reduction. Before testing, $\delta r_{j,j}^2 > \sum_{i=j}^{l} r_{i,l}^2 = \|\pi_j(\mathbf{r}_l)\|^2$ shorten $\pi_j(\mathbf{b}_l)$ as follows

2.2.b *additional size-reduction of* $\pi_j(\mathbf{b}_l)$:

```
WHILE ∃h : j ≤ h < l such that μ'_{j,h} := Σ_{i=j}^{h} r_{i,l} r_{i,h} / Σ_{i=j}^{h} r²_{i,h} satisfies δ|μ'_{j,h}| ≥ 1/2
  DO b_l := b_l − ⌈μ'_{j,h}⌋ b_h,  r_l := r_l − ⌈μ'_{j,h}⌋ r_h.
```

Obviously, deep insertion with additional size-reduction remains polynomial time per round. This makes the improved deep insertion quite attractive and more efficient than the algorithms of Sections "Semi Block 2k-Reduction Revisited" FF. In addition, one can perform deep insertion of $\mathbf{b}_l$ and of several combinations of $\mathbf{b}_l$ with other lattice vectors. This leads to random sampling reduction of [69] to be studied in Section "Analysis of the Lattice Reduction in two Dimensions".

Decreasing α *by deep insertion of depth* ≤ 2. The bound $\|\mathbf{b}_1\| \leq \alpha^{\frac{n-1}{2}} (\det \mathcal{L})^{\frac{2}{n}}$ of Theorem 1 is sharp for the LLL-GNF $R = [r_{i,j}] \in \mathbb{R}^{n \times n}$: $r_{i,i} = \alpha^{(-i+1)/2}$, $r_{i,i+1} = \frac{1}{2} r_{i,i}$, $r_{i,j} = 0$ for $j \geq i + 2$. (Note that deep insertion at $(1, n)$ results in

$r_{1,1} = \lambda_1$ right-away.) While this GNF satisfies $r_{i,i}^2/r_{i+1,i+1}^2 = \alpha$, the maximum of $r_{i,i}^2/r_{i+2,i+2}^2$ under LLL-reduction with deep insertion of depth ≤ 2 for $\delta = 1$ is $\frac{3}{2}$, $\frac{3}{2} < \alpha^2 = \frac{16}{9}$, since the HKZ-reduced GNF of the critical lattice of dimension 3 satisfies $r_{1,1}^2/r_{3,3}^2 = \frac{3}{2}$. This shows that deep insertion of depth ≤ 2 decreases α in clauses **1**, **2** of Theorem 1 from $\frac{4}{3}$ to $(\frac{3}{2})^{1/2}$.

LLL-reduction of bases B with large entries. Similar to Lehmer's version of Euclid's algorithm for large numbers [37] Sect. 4.5.2, p. 342, most of the LLL-work can be done in single precision arithmetic:

Pick a random integer $\rho' \in_R [1, 2^{43}]$ and truncate $B = [b_{i,j}]$ into the matrix B' with entries $b'_{i,j} := \lceil b_{i,j}\rho'/2^\tau \rfloor$ such that $|b_{i,j}\rho'/2^\tau - b'_{i,j}| \leq 1/2$ and $|b'_{i,j}| \leq \rho' + 1/2$, where $\tau := \max_{i,j} \lceil \log_2 |b_{i,j}| \rceil \gg 53$.

LLL-reduce B' into $B'T'$ working mostly in single precision, e.g. with 53 precision bits, and transform B into BT'. Iterate this process as long as it shortens B.

Note that $(2^\tau/\rho')B'$ and $(2^\tau/\rho')^n \det B'$ well approximate B and $\det B$. In particular, $|\det B| \ll (2^\tau/\rho')^n$ implies that $\det B' = 0$, and thus LLL-reduction of B' produces zero vectors of $B'T'$ and highly shortened vectors of BT'.

LLL-Reduction of Quadratic and Indefinite Forms

We present extensions of the LLL-algorithm to domains other than $\mathbb{Z}$ and general quadratic forms.

Rational, algebraic and real numbers. It is essential for the LLL-algorithm that the input basis $B \in \mathcal{R}^{m\times n}$ has entries in an euclidean domain $\mathcal{R}$ with efficient, exact arithmetic such as $\mathcal{R} = \mathbb{Z}$. Using rational arithmetic the LLL-algorithm directly extends from $\mathbb{Z}$ to the field of rational numbers $\mathbb{Q}$ and to rings $\mathcal{R}$ of algebraic numbers. However, the bit length of the numbers occuring within the reduction requires further care. If the rational basis matrix B has an integer multiple $\mathrm{d}B \in \mathbb{Z}^{m\times n}$, $\mathrm{d} \in \mathbb{Z}$ of moderate size, then the LLL-algorithm applied to the integer matrix $\mathrm{d}B \in \mathbb{Z}^{m\times n}$ yields an LLL-basis of $\mathcal{L}(\mathrm{d}B) \in \mathbb{Z}^m$ which is the d-multiple of an LLL-basis of the lattice $\mathcal{L}(B)$.

Gram-matrices, symmetric matrices, quadratic forms. The LLL-algorithm directly extends to real bases $B \in \mathbb{R}^{m\times n}$ of rank n that have an integer Gram-matrix $B^t B \in \mathbb{Z}^{n\times n}$. It transforms $A := B^t B$ into $A' = T^t A T$ such that the GNF R' of $A' = R'^t R'$ is LLL-reduced.

We identify symmetric matrices $A = A^t = [a_{i,j}]_{1\leq i,j\leq n} \in \mathbb{R}^{n\times n}$ with n-ary quadratic forms $\mathbf{x}^t A\mathbf{x} \in \mathbb{R}[x_1, \ldots, x_n]$. The forms A, A' are *equivalent* if $A' = T^t A T$ holds for some $T \in \mathrm{GL}_n(\mathbb{Z})$.

The form $A \in \mathbb{R}^{n\times n}$ with $\det(A) \neq 0$ is *indefinite* if $\mathbf{x}^t A\mathbf{x}$ takes positive and negative values; otherwise, A is either *positive* or *negative* (definite). The form A is *regular* if $\det(A) \neq 0$. We call $A = A^t = [a_{i,j}]$ *strongly regular* (s.r.) if $\det([a_{i,j}]_{1\leq i,j\leq \ell}) \neq 0$ for $\ell = 1, \ldots, n$. Let $D_{\boldsymbol{\sigma}} \in \{0, \pm 1\}^{n\times n}$ denote the diagonal matrix with diagonal $\boldsymbol{\sigma} = (\sigma_1, \ldots, \sigma_n) \in \{\pm 1\}^n$. An easy proof shows

Proposition 1. *Every s.r. form* $A = A^t \in \mathbb{R}^{n\times n}$ *has a unique decomposition* $A = R^t D_\sigma R \in \mathbb{R}^{n\times n}$ *with a GNF* $R \in \mathbb{R}^{n\times n}$ *and a diagonal matrix* D_σ *with diagonal* $\sigma \in \{\pm 1\}^n$.

We call R the geometric normal form (GNF) of A, $R = \mathrm{GNF}(A)$. The *signature* $\#\{i \mid \sigma_i = 1\}$ is invariant under equivalence. The form $A = R^t D_\sigma R$ is positive, respectively, negative if all entries σ_i of $\boldsymbol{\sigma}$ are $+1$, respectively, -1. The form is indefinite if $\boldsymbol{\sigma}$ has -1 and $+1$ entries.

Definition 1. A s.r. form $A = RD_\sigma R$ is an LLL-form if its GNF R is LLL-reduced.

The LLL-form $A = [a_{i,j}] = B^t B$ of an LLL-basis $B = (\mathbf{b}_1, \ldots, \mathbf{b}_n)$ satisfies the classical bound $a_{1,1}^2 = \|\mathbf{b}_1\|^2 \leq \alpha^{\frac{n-1}{2}} |\det(A)|^{\frac{2}{n}}$ of Theorem 1.

If $B \in \mathbb{R}^{m\times n}$ generates a lattice $\mathcal{L} = \{B\mathbf{x} \mid \mathbf{x} \in \mathbb{Z}\}$ of dimension $n' \leq n$, the LLL-algorithm transforms the input basis B into an output basis $[\mathbf{0}, \ldots, \mathbf{0}, \mathbf{b}'_1, \ldots, \mathbf{b}'_{n'}] \in \mathbb{R}^{m\times n}$ such that $[\mathbf{b}'_1, \ldots, \mathbf{b}'_{n'}]$ is an LLL-basis. This generalizes to

Corollary 1. *An adjusted LLL transforms an input form* $A \in \mathbb{Z}^{n\times n}$ *of rank* n' *into an equivalent regular form* $A' \in \mathbb{Z}^{n'\times n'}$ *with* $\det(A') \neq 0$.

LLL-reduction easily extends to s.r. forms $A \in \mathbb{Z}^{n\times n}$: compute $R = \mathrm{GNF}(A)$; LLL-reduce R into RT and transform A into the equivalent form $A' = T^t AT$. By Lemma 1, the LLL-reduction of non s.r. indefinite forms reduces to LLL-reduction in dimension $n-2$, see also Simon [72].

Lemma 1. *An adjusted LLL-algorithm transforms a non s.r. input form* $A \in \mathbb{Z}^{n\times n}$ *in polynomial time into an equivalent* $A' = [a'_{i,j}]$ *such that* $a'_{1,i} = a'_{2,j} = 0$ *for all* $i \neq 2$, $j \geq 3$ *and* $0 \leq a'_{2,2} \leq 2a'_{1,2}$. *Such* A' *is a direct sum* $\begin{bmatrix} 0 & a'_{1,2} \\ a'_{1,2} & a'_{2,2} \end{bmatrix} \oplus [a'_{i,j}]_{3\leq i,j\leq n}$. *Moreover* $a'_{1,2} = 1$ *if* $\det(A) \neq 0$ *is square-free.*

Proof. If $\det([a_{i,j}]_{1\leq i,j\leq \ell}) = 0$ for some $\ell \leq n$, then the LLL-algorithm achieves in polynomial time that $a_{1,1} = 0$. The corresponding A can be transformed into A' such that $a'_{1,2} = \gcd(a_{1,2}, \ldots, a_{1,n})$, $a'_{1,3} = \cdots = a'_{1,n} = 0 = a'_{2,3} = \cdots = a'_{2,n}$. Moreover $a'_{2,2}$ can be reduced modulo $2a'_{1,2}$. Doing all transforms symmetrically on rows and columns, the transformed $A' = T^t AT$ is symmetric $a'_{1,2} = a'_{2,1}$, and thus $(a'_{1,2})^2$ divides $\det(A)$. If $\det(A)$ is square-free, then $|a'_{1,2}| = 1$ and $a'_{1,2} = 1$ is easy to achieve. □

Lemma 1 shows that the LLL-algorithm can be adjusted to satisfy

Theorem 4. *An adjusted LLL-algorithm transforms a given form* $A \in \mathbb{Z}^{n\times n}$ *in polynomial time into a direct sum* $\oplus_{i=1}^{k} A^{(i)}$ *of an LLL-form* $A^{(k)}$ *and binary forms* $A^{(i)} = \begin{bmatrix} 0 & a_i \\ a_i & b_i \end{bmatrix}$ *for* $i = 1, \ldots, k-1$, *where* $0 \leq b_i < 2a_i$ *and* $a_i = 1$ *if* $\det(A) \neq 0$ *is square-free. If* A *is positive definite and* $\det A \neq 0$, *then* $k = 1$.

LLL-reduction of indefinite forms is used in cryptographic schemes based hard problems of indefinite forms. In particular, the equivalence problem is NP-hard for ternary indefinite forms [29]. Hartung and Schnorr [30] presents public key identification and signatures based on the equivalence problem of quadratic forms.

LLL Using Householder Orthogonalization

In practice GNF$(B) = R$ is computed in *fpa* within **LLL** whereas B is transformed in exact integer arithmetic. For better accuracy, we replace steps 2.1, 2.2 of **LLL** by the procedure $\texttt{TriCol}_l$ below, denoting the resulting LLL-algorithm by $\mathbf{LLL}_H$. $\texttt{TriCol}_l$ performs HO via *reflections*; these isometric transforms preserve the length of vectors and of error vectors. $\mathbf{LLL}_H$ improves the accuracy of the LLL of [66]. All subsequent reduction algorithms are based on $\mathbf{LLL}_H$. We streamline the $\mathbf{LLL}_H$-analysis of [70].

Computing the GNF of B. Numerical algorithms for computing the GNF $R = [r_{i,j}] \in \mathbb{R}^{n\times n}$ of $B = QR$ have been well analyzed, see [34] Chap. 19. It has been known for quite some time that HO is very stable. Wilkinson showed that the computation of a Householder vector and the transform of a given matrix by a reflection are both normwise stable in the sense that the computed Householder vector is very close to the exact one, and the computed transform is the update of a tiny normwise perturbation of the original matrix. Wilkinson also showed that the QR factorization algorithmn is normwise backwards stable [76] pp.153–162, p. 236. For a componentwise and normwise error analysis see [34] Chap. 19.3.

To simplify the analysis, let all basis vectors $\mathbf{b}_j$ start with n zero entries $\mathbf{b}_j = (0^n, \mathbf{b}'_j) \in 0^n\mathbb{Z}^{m-n}$. The bases $\mathbf{b}'_1, \dots, \mathbf{b}'_n$ and $\mathbf{b}_1, \dots, \mathbf{b}_n$ have the same GNF. The padding with initial zeros increases $\texttt{TriCol}$'s number of steps by the factor $\frac{4}{3}$; it is not required in practice. We compute an orthogonal matrix $Q' \in \mathbb{R}^{m\times m}$ that extends $Q \in \mathbb{R}^{m\times n}$ by $m - n$ columns and a matrix $R' \in \mathbb{R}^{m\times n}$ that extends $R \in \mathbb{R}^{n\times n}$ by final zero rows. In ideal arithmetic, we get R' by a sequence of Householder reflections $Q_j \in \mathbb{R}^{m\times m}$ as

$$R'_0 := B, \quad R'_j := Q_j R'_{j-1} \quad \text{for } j = 1, \dots, n,$$
$$R' := R'_n, \quad Q' := Q_1 \cdots Q_n = Q_1^t \cdots Q_n^t,$$

where $Q_j := I_m - 2\|\mathbf{h}_j\|^{-2}\mathbf{h}_j\mathbf{h}_j^t$ is orthogonal and symmetric, and $\mathbf{h}_j \in \mathbb{R}^m$.

The transform $R'_j \mapsto Q_j R'_{j-1}$ zeroes the entries in positions $j + 1$ through m of col(j, R'_{j-1}), it *triangulates* $\mathbf{r} := (r_1, \dots, r_m)^t :=$ col(j, R'_{j-1}) so that $R'_j \in \mathbb{R}^{m\times n}$ is upper-triangular for the first j columns. The reflection Q_j reflects about the hyperplane span$(\mathbf{h}_j)^\perp$:

$$Q_j\mathbf{h}_j = -\mathbf{h}_j, \qquad Q_j\mathbf{x} = \mathbf{x} \quad \text{for } \langle \mathbf{h}_j, \mathbf{x}\rangle = 0.$$

Note that $(r_j, \dots, r_n)^t = 0^{n-j+1}$ due to $\mathbf{b}_1, \dots, \mathbf{b}_j \in 0^n \mathbb{Z}^{m-n}$.

We set $r_{j,j} := (\sum_{i=j}^m r_i^2)^{1/2}$, $\mathbf{h}_j := (0^{j-1}, -r_{j,j}, 0^{n-j}, r_{n+1}, \dots, r_m)^t$.

Correctness of $\mathbf{h}_j$. We have $2\langle \mathbf{h}_j, \mathbf{r}\rangle \|\mathbf{h}_j\|^{-2} = 1$ and $\|\mathbf{h}_j\|^2 = 2r_{j,j}^2$ and thus

$$Q_j \mathbf{r} = \mathbf{r} - \mathbf{h}_j = (r_1, \dots, r_{j-1}, r_{j,j}, 0^{m-j})^t \in \mathbb{R}^m.$$

Hence $Q_j \mathbf{r}$ is correctly triangulated and the *Householder vector* $\mathbf{h}_j$ is well chosen.

`TriCol` $(\mathbf{b}_1, \dots, \mathbf{b}_l, \mathbf{h}_1, \dots, \mathbf{h}_{l-1}, \mathbf{r}_1, \dots, \mathbf{r}_{l-1})$ (`TriCol`$_l$ for short)
`TriCol`$_l$ *computes* $\mathbf{h}_l$ *and* $\mathbf{r}_l := \mathrm{col}(l, R)$ *and size-reduces* $\mathbf{b}_l, \mathbf{r}_l$.
1. $\mathbf{r}_{0,l} := \mathbf{b}_l$, `FOR` $j = 1, \dots, l-1$ `DO` $\mathbf{r}_{j,l} := \mathbf{r}_{j-1,l} - \langle \mathbf{h}_j, \mathbf{r}_{j-1,l}\rangle \mathbf{h}_j / r_{j,j}^2$.
2. $(r_1, \dots, r_m)^t := \mathbf{r}_{l-1,l}$, $\zeta := \max_i |r_i|$, $r_{l,l} := \zeta(\sum_{i=n+1}^m (r_i/\zeta)^2)^{1/2}$,
   # ζ *prevents under/overflow;* $r_i = 0$ *holds for* $i = l, \dots, n$
3. $\mathbf{h}_l := (0^{l-1}, -r_{l,l}, 0^{n-l}, r_{n+1}, \dots, r_m)^t$, # *note that* $\|\mathbf{h}_l\|^2 = 2r_{l,l}^2$.
4. $\mathbf{r}_l := (r_1, \dots, r_{l-1}, r_{l,l}, 0^{m-l})^t \in \mathbb{R}^m$.
5. # *size-reduce* $\mathbf{b}_l$ *and* $\mathbf{r}_l$: `FOR` $i = l-1, \dots, 1$ `DO`
   `IF` $|r_{i,l}/r_{i,i}| \le \frac{1}{2} + \varepsilon$
   `THEN` $\mu_i := 0$ `ELSE` $\mu_i := \lceil r_{i,l}/r_{i,i} \rfloor$, $\mathbf{b}_l := \mathbf{b}_l - \mu_i \mathbf{b}_i$, $\mathbf{r}_l :=$
   $\mathbf{r}_l - \mu_i \mathbf{r}_i$.
6. `IF` $\sum_{i=1}^{l-1} |\mu_i| \neq 0$ `THEN GO TO` 1 `ELSE output` $\mathbf{b}_l, \mathbf{r}_l, \mathbf{h}_l$.

`TriCol`$_l$ *under* fpa *with t precision bits*. Zeroing μ_i in case $|r_{i,l}/r_{i,i}| <\le \frac{1}{2} + \varepsilon$ in step 5 cancels a size-reduction step and prevents cycling through steps 1–6. In `TriCol`$_l$'s last round size-reduction is void. Zeroing μ_i, the use of ζ and the loop through steps 1–6 are designed for *fpa*. The μ_i in steps 5, 6 consist of the leading $\Theta(t)$ bits of the $\mu_{l,i}$. `TriCol`$_l$ replaces in the Schnorr-Euchner-LLL [66] classical Gram–Schmidt by an economic, modified Gram–Schmidt, where $\mathrm{col}(l, R)$ gets merely transformed by the $l-1$ actual reflections.

Accurate floating point summation. Compute the scalar product $\langle \mathbf{h}_j, \mathbf{r}_{j-1,l}\rangle = \langle \mathbf{x}, \mathbf{y}\rangle$ in step 1 by summing up positive and negative terms $x_i y_i$ separately, both in increasing order to $\sum_{>0}$ and $\sum_{<0}$. The increasing order minimizes the apriori forward error [34] chapter 4, see also [19]. If both $\sum_{>0}$ and $-\sum_{<0}$ are larger than $2^{-t/2}$ and nearly opposite, $|\sum_{>0} + \sum_{<0}| < 2^{-t/2}(\sum_{>0} - \sum_{<0})$, then compute $\langle \mathbf{x}, \mathbf{y}\rangle$ exactly. This provision is far more economic than that of [66] to compute $\langle \mathbf{x}, \mathbf{y}\rangle$ exactly if $|\langle \mathbf{x}, \mathbf{y}\rangle| < 2^{t/2}\|\mathbf{x}\|\|\mathbf{y}\|$. Proposition 2 and Theorem 5 do not require occasional exact computations of $\langle \mathbf{x}, \mathbf{y}\rangle$.

Efficiency. `TriCol`$_l$ performs $4ml + \frac{3}{2}l^2 + O(l)$ arithmetic steps and one sqrt per round, the $\frac{3}{2}l^2$ steps cover step 1. `TriCol`$_l$ is more economic than the fully modified GSO of [34] as only the reflections of the current Householder vectors $\mathbf{h}_1, \dots, \mathbf{h}_{l-1}$ are applied to $\mathbf{b}_l$. The contribution of step 1 to the overall costs is negligible for reasonably long input bases since on average $l \le n/2$, and there are no long integer steps. `TriCol`$_l$ performs at most $\log_2(2\|\mathbf{b}_l\|/2^t)$ rounds; each round shortens $\mathbf{b}_l$ by a factor $\le 1/2^{t-1}$.

fpa-Heuristics. There is room for heuristics in speeding up the LLL as the correctness of the output can most likely be efficiently verified [75]. We want to catch the typical behaviour of *fpa*-errors knowing that worst case error bounds are to pessimistic. Note that error vectors $\mathbf{x} - \bar{\mathbf{x}}$ are rarely near parallel to $\mathbf{x}$ in high dimension. Small errors $\|\mathbf{x} - \bar{\mathbf{x}}\| \le \varepsilon\|\mathbf{x}\|$ with expected value $\mathbf{E}[\langle \mathbf{x}, \bar{\mathbf{x}} - \mathbf{x}\rangle] = 0$ satisfy $\mathbf{E}[\,\|\bar{\mathbf{x}}\|^2] = \mathbf{E}[\,\|\mathbf{x} + (\bar{x} - \mathbf{x})\|^2] = \mathbf{E}[\,\|\mathbf{x}\|^2 + \|\bar{\mathbf{x}} - \mathbf{x}\|^2] \le (1 + \varepsilon^2)\mathbf{E}[\,\|\mathbf{x}\|^2]$. Hence, the relative error $|\,\|\bar{\mathbf{x}}\| - \|\mathbf{x}\|\,|/\|\mathbf{x}\|$ is for $\varepsilon \ll 1$, on average smaller than ε^2 and can be neglected.

We let the projection $\pi'_n : \mathbb{R}^m \to \mathbb{R}^{m-n}$ remove the first n coordinates so that $\pi'_n(\mathbf{b}_j) = \mathbf{b}'_j$ for $\mathbf{b}_j = (0^n, \mathbf{b}'_j)$. Recall that $\texttt{TriCol}_l$ computes in step 2 $\mathbf{r}_{l-1,l} = \prod_{j=1}^{l-1} Q_j \mathbf{b}_l$, as $\bar{\mathbf{r}}_{0,l} := \mathbf{b}_l$, $\bar{\mathbf{r}}_{j,l} := fl(\bar{Q}_j \bar{\mathbf{r}}_{j-1,l})$ for $1 \le j < l$ and $\bar{Q}_j = I_m - \|\bar{\mathbf{h}}_j\|^{-2}\bar{\mathbf{h}}_j\bar{\mathbf{h}}_j^t$.

Proposition 2. [fpa-Heur.] $\texttt{TriCol}_l$ *applied to an LLL-basis* $\mathbf{b}_1, \ldots, \mathbf{b}_l \in 0^n \mathbb{Z}^{m-n}$ *approximates the GNF* $R = [r_{i,j}] = [\mathbf{r}_1, \ldots, \mathbf{r}_l]$ *such that for* $j = 0, \ldots, l-1$

1. $|\bar{r}_{j+1,l} - r_{j+1,l}| \le \|\pi'_n(\bar{\mathbf{r}}_{j,l} - \mathbf{r}_{j,l})\| \le 10\,m\,(\frac{3}{2})^{j-1} \max_{1 \le i \le l} r_{i,i}\, 2^{-t}$,
2. $\mathbf{E}[\|\pi'_n(\bar{\mathbf{r}}_{j,l} - \mathbf{r}_{j,l})\|] \le 10\,m(\frac{5}{4})^{(j-1)/2} \max_{1 \le i \le l} r_{i,i}\, 2^{-t}$ *holds for random* fpa-*error vectors.*

LLL-bases $\mathbf{b}_1, \ldots, \mathbf{b}_l$ satisfy $r_{i,i}^2 \le \alpha^{l-i} r_{l,l}^2$ and $\|\mathbf{b}_l\| \le \alpha^{\frac{l-1}{2}} r_{l,l}$.Clause 1. of Proposition 2 shows for $j = i - 1$ that $\texttt{TriCol}_l$ achieves for $i = 1, \ldots, l$

$$|\bar{\mu}_{l,i} - \mu_{l,i}| \approx |\bar{r}_{i,l} - r_{i,l}|/r_{i,i} \le 10\,m\,\alpha^{\frac{l-1}{2}} \left(\frac{3}{2}\right)^{i-2} \frac{r_{l,l}}{r_{i,i}} 2^{-t}. \tag{4.1}$$

This bound can easily be adjusted to the case that merely $\mathbf{b}_1, \ldots, \mathbf{b}_{l-1}$ is LLL-reduced. Thus, it covers for $i = l - 1$ the critical situation of swapping $\mathbf{b}_{l-1}, \mathbf{b}_l$ within $\mathbf{LLL}_H$. It guarantees correct swapping of $\mathbf{b}_{l-1}, \mathbf{b}_l$ if $2^t \ge 100\,m\sqrt{3}^{\,l-1}$, as $\frac{3}{2}\sqrt{\alpha} \approx \sqrt{3}$ holds for $\alpha \approx \frac{4}{3}$. On average the error bounds are much smaller for random *fpa*-error vectors. Clause 2. of Proposition 2 reduces $\frac{3}{2}$ in (4.1) on average to $(\frac{5}{4})^{1/2} \approx 1.12$. Moreover, the constant α in (1) reduces considerably by stronger lattice reduction. Sections "The LLL algorithm–Analysis of Lattice Reduction in Two-Dimensions" show that α can be decreased by feasible lattice reduction to about 1.025. This reduces our correctness condition $2^t \ge 100\,m\sqrt{3}^{\,l-1}$, for LLL-swapping of $\mathbf{b}_{l-1}, \mathbf{b}_l$ to $2^t \ge 100\,m\,1.132^{l-1}$.

Proof of Proposition 2. Induction on j; $j = 0$: Consider the last round of $\texttt{TriCol}_l$ where size-reduction in step 5 is void. So let $\mathbf{b}_l$ and $\mathbf{r}_l$ be size-reduced. We have that $\mathbf{r}_{0,l} = \mathbf{b}_l$, $\mathbf{r}_{1,l} = \mathbf{b}_l - \langle \mathbf{h}_1, \mathbf{b}_l\rangle \mathbf{h}_1/r_{1,1}^2$, $\mathbf{h}_1 = (-r_{1,1}, 0^{n-1}, \mathbf{b}'_1)$, $r_{1,1} = \|\mathbf{b}_1\|$. A lengthy proof shows $\|\bar{\mathbf{r}}_{1,l} - \mathbf{r}_{1,l}\|/\|\mathbf{r}_{1,l}\| \le (\frac{d}{2} + 3)2^{-t} + O(2^{-2t})$, see [44] pp. 84, 85, (15.21). We disregard all $O(2^{-2t})$-terms. The claim holds for $j = 0$ and arbitrary l since the size-reduced $\mathbf{b}_l$ satisfies $\|\mathbf{b}_l\| = \|\mathbf{r}_{1,l}\| \le (\sum_{1 \le i \le l} r_{i,i}^2)^{1/2}$. The constant factor 10 in the claim is a crude upper bound.

Induction step $j - 1 \to j$: Clearly $\mathbf{r}_{j,l} = Q_j \mathbf{r}_{j-1,l}$, $\|\mathbf{h}_j\|^2 = 2r_{j,j}^2$, $\|\pi'_n(\mathbf{r}_{j-1,j})\| = r_{j,j}$, $\pi'_n(\mathbf{h}_j) = \pi'_n(\mathbf{r}_{j-1,j})$, the jth entries of $\mathbf{h}_j, \mathbf{r}_{j,l}$ are $-r_{j,j}, r_{j,l}$. Hence,

$$r_{j,l} = \langle \mathbf{h}_j, \pi'_n(\mathbf{r}_{j-1,l}) \rangle / r_{j,j}, \tag{4.2}$$

$$\mathbf{r}_{j,l} - \mathbf{r}_{j-1,l} = -\langle \mathbf{h}_j, \pi_n(\mathbf{r}_{j-1,l}) \rangle \, \mathbf{h}_j / r_{j,j}^2 = -r_{j,l}\mathbf{h}_j / r_{j,j},$$

$$\pi'_n(\mathbf{r}_{j,l}) = \pi'_n\left(\mathbf{r}_{j-1,l} - \frac{r_{j,l}}{r_{j,j}}\mathbf{r}_{j-1,j}\right). \tag{4.3}$$

Consider the part of $\|\pi'_n(\bar{\mathbf{r}}_{j,l} - \mathbf{r}_{j,l})\|$ induced via (5) from $|\bar{r}_{j,l} - r_{j,l}|$. We neglect *fpa*-errors from $\bar{\mathbf{r}}_{j-1,l} \mapsto fl(\bar{Q}_j \bar{\mathbf{r}}_{j-1,l}) = \bar{\mathbf{r}}_{j,l}$ as they are minor. (4.2) shows that

$$\bar{r}_{j,l} - r_{j,l} r_{j,j} = \langle \bar{\mathbf{h}}_j, \pi'_n(\bar{\mathbf{r}}_{j-1,l}) \rangle - \langle \mathbf{h}_j, \pi'_n(\mathbf{r}_{j-1,l}) \rangle = \langle \bar{\mathbf{h}}_j - \mathbf{h}_j, \pi'_n(\mathbf{r}_{j-1,l}) \rangle$$
$$+ \langle \bar{\mathbf{h}}_j, \pi'_n(\bar{\mathbf{r}}_{j-1,l} - \mathbf{r}_{j-1,l}) \rangle.$$

The contribution of $\langle \bar{\mathbf{h}}_j, \pi'_n(\bar{\mathbf{r}}_{j-1,l} - \mathbf{r}_{j-1,l}) \rangle$ via $\bar{r}_{j,l} - r_{j,l}$ to $\|\pi'_n(\bar{\mathbf{r}}_{j,l} - \mathbf{r}_{j,l})\|$ is part of $\bar{Q}_j \pi'_n(\mathbf{r}_{j-1,l} - \bar{\mathbf{r}}_{j-1,l})$. As the orthogonal $\bar{Q}_j$ preserves the length of error vectors, that contribution is covered by $\|\pi'_n(\bar{\mathbf{r}}_{j-1,l} - \mathbf{r}_{j-1,l})\|$. We neglect the contribution of $\langle \bar{\mathbf{h}}_j - \mathbf{h}_j, \pi'_n(\mathbf{r}_{j-1,l}) \rangle$ to $\|\pi'_n(\bar{\mathbf{r}}_{j,l} - \mathbf{r}_{j,l})\|$ assuming that the error $\bar{\mathbf{h}}_j - \mathbf{h}_j$ is random. Therefore, (4.3) shows up to minor errors that

$$\pi'_n(\bar{\mathbf{r}}_{j,l} - \mathbf{r}_{j,l}) \approx \pi'_n(\bar{\mathbf{r}}_{j-1,l} - \mathbf{r}_{j-1,l}) + \frac{r_{j,l}}{r_{j,j}} \pi'_n(\bar{\mathbf{r}}_{j-1,j} - \mathbf{r}_{j-1,j}). \tag{4.4}$$

Applying the induction hypothesis for $j-1$ and size-reducedness, $|r_{j,l}/r_{j,j}| \le \frac{1}{2}$, we get the second part of the induction claim:

$$\|\pi'_n(\bar{\mathbf{r}}_{j,l} - \mathbf{r}_{j,l})\| \le 10\, m \left(\frac{3}{2}\right)^{j-2} \left(\frac{3}{2}\right) \max_{1 \le i \le l} r_{i,i}\, 2^{-t}.$$

The first part $|\bar{r}_{j+1,l} - r_{j+1,l}| \le \|\pi'_n(\bar{\mathbf{r}}_{j,l} - \mathbf{r}_{j,l}\|$ holds since $r_{j+1,l}$ is an entry of $\mathbf{r}_{j+1,l} = Q_j \mathbf{r}_{j,l}$. On average, the error bounds are much smaller since independent random error vectors are nearly orthogonal with high probability. Thus (4.4) shows

$$\mathbf{E}[\|\pi'_n(\bar{\mathbf{r}}_{j,l} - \mathbf{r}_{j,l})\|^2] \approx \mathbf{E}[\|\pi'_n(\bar{\mathbf{r}}_{j-1,l} - \mathbf{r}_{j-1,l})\|^2]$$
$$+\mathbf{E}\left[|\frac{r_{j,l}}{r_{j,j}}|^2 \|\pi'_n(\bar{\mathbf{r}}_{j-1,j} - \mathbf{r}_{j-1,j})\|^2 \right].$$

This proves by induction on j clause 2 of Proposition 2.

We set $\delta := 0.98$, $\delta_- := 0.97$, $\delta_+ := 0.99$, $\alpha := 1/0.73 < 1.37$, $\rho := \frac{3}{2}\sqrt{\alpha} \approx \sqrt{3}$, $\varepsilon := 0.01$ and $\alpha_\varepsilon := (1 + \varepsilon^2\alpha) / \left(\frac{3}{4} - 4\varepsilon - \varepsilon^2/4 - (1 + 2\varepsilon)\varepsilon\alpha^{1/2}\right) < 1.44$.

Recall that $\mathbf{LLL}_H$ is obtained by replacing steps 2.1, .2.2 of **LLL** by $\texttt{TriCol}_l$; let $M = \max(d_1, \ldots, d_n, 2^n)$ for the input basis and $M_0 = \max(\|\mathbf{b}_1\|, \ldots, \|\mathbf{b}_n\|)$.

Theorem 5. [Theorem 2 of [70] using fpa-Heur.] $\mathbf{LLL}_H$ *transforms with* fpa *of precision* $2^t \ge 2^{10}\, m\, \rho^n$ *a basis* $\mathbf{b}_1, \ldots, \mathbf{b}_n \in \mathbb{Z}^m$ *into an approximate LLL-basis satisfying*

1. $|\mu_{j,i}| < \frac{1}{2} + \varepsilon\alpha_\varepsilon^{\frac{j-i}{2}} r_{j,j}/r_{i,i} + \varepsilon$ *for* $1 \le i < j \le n$,
2. $\delta\_r_{i,i}^2 \le \mu_{i+1,i}^2 r_{i,i}^2 + r_{i+1,i+1}^2$ *for* $i = 1, \ldots, n-1$.
3. *Clauses 1.-3. of Theorem 1 hold with α replaced by α_ε.*

LLL$_H$ *runs in* $O(n^2 m \log_{1/\delta} M)$ *arithmetic steps using* $n + \log_2 M_0$ *bit integers and* fpa *numbers.*

We call a basis *size-reduced under* fpa if it satisfies clause 1 of Theorem 5. The term $\varepsilon\alpha_\varepsilon^{\frac{j-i}{2}} r_{j,j}/r_{i,i} \ge \varepsilon$ covers the error $|\bar{\mu}_{j,i} - \mu_{j,i}|$ from (1).

In particular, **LLL**$_H$ runs for $M_0 = 2^{O(n)}$, $m = O(n)$ in $O(n^5)$ arithmetic steps and in $O(n^6)/O(n^{5+\varepsilon})$ bit operations under school-/FFT-multiplication.

Similar to L^2 of [56], **LLL**$_H$ should be quadratic in $\log_2 M_0$ since size-reduction in step 5 of $\texttt{TriCol}_l$ is done in *fpa* using the t most significant bits of $r_{i,l}, r_{l,l}, \mu_{l,i}$. Thus, at least one argument of multiplication/division has bit length t. The maximal bit length of the $\mu_{l,i}$ decreases in practice by $\Theta(t)$ per round of $\texttt{TriCol}_l$, e.g. by about 30 for $t = 53$, see Fig. 1 [41]. In our experience, **LLL**$_H$ is most likely correct up to dimension $n = 350$ under *fpa* with $t = 53$ precision bits for *arbitrary* M_0, and not just for $t \ge \Omega(n + \log_2 M_0)$ as shown in Theorem 5. **LLL**$_H$ computes minimal, near zero errors for the LLL-bases given in [56], where the *fpa*-LLL-code of NTL fails in dim. 55, and NTL orthogonalizes the Gram-matrix by classical GSO and **LLL**$_H$ by HO.

Givens rotations zero out a single entry of col(j, R_{j-1}) and provide a slightly better *fpa*-error bound than HO [34] Chap. 18.5. Givens rotations have been used in parallel LLL-algorithms of Heckler, Thiele and Joux.

The L^2 of Nguyen, Stehlé [56] uses GSO in *fpa* for the Gram matrix $B^t B$. Theorem 2 of [56] essentially replaces $\sqrt{3}$ by 3 in our condition $2^t \ge 100\, m \sqrt{3}^{l-1}$ for correct LLL-swapping. **LLL**$_H$ performs correct LLL-swaps in about twice the dimension compared to the L^2 of [56]. This is because replacing the basis B by the Gram matrix $B^t B$ squares the matrix condition number $\mathcal{K}(B) = \|B\| \|B^{-1}\|$ which characterizes the sensitivity of Q, R to small perturbations of B, see [34] Chap. 18.8. As *fpa*-errors during the QR-factorization act in the same way as perturbations of B, the squaring $\mathcal{K}(B^t B) = O(\mathcal{K}(B)^2)$ essentially halves the accurate bits for R, Q and the dimension for which L^2 is correct. The squaring also doubles the bit length of B and more than doubles the running time. While [74] reports that L^2 is most likely correct with 53 precision bits up to dimension 170, **LLL**$_H$ is most likely correct up to dimension 350. The strength of L^2 is its proven accuracy; moreover, L^2 is quadratic in the bit length $\log_2 M_0$. While the proven accuracy bound of [56] is rather modest for 53 precision bits, the Magma-code of L^2 uses intermediary heuristics and provides proven accuracy of the output via multiprecision *fpa* [74].

Scaled LLL-reduction. Scaling is a useful concept of numerical analysis for reducing *fpa*-errors. Scaled LLL-reduction of [41] associates with a given lattice basis a scaled basis of a sublattice of the given lattice. The scaled basis satisfies $\frac{1}{2} \le |r_{1,1}^2/r_{j,j}^2| \le 2$ for all j. Scaled LLL-reduction performs a relaxed size-reduction, reducing relative to an associated scaled basis. The relaxed size-reduction is very accurate, independent of *fpa*-errors, and its relaxation is negligible. Scaled LLL-reduction is useful in dimension $n > 350$ where $\mathbf{LLL}_H$ becomes inaccurate. This way, we reduced in 2002 a lattice bases of dimension 1,000 consisting of integers of bit length 400, in 10 h on a 800 MHz PC.

Comparison with [65] and the modular LLL of [73]. Theorem 7 improves the time bound and the accuracy of the theoretic method of [S88], which uses approximate rational arithmetic.

The modular LLL [73] performs $O(nm \log_{1/\delta} M)$ arithmetic steps on integers of bit length $\log_2(M_0 M)$ using standard matrix multiplication, where M denotes $\max(d_1, \ldots, d_n, 2^n)$. If $M_0 = 2^{\Omega(n)}$, the LLL's of [65, 73] match asymptotically the bound for the number of bit operations of $\mathbf{LLL}_H$. Neither of these LLL's is quadratic in M_0. The practicability of $\mathbf{LLL}_H$ rests on the use of small integers of bit length $1.11\, n + \log_2 M_0$ whereas [73] uses long integers of bit length $\log_2(M_0 M) = O(n \log M_0)$.

Semi Block $2k$-Reduction Revisited

Survey, background and perspectives this and the next two sections. We survey feasible basis reduction algorithms that decrease α in clauses 1, 2 of Theorem 1 to $\bar{\alpha} < \alpha$ for $n \gg 2$. The factors $\alpha^{\frac{n-1}{2}}, \alpha^{n-1}$ of Theorem 1 decrease within polynomial time reduction to $2^{O((n \log\log n)^2/\log n)}$ [64] and combined with [4] to $2^{O(n \log\log n/\log n)}$. In this survey, we focus on reductions of α achievable in feasible lattice reduction time. Some reductions are proven by heuristics to be feasible on the average.

For the rest of the paper, let $\delta \approx 1$ so that $\alpha \approx 4/3$. LLL-bases approximate λ_1 up to a factor $\alpha^{\frac{n-1}{2}} \lesssim 1.155^n$. They approximate λ_1 much better for lattices of *high density* where $\lambda_1^2 \approx \gamma_n (\det \mathcal{L})^{2/n}$, namely up to a factor $\lesssim \alpha^{\frac{n-1}{4}} / \sqrt{\gamma_n} \lesssim 1.075^n$ as a result of part **1** of Theorem 1. Moreover, [57] reports that α decreases on average to about $1.02^4 \approx 1.08$ for the random lattices of [57].

The constant α can be further decreased within polynomial reduction time by blockwise basis reduction. We compare Schnorr's algorithm for semi block $2k$-reduction [64] and Koy's primal–dual reduction [39] with blocksize $2k$. Both algorithms perform HKZ-reductions in dimension $2k$ and have similar polynomial time bounds. They are feasible for $2k \leq 50$. Koy's algorithm guarantees within the same time bound under known proofs better approximations of the shortest lattice vector within the same time bound, under known proofs. Under reasonable heuristics, both algorithms are equally strong and much better than proven in worst-case. We combine primal–dual reduction with Schnorr's random sampling reduction (RSR) with a highly parallel reduction algorithm, which is on the average more efficient than previous algorithms. It reduces the approximation factor $\left(\frac{4}{3}\right)^{n/2}$ guaranteed by the LLL-algorithm on average to $1.025^{n/2}$ using feasible lattice reduction.

Semi block $2k$-reduced bases of [64] satisfy by Theorem 6 the inequalities of Theorem 1, with α replaced by $(\beta_k/\delta)^{1/k}$, for a lattice constant β_k such that $\lim_{k\to\infty}\beta_k^{1/k} = 1$. The best known bounds on β_k are $k/12 < \beta_k < \left(1+\frac{k}{2}\right)^{2\ln 2+1/k}$ [23]. Primal–dual reduction (**Algorithm 3**) replaces α by $(\alpha\gamma_{2k}^2)^{1/2k}$ (Theorem 7). The second bound outperforms the first, unless β_k is close to its lower bound $k/12$. Primal–dual reduction for blocks of length 48 replaces α in Theorem 1 within feasible reduction time by $(\alpha\gamma_{48}^2)^{1/48} \approx 1.084$. The algorithms **Algorithms 2 and 3** for semi block $2k$ reduction and primal–dual reduction are equally powerful in approximating λ_1 under the **GSA**-heuristic of [69]. Under **GSA**, they perform much better than proven in a worst case.

Section "Primal-Dual Random Sampling Reduction" surveys some basis reduction algorithms that are effcient on average but not proven in polynomial time. BKZ-reduction of [66] runs in practice for blocksize 10 in less than twice the LLL-time. The LLL with the *deep insertion* step of [66] seems to be in polynomial time on average and greatly improves the approximational power of the LLL. Based on experiments, [57] reports that LLL with deep insertions decreases α for random lattices on average to $1.012^4 \approx 1.05 \approx \alpha^{1/6}$.

In Section "Primal-Dual Random Sampling Reduction", we replace HKZ-reduction within primal–dual reduction by *random sampling reduction* (RSR) of [69], a parallel extension of the deep insertion step of [66]. RSR is nearly feasible up to blocksize $k = 80$. The new algorithm, *primal–dual RSR* (**Algorithm 4**) replaces under the worst-case **GSA**-heuristics α in Theorem 1 by $(80/11)^{1/80} \approx 1.025$. **Algorithm 4** is highly parallel and polynomial time on the average but not proven polynomial time.

For Table 1, we assume that the densest known lattice packings P_{48p}, P_{48q} in dimension 48 [16] Table 1.3, have nearly maximal density; then $\gamma_{48} \approx 6.01$. For the assumptions **GSA, RA** see Section "Returning to the Gauss Algorithm". **GSA** is a worst case heuristics in the sense that bases $B = QR$ having a large spread of the values $r_{i+1,i+1}^2/r_{i,i}^2$, are in general, easier to reduce.

Table 4.1 Reductions $\bar{\alpha}$ of $\alpha \approx \frac{4}{3}$ in Theorem 1 under feasible lattice basis reduction for $n \gg 2$

	$\bar{\alpha}$
1. Semi block $2k$-reduction [64], $k = 24$	
proven [23]	$(\beta_{24}/\delta)^{1/24} < 1.165$
by heuristic, **GSA**	$\gamma_{47}^{1/47} \approx 1.039$
2. Primal–dual reduction, Koy 2004, $k = 48$	
proven	$(\alpha\gamma_{48}^2)^{1/48} \approx 1.084$
by heuristic, **GSA**	$\gamma_{48}^{1/47} \approx 1.039$
3. Primal–dual RSR, $k = 80$, under **GSA, RA**	1.025
4. LLL on the average for random lattices, experimental [57]:	1.08
5. LLL with deep insertion [66] on the average for random lattices, experimental [7,57]:	$1.012^4 \approx 1.05$

Notation. For a basis $B = QR \in \mathbb{R}^{m\times n}$, $R = [r_{i,j}]_{1\le i,j\le n}$ with $n = hk$, let

$$R_\ell := [r_{i,j}]_{k\ell-k<i,j\le k\ell} \in \mathbb{R}^{k\times k} \quad \text{for } \ell \le h$$
$$R_{\ell,\ell+1} = [r_{i,j}]_{k\ell-k<i,j\le k\ell+k} = \begin{bmatrix} R_\ell & R'_\ell \\ O & R_{\ell+1} \end{bmatrix} \in \mathbb{R}^{2k\times 2k} \quad \text{for } \ell < h$$

denote the principal submatrices of the GNF R corresponding to the segments $B_\ell = [\mathbf{b}_{k\ell-k+1},\ldots,\mathbf{b}_{k\ell}]$ and $[B_\ell, B_{\ell+1}]$ of B. We denote

$$\mathcal{D}_\ell =_{def} (\det R_\ell)^2 = d_{k\ell}/d_{k\ell-k},$$
$$\mathcal{D}_k^{(1)} = \mathcal{D} =_{def} \prod_{\ell=1}^{h-1} d_{k\ell} = \prod_{\ell=1}^{h-1} \mathcal{D}_\ell^{h-\ell}.$$

The lattice constant β_k. Let $\beta_k =_{def} \max(\det R_1/\det R_2)^{1/k}$ maximized over all HKZ-reduced GNF's $R = R_{1,2} = \begin{bmatrix} R_1 & R'_1 \\ O & R_2 \end{bmatrix} \in \mathbb{R}^{2k\times 2k}$.

Note that $\beta_1 = \max r_{1,1}^2/r_{2,2}^2$ over all GNF's $R = \begin{bmatrix} r_{1,1} & r_{1,2} \\ 0 & r_{2,2} \end{bmatrix} \in \mathbb{R}^{2\times 2}$ satisfying $|r_{1,2}| \le r_{1,1}/2$, $r_{1,1}^2 \le r_{1,2}^2 + r_{2,2}^2$ and thus $\beta_1 = \frac{4}{3} = \alpha$ holds for $\delta = 1$, $\eta = \frac{1}{2}$. Note that $\beta_k \le (1+\frac{k}{2})^{2\ln 2+1/k}$ [23].

Definition 2. [64] A basis $B = QR \in \mathbb{R}^{m\times n}$, $n = hk$, is *semi block $2k$-reduced* for $\delta \in (\eta^2, 1]$ and $\alpha = 1/(\delta-\eta^2)$ if the *GNF* $R = [r_{i,j}]$ satisfies

1. $R_1,\ldots,R_h \subset R$ are HKZ-reduced,
2. $r_{k\ell,k\ell}^2 \le \alpha\, r_{k\ell+1,k\ell+1}^2$ for $\ell = 1,\ldots,h-1$,
3. $\delta^k\, \mathcal{D}_\ell \le \beta_k^k\, \mathcal{D}_{\ell+1}$ for $\ell = 1,\ldots,h-1$.

In [64], α in clause 2 has been set to 2 and δ^k in clause 3 has been set to $\frac{3}{4}$. For $k = 1$, clause 3 means that $r_{\ell,\ell}^2 \le \alpha r_{\ell+1,\ell+1}^2$. LLL-bases for δ are semi block $2k$-reduced for $k = 1$.

Theorem 6. [64]. *A semi block $2k$-reduced basis $B = QR \in \mathbb{R}^{m\times n}$, $n = hk$, satisfies*

$$\|\mathbf{b}_1\|^2 \le \gamma_k(\beta_k/\delta)^{\frac{n/k-1}{2}}(\det\mathcal{L}(B))^{2/n}.$$

Proof. We have that $\|\mathbf{b}_1\|^2 \le \gamma_k \mathcal{D}_1^{1/k} = \gamma_k(\det R_1)^{2/k}$ since R_1 is HKZ-reduced. Clause 3 of Definition 3 shows $\mathcal{D}_\ell^{1/k} \le (\beta_k/\delta)\mathcal{D}_{\ell+1}^{1/k}$ and yields

$$\|\mathbf{b}_1\|^2 \le \gamma_k \mathcal{D}_1^{1/k} \le \gamma_k(\beta_k/\delta)^{\ell-1}\mathcal{D}_\ell^{1/k} \quad \text{for } \ell = 1,\ldots,h = n/k.$$

Multiplying these inequalities and taking hth roots yields the claim. □

Moreover, $\|\mathbf{b}_1\|^2\lambda_1^{-2} \le k^{\ln k+2}\,(\beta_k/\delta)^{n/k-2}$ holds for $k \ge 3$ [64] Theorem 3.1, Corollary 3.5. Ajtai [3] proved these bounds to be optimal up to a constant factor in the exponent. There exist semi block $2k$-reduced bases of arbitrary dimension n satisfying $\|\mathbf{b}_1\|^2 \ge \gamma_k(\beta_k/\delta)^{\Omega(n/k)}(\det \mathcal{L}(B))^{2/n}$.

Algorithm 2 rephrases the algorithm of [64] without using the unknown constant β_k. For $k = 1$, **Algorithm 2** essentially coincides with LLL-reduction.

The transforms T of $R_{\ell,\ell+1}$ that perform LLL-, respectively HKZ-reduction are only transported to the basis B if this decreases $\mathcal{D}_\ell$ by the factor δ, respectively $\delta^{k/2}$.

Algorithm 2: Semi block $2k$-reduction

INPUT basis $B = [\mathbf{b}_1, \dots, \mathbf{b}_n] \in \mathbb{Z}^{m\times n}$, $\delta \in [\eta^2, 1)$, $\eta = \frac{1}{2} + \varepsilon$, $n = hk$.

OUTPUT semi block $2k$-reduced basis B.

1. LLL-reduce B, HKZ-reduce $[B_1, B_2] = [\mathbf{b}_1, \dots, \mathbf{b}_{2k}]$, compute $R = [r_{i,j}] \in \mathbb{R}^{n\times n}$, $\ell := 2$.
2. HKZ-reduce $R_{\ell+1}$ into $R_{\ell+1}T'$ for some $T' \in \mathrm{GL}_k(\mathbb{Z})$, $B_{\ell+1} := B_{\ell+1}T'$,
   LLL-reduce $R_{\ell,\ell+1}$ into $R_{\ell,\ell+1}T$.
   If an LLL-swap bridging R_ℓ and $R_{\ell+1}$ occured THEN
   $[B_\ell, B_{\ell+1}] := [B_\ell, B_{\ell+1}]\,T$, $\ell := \max(\ell-1, 1)$ GO TO 2
   HKZ-reduce $R_{\ell,\ell+1}$ into $R_{\ell,\ell+1}T$ for some $T \in \mathrm{GL}_{2k}(\mathbb{Z})$.
3. Compute $\mathcal{D}_\ell^{new} := (\det R_\ell^{new})^2$ for $\begin{bmatrix} R_\ell^{new} & R_\ell'^{new} \\ O & R_{\ell+1}^{new} \end{bmatrix} := \mathrm{GNF}(R_{\ell,\ell+1}T)$,
   IF $\mathcal{D}_\ell^{new} \le \delta^{k/2}\mathcal{D}_\ell$ THEN $[B_\ell, B_{\ell+1}] := [B_\ell, B_{\ell+1}]\,T$, recomputr $R_\ell, R_{\ell+1}, \ell := \ell - 1$
   ELSE $\ell := \ell + 1$.
4. IF $\ell = 1$ THEN GO TO 1, IF $1 < \ell < h$ THEN GO TO 2
   ELSE terminate.

Correctness. Induction over the rounds of the algorithm shows that the basis $\mathbf{b}_1, \dots, \mathbf{b}_{k\ell}$ is always semi block $2k$-reduced for the current ℓ. We show that clauses 2, 3 of Definition 2 hold; clause 1 obviously holds.

Clause 2: LLL-reduction of $R_{\ell,\ell+1}$ in step 2 guarantees clause 2. In particular, if an LLL-swap bridging R_ℓ, $R_{\ell+1}$ occured the blocks B_ℓ, $B_{\ell+1}$ get transformed.

Clause 3: After HKZ-reduction of $R_{\ell,\ell+1}$, we have $\mathcal{D}_\ell^{new} \le \beta_k^k\,\mathcal{D}_{\ell+1}^{new}$. Before increasing ℓ, we have $\mathcal{D}_\ell^{new} > \delta^{k/2}\,\mathcal{D}_\ell$ and thus, $\mathcal{D}_{\ell+1}^{new} < \delta^{-k/2}\,\mathcal{D}_{\ell+1}$ resulting in $\delta^{k/2}\,\mathcal{D}_\ell < \mathcal{D}_\ell^{new} \le \beta_k^k\,\mathcal{D}_{\ell+1}^{new} < \delta^{-k/2}\beta_k^k\mathcal{D}_{\ell+1}$, and therefore, $\delta^k\,\mathcal{D}_\ell < \beta_k^k\,\mathcal{D}_{\ell+1}$. □

Lemma 2. *Semi block 2k-reduction performs at most* $h - 1 + 2n(h-1)\log_{1/\delta} M_0$ *rounds, i.e., passes of step 2.*

Proof. Let $k \ge 2$. Semi block $2k$-reduction iteratively decreases $\mathcal{D}_\ell$ either by LLL-reduction or by HKZ-reduction of $R_{\ell,\ell+1}$ Each pass of steps 2, 3 either decreases $\mathcal{D}_\ell$ and $\mathcal{D} = \prod_{\ell=1}^{h-1}\mathcal{D}_\ell^{h-\ell}$ by the factor δ, respectively $\delta^{k/2}$, or else increments ℓ. Since initially $\mathcal{D} = \prod_{\ell=1}^{h-1} d_{k\ell} \le M_0^{2k\binom{h}{2}} = M_0^{n(h-1)}$, the integer $\mathcal{D}$ can be decreased at most $2n(h-1)\log_{1/\delta} M_0$ times by the factor δ. Hence, there are at most $n(h-1)$

$\log_{1/\delta} M_0$ passes of steps 2, 3 that decrease $\mathcal{D}$ by the factor δ, respectively $\delta^{k/2}$, and at most $h - 1 + n(h-1)\log_{1/\delta} M_0$ passes of step 2 that do not change $\mathcal{D}$ but increment ℓ in step 3. □

The proof of [64] Theorem 3.2 shows that an HKZ-reduction of $R_{\ell,\ell+1}$ performs $O(n^2 k + k^4 \log M_0) + (2k)^{k+o(k)}$ arithmetic steps using integers of bit length $O(n \log M_0)$. Following [S87], semi block $2k$-reduction performs $O((n^4 + n^2(2k)^{k+o(k)}) \log_{1/\delta} M_0)$ arithmetic steps.

Primal–Dual Reduction

Koy's primal–dual reduction [39] decreases $\mathcal{D}_\ell = (\det R_\ell)^2$ as follows. It maximizes $r_{k\ell,k\ell}$ over the GNF's of $R_\ell T_\ell$ and minimizes $r_{k\ell+1,k\ell+1}$ over the GNF's of $R_{\ell+1} T_{\ell+1}$ for all $T_\ell, T_{\ell+1} \in \mathrm{GL}_k(\mathbb{Z})$, and then swaps $\mathbf{b}_{k\ell}, \mathbf{b}_{k\ell+1}$ if this decreases $r_{k\ell,k\ell}$ and $\mathcal{D}_\ell$. Primal–dual reduction with double blocksize $2k$ replaces the constant β_k/δ in Theorem 6 by $\sqrt{\alpha}\,\gamma_{2k}$, which is better understood than β_k/δ since $\gamma_k = \Theta(k)$.

Dual lattice and dual basis. The *dual* of lattice $\mathcal{L} = \mathcal{L}(QR)$ is the lattice

$$\mathcal{L}^* = \{\mathbf{z} \in \mathrm{span}(\mathcal{L}) \mid \mathbf{z}^t \mathbf{y} \in \mathbb{Z} \quad \text{for all } \mathbf{y} \in \mathcal{L}\}.$$

$\mathcal{L}^* = \mathcal{L}(QR^{-t})$ holds because $(QR^{-t})^t QR = R^{-1} Q^t QR = R^{-1} R = I_n$. $QR^{-t} = B^{-t}$ holds for $m = n$.

R^{-t} is a lower triangular matrix and $U_n R^{-t} U_n$ isupper-triangular with positive diagonal entries. Clearly, $\mathcal{L}^* = \mathcal{L}(QR^{-t} U_n)$, the basis $QR^{-t} U_n$ has QR-decomposition $QR^{-t} U_n = (QU_n)(U_n R^{-t} U_n)$ because QU_n is isometric and $U_n R^{-t} U_n$ is upper-triangular. $B^* := QR^{-t} U_n$ is the (reversed) *dual basis* of $B = QR$. Note that $(B^*)^* = B$. B^* has the *dual GNF* $R^* := U_n R^{-t} U_n$. The (reversed) dual basis $B^* = [\mathbf{b}_1^*, \ldots, \mathbf{b}_n^*]$ of $B = [\mathbf{b}_1, \ldots, \mathbf{b}_n]$ is characterized by

$$\langle \mathbf{b}_i^*, \mathbf{b}_{n-j+1} \rangle = \delta_{i,j} = \langle \mathbf{b}_i, \mathbf{b}_{n-j+1}^* \rangle,$$

where $\delta_{i,j} \in \{0, 1\}$ is 1 iff $i = j$. The dual basis B^* satisfies $B^* = [\mathbf{b}_1^*, \ldots, \mathbf{b}_n^*] = B^{-t}$ for $m = n$. (The $\mathbf{b}_i^*$ denote the dual basis vectors and not the orthogonal vectors $\mathbf{q}_i = \pi_i(\mathbf{b}_i)$ as in [47]. The diagonal entries of $R = [r_{i,j}]$ and $R^* = [r_{i,j}^*]$ satisfy

$$r_{i,i} = 1/r_{n-i+1,n-i+1}^* \quad \text{for } i = 1, \ldots, n. \tag{4.5}$$

HKZ-reduction of R^* minimizes $r_{1,1}^* = \|\mathbf{b}_1^*\|$ and maximizes $r_{n,n} = 1/r_{1,1}^*$.

Notation. For a basis $B = QR \in \mathbb{R}^{m \times n}$, we let $\bar{r}_{k\ell,k\ell}$ for $k\ell \le n$ denote the maximum of $\widetilde{r}_{k\ell,k\ell}$ over the GNF's $[\widetilde{r}_{i,j}]_{k\ell-k<i,j\le k\ell} = \mathrm{GNF}(R_\ell T)$ for all $T \in \mathrm{GL}_k(\mathbb{Z})$. Shortly, $\bar{r}_{k\ell,k\ell}$ is the maximum of $r_{k\ell,k\ell}$ over the transforms of $R_\ell \subset R$. If $R_\ell^* = U_k R_\ell^{-t} U_k$ is HKZ-reduced, then $r_{k\ell,k\ell} = \bar{r}_{k\ell,k\ell}$. We compute $\bar{r}_{k\ell,k\ell}$ by

HKZ-reducing R_ℓ^* into $R_\ell^* T$, then $[\widetilde{r}_{i,j}] := \mathrm{GNF}(R_\ell U_k T^{-t})$ satisfies $\bar{r}_{k\ell,k\ell} = \widetilde{r}_{k\ell,k\ell}$.

Definition 3. A basis $B = QR \in \mathbb{R}^{m\times n}$, $n = hk$ is a *primal–dual basis* for k and $\delta \in (\eta^2, 1]$, $\alpha = 1/(\delta - \frac{1}{4})$ if its *GNF* $R = [r_{i,j}]$ satisfies

1. $R_1, \dots, R_h \subset R$ are HKZ-reduced,
2. $\bar{r}^2_{k\ell,k\ell} \le \alpha\, r^2_{k\ell+1,k\ell+1}$ for $\ell = 1, \dots, h-1$.

We see from (4.5) that clause 2 of Definition 3 also holds for the dual B^* of a primal–dual basis B. Therefore, such B^* can be transformed into a primal–dual basis by HKZ-reducing $B_\ell^* = [\mathbf{b}^*_{k\ell+1}, \dots, \mathbf{b}^*_{k\ell+k}]$ into $B_\ell^* T_\ell$ for $\ell = 1, \dots, h$. Moreover, clauses 2 and 3 of Definition 1 are preserved under duality, they hold for the dual of a semi block $2k$-reduced basis. Theorem 7 replaces α in Theorem 1 by $(\alpha\gamma_k^2)^{1/k}$.

Theorem 7. [23,39]. *A primal–dual basis $B = QR \in \mathbb{R}^{m\times n}$, $n = hk$ of the lattice $\mathcal{L}$ satisfies* **1.** $\|\mathbf{b}_1\|^2 \le \gamma_k (\alpha\gamma_k^2)^{\frac{h-1}{2}} (\det \mathcal{L})^{2/n}$, **2.** $\|\mathbf{b}_1\|^2 \le (\alpha\gamma_k^2)^{h-1}\lambda_1^2$.

Proof. 1. The maximum $\bar{r}^2_{k\ell,k\ell}$ of $r^2_{k\ell,k\ell}$ over the $R_\ell T$ satisfies by clause 2 of Definition 3 $\bar{r}^2_{k\ell,k\ell} \le \alpha\, r^2_{k\ell+1,k\ell+1}$.
Moreover, we have $\mathcal{D}_\ell^{1/k} \le \gamma_k \bar{r}^2_{k\ell,k\ell} = \gamma_k / \lambda_1^2(\mathcal{L}(R_\ell^*))$
since $\bar{r}^2_{k\ell,k\ell}$ is computed by HKZ-reduction of R_ℓ^*, and

$$\lambda_1^2(\mathcal{L}(R_{\ell+1})) = r^2_{k\ell+1,k\ell+1} \le \gamma_k \mathcal{D}_{\ell+1}^{1/k}$$

since $R_{\ell+1}$ is HKZ-reduced. Combining these inequalities, we get

$$\mathcal{D}_\ell^{1/k} \le \gamma_k \bar{r}^2_{k\ell,k\ell} \le \alpha\gamma_k r^2_{k\ell+1,k\ell+1} \le \alpha\gamma_k^2 \mathcal{D}_{\ell+1}^{1/k}. \tag{4.6}$$

Since R_1 is HKZ-reduced this yields

$$\|\mathbf{b}_1\|^2 \le \gamma_k \mathcal{D}_1^{1/k} \le \gamma_k (\alpha\gamma_k^2)^{\ell-1} \mathcal{D}_\ell^{1/k} \quad \text{for } \ell = 1, \dots, h.$$

Multiplying these h inequalities and taking hth roots yields the claim.
2. Note that the inequality (4.6) also holds for the dual basis B^*, i.e., $(\mathcal{D}_\ell^*)^{1/k} \le \alpha\gamma_k^2 (\mathcal{D}^*_{\ell+1})^{1/k}$ holds for $\mathcal{D}_\ell^* = (\det R_\ell^*)^2 = \mathcal{D}^{-1}_{h-\ell+1}$. Hence, the dual **1*.** of part **1.** of Theorem 7 also holds:

$$\mathbf{1}*.\ \bar{r}^2_{n,n} \ge \gamma_k^{-1} (\alpha\gamma_k^2)^{\frac{-h+1}{2}} (\det \mathcal{L})^{2/n}.$$

1. and 1.* yield $\|\mathbf{b}_1\|^2 \le \gamma_k^2 (\alpha\gamma_k^2)^{h-1} \bar{r}^2_{n,n}$. By clause 2 of Definition 3, we get

$$\|\mathbf{b}_1\|^2 \le \gamma_k^2 (\alpha\gamma_k^2)^{\ell-1} \bar{r}^2_{k\ell,k\ell} \le (\alpha\gamma_k^2)^{\ell} r^2_{k\ell+1,k\ell+1} \quad \text{for } \ell = 0, \dots, h-1.$$

Therefore, $r_{k\ell+1,k\ell+1} \le \lambda_1$ yields the claim. In fact $r_{k\ell+1,k\ell+1} \le \|\pi_{k\ell+1}(\mathbf{b})\| \le \lambda_1$ holds if the shortest lattice vector $\mathbf{b} = \sum_{j=1}^n r_j \mathbf{b}_j \neq \mathbf{0}$ satisfies $k\ell < \mu \le k\ell + k$ for $\mu := \max\{j \mid r_j \neq 0\}$, because $R_{\ell+1}$ is HKZ-reduced. □

Algorithm 3: Koy's algorithm for primal–dual reduction
INPUT basis $[\mathbf{b}_1, \dots, \mathbf{b}_n] = B = QR \in \mathbb{Z}^{m \times n}$, $\delta \in ((\frac{1}{2} + \varepsilon)^2, 1)$, $n = hk$.
OUTPUT primal–dual reduced basis B for k, δ.

1. LLL-reduce B, HKZ-reduce $B_1 = [\mathbf{b}_1, \dots, \mathbf{b}_k]$ and compute $R = \mathrm{GNF}(B)$, $\ell := 1$.
2. *#reduce $R_{\ell,\ell+1}$ by primal and dual HKZ-reduction of blocksize k:*
   HKZ-reduce $R_{\ell+1}$ into $R_{\ell+1}T'$, $B_{\ell+1} := B_{\ell+1}T'$.
   HKZ-reduce $R_\ell^* = U_k R_\ell^{-t} U_k$ into $R_\ell^* T_\ell$, set $\bar{T} := \begin{bmatrix} U_k T_\ell^{-t} & O \\ O & I_k \end{bmatrix}$.
   LLL-reduce $R_{\ell,\ell+1}\bar{T}$ into $R_{\ell,\ell+1}T$ with δ.
3. IF an LLL-swap bridging R_ℓ and $R_{\ell+1}$ occured THEN
   $[B_\ell, B_{\ell+1}] := [B_\ell, B_{\ell+1}]T$, $\ell := \max(\ell - 1, 1)$ ELSE $\ell := \ell + 1$.
4. IF $\ell < h$ THEN GO TO 2 ELSE output B and terminate.

Correctness. Induction over the rounds of the algorithm shows that the basis $\mathbf{b}_1, \dots, \mathbf{b}_{k\ell}$ is always a primal–dual basis for the current ℓ.

The algorithm deviates from semi block $2k$-reduction in the steps 2, 3. Step 2 maximizes $r_{k\ell,k\ell}$ within R_ℓ by HKZ-reduction of R_ℓ^* and minimizes $r_{k\ell+1,k\ell+1}$ within $R_{\ell+1}$ by HKZ-reduction of $R_{\ell+1}$, and then LLL-reduces $R_{\ell,\ell+1}$. If no LLL-swap bridging R_ℓ and $R_{\ell+1}$ occured in step 2, then clause 2 of Definition 2 was previously satisfied for ℓ.

Lemma 3. *Primal–dual reduction performs at most $h - 1 + 2n(h-1)\log_{1/\delta} M_0$ passes of step 2.*

Proof. Initially, we have $\mathcal{D} = \prod_{\ell=1}^{h-1} d_{\ell k} \le M_0^{n(h-1)}$. Steps 2, 3 either decrease $\mathcal{D}_\ell$ by a factor δ or else increment ℓ. Thus the proof of Lemma 2 applies. □

The number of passes of step 2 are in the worst case $h-1+2n(h-1)\log_{1/\delta} M_0$. The actual number of passes may be smaller, but on average, it should be proportional to $h - 1 + 2n(h-1)\log_{1/\delta} M_0$.

The dual clause 2 of Definition 3 has been strengthened in [24] for arbitrary small $\varepsilon > 0$ to

$$2^+ .\ \bar{r}_{kl+1,kl+1} \le (1+\varepsilon) r_{kl+1,kl+1} \quad \text{for } l = 1, \dots, h-1$$

denoting $\bar{r}_{kl+1,kl+1} := \max_T r'_{kl+1,kl+1}$ of $[r'_{i,j}] := \mathrm{GNF}([r_{i,j}]_{kl-k+2 \le i,j \le kl+1} T)$ over all $T \in \mathrm{GL}_k(\mathbb{Z})$.

This improves the inequalities of Theorem 7 to hold with $\alpha\gamma_k^2$ replaced by $\big((1+\varepsilon)\gamma_k\big)^{\frac{2k}{k-1}}$ [24]:

1. $\|\mathbf{b}_1\|^2 \le \gamma_k \big((1+\varepsilon)\gamma_k\big)^{\frac{n-k}{k-1}} (\det \mathcal{L})^{2/n}$, **2.** $\|\mathbf{b}_1\| \le \big((1+\varepsilon)\gamma_k\big)^{\frac{n-k}{k-1}} \lambda_1$.

In adjusting **Algorithm 3** to the new clause 2^+, it is crucial that increasing $r_{kl+1,kl+1}$ by a factor $1+\varepsilon$ to satisfy clause 2^+ decreases $\mathcal{D}_l = \det R_l^2$ by the factor $(1+\varepsilon)^{-2}$ and preserves $\mathcal{D}_l\mathcal{D}_{l+1}$ and all $\mathcal{D}_i$ for $i \neq l, l+1$. The adjusted **Algorithm 3** performs at most $h + 2nh \log_{1+\varepsilon} M_0$ HKZ-reductions in dimension k.

Algorithms 2 and **3** (for blocksize $2k$) have by Lemma 2 and 3 the same time bound, and both algorithms do HKZ-reductions in dimension $2k$.

For double blocksize $2k$ Theorem 7 shows

1. $\|\mathbf{b}_1\|^2 \leq \gamma_{2k}(\alpha\gamma_{2k}^2)^{n/4k-1/2}(\det \mathcal{L})^{2/n}$, **2.** $\|\mathbf{b}_1\|^2 \leq (\alpha\gamma_{2k}^2)^{n/2k-1}\lambda_1^2$.

These bounds are better than the bounds of Theorem 6 unless β_k is close to the lower bound $\beta_k > k/12$ of [23] so that $\beta_k/\delta \leq \sqrt{\alpha}\,\gamma_{2k}$. Because of the unknown values β_k semi block $2k$-reduction can still be competitive.

Theorems 6 and 7 under the **GSA***-heuristics.* We associate the quotients $q_i := r_{i+1,i+1}^2/r_{i,i}^2$ with the GNF $R = [r_{i,j}] \in \mathbb{R}^{n\times n}$. In practice, the q_i are all very close for typical reduced bases. For simplicity, we assume the geometric series assumption (**GSA**) of [69]:

GSA $\qquad q =_{def} q_1 = q_2 = \cdots = q_{n-1}.$

GSA is a worst-case property. Bases that do not satisfy **GSA** are easier to reduce, see [69]. The worst case bases of Ajtai [3] also satisfy **GSA**.

We next show that the bounds of Theorem 6 improve under **GSA** to those of Theorem 7 while the bounds of Theorem 7 are mainly preserved under **GSA**.

Under **GSA**, Theorem 1 holds with α replaced by $1/q$. Note that $(\det \mathcal{L})^{2/n}/r_{1,1}^2 = q^{\binom{n}{2}\frac{1}{n}} = q^{\frac{n-1}{2}}$ holds under **GSA** and thus, $r_{1,1}^2 = q^{\frac{1-n}{2}}(\det \mathcal{L})^{2/n}$. Hence, part **1.** of Theorem 1 holds with α replaced by $1/q$. Part **2.** also holds because of the duality argument used in the proof of Theorem 7.

HKZ-bases R_ℓ satisfy, under **GSA**,

$$\|\mathbf{b}_1\|^2 = \lambda_1^2(\mathcal{L}(R_\ell)) \leq \gamma_k \det \mathcal{L}(R_\ell)^{\frac{2}{k}} = \gamma_k\|\mathbf{b}_1\|^2 q^{\binom{k}{2}\frac{1}{k}} = \gamma_k\|\mathbf{b}_1\|^2 q^{\frac{k-1}{2}}.$$

This shows that $1/q \leq \gamma_k^{\frac{2}{k-1}}$ holds, under **GSA**. Replacing in Theorem 1 α by $1/q \leq \gamma_k^{\frac{2}{k-1}}$ we get

Corollary 2. *Primal–dual bases of blocksize k and $n = hk$ satisfy, under* **GSA**, *the bounds of Theorem 1 with α replaced by $\gamma^{\frac{2}{k-1}}$, in particular* $\|\mathbf{b}_1\|^2 \leq \gamma_k^{\frac{n-1}{k-1}}(\det \mathcal{L})^{2/n}$, $\|\mathbf{b}_1\|^2 \leq \gamma_k^{2\frac{n-1}{k-1}}\lambda_1^2$.

The bounds of Corollary 2 and those of Theorem 7 nearly coincide. Corollary 2 eliminates α from Theorem 7 and replaces $h = \frac{n}{k}$ by the slightly larger value $\frac{n-1}{k-1}$. Interestingly , the bounds of Corollary 2 coincide for $k = 2$ with clauses 1, 2 of Theorem 1 for LLL-bases with $\delta = 1, \alpha = \frac{4}{3}$ because $\gamma_2^2 = \frac{4}{3}$.

Similarly, $1/q \le \gamma_{2k}^{\frac{2}{2k-1}}$ holds, under **GSA**, for any HKZ-reduced GNF $R = R_{1,2} \in \mathbb{R}^{2k \times 2k}$. Hence,

Corollary 3. *Semi block $2k$-reduced bases with $n = hk$ satisfy under* **GSA** *the inequalities of Theorem 1 with α replaced by $\gamma_{2k}^{\frac{2}{2k-1}}$, in particular* $\|\mathbf{b}_1\|^2 \le \gamma_{2k}^{\frac{n-1}{2k-1}} (\det \mathcal{L})^{2/n}$, $\|\mathbf{b}_1\|^2 \le \gamma_{2k}^{2\frac{n-1}{2k-1}} \lambda_1^2$.

Primal–dual bases of blocksize $2k$ and semi block $2k$-reduced bases are by Corollarys 2 and 3 almost equally strong under **GSA**. This suggests that **Algorithm 2** for semi block $2k$-reduction can be strengthened by working towards **GSA** (similar to **Algorithm 4**) so that **GSA** holds approximately for the output basis.

Practical bounds for $\|\mathbf{b}_1\|^2 \lambda_1^{-2}$. We compare the bounds of Theorem 6 for $2k = 48$ to those of Theorem 7 for double blocksize 48. Note that $\gamma_{24} = 4$ [15]. Assuming that the densest known lattice packings P_{48p}, P_{48q} in dimension 48 [16] Table 1.3, is nearly maximal, we have $\gamma_{48} \approx 6.01$. HKZ-reduction in dimension 48 is nearly feasible. Let $\delta = 0.99$.

Semi block $2k$-reduction for $k = 24$. Using $\beta_{24} \le 13^{2\ln 2 + 1/24}$ Theorem 6 proves $\|\mathbf{b}_1\|^2/(\det \mathcal{L})^{2/n} \le \gamma_{24}(\beta_{24}/\delta)^{n/48-1/2} < \gamma_{24}\, 1.165^{n/2}$. Moreover

$$\|\mathbf{b}_1\|^2/(\det \mathcal{L})^{2/n} \le \gamma_{48}^{\frac{1}{47}\frac{n-1}{2}}$$

holds, under **GSA**. This replaces α in Theorem 1 by $\gamma_{48}^{1/47} < 1.039$.

Primal–dual bases of blocksize 48 satisfy, by Theorem 7,

$$\|\mathbf{b}_1\|^2/(\det \mathcal{L})^{2/n} < \gamma_{48}(\alpha\gamma_{48}^2)^{\frac{n/48-1}{2}} \lesssim 1.075^{n/2}/\sqrt{\alpha}.$$

This replaces α in Theorem 1 by $(\alpha\gamma_{48}^2)^{1/48} \approx 1.084$. Moreover,

$$\|\mathbf{b}_1\|^2/(\det \mathcal{L})^{2/n} \le \gamma_{48}^{\frac{1}{47}\frac{n-1}{2}}$$

holds under **GSA**, which replaces α in Theorem 1 by $\gamma_{48}^{1/47} < 1.039$.

While γ_{48} is relatively large, **Algorithms 2** and **3** perform better when γ_k is relatively small. This suggests the choice of k in practice, clearly apart from multiples of 24, since γ_k is relatively large for $k = 0 \mod 24$.

Primal–Dual Random Sampling Reduction

We replace HKZ-reduction in primal–dual reduction by random sampling reduction (RSR) of [69], a method that shortens the first basis vector. RSR extends the deep insertion step of [66] to a highly parallel algorithm. We use local RSR to approximate the **GSA**-property, which has been used in the analysis of [69]. Moreover, global RSR breaks the worst-case bases of [Aj03] against semi-block k-reduction and those of [24] against primal–dual reduction. These worst-case bases $B = QR$ satisfy $r_{i,j} = 0$ for $j \geq i + k$, which results in a very short last vector $\mathbf{b}_n$. Primal–dual RSR (**ALG. 4**) runs for $k = 80$ in expected feasible time under reasonable heuristics. It reduces α in Theorem 1 to less than $(80/11)^{1/80} \approx 1.025$, which so far is the smallest feasible reduction of α proven under heuristics.

Notation. We associate with a basis $B = QR = [\mathbf{b}_1, \ldots, \mathbf{b}_n] \in \mathbb{R}^{m \times n}$, $R = [r_{i,j}]_{1 \leq i,j \leq n}$ the submatrix $R_{\nu,k} := [r_{i,j}]_{\nu < i,j \leq \nu+k} \subset R$ corresponding to

$$B_{\nu,k} := [\mathbf{b}_{\nu+1}, \ldots, \mathbf{b}_{\nu+k}] \subset B. \text{ Let } T_{\mathbf{a},k} = \begin{bmatrix} a_1 & 1 & & \\ \vdots & 0 & \ddots & \\ a_{k-1} & & \ddots & 1 \\ 1 & 0 & \cdots & 0 \end{bmatrix}.$$

We replace in **Algorithm 4.** the HKZ-reductions of R_ℓ, R_ℓ^* occuring in **Algorithm 3** by RSR of suitable $R_{\nu,k}, R_{\nu,k}^*$.

RSR of $R_{\nu,k}$. Let $R_{\nu,k} = [r_{i,j}]_{\nu < i,j \leq \nu+k} \subset R = [\mathbf{r}_1, \ldots, \mathbf{r}_n]$. Enumerate in parallel all vectors $\mathbf{r} := \sum_{j=k/2+1}^{k} a_j \mathbf{r}_{\nu+j} \in \mathcal{L}(R)$ for $(a_{k/2+1}, \ldots, a_k) \in \mathbb{Z}^{k/2}, a_k = 1$ that satisfy $\|\pi_{\nu+k/2+1}(\mathbf{r})\| \leq \eta$ for some η having about $(k/11)^{k/4}$ such vectors $\mathbf{r}$. Extend each sum to a short vector $\sum_{j=1}^{k} a_j \mathbf{r}_{\nu+j}$ using $n^2 k^{o(k)}$ bit operations for minimization of $\| \sum_{j=1}^{k} a_j \mathbf{r}_{\nu+j} \|$. We can use the additional size-reduction step 2.3 b of Section "The Lattice Reduction Algorithm in the Two-Dimensional Case" and the Schnorr-Hörner heuristics [68]; full exhaustive search minimization, requiring $k^{k/4+o(k)} \log_2 M_0$ bit operations, is too expensive. RSR extends the Schnorr-Euchner *deep insertion* step [66] of depth k. This step coincides with RSR for $a_1, \ldots, a_{k-1}$ set to zero. RSR tries the zero choice and about $(k/11)^{k/4}$ more instances $(a_1, \ldots, a_{k-1}) \in \mathbb{Z}^{k-1}$.

Analysis of RSR on $R_{\nu,k}$. Under the assumptions
RA $\quad r_{\nu+j,\nu+j'} \in_R [-\frac{1}{2}, \frac{1}{2}]$ is random for $j' > j$,
GSA $\quad r_{i+1,i+1}/r_{i,i} = q_\nu$ for all i, $\nu < i < \nu + k$,
[S03, Theorem 1, Remark 2] shows that RSR of $R_{\nu,k}$ and $R_{\nu,k}^*$ succeeds in steps 3, 4 as long as $q_\nu < (11/k)^{1/k}$. Primal–dual RSR uses all indices $\nu = 1, \ldots, n-k$ in a uniform way; this helps to approximate the **GSA**-property.

Theorem 8. [69] RA, GSA. *Primal–dual RSR transforms a basis $B \in \mathbb{R}^{m \times n}$ of $\mathcal{L} = \mathcal{L}(B)$ such that*

1. $\|\mathbf{b}_1\|^2 \leq (k/11)^{\frac{n-1}{2k}} (\det \mathcal{L})^{2/n}$, $\qquad$ **2.** $\|\mathbf{b}_1\|^2 \leq (k/11)^{\frac{n-1}{k}} \lambda_1^2$.

Theorem 8 replaces α in Theorem 1 by $(k/11)^{1/k}$ with $(80/11)^{1/80} \approx 1.025$ for $k = 80$.

Algorithm 4. Primal–dual RSR

INPUT basis $B = QR \in \mathbb{Z}^{m\times n}$, $\delta \in [\eta^2, 1)$, $\eta = \frac{1}{2} + \varepsilon$, k

OUTPUT reduced basis B.

1. LLL-reduce B with δ and compute the GNF R of B, $\ell := 0$.
2. IF $\ell = 0 \mod \lfloor n/k \rfloor$ THEN [BKZ-reduce B into BT with δ and blocksize 20, compute the new GNF R, $\ell := \ell + 1$.] # *this approximates the* **GSA**-*property.*
3. *Primal RSR-step.* Randomly select $0 \le \nu \le n-k$ that nearly maximizes $r_{\nu+1,\nu+1}/|\det R_{\nu,k}|^{1/k}$. Try to decrease $r_{\nu+1,\nu+1}$ by the factor δ through RSR of $R_{\nu,k} \subset R$, i.e., compute some $T_{\mathbf{a},k}$ in $\mathrm{GL}_k(\mathbb{Z})$ such that the GNF $\widetilde{R}_{\nu,k} = [\widetilde{r}_{i,j}]$ of $R_{\nu,k}T_{\mathbf{a},k}$ satisfies $\widetilde{r}_{\nu+1,\nu+1} \le \delta\, r_{\nu+1,\nu+1}$. Transform $B_{\nu,k} := B_{\nu,k}T_{\mathbf{a},k}$. Recompute $R_{\nu,k}$.
4. *Dual RSR-step.* Randomly select $0 \le \nu \le n-k$ that nearly minimizes $r_{\nu+k,\nu+k}/|\det R_{\nu,k}|^{1/k}$. Try to increase $r_{\nu+k,\nu+k}$ by the factor $1/\delta$ through RSR of the dual GNF $R^*_{\nu,k} = U_k R^{-t}_{\nu,k} U_k$, i.e., compute by RSR of $R^*_{\nu,k} = [r^*_{i,j}]$ some $T_{\mathbf{a},k} \in \mathrm{GL}_k(\mathbb{Z})$ such that the GNF $\widetilde{R}^*_{\nu,k} = [\widetilde{r}^*_{i,j}]$ of $R^*_{\nu,k}T_{\mathbf{a},k}$ satisfies $\widetilde{r}^*_{\nu+1,\nu+1} \le \delta r_{\nu+1,\nu+1}$. # *Hence the GNF* $\widetilde{R}_{\nu,k} = [\widetilde{r}_{i,j}]$ *of* $R_{\nu,k}U_kT^{-t}_{\mathbf{a},k}$ *satisfies* $\widetilde{r}_{\nu+k,\nu+k} \ge r_{\nu+k,\nu+k}/\delta$. Transform $B_{\nu,k} := B_{\nu,k}U_kT^{-t}_{\mathbf{a},\mathbf{k}}$ and recompute $R_{\nu,k}$.
5. *Global RSR-step.* Try to decrease $r_{1,1}$ by the factor$1/\delta$ through RSR of R, i.e., compute some $T_{\mathbf{a},n}$ in $\mathrm{GL}_n(\mathbb{Z})$ such that the GNF $\widetilde{R} = [\widetilde{r}_{i,j}]$ of $RT_{\mathbf{a},n}$ satisfies $\widetilde{r}_{1,1} \le \delta\, r_{1,1}$. Transform $B := BT_{\mathbf{a},n}$ and recompute R.
6. IF either of steps 3,4,5 succeeds, or $\ell \ne 0 \mod \lfloor n/k \rfloor$ THEN GOTO 2 ELSE output B and terminate.

Primal–dual RSR time bound. RSR succeeds under **RA, GSA** in steps 3 and 4 using $(k/11)^{k/4+o(k)}$ arithm. steps, provided that $q_\nu < (11/k)^{1/k}$ [69] Theorem 1, ff]. For **RA**, see [57] Figs. 4 and 5 (Randomness of $r_{i,i+1}$ is irrelevant; $r_{i,i+1}$ is in practice, nearly random in $[-\frac{1}{2}, \frac{1}{2}]$, under the condition that $r^2_{i,i} \approx r^2_{i,i+1} + r^2_{i+1,i+1}$; and this improves by deep insertion.) **GSA** is a worst-case assumption; in practice, **GSA** is approximately satisfied. Ludwig [51] analyses an approximate version of **GSA**.

On average, one round of **Algorithm 4** decreases the integer $\mathcal{D}^{(1)} := \prod_{i=1}^{n-1} d_i$ by the factor δ^2. This bounds the average number of rounds by about $\frac{1}{2}n^2 \log_{1/\delta} M_0$ since initially $\mathcal{D}^{(1)} \le M_0^{n(n-1)}$. In worst case, however, $\mathcal{D}^{(1)}$ can even increase per round, and **Algorithm 4** must not terminate.

Comparing **Algorithms 3** *and* **4** *for* $k = 80$. We assume that $\gamma_{80} \approx 4 \cdot 2^{40.14/40} \approx 8.02$, i.e., that the Mordell-Weil lattice MW_n in dimension $n = 80$ has near maximal density, [16] Table 1.3; similarly, we assume $\gamma_{400} \approx 24$.

Algorithm 3 would under **GSA** reduce α in Theorem 1 for $k = 80$ to $\gamma_{80}^{1/79} \approx 1.027$, but **Algorithm 3** does not work towards **GSA**. Moreover, primal–dual reduction with full HKZ-reduction is infeasible in dimension $k = 80$ requiring $80^{40+o(1)}$ steps, whereas RSR is nearly feasible.

Algorithm 4 for primal–dual RSR reduces α in Theorem 1 to $(80/11)^{1/80} \approx 1.025$ (by Theorem 8), and thus achieves $\|\mathbf{b}_1\|/(\det\mathcal{L})^{1/n} \lesssim 1.025^{\frac{n-1}{4}}$.

For lattices of high density, $\lambda_1 \approx \gamma_n \det(\mathcal{L})^{2/n}$, and $n = 400$, $k = 80$, this yields

$$\|\mathbf{b}_1\|/\lambda_1 \lesssim 1.025^{99.75}/\sqrt{\gamma_{400}} \lesssim 2.4.$$

If this bound further decreases on the average case, this might endanger the NTRU schemes for parameters $N \leq 200$. The secret key is a sufficiently short lattice vector of a lattice of dimension $2N$. Ludwig [51] reports on attacks to NTRU via RSR.

HKZ-reduction via the sieve algorithm of [4] reduces all asymptotic time bounds at the expense of superpolynomial space. Regev [62] and Nguyen and Vidick [60] improve the AKS-algorithm and its analysis. The experimental comparison of AKS in [60] with the Schnorr, Euchner version [66] of [64] shows that the SE-algorithm with BKZ for block size 20 outperforms the improved AKS for dimension $n \leq 50$.

Basic Segment LLL

Segment LLL uses an idea of Schönhage [71] to do most of the LLL-reduction locally in segments of low dimension k using k local coordinates. It guarantees that the determinants of the segments do not decrease too fast, see Definition 5. Here, we present the basic algorithm $\mathbf{SLLL_0}$. Theorem 12 bounds the number of local LLL-reductions within $\mathbf{SLLL_0}$. Lemma 4 and Corollary 5 bound the norm of and the *fpa*-errors induced by local LLL-transforms. The algorithm $\mathbf{SLLL_0}$ is faster by a factor n in the number of arithmetic steps compared to $\mathbf{LLL}_H$, but uses longer integers and *fpa* numbers, a drawback that will be repaired by **SLLL**.

Segments and local coordinates. Let the basis $B = [\mathbf{b}_1, \ldots, \mathbf{b}_n] \in \mathbb{Z}^{m \times n}$ have dimension $n = k\,h$ and GNF $R \in \mathbb{R}^{n \times n}$. We partition B into m *segments* $B_{l,k} = [\mathbf{b}_{lk-k+1}, \ldots, \mathbf{b}_{lk}]$ for $l = 1, \ldots, h$. Local LLL-reduction of two consecutive segments $B_{l,k}, B_{l+1,k}$ is done in local coordinates of the principal submatrix

$$R_{l,k} := [r_{lk+i,lk+j}]_{-k<i,j\leq k} \in \mathbb{R}^{2k \times 2k}$$

of R. Let $H = [\mathbf{h}_1, \ldots, \mathbf{h}_n] = [h_{i,j}] \in \mathbb{R}^{m \times n}$ be the lower triangular matrix of Householder vectors and $H_{l,k} = [h_{lk+i,lk+j}]_{-k<i,j\leq k} \subset H$, the submatrix for $R_{l,k}$. We control the calls, and minimize the number of local LLL-reductions of the $R_{l,k}$ by means of the *local squared determinant* of $B_{l,k}$

$$D_{l,k} =_{\text{def}} \|\mathbf{q}_{lk-k+1}\|^2 \cdots \|\mathbf{q}_{lk}\|^2.$$

We have $d_{lk} = \|\mathbf{q}_1\|^2 \cdots \|\mathbf{q}_{lk}\|^2 = D_{1,k} \cdots D_{l,k}$. Moreover, we will use

$$\mathcal{D}^{(k)} =_{\text{def}} \prod_{l=1}^{h-1} d_{lk} = \prod_{l=1}^{h-1} D_{l,k}^{h-l},$$

$$M_{l,k} =_{\text{def}} \max_{lk-k<i\leq j\leq lk+k} \|\mathbf{q}_i\| / \|\mathbf{q}_j\|.$$

For the input basis $B = QR$, we denote $M_1 := \max_{1 \le i \le j \le n} \|\mathbf{q}_i\|/\|\mathbf{q}_j\|$. $M_{l,k}$ is the M_1-value of $R_{l,k}$ when calling locLLL($R_{l,k}$); obviously $M_{l,k} \le M_1$. Recall that $M = \max(d_1, \ldots, d_n, 2^n)$.

Definition 4. A basis $\mathbf{b}_1, \ldots, \mathbf{b}_n \in \mathbb{Z}^m$, $n = kh$, is an *SLLL$_0$-basis (or SLLL$_0$-reduced)* for given $k, \delta \ge \eta^2, \alpha = 1/(\delta - \frac{3}{4})$ if it is size-reduced and

1. $\delta \|\mathbf{q}_i\|^2 \le \mu_{i+1,i}^2 \|\mathbf{q}_i\|^2 + \|\mathbf{q}_{i+1}\|^2$ for $i \in [1, n-1] \setminus k\mathbb{Z}$,
2. $D_{l,k} \le (\alpha/\delta)^{k^2} D_{l+1,k}$ for $l = 1, \ldots, h-1$.

Size-reducedness under *fpa* is defined by clause 1 of Theorem 5. Segment $B_{l,k}$ of an SLLL$_0$-basis is LLL-reduced in the sense that the $k \times k$-submatrix $[r_{lk+i,lk+j}]_{-k<i,j\le 0} \subset R$ is LLL-reduced. Clause 1 does not bridge distinct segments since the $i \in k\mathbb{Z}$ are excepted. Clause 2 relaxes the inequality $D_{l,k} \le \alpha^{k^2} D_{l+1,k}$ of LLL-bases, and this makes it possible to bound the number of local LLL-reductions, see Theorem 12.

We could have used two independent δ-values for the two clauses of Definition 4. Theorem 9 shows that the first vector of an SLLL$_0$-basis of lattice $\mathcal{L}$ is almost as short relative to $(\det \mathcal{L})^{1/n}$ as for LLL-bases.

Theorem 9. Theorem 3 of [70]. $\|\mathbf{b}_1\| \le (\alpha/\delta)^{\frac{n-1}{4}} (\det \mathcal{L})^{\frac{1}{n}}$ *holds for all SLLL$_0$-bases* $\mathbf{b}_1, \ldots, \mathbf{b}_n$.

The dual of Theorem 9. Clause 2 of Definition 4 is preserved under duality. If it holds for a basis $\mathbf{b}_1, \ldots, \mathbf{b}_n$, it also holds for the dual basis $\mathbf{b}_1^*, \ldots, \mathbf{b}_n^*$ of the lattice $\mathcal{L}^*$. We have $\|\mathbf{b}_1^*\| = \|\mathbf{q}_n\|^{-1}$ and $\det(\mathcal{L}^*) = (\det \mathcal{L})^{-1}$. Hence, Theorem 9 implies that every SLLL$_0$-basis satisfies $\|\mathbf{q}_n\| \ge (\delta/\alpha)^{\frac{n-1}{4}} (\det \mathcal{L})^{\frac{1}{n}}$.

Local LLL-reduction. The procedure locLLL($R_{l,k}$)of [S06] locally LLL-reduces $R_{l,k} \subset R$ given $H_{l,k} \subset H$. Initially, it produces a copy $[\mathbf{b}'_1, \ldots, \mathbf{b}'_{2k}]$ of $R_{l,k}$. It LLL-reduces the local basis $[\mathbf{b}'_1, \ldots, \mathbf{b}'_{2k}]$ consisting of *fpa*-vectors. It updates and stores the local transform $T_{l,k} \in \mathbb{Z}^{2k \times 2k}$ so that $[\mathbf{b}'_1, \ldots, \mathbf{b}'_{2k}] = R_{l,k} T_{l,k}$ always holds for the current local basis $[\mathbf{b}'_1, \ldots, \mathbf{b}'_{2k}]$ and the initial $R_{l,k}$, e.g., it does $\mathrm{col}(l', T_{l,k}) := \mathrm{col}(l', T_{l,k}) - \mu\, \mathrm{col}(i, T_{l,k})$ along with $\mathbf{b}'_{l'} := \mathbf{b}'_{l'} - \mu\, \mathbf{b}'_i$ within TriCol$_l$. It freshly computes $\mathbf{b}'_{l'}$ from the updated $T_{l,k}$. Using a correct $T_{l,k}$ this correction of $\mathbf{b}'_{l'}$ limits *fpa*-errors of the local basis, see Corollary 5. Local LLL-reduction of $R_{l,k}$ is done in local coordinates of dimension $2k$. A local LLL-swap merely requires $O(k^2)$ arithmetic steps and update of $R_{l,k}$, local triangulation and size-reduction via TriCol$_l$ included, compared to $O(nm)$ arithmetic steps for an LLL-swap in global coordinates.

SLLL$_0$-algorithm. **SLLL$_0$** transforms a given basis into an SLLL$_0$-basis. It iterates locLLL($R_{l,k}$) for submatrices $R_{l,k} \subset R$, followed by a global update that *transports* $T_{l,k}$ to B and triangulates $B_{l,k}, B_{l,k+1}$ via TriSeg$_{l,k}$. *Transporting* $T_{l,k}$ to $B, R, T_{1,n/2}$ and so on means multiplying the submatrix consisting of $2k$ columns of $B, R, T_{1,n/2}$ corresponding to $R_{l,k}$ from the right by $T_{l,k}$.

SLLL$_0$
INPUT $\mathbf{b}_1, \ldots, \mathbf{b}_n \in \mathbb{Z}^d$ (a basis with M_0, M_1, M), k, m, δ
OUTPUT $\mathbf{b}_1, \ldots, \mathbf{b}_n$ SLLL$_0$-basis for k, δ
WHILE $\exists l, 1 \leq l < m$ such that either $D_{l,k} > (\alpha/\delta)^{k^2} D_{l+1,k}$
or $\texttt{TriSeg}_{l,k}$ has not yet been executed
DO for the minimal such l: $\texttt{TriSeg}_{l,k}$, $\texttt{locLLL}(R_{l,k})$
\# *global update*: $[B_{l,k}, B_{l+1,k}] := [B_{l,k}, B_{l+1,k}]\, T_{l,k}$, $\texttt{TriSeg}_{l,k}$.

The procedure $\texttt{TriSeg}_{l,k}$ *triangulates* and size-reduces two adjacent segments $B_{l,k}, B_{l+1,k}$. Given that $B_{l,k}, B_{l+1,k}$ and $\mathbf{h}_1, \ldots, \mathbf{h}_{lk-k}$, it computes $[\mathbf{r}_{lk-k+1}, \ldots, \mathbf{r}_{lk+k}] \subset R$ and $[\mathbf{h}_{lk-k+1}, \ldots, \mathbf{h}_{lk+k}] \subset H$.

$\texttt{TriSeg}_{l,k}$
1. FOR $l' = lk - k + 1, \ldots, lk + k$ DO $\texttt{TriCol}_{l'}$ (including updates of $T_{l,k}$)
2. $D_{j,k} := \prod_{i=0}^{k-1} r^2_{kj-i,kj-i}$ for $j = l, l+1$.

Correctness in ideal arithmetic. All inequalities $D_{l,k} \leq (\alpha/\delta)^{k^2} D_{l+1,k}$ hold upon termination of **SLLL$_0$**. All segments $B_{l,k}$ are locally LLL-reduced and globally size-reduced, and thus, the terminal basis is SLLL$_0$-reduced.

The number of rounds of **SLLL$_0$**. Let $\#_k$ denote the number of $\texttt{loclll}(R_{l,k})$-executions as a result of $D_{l,k} > (\alpha/\delta)^{k^2} D_{l+1,k}$ for all l. The first $\texttt{loclll}(R_{l,k})$-executions for each l are possibly not counted in $\#_k$; this yields at most $n/k - 1$ additional rounds.

$\#_k$ can be bounded by the Lovász volume argument.

Theorem 10. Theorem 4 of [70]. $\#_k \leq 2\, n\, k^{-3} \log_{1/\delta} M$.

All intermediate $M_{l,k}$-values within **SLLL$_0$** are bounded by the M_1-value of the input basis of **SLLL$_0$**. Consider the local transform $T_{l,k} \in \mathbb{Z}^{2k \times 2k}$ within $\texttt{locLLL}(R_{l,k})$. Let $\|T_{l,k}\|_1$ denote the maximal $\| \ \|_1$-norm of the columns of $T_{l,k}$.

Lemma 4. [70] *Within* $\texttt{locLLL}(R_{l,k})$ *we have* $\|T_{l,k}\|_1 \leq 6k(\frac{3}{2})^{2k} M_{l,k}$.

Next consider $\texttt{locLLL}(R_{l,k})$ under *fpa*, based on the iterative *fpa*-version of $\texttt{TriCol}_l$. Let $\|[r_{i,j}]\|_F = (\sum_{i,j} r^2_{i,j})^{1/2}$ denote the Frobenius *norm*. [S06] shows

Corollary 4. [fpa-Heur.]

1. *Within* $\texttt{locLLL}(R_{l,k})$ *the current* $R'_{l,k} := R_{l,k} T_{l,k}$ *and its approximation* $\bar{R}'_{l,k}$ *satisfy* $\|\bar{R}'_{l,k} - R'_{l,k}\|_F \leq \|\bar{R}_{l,k} - R_{l,k}\|_F\, 2^{2k} M_{l,k} + 7n\|R_{l,k}\|_F 2^{-t}$.
2. *Let* $\texttt{TriSeg}_{l,k}$ *and* $\texttt{locLLL}$ *use* fpa *with precision* $2^t \geq 2^{10} d\, \rho^n M_1^2$. *If* $\bar{R}_{l,k}$ *is computed by* $\texttt{TriSeg}_{l,k}$ *then* $\texttt{locLLL}(\bar{R}_{l,k})$ *computes a correct* $T_{l,k}$ *so that* $R_{l,k} T_{l,k}$ *is LLL-reduced with* $\delta\_$.

Theorem 11. Theorem 5 of [70] using fpa-Heur. *Let $k = \Theta(\sqrt{n})$. Given a basis with M_0, M_1, M,* **SLLL$_0$** *computes under* fpa *with precision $2^t \geq 2^{10} m \, \rho^n M_1^2$ an SLLL$_0$-basis for δ_-. It runs in $O(nm \log_{1/\delta} M)$ arithmetic steps using $2n + \log_2(M_0 M_1^2)$-bit integers.*

SLLL$_0$ saves a factor n in the number of arithmetic steps compared to **LLL$_H$** but uses longer integers and *fpa* numbers. **SLLL$_0$** runs for $M_0 = 2^{O(n)}$, and thus for $M = 2^{O(n^2)}$, in $O(n^3 m)$ arithmetic steps using $O(n^2)$ bit integers. Algorithm **SLLL** of Section "First steps in the Probabilistic Analysis of the LLL Algorithm" reduces the bit length $O(n^2)$ to $O(n)$.

Gradual SLLL Using Short *fpa*-Numbers

SLLL reduces the required precision and the bit length of integers and *fpa* numbers compared to **SLLL$_0$**. This results from limiting the norm of local transforms to $O(2^n)$. Theorem 14 shows that **SLLL**-bases are as strong as LLL-bases. For input bases of length $2^{O(n)}$ and $d = O(n)$, **SLLL** performs $O(n^{5+o(1)})$ bit operations compared to $O(n^{6+o(1)})$ bit operations for **LLL$_H$**, **SLLL$_0$**, and the LLL-algorithms of [65, 73]. The advantage of **SLLL** is the use of small integers, which is crucial in practice.

The use of small integers and short intermediate bases within **SLLL** rests on a gradual LLL-type reduction so that all local LLL-transforms $T_{l,2^\sigma}$ of $R_{l,2^\sigma}$ have norm $O(2^n)$. For this, we work with segments of all sizes 2^σ and also to perform LLL-reduction on $R_{l,2^\sigma}$ with a measured strength, i.e., SLLL-reduction according to Definition 6. If the submatrices $R_{2l,2^{\sigma-1}}, R_{2l+1,2^{\sigma-1}} \subset R_{l,2^\sigma}$ are already SLLL-reduced, then `locLLL`$(R_{l,k})$ performs a transform $T_{l,2^\sigma}$ bounded as $\|T_{l,2^\sigma}\|_F = O(2^n)$. This is the core of *fpa*-correctness of **SLLL**.

Comparison with Schönhage's semi-reduction [71]. Semi-reduction also uses segments but proceeds without adjusting LLL-reduction according to Definition 5 and does not satisfy Theorems 11 and 12. While **SLLL** achieves length defect $\|\mathbf{b}_1\|/\lambda_1 \leq (\frac{4}{3} + \varepsilon)^{n/2}$, semi-reduction achieves merely $\|\mathbf{b}_1\|/\lambda_1 \leq 2^n$. **SLLL** is practical even for small n, where all O-constants and n_0-values are small.

We let n be a power of 2. We set $s := \lceil \frac{1}{2} \log_2 n \rceil$ so that $\sqrt{n} \leq 2^s < 2\sqrt{n}$.

Definition 5. A basis $\mathbf{b}_1, \ldots, \mathbf{b}_n \in \mathbb{R}^m$ is an *SLLL-basis (or SLLL-reduced)* for $\delta \geq \frac{1}{2}$ *if it satisfies for $\sigma = 0, \ldots, s = \lceil \frac{1}{2} \log_2 n \rceil$ and all l, $1 \leq l < n/2^\sigma$:*

$$D_{l,2^\sigma} \leq \alpha^{4^\sigma} \delta^{-n} D_{l+1,2^\sigma}.$$

If the inequalities of Definition 6 hold for a basis, they also hold for the dual basis. Thus, the dual of an SLLL-basis is again an SLLL-basis. To preserve SLLL-reducedness by duality, we do not require SLLL-bases to be size-reduced.

The inequalities of Definition 6 for $\sigma = 0$ mean that $\|\mathbf{q}_l\|^2 \le \alpha\delta^{-n}\|\mathbf{q}_{l+1}\|^2$ holds for all l. The inequalities of Definition 6 are merely required for $2^\sigma \le 2\sqrt{n}$. Therefore, **SLLL** locally LLL-reduces $R_{l,2^\sigma}$ via $\texttt{locLLL}(R_{l,2^\sigma})$ merely for segment sizes $2^\sigma < 2\sqrt{n}$, where size-reduction of a vector requires $O(2^{2\sigma}) = O(n)$ arithmetic steps.

The inequalities of Definition 6 and $D_{l,k} \le (\alpha/\delta)^{k^2} D_{l+1,k}$ of Definition 5 coincide for $k = 2^\sigma$ when setting $\delta := \delta_\sigma$ in Definition 6, and $\delta_\sigma := \delta^{n4^{-\sigma}}$ for the δ of Definition 6. Note that δ_σ can be arbitrarily small, e.g. $\delta_\sigma \ll \frac{1}{4}$, δ_σ decreases with σ. In particular, for $2^\sigma = k \ge \sqrt{n}$ we have $\alpha^{4^\sigma}\delta^{-n} \le (\alpha/\delta)^{k^2}$ and thus, the inequalities of Definition 6 are stronger than the ones of Definition 5. Theorem 12 shows that the vectors of SLLL-bases approximate the successive minima in nearly the same way as for LLL-bases.

Theorem 12. Theorem 6 of [70]. *Every size-reduced SLLL-basis satisfies*

1. $\lambda_j^2 \le \alpha^{j-1}\delta^{-7n} r_{j,j}^2$ *for* $j = 1, \ldots, n,$
2. $\|\mathbf{b}_l\|^2 \le \alpha^{j-1}\delta^{-7n} r_{j,j}^2$ *for* $l \le j,$
3. $\|\mathbf{b}_j\|^2 \le \alpha^{n-1}\delta^{-7n}\lambda_j^2$ *for* $j = 1, \ldots, n.$

$\texttt{LLLSeg}_{l,1}$

\# *Given* $R_{l,1}, \mathbf{b}_1, \ldots, \mathbf{b}_{l+1}, \mathbf{h}_1, \ldots, \mathbf{h}_l, \mathbf{r}_1, \ldots, \mathbf{r}_l$, $\texttt{LLLSeg}_{l,1}$ *LLL-reduces* $R_{l,1}$.

1. IF $r_{l,l}/r_{l+1,l+1} > 2^{n+1}$ THEN [$R'_{l,1} := R_{l,1}$,
   $\mathrm{row}(2, R'_{l,1}) := \mathrm{row}(2, R'_{l,1})\, 2^{-n-1}\, r_{l,l}/r_{l+1,l+1}$ $\texttt{locLLL}(R'_{l,1})$,
   \# *global update*: $[\mathbf{b}_l, \mathbf{b}_{l+1}] := [\mathbf{b}_l, \mathbf{b}_{l+1}]\, T_{l,1}$, $\texttt{TriCol}_l$, $\texttt{TriCol}_{l+1}$]
2. $\texttt{locLLL}(R_{l,1})$.

SLLL uses the procedure $\texttt{LLLSeg}_{l,1}$ that breaks $\texttt{locLLL}(R_{l,1})$ up into parts, each with a bounded transform $\|T_{l,1}\|_1 \le 9 \cdot 2^{n+1}$. This keeps intermediate bases of length $O(4^n M_0)$ and limits *fpa*-errors within $\texttt{LLLSeg}_{l,1}$. $\texttt{LLLSeg}_{l,1}$ LLL-reduces the basis $R_{l,1} = \begin{bmatrix} r_{l,l} & r_{l,l+1} \\ 0 & r_{l+1,l+1} \end{bmatrix} \subset R$ after dilating $\mathrm{row}(2, R_{l,1})$ so that $r_{l,l}/r_{l+1,l+1} \le 2^{n+1}$. After the LLL-reduction of the dilated $R_{l,1}$, we undo the dilation by transporting the local transform $T_{l,1} \in \mathbb{Z}^{2\times 2}$ to B. $\texttt{LLLSeg}_{l,1}$ includes global updates between local rounds.

$\texttt{LLLSeg}_{l,1}$ performs $O(nm)$ arithmetic steps [70] Lemma 3. An effectual step 1 decreases $\mathcal{D}^{(1)}$ by a factor $2^{-n/2}$ via a transform $T_{l,1}$ satisfying $\|T_{l,1}\|_1 \le 9 \cdot 2^{n+1}$.

SLLL
INPUT $\mathbf{b}_1, \ldots, \mathbf{b}_n \in \mathbb{Z}^m$ (a basis with M_0, M_1, M), δ, α, ε
OUTPUT $\mathbf{b}_1, \ldots, \mathbf{b}_n$ size-reduced SLLL-basis for δ, ε

```
1. TriCol₁, TriCol₂, l' := 2, s := ⌈½ log₂ n⌉
       # TriCol_l' has always been executed for the current l'
2. WHILE ∃σ ≤ s, l, 2^σ(l + 1) ≤ l' such that D_{l,2^σ} > α^{4^σ} δ^{-n} D_{l+1,2^σ}
             # Note that r_{1,1}, ..., r_{l',l'} and thus D_{l,2^σ}, D_{l+1,2^σ} are given
             DO for the minimal such σ and the minimal l:
                 IF σ = 0 THEN LLLSeg_{l,1} ELSE locLLL(R_{l,2^σ})
                 #global update: transport T_{l,2^σ} to B, TriSeg_{l,2^σ}
3. IF l' < n THEN [l' := l' + 1, TriCol_l', GOTO 2.]
```

Correctness in ideal arithmetic. All inequalities $D_{l,2^\sigma} \le \alpha^{4^\sigma} \delta^{-n} D_{l+1,2^\sigma}$ hold upon termination of **SLLL**. As $\texttt{TriSeg}_{l,2^\sigma}$ results in size-reduced segments $B_{l,2^\sigma}$, $B_{l+1,2^\sigma}$ the terminal basis is size-reduced.

Theorem 13. Theorem 7 of [70] using fpa-Heur. *Given a basis with* M_0, M, **SLLL** *finds an SLLL-basis for* $\delta\_$ *under* fpa *of precision* $t = 3n + O(\log m)$. *It runs in* $O(nm \log_2 n \log_{1/\delta} M)$ *arithmetic steps using integers of bit length* $2n + \log_2 M_0$.

For $M_0 = 2^{O(n)}$ and $m = O(n)$ **SLLL** runs in $O(n^4 \log n)$ arithmetic steps, and in $O(n^{6+\varepsilon})/O(n^{5+\varepsilon})$ bit operations under school-/FFT-multiplication.

SLLL-bases versus LLL-bases. LLL-bases with δ satisfy the inequalities of Theorem 14 with δ replaced by 1. Thus, $\|\mathbf{b}_j\|$ approximates λ_j to within a factor $\alpha^{\frac{n-1}{2}}$ for LLL-bases, respectively, within a factor $(\alpha/\delta^7)^{\frac{n-1}{2}}$ for SLLL-bases. But SLLL-bases for $\delta' = \delta^{1/8}$ are "better" than LLL-bases for δ, in the sense that they guarantee a smaller length defect, because $\alpha'/\delta'^7 = \frac{1}{\delta'^8 - \delta'^7/4} = \frac{1}{\delta - \delta'^7/4} < \frac{1}{\delta - 1/4} = \alpha$.

Dependence of time bounds on δ. The time bounds contain a factor $\log_{1/\delta} 2$,

$$\log_{1/\delta} 2 = \log_2(e)/\ln(1/\delta) \le \log_2(e)\frac{\delta}{1-\delta},$$

since $\ln(1/\delta) \ge 1/\delta - 1$. We see that replacing δ by $\sqrt{\delta}$ essentially halves $1-\delta$ and doubles the SLLL-time bound. Hence, replacing δ by $\delta^{1/8}$ increases the **SLLL**-time bound at most by a factor 3. In practice, the LLL-time may increase slower than by the factor $\frac{\delta}{1-\delta}$ as δ approaches 1, see [41] Fig. 3.

Reducing a generator system. There is an algorithm that, given a generator matrix $B \in \mathbb{Z}^{m \times n}$ of arbitrary rank $\le n$, transforms B with the performance of **SLLL**, into an SLLL-basis for $\delta\_$ of the lattice generated by the columns of B.

SLLL-Reduction via iterated subsegments. **SLLL**$^+$ of [70] is a variant of **SLLL** that extends LLL-operations stepwise to increasingly larger submatrices $R_{l,2^\sigma} \subset R$ by transporting local transforms from level $\sigma - 1$ to level σ recursively for $\sigma = 1, \ldots, s = \log_2 n$. Local LLL-reduction and the transport of local LLL-transforms is done by a local procedure $\texttt{locSLLL}(R_{l,2^\sigma})$ that recursively executes `locSLLL`$(R_{l',2^\sigma-1})$ for $l' = 2l-1, 2l, 2l+1$. **SLLL$^+$** does not iterate the global procedure `TriSeg` but a faster local one.

Definition 6. A basis $\mathbf{b}_1, \ldots, \mathbf{b}_n \in \mathbb{Z}^m$ with $n = 2^s$ is an *SLLL$^+$-basis (or SLLL$^+$-reduced)* for δ if it satisfies for $\sigma = 0, \ldots, s = \log_2 n$

$$D_{l,2^\sigma} \le (\alpha/\delta)^{4^\sigma} D_{l+1,2^\sigma} \quad \text{for } odd\ l \in [1, n/2^\sigma]. \tag{4.7}$$

Unlike in Definitions 5 and 6, the inequalities (4.7) are not required for *even* l; this opens new efficiencies for SLLL$^+$-reduction. The inequalities (7) hold for each σ and odd l locally in double segments $[B_{l,2^\sigma}, B_{l+1,2^\sigma}]$; they do not bridge these pairwise disjoint double segments. For $\sigma = 0$, the inequalities (7) mean that $\|\mathbf{q}_l\|^2 \le \alpha/\delta\, \|\mathbf{q}_{l+1}\|^2$ holds for odd l.

The inequalities (7) are preserved under duality. If $\mathbf{b}_1, \ldots, \mathbf{b}_n$ is an SLLL$^+$-basis, then so is the dual basis $\mathbf{b}_1^*, \ldots, \mathbf{b}_n^*$. Theorem 16 extends Theorem 11 and shows that the first vector of an SLLL$^+$-basis is almost as short relative to $(\det \mathcal{L})^{\frac{2}{n}}$ as for LLL-bases.

Theorem 14. Theorem 8 of [70]. *Every SLLL$^+$-basis $\mathbf{b}_1, \ldots, \mathbf{b}_n$, where n is a power of 2 satisfies*

1. $\|\mathbf{b}_1\| \le (\alpha/\delta)^{\frac{n-1}{4}} (\det \mathcal{L})^{\frac{1}{n}}$ **2.** $\|\mathbf{q}_n\| \ge (\delta/\alpha)^{\frac{n-1}{4}} (\det \mathcal{L})^{\frac{1}{n}}$.

Theorem 15. Theorem 9 of [70]. *In ideal arithmetic, algorithm* **SLLL$^+$** *of [S06] computes a size-reduced SLLL$^+$-basis for δ and runs in $O(n^2 m + n \log_2 n \log_{1/\delta} M)$ arithmetic steps.*

SLLL$^+$ requires $t = O(\log(M_0 M_1)) = O(n \log M_0)$ precision bits to cover the *fpa*-errors that get accumulated by the initial `TriSeg` and by iterating `locTri`. For $M_0 = 2^{O(n)}$ and $m = O(n)$, **SLLL$^+$** saves a factor n in the number of arithmetic steps compared to **SLLL** but requires n-times longer *fpa*-numbers.

Acknowledgements I would thank P. Nguyen and D. Stehlé for their useful comments.

References

1. *E. Agrell, T. Eriksson, A. Vardy, and K. Zeger*, Closest Point Search in Lattices. *IEEE Trans. Inform. Theor.*, **48**(8), pp. 2201–2214, 2002.
2. *M. Ajtai*, The Shortest Vector Problem in L_2 is NP-hard for Randomized Reductions. In Proc. 30th STOC, ACM, New York, pp. 10–19, 1998.
3. *M. Ajtai*, The Worst-case Behavior of Schnorr's Algorithm Approximating the Shortest Nonzero Vector in a Lattice. In Proc. 35th STOC, ACM, New York, pp. 396–406, 2003.
4. *M. Ajtai, R. Kumar, and D. Sivakumar*, A Sieve Algorithm for the Shortest Lattice Vector Problem. In Proc. 33th STOC, ACM, New York, pp. 601–610, 2001.
5. *A. Akhavi*, Worst Case Complexity of the Optimal LLL. In Proc. of LATIN 2000, LNCS 1776, Springer, Berlin, 2000.
6. *A. Akhavi and D.Stehlé*, Speeding-up Lattice Reduction with Random Projections. In Proc. of LATIN 2008, to be published in LNCS by Springer, Berlin, 2008.

7. *W. Backes and S. Wetzel*, Heuristicsc on Lattice Reduction in Practice. *J. Exp. Algorithm*, ACM **7**(1), 2002.
8. *G. Bergman*, Notes on Ferguson and Forcade's Generalized Euclidean Algorithm. TR. Dep. of Mathematics, University of Berkeley, CA, 1980.
9. *D. Bleichenbacher and A. May*, New Attacks on RSA with Small Secret CRT-Exponents. In Proc. PKC 2006, LNCS 3958, Springer, Berlin, pp. 1–13, 2006.
10. *J. Blömer and A. May*, New Partial Key Exposure Attacks on RSA. In Proc. Crypto 2003, LNCS 2729, Springer, Berlin, pp. 27–43, 2003.
11. *J. Blömer and J.P. Seifert*, On the Complexity of Computing Short Linearly Independent Vectors and Short Bases in a Lattice. In Proc. 31th STOC, ACM, New York, pp. 711–720, 1999.
12. *J. Cai*, The Complexity of some Lattice Problems. In Proc. Algorithmic- Number Theory, LNCS 1838, Springer, Berlin, pp. 1–32, 2000.
13. *J.W.S. Cassels*, Rational Quadratic Forms. London Mathematical Society Monographs, 13, Academic, London, 1978.
14. *H. Cohen*, A Course in Computational Number Theory, Second edition, Springer, Berlin, 2001.
15. *H. Cohn and A. Kumar*, Optimality and Uniqueness of the Leech Lattice Among Lattices. arXiv:math.MG/04 03263v1, 16 Mar 2004.
16. *J.H. Conway and N.J.A. Sloane*, Sphere Packings, Lattices and Groups. Third edition, Springer, Berlin, 1998.
17. *D. Coppersmith*, Small Solutions to Polynomial Equations, and Low Exponent RSA Vulnerabilities. *J. Crypt.*, **10**, pp. 233–260, 1997.
18. *D. Coppersmith*, Finding Small Solutions to Small Degree Polynomials. In Proc. CaLC 2001, LNCS 2146, Springer, Berlin, pp. 20–1, 2001.
19. *J.Demmel and Y. Hida*, Accurate floating point summation. TR University Berkeley, 2002, http://www.cs.berkeley.edu/ demmel/AccurateSummation.ps.
20. *P. van Emde Boas*, Another NP-Complete Partition Problem and the Complexity of Computing Short Vectors in a Lattice. Mathematics Department, University of Amsterdam, TR 81-04, 1981.
21. *U. Fincke and M. Pohst*, A Procedure for Determining Algebraic Integers of a Given Norm. In Proc. EUCAL, LNCS 162, Springer, Berlin, pp. 194–202, 1983.
22. *C.F. Gauss*, Disquisitiones Arithmeticae. 1801; English transl., Yale University Press, New Haven, Conn. 1966.
23. *N. Gama, N. How-Grave-Graham, H. Koy, and P. Nguyen*, Rankin's Constant and Blockwise Lattice Reduction. In Proc. CRYPTO 2006, LNCS 4117, Springer, Berlin, pp. 112–139, 2006.
24. *N. Gama and P. Q. Nguyen*, Finding short lattice vectors within Mordell's inequality. In Proc. of the 2008 ACM Symposium on Theory of Computing, pp. 207–216, 2008.
25. *O. Goldreich and S. Goldwasser*, On the Limits of Nonapproximability of Lattice Problems. *J. Comput. Syst. Sci.*, **60**(3), pp. 540–563, 2000. Preliminary version in STOC '98.
26. *G. Golub and C. van Loan*, Matrix Computations. John Hopkins University Press, Baltimore, 1996.
27. *M. Grötschel, L. Lovász, and A. Schrijver*, Geometric Algorithms and Combinatorial Optimization, Springer, Berlin, 1988.
28. *G. Hanrot and D. Stehlé*, Improved Analysis of Kannan's Shortest Lattice Vector Algorithm. In Proc. CRYPTO 2007, LNCS 4622, Springer, Berlin, pp. 170–186, 2007.
29. *H.J. Hartung and C.P. Schnorr*, Public Key Identification Based on the Equivalence of Quadratic Forms. In Proc. of Math. Found. of Comp. Sci., Aug. 26–31, Český Krumlov, Czech Republic, LNCS 4708, Springer, Berlin, pp. 333–345, 2007.
30. *R.J. Hartung and C.P. Schnorr*, Identification and Signatures Based on NP-hard Problems of Indefinite Quadratic Forms. *J. Math. Crypt.* **2**, pp. 327–341, 2008. Preprint University Frankfurt, 2007,//www.mi.informatik.uni-frankfurt.de/research/papers.html.
31. *J. Håstad, B. Just, J.C. Lagarias, und C.P. Schnorr*, Polynomial Time Algorithms for Finding Integer Relations Among Real Numbers, *SIAM J. Comput.*, **18**, pp. 859–881, 1989.
32. *B. Helfrich*, Algorithms to Construct Minkowski Reduced and Hermite Reduced Bases. *Theor. Comput. Sci.*, **41**, pp. 125–139, 1985.

33. *C. Hermite*, Extraits de lèttres de M. Ch. Hermiteà Jacobi sur differents objects de la théorie des nombres, deuxième lettre, *J. Reine Angew. Math.*, **40**, pp. 279–290, 1850.
34. *N.J. Higham*, Accuracy and Stability of Numerical Algorithms. SIAM, Philadelphia, Second edition, 2002.
35. *R. Kannan*, Minkowski's Convex Body Theorem and Integer Programming. *Math. Oper. Res.*, **12**, pp. 415–440, 1987.
36. *S. Khot*, Hardness of Approximating the Shortest Vector Problem in Lattices. In Proc. FOCS, 2004.
37. *D.E. Knuth*, The Art of Computer Programming, 2, Addison-Wesley, Boston, Third edition, 1997.
38. *A. Korkine und G. Zolotareff*, Sur les Formes Quadratiques, *Math. Ann.*, **6**, pp. 366–389, 1873.
39. *H. Koy*, Primal/duale Segment-Reduktion von Gitterbasen, Lecture Universität Franfurt 2000, files from Mai 2004. //www.mi.informatik.uni-frankfurt.de/research/papers.html
40. *H. Koy and C.P. Schnorr*, Segment LLL-Reduction. In Proc. CaLC 2001, LNCS 2146, Springer, Berlin, pp.67–80, 2001. //www.mi.informatik.uni-frankfurt.de/research/papers.html
41. *H. Koy and C.P. Schnorr*, Segment LLL-Reduction with Floating Point Orthogonalization. In Proc. CaLC 2001, LNCS 2146, Springer, Berlin, pp. 81–96, 2001. //www.mi.informatik.uni-frankfurt.de/research/papers.html
42. *H. Koy and C.P. Schnorr*, Segment and Strong Segment LLL-Reduction of Lattice Bases. TR Universität Franfurt, April 2002, //www.mi.informatik.uni-frankfurt.de/research/papers.html
43. *J.L.Lagrange*, Recherches d'arithmétique. Nouveaux Mémoires de l'Académie de Berlin, 1773.
44. *C.L. Lawson and R.J. Hanson*, Solving Least Squares Problems. SIAM, Philadelphia, 1995.
45. *H.W. Lenstra, Jr.*, Integer Programming With a Fixed Number of Variables. *Math. Oper. Res.* ,8, pp. 538–548, 1983.
46. LIDIA, a C++ Library for computational number theory. The LIDIA Group, http://www.informatik.tu-darmstadt.de/TI/LiDIA/
47. *A. K. Lenstra, H. W. Lenstra Jr., and L. Lovász*, Factoring Polynomials with Rational Coefficients. *Math. Ann.*, **261**, pp. 515–534, 1982.
48. *J.C. Lagarias, H.W. Lenstra Jr., and C.P. Schnorr*, Korkin-Zolotarev Bases and Successive Minima of a Lattice and its Reciprocal Lattice. *Combinatorica*, **10**, pp. 333–348, 1990.
49. *L. Lovász*, An Algorithmic Theory of Numbers, Graphs and Convexity, CBMS-NSF Regional Conference Series in Applied Mathematics, **50**, SIAM, Philadelphia, 1986.
50. *L. Lovász and H. Scarf*, The Generalized Basis Reduction Algorithm. *Math. Oper. Res.*, **17** (3), pp. 754–764, 1992.
51. *C. Ludwig*, Practical Lattice Basis Reduction. Dissertation, TU-Darmstadt, December 2005, http://elib.tu-darmstadt.de/diss/000640 and http://deposit.ddb.de/cgi-bin/dokserv?idn=978166493.
52. *J. Martinet*, Perfect Lattices in Euclidean Spaces. Springer, Berlin, 2002.
53. *A. May*, Computing the RSA Secret Key is Deterministic Polynomial Time Equivalent to Factoring. In Proc. CRYPTO 2004, LNCS 3152, Springer, Berlin, pp. 213–219, 2004.
54. *D. Micciancio and S. Goldwasser*, Complexity of Lattice Problems, A Cryptographic Perspective. Kluwer, London, 2002.
55. *V. Shoup*, Number Theory Library. http: //www.shoup.net/ntl/
56. *P.Q. Nguyen and D. Stehlé*, Floating-Point LLL Revisited. In Proc. Eurocrypt'05, LNCS 3494, Springer, Berlin, pp. 215–233, 2005.
57. *P. Nguyen and D. Stehlé*, LLL on the Average. In Proc. ANTS-VII, LNCS 4076, Springer, Berlin, pp. 238–356, 2006.
58. *P.Q. Nguyen and J. Stern*, Lattice Reduction in Cryptology, An Update. Algorithmic Number Theory, LNCS 1838, Springer, Berlin, pp. 85–112, 2000. full version http://www.di.ens.fr/pnguyen,stern/
59. *P.Q. Nguyen and J. Stern*, The Two Faces of Cryptology, In Proc. CaLC'01, LNCS 2139, Springer, Berlin, pp. 260–274, 2001.
60. *P.Q. Nguyen and T. Vidick*, Sieve Algorithms for the Shortest Vector Problem are Practical. Preprint, 2007.

61. *A.M.Odlyzko*, The Rise and the Fall of Knapsack Cryptosystems, In Proc. of Cryptology and Computational Number Theory, vol. 42 of Proc. of Symposia in Applied Mathematics, pp. 75–88, 1989.
62. *O. Regev*, Lecture Notes of Lattices in Computer Science, taught at the Computer Science Tel Aviv University, 2004; available at http://www.cs.tau.il/ odedr.
63. *C. Rössner and C.P. Schnorr*, An Optimal, Stable Continued Fraction Algorithm for Arbitrary Dimension. In Proc. 5-th IPCO 1996, LNCS 1084, Springer, Berlin, pp. 31–43, 1996.
64. *C.P. Schnorr*, A Hierarchy of Polynomial Time Lattice Basis Reduction Algorithms. *Theoret. Comput. Sci.*, **53**, pp. 201–224, 1987.
65. *C.P. Schnorr*, A More Efficient Algorithm for Lattice Reduction. *J. Algorithm* **9**, 47–62, 1988.
66. *C.P. Schnorr and M. Euchner*, Lattice Basis Reduction and Solving Subset Sum Problems. In Proc. Fundamentals of Comput. Theory, LNCS 591, Springer, Berlin, pp. 68–85, 1991. Complete paper in *Math. Program. Stud.*, **66A**, 2, pp. 181–199, 1994.
67. *C.P. Schnorr*, Block Reduced Lattice Bases and Successive Minima. *Combin. Probab. and Comput.*, **3**, pp. 507–522, 1994.
68. *C.P. Schnorr and H.H. Hörner*, Attacking the Chor-Rivest cryptosystem by improved lattice reduction. In Proc. Eurocrypt 1995, LNCS 921, Springer, Berlin, pp. 1–12, 1995.
69. *C.P. Schnorr*, Lattice Reduction by Random Sampling and Birthday Methods. In Proc. STACS 2003, H. Alt and M. Habib (Eds.), LNCS 2607, Springer, Berlin, pp. 145–156, 2003.
70. *C.P. Schnorr*, Fast LLL-type lattice reduction. Information and Computation, **204**, pp. 1–25, 2006. //www.mi.informatik.uni-frankfurt.de
71. *A. Schönhage*, Factorization of Univariate Integer Polynomials by Diophantine Approximation and Improved Lattice Basis Reduction Algorithm. In Proc. ICALP 1984, LNCS 172, Springer, Berlin, pp. 436–447, 1984.
72. *D.Simon*, Solving Quadratic Equations Using Reduced Unimodular Quadratic Forms. *Math. Comp.* **74** (251), pp. 1531–1543, 2005.
73. *A. Storjohann*, Faster Algorithms for Integer Lattice Basis Reduction. TR 249, Swiss Federal Institute of Technology, ETH-Zurich, July 1996. //www.inf.ethz.ch/research/publications/html.
74. *D. Stehlé*, Floating Point LLL: Theoretical and Practical Aspects. This Proceedings.
75. *G. Villard*, Certification of the QR Factor R, and of Lattice Basis Reducedness. LIP Research Report RR2007-03, ENS de Lyon, 2007.
76. *J.H. Wilkinson*, The Algebraic Eigenvalue Problem. Oxford University Press, New York, 1965, reprinted 1999.

Chapter 5
Floating-Point LLL: Theoretical and Practical Aspects

Damien Stehlé

Abstract The text-book LLL algorithm can be sped up considerably by replacing the underlying rational arithmetic used for the Gram–Schmidt orthogonalisation by floating-point approximations. We review how this modification has been and is currently implemented, both in theory and in practice. Using floating-point approximations seems to be natural for LLL even from the theoretical point of view: it is the key to reach a bit-complexity which is quadratic with respect to the bit-length of the input vectors entries, without fast integer multiplication. The latter bit-complexity strengthens the connection between LLL and Euclid's gcd algorithm. On the practical side, the LLL implementer may weaken the provable variants in order to further improve their efficiency: we emphasise on these techniques. We also consider the practical behaviour of the floating-point LLL algorithms, in particular their output distribution, their running-time and their numerical behaviour. After 25 years of implementation, many questions motivated by the practical side of LLL remain open.

Introduction

The LLL lattice reduction algorithm was published in 1982 [35], and was immediately heartily welcome in the algorithmic community because of the numerous barriers it broke. Thanks to promising applications, the algorithm was promptly implemented. For example, as soon as the Summer 1982, Erich Kaltofen implemented LLL in the Macsyma computer algebra software to factor polynomials. This implementation followed very closely the original description of the algorithm. Andrew Odlyzko tried to use it to attack knapsack-based cryptosystems, but the costly rational arithmetic and limited power of computers available at that time limited him to working with lattices of small dimensions. This led him to replace the rationals by floating-point approximations, which he was one of the very first

D. Stehlé
CNRS/Universities of Macquarie, Sydney and Lyon/INRIA/ÉNS Lyon
Department of Mathematics and Statistics, University of Sydney, NSW 2008, Australia,
e-mail: damien.stehle@gmail.com, http://perso.ens-lyon.fr/damien.stehle

P.Q. Nguyen and B. Vallée (eds.), *The LLL Algorithm*, Information Security and Cryptography, DOI 10.1007/978-3-642-02295-1_5,

persons to do, at the end of 1982. This enabled him to reduce lattices of dimensions higher than 20.

The reachable dimension was then significantly increased by the use of a Cray-1, which helped solving low-density knapsacks [34] and disproving the famous Mertens conjecture [46]. The use of the floating-point arithmetic in those implementations was heuristic, as no precautions were taken (and no theoretical methods were available) to assure correctness of the computation when dealing with approximations to the numbers used within the LLL algorithm.

In the 1990's, it progressively became important to reduce lattice bases of higher dimensions. Lattice reduction gained tremendous popularity in the field of public-key cryptography, thanks to the birth of lattice-based cryptosystems [5, 21, 27], and to the methods of Coppersmith to find small roots of polynomials [14–16], a very powerful tool in public-key cryptanalysis (see the survey [38] contained in this book). Lattice-based cryptography involves lattice bases of huge dimensions (502 for the first NTRU challenge and several hundreds in the case of the GGH challenges), and in Coppersmith's method lattices of dimensions between 40 and 100 are quite frequent [10,39]. Nowadays, very large dimensional lattices are also being reduced in computational group theory or for factoring univariate polynomials.

All competitive implementations of the LLL algorithm rely on floating-point arithmetic. Sometimes, however, one wants to be sure of the quality of the output, to obtain mathematical results, e.g., to prove there is no small linear integer relation between given numbers. More often, one wants to be sure that the started execution will terminate. This motivates the study of the reliability of floating-point operations within LLL, which is the line of research we are going to survey below. In 1988, Schnorr described the first provable floating-point variant of the LLL algorithm [50]. This was followed by a series of heuristic and provable variants [33, 42, 51, 53]. Apart from these references, most of the practical issues we are to describe are derived from the codes of today's fastest floating-point LLL routines: LiDIa's [1], Magma's [11], NTL's [59], as well as in `fplll-2.0` [12].

The practical behaviour of the LLL algorithm is often considered as mysterious. Though many natural questions concerning the average behaviour of LLL remain open in growing dimensions (see the survey [62]), a few important properties can be obtained by experimentation. It appears then that there is a general shape for bases output by LLL, that the running-time behaves predictively for some families of inputs, and that there exists a generic numerical behaviour. We explain these phenomena and describe how some of the observed properties may be used to improve the code: for example, guessing accurately the causes of an undesirable behaviour that occurs during the execution of a heuristic variant helps selecting another variant which is more likely to work.

Road-Map of the Survey

In Section "Background Definitions and Results", we give some necessary background on LLL and floating-point arithmetic. We then describe the provable floating-point algorithms in Section "The Provable Floating-Point LLL Algorithms", as well as the heuristic practice-oriented ones in Section "Heuristic Variants and Implementations of the Floating-Point LLL". In Section "Practical Observations on LLL", we report observations on the practical behaviour of LLL, and finally, in Section "Open Problems", we draw a list of open problems and ongoing research topics related to floating-point arithmetic within the LLL algorithm.

Model of Computation

In the paper, we consider the usual bit-complexity model. For the integer and arbitrary precision floating-point arithmetics, unless stated otherwise, we restrict ourselves to naive algorithms, i.e., we do not use any fast multiplication algorithm [20]. This choice is motivated by two main reasons. First, the integer arithmetic operations dominating the overall cost of the described floating-point LLL algorithms are multiplications of large integers by small integers (most of the time, linear in the lattice dimension): using fast multiplication algorithms here is meaningless in practice, since the lattice dimensions remain far below the efficiency threshold between naive and fast integer multiplications. Second, we probably do not know, yet, how to fully exploit fast multiplication algorithms in LLL-type algorithms: having a quadratic cost with naive integer arithmetic suggests that a quasi-linear cost with fast integer arithmetic may be reachable (see Section "Open Problems" for more details).

Other LLL-Reduction Algorithms

Some LLL-type algorithms have lower complexity upper bounds than the ones described below, with respect to the lattice dimension [32,33,51,55,60]. However, their complexity upper bounds are worse than the ones below with respect to the bit-sizes of the input matrix entries. Improving the linear algebra cost and the arithmetic cost can be thought of as independent strategies to speed up lattice reduction algorithms. Ideally, one would like to be able to combine these improvements into one single algorithm. Improving the linear algebra cost of LLL is not the scope of the present survey, and for this topic, we refer to [48].

Notation

During the survey, vectors will be denoted in bold. If $\mathbf{b}$ is an n-dimensional vector, we denote its i-th coordinate by $\mathbf{b}[i]$, for $i \leq n$. Its length $\sqrt{\sum_{i=1}^{n} \mathbf{b}[i]^2}$ is denoted by $\|\mathbf{b}\|$. If $\mathbf{b}_1$ and $\mathbf{b}_2$ are two n-dimensional vectors, their scalar product $\sum_{i=1}^{n} \mathbf{b}_1[i] \cdot \mathbf{b}_2[i]$ is denoted by $\langle \mathbf{b}_1, \mathbf{b}_2 \rangle$. If x is a real number, we define $\lfloor x \rceil$ as the closest integer to x (the even one if x is equally distant from two consecutive integers).

We use bars to denote approximations: for example $\bar{\mu}_{i,j}$ is an approximation to $\mu_{i,j}$. By default, the function log will be the base-2 logarithm.

Background Definitions and Results

An introduction to the geometry of numbers can be found in [37]. The algorithmic aspects of lattices are described in the present book, in particular, in the survey [48], and therefore, we only give the definitions and results that are specific to the use of floating-point arithmetic within LLL. In particular, we briefly describe floating-point arithmetic. We refer to the first chapters of [26] and [40] for more details.

Floating-Point Arithmetic

Floating-point arithmetic is the most frequent way to simulate real numbers in a computer. Contrary to a common belief, floating-point numbers and arithmetic operations on floating-point numbers are rigorously specified, and mathematical proofs, most often in the shape of error analysis, can be built upon these specifications.

The most common floating-point numbers are the binary double precision floating-point numbers (doubles for short). They are formally defined in the IEEE-754 standard [2]. The following definition is incomplete with respect to the IEEE-754 standard, but will suffice for our needs.

Definition 1. A double consists of 53 bits of mantissa m which are interpreted as a number in $\{1, 1+2^{-52}, 1+2\cdot 2^{-52}, \ldots, 2-2^{-52}\}$; a bit of sign s; and 11 bits of exponent e which are interpreted as an integer in $[-1022, 1023]$. The real number represented that way is $(-1)^s \cdot m \cdot 2^e$.

If x is a real number, we define $\diamond(x)$ as the closest double to x, choosing the one with an even mantissa in case there are two possibilities. Other rounding modes are defined in the standard, but here we will only use the rounding to nearest. We will implicitly extend the notation $\diamond(\cdot)$ to extensions of the double precision. The IEEE-754 standard also dictates how arithmetic operations must be performed on doubles. If $op \in \{+, -, \times, \div\}$, the result of $(a\ op\ b)$ where a and b are doubles is the double corresponding to the rounding of the real number $(a\ op\ b)$, i.e., $\diamond(a\ op\ b)$. Similarly, the result of $\sqrt{a}$ is $\diamond(\sqrt{a})$.

Doubles are very convenient because they are widely available, they are normalised, they are extremely efficient since most often implemented at the processor level, and they suffice in many applications. They nevertheless have two major limitations: the exponent is limited (only 11 bits) and the precision is limited (only 53 bits).

A classical way to work around the exponent limitation is to batch an integer (most often a 32-bit integer suffices) to each double, in order to extend the exponent.

For example, the pair (x, e), where x is a double and e is an integer, could encode the number $x \cdot 2^e$. One must be careful that a given number may have several representations, because of the presence of two exponents (the one of the double and the additional one), and it may thus prove useful to restrict the range of the double to $[1, 2)$ or to any other binade. An implementation of such a double plus exponent arithmetic is the `dpe`[1] library written by Patrick Pélissier and Paul Zimmermann, which satisfies specifications similar to the IEEE-754 standard. We will use the term dpe to denote this extension of the standard doubles.

If for some application a larger precision is needed, one may use arbitrary precision real numbers. In this case, a number comes along with its precision, which may vary. It is usually implemented from an arbitrary precision integer package. An example is MPFR [47], which is based on GNU MP [23]. It is a smooth extension of the IEEE-754 standardised doubles. Another such implementation is the `RR`-class of Shoup's Number Theory Library [59], which can be based, at compilation time, either on GNU MP or on NTL's arbitrary precision integers. Arbitrary precision floating-point numbers are semantically very convenient, but one should try to limit their use in practice, since they are significantly slower than the processor-based doubles: even if the precision is chosen to be 53 bits, the speed ratio for the basic arithmetic operations can be larger than 15.

Lattices

A lattice L is a discrete subgroup of some $\mathbb{R}^n$. Such an object can always be represented as the set of integer linear combinations of some vectors $\mathbf{b}_1, \ldots, \mathbf{b}_d \in \mathbb{R}^n$ with $d \leq n$. If these vectors are linearly independent, we say that they form a basis of the lattice L. A given lattice may have an infinity of bases, related to one another by unimodular transforms, i.e., by multiplying on the right the column expressions of the basis vectors by a square integral matrix of determinant ± 1. The cardinalities d of the bases of a given lattice match and are called the lattice dimension, whereas n is called the embedding dimension. Both are lattice invariants: they depend on the lattice but not on the chosen basis of the lattice. There are two other important lattice invariants: the volume $\mathrm{vol}(L) := \sqrt{\det(B^t \cdot B)}$ where B is the matrix whose columns are any basis of L, and the minimum $\lambda(L)$ which is the length of a shortest nonzero lattice vector.

Gram–Schmidt Orthogonalisation

Let $\mathbf{b}_1, \ldots, \mathbf{b}_d$ be linearly independent vectors. Their *Gram–Schmidt orthogonalisation* (GSO for short) $\mathbf{b}_1^*, \ldots, \mathbf{b}_d^*$ is the orthogonal family defined recursively as follows: the vector $\mathbf{b}_i^*$ is the component of the vector $\mathbf{b}_i$ which is orthogonal to

[1] http://www.loria.fr/~zimmerma/free/dpe-1.4.tar.gz

the linear span of the vectors $\mathbf{b}_1, \ldots, \mathbf{b}_{i-1}$. We have $\mathbf{b}_i^* = \mathbf{b}_i - \sum_{j=1}^{i-1} \mu_{i,j} \mathbf{b}_j^*$ where $\mu_{i,j} = \frac{\langle \mathbf{b}_i, \mathbf{b}_j^* \rangle}{\|\mathbf{b}_j^*\|^2}$. For $i \leq d$ we let $\mu_{i,i} = 1$. The quantity $\mu_{i,j}$ is the component of the vector $\mathbf{b}_i$ on the vector $\mathbf{b}_j^*$ when written as a linear combination of the $\mathbf{b}_k^*$'s. The Gram–Schmidt orthogonalisation is widely used in lattice reduction because a reduced basis is somehow close to being orthogonal, which can be rephrased conveniently in terms of the GSO coefficients: the $\|\mathbf{b}_i^*\|$'s must not decrease too fast, and the $\mu_{i,j}$'s must be relatively small. Another interesting property of the GSO is that the volume of the lattice L spanned by the $\mathbf{b}_i$'s satisfies $\mathrm{vol}(L) = \prod_{i \leq d} \|\mathbf{b}_i^*\|$.

Notice that the GSO family depends on the order of the vectors. Furthermore, if the $\mathbf{b}_i$'s are integer vectors, the $\mathbf{b}_i^*$'s and the $\mu_{i,j}$'s are rational numbers. We also define the variables $r_{i,j}$ for $i \geq j$ as follows: for any $i \in [1, d]$, we let $r_{i,i} = \|\mathbf{b}_i^*\|^2$, and for any $i \geq j$, we let $r_{i,j} = \mu_{i,j} r_{j,j} = \langle \mathbf{b}_i, \mathbf{b}_j^* \rangle$. We have the relation $r_{i,j} = \langle \mathbf{b}_i, \mathbf{b}_j \rangle - \sum_{k<j} r_{i,k} \mu_{j,k}$, for any $i \geq j$. In what follows, the *GSO family* denotes the $r_{i,j}$'s and $\mu_{i,j}$'s. Some information is redundant in rational arithmetic, but in the context of our floating-point calculations, it is useful to have all these variables.

QR and Cholesky Factorisations

The GSO coefficients are closely related to the Q and R factors of the QR-factorisation of the basis matrix. Suppose that the linearly independent vectors $\mathbf{b}_1, \ldots, \mathbf{b}_d$ are given by the columns of an $n \times d$ matrix B, then one can write $B = Q \cdot \begin{bmatrix} R \\ 0 \end{bmatrix}$, where Q is an $n \times n$ orthogonal matrix and R is a $d \times d$ upper triangular matrix with positive diagonal entries. The first d columns of Q and the matrix R are unique and one has the following relations with the GSO family:

- For $i \leq d$, the i-th column Q_i of the matrix Q is the vector $\frac{1}{\|\mathbf{b}_i^*\|} \mathbf{b}_i^*$.
- The diagonal coefficient $R_{i,i}$ is $\|\mathbf{b}_i^*\|$.
- If $i < j$, the coefficient $R_{i,j}$ is $\frac{\langle \mathbf{b}_j, \mathbf{b}_i^* \rangle}{\|\mathbf{b}_i^*\|} = \frac{r_{j,i}}{\sqrt{r_{i,i}}}$.

In the rest of the survey, in order to avoid any confusion between the matrices $(R_{i,j})_{i \geq j}$ and $(r_{i,j})_{i \leq j}$, we will only use the $r_{i,j}$'s.

The Cholesky factorisation applies to a symmetric definite positive matrix. If A is such a matrix, its Cholesky factorisation is $A = R^t \cdot R$, where R is upper triangular with positive diagonal entries. Suppose now that A is the Gram matrix $B^t \cdot B$ of a basis matrix B, then the R-matrix of the Cholesky factorisation of A is exactly the R-factor of the QR-factorisation of B. The QR and Cholesky factorisations have been extensively studied in numerical analysis, and we refer to [26] for a general overview.

Size-reduction

A basis $(\mathbf{b}_1, \ldots, \mathbf{b}_d)$ is called *size-reduced* with factor $\eta \geq 1/2$ if its GSO family satisfies $|\mu_{i,j}| \leq \eta$ for all $1 \leq j < i \leq d$. The i-th vector $\mathbf{b}_i$ is *size-reduced* if $|\mu_{i,j}| \leq \eta$ for all $j \in [1, i-1]$. Size-reduction usually refers to $\eta = 1/2$, but it is essential for the floating-point LLLs to allow one to take at least slightly larger factors η, since the $\mu_{i,j}$'s will be known only approximately.

The Lenstra-Lenstra-Lovász Reduction

A basis $(\mathbf{b}_1, \ldots, \mathbf{b}_d)$ is called *LLL-reduced* with factor (δ, η), where $\delta \in (1/4, 1]$ and $\eta \in [1/2, \sqrt{\delta})$, if the basis is size-reduced with factor η and if its GSO satisfies the $(d-1)$ following conditions, often called Lovász conditions:

$$\delta \cdot \|\mathbf{b}^*_{\kappa-1}\|^2 \leq \|\mathbf{b}^*_\kappa + \mu_{\kappa,\kappa-1}\mathbf{b}^*_{\kappa-1}\|^2,$$

or equivalently $\left(\delta - \mu^2_{\kappa,\kappa-1}\right) \cdot r_{\kappa-1,\kappa-1} \leq r_{\kappa,\kappa}$. This implies that the norms of the GSO vectors $\mathbf{b}^*_1, \ldots, \mathbf{b}^*_d$ never drop too much: intuitively, the vectors are not far from being orthogonal. Such bases have useful properties. In particular, their first vector is relatively short. Theorem 1 is an adaptation of [35, (1.8) and (1.9)].

Theorem 1. *Let $\delta \in (1/4, 1]$ and $\eta \in [1/2, \sqrt{\delta})$. Let $(\mathbf{b}_1, \ldots, \mathbf{b}_d)$ be a (δ, η)-LLL-reduced basis of a lattice L. Then:*

$$\|\mathbf{b}_1\| \leq \left(\frac{1}{\delta - \eta^2}\right)^{\frac{d-1}{4}} \cdot \mathrm{vol}(L)^{\frac{1}{d}},$$

$$\prod_{i=1}^{d} \|\mathbf{b}_i\| \leq \left(\frac{1}{\delta - \eta^2}\right)^{\frac{d(d-1)}{4}} \cdot \mathrm{vol}(L).$$

LLL-reduction classically refers to the factor pair $(3/4, 1/2)$ initially chosen in [35], in which case the quantity $\frac{1}{\delta - \eta^2}$ is conveniently equal to 2. But the closer δ and η, respectively, to 1 and $1/2$, the smaller the upper bounds in Theorem 1. In practice, one often selects $\delta \approx 1$ and $\eta \approx 1/2$, so that we almost have $\|\mathbf{b}_1\| \leq (4/3)^{\frac{d-1}{4}} \cdot \mathrm{vol}(L)^{\frac{1}{d}}$. It also happens that one selects weaker factors in order to speed up the execution of the algorithm (we discuss this strategy in Section "A Thoughtful Wrapper").

The LLL Algorithm

We give in Fig. 5.1 a description of LLL that we will use to explain its floating-point variants. The LLL algorithm obtains, in polynomial time, a (δ, η)-reduced basis,

Input: A basis $(\mathbf{b}_1, \ldots, \mathbf{b}_d)$ and a valid pair of factors (δ, η).
Output: A (δ, η)-LLL-reduced basis of $L[\mathbf{b}_1, \ldots, \mathbf{b}_d]$.

1. $r_{1,1}{:=}\|\mathbf{b}_1\|^2$, $\kappa{:=}2$. While $\kappa \leq d$ do
2. η-size-reduce $\mathbf{b}_\kappa$:
3. Compute $\mu_{\kappa,1}, \ldots, \mu_{\kappa,\kappa-1}$ and $r_{\kappa,\kappa}$, using the previous GSO coefficients.
4. For $i = \kappa - 1$ down to 1 do, if $|\mu_{\kappa,i}| > \eta$:
5. $\mathbf{b}_\kappa{:=}\mathbf{b}_\kappa - \lfloor \mu_{\kappa,i} \rceil \mathbf{b}_i$, update $\mu_{\kappa,1}, \ldots, \mu_{\kappa,i}$ accordingly.
6. If $(\delta - \mu_{\kappa,\kappa-1}^2) \cdot r_{\kappa-1,\kappa-1} \leq r_{\kappa,\kappa}$ then $\kappa{:=}\kappa + 1$.
7. Else swap $\mathbf{b}_\kappa$ and $\mathbf{b}_{\kappa-1}$ and set $\kappa{:=}\max(2, \kappa - 1)$.
8. Return $(\mathbf{b}_1, \ldots, \mathbf{b}_d)$.

Fig. 5.1 The LLL algorithm

even if one chooses $\eta = 1/2$. The factor $\delta < 1$ can be chosen arbitrarily close to 1. It is unknown whether polynomial time complexity can be achieved or not for $\delta = 1$ (partial results can be found in [6] and [36]).

The floating-point LLL algorithms do not achieve $\eta = 1/2$, because the GSO coefficients are known only approximately. Choosing $\eta = 1/2$ in these algorithms may make them loop forever. Similarly, one has to relax the LLL factor δ, but this relaxation only adds up with the already necessary relaxation of δ in the classical LLL algorithm. The LLL factor η can be chosen arbitrarily close to $1/2$ in the provable floating-point L^2 algorithm of Nguyen and Stehlé [42] (to be described in Section "The Provable Floating-Point LLL Algorithms") which terminates in quadratic time (without fast integer multiplication) with respect to the bit-size of the matrix entries. Finally, $\eta = 1/2$ can also be achieved within the same complexity: first, run the L^2 algorithm on the given input basis with $\delta' = \delta + 2(\eta - 1/2)$ and a factor $\eta > 1/2$; and second, run the LLL algorithm on the output basis. One can notice that the second reduction is simply a size-reduction and can be performed in the prescribed time.

Remarkable Variables in the LLL Algorithm

The LLL index $\kappa(t)$ denotes the vector under investigation at the t-th loop iteration of the algorithm. Its initial value is 2 and at the end of the execution, one has $\kappa(\tau + 1) = d + 1$, where τ is the number of loop iterations and $\kappa(\tau + 1)$ is the value of κ at the end of the last loop iteration. We will also use the index $\alpha(t)$ (introduced in [42]), which we define below and illustrate in Fig. 5.2. It is essentially the smallest swapping index since the last time the index κ was at least $\kappa(t)$ (this last time is rigorously defined below as $\phi(t)$).

Definition 2. Let t be a loop iteration. Let $\phi(t) = \max(t' < t, \kappa(t') \geq \kappa(t))$ if it exists and 1 otherwise, and let $\alpha(t) = \min(\kappa(t'), t' \in [\phi(t), t-1]) - 1$.

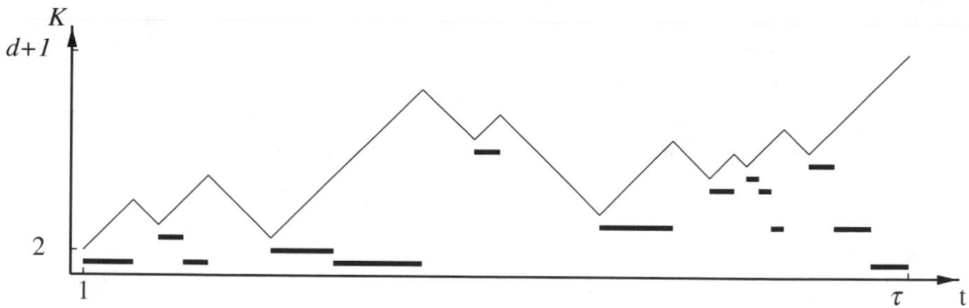

Fig. 5.2 A possible curve for $\kappa(t)$ (*thin continuous line*), with the corresponding curve for $\alpha(t)$ (*thick line* when κ increases, and same as κ otherwise)

The index $\alpha(t)$ has the remarkable property that between the loop iterations $\phi(t)$ and t, the vectors $\mathbf{b}_1, \ldots, \mathbf{b}_{\alpha(t)}$ remain unchanged: because κ remains larger than $\alpha(t)$, these first vectors are not swapped nor size-reduced between these iterations.

The Provable Floating-Point LLL Algorithms

When floating-point calculations are mentioned in the context of the LLL algorithm, this systematically refers to the underlying Gram–Schmidt orthogonalisation. The transformations on the basis and the basis itself remain exact, because one wants to preserve the lattice while reducing it. The LLL algorithm heavily relies on the GSO. For example, the LLL output conditions involve all the quantities $\mu_{i,j}$ for $j < i \leq d$ and $\|\mathbf{b}_i^*\|^2$ for $i \leq d$. The floating-point arithmetic is used on these GSO quantities $\mu_{i,j}$ and $\|\mathbf{b}_i^*\|^2$.

In this section, we are to describe three ways of implementing this idea: the first way is the most natural solution, but fails for different reasons, that we emphasise because they give some intuition about the provable variants; the second one, due to Schnorr [50], is provable but suffers from a number of practical drawbacks; and the last one, due to Nguyen and Stehlé [42], was introduced recently and seems more tractable in practice.

The following table summarises the complexities of two rational LLL algorithms and the two provable floating-point LLLs described in this section. The second line contains the required precisions, whereas the last line consists of the best known complexity upper bounds. The variant of Kaltofen [28] differs only slightly from LLL. The main improvement of the latter is to analyse more tightly the cost of the size-reductions, providing a complexity bound of total degree 8 instead of 9. This bound also holds for the LLL algorithm. On the floating-point side, both Schnorr's algorithm and L^2 have complexities of total degree 7, but the complexity bound of L^2 is always better and is quadratic with respect to $\log B$, the bit-size of the input matrix entries.

LLL [35]	Kaltofen [28]	Schnorr [50]	L^2 [42]
$O(d \log B)$	$O(d \log B)$	$\geq 12d + 7\log_2 B$	$d \log_2 3 \approx 1.58d$
$O(d^5 n \log^3 B)$	$O(d^4 n(d + \log B) \log^2 B)$	$O(d^3 n(d + \log B)^2 \log B)$	$O(d^4 n \log B(d + \log B))$

Fig. 5.3 Complexity bounds of the original LLL and the provable floating-point LLL algorithms

A First Attempt

A natural attempt to define a floating-point LLL algorithm is as follows: one keeps the general structure of LLL as described in Fig. 5.1, and computes approximations to the GSO quantities, by converting into floating-point arithmetic the formulas that define them (as given in Section "Lattices"). The scalar product $\langle \mathbf{b}_i, \mathbf{b}_j \rangle$ is approximated by the quantity $\bar{g}_{i,j}$, computed as follows:

$$\bar{g}_{i,j} := 0. \texttt{ For } k \texttt{ from } 1 \texttt{ to } n, \texttt{ do}$$
$$\bar{g}_{i,j} := \diamond\left(\bar{g}_{i,j} + \diamond\left(\diamond(\mathbf{b}_i[k]) \cdot \diamond(\mathbf{b}_j[k])\right)\right).$$

Similarly, the quantities $r_{i,j}$ and $\mu_{i,j}$ are approximated respectively by the $\bar{r}_{i,j}$'s and $\bar{\mu}_{i,j}$'s, computed as follows:

$$\bar{r}_{i,j} := \bar{g}_{i,j}. \texttt{ For } k \texttt{ from } 1 \texttt{ to } j-1, \texttt{ do } \bar{r}_{i,j} := \diamond\left(\bar{r}_{i,j} - \diamond\left(\bar{r}_{i,k} \cdot \bar{\mu}_{j,k}\right)\right).$$
$$\bar{\mu}_{i,j} := \diamond\left(\bar{r}_{i,j} / \bar{r}_{j,j}\right).$$

As a first consequence, since the $\mu_{i,j}$'s are known only approximately, one cannot ensure ideal size-reduction anymore. One has to relax the condition $|\mu_{i,j}| \leq 1/2$ into the condition $|\mu_{i,j}| \leq \eta$ for some $\eta > 1/2$ that takes into account the inaccuracy of the $\bar{\mu}_{i,j}$'s.

This first attempt suffers from three major drawbacks. First, the scalar products can be miscalculated. More precisely, the quantity $\bar{g}_{i,j}$ is a sum of floating-point numbers and the classical phenomena of cancellation and loss of precision can occur. We do not have better than the following error bound:

$$\left|\bar{g}_{i,j} - \langle \mathbf{b}_i, \mathbf{b}_j \rangle\right| \leq f(n, \ell) \cdot \sum_{k \leq n} |\mathbf{b}_i[k]| \cdot \left|\mathbf{b}_j[k]\right|,$$

where the function f depends on the precision ℓ and the number n of elements to sum. Unfortunately, such a summation prevents us from getting absolute error bounds on the $\mu_{i,j}$'s. In order to obtain an absolute error bound on $\mu_{i,j} = \frac{\langle \mathbf{b}_i, \mathbf{b}_j \rangle}{r_{i,i}}$, one would like an error on $\langle \mathbf{b}_i, \mathbf{b}_j \rangle$ which is no more than proportional to $r_{i,i} = \|\mathbf{b}_i^*\|^2$. We illustrate this with an example in dimension 2 and double precision. Consider the columns of the following matrix:

$$\begin{bmatrix} 1 & 2^{100} + 2^{40} \\ -1 & 2^{100} - 2^{40} \end{bmatrix}.$$

Here, $r_{1,1} = 2$ and we would like the error on $\langle \mathbf{b}_2, \mathbf{b}_1 \rangle$ to be small compared to that quantity. If the scalar product of the two vectors is computed by first rounding the matrix entries to double precision, then it is estimated to 0. This implies that the computed $\bar{\mu}_{2,1}$ is 0, and the basis is deemed LLL-reduced. But it is not, since, in fact, $\mu_{2,1} = 2^{40}$, which contradicts the size-reduction condition. If one changes the values 2^{40} by any values below 2^{100}, one sees that the less significant bits are simply ignored though they may still be contributing significantly to $\mu_{2,1}$. In order to test the size-reduction conditions from the basis matrix, it seems necessary to use a precision which is at least as large as the bit-length of the input matrix entries, which may be very expensive.

Second, the precision may not be sufficient to perform the size-reductions completely. It can easily be illustrated by an example. Consider the following lattice basis:

$$\begin{bmatrix} 1 & 2^{54}+1 \\ 0 & 1 \end{bmatrix}.$$

The algorithm will compute $\bar{\mu}_{2,1} = 2^{54}$: the bit-length of the true quantity is too large to be stored in a double precision floating-point number. Then it will try to size-reduce the second vector by performing the operation $\mathbf{b}_2 := \mathbf{b}_2 - 2^{54}\mathbf{b}_1 = (1,1)^t$. It will then check that Lovász's condition is satisfied, and terminate. Unfortunately, the output basis is still not size-reduced, because $\mu_{2,1} = 1$. One can change the example to make $\mu_{2,1}$ as large as desired. The trouble here is that the mantissa size is too small to handle the size-reduction. Either more precision or a reparation routine seems necessary. Such a reparation process will be described in Section "The L^2 Algorithm".

The third weakness of the first attempt is the degradation of precision while computing the GSO coefficients. Indeed, a given $\bar{r}_{i,j}$ is computed from previously computed and already erroneous quantities $\bar{r}_{i,k}$ and $\bar{\mu}_{j,k}$, for $k \leq j$. The floating-point errors not only add up, but also get amplified. No method to prevent this amplification is known, but it is known how to bound and work around the phenomenon: such techniques come from the field of numerical analysis. In particular, it seems essential for the good numerical behaviour of the LLL algorithm to always consider a vector $\mathbf{b}_k$, such that all previous vectors $\mathbf{b}_i$ for $i < k$ are LLL-reduced: this means that when one is computing the orthogonalisation of a vector with respect to previous vectors, the latter are always LLL-reduced and therefore fairly orthogonal, which is good for the numerical behaviour. The structure of the LLL algorithm as described in Fig. 5.1 guarantees this property.

Schnorr's Algorithm

Schnorr [50] described the first provable variant of the LLL algorithm relying on floating-point arithmetic. Instead of using GSO coefficients represented as rational numbers of bit-lengths $O(d \log B)$, Schnorr's algorithm approximates them by arbitrary precision floating-point numbers, of mantissa size $\ell = O(d + \log B)$.

Input: A basis $(\mathbf{b}_1, \ldots, \mathbf{b}_d)$, a precision ℓ.
Output: A $(0.95, 0.55)$-reduced basis of the lattice spanned by the $\mathbf{b}_i$'s.

1. $\kappa := 2$, $\mathbf{b}_1^* := \mathbf{b}_1$.
2. $r := \lfloor \langle \mathbf{b}_\kappa, \bar{\mathbf{b}}^*_{\kappa-1} \rangle / \|\bar{\mathbf{b}}^*_{\kappa-1}\|^2 \rceil$, $\mathbf{b}_\kappa := \mathbf{b}_\kappa - r\mathbf{b}_{\kappa-1}$.
3. If $(\|\mathbf{b}_\kappa\|^2 - \sum_{j<\kappa-1} \langle \mathbf{b}_\kappa, \bar{\mathbf{b}}^*_j \rangle / \|\bar{\mathbf{b}}^*_j\|^2) \cdot 1.025 \geq \|\bar{\mathbf{b}}^*_{\kappa-1}\|^2$, go to Step 5. Otherwise:
4. Exchange $\mathbf{b}_\kappa$ and $\mathbf{b}_{\kappa-1}$, $\kappa := \max(2, \kappa - 1)$. Update $\overline{G_\kappa^{-1}}$ and go to Step 2.
5. For j from $\kappa - 2$ down to 1, do $r := \lfloor \langle \mathbf{b}_\kappa, \bar{\mathbf{b}}^*_j \rangle / \|\bar{\mathbf{b}}^*_j\|^2 \rceil$, $\mathbf{b}_\kappa := \mathbf{b}_\kappa - r\mathbf{b}_j$.
6. Compute a first approximation of $\nu_{\kappa,1}, \ldots, \nu_{\kappa,\kappa-1}, \mathbf{b}^*_\kappa, G_\kappa^{-1}$, from the $\mathbf{b}_i$'s and the matrix $\overline{G_{\kappa-1}^{-1}}$.
7. Use a finite number of iterations of Schulz's method on $\overline{G_k^{-1}}$ using G_k. This helps improving the approximations of $\nu_{\kappa,1}, \ldots, \nu_{\kappa,\kappa-1}, \mathbf{b}^*_\kappa$ and G_κ^{-1}.
8. Truncate the $\bar{\nu}_{\kappa,i}$'s to ℓ bits after the point. Compute the corresponding vectors $\bar{\mathbf{b}}^*_\kappa$ and $\overline{G_\kappa^{-1}}$.
9. $\kappa := \kappa + 1$. If $\kappa \leq n$, go to Step 2.

Fig. 5.4 Schnorr's algorithm

This provides a gain of 2 in the total degree of the polynomial complexity of LLL: from $O(d^5 n \log^3 B)$ to $O(d^3 n(d + \log B)^2 \log B)$.

In Fig. 5.4, we give a description of this algorithm. It uses exact integer operations on the basis vectors (Steps 2 and 5), and approximate operations on the inverses of the partial Gram matrices $G_k = (\langle \mathbf{b}_i, \mathbf{b}_j \rangle)_{i,j \leq k}$ and the inverse $(\nu_{i,j})_{i,j \leq d}$ of the lower triangular matrix made of the $\mu_{i,j}$'s. These operations are not standard floating-point operations, since most of them are, in fact, exact operations on approximate values: in floating-point arithmetic, the result of a basic arithmetic operation is a floating-point number closest to the true result, whereas here the true result is kept, without any rounding. This is the case everywhere, except at Step 8, where the quantities are truncated in order to avoid a length blow-up. The truncation itself is similar to fixed-point arithmetic since it keeps a given number of bits after the point instead of keeping a given number of most significant bits. It can be checked that all quantities computed never have more than $c \cdot \ell$ bits after the point, for some small constant c depending on the chosen number of iterations at Step 7. At Step 7, a few steps of Schulz's iteration are performed. Schulz's iteration [57] is a classical way to improve the accuracy of an approximate inverse (here $\overline{G_k^{-1}}$) of a known matrix (here G_k). This is a matrix generalisation of Newton's iteration for computing the inverse of a real number.

Schnorr proved that by taking a precision $\ell = c_1 \cdot d + c_2 \cdot \log B$ for some explicitly computable constants c_1 and c_2, the algorithm terminates and returns a $(0.95, 0.55)$-LLL-reduced basis. The constants 0.95 and 0.55 can be chosen arbitrarily close, but different to respectively 1 and 0.5, by changing the constants c_1 and c_2 as well as the constant 1.025 from Step 3. Finally, it can be checked that the bit-cost of the algorithm is $O\left(d^3 n(d + \log B)^2 \log B\right)$.

This algorithm prevents the three problems of the naive floating-point LLL from occurring: the inaccuracy of the scalar products is avoided because they are always computed exactly; the incomplete size-reductions cannot occur because the precision is set large enough to guarantee that any size-reduction is performed correctly and fully at once; the accumulation of inaccuracies is restrained because most of the operations performed on approximations are done exactly, so that few errors may add up, and the amplification of the errors (due to a bad conditioning of the problem) is compensated by the large precision.

Schnorr's algorithm gives the first answer to the question of using approximate GSO quantities within the LLL algorithm, but:

- The constants c_1 and c_2, on which the precision depends, may be large. What is most annoying is that the precision actually depends on $\log B$. This means that the approximate operations on the GSO still dominate the integer operations on the basis matrix.
- As explained above, it is not using standard floating-point arithmetic, but rather a mix between exact computations on approximate values and arbitrary precision fixed-point arithmetic.

The L^2 Algorithm

The L^2 algorithm was introduced by Nguyen and Stehlé [42] in 2005. It is described in Fig. 5.5. L^2 is a variant of the LLL algorithm relying on arbitrary precision floating-point arithmetic for the underlying Gram–Schmidt orthogonalisation, in a provable way. Apart from giving a sound basis for floating-point calculations within LLL, it is also the sole variant of LLL that has been proven to admit a quadratic bit-complexity with respect to the bit-size of the input matrix entries. This latter property is very convenient since LLL can be seen as a multi-dimensional generalisation of Euclid's gcd algorithm, Gauss' two-dimensional lattice reduction algorithm and the three and four dimensional greedy algorithm of Semaev et al. [41, 58], all of which admit quadratic complexity bounds. This property, from which the name of the algorithm comes, arguably makes it a natural variant of LLL.

In L^2, the problem of scalar product cancellations is handled very simply, since all the scalar products are known exactly during the whole execution of the algorithm. Indeed, the Gram matrix of the initial basis matrix is computed at the beginning of the algorithm and updated for each change of the basis vectors. In fact, the algorithm operates on the Gram matrix and the computed transformations are forwarded to the basis matrix. It can be seen that this can be done with only a constant factor overhead in the overall complexity. Second, the size-reduction procedure is modified into a lazy size-reduction. One size-reduces as much as possible, given the current knowledge of the Gram–Schmidt orthogonalisation, then recomputes the corresponding Gram–Schmidt coefficients from the exact Gram matrix, and restarts the lazy size-reduction until the vector under question stays the same. When this happens, the vector is size-reduced and the corresponding Gram–Schmidt coefficients are well approximated. This lazy size-reduction was already contained inside NTL's LLL, and described in [33], in the context of a heuristic

Input: A valid pair (δ, η) with $\eta > 1/2$, a basis $(\mathbf{b}_1, \ldots, \mathbf{b}_d)$ and a precision ℓ.
Output: A (δ, η)-LLL-reduced basis.

1. Compute exactly $G = G(\mathbf{b}_1, \ldots, \mathbf{b}_d)$, $\eta^- := \frac{\eta + 1/2}{2}$, $\delta^+ := \frac{\delta+1}{2}$.
2. $\bar{r}_{1,1} := \diamond\left(\langle \mathbf{b}_1, \mathbf{b}_1 \rangle\right)$, $\kappa := 2$. While $\kappa \leq d$, do
3. $\quad\eta$-size-reduce $\mathbf{b}_\kappa$:
4. $\quad$Compute the $\bar{r}_{\kappa,j}$'s and $\bar{\mu}_{\kappa,j}$'s from G and the previous $\bar{r}_{i,j}$'s and $\bar{\mu}_{i,j}$'s.
5. $\quad$If $\max_{i<\kappa} |\bar{\mu}_{\kappa,i}| > \eta^-$, then, for $i = \kappa - 1$ down to 1, do:
6. $\quad\quad X := \lfloor \bar{\mu}_{\kappa,i} \rceil$, $\mathbf{b}_\kappa := \mathbf{b}_\kappa - X \cdot \mathbf{b}_i$ and update $G(\mathbf{b}_1, \ldots, \mathbf{b}_d)$ accordingly.
7. $\quad\quad$For $j = 1$ to $i - 1$, $\bar{\mu}_{\kappa,j} := \diamond\left(\bar{\mu}_{\kappa,j} - \diamond(X \cdot \bar{\mu}_{i,j})\right)$.
8. $\quad\quad$Go to Step 4.
9. $\quad$If $\delta^+ \cdot \bar{r}_{\kappa-1,\kappa-1} < \bar{r}_{\kappa,\kappa} + \bar{\mu}^2_{\kappa,\kappa-1} \bar{r}_{\kappa-1,\kappa-1}$, $\kappa := \kappa + 1$.
10. $\quad$Else, swap $\mathbf{b}_{\kappa-1}$ and $\mathbf{b}_\kappa$, update G, $\bar{\mu}$ and $\bar{r}$ accordingly, $\kappa := \max(2, \kappa - 1)$.
11. Return $(\mathbf{b}_1, \ldots, \mathbf{b}_d)$.

Fig. 5.5 The L^2 algorithm

floating-point LLL algorithm based on Householder transformations. In this context, fixing $\eta = 1/2$ can have dramatic consequences: apart from asking for something which is not reachable with floating-point computations, the lazy size-reduction (i.e., the inner loop between Steps 4 and 8 in Fig. 5.5) may loop forever. Finally, an a priori error analysis provides a bound on the loss of accuracy, which provides the provably sufficient precision.

Theorem 2 ([42, Theorem 1]). *Let* (δ, η) *such that* $1/4 < \delta < 1$ *and* $1/2 < \eta < \sqrt{\delta}$. *Let* $c = \log \frac{(1+\eta)^2 + \varepsilon}{\delta - \eta^2} + C$, *for some arbitrary* $\varepsilon \in (0, 1/2)$ *and* $C > 0$. *Given as input a* d*-dimensional lattice basis* $(\mathbf{b}_1, \ldots, \mathbf{b}_d)$ *in* $\mathbb{Z}^n$ *with* $\max_i \|\mathbf{b}_i\| \leq B$, *the* L^2 *algorithm of Fig. 5.5 with precision* $\ell = cd + o(d)$ *outputs a* (δ, η)*-LLL-reduced basis in time* $O\left(d^4 n(d + \log B) \log B\right)$. *More precisely, if* τ *denotes the number of iterations of the loop between Steps 3 and 10, then the running time is* $O\left(d^2 n(\tau + d \log(dB))(d + \log B)\right)$.

The precision $\ell = cd + o(d)$ can be made explicit from the correctness proof of [42]. It suffices that the following inequality holds, for some arbitrary $C > 0$:

$$d^2 \left(\frac{(1+\eta)^2 + \varepsilon}{\delta - \eta^2} \right)^d 2^{-\ell + 10 + Cd} \leq \min\left(\varepsilon, \eta - \frac{1}{2}, 1 - \delta \right).$$

Notice that with double precision (i.e., $\ell = 53$), the dimension up to which the above bound guarantees that L^2 will work correctly is very small. Nevertheless, the bound is likely to be loose: in the proof of [42], the asymptotically negligible components are chosen to simplify the error analysis. Obtaining tighter bounds for the particular case of the double precision would be interesting in practice for small dimensions. For larger dimensions, the non-dominating components become meaningless. Asymptotically, in the case of LLL-factors (δ, η) that are close to $(1, 1/2)$, a floating-point precision $\ell = 1.6 \cdot d$ suffices.

Here is a sketch of the complexity analysis of the L^2 algorithm. We refer to [44] for more details.

1. There are $\tau = O(d^2 \log B)$ loop iterations.
2. In a given loop iteration, there can be up to $O\left(1 + \frac{\log B}{d}\right)$ iterations within the lazy size-reduction. However, most of the time there are only $O(1)$ such loop iterations. The lengthy size-reductions cannot occur often during a given execution of L^2, and are compensated by the other ones. In the rigorous complexity analysis, this is formalised by an amortised analysis (see below for more details). In practice, one can observe that there are usually two iterations within the lazy size-reduction: the first one makes the $|\mu_{\kappa,i}|$'s smaller than η and the second one recomputes the $\mu_{\kappa,i}$'s and $r_{\kappa,i}$'s with better accuracy. This is incorrect in full generality, especially when the initial $\mu_{\kappa,i}$'s are very large.
3. In each iteration of the lazy size-reduction, there are $O(dn)$ arithmetic operations.
4. Among these arithmetic operations, the most expensive ones are those related to the coefficients of the basis and Gram matrices: these are essentially multiplications between integers of lengths $O(\log B)$ and the computed X's, which can be represented on $O(d)$ bits.

The proof of the quadratic complexity bound generalises the complexity analysis of Euclid's gcd algorithm. In Euclid's algorithm, one computes the gcd of two integers $r_0 > r_1 > 0$, by performing successive euclidean divisions: $r_{i+1} = r_{i-1} - q_i r_i$, with $|r_{i+1}| < |r_i|$, until one gets 0. Standard arguments show that the number of divisions is $O(\log r_0)$. To obtain a quadratic complexity bound for Euclid's algorithm, one has to compute q_i by using only some (essentially $\log q_i \approx \log |r_{i-1}| - \log |r_i|$) of the most significant bits of r_{i-1} and r_i, to get r_{i+1} with a bit-complexity $O(\log r_0 \cdot (1 + \log |r_{i-1}| - \log |r_i|))$. It is crucial to consider this bound, instead of the weaker $O\left(\log^2 |r_{i-1}|\right)$, to be able to use an amortised cost analysis: the worst-case cost of a sequence of steps can be much lower than the sum of the worst cases of each step of the sequence. In the quadratic complexity bound of the L^2 algorithm, the Euclidean division becomes the lazy size-reduction and the term $O\left(\log r_0 \cdot (1 + \log |r_{i-1}| - \log |r_i|)\right)$ is replaced by the new term $O(\log B \cdot (d + \log \|\mathbf{b}_{\kappa(t)}\| - \log \|\mathbf{b}_{\alpha(t)}\|))$ for the t-th loop iteration: intuitively, the cost of the size-reduction does not depend on the $\alpha(t) - 1$ first vectors, since the vector $\mathbf{b}_{\kappa(t)}$ is already size-reduced with respect to them. In the analysis of Euclid's algorithm, terms cancel out as soon as two consecutive steps are considered, but in the case of L^2, one may need significantly more than two steps to observe a possible cancellation. The following lemma handles this difficulty.

Lemma 1 ([42, Lemma 2]). *Let $k \in [2, d]$ and $t_1 < \ldots < t_k$ be loop iterations of the L^2 algorithm such that for any $j \leq k$, we have $\kappa(t_j) = k$. For any loop iteration t and any $i \leq d$, we define $\mathbf{b}_i^{(t)}$ as the i-th basis vector at the beginning of the t-th loop iteration. Then there exists $j < k$ such that:*

$$d(\delta - \eta^2)^{-d} \cdot \left\|\mathbf{b}_{\alpha(t_j)}^{(t_j)}\right\| \geq \left\|\mathbf{b}_k^{(t_k)}\right\|.$$

This result means that when summing all the bounds of the costs of the successive loop iterations, i.e., $O\left(\log B \cdot (d + \log \|\mathbf{b}_{\kappa(t)}\| - \log \|\mathbf{b}_{\alpha(t)}\|)\right)$, some quasi-cancellations of the following form occur: a term $\log \|\mathbf{b}_{\kappa(t)}\|$ can be cancelled out with a term $\log \|\mathbf{b}_{\alpha(t')}\|$, where the relationship between t' and t is described in the lemma. This is not exactly a cancellation, since the difference of the two terms is replaced by $O(d)$ (which does not involve the size of the entries).

The proof of correctness of the L^2 algorithm relies on a forward error analysis of the Cholesky factorisation algorithm while applied to a Gram matrix of a basis whose first vectors are already LLL-reduced. We give here a sketch of the error analysis in the context of a fully LLL-reduced basis (i.e., the whole basis is LLL-reduced). This shows the origin of the term $\frac{(1+\eta)^2}{\delta-\eta^2}$ in Theorem 2.

We define $err_j = \max_{i \in [j,d]} \frac{|\bar{r}_{i,j} - r_{i,j}|}{r_{j,j}}$, i.e., the approximation error on the $r_{i,j}$'s relatively to $r_{j,j}$, and we bound its growth as j increases. We have:

$$err_1 = \max_{i \leq d} \frac{|\diamond \langle \mathbf{b}_i, \mathbf{b}_1 \rangle - \langle \mathbf{b}_i, \mathbf{b}_1 \rangle|}{\|\mathbf{b}_1\|^2} \leq 2^{-\ell} \cdot \max_{i \leq d} \frac{|\langle \mathbf{b}_i, \mathbf{b}_1 \rangle|}{\|\mathbf{b}_1\|^2} = 2^{-\ell} \cdot \max_{i \leq d} |\mu_{i,1}| \leq 2^{-\ell},$$

because of the size-reduction condition. We now choose $j \in [2, d]$. We have, for any $i \leq d$ and any $k < j$:

$$|\bar{\mu}_{i,k} - \mu_{i,k}| \lesssim \left| \frac{r_{k,k}}{\bar{r}_{k,k}} \right| err_k + |r_{i,k}| \left| \frac{1}{\bar{r}_{k,k}} - \frac{1}{r_{k,k}} \right| \lesssim (\eta + 1) \cdot err_k,$$

where we neglected low-order terms and used the fact that $|r_{i,k}| \leq \eta \cdot r_{k,k}$, which comes from the size-reduction condition. This implies that:

$$\begin{aligned} |\diamond(\bar{\mu}_{j,k} \cdot \bar{r}_{i,k}) - \mu_{j,k} r_{i,k}| &\lesssim |\bar{\mu}_{j,k} - \mu_{j,k}| \cdot |\bar{r}_{i,k}| + |\mu_{j,k}| \cdot |\bar{r}_{i,k} - r_{i,k}| \\ &\lesssim \eta(\eta + 2) \cdot err_k \cdot \|\mathbf{b}_k^*\|^2, \end{aligned}$$

where we also neglected low-order terms and used the size-reduction condition twice. Thus,

$$err_j \lesssim \eta(\eta + 2) \sum_{k<j} \frac{\|\mathbf{b}_k^*\|^2}{\|\mathbf{b}_j^*\|^2} err_k \lesssim \eta(\eta + 2) \sum_{k<j} (\delta - \eta^2)^{k-j} \cdot err_k,$$

by using the fact that Lovász's conditions are satisfied. This finally gives

$$err_j \lesssim \left(\frac{(1+\eta)^2}{\delta - \eta^2} \right)^j \cdot err_1.$$

Heuristic Variants and Implementations of the Floating-Point LLL

Floating-point arithmetic has been used in the LLL implementations since the early 1980's, but only very few articles describe how this should be done in order to balance efficiency and correctness. The reference for LLL implementers is the article by Schnorr and Euchner on practical lattice reduction [52, 53]. Until very recently, all the fastest LLL implementations were relying on it, including the one in Victor Shoup's NTL, Allan Steel's LLL in Magma, and LiDIA's LLL (written by Werner Backes, Thorsten Lauer, Oliver van Sprang and Susanne Wetzel). Magma's LLL is now relying on the L^2 algorithm. In this section, we describe the Schnorr-Euchner heuristic floating-point LLL, and explain how to turn the L^2 algorithm into an efficient and reliable code.

The Schnorr-Euchner Heuristic LLL

Schnorr-Euchner's floating-point LLL follows very closely the classical description of LLL. It mimics the rational LLL while trying to work around the three pitfalls of the naive strategy (see Section "The Provable Floating-Point LLL Algorithms"). Let us consider these three difficulties separately.

It detects cancellations occurring during the computation of scalar products (at Step 3 of the algorithm of Fig. 5.1), by comparing their computed approximations with the (approximate) product of the norms of the corresponding vectors. Since norms consist in summing positive values, no cancellation occurs while computing them approximately, and the computed values are, therefore, very reliable. If more than half the precision within the scalar product is likely to be lost (i.e., the ratio between the magnitude of the computed value and the product of the norms is smaller than $2^{-\ell/2}$ where ℓ is the precision), the scalar product is computed exactly (with the integer vectors and integer arithmetic) and then rounded to a closest double. As a consequence, not significantly more than half the precision can be lost while computing a scalar product. In NTL's LLL (which implements the Schnorr-Euchner variant), Victor Shoup replaced the 50% loss of precision test by a stronger requirement of not losing more than 15% of the precision.

Second, if some coefficient $\mu_{i,j}$ is detected to be large (between Steps 3 and 4 of the algorithm of Fig. 5.1), i.e., more than $2^{\ell/2}$ where ℓ is the precision, then another size-reduction will be executed after the current one. This prevents incomplete size-reductions from occurring.

Finally, the algorithm does not tackle the error amplification due to the Gram–Schmidt orthogonalisation process: one selects the double precision and hopes for the best.

Let us now discuss these heuristics. If the scalar products are detected to be cancelling frequently, they will often be computed exactly with integer arithmetic. In that situation, one should rather keep the Gram matrix and update it. On the

Number of unknown bits of p	220	230	240	245
Dimension of the lattice	17	22	34	50
NTL's LLL_XD	13.1	78.6	1180	13800
Time to compute the scalar products exactly	8.05	51.6	914	11000
Magma's LLL	8.47	44.5	710	10000

Fig. 5.6 Comparison between NTL's LLL_XD and Magma's LLL for lattice bases arising in Coppersmith's method applied to the problem of factoring with high bits known

theoretical side, the Schnorr-Euchner strategy for scalar products prevents one from getting a quadratic bit complexity. In practice, it may slow down the computation significantly. In particular, this occurs when two vectors have much different lengths and are nearly orthogonal. This may happen quite frequently in some applications of LLL, one of them being Coppersmith's method [14]. In the table of Fig. 5.6, we compare NTL's LLL_XD with Magma's LLL for input lattice bases that correspond to the use of Coppersmith's method for the problem of factoring with high bits known, such as described in [39], for a 1024 bit RSA modulus $p \cdot q$ and for different numbers of most significant bits of p known. The experiments were performed with NTL-5.4 and Magma-2.13, both using GNU MP for the integer arithmetic, on a Pentium double-core 3.00 GHz. In both cases, the chosen parameters were $\delta = 0.75$ and η very close to $1/2$, and the transformation matrix was not computed. The timings are given in seconds. Here Magma's LLL uses the Gram matrix, whereas NTL's LLL_XD recomputes the scalar products from the basis vectors if a large cancellation is detected. In these examples, NTL spends more than 60% of the time recomputing the scalar products from the basis vectors.

Second, when the $\mu_{i,j}$'s are small enough, they are never recomputed after the size-reduction. This means that they are known with a possibly worse accuracy. NTL's LLL is very close to Schnorr-Euchner's heuristic variant but differs on this point: a routine similar to the lazy size-reduction of the L^2 algorithm is used. Shoup's strategy consists in recomputing the $\mu_{k,i}$'s as long as one of them seems (we know them only approximately) larger than η, where η is extremely close to $1/2$ (the actual initial value being $1/2 + 2^{-26}$), and call the size-reduction again. When unexpectedly long lazy size-reductions are encountered (the precise condition being more than ten iterations), the accuracy of the GSO coefficients is deemed very poor, and η is increased slightly to take into account larger errors. This is a good strategy on the short term, since it may accept a larger but manageable error. However, on the long term, weakening the size-reduction condition may worsen the numerical behaviour (see Theorem 2 and Section "Practical Observations on LLL"), and thus even larger errors, and, therefore, stronger misbehaviours are likely to occur.

The fact that the error amplification is not dealt with would not be a problem if there was a way to detect misbehaviours and to handle them. This amplification may cause meaningless calculations: if the current GSO coefficients are very badly approximated, then the performed Lovász tests are meaningless with respect to the basis; it implies that the performed operations may be irrelevant, and not reducing the basis at all; nothing ensures then that the execution will terminate, since it is too different from the execution of the rational LLL. In NTL's LLL, one tries a given precision. If the execution seems too long, the user has to stop it, and restart with some higher precision or some more reliable variant, without knowing if the algorithm was misbehaving (in which case increasing the precision may help), or just long to finish (in which case increasing the precision will slow down the process even more). The lack of error detection and interpretation can be quite annoying from the user point of view: in NTL and LiDia, one may have to try several variants before succeeding.

Implementing the L^2 Algorithm

We now consider the task of implementing the L^2 algorithm (described in Fig. 5.5). In practice, one should obviously try to use heuristic variants before falling down to the guaranteed L^2 algorithm. To do this, we allow ourselves to weaken the L^2 algorithm in two ways: we may try not to use the Gram matrix but the basis matrix only, and we may try to use a floating-point precision which is much lower than the provably sufficient one. We describe here such a possible implementation.

We consider four layers for the underlying floating-point arithmetic:

- Double precision: it is extremely fast, but has a limited exponent (11 bits) and a limited precision (53 bits). The exponent limit allows one to convert integers that have less than 1022 bits (approximately half if one wants to convert the Gram matrix as well). The limited precision is less annoying, but prevents from considering high dimensions.
- Doubles with additional exponents (dpes): it is still quite fast, but the precision limit remains.
- Heuristic extended precision: if more precision seems to be needed, then one will have to use arbitrary precision floating-point numbers. According to the analysis of the L^2 algorithm, a precision $\ell \approx \log \frac{(1+\eta)^2}{\delta-\eta^2} \cdot d$ always suffices. Nevertheless, one is allowed to try a heuristic lower precision first.
- Provable extended precision: use arbitrary precision floating-point numbers with a provably sufficient mantissa size of $\ell \approx \log \frac{(1+\eta)^2}{\delta-\eta^2} \cdot d$ bits.

In [33], Koy and Schnorr suggest to extend the 53 bit long double precision to a precision of 106 bits. This is an interesting additional layer between the double precision and the arbitrary precision, since it can be implemented in an efficient way with a pair of doubles (see [31, Chap. 4.2.2, Exercise 21]).

One can also perform the computations with or without the Gram matrix. If it is decided not to consider the Gram matrix, then the scalar products are computed from floating-point approximations of the basis matrix entries. As mentioned previously, cancellations may occur and the computed scalar products may be completely incorrect. Such misbehaviours will have to be detected and handled. If it is decided to consider the Gram matrix, then there are more operations involving possibly long integers, since both the Gram and basis matrices have to be updated. One may, however, forget about the basis matrix during the execution of the algorithm, by computing the transformation instead, and applying the overall transformation to the initial basis: it then has to be determined from the input which would be cheaper between computing with the transformation matrix and computing with the basis matrix.

So far, we have eight possibilities. A very frequent one is dpes without the Gram matrix, which corresponds to NTL's `LLL_XD` routine. This choice can be sped up by factoring the exponents of the dpes: the idea is to have one common exponent per vector, instead of n exponents. To do this, we consider $e_i = \lfloor 1 + \log \max_{j \leq n} |\mathbf{b}_i[j]| \rfloor$ together with the vector $2^{-\ell} \lfloor \mathbf{b}_i \cdot 2^{\ell - e_i} \rceil$. More summing cancellations are likely to occur than without factoring the exponents (since we may lose some information by doing so), but we obtain a variant which is essentially as fast as using the processor double precision only, while remaining usable for large matrix inputs.

A Thoughtful Wrapper

In a complete implementation of LLL, the choice of the variant and the transitions between variants should be oblivious to the user. When calling the LLL routine, the user expects the execution to terminate and to return a guaranteed answer. At the time of the publishing of this survey, such a routine is available only in Magma and `fplll`: using the LLL routines in the other libraries requires, to some extent, some understanding of the algorithms used. To obtain an LLL routine which is guaranteed, but also makes use of heuristics, misbehaviours should be detected and interpreted in such a way that the cheapest variant that is likely to work is chosen. In the Schnorr-Euchner algorithm, two such detections already exist: scalar product cancellations and too large GSO coefficients.

When considering floating-point LLL algorithms, the main source of infinite looping is the lazy size-reduction (Steps 4–8 in Fig. 5.5). It is detected by watching if the $\mu_{i,j}$'s appearing are indeed decreasing at each loop iteration of the size-reduction. If this stops being the case, then something incorrect is happening. The other source of infinite looping is the succession of incorrect Lovász tests. Fortunately, the proof of the LLL algorithm provides an upper bound to the number of Lovász tests performed during the execution, as a function of the input basis. One can test whether the current number of Lovász tests is higher than this upper bound. This is a crude upper bound, but this malfunction seems to be much less frequent than the incomplete size-reduction.

In Fig. 5.7, we give an overview of the reduction strategy in the LLL routine of Magma. Each box corresponds to a floating-point LLL using the Gram matrix or not, and using one of the aforementioned floating-point arithmetics. When a variant fails, another is tried, following one of the arrows. In addition to this graph, one should rerun a provable variant at the end of the execution if it succeeded with a heuristic one, since the output might then be incorrect. Other boxes and arrows than

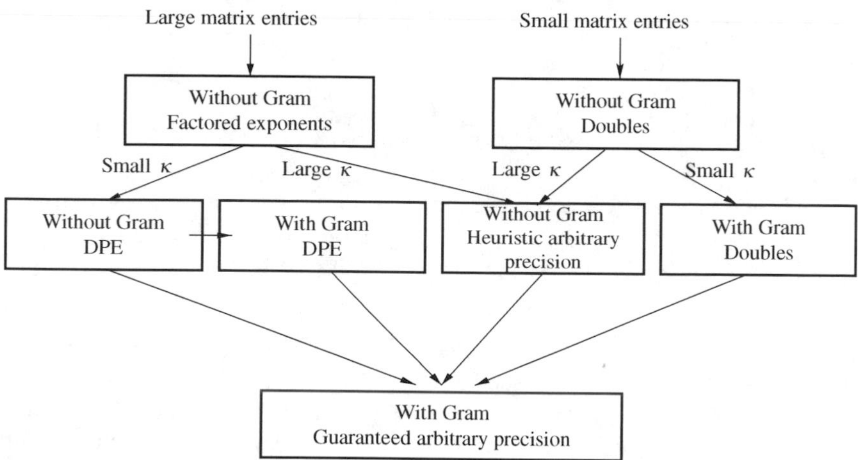

Fig. 5.7 Overview of the LLL reduction strategy in Magma

the ones displayed may be added. For example, one may stop using the factored exponents variants if the entries of the basis matrix start being small: in this case, the doubles without the Gram matrix will be more efficient.

When a malfunction is detected (by a non-decrease of the GSO coefficients during a size-reduction or by a too large number of Lovász tests), another variant must be selected. Essentially, two problems can occur: cancellations of scalar products and lack of precision for the GSO calculations. The first trouble may occur in any dimension, while the second one can only occur when the dimension increases: around $d = 30$ in the worst case and around $d = 180$ on the average, for close to optimal LLL parameters δ and η (for a heuristic explanation of the last figure, see Section "Practical Observations on LLL"). As a consequence, if a misbehaviour is detected in a low dimension or for a small LLL index κ (the magnitude of the floating-point errors essentially depends on κ, see the first-order analysis of the end of Section "The Provable Floating-Point LLL Algorithms"), cancellations in scalar products are likely to be the cause of the problem, and one should start using the Gram matrix. Otherwise, it is likely that the mantissa size is not sufficient. In Fig. 5.7, this choice is represented by the arrows with the labels "Small κ" and "Large κ".

The labels "Large matrix entries" and "Small matrix entries" denote the possibility of converting the Gram matrix coefficients to double precision floating-point numbers: the top boxes do not involve the Gram matrices, but those matrices may be needed later on if misbehaviours occur.

As mentioned earlier, in order to guarantee the correctness of the output, one has to run the most reliable (and thus slower) variant on the output. This can dominate the overall cost, especially if we are given an already reduced basis. Villard [63] recently introduced a method to certify that a given basis is reduced. It will not always work, but if it does the result is guaranteed. It can be made very efficient (for example by using double precision floating-point numbers), and indeed much faster than using the provable precision in the L^2 algorithm. The general principle is as follows:

1. Compute an approximation $\bar{R}$ of the R-factor R of the QR-factorisation of the basis matrix.
2. Certify that the approximation $\bar{R}$ is indeed close to R, by using a result of [61], showing that it suffices to bound the spectral radius of some related matrix.
3. Check the LLL conditions in a certified way from the certified approximation $\bar{R}$ of R.

Another drawback of the general strategy is that it always goes towards a more reliable reduction. It may be that such a reliable variant is needed at some moment and becomes superfluous after some time within an execution: generally speaking, the reduction of the basis improves the accuracy of the computations and therefore some precautions may become superfluous. One would thus have to devise heuristic tests to decide if one should change for a more heuristic but faster variant. For example, suppose we did start using the Gram matrix before scalar product cancellations were detected. The most annoying scalar product cancellations occur when some vectors have very unbalanced lengths and are at the same time fairly orthogonal. One can check with the Gram matrix if it remains the case during the execution of the chosen variant. Suppose now that we did increase the floating-point precision. This was done in particular because the basis was not orthogonal enough. It may happen that it becomes significantly more orthogonal, later within the LLL-reduction: this can be detected by looking at the decrease of the $\|\mathbf{b}_i^*\|$'s.

Finally, one may try to adapt the η and δ parameters in order to speed up the LLL reduction. If one is only interested in a reduced basis without paying attention to the LLL factors δ and η, then one should try the fastest pair, while still requiring only double precision. The first requirement usually implies a weakening of the pair (δ further away from 1 and η further away from $1/2$), whereas the second one involves a strengthening, so that there is a trade-off to be determined. Furthermore, one may also try to change the LLL factors during the LLL reduction itself, for example starting with weak LLL factors to perform most of the reduction efficiently and strengthen the factors afterwards to provide a basis of a better quality.

Adapting the Algorithm to Particular Inputs

It is possible to adapt the algorithm to particular lattice basis inputs that occur frequently. We give here an example of a dedicated strategy called early size-reduction, which was initially introduced by Allan Steel. The computational saving of this method can easily be explained for input lattice bases of the following shape (they arise for example for detecting small integer relations between numbers):

$$\begin{pmatrix} a_1 & a_2 & a_3 & \dots & a_d \\ 0 & 1 & 0 & \dots & 0 \\ 0 & 0 & 1 & \dots & 0 \\ \vdots & \vdots & \vdots & \ddots & \vdots \\ 0 & 0 & 0 & \dots & 1 \end{pmatrix},$$

where the a_i's have large magnitudes. Let $A = \max_i |a_i|$. The idea of the early size-reduction is as follows: when the LLL index κ reaches a new value for the first time, instead of only size-reducing the vector $\mathbf{b}_\kappa$ with respect to the vectors $\mathbf{b}_1, \dots, \mathbf{b}_{\kappa-1}$, reduce the $\mathbf{b}_i$'s with respect to the vectors $\mathbf{b}_1, \dots, \mathbf{b}_{\kappa-1}$ for all $i \geq \kappa$. The speed-up is higher for the longest $\mathbf{b}_i$'s, so that it may be worth restricting the strategy to these ones.

One may think, at first sight, that this variant is going to be more expensive: in fact, the overall size-reduction of any input vector $\mathbf{b}_i$ will be much cheaper. In the first situation, if the first $i-1$ vectors behave fairly randomly, we will reduce in dimension i a vector of length $\approx A$ with respect to $i-1$ vectors of length $\approx A^{\frac{1}{i-1}}$: if the first $i-1$ vectors behave randomly, the lengths of the reduced vectors are all approximately the $(i-1)$-th root of the determinant of the lattice they span, which is itself approximately A. In the second situation, we will:

- Reduce in dimension 3 a vector of length $\approx A$ with respect to 2 vectors of length $\approx A^{\frac{1}{2}}$, when κ reaches 3 for the first time.
- Reduce in dimension 4 a vector of length $\approx A^{\frac{1}{2}}$ with respect to 3 vectors of length $\approx A^{\frac{1}{3}}$, when κ reaches 4 for the first time.
- ...
- Reduce in dimension i a vector of length $\approx A^{\frac{1}{i-2}}$ with respect to $i-1$ vectors of length $\approx A^{\frac{1}{i-1}}$, when κ reaches i for the first time.

We gain much because most of the time the number of nonzero coordinates is less than i.

We now describe a very simple dedicated strategy for lattice bases occurring in Coppersmith's method for finding the small roots of a polynomial modulo an integer [16]. We consider the univariate case for the sake of simplicity. In this application of LLL, the input basis vectors are made of the weighted coefficients of polynomials $(P_i(x))_i$: the i-th basis vector is made of the coordinates of $P_i(xX)$, where X is the weight. This implies that the $(j+1)$-th coordinates of all vectors are multiples of X^j. Rather than reducing a basis where the coordinates share large factors, one may consider the coordinates of the $P_i(x)$'s themselves and modify the scalar product by giving a weight X^j to the $(j+1)$-th coordinate. This decreases the size of the input basis with a negligible overhead on the computation of the scalar products. If X is a power of 2, then this overhead can be made extremely small.

Practical Observations on LLL

The LLL algorithm has been widely reported to perform much better in practice than in theory. In this section, we describe some experiments whose purpose is to measure this statement. These systematic observations were made more tractable thanks to the faster and more reliable floating-point LLLs based on L^2. Conversely, they also help improving the codes:

- They provide heuristics on what to expect from the bases output by LLL. For example, when LLL is needed for an application, these heuristic bounds may be used rather than the provable ones, which may decrease the overall cost. For example, in the cases of Coppersmith's method (see [38]) and the reconstruction of algebraic numbers (see [24]), the bases to be reduced will have smaller bit-lengths.
- They explain precisely which steps are expensive during the execution, so that the coder may be performing relevant code optimisations.
- They also help guessing which precision is likely to work in practice if no scalar product cancellation occurs, which helps choosing a stronger variant in case a malfunction is detected (see Section "A Thoughtful Wrapper").

Overall, LLL performs quite well compared to the worst-case bounds with respect to the quality of the output: the practical approximation factor between the first basis vector and a shortest lattice vector remains exponential, but the involved constant is significantly smaller. Moreover, the floating-point LLLs also seem to outperform the worst-case bounds with respect to their running-time and the floating-point precision they require. We refer to [43] for more details about the content of this section. Further and more rigorous explanations of the observations can be found in [62].

The Lattice Bases Under Study

The behaviour of LLL can vary much with the type of lattice and the type of input basis considered. For instance, if the lattice minimum is extremely small compared to the other lattice minima (the k-th minimum being the smallest R such that there are $\geq k$ linearly independent lattice vectors of length $\leq R$), the LLL algorithm will find a vector whose length reaches it (which is of course not the case in general). If the basis is to be reduced or is close to being LLL-reduced, the LLL algorithm will not behave generically. For instance, if one selects vectors uniformly and independently in the d-dimensional hypersphere, they are close to be reduced with high probability (see [6, 7] and the survey [62] describing these results in a more general probabilistic setup). We must, therefore, define precisely what we will consider as input.

First of all, there exists a natural notion of random lattice. A full-rank lattice (up to scaling) can be seen as an element of $SL_d(\mathbb{R})/SL_d(\mathbb{Z})$. The space $SL_d(\mathbb{R})$

inherits a Haar measure from $\mathbb{R}^{d^2}$, which projects to a finite measure when taking the quotient by $SL_d(\mathbb{Z})$ (see [3]). One can, therefore, define a probability measure on real-valued lattices. There are ways to generate integer valued lattices so that they converge to the uniform distribution (with respect to the Haar measure) when the integer parameters grow to infinity. For example, Goldstein and Mayer [22] consider the following random family of lattices: take a large prime p, choose $d-1$ integers $x_2, \ldots, x_d$ randomly, independently and uniformly in $[0, p-1]$, and consider the lattice spanned by the columns of the following $d \times d$ matrix:

$$\begin{pmatrix} p & x_2 & x_3 & \ldots & x_d \\ 0 & 1 & 0 & \ldots & 0 \\ 0 & 0 & 1 & \ldots & 0 \\ \vdots & \vdots & \vdots & \ddots & \vdots \\ 0 & 0 & \ldots & 0 & 1 \end{pmatrix}.$$

Amazingly, these lattice bases resemble those arising from knapsack-type problems, the algebraic reconstruction problem (finding the minimal polynomial of an algebraic number given a complex approximation to it), and the problem of detecting integer relations between real numbers [24]. We define knapsack-type lattice bases as follows: take a bound B, choose d integers $x_1, \ldots, x_d$ randomly, independently and uniformly in $[0, B-1]$ and consider the lattice spanned by the columns of the following $(d+1) \times d$ matrix:

$$\begin{pmatrix} x_1 & x_2 & x_3 & \ldots & x_d \\ 1 & 0 & 0 & \ldots & 0 \\ 0 & 1 & 0 & \ldots & 0 \\ 0 & 0 & 1 & \ldots & 0 \\ \vdots & \vdots & \vdots & \ddots & \vdots \\ 0 & 0 & 0 & \ldots & 1 \end{pmatrix}.$$

In our experiments, we did not notice any difference of behaviour between these random bases and the random bases of Goldstein and Mayer. Similarly, removing the second row or adding another row of random numbers do not seem to change the observations either.

We will also describe experiments based on what we call Ajtai-type bases. Similar bases were introduced by Ajtai [4] to prove a lower-bound on the quality of Schnorr's block-type algorithms [49]. Select a parameter a. The basis is given by the columns of the $d \times d$ upper-triangular matrix B such that $B_{i,i} = \lfloor 2^{(2d-i+1)^a} \rceil$ and the $B_{j,i}$'s (for $i > j$) are randomly, independently and uniformly chosen in $\mathbb{Z} \cap [-B_{j,j}/2, \ldots, B_{j,j}/2)$. The choice of the function $2^{(2d-i+1)^a}$ is arbitrary: one may generalise this family by considering a real-valued function $f(i,d)$ and by taking $B_{i,i} = \lfloor f(i,d) \rceil$. One advantage of choosing $f(i,d) = 2^{(2d-i+1)^a}$ is that the $\|\mathbf{b}_i^*\|$'s are decreasing very quickly, so that the basis is far from being reduced.

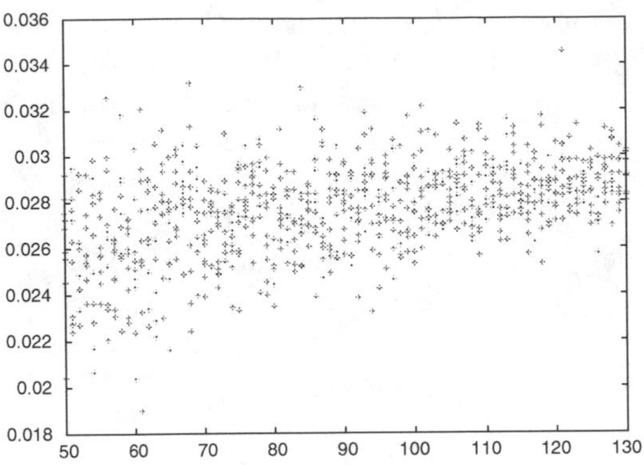

Fig. 5.8 Samples of $\frac{1}{d}\log_2 \frac{\|\mathbf{b}_1\|}{\mathrm{vol}(L)^{1/d}}$ for increasing dimensions d

The Output Quality

In low dimensions, it has been observed for quite a while that the LLL algorithm computes vectors whose lengths are close (if not equal) to the lattice minimum [45]. Hopefully for lattice-based cryptosystems, this does not remain the case when the dimension increases.

By experimenting, one can observe that the quality of the output of LLL is similar for all input lattice bases generated from the different families mentioned above. For example, in Fig. 5.8, each point corresponds to the following experiment: generate a random knapsack-type basis with $B = 2^{100 \cdot d}$ and reduce it with the L^2 algorithm, with $(\delta, \eta) = (0.999, 0.501)$; a point corresponds to the value of $\frac{1}{d}\log_2 \frac{\|\mathbf{b}_1\|}{\mathrm{vol}(L)^{1/d}}$ for the corresponding returned basis. We conclude that experimentally, it seems that for a growing dimension d, the first output vector is such that:

$$\|\mathbf{b}_1\| \approx c^d \cdot \mathrm{vol}(L)^{1/d},$$

where $c \approx 2^{0.03} \approx 1.02$. The exponential factor 1.02^d remains tiny even in moderate dimensions: e.g., $(1.02)^{50} \approx 2.7$ and $(1.02)^{100} \approx 7.2$.

One may explain this global phenomenon on the basis, by looking at the local two-dimensional bases, i.e., the pairs $(\mathbf{b}_{i-1}^*, \mathbf{b}_i^* + \mu_{i,i-1}\mathbf{b}_{i-1}^*)$. If we disregard some first and last local pairs, then all the others seem to behave quite similarly. In Fig. 5.9, each point corresponds to a local pair (its coordinates being $\left(\mu_{i,i-1}, \|\mathbf{b}_i^*\|/\|\mathbf{b}_{i-1}^*\|\right)$) of a basis that was reduced with `fplll` with parameters $\delta = 0.999$ and $\eta = 0.501$, starting from a knapsack-type basis with $B = 2^{100 \cdot d}$. These observations seem to stabilise between the dimensions 40 and 50: the behaviour differs in low dimensions (in particular, the quantity $\frac{1}{d}\log_2 \frac{\|\mathbf{b}_1\|}{\mathrm{vol}(L)^{1/d}}$ is lower), but converges to a limit when the dimension increases. The mean value of the $|\mu_{i,i-1}|$'s is close to 0.38, and the mean value of $\frac{\|\mathbf{b}_{i-1}^*\|}{\|\mathbf{b}_i^*\|}$ is close to 1.04, which matches the above constant 1.02. One may wonder if the geometry of "average" LLL-reduced bases is due to the fact that

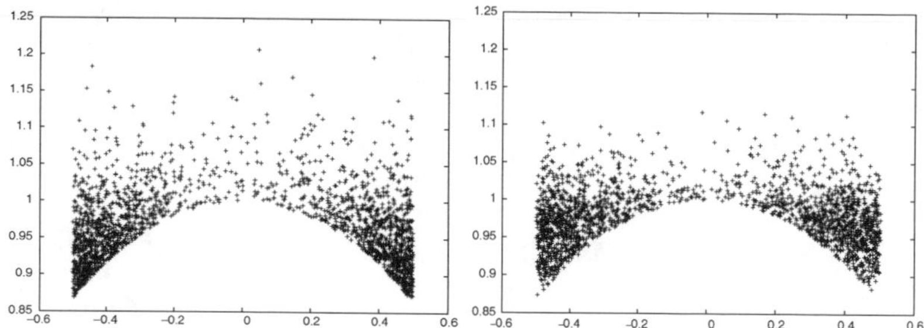

Fig. 5.9 Distribution of the local bases after LLL (*left*) and deep-LLL (*right*)

most LLL-reduced bases are indeed of this shape, or if the LLL algorithm biases the distribution. It is hard to decide between both possibilities: one would like to generate randomly and uniformly LLL-reduced bases of a given lattice, but it is unknown how to do it efficiently; for example, the number of LLL-reduced bases of a given lattice grows far too quickly when the dimension increases.

On the right hand-side of Fig. 5.9, we did the same experiment except that we replaced LLL by the Schnorr-Euchner deep insertion algorithm [53] (see also [48]), which is a variant of the LLL algorithm where the Lovász condition is changed into the stronger requirement:

$$\forall \kappa \leq d, \forall i < \kappa, \;\; \delta \cdot \|\mathbf{b}_i^*\|^2 \leq \left\| \mathbf{b}_\kappa^* + \sum_{j=i}^{\kappa-1} \mu_{\kappa,j} \mathbf{b}_j^* \right\|^2 .$$

The quality of the local bases improves by considering the deep insertion algorithm, the constant 1.04 becoming ≈ 1.025, for close to optimal parameters δ and η. These data match the observations of [9] on the output quality improvement obtained by considering the deep insertion algorithm.

Practical Running-Time

The floating-point LLL algorithms seem to run much faster in practice than the worst-case theoretical bounds. We argue below that these bounds should be reached asymptotically for some families of inputs. We also heuristically explain why the algorithms terminate significantly faster in practice. We will consider bases for which $n = \Theta(d) = O(\log B)$, so that the worst-case bound given in Theorem 2 is simplified to $O(d^5 \log^2 B)$.

The worst-case analysis of the L^2 algorithm given in Section "The Provable Floating-Point LLL Algorithms" seems to be tight for Ajtai-type random bases. More precisely: if $a > 1$ is fixed and d grows to infinity, the average bit-complexity

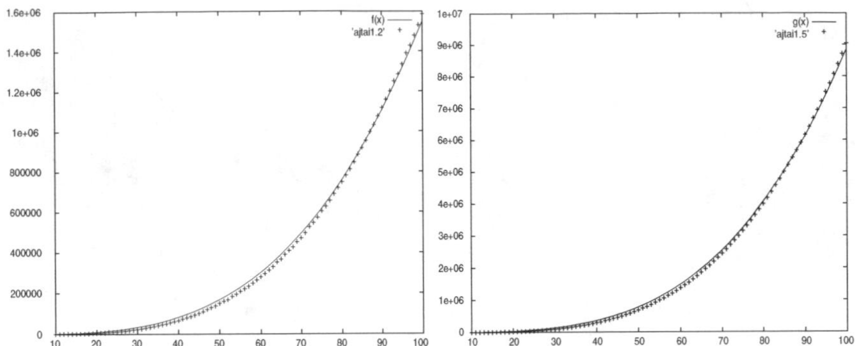

Fig. 5.10 Number of loop iterations of L^2 for Ajtai-type random bases

of the L^2 algorithm given as input a randomly and uniformly chosen d-dimensional Ajtai-type basis with parameter a seems to be $\Theta(d^{5+2a})$ (in this context, we have $\log B \approx d^a$).

When L^2 is run on these input bases, all the bounds of the heuristic analysis but one seem tight, the exception being the $O(d)$ bound on the size of the X's (computed at Step 6 of the L^2 algorithm, as described in Fig. 5.5). First, the $O(d^2 \log B)$ bound on the loop iterations seems to be tight in practice, as suggested by Fig. 5.10. The left side of the figure corresponds to Ajtai-type random bases with $a = 1.2$: the points are the experimental data and the continuous line is the `gnuplot` interpolation of the form $f(d) = c_1 \cdot d^{3.2}$. The right side of the figure has been obtained similarly, for $a = 1.5$, and $g(d) = c_2 \cdot d^{3.5}$. With Ajtai-type bases, size-reductions rarely contain more than two iterations. For example, for $d \leq 75$ and $a = 1.5$, less than 0.01% of the size-reductions involve more than two iterations. The third bound of the heuristic worst-case analysis, i.e., the number of arithmetic operations within each loop iteration of the lazy size-reduction, is also reached.

These similarities between the worst and average cases do not go on for the size of the integers involved in the arithmetic operations. The X's computed during the size-reductions are most often shorter than a machine word, which makes it difficult to observe the $O(d)$ factor in the complexity bound coming from their sizes. For an Ajtai-type basis with $d \leq 75$ and $a = 1.5$, less than 0.2% of the nonzero X's are longer than 64 bits. In the worst case [42] and for close to optimal parameters δ and η, we have $|X| \lesssim (3/2)^{\kappa} M$, where M is the maximum of the $|\mu_{\kappa,j}|$'s before the lazy size-reduction starts, and κ is the current LLL index. In practice, M happens to be small most of the time. It is essentially the length ratio between $\mathbf{b}_\kappa$ and the smallest of the $\mathbf{b}_i$'s for $i \leq \kappa$: it is very rare that the lengths of the basis vectors differ significantly in a nonnegligible number of loop iterations during an execution of the LLL algorithm. It can be argued (see [43] for more details), using the generic geometry of LLL-reduced bases described previously, that the average situation is $|X| \approx (1.04)^{\kappa-i} M$, if X is derived from $\mu_{\kappa,i}$. This bound remains exponential, but for a small M, the integer X becomes larger than a machine word only in dimensions higher than several hundreds. We thus expect the $|X|$'s to be of length $\lesssim (\log_2 1.04) \cdot d \approx 0.057 \cdot d$. For example, the quantity $(1.04)^d$ becomes larger than 2^{64} for $d \geq 1100$. Since it is not known how to reduce lattice bases which simultaneously have such huge dimensions and reach the other bounds of the heuristic worst-case complexity analysis, it is, at the moment, impossible to observe the asymptotic behaviour. The practical running time is rather $O(d^4 \log^2 B)$.

One can take advantage of the fact that most X's are small by optimising the operation $\mathbf{b}_\kappa := \mathbf{b}_\kappa + X\mathbf{b}_i$ for small X's. For example, one may consider the cases $X \in \{-2, -1, 1, 2\}$ separately. One may also use fast multiply-and-add operations, as available in Pierrick Gaudry's GNU MP patch for AMD 64 processors[2].

Furthermore, in many situations, a much better running-time can be observed. Of course, it highly depends on the input basis: if it is already reduced, it will terminate very quickly since there will be no more than d loop iterations. But this can also happen for bases that are very far from being reduced, such as knapsack-type bases. In this case, two facts improve the running-time of LLL. First, the number of loop iterations is only $O(d \log B)$ instead of $O(d^2 \log B)$ in the worst-case: this provably provides a $O(d^4 \log^2 B)$ worst-case complexity bound for these lattice bases (from Theorem 2). Second, the basis matrix entries become very short much sooner than usual: if the maximum value of the index κ so far is some j, one can heuristically expect the entries of the vectors $\mathbf{b}_i$ for $i < j$ to be of length $O\left(\frac{1}{j-1} \log B\right)$ (see [43] for more details). It is not known, yet, how to use this second remark to decrease the complexity bound in a rigorous way, but one can heuristically expect the following worst-case bound:

$$O\left(d \log B \cdot d^2 \cdot d \cdot \frac{\log B}{d}\right) = O(d^3 \log^2 B).$$

Finally, by also considering the $\Theta(d)$ gain due to the fact that the size-reduction X's do not blow up sufficiently for dimensions that can be handled with today's LLL codes, we obtain a $O(d^2 \log^2 B)$ bound in practice (for a very large $\log B$).

Numerical Behaviour

We now describe the practical error amplification of the GSO coefficients in the L^2 algorithm. To build the wrapper described in Section "A Thoughtful Wrapper", it is important to be able to guess up to which dimension a given precision will work, for example the double precision, which is much faster than arbitrary precision. Is it possible to predict the chance of success when using double precision? We suppose here that we work with close to optimal parameters, i.e., δ close to 1 and η close to $1/2$, and with some precision ℓ lower than the provably sufficient $\approx \log_2(3) \cdot d$

[2] http://www.loria.fr/~gaudry/mpn_AMD64/

precision. We do not consider the cancellations that may arise in the scalar product computations. Two problems may occur: some lazy size-reduction or some consecutive Lovász tests may be looping forever. In both situations, the misbehaviour is due to the incorrectness of the involved approximate Gram–Schmidt coefficients. The output basis may also be incorrect, but most often if something goes wrong, the execution loops within a size-reduction.

In practice, the algorithm seems to work correctly with a given precision for much larger dimensions than guaranteed by the worst-case analysis: for example, double precision seems sufficient most of the time up to dimension 170. This figure depends on the number of loop iterations performed: if there are fewer loop iterations, one can expect fewer large floating-point errors since there are fewer floating-point calculations. It can be argued that the average required precision grows linearly with the dimension, but with a constant factor significantly smaller than the worst-case one: for close to optimal LLL parameters and for most input lattice bases, L^2 behaves correctly with a precision of $\approx 0.25 \cdot d$ bits. The heuristic argument, like in the previous subsection, relies on the generic geometry of LLL-reduced bases.

Open Problems

Though studied and used extensively since 25 years, many questions remain open about how to implement the LLL algorithm as efficiently as possible and about its practical behaviour. Some open problems have been suggested along this survey. For example, Section "Practical Observations on LLL" is essentially descriptive (though some relations between the diverse observations are conjectured), and obtaining proven precise results would help to understand more clearly what happens in practice and how to take advantage of it: the survey [62] formalises more precisely these questions and answers some of them. We suggest here a few other lines of research related to the topics that have been presented.

Decreasing the Required Precision in Floating-Point LLLs

Since processor double precision floating-point numbers are drastically faster than other floating-point arithmetics (in particular arbitrary precision), it is tempting to extend the family of inputs for which double precision will suffice. One way to do this, undertaken by Schnorr [33,48,51], is to use other orthogonalisation techniques like the Givens and Householder algorithms. These algorithms compute the Q (as a product of matrices) and R factors of the QR-factorisation of the basis matrix. The L^2 algorithm relies on the Cholesky factorisation (computing the R factor from the Gram matrix). Unfortunately, the condition number of the Cholesky factorisation is essentially the square of the condition number of the QR-factorisation (see [26] for more details). With fully general input matrices, this heuristically means that one needs approximately twice the precision with Cholesky's factorisation than with the QR-factorisation. Another significant advantage of relying on the QR-factorisation rather than Cholesky's is that the Gram matrix becomes superfluous: a large ratio of the integer operations can thus be avoided, which should provide better running-times, at least for input matrices having dimensions that are small compared to the bit-sizes of their entries. Nevertheless, the Householder and Givens algorithms have at least two drawbacks. First, they require more floating-point operations: for $d \times d$ matrices, the Householder algorithm requires $\frac{4}{3}d^3 + o(d^3)$ floating-point operations whereas the Cholesky algorithm requires only $\frac{1}{3}d^3 + o(d^3)$ floating-point operations (see [26]). And second, they suffer from potential cancellations while computing scalar products (the first of the three drawbacks of the naive floating-point LLL of Section "The Provable Floating-Point LLL Algorithms"). A reduction satisfying our definition of LLL-reduction seems unreachable with these orthogonalisation techniques. In [51], Schnorr suggests to replace the size-reduction condition $|\mu_{i,j}| \leq \eta$ by $|\mu_{i,j}| \leq \eta + \varepsilon \frac{\|\mathbf{b}_i^*\|}{\|\mathbf{b}_j^*\|}$ for some small $\varepsilon > 0$. So far, the best results in this direction remain heuristic [48,51]: making them fully provable would be a significant achievement. It would prove that one can double the dimension up to which the double precision rigorously suffices, and provide a sounder insight on the possibilities of such orthogonalisation techniques in practice.

To decrease the precision even further, one could strengthen the orthogonality of the bases that we are reducing. To do this, deep insertions [53] (see also Section "Practical Observations on LLL") may be used, but this may become slow when the dimension increases. Another alternative would be to perform a block-type reduction (such as described in [19, 49]), for some small size of block: one performs strong reductions such as Hermite-Korkine-Zolotarev (HKZ for short) or dual-HKZ to make these small blocks extremely reduced and thus extremely orthogonal. Indeed, a small size of block is sufficient to strengthen the overall orthogonality of the basis, and if the block-size is small enough, the actual cost of HKZ-reducing for this block-size remains dominated by the size-reduction step. Asymptotically, a block-size $k = \Theta\left(\frac{\log d}{\log \log d}\right)$ would satisfy these requirements. In practice, a block-size below 15 does not seem to create a large running-time overhead.

Using Floating-point Arithmetic in Other Lattice Algorithms

Replacing the text-book rational arithmetic by an approximate floating-point arithmetic can lead to drastic theoretical and practical speed-ups. The counterpart is that the correctness proofs become more intricate. One may extend the error analysis strategy of the L^2 algorithm to derive complete (without neglected error terms) explicit error bounds for modifications of the LLL algorithm such as the algorithm of Schönhage [55] and the Strong Segment-LLL of Koy and Schnorr [32]. Adapting these algorithms to floating-point arithmetic has already been considered [33, 51], but often the provably sufficient precision is quite large in the worst case (linear in the bit-size of the matrix entries), though better heuristic bounds outperform those of L^2 (see [51] and the survey [48] in this book). Developing high-level techniques to prove such bounds would be helpful. Second, some lattice reduction algorithms such as short lattice point enumeration, HKZ reduction and block-type reductions [18, 19, 25, 29, 53] are usually implemented with floating-point numbers, though no analysis at all has been made. This simply means that the outputs of these codes come with no correctness guarantee. This fact is particularly annoying, since checking the solutions of these problems is often very hard. Amazingly, devising strong reduction algorithms based on floating-point arithmetic may help decreasing the precision required for the LLL-reduction, as mentioned above.

Decreasing the Linear Algebra Cost

In all known LLL algorithms, the embedding dimension n is a factor of the overall cost. This comes from the fact that operations are performed on the basis vectors, which are made of n coordinates. This may not seem natural, since one could reduce the underlying quadratic form (i.e., LLL-reduce by using only the Gram matrix), store the transformation matrix, and finally apply it to the initial basis. Then the cost would be a smaller function of n. We describe here a possible way to reduce a lattice basis whose embedding dimension n is much larger than its rank d. It consists in applying a random projection (multiplying the embedding space by a random $d \times n$ matrix), reducing the projected lattice, and applying the obtained transformation to the initial basis: one then hopes that the obtained lattice basis is somehow close to being reduced, with high probability. Results in that direction are proved in [8]. This strategy can be seen as a dual of the probabilistic technique recently introduced by Chen and Storjohann [13] to decrease the number of input vectors, when they are linearly dependent: their technique decreases the number of input vectors while the one above decreases the number of coordinates of the input vectors.

Decreasing the Integer Arithmetic Cost

Finally, when the size of the matrix entries is huge and the dimension is small, one would like to have an algorithm with a sub-quadratic bit-complexity (with respect to the size of the entries). Both Euclid's and Gauss' algorithms have quasi-linear variants (see [30, 54, 56, 64]): is it possible to devise an LLL algorithm which is quasi-linear in any fixed dimension? Eisenbrand and Rote [17] answered the question positively, but the cost of their algorithm is more than exponential with respect to d. So we may restate the question as follows: is it possible to devise an LLL algorithm whose bit-complexity grows quasi-linearly with the size of the entries and polynomially with the dimension?

Acknowledgements The author gratefully thanks John Cannon, Claude-Pierre Jeannerod, Erich Kaltofen, Phong Nguyen, Andrew Odlyzko, Peter Pearson, Claus Schnorr, Victor Shoup, Allan Steel, Brigitte Vallée and Gilles Villard for helpful discussions and for pointing out errors on drafts of this work.

References

1. LIDIA 2.1.3. A C++ library for computational number theory. Available at http://www.informatik.tu-darmstadt.de/TI/LiDIA/.
2. IEEE Standards Committee 754. ANSI/IEEE standard 754-1985 for binary floating-point arithmetic. Reprinted in SIGPLAN Notices, 22(2):9–25, 1987.
3. M. Ajtai. Random lattices and a conjectured 0-1 law about their polynomial time computable properties. In *Proceedings of the 2002 Symposium on Foundations of Computer Science (FOCS 2002)*, pages 13–39. IEEE Computer Society Press, 2002.
4. M. Ajtai. The worst-case behavior of Schnorr's algorithm approximating the shortest nonzero vector in a lattice. In *Proceedings of the 35th Symposium on the Theory of Computing (STOC 2003)*, pages 396–406. ACM, 2003.
5. M. Ajtai and C. Dwork. A public-key cryptosystem with worst-case/average-case equivalence. In *Proceedings of the 29th Symposium on the Theory of Computing (STOC 1997)*, pages 284–293. ACM, 1997.
6. A. Akhavi. Worst-case complexity of the optimal LLL algorithm. In *Proceedings of the 2000 Latin American Theoretical Informatics conference (LATIN 2000)*, volume 1776 of *Lecture Notes in Computer Science*, pages 355–366. Springer, 2000.
7. A. Akhavi, J.-F. Marckert, and A. Rouault. On the reduction of a random basis. In *Proceedings of the 4th Workshop on Analytic Algorithmics and Combinatorics*. SIAM, 2007.
8. A. Akhavi and D. Stehlé. Speeding-up lattice reduction with random projections (extended abstract). In *Proceedings of the 2008 Latin American Theoretical Informatics conference (LATIN'08)*, volume 4957 of *Lecture Notes in Computer Science*, pages 293–305. Springer, 2008.
9. W. Backes and S. Wetzel. Heuristics on lattice reduction in practice. *ACM Journal of Experimental Algorithms*, 7:1, 2002.
10. D. Boneh and G. Durfee. Cryptanalysis of RSA with private key d less than $N^{0.292}$. *IEEE Transactions on Information Theory*, 46(4):233–260, 2000.
11. W. Bosma, J. Cannon, and C. Playoust. The Magma algebra system. I. The user language. *Journal of Symbolic Computation*, 24(3–4):235–265, 1997.
12. D. Cadé and D. Stehlé. fplll-2.0, a floating-point LLL implementation. Available at http://perso.ens-lyon.fr/damien.stehle.
13. Z. Chen and A. Storjohann. A BLAS based C library for exact linear algebra on integer matrices. In *Proceedings of the 2005 International Symposium on Symbolic and Algebraic Computation (ISSAC'02)*, pages 92–99. ACM, 2005.
14. D. Coppersmith. Finding a small root of a bivariate integer equation. In *Proceedings of Eurocrypt 1996*, volume 1070 of *Lecture Notes in Computer Science*, pages 178–189. Springer, 1996.
15. D. Coppersmith. Finding a small root of a univariate modular equation. In *Proceedings of Eurocrypt 1996*, volume 1070 of *Lecture Notes in Computer Science*, pages 155–165. Springer, 1996.
16. D. Coppersmith. Small solutions to polynomial equations, and low exponent RSA vulnerabilities. *Journal of Cryptology*, 10(4):233–260, 1997.

17. F. Eisenbrand and G. Rote. Fast reduction of ternary quadratic forms. In *Proceedings of the 2001 Cryptography and Lattices Conference (CALC'01)*, volume 2146 of *Lecture Notes in Computer Science*, pages 32–44. Springer, 2001.
18. U. Fincke and M. Pohst. A procedure for determining algebraic integers of given norm. In *Proceedings of EUROCAL*, volume 162 of *Lecture Notes in Computer Science*, pages 194–202, 1983.
19. N. Gama and P. Q. Nguyen. Finding short lattice vectors within Mordell's inequality. In *Proceedings of the 40th Symposium on the Theory of Computing (STOC'08)*. ACM, 2008.
20. J. von zur Gathen and J. Gerhardt. *Modern Computer Algebra, 2nd edition*. Cambridge University Press, Cambridge, 2003.
21. O. Goldreich, S. Goldwasser, and S. Halevi. Public-key cryptosystems from lattice reduction problems. In *Proceedings of Crypto 1997*, volume 1294 of *Lecture Notes in Computer Science*, pages 112–131. Springer, 1997.
22. D. Goldstein and A. Mayer. On the equidistribution of Hecke points. *Forum Mathematicum*, 15:165–189, 2003.
23. T. Granlund. The GNU MP Bignum Library. Available at http://gmplib.org/.
24. G. Hanrot. LLL: a tool for effective diophantine approximation. This book.
25. B. Helfrich. Algorithms to construct Minkowski reduced and Hermite reduced lattice bases. *Theoretical Computer Science*, 41:125–139, 1985.
26. N. Higham. *Accuracy and Stability of Numerical Algorithms*. SIAM, 2002.
27. J. Hoffstein, J. Pipher, and J. H. Silverman. NTRU: a ring based public key cryptosystem. In *Proceedings of the 3rd Algorithmic Number Theory Symposium (ANTS III)*, volume 1423 of *Lecture Notes in Computer Science*, pages 267–288. Springer, 1998.
28. E. Kaltofen. On the complexity of finding short vectors in integer lattices. In *Proceedings of EUROCAL'83*, volume 162 of *Lecture Notes in Computer Science*, pages 236–244. Springer, 1983.
29. R. Kannan. Improved algorithms for integer programming and related lattice problems. In *Proceedings of the 15th Symposium on the Theory of Computing (STOC 1983)*, pages 99–108. ACM, 1983.
30. D. Knuth. The analysis of algorithms. In *Actes du Congrès International des Mathématiciens de 1970*, volume 3, pages 269–274. Gauthiers-Villars, 1971.
31. D. Knuth. *The Art of Computer Programming, vol. 2, third edition*. Addison-Wesley, Reading, MA, 1997.
32. H. Koy and C. P. Schnorr. Segment LLL-reduction of lattice bases. In *Proceedings of the 2001 Cryptography and Lattices Conference (CALC'01)*, volume 2146 of *Lecture Notes in Computer Science*, pages 67–80. Springer, 2001.
33. H. Koy and C. P. Schnorr. Segment LLL-reduction of lattice bases with floating-point orthogonalization. In *Proceedings of the 2001 Cryptography and Lattices Conference (CALC'01)*, volume 2146 of *Lecture Notes in Computer Science*, pages 81–96. Springer, 2001.
34. J. C. Lagarias and A. M. Odlyzko. Solving low-density subset sum problems. *Journal of the ACM*, 32:229–246, 1985.
35. A. K. Lenstra, H. W. Lenstra, Jr., and L. Lovász. Factoring polynomials with rational coefficients. *Mathematische Annalen*, 261:515–534, 1982.
36. H. W. Lenstra, Jr. Flags and lattice basis reduction. In *Proceedings of the third European congress of mathematics, volume 1*. Birkhäuser, 2001.
37. J. Martinet. *Perfect Lattices in Euclidean Spaces*. Springer, Berlin, 2002.
38. A. May. Using LLL-reduction for solving RSA and factorization problems: a survey. This book.
39. A. May. *New RSA Vulnerabilities Using Lattice Reduction Methods*. PhD thesis, University of Paderborn, 2003.
40. J.-M. Muller. *Elementary Functions, Algorithms and Implementation*. Birkhäuser, 1997.
41. P. Nguyen and D. Stehlé. Low-dimensional lattice basis reduction revisited (extended abstract). In *Proceedings of the 6th Algorithmic Number Theory Symposium (ANTS VI)*, volume 3076 of *Lecture Notes in Computer Science*, pages 338–357. Springer, 2004.

42. P. Nguyen and D. Stehlé. Floating-point LLL revisited. In *Proceedings of Eurocrypt 2005*, volume 3494 of *Lecture Notes in Computer Science*, pages 215–233. Springer, 2005.
43. P. Nguyen and D. Stehlé. LLL on the average. In *Proceedings of the 7th Algorithmic Number Theory Symposium (ANTS VII)*, volume 4076 of *Lecture Notes in Computer Science*, pages 238–256. Springer, 2006.
44. P. Nguyen and D. Stehlé. An LLL algorithm with quadratic complexity. *SIAM Journal on Computing*, 39(3):874–903, 2009.
45. A. M. Odlyzko. The rise and fall of knapsack cryptosystems. In *Proceedings of Cryptology and Computational Number Theory*, volume 42 of *Proceedings of Symposia in Applied Mathematics*, pages 75–88. American Mathematical Society, 1989.
46. A. M. Odlyzko and H. J. J. te Riele. Disproof of Mertens conjecture. *Journal für die reine und angewandte Mathematik*, 357:138–160, 1985.
47. The SPACES Project. MPFR, a LGPL-library for multiple-precision floating-point computations with exact rounding. Available at http://www.mpfr.org/.
48. C. P. Schnorr. Hot topics of LLL and lattice reduction. This book.
49. C. P. Schnorr. A hierarchy of polynomial lattice basis reduction algorithms. *Theoretical Computer Science*, 53:201–224, 1987.
50. C. P. Schnorr. A more efficient algorithm for lattice basis reduction. *Journal of Algorithms*, 9(1):47–62, 1988.
51. C. P. Schnorr. Fast LLL-type lattice reduction. *Information and Computation*, 204:1–25, 2006.
52. C. P. Schnorr and M. Euchner. Lattice basis reduction: Improved practical algorithms and solving subset sum problems. In *Proceedings of the 1991 Symposium on the Fundamentals of Computation Theory (FCT'91)*, volume 529 of *Lecture Notes in Computer Science*, pages 68–85. Springer, 1991.
53. C. P. Schnorr and M. Euchner. Lattice basis reduction: improved practical algorithms and solving subset sum problems. *Mathematics of Programming*, 66:181–199, 1994.
54. A. Schönhage. Schnelle Berechnung von Kettenbruchentwicklungen. *Acta Informatica*, 1: 139–144, 1971.
55. A. Schönhage. Factorization of univariate integer polynomials by Diophantine approximation and improved basis reduction algorithm. In *Proceedings of the 1984 International Colloquium on Automata, Languages and Programming (ICALP 1984)*, volume 172 of *Lecture Notes in Computer Science*, pages 436–447. Springer, 1984.
56. A. Schönhage. Fast reduction and composition of binary quadratic forms. In *Proceedings of the 1991 International Symposium on Symbolic and Algebraic Computation (ISSAC'91)*, pages 128–133. ACM, 1991.
57. G. Schulz. Iterative Berechnung der reziproken Matrix. *Zeitschrift für Angewandte Mathematik und Mechanik*, 13:57–59, 1933.
58. I. Semaev. A 3-dimensional lattice reduction algorithm. In *Proceedings of the 2001 Cryptography and Lattices Conference (CALC'01)*, volume 2146 of *Lecture Notes in Computer Science*, pages 181–193. Springer, 2001.
59. V. Shoup. NTL, Number Theory C++ Library. Available at http://www.shoup.net/ntl/.
60. A. Storjohann. Faster algorithms for integer lattice basis reduction. Technical report, ETH Zürich, 1996.
61. J.-G. Sun. Componentwise perturbation bounds for some matrix decompositions. *BIT Numerical Mathematics*, 31:341–352, 1992.
62. B. Vallée and A. Vera. Probabilistic analyses of lattice reduction algorithms. This book.
63. G. Villard. Certification of the QR factor R, and of lattice basis reducedness. In *Proceedings of the 2007 International Symposium on Symbolic and Algebraic Computation (ISSAC'07)*, pages 361–368. ACM, 2007.
64. C. K. Yap. Fast unimodular reduction: planar integer lattices. In *Proceedings of the 1992 Symposium on the Foundations of Computer Science (FOCS 1992)*, pages 437–446. IEEE Computer Society Press, 1992.

Chapter 6
LLL: A Tool for Effective Diophantine Approximation

Guillaume Hanrot

Abstract The purpose of this paper is to survey in a unified setting some of the results in diophantine approximation that the LLL algorithm can make effective in an efficient way. We mostly study the problems of finding good rational approximations to vectors of real and p-adic numbers, and of finding approximate linear relations between vectors of real numbers. We also discuss classical applications of those effective versions, among which Mertens' conjecture and the effective solution of diophantine equations.

Introduction

The invention of the LLL lattice basis reduction algorithm in 1982 [1] was a key advancement in the development of fast algorithms in number theory. Its main feature was to provide the first notion of lattice basis reduction which was at the same time sufficiently sharp and could be computed efficiently. Since then, it has been used in a variety of situations for algorithmic purposes. Its *constructive facet* is the fact that it can be used to "make things small." We shall mostly use its ability to find "small" linear combinations of integers or real numbers, or good rational approximations to real or p-adic numbers. Its *negative facet* is the fact that it makes things "almost smallest possible," which means that nothing much smaller can exist, in a precise and quantitative way.

In this survey, we will stress out the links of LLL with the theory of diophantine approximation. Many results in number theory, especially (but not only) in the field of diophantine approximation and diophantine equations are proved through the use of some kind of a pigeonhole principle, and especially through the use of Minkowski's first theorem. It thus comes as no surprise that the LLL algorithm can be used to provide one with "effective" versions of those theorems. We review some of those results and their LLL-based effective versions in the present survey.

G. Hanrot
INRIA/LORIA, Projet CACAO – Bâtiment A, 615 rue du jardin botanique,
F-54602 Villers-lès-Nancy Cedex, France,
e-mail: hanrot@loria.fr

P.Q. Nguyen and B. Vallée (eds.), *The LLL Algorithm*, Information Security and Cryptography, DOI 10.1007/978-3-642-02295-1_6,

Most of the results presented in this survey belong to algorithmic number theory community folklore, but they might not always be written in a precise form. It is a widely known fact that LLL finds small linear relations, good simultaneous approximations, and can be used to get lower bounds for (in)homogeneous linear forms with bounded coefficients. As such, we do not claim any original result in this survey. Our goal is just to present all those results in a somewhat unified setting, and to discuss some striking applications in number theory. We might have forgotten to give due credit for some of the results described above; in that case, it is simply due to ignorance.

We have tried to give precise results whenever possible for the "building blocks" of all our applications. The fact that these building blocks can be stated in simple terms and that the proofs are short and simple appeals for completeness. On the other hand, the applications that we discuss usually involve deep and technical number theory. For those aspects, hence, we decided to rather adopt a more informal style, trying to emphasize upon the link with LLL, and giving only sufficient hints about the theory to allow the reader to understand where and how LLL comes into play.

In a nutshell, the results we discuss split into two parts, the *constructive* results, and the *negative* ones. This comes from the fact that the main result about LLL can be seen under two different viewpoints: LLL builds a *small vector* or LLL builds an *almost smallest vector*. In other words, LLL shall be used either to construct a small object (relation, approximation) or to prove that no very small object exists.

In the present survey, we focus on two types of results, simultaneous diophantine approximation and small values of linear forms, and on some of their applications.

(Simultaneous) approximation. A classical theorem by Dirichlet asserts that for $x_1, \ldots, x_n$ real numbers, and for Q a parameter, one can find $q < Q^n$ such that $|qx_i - p_i| \leqslant 1/Q$ for some integers p_i. In the case $n = 1$, the very classical theory of continued fractions provides one with an effective version of this result, i.e., a simple way to construct the integer q.

In an higher dimensional setting, however, the effective side is much poorer. We shall show how Dirichlet's theorem can be made effective, though we shall lose some factors with respect to the (optimal) result stated above.

We shall discuss two applications of this technique. The first one is the striking application to the study of the Mertens' conjecture initiated by Odlyzko and te Riele in the mid-80s. The second one is a *negative* result (see below for discussion on constructive and negative aspects of the applications of LLL) on inhomogeneous approximation.

Small values of linear forms. The dual problem of the simultaneous approximation problem is, given one linear form $\sum_{i=1}^n \lambda_i x_i$, where the x_i are fixed real (or p-adic) numbers, to find integers λ_i such that $\sum_{i=1}^n \lambda_i x_i$ is as small as possible.

Exact linear relations. In the case where the x_i are linearly dependent over $\mathbb{Z}$, or for instance when the x_i are integers, one can find *exact* linear relations. The size of those relations is controlled by a result called Siegel's lemma, which is at the core of the construction of many "small" auxiliary objects; for instance, constructing a polynomial of small height with roots at some given points. If one sees the set of the relations between the x_i as a $\mathbb{Z}$-module, then, one can use LLL on the basis of this (discrete) module to build small relations between x_i. This is at the core of a recent approach of Dokchitser to build extremal cases for Masser–Oesterlé's ABC conjecture.

Approximate linear relations. In the case where the x_i are linearly independent over $\mathbb{Z}$, we can no longer hope for an exact linear relation; we shall thus content ourselves with approximate relations, ie n-uples of "not too large" integers $(\lambda_1, \ldots, \lambda_n)$ such that $\sum_{i=1}^{n} \lambda_i x_i$ is as small as possible. We shall also discuss the inhomogeneous case $\sum_{i=1}^{n} \lambda_i x_i + x_{n+1}$.

In that case, both *constructive* results and *negative* ones have important applications. On the constructive side, let us quote the ability to find the minimal polynomial of an algebraic number. On the negative side, those results are at the core of algorithms based on Baker's method for solving certain families of Diophantine equations, since in that setting solutions correspond to extremely small values of linear forms; those values must be so small that LLL can be used to prove that they do not exist.

Roadmap of the paper. Section "Facts and Notations About the LLL Algorithm" reviews the basic definitions and the main results about LLL-reduced bases. The next two sections are the most technical of this survey, and develop our building blocks. Section "Approximation by Rationals" studies approximation of real numbers by rationals and its applications, whereas Section "Linear Relations" studies small values of homogeneous and inhomogeneous linear forms. Then, we apply the material of these two technical sections to the solution of Diophantine equations in Section "Applications to Diophantine Equations", to the approximation by algebraic numbers in Section "Approximation by Algebraic Numbers". Finally, we briefly discuss what we might call solving diophantine inequalities, i.e., algorithms to find integer or rational points close to a variety.

Notations and conventions. If x, y are two real numbers, we define $x \operatorname{cmod} y$ as being the element of $x + \mathbb{Z}y$ lying in the interval $(-y/2, y/2]$, and $\lfloor x \rceil$ as $x - (x \operatorname{cmod} 1) \in \mathbb{Z}$. Finally, we note $d(x, \mathbb{Z}) = |x \operatorname{cmod} 1|$. Similarly, if x is a p-adic integer, we define $x \operatorname{cmod} p^\ell$ as being the element of $(-p^\ell/2, \ldots, p^\ell/2]$ congruent to x modulo p^ℓ.

We use boldface letters to denote vectors. We use the notations $f = O(g)$ or $f \ll g$ with their classical meaning, i.e., $|f(x)| \leqslant c|g(x)|$ for some constant c and x large enough.

Finally, when describing an algorithm, we assume that each real, complex or p-adic number is an oracle which returns one bit of the corresponding number at each call.

Facts and Notations About the LLL Algorithm

In order to make this survey as self-contained as possible, we review some basic facts about the LLL algorithm. We refer to other texts in this volume for fine algorithmic and complexity details concerning LLL.

Lattices

In all the sequel, $\mathbb{R}^n$ is endowed with its canonical euclidian structure, i.e., if $\mathbf{x} = (x_i)_{1 \leqslant i \leqslant n} \in \mathbb{R}^n$ and $\mathbf{y} = (y_i)_{1 \leqslant i \leqslant n} \in \mathbb{R}^n$, we note $(\mathbf{x}|\mathbf{y}) = \sum_{i=1}^n x_i y_i$, and we put $\|\mathbf{x}\| = (x|x)^{1/2}$.

A d-dimensional lattice L of $\mathbb{R}^n$ is a discrete subgroup of $\mathbb{R}^n$ of rank d, which means that there exists vectors $\mathbf{b}_1, \ldots, \mathbf{b}_d \in \mathbb{R}^n$, linearly independent over $\mathbb{R}$, such that

$$L = \oplus_{i=1}^d \mathbb{Z}\mathbf{b}_i.$$

In practice, we shall always consider rational, and almost always integral lattices, i.e., assume that $L \subset \mathbb{Q}^d$ or $\mathbb{Z}^d$, or equivalently that $\mathbf{b}_i \in \mathbb{Q}^d$ or $\mathbb{Z}^d$ for all i. This is critical for computational purposes, since this allows one to avoid the difficult problem of numerical stability of lattice basis reduction and of the corresponding algorithms.

The *volume*, or *discriminant* of L is the quantity $\mathrm{vol}(L) = \sqrt{\det((\mathbf{b}_i|\mathbf{b}_j))}$. In the case $d = n$, this volume can also be written as $\mathrm{vol}(L) = |\det(\mathbf{b}_i)|$.

We start by recalling Minkowski's first theorem:

Theorem 1. *Let L be a d-dimensional lattice, and C a compact, convex set, symmetric with respect to* 0*, with* $\mathrm{vol}(C) \geqslant 2^d \mathrm{vol}(L)$. *Then* $C \cap L \neq \{0\}$.

Applied to the d-dimensional sphere, this yields the following corollary:

Corollary 1. *There exists a constant γ_d such that*

$$\min_{\mathbf{x} \in L - \{0\}} \|\mathbf{x}\| \leqslant \sqrt{\gamma_d} (\mathrm{vol}(L))^{1/d}.$$

The optimal value of the constant γ_d is known for $d \leqslant 8$ and $d = 24$, and one has $\gamma_d \leqslant (d+4)/4$ for all d.

Proof. See e.g., Martinet [2] and Cohn and Kumar [3]. □

LLL Reduced Bases

Let $\mathbf{b}_1, \ldots, \mathbf{b}_d \in \mathbb{R}^n$ be linearly independent vectors. The Gram–Schmidt orthogonalization procedure constructs orthogonal vectors $\mathbf{b}_i^*$ defined by the fact that $\mathbf{b}_i^*$ is the orthogonal projection of $\mathbf{b}_i$ on $(\mathbb{R}\mathbf{b}_1 \oplus \cdots \oplus \mathbb{R}\mathbf{b}_{i-1})^\perp$.

More explicitly, one has

$$\mathbf{b}_i^* = \mathbf{b}_i - \sum_{k=1}^{i-1} \mu_{ik} \mathbf{b}_k^*,$$

with $\mu_{ik} = (\mathbf{b}_i|\mathbf{b}_k^*)/(\mathbf{b}_k^*|\mathbf{b}_k^*)$ for $k < i$.

We say that a basis $(\mathbf{b}_1, \ldots, \mathbf{b}_d)$ is LLL-reduced if

- For all $k < i$, $|\mu_{ik}| \leqslant 1/2$;
- For all i, $\frac{3}{4}\|\mathbf{b}_i^*\| \leqslant \|\mu_{i+1,i}\mathbf{b}_i^* + \mathbf{b}_{i+1}^*\|$.

From a practical viewpoint, the second condition (often called Lovasz' condition) implies, using the first, that $\|\mathbf{b}_{i+1}^*\| \geqslant \|\mathbf{b}_i^*\|/2$. Good bases are almost orthogonal, i.e., that the sequence $\|\mathbf{b}_i^*\|$ should not decrease too quickly is one of the ideas of lattice basis reduction.

Remark 1. One can define the notion of t-LLL reduction, where the factor $3/4$ of the second (Lovász's) condition is replaced by $1 > t > 1/4$.

SVP and CVP

One of the main features of the LLL algorithm is to be a polynomial-time approximation algorithm for the shortest vector problem (henceforth SVP), and to yield a polynomial-time approximation algorithm for the closest vector problem (henceforth CVP). We shall review this in Theorems 2 and 3.

The SVP and the CVP problems, especially their computational versions, are at the core of the questions studied in this survey. The computational version of the SVP problem is defined as follows:

Given a basis $b_1, \ldots, b_d \in \mathbb{Z}^n$ of a lattice L, compute an x such that

$$\|\mathbf{x}\| = \min_{y \in L - \{0\}} \|\mathbf{y}\|.$$

The computational version of the CVP problem is defined by:

Given a basis $b_1, \ldots, b_d \in \mathbb{Z}^n$ of a lattice L and a vector $t \in \mathbb{Z}^n$, compute an x such that

$$\|\mathbf{x} - \mathbf{t}\| = \min_{y \in L} \|\mathbf{y} - \mathbf{t}\|.$$

Both of these problems are considered to be hard ones, and precise hardness results exist on both problems, even if one allows rather large approximation factors (e.g., $2^{(\log d)^{1-\varepsilon}}$ for SVP [4]). We shall not discuss this subject at length, and rather refer to Regev's survey in this volume. We refer to [5,6] for algorithms for SVP and CVP.

From our point of view, LLL (together with Babai's nearest plane algorithm in the case of CVP) is an approximation algorithm, with an approximation factor exponential in the dimension only. Stronger notions of reduction improve very slightly on this in polynomial time [7], but one is still very far away from a polynomial approximation factor.

LLL as an Approximation to SVP

The main properties of LLL-reduced bases that we shall use in the sequel are summed up in the following theorem.

Theorem 2. *Let* $\mathbf{u}_1, \ldots, \mathbf{u}_d \in \mathbb{Q}^n$ *be* $\mathbb{R}$*-linearly independent vectors. One can compute in polynomial time in* d, n, *and* $\log \max_i \|\mathbf{u}_i\|$ *an LLL-reduced basis* $(\mathbf{b}_1, \ldots, \mathbf{b}_d)$. *Further, in that case,*

$$\|\mathbf{b}_1\| \leqslant 2^{(d-1)/4}\mathrm{vol}(L), \tag{6.1}$$

$$\min_{\mathbf{x} \in L - \{0\}} \|\mathbf{x}\| \geqslant 2^{(1-d)/2}\|\mathbf{b}_1\|. \tag{6.2}$$

More generally, for any $x_1, \ldots, x_t$ *linearly independent vectors in L, we have*

$$\max(\|\mathbf{x}_1\|, \|\mathbf{x}_2\|, \ldots, \|\mathbf{x}_t\|) \geqslant 2^{(1-d)/2}\|\mathbf{b}_j\|, \quad 1 \leqslant j \leqslant t. \tag{6.3}$$

For detailed proofs, we refer to the original paper [1] or Cohen's book [8]. The first part of the result shall be used on the *constructive* side, where we want to compute a short vector in a lattice, the volume of which is known. The second part shall be used on the *negative* side, where we compute an LLL-reduced basis of a lattice L and want a lower bound on $\min_{\mathbf{x} \in L - \{0\}} \|\mathbf{x}\|$.

For the subjects discussed in this paper, stronger notions of reduction might actually yield better results than LLL algorithm.

Remark 2. The assumption that the vectors are linearly independent is not necessary, as was shown by Pohst [9].

Remark 3. In the case of a t-LLL reduced basis (see Remark 1), the approximation factor $2^{(d-1)/2}$ is replaced by $(2/\sqrt{4t-1})^{d-1}$, and $2^{(d-1)/4}$ becomes $(2/\sqrt{4t-1})^{(d-1)/2}$. In the sequel, we shall state our results for LLL-reduced basis, i.e., $t = 3/4$, but one can obtain the same results in polynomial time for any $t < 1$. In practice, some "black-box" implementations use $t = 0.99$ or 0.999.

Remark 4. The bound $\min_{x \in L - \{0\}} \|\mathbf{x}\| \geqslant 2^{(1-d)/2}\|\mathbf{b}_1\|$ can often be somewhat improved in a specific example by using the bound

$$\min_{\mathbf{x} \in L - \{0\}} \|\mathbf{x}\| \geqslant \min_{1 \leqslant i \leqslant d} \|\mathbf{b}_i^*\|.$$

Babai's Algorithm: LLL as an Approximation to CVP

Babai's Algorithm

In the Euclidean setting, one finds the coordinates of a given vector $\mathbf{x} \in \mathbb{R}^n$ over a basis $M = (\mathbf{b}_1, \ldots, \mathbf{b}_d)$, for example, by computing the vector $M^{-1}\mathbf{x}$.

Alternatively, the last coefficient x_d can be computed as $(\mathbf{b}_d^*|\mathbf{x})/\|\mathbf{b}_d^*\|^2$, where $(\mathbf{b}_1^*, \ldots, \mathbf{b}_d^*)$ is the Gram–Schmidt basis associated to $(\mathbf{b}_1, \ldots, \mathbf{b}_d)$.

A natural idea (sometimes called Babai's first algorithm) when dealing with the CVP problem is to round the exact solution $M^{-1}\mathbf{x}$. Babai's nearest plane algorithm [10] is also based on this, but refines the idea neatly. Indeed, rounding for each coordinate introduces an error. This error should definitely not be ignored, but instead be reinjected into $\mathbf{x}$ at each step, so that it is taken into account when computing the next coordinates. This gives Algorithm 2.5.1.

Algorithm: ApproximatedCVP
Data: An LLL-reduced basis B ; a vector $\mathbf{v}$
Result: An approximation to the factor $2^{d/2}$ of the CVP of $\mathbf{v}$
begin
  $\mathbf{t} = \mathbf{v}$
  for $(j = d\ ;\ j \geq 1\ ;\ j\text{-}\text{-})$ **do**
    $\mathbf{t} = \mathbf{t} - \left\lfloor \frac{\langle \mathbf{t}, \mathbf{b}_j^* \rangle}{\langle \mathbf{b}_j^*, \mathbf{b}_j^* \rangle} \right\rceil \mathbf{b}_j$
  end
  return $\mathbf{v} - \mathbf{t}$
end

Algorithm 1: Babai nearest Plane algorithm

This algorithm is independent of the basis $(\mathbf{b}_i)$ irrespective of the fact that it is reduced or not. However, as pointed out, it performs much better when the $(\mathbf{b}_i^*)$ are not too small, for instance when the basis is LLL-reduced.

Theorem 3. *Let $\mathbf{b}_1, \ldots, \mathbf{b}_d$ be an LLL-reduced basis of a lattice. For any $\mathbf{t} \in \oplus_{i=1}^d \mathbb{R}\mathbf{b}_i$, Babai's algorithm outputs a vector $\mathbf{x}$ of the lattice such that*

$$\|\mathbf{x} - \mathbf{t}\| \leqslant \frac{1}{2}\left(\sum_{i=1}^{d} \|\mathbf{b}_i^*\|^2\right)^{1/2} \leqslant 2^{d/2} \min_{y \in L} \|\mathbf{y} - \mathbf{t}\|.$$

For a proof, one can for instance consult Lovasz' book [11].

Remark 5. We can give an analog of Remark 4:

Proposition 1. *Let $b_1, \ldots, b_d$ be d vectors in $\mathbb{R}^n$, and write $\mathbf{t} = \sum_{i=1}^d t_i \mathbf{b}_i^*$. Then, Babai's algorithm outputs a vector $\mathbf{x} = \sum_{i=1}^d x_i \mathbf{b}_i^*$ of the lattice such that, for all $\mathbf{y} \in L$,*

$$\|\mathbf{y} - \mathbf{t}\| \geqslant \min_{1 \leqslant j \leqslant d} \left(\frac{1}{4}\|\mathbf{b}_j^*\|^2 + \sum_{i=j+1}^{d} (x_j - t_j)^2 \|\mathbf{b}_j^*\|^2\right)^{1/2}.$$

Proof. If $\mathbf{x}$ is the output of Babai's algorithm, write $\mathbf{x} = \sum_{i=1}^d x_i \mathbf{b}_i^*$, and for any vector $\mathbf{y} \in L$, write $\mathbf{y} = \sum_{i=1}^d y_i \mathbf{b}_i^*$. Let j be the largest index such that $y_j \neq x_j$.

Then,

$$\|\mathbf{y}-\mathbf{t}\|^2 = \sum_{i=1}^{j}(y_j - t_j)^2\|\mathbf{b}_j^*\|^2 + \sum_{i=j+1}^{d}(x_j - t_j)^2\|\mathbf{b}_j^*\|^2.$$

The construction of $\mathbf{x}$ in Babai's algorithms shows that $|y_j - t_j| \geqslant 1/2$; hence, we have

$$\|\mathbf{y}-\mathbf{t}\|^2 \geqslant \frac{1}{4}\|\mathbf{b}_j^*\|^2 + \sum_{i=j+1}^{d}(x_j - t_j)^2\|\mathbf{b}_j^*\|^2.$$

□

In the literature, Babai's nearest plane algorithm is often replaced by the conceptually simpler Babai's first algorithm, i.e., rounding $M^{-1}\mathbf{v}$. In that case, de Weger [12] proved the lower bound

$$\min_{\mathbf{y}\in L}\|\mathbf{y}-\mathbf{t}\| \geqslant d(x_{i_0},\mathbb{Z})\|\mathbf{b}_{i_0}^*\|$$

where $i_0 = \max\{1 \leqslant i \leqslant d\,;\, x_i \notin \mathbb{Z}\}$. If $i_0 = -\infty$, the lower bound should be understood as 0.

Overall Strategy

Since many results in diophantine approximation make use of Minkowski's theorem, LLL has a natural role to play as an (approximate) effective counterpart. We shall obtain two types of results. The "constructive" type is the existence of certain small linear relations with not too large coefficients. It follows from inequality. The "negative" results rest only on the optimality of LLL, and are used to get lower bounds on linear forms with bounded coefficients.

General Comments on the Positive Side

On the positive side, there is usually a double loss of quality with respect to the classical, more or less optimal results known in the field of diophantine approximation.

First, we are doomed to lose an exponential factor in the dimension due to the fact that LLL only yields an approximation to SVP/CVP (via Babai's algorithm described in Section "Babai's Algorithm". This approximation factor involves only the dimension, not the size of the defining matrix. For lattices of fixed dimension and size of the matrix tending to infinity, LLL thus often performs quite well; when the dimension varies, a detailed heuristic analysis was undertaken by Nguyen and Stehlé [13], who proved that one should indeed expect to lose a small exponential factor.

This factor can be removed if one is ready to pay the (exponential time at least, see [5, 14]) computing price of an exact CVP/SVP. In that case, one will still lose a factor $\sqrt{\gamma_k} \approx \sqrt{k}$, which corresponds to going from an L^2-type result to an L^∞-type result.

Second, in the archimedean case, the restriction that we are bound to work with integers will often make us replace a real number x by the integer round $\lfloor Cx \rceil$ for a large constant C. We then have to go back to the original problem. This accounts for the loss of a small constant factor, which can go up to $\sqrt{n}$ in the case of the search for linear relations. In the non-archimedean case, this problem does not occur.

General Comments on the Negative Side

On the negative side, we are trying to prove that certain linear relations cannot exist with small coefficients. However, whereas in the positive side we have guarantees that the object we are looking for exists, on the negative side, no such guarantee exists. Indeed, in general, we are trying to prove that relations of a certain kind between real numbers do not exist. But it may happen that such a relation exists in some situations, and in those situations, a direct application of the theorems stated there is bound to fail.

In that case, what one can do and should try to do is assume that the very small relation found is a true relation, and modify the theorems stated there so as to prove that the only approximate linear relations that exist are those colinear to the true relation. This is done by replacing the Inequality (6.2) by Inequality (6.3) in the proofs.

Approximation by Rationals

An old problem in number theory is to find good rational approximations to given real numbers, and the classical case of *one* real number is addressed by the theory of continued fractions. Apart from its own interest, being able to get good simultaneous approximation is a very useful building block for other algorithms. This application of LLL was already discussed in the original paper [1], in [15] and in [11].

The Archimedean Case

In this section, we are given n real numbers $x_1, \dots, x_n$, and we are looking for approximations to the x_i with a common denominator q, i.e., such that $|qx_i - \lfloor qx_i \rceil|$ is small for all i. We first review the theory of the problem and then discuss how the LLL algorithm can be used to make it constructive.

Theoretical Results

These results go back to Dirichlet, and are based on the pigeonhole principle, which is in some sense constructive but yields algorithms inherently exhaustive; though one can find similarities between this approach and randomized algorithms for the SVP and CVP problems due to Ajtai et al. [5].

Theorem 4 (Dirichlet). *Let $(x_i)_{1\leqslant i\leqslant n}$ be real numbers. For any integer number Q, there exists $0 < q \leqslant Q^n$ such that for all i,*

$$\max_{1\leqslant i\leqslant n} |qx_i - \lfloor qx_i \rceil| < \frac{1}{Q}.$$

Proof. Consider the set $S := \{(qx_i \operatorname{cmod} 1)_{1\leqslant i\leqslant n}, q \in [0, Q^n]\} \subset U := (-1/2, 1/2]^n$. If one splits U into the following Q^n parts, for $\mathbf{v} \in \{0, \ldots, Q-1\}^n$,

$$U_{\mathbf{v}} = \prod_{i=1}^{n} \left(-\frac{1}{2} + \frac{v_i}{Q}, -\frac{1}{2} + \frac{v_i + 1}{Q} \right],$$

then there exists a $\mathbf{v}$ such that two points of S are in the same $U_{\mathbf{v}}$. This means that there exists $q_1 \neq q_2$ such that for all i,

$$|(q_1 x_i \operatorname{cmod} 1) - (q_2 x_i \operatorname{cmod} 1)| < 1/Q,$$

which proves our claim by taking $q = |q_1 - q_2|$. □

Recall that the exponent $1/n$ is best possible for almost all n-uples $(x_1, \ldots, x_n) \in [0, 1)^n$ (i.e., for all, except maybe a set of Lebesgue measure 0). The following weighted version is a straightforward generalization.

Theorem 5. *Let $x_1, \ldots, x_n$ be real numbers, and $w_1, \ldots, w_n$ be positive real numbers such that $\prod_{i=1}^{n} w_i = 1$.*

For any integer Q, there exists $0 < q \leqslant Q^n$ such that

$$\max_{1\leqslant i\leqslant n} w_i\, |qx_i - \lfloor qx_i \rceil| < \frac{1}{Q}.$$

We now study the very classical case $n = 1$, which is based on the theory of continued fractions; we shall see how part of the results of this theory can be recovered using lattice basis reduction. Then, we shall generalize the LLL approach to the general case.

Algorithms – Univariate Case

In the univariate case, we are simply trying to get good approximations of a real number x. This case is settled by the classical continued fraction theory, but we shall see that it can also be dealt with using lattice basis reduction, the main advantage being that the latter approach generalizes to higher dimensions.

For completeness, review shortly the theory of continued fractions. A complete account can be found in [16–18].

Recall that if $\xi = \xi_0$ is a real number, we can define a sequence a_i of integers by

$$a_i = \lfloor \xi_i \rfloor, \xi_{i+1} = (\xi_i - a_i)^{-1},$$

where the process stops if ξ_i is an integer, which happens for some i iff. ξ is a rational number.

Using the sequence a_i, we can build sequences p_i, q_i of integers by $p_{-1} = q_0 = 0$, $p_0 = q_{-1} = 1$,

$$\begin{cases} p_{i+1} = a_i p_i + p_{i-1} \\ q_{i+1} = a_i q_i + q_{i-1}. \end{cases} \tag{6.4}$$

The sequences $(p_i/q_i)_{i \geqslant 1}$ define a sequence of best approximations to ξ in a very precise sense, which we describe in the following theorem.

Theorem 6. *Let ξ be a real number, and $(p_i), (q_i)$ be the integer sequences defined above. Then, for all i, one has*

$$\left| \xi - \frac{p_i}{q_i} \right| \leqslant \frac{1}{q_i q_{i+1}},$$

and for all $q < q_{i+1}$,

$$|q_i x - p_i| \leqslant |qx \operatorname{cmod} 1|.$$

The strength of continued fractions is to give an optimal answer to an important problem, and to do it in a very efficient way. One can compute the largest $q_i \leqslant Q$ in time $O(M(\log Q) \log\log Q)$, where $M(n)$ is the complexity of the multiplication of two numbers of size n, i.e., $M(n) = O(n(\log n)2^{\log^*(n)})$ [19], at the time this survey is written. The corresponding algorithm is described e.g., in [20].

The generalizations of this theory are far from being satisfactory. Either they give a more or less satisfactory theoretical answer, but do not give any practical algorithm; or we have an efficient algorithm, but the result is not satisfactory from a theoretical point of view.

The Two-Dimensional LLL Algorithm

An often better way to understand the continued fraction algorithm is to write it down in matrix form. The iteration (6.4) can be rewritten as:

$$M(i) := \begin{pmatrix} q_i & q_{i+1} \\ p_i & p_{i+1} \end{pmatrix} = M(i-1) \cdot \begin{pmatrix} 0 & 1 \\ 1 & a_i \end{pmatrix},$$

with $M(-1) = I_2$.

In this form, we see that the matrix $M(i)$ is in $GL_n(\mathbb{Z})$, and also that

$$\begin{pmatrix} 1 & 0 \\ -C\xi & C \end{pmatrix} M(i) = \begin{pmatrix} q_i & q_{i+1} \\ C(p_i - q_i\xi) & C(p_{i+1} - q_{i+1}\xi) \end{pmatrix}.$$

If C is chosen so that $q_i \approx C(p_i - q_i\xi)$, the right hand side matrix looks very much like an LLL-reduced matrix. We can thus try to proceed in the reverse way, and recover good approximations of ξ by reducing the lattice $L_C(\xi)$ generated by the columns of the matrix

$$\begin{pmatrix} 1 & 0 \\ -C\xi & C \end{pmatrix}.$$

Theorem 7. *Let ξ be a real number, and C a positive real number. If (u, v) is the shortest vector of $L_C(\xi)$, then u is the denominator of a convergent of the continued fraction expansion of ξ, with $|u| \leqslant \frac{2}{\sqrt{3}}\sqrt{C}$.*

Proof. Assume that this is not the case, and let (p_n, q_n) be the largest convergent with $p_n < u$. Then, it is known that $|p_n - q_n\xi| \leqslant |u\xi - \lfloor u\xi \rceil| \leqslant |u\xi - v|$, and hence

$$p_n^2 + C|p_n - q_n\xi|^2 < u^2 + C|u\xi - v|^2,$$

contradicting the fact that (u, v) is minimal.

The remaining part comes from Minkowski's first theorem applied to L_C, and the fact that the Hermite constant in dimension 2 is $4/3$. □

Note that in dimension 2, we have an efficient algorithm (Gauß' algorithm [21]) to solve the SVP. Unfortunately, we have little control on the convergent which is returned; in particular, this is *not* the largest convergent with denominator less than $2\sqrt{C/3}$.

Example 1. The very good approximation 355/113 of π is obtained by this process for C up to 1.19×10^9. In contrast, the denominator of the next convergent is $33102 < 2/\sqrt{3}\sqrt{1.19 \times 10^9}$. In the same spirit, one can build examples where the second vector of a (Gauß-)reduced basis is not a convergent, for example, for $\xi = \pi$ and $C = 10^3$.

As a final remark, if one wants to apply this method, in practice one will replace $C\xi$ by $\lfloor C\xi \rceil$ to deal with an integral lattice. In that case, the first vector is a convergent of the continued fraction expansion of $\lfloor C\xi \rceil / C$, which we can expect to coincide with the continued fraction expansion of ξ up to denominators of the order of $\sqrt{C}$. Heuristically, thus, we still should get a convergent of ξ, or something quite close. Formally, if $(q, p - q\lfloor C\xi \rceil)$ is the first vector of a Gauss-reduced basis of the lattice generated by the columns of

$$L'_C(\xi) = \begin{pmatrix} 1 & 0 \\ -\lfloor C\xi \rceil & C \end{pmatrix},$$

we have, for $q' < q$,

$$|q' \lfloor C\xi \rceil - Cp'| \geqslant |q \lfloor C\xi \rceil - C\lfloor q\xi \rceil|,$$

from which we deduce

$$|q'\xi - p'| \geqslant |q\xi - \lfloor q\xi \rceil| - \frac{q}{C} \geqslant |q\xi - \lfloor q\xi \rceil| - \frac{2}{\sqrt{3C}},$$

so that $\lfloor q\xi \rceil / q$ is almost a best approximation of ξ itself. Note that we still have $q \leqslant 2\sqrt{C/3}$ in that case.

Algorithms: Multivariate Case

Having somehow reformulated the problem in terms of lattices in the univariate case, we can generalize the results to the multivariate case. Since the general theory of simultaneous approximation is much poorer than the univariate case, the optimality assertion below is by far not as good as previously. Further, in higher dimension, we shall need to replace SVP by the LLL algorithm in order to get efficient algorithms, thus losing an exponential factor.

Theorem 8. *Let Q be a positive integer; let $(x_1, \ldots, x_n)$ be n real numbers. There is a deterministic polynomial time algorithm which finds an integer $q \leqslant 2^{n/4} Q^n$ such that*

$$\max_{1 \leqslant i \leqslant n} |qx_i - (qx_i \operatorname{cmod} 1)| \leqslant \sqrt{5} \times 2^{(n-4)/4} Q^{-1}.$$

Proof. Let $C_1, \ldots, C_n$ be positive integers to be chosen in the course of the proof, and consider the lattice generated by the columns of the matrix

$$\begin{pmatrix} 1 & 0 & 0 & \ldots & 0 \\ \lfloor C_1 x_1 \rceil & C_1 & 0 & \ldots & 0 \\ \lfloor C_2 x_2 \rceil & 0 & C_2 & \ldots & 0 \\ \vdots & 0 & 0 & \ddots & \vdots \\ \lfloor C_n x_n \rceil & 0 & 0 & \ldots & C_n \end{pmatrix}.$$

The dimension of this lattice is $n+1$; its volume is $C_1 \ldots C_n$. Thus, in view of Inequality (6.1), the LLL algorithm finds a vector $(q, r_1, \ldots, r_n)^t$ of L of size at most

$$\Lambda := 2^{n/4} (C_1 \ldots C_n)^{1/(n+1)}.$$

Define $p_i = \lfloor qx_i \rceil$. We have $|C_i r_i - q \lfloor C_i x_i \rceil| \geqslant |C_i p_i - q \lfloor C_i x_i \rceil|$, so that

$$q^2 + \sum_{i=1}^{n} (C_i p_i - q \lfloor C_i x_i \rceil)^2 \leqslant q^2 + \sum_{i=1}^{n} (C_i r_i - q \lfloor C_i x_i \rceil)^2 \leqslant \Lambda^2.$$

In particular, $q \leqslant \Lambda$, and $\max_{1 \leqslant i \leqslant n} |C_i p_i - q \lfloor C_i x_i \rceil| \leqslant \Lambda$. Since $|C_i p_i - q \lfloor C_i x_i \rceil| \geqslant C_i |p_i - qx_i| - q/2$, we see that

$$|p_i - qx_i| \leqslant C_i^{-1}\left(|C_i p_i - q\lfloor C_i x_i \rceil| + q/2\right).$$

Furthermore, when $u^2 + v^2 \leqslant A$, one has $u + v/2 \leqslant \sqrt{5A}/2$, so that $q \leqslant \Lambda$ and, for all i, $|p_i - qx_i| \leqslant \frac{\sqrt{5}}{2C_i}\Lambda$.

Taking $C_i = C$ for $i \geqslant 1$, we see that it suffices to choose $C = Q^{n+1}$ to get q as claimed and

$$|p_i - qx_i| \leqslant \frac{\sqrt{5}}{2} 2^{n/4} Q^{-1}.$$

□

We can also somehow prove some kind of semi-optimality assertion on the result. By Inequality (6.2), for all q', the corresponding vector $(q', \lfloor q'x_1 \rceil, \ldots, \lfloor q'x_n \rceil)$ has length

$$q'^2 + \sum_{i=1}^{n} (Cp'_i - q'\lfloor Cx_i \rceil)^2 \geqslant 2^{-n}\left(q^2 + \sum_{i=1}^{n}(Cp_i - q\lfloor Cx_i \rceil)^2\right).$$

Assume that $q' \leqslant q$. If one wants a more tractable bound, for instance in the L^∞ form, this can be rewritten as

$$\max_{1 \leqslant i \leqslant n} \left(C|p'_i - q'x_i| + \left(\frac{1}{2} + \frac{1}{\sqrt{n}}\right)q'\right)^2 \geqslant \frac{2^{-1-n}}{n} \max_{1 \leqslant i \leqslant n} (C|p_i - qx_i| + \frac{1}{2}q)^2,$$

or, finally

$$\max_{1 \leqslant i \leqslant n} |p'_i - q'x_i| \geqslant \frac{2^{(-1-n)/2}}{\sqrt{n}} \max_{1 \leqslant i \leqslant n} |p_i - qx_i| - \frac{q}{C}\frac{\sqrt{n} + 2 - 2^{(-3-n)/2}}{2\sqrt{n}}$$

Lagarias [15] has obtained a similar result in a stronger sense of optimality, i.e., comparing q with the best q' smaller than Q^n. In order to achieve this, however, he has to reduce $O(n \log Q)$ to different matrices, replacing the 1 in the upper left corner by powers of 2 less than Q^n.

The previous theorem extends, by modifying the choice of the C_i, to the weighted case:

Theorem 9. *Let Q be a positive integer, and w_i be positive rational numbers normalized so that*

$$\prod_{i=1}^{n} w_i = 1.$$

One can find in deterministic polynomial time in the bitsizes of Q, w_i, an integer $q \leqslant 2^{n/4}Q^n$ such that

$$\max_{1 \leqslant i \leqslant n} w_i\, d(qx_i, \mathbb{Z}) \leqslant \sqrt{5} \times 2^{(n-4)/4} Q^{-1}.$$

Proof. Let z be the least common multiple of the denominators of the w_i. Then, use the same matrix as above but take $C_i = zw_i Q^{n+1}$ and replace the top coefficient 1 by z. □

In this last statement, one should be careful that the complexity also depends on the bitsize of the w_i. Usually, this problem occurs with real weights w_i; what this means is that one should replace the w_i by good rational approximations with not too large denominators.

Remark 6. In the last two theorems, we have chosen to lose a factor of $2^{n/4}$ on both sides, i.e., on the size of q and on the quality of the approximation, in order to get a "symmetric" result. Of course, by varying C, one can get a factor of that type on one side only, as in [1].

The p-Adic Case

Usually, in diophantine approximation, any result valid on approximation of a real or complex number extends to the p-adic case. Here, the above discussion generalizes *mutatis mutandis* to the p-adic case as indicated briefly in this subsection. The corresponding question is sometimes (especially in the univariate case) called *rational reconstruction*. One can find a complete discussion of this problem for example in [20].

Good Rational Approximations

Finding a good rational approximation of small height of a given p-adic number x is finding two small integers ℓ and q such that the p-adic valuation of $qx - \ell$ is as large as possible, or equivalently that

$$qx - \ell = 0 \bmod p^k$$

for k as large as possible. From an algorithmic point of view, we shall fix k and try to find a pair (q, ℓ) as small as possible.

We shall assume that x is a p-adic integer, i.e., $v_p(x) \geqslant 0$; otherwise, one replaces x by $xp^{-v_p(x)}$, and there is a one-to-one correspondence between rational approximations of x and of $xp^{-v_p(x)}$.

In the sequel, we fix a prime number p and p-adic numbers $x_1, \ldots, x_n$. The analog of Dirichlet's theorem is as follows.

Theorem 10. *Let $x_1, \ldots, x_n$ be p-adic integers and k a positive integer. For all integer Q, there is a $0 < q \leqslant Q^n$ such that $|qx \operatorname{cmod} p^k| < p^k/Q$.*

As in the archimedean case, the univariate case has been settled by using the continued fraction algorithm. In that case, this is equivalent to using the extended

Euclidean algorithm on x mod p^k and p^k, and stopping as soon as the coefficients of Bézout's relation get larger than the size we want to achieve. The following theorem is sometimes described as "rational reconstruction."

Theorem 11. *Let Q be a positive integer and x be a p-adic number. Let q be the largest denominator of a convergent of $(x \operatorname{cmod} p^k)/p^k$ smaller than Q. Then, $|qx \operatorname{cmod} p^k| \leqslant p^k/Q$.*

Proof. Let ℓ be the numerator of the corresponding convergent. By Theorem 6, we have

$$\left|\frac{qx \operatorname{cmod} p^k}{p^k}\right| \leqslant \left|q\frac{x}{p^k} - \ell\right| < \frac{1}{Q},$$

which gives the result after multiplication by p^k. □

The multivariate case is settled by a method very close to the real case:

Theorem 12. *Given a prime p, an integer k, n p-adic integers $(x_1, \ldots, x_n)$, and a positive integer Q, there is a deterministic polynomial time algorithm which finds an integer $q \leqslant 2^{n/4}Q^n$ such that*

$$\max_{1\leqslant i\leqslant n} |qx_i \operatorname{cmod} p^k| \leqslant 2^{n/4}Q^{-1}.$$

Proof. Consider the lattice L generated by the columns of the matrix

$$\begin{pmatrix} p^k & 0 & 0 & \ldots & 0 \\ Q^{n+1}(x_1 \operatorname{cmod} p^k) & Q^{n+1}p^k & 0 & \ldots & 0 \\ Q^{n+1}(x_2 \operatorname{cmod} p^k) & 0 & Q^{n+1}p^k & \ldots & 0 \\ Q^{n+1}(x_3 \operatorname{cmod} p^k) & 0 & 0 & \ddots & \vdots \\ Q^{n+1}(x_k \operatorname{cmod} p^k) & 0 & 0 & \ldots & Q^{n+1}p^k \end{pmatrix}.$$

The volume of this lattice is $(Q^n p^k)^{n+1}$. Thus, the LLL algorithm finds a vector $(q, l_1, \ldots, l_k)^t$ of L of size at most

$$\Lambda := 2^{n/4}p^kQ^n.$$

As such, we have

$$q \leqslant \frac{\Lambda}{p^k}, \quad |qx_i \operatorname{cmod} p^k| \leqslant \frac{\Lambda}{Q^{n+1}}.$$

Thus, we get $q \leqslant 2^{n/4}Q^n$, and

$$|qx_i - l_ip^k| \leqslant 2^{n/4}\frac{p^k}{Q}.$$

□

As a final remark, note that we have been working with p-adic numbers, which are the right objects for the applications we shall discuss. However, in practice, with fixed finite precision, a p-adic number is simply a number modulo p^n for some n, and all results stated here remain valid modulo a general integer modulus N, or even a different modulus N_i for each x_i.

Application: The Mertens Conjecture

One of the most striking applications of LLL in number theory was the disproof by Odlyzko and te Riele [22] of long-standing Mertens' conjecture [23].

Let $\mu : \mathbb{Z}_{\geqslant 0} \to \mathbb{R}$ be the Möbius function defined by

$$\mu(n) = \begin{cases} 0, & n \text{ is not squarefree} \\ (-1)^{\omega(n)} & \text{otherwise,} \end{cases}$$

where $\omega(n)$ is the number of prime factors of n.

Heuristically, one can consider that, on a set of integers of positive density (the squarefree numbers), μ is a random variable which takes the values 1 and -1 with equal probabilities. Thus, if we take the heuristics to its limits, a very crude application of the central limit theorem might suggest that

$$M(N) := \sum_{n \leqslant N} \mu(n) \ll \sqrt{N}. \tag{6.5}$$

A stronger form of this heuristic, namely

$$|M(N)| \leqslant \sqrt{N}, \tag{6.6}$$

is called Mertens' conjecture. It has been checked numerically that this inequality holds up to very large values of N, and the estimates for the smallest counterexample for that inequality seem beyond reach, unless one comes up with an algorithm having complexity polynomial in $\log N$ to compute $M(N)$.

Conjecture (6.5), and a fortiori Conjecture (6.6) would imply Riemann's hypothesis; this is even the case for the weaker conjecture (actually equivalent to RH)

$$M(N) \ll_{\varepsilon} N^{1/2+\varepsilon}, \qquad \forall \varepsilon > 0. \tag{6.7}$$

Indeed, any of those forms implies, by partial summation, that the Dirichlet series

$$\zeta(s)^{-1} = \sum_{n \geqslant 1} \frac{\mu(n)}{n^s}$$

defines a holomorphic function on the open set $\mathrm{Re}(s) > 1/2$, which is exactly the Riemann hypothesis.

Conjecture (6.5) is expected to be false. However, it was some kind of a surprise when Odlyzko and te Riele, in 1985 [22], were able to use computational arguments based on the LLL algorithm to disprove the Conjecture (6.6).

Nowadays, it is common belief that LLL can be used to prove that $\limsup M(N)/\sqrt{N} \geqslant c$ for any $c > 0$, at least in principle. Understanding what the limits of this LLL-based approach are might even suggest reasonable orders of magnitude for the N^{ε} term in (6.7).

Explicit Formula

In order to explain the link between Mertens' conjecture and the LLL algorithm, the best starting point is the classical "explicit formula." Assume that all the zeros of ζ are simple (which we can do, since it is implied by Mertens' conjecture). This formula is obtained by applying Perron's formula to the function $\zeta(s)^{-1}$, and pushing the integration line to the left to get the contribution corresponding to the poles of $\zeta(s)^{-1}$:

$$\begin{aligned} M(x) &= \frac{1}{2i\pi}\int_{2-i\infty}^{2+i\infty} \frac{x^s}{s\zeta(s)} \mathrm{d}s + O(1), \\ &= \sum_{\rho} \frac{x^{\rho}}{\rho\zeta'(\rho)} + O(1). \end{aligned}$$

The series is over the nontrivial zeros of ζ, and each zero should be paired with its conjugate in order to guarantee the convergence of the series.

Then, under the Riemann hypothesis,

$$M(x)x^{-1/2} = \sum_{\rho} \frac{x^{i\,\mathrm{Im}(\rho)}}{\rho\zeta'(\rho)} + O(x^{-1/2}).$$

An important point there is that the series $\sum 1/|\rho\zeta'(\rho)|$ diverges; this means, in particular, that for $M(x)x^{-1/2}$ to remain bounded, there must be compensations coming from the oscillating terms

$$\exp\left(i\ \mathrm{Im}\rho \log x - i\,\mathrm{Arg}(\rho\zeta'(\rho))\right),$$

for all positive real number x. Let $y = \log x$, write a zero $\rho = 1/2 + i\gamma$, and write $\mathrm{Arg}(\rho\zeta'(\rho)) = \theta_\gamma$.

The assumption about the oscillating terms means that the terms $y\,\gamma - \theta_\gamma \operatorname{cmod} 2\pi$ cannot be all close to each other. This heuristics, which involves in that form infinitely many terms, can be made precise and practical, via the following theorem (an explicit formula with a kernel, proved in [22]):

Theorem 13. *Assume the Riemann hypothesis and that all zeros of ζ are simple, and let us denote by $\rho = 1/2 + i\gamma$ the zeros of $\zeta(s)$. Then,*

$$h(y,T) := 2 \sum_{0<\gamma<T} ((1-\gamma/T)\cos(\pi\gamma/T) + \sin(\pi\gamma/T)/\pi) \frac{\cos(\gamma y - \theta_\gamma)}{|\rho\zeta'(\rho)|}.$$

Then, for all y, T,

$$\liminf_x M(x)x^{-1/2} \leqslant h(y,T) \leqslant \limsup_x M(x)x^{-1/2}.$$

In fact, it can be shown that the theorem is optimal, i.e., $\{h(y,T)\}$ is exactly the set of accumulation points of the sequence $M(n)n^{-1/2}$.

Disproof of Mertens' Conjecture

We are, thus, left with the problem of finding y and T such that $|h(y,T)|$ is large. As previously, this amounts to finding a real y such that all the values $\gamma y - \theta_\gamma - 2k_\gamma\pi$ are simultaneously small, or simultaneously close to π. If we restrict y to be an integer, this is exactly a weighted (to take into account the coefficient in front of each cosine) simultaneous inhomogeneous approximation problem.

Theorem 9 allows us to compute a good value of y, and even with a modest number of zeros, this is sufficient to prove that Mertens' conjecture is false.

This approach was introduced by Odlyzko and te Riele in a spectacular paper in 1984, where they disproved Mertens' conjecture by using 70 zeros of the ζ function. Since then, there have been progresses, and the last paper on the subject, due to Kotnick and te Riele [24], shows that

$$\liminf_x M(x)x^{-1/2} \leqslant -1.229 \leqslant 1.218 \leqslant \limsup_x M(x)x^{-1/2},$$

using lattice basis reduction on up to 200 zeros of ζ, and using then local optimization techniques to take into account the next 9800 zeros. Some unpublished computations by Stehlé and the author, using faster lattice basis reduction algorithms [25] and a more careful analysis of the precision needed, on a dimension 230 lattice, allowed them to prove that $\limsup M(x)x^{-1/2} \geqslant 1.4812$; a similar result had also been obtained (but not published) by te Riele (private communication) at the same time.

Let us finish by mentioning that the LLL algorithm may also be used to give an upper bound on the smallest x such that $M(x)x^{-1/2} \geqslant A$ for given A. We refer to [24] for the details.

Baker–Davenport Lemma

LLL-based algorithms for finding simultaneous Diophantine approximation also have an application to inhomogeneous linear approximation, the so-called Generalized Baker–Davenport lemma [26]. This duality is not surprising, since, as we shall see shortly, there is a strong relationship between small values of linear forms and simultaneous approximation. This Lemma gives a result of negative type for inhomogeneous linear approximation.

Let $(x_i)_{1\leqslant i\leqslant n+1}$ be real numbers, and $(\lambda_i)_{1\leqslant i\leqslant n}$ be integers. Consider the inhomogeneous linear form

$$\Lambda := \sum_{i=1}^{n} \lambda_i x_i + x_{n+1}.$$

We want to find a lower bound for $|\Lambda|$ under the assumption that $|\lambda_i| \leqslant B$. The idea is the following: if Q is an integer such that $d(Qx_i, \mathbb{Z})$ is small for $1 \leqslant i \leqslant n$ and Q large enough, then the numbers $(x_i)_{1\leqslant i\leqslant n}$ and hence, when the $|\lambda_i|$ are much smaller than B, the linear combination $\sum_{i=1}^{n} \lambda_i x_i$ behave almost as rationals with denominator Q. This means that

$$\left|\sum_{i=1}^{n} \lambda_i x_i + x_{n+1}\right| \geqslant d\left(\sum_{i=1}^{n} \lambda_i x_i + x_{n+1}, \mathbb{Z}\right) \approx d(x_{n+1}, \frac{1}{Q}\mathbb{Z}).$$

More precisely, we have the following:

Lemma 1. *Let $(x_1, \ldots, x_n)$ be real numbers, ε a positive real number and let Q be an integer such that $d(Qx_i, \mathbb{Z}) \leqslant \varepsilon$ for $i = 1, \ldots, n$. Then, for any $\lambda_1, \ldots, \lambda_n \in [-B, B]^n$, one has*

$$\left|\sum_{i=1}^{n} \lambda_i x_i + x_{n+1}\right| \geqslant Q^{-1}\left(d(Qx_{n+1}, \mathbb{Z}) - nB\varepsilon\right).$$

Proof. Follows from the chain of inequalities:

$$\begin{aligned} Q\left|\sum_{i=1}^{n} \lambda_i x_i + x_{n+1}\right| &\geqslant d\left(\sum_{i=1}^{n} Q\lambda_i x_i + Qx_{n+1}, \mathbb{Z}\right) \\ &\geqslant d(Qx_{n+1}, \mathbb{Z}) - \left(\max_{1\leqslant i\leqslant n} |\lambda_i|\right)\sum_{i=1}^{n} d(Qx_i, \mathbb{Z}) \\ &\geqslant d(Qx_{n+1}, \mathbb{Z}) - nB\varepsilon. \end{aligned}$$

□

In practice, Q is found by solving a simultaneous diophantine approximation problem, by using LLL. In practice, we expect that $\varepsilon \approx Q^{-1/n}$, and if x_{n+1} is linearly independent from the $(x_i)_{1\leqslant i\leqslant n}$, then $d(Qx_{n+1}, \mathbb{Z})$ is "on average" 1/4. Thus, Q should be chosen to be slightly larger than $(nB)^n$, and the corresponding lower bound is of the order of $(nB)^{-n}$, which is the expected order of magnitude, up to the factor n^n.

Linear Relations

In this section, we survey applications of the LLL algorithm to the situation that is dual to the previous section. Indeed, in practice, the matrix defining the underlying lattice is the transpose of the kind of matrices we encountered in the previous part. It is also a well-known fact that goes back to Khinchin that results on simultaneous approximation are closely related to small values of linear forms through transference arguments [27], i.e., relating the minima of a lattice and of its dual lattice.

In the sequel, we are given n real numbers $x_1, \ldots, x_n$ and want to find n real numbers $\lambda_1, \ldots, \lambda_n$ such that $\sum_{i=1}^{n} \lambda_i x_i$ is small. We shall again derive

- *Constructive results*: We are really looking for small values of the linear form, with a bound on $\max_i |\lambda_i|$;
- *Negative results*: We expect that a very small relation does not exist, and want to prove that it is indeed the case for all (λ_i) such that $\max_i |\lambda_i| \leqslant B$.

For completeness, we mention that the popular algorithm PSLQ can also be used for those two tasks. See [28, 29] for details.

Effective Siegel's Lemma

We start with the easiest case of an underdetermined system of linear equations over the integers, where an exact relation is searched. This is what is called Siegel's Lemma [30, Ges. Abh. I, p. 213, Hilfssatz] in diophantine approximation, and is a basic construction used in many places. The typical application is to build an auxiliary polynomial of given degree in several variables, with roots of given order at given points, and with small coefficients. One writes the linear system that the coefficients of such a polynomial must verify in order to get roots at the right points with the right order, and can apply Siegel's lemma if the degree was chosen large enough.

Theorem 14 (Siegel's Lemma). *Let $r < n$ two integers and $(x_{ij})_{1 \leqslant i \leqslant r, 1 \leqslant j \leqslant n} \in \mathbb{Z}^{r \times n}$ be a rank r matrix. Let $X = \max_{ij} |x_{ij}|$. If $(\lambda_1, \ldots, \lambda_n)$ is the first vector of an LLL-reduced basis of the integral kernel of the matrix (x_{ij}), then one has*

$$\sum_{i=1}^{r} \lambda_i^2 \leqslant 2^{(n-r-1)/2} \prod_{i=1}^{r} \left(\sum_{j=1}^{n} x_{ij}^2 \right)^{1/(n-r)}$$
$$\leqslant 2^{(n-r-1)/2} \left(nX^2\right)^{1/(n-r)}.$$

Proof. [31, Lemma 4C] shows that the volume of the kernel of the matrix $(x_{ij})_{i,j}$, seen as a lattice, is $(\det(x_{ij}))^{1/(n-r)}$. We then apply Hadamard's inequality and Inequality (6.1). □

Approximate Linear Relations

In the previous paragraph, we insisted on finding an exact relation between integers. In practice, we shall be interested by finding relations between real numbers. In order to do this, we shall use an approximation, namely $\lfloor Cx \rceil$, where C is a large integer constant, as we did for simultaneous rational approximation. By doing this, we introduce a perturbation to the initial problem, which means that looking for an *exact* relation is usually no longer relevant: the fact that the problem is approximate means that we need to balance the order of magnitude of the relation and the order of magnitude of the coefficients in some way.

This is at the core of many applications of the LLL algorithm. Among those that we shall not discuss, we mention solving knapsack-type problems in cryptology [32] or in computer algebra, such as the beautiful algorithm by van Hoeij [33] for polynomial factoring over $\mathbb{Q}$, see Kluners' survey in this volume.

If $1 \leqslant r \leqslant n$ and C are integers, and $\mathbf{x}_1, \ldots, \mathbf{x}_r$ are vectors in $\mathbb{R}^n$, we define

$$M(C, \mathbf{x}_1, \ldots, \mathbf{x}_r) := \begin{pmatrix} 1 & 0 & \ldots & 0 \\ 0 & 1 & \ldots & 0 \\ \vdots & \vdots & \ddots & \vdots \\ 0 & 0 & \ldots & 1 \\ \lfloor Cx_{11} \rceil & \lfloor Cx_{12} \rceil & \ldots & \lfloor Cx_{1n} \rceil \\ \vdots & \vdots & \vdots & \vdots \\ \lfloor Cx_{r1} \rceil & \lfloor Cx_{r2} \rceil & \ldots & \lfloor Cx_{rn} \rceil \end{pmatrix},$$

Lemma 2. *Let* $\mathbf{x}_1, \ldots, \mathbf{x}_r$ *be vectors in* $\mathbb{R}^n$. *One has*

$$\det\left(M(C, \mathbf{x}_1, \ldots, \mathbf{x}_r)^t M(C, \mathbf{x}_1, \ldots, \mathbf{x}_r)\right) \leqslant \left(1 + \frac{1}{r} \sum_{i,j} \lfloor Cx_{ij} \rceil^2\right)^r,$$

with equality for $r = 1$.

Proof. For simplicity, we write M instead of $M(C, \mathbf{x}_1, \ldots, \mathbf{x}_r)$ in this proof. First notice that 1 is an eigenvalue of $M^t M$ with multiplicity at least $n - r$. Indeed, if $\mathbf{u} = (u_1, \ldots, u_n)^t \in K := \mathrm{Ker}\left(\lfloor Cx_{ij} \rceil_{1 \leqslant i \leqslant r, 1 \leqslant j \leqslant n}\right)$, we have

$$M^t M \mathbf{u} = M^t (u_1, \ldots, u_n, 0, \ldots, 0)^t = \mathbf{u}.$$

Since K has dimension at least $n - r$, this proves our claim.

Now, if $\alpha_1, \ldots, \alpha_r$ are the other eigenvalues of $M^t M$, we have

$$\operatorname{Tr}(M^t M) = n + \sum_{i,j} \lfloor C x_{ij} \rceil^2 = (n - r) + \sum_{l=1}^{r} \alpha_l, \tag{6.8}$$

or equivalently $\sum_{l=1}^{r} \alpha_l = r + \sum_{i,j} \lfloor C x_{ij} \rceil^2$. By the arithmetico-geometrical inequality, we deduce

$$\det M^t M = \prod_{l=1}^{r} \alpha_l \leqslant \left(\frac{\sum_{l=1}^{r} \alpha_l}{r} \right)^r \leqslant \left(1 + \frac{1}{r} \sum_{i,j} \lfloor C x_{ij} \rceil^2 \right)^r.$$

The fact that this is an equality for $r = 1$ comes from the fact that (6.8) gives in that case the exact value of α_1. □

We start with a constructive result: one can always find linear relations with coefficients $\approx B$ such that r simultaneous linear forms are of size $\approx B^{r/n-1}$.

Theorem 15. *With the same notations as Lemma 2, the first vector* $(\lambda_1, \ldots, \lambda_n, \mu_1, \ldots, \mu_r)^t$ *of an LLL-reduced basis of the lattice generated by the columns of* $M(C, \mathbf{x}_1, \ldots, \mathbf{x}_r)$ *verifies*

$$\max_{1 \leqslant i \leqslant n} |\lambda_i| \leqslant 2^{(n-1)/4} \left(1 + \frac{1}{r} \sum_{i,j} \lfloor C x_{ij} \rceil^2 \right)^{r/2n}$$

and

$$\max_{1 \leqslant j \leqslant r} \left| \sum_{i=1}^{n} \lambda_i x_{ji} \right| \leqslant 2^{(n-5)/4} \frac{\sqrt{n+4}}{C} \left(1 + \frac{1}{r} \sum_{i,j} \lfloor C x_{ij} \rceil^2 \right)^{r/2n}.$$

Proof. Inequality 6.1 implies that the first vector of an LLL-reduced basis of L has length at most

$$\Lambda := 2^{(n-1)/4} \left(\frac{1}{r} \sum_{i,j} \lfloor C x_{ij} \rceil^2 \right)^{r/2n}$$

and coordinates $(\lambda_1, \ldots, \lambda_{n-1}, \lambda_n, \sum_{i=1}^{n} \lambda_i \lfloor C x_{1i} \rceil, \ldots, \sum_{i=1}^{n} \lambda_i \lfloor C x_{ri} \rceil)$. In particular, for all j,

$$\sum_{i=1}^{n} \lambda_i^2 + \sum_{j=1}^{r} \left(\sum_{i=1}^{n} \lambda_i \lfloor C x_{ji} \rceil \right)^2 \leqslant \Lambda^2$$

Hence,

$$\max_{1\leqslant j\leqslant r}\left|\sum_{i=1}^{n}\lambda_i \lfloor Cx_{ji}\rceil\right| \leqslant \left(\Lambda^2-\sum_{i=1}^{n}\lambda_i^2\right)^{1/2}.$$

Since

$$\left|\sum_{i=1}^{n}\lambda_i \lfloor Cx_{ji}\rceil - C\sum_{i=1}^{n}\lambda_i x_{ji}\right| \leqslant \sum_{i=1}^{n}\frac{|\lambda_i|}{2}, \tag{6.9}$$

this can be rewritten as

$$C\max_{1\leqslant j\leqslant r}\left|\sum_{i=1}^{n}\lambda_i x_{ji}\right| \leqslant \left(\Lambda^2-\sum_{i=1}^{n}\lambda_i^2\right)^{1/2}+\frac{1}{2}\sum_{i=1}^{n}|\lambda_i|.$$

Finally, the right hand side is maximal when $\lambda_i = \Lambda/\sqrt{n+4}$ for all i, and we get, for all j,

$$\max_{1\leqslant j\leqslant r}\left|\sum_{i=1}^{n}\lambda_i x_{ji}\right| \leqslant \frac{\sqrt{n+4}}{2C}\Lambda,$$

as claimed. □

To sum up, putting $B = C^{r/n} \approx \Lambda$, we get a relation with coefficients of size roughly B, the relation itself being of size $B^{1-n/r}$, as predicted by the pigeonhole principle.

The following theorem is the translation of the fact that LLL solves the SVP in an almost optimal way, i.e., the corresponding *negative* type result.

Theorem 16. *Let $\tilde{M}(C, \mathbf{x}_1, \ldots, \mathbf{x}_r)$ be the matrix obtained by removing rows $n-r+1$ to n from $M(C, \mathbf{x}_1, \ldots, \mathbf{x}_r)$. Let $\boldsymbol{v}$ be the first vector of an LLL-reduced basis of the lattice L'_C generated by the columns of this matrix. Then, for all positive real number B we have, as soon as $\|\boldsymbol{v}\| \geqslant 2^{(n-1)/2}\sqrt{n-r}B$,*

$$\min_{(\lambda_i)\in[-B,B]^n-\{0\}}\ \max_{1\leqslant j\leqslant r}\left|\sum_{i=1}^{n}\lambda_i x_{ji}\right| \geqslant C^{-1}\left\{\left(\frac{2^{(1-n)}\|\boldsymbol{v}\|^2-(n-r)B^2}{r}\right)^{1/2}-\frac{nB}{2}\right\}$$

Proof. By Theorem 2, we have

$$\min_{(\lambda_i)\neq 0}\lambda_1^2+\cdots+\lambda_{n-r}^2+\sum_{j=1}^{r}\left(\sum_{i=1}^{n}\lambda_i\lfloor Cx_{ji}\rceil\right)^2 \geqslant 2^{(1-n)}\|\boldsymbol{v}\|^2.$$

This implies

$$\min_{(\lambda_i)\in[-B,B]^n-\{0\}}\ \max_{1\leqslant j\leqslant r}\left|\sum_{i=1}^{n}\lambda_i\lfloor Cx_{ji}\rceil\right| \geqslant \left(\frac{2^{(1-n)}\|\boldsymbol{v}\|^2-(n-r)B^2}{r}\right)^{1/2},$$

or, using (6.9),

$$\min_{(\lambda_i)\in[-B,B]^n-\{0\}} \max_{1\leqslant j\leqslant r} \left|\sum_{i=1}^{n-1} \lambda_i x_{ji}\right| \geqslant \frac{1}{C}\left(\frac{2^{(1-n)}\|v\|^2-(n-r)B^2}{r}\right)^{1/2} - \frac{nB}{2C}.$$

□

When applying this theorem, the main difficulty is to ensure that the lower bound is meaningful and positive, which means that $\|v\|^2$ is at least $(rn^2/4+n-r)2^{n-1}B^2$. In practice, unless an exact linear relation exists, we expect that for C large enough, the shortest vector of L'_C attains the order of magnitude of Minkowski's bound. This means that we expect $\|v\| \approx C^{r/n}$, and that C should be chosen somewhat larger than B^n in practice. If this heuristic does not work, one should increase C. Otherwise, we will find a lower bound of the order of $B/C \approx B^{1-n/r}$.

Remark 7. There is a second way to see this problem, which has proved to be quite useful when $n-r$ is small. Assume that the matrix $X=(x_{ji})$ has maximal rank r, and without loss of generality that the minor $\tilde{X}=(x_{ji})_{1\leqslant j\leqslant r,n-r+1\leqslant i\leqslant n}$ is invertible. Put $(y_{ji})=\tilde{X}^{-1}(x_{ji})_{1\leqslant j\leqslant r,1\leqslant i\leqslant n-r}$. Then, from any simultaneous approximate linear relation

$$\left|\sum_{i=1}^{n}\lambda_i x_{ji}\right| \leqslant \varepsilon,$$

we can deduce a simultaneous approximate relation

$$\left|\sum_{i=1}^{n-r}\lambda_i y_{ji}+\lambda_{n-r+j}\right| \leqslant n\|\tilde{X}^{-1}\|_\infty\varepsilon.$$

As far as finding linear relations is concerned, this is not much of a progress: what we did just amounts to making the initial matrix sparser, which is not really useful. But, if one wants to get a lower bound on any linear relation, then this is a quite different story: finding a lower bound on one single $\left|\sum_{i=1}^{n-r}\lambda_i y_{ji}+\lambda_{n-r+j}\right|$ yields a lower bound for simultaneous approximate relations on the x_{ji}.

We are thus reduced to a one relation problem in dimension $n-r+1$, which is algorithmically much easier if n, r are huge but $n-r$ is small. This trick is described in [34], and is especially well suited to Baker's method for some types of Diophantine equations, mostly the absolute Thue equation; it gives worse lower bounds (since it finally takes into account only one linear constraint), but at a lower computational cost: an LLL reduction in dimension n is replaced by an inversion in dimension r and an LLL reduction in dimension $n-r$.

p-Adic Version

In that case again, we have an almost direct analog of the archimedean case with p-adic numbers. Let $\mathbf{x}_1, \ldots, \mathbf{x}_n$ be in $\mathbb{Z}_p^r$. If the $\mathbf{x}_i$ are in $\mathbb{Q}_p^r$ instead, one should first remove denominators before applying what follows. We also fix an integer ℓ, and for $x \in \mathbb{Z}_p$, denote by $\phi_\ell(x)$ the representative of $x \bmod p^\ell$ which is in $\{0, \ldots, p^\ell - 1\}$.

We define the $(n+r) \times (n+r)$ matrix

$$M_{\ell,C_1,C_2}(\mathbf{x}_1, \ldots, \mathbf{x}_n) := \begin{pmatrix} C_1 & 0 & \ldots & 0 & 0 & \ldots & 0 \\ 0 & C_1 & \ldots & 0 & 0 & \ldots & 0 \\ \vdots & \vdots & \ddots & \vdots & \vdots & \vdots & \vdots \\ 0 & 0 & \ldots & C_1 & 0 & \ldots & 0 \\ C_2\phi_\ell(x_{11}) & C_2\phi_\ell(x_{12}) & \ldots & C_2\phi_\ell(x_{1n}) & C_2 p^\ell & \ldots & 0 \\ \vdots & \vdots & \vdots & \vdots & 0 & \ddots & 0 \\ C_2\phi(x_{r1}) & C_2\phi(x_{r2}) & \ldots & C_2\phi(x_{rn}) & 0 & \ldots & C_2 p^\ell \end{pmatrix}.$$

Applying the same techniques as in the previous section (with the difference that we no longer have to control rounding errors) yield

Theorem 17. *Let $\mathbf{x}_1, \ldots, \mathbf{x}_n$ be vectors of p-adic integers, ℓ an integer and C_1, C_2 two integer parameters. Let $(\lambda_1, \ldots, \lambda_n, \mu_1, \ldots, \mu_r)$ the first vector of an LLL-reduced basis, of the lattice generated by the columns of the matrix M_{ℓ,C_1,C_2} $(\mathbf{x}_1, \ldots, \mathbf{x}_n)$. Then, one has*

$$|\lambda_i| \leqslant 2^{(n+r-1)/4} p^{\frac{r\ell}{n+r}} \left(\frac{C_2}{C_1}\right)^{\frac{r}{n+r}}$$

and, for all $1 \leqslant j \leqslant r$,

$$\left| \sum_{i=1}^{r} \lambda_i x_{ji} \operatorname{cmod} p^\ell \right| \leqslant 2^{(n+r-1)/4} p^{\frac{r\ell}{n+r}} \left(\frac{C_1}{C_2}\right)^{\frac{n}{n+r}}.$$

Proof. Follows from the fact that the determinant of the lattice generated by the columns of $M_{\ell,C_1,C_2}(\mathbf{x}_1, \ldots, \mathbf{x}_n)$ is $p^{r\ell} C_1^n C_2^r$. □

As a negative counterpart of this result, we have

Theorem 18. *Let $\mathbf{v}$ the first vector of an LLL-reduced basis the lattice generated by the columns of $M_{\ell,1,1}(\mathbf{x}_1, \ldots, \mathbf{x}_n)$. Then, for all $\lambda_1, \ldots, \lambda_n \in [-B, B]^n - \{0\}$, one has*

$$\max_{1 \leqslant j \leqslant r} \left| \left(\sum_{i=1}^{n} \lambda_i x_{ji} \right) \operatorname{cmod} p^\ell \right| \geqslant \left(\frac{2^{(1-n-r)}}{r} \|\mathbf{v}\|^2 - nB^2 \right)^{1/2}.$$

In practice, this negative counterpart is often used for proving that there is no small linear combination of the x_{ij} with large p-adic valuation. Indeed, if the lower bound is positive, the statement of the theorem implies that

$$\min_{1\leqslant j\leqslant r} v_p\left(\sum_{i=1}^{n}\lambda_i x_{ji}\right) < \ell$$

for all $\lambda_1,\ldots,\lambda_n \in [-B,B]^n - \{0\}$. Thus, the goal is to choose the smallest ℓ such that the lower bound of the theorem is positive. If one assumes that the right order of magnitude for $\|\boldsymbol{v}\|$ is that of Minkowski's theorem, we see that one should choose ℓ such that

$$\frac{r\ell}{n+r}\log p \approx \log B,$$

or equivalently

$$\ell \approx \frac{(n+r)}{r}\frac{\log B}{\log p},$$

or, in practice, slightly larger.

Remark 8. Though we have treated in this part only the case where x_{ji} live in $\mathbb{Z}_p$, we can also use the same technique to treat the case where x_{ji} are integral elements in some degree d algebraic extension of $\mathbb{Q}_p$, say $x_{ji} \in \mathbb{Z}_p[\theta]$, for some θ integral over $\mathbb{Z}_p$. To do this, if $x_{ji} = \sum_{k=0}^{d-1} x_{kji}\theta^k$, we associate to each linear form $\sum_{i=1}^{n}\lambda_i x_{ji}$ over $\mathbb{Z}_p[\theta]$ the d linear forms $\sum_{i=1}^{n}\lambda_i x_{kji}$ over $\mathbb{Z}_p$.

The Inhomogeneous Case

It is sometimes also useful to find an approximation of a given real number by a linear combination of other real numbers. This amounts to searching for a linear combination with one of the coefficients equal to 1. When one is looking for negative results, this can be done in a similar way to the previous section, by looking for a general relation. If no relation of small enough size exists, then a fortiori no relation exists with $x_{n+1} = 1$, and this is often sufficient. When one is interested in a lower bound, Inequality (6.2) applies.

However, a more refined tactic is possible. The main remark is the fact that, as the homogeneous case is closely approximated by SVP, itself closely approximated by the first vector of an LLL-reduced basis, the inhomogeneous case is closely approximated by CVP, itself closely approximated by Babai's nearest plane algorithm.

Applications to Inhomogeneous Diophantine Approximation

The question of inhomogeneous diophantine approximation is a much more complicated question than the homogeneous case. Affirmative results (i.e., existence of small values of $\sum_{j=1}^{n} \lambda_j x_j - x_0$) of effective type lie much deeper than the homogeneous case.

Indeed, even in the one-dimensional case, understanding what happens is closely related with the theory of uniform distribution. For instance, for α, γ real numbers, the inhomogeneous form $\alpha x + y + \gamma$ can only be arbitrarily small for (x, y) integers,

- Either if α is irrational, where it follows from the fact that $\alpha\mathbb{Z} + \mathbb{Z}$ is dense;
- Or if $\text{denom}(\alpha)\gamma \in \mathbb{Z}$ if α is rational.

More generally, Kronecker's theorem shows that for $\alpha \in \mathbb{R}^n$, the set $\{\sum_{i=1}^{n} \lambda_i \alpha_i \bmod 1\}$ is dense iff. $1, \alpha_1, \ldots, \alpha_n$ are $\mathbb{Q}$-linearly independent.

From a quantitative viewpoint then, measuring the distance between the set $\alpha x + y$ for $x, y \leqslant X$ and an arbitrary $\gamma \in [0, 1]$ is related with the discrepancy of the sequence $\alpha x \bmod 1$, itself related to the quality of the convergents of the continued fraction of α.

Example 2. For instance, if $\alpha = 1/2 + 10^{-100}$, $\gamma = 1/3$, for $x \leqslant 10^{99}$, one has $|\alpha x - y + \gamma| \geqslant 1/15$. This means that it is quite difficult to guarantee anything from what we called the positive point of view on the quality of the approximation.

Example 3. We can also argue on the lattice side that it is difficult to get constructive results. Let L be a $(2d + 1)$-dimensional lattice with an orthogonal basis $(\mathbf{b}_i)_{0 \leqslant i \leqslant 2d}$, with $\|\mathbf{b}_i\| = C^{d-i}$ for some real constant C. Then $\text{vol}(L) = 1$. Still, if we take $\mathbf{t} = \sum \mathbf{b}_i/2$, we see that $d(\mathbf{t}, L) \geqslant C^d/2$. Hence, it is not possible to bound $d(\mathbf{t}, L)$ in terms of the volume of the lattice only.

However, we can still try to construct good (though we cannot control their quality in a quantitative manner) approximations by using Babai's nearest plane algorithm, and also obtain negative results. Indeed, according to Theorem 3, using Babai's nearest plane algorithm gives us an almost optimal result.

Theorem 19. *Let* $\mathbf{x}_1, \ldots, \mathbf{x}_n, \mathbf{x}_{n+1} \in \mathbb{R}^r$ *be vectors of real numbers. Let* $\tilde{M}_C(\mathbf{x}_1, \ldots, \mathbf{x}_r)$ *be the matrix defined in Theorem 16, and* $\mathbf{v}_C$ *be the vector* $(0, \ldots, \lfloor -Cx_{1,n+1} \rceil, \lfloor -Cx_{r,n+1} \rceil)^t$. *Let* $\mathbf{v}$ *be the output of Babai's nearest plane algorithm applied to the lattice generated by the columns of* M_C *and the vector* $\mathbf{v}_C$. *Then, if* $\|\mathbf{v}\| \geqslant 2^{n/2}\sqrt{n-r}B$, *one has, for* $(\lambda_i)_{1 \leqslant i \leqslant n} \in [-B, B]^n$,

$$\left|\sum_{i=1}^{n} \lambda_i x_{ji} + x_{j,n+1}\right| \geqslant \frac{1}{C}\left(\sqrt{\frac{2^{-n}\|\mathbf{v}\|^2 - (n-r)B^2}{r}} - \frac{nB+1}{2}\right)$$

Proof. Similar to the proof of Theorem 16, replacing Theorem 2 by Theorem 3. □

Note that this result has important applications when solving Diophantine equations by Baker's method. In practice, one should rather use Proposition 1 than Theorem 3.

Remark 9. Another way to find lower bounds for that kind of quantities is to use the generalized Baker–Davenport Lemma 1.

In the p-adic case, we have very similar results:

Theorem 20. *Let* $\mathbf{x}_1, \ldots, \mathbf{x}_n \in \mathbb{Z}_p^r$ *be vectors of* p*-adic integers. Let* $\mathbf{v}$ *be the output of Babai's algorithm applied to an LLL-reduced basis of the lattice generated by the columns of the matrix* $M_{\ell,1,1}(\mathbf{x}_1, \ldots, \mathbf{x}_n)$ *and*

$$\mathbf{v}_\ell := (0, \ldots, 0, x_{1,n+1} \operatorname{cmod} p^\ell, \ldots x_{r,n+1} \operatorname{cmod} p^\ell)^t.$$

Then, for all $\lambda_1, \ldots, \lambda_n \in [-B, B]^n - \{0\}$*, one has*

$$\max \left| \left(\sum_{i=1}^{n} \lambda_i x_{ji} + x_{j,n+1} \right) \operatorname{cmod} p^\ell \right| \geqslant \left(\frac{2^{(-n-r)} \|\boldsymbol{v}\|^2 - nB^2}{r} \right)^{1/2}.$$

Schnorr's Factoring and Discrete Logarithm Algorithm

Though the behavior of the constructive variant of the inhomogeneous case described above is difficult to estimate precisely, there is still a beautiful application of it, described by Schnorr [35]. This application concerns both factoring and discrete logarithm, but we shall only describe the "factoring" part.

As many modern discrete log and factoring algorithms, Schnorr's method is based on constructing relations and recombining them. Let us point that our description and analysis are highly heuristic. We refer the interested reader to Schnorr's original paper [35] or Adleman's follow-up [36] for more details.

Building Relations

Let $p_1 < p_2 \cdots < p_t$ be the first t prime numbers. Numbers of the form $p_1^{e_1} \ldots p_t^{e_t}$ are said to be p_t-smooth, or sometimes p_t-friable. Our goal is to produce many non-trivial (i.e., $u_i \neq v_i$) relations of the form

$$\prod_{i=1}^{t} p_i^{u_i} = \pm \prod_{i=1}^{t} p_i^{w_i} \bmod N.$$

In order to obtain a relation, we shall try to find u of the form $\prod_{i=1}^{t} p_i^{u_i}$ such that $u \operatorname{cmod} N$ is very small. If $|u \operatorname{cmod} N|$ is smaller than p_t^α, a theorem by Dickman [37] shows that the "probability" that $|u \operatorname{cmod} N|$ is p_t-smooth is $\rho(1/\alpha)$, where ρ is the Dickman–de Bruijn function defined to be the continuous solution of

$$\rho(x) = 1, 0 \leqslant x \leqslant 1, \quad x\rho'(x) - x\rho(x-1) = 0, x \geqslant 1.$$

As a consequence, the problem of producing approximately t relations is reduced to the problem of computing $\approx t/\rho(1/\alpha)$ different values of u such that $|u \,\mathrm{cmod}\, N|$ is smaller than p_t^{α}.

Write $u \,\mathrm{cmod}\, N$ as $u - vN$. The fact that $u - vN$ is very small means that $\frac{u}{vN} \approx 1$, that

$$\log \frac{u}{vN} \approx \frac{u - vN}{vN}.$$

Thus, we want $\log \frac{u}{vN}$ to be very small. If we also request for v to be p_t-smooth, we find that this means that

$$\sum_{i=1}^{t} u_i \log p_i - \sum_{i=1}^{t} v_i \log p_i - \log N$$

is very small. In other words, such relations correspond to small values of inhomogeneous linear forms!

Consider the t-dimensional lattice L generated by the columns of the matrix

$$\begin{pmatrix} \log p_1 & 0 & \dots & 0 \\ 0 & \log p_2 & \dots & 0 \\ 0 & 0 & \dots & \log p_t \\ C \log p_1 & C \log p_2 & \dots & C \log p_t \end{pmatrix}$$

and the vector $\mathbf{v}_N = (0, \dots, 0, -C \log N)^t$. The matrix above has real coefficients; in practice, one should replace the terms $\log p_i$ by $\lfloor C' \log p_i \rceil$ and replace C by CC'. This variant is analyzed in [36]. We stick to the matrix above in order to simplify the analysis.

For $\boldsymbol{e} \in \mathbb{Z}^n$, we shall define $\Lambda(\boldsymbol{e}) := \sum_{i=1}^{n} e_i \log p_i - \log N$.

A vector of L is of the form

$$\mathbf{v} = (e_1 \log p_1, \dots, e_t \log p_t, \sum_{i=1}^{t} e_i C \log p_i),$$

and we can associate to it two integers $u = \prod_{e_i > 0} p_i^{e_i}$, $v = \prod_{e_i < 0} p_i^{-e_i}$ such that $\Lambda(\boldsymbol{e}) = \log u - \log vN$.

Let us evaluate $\|\mathbf{v} - \mathbf{v}_N\|$.

$$\begin{aligned} \|\mathbf{v} - \mathbf{v}_N\|^2 &= \sum_{i=1}^{t} (e_i \log p_i)^2 + \left(\sum_{i=1}^{t} e_i C \log p_i - C \log N \right)^2, \\ &= \sum_{i=1}^{t} (e_i \log p_i)^2 + C^2 \Lambda(\boldsymbol{e})^2 \\ &\geqslant \frac{1}{t} (\sum_{i=1}^{t} |e_i| \log p_i)^2 + C^2 \Lambda(\boldsymbol{e})^2. \end{aligned}$$

Thus

$$\log u + \log v \le \sqrt{t}\|\mathbf{v} - \mathbf{v}_N\|, \quad |\log u - \log v - \log N| \le \frac{\|\mathbf{v} - \mathbf{v}_N\|}{C}.$$

Since C is large, for $\mathbf{v}$ close to $\mathbf{v}_N$ this means that $u \approx vN$, as expected, and that

$$\begin{aligned}\log v &\le \log u - \log N + \frac{\|\mathbf{v} - \mathbf{v}_N\|}{C} \\ &\le -\log N - \log v + \|\mathbf{v} - \mathbf{v}_N\|(\sqrt{t} + C^{-1}).\end{aligned}$$

so that

$$\log v \le -\frac{\log N}{2} + \|\mathbf{v} - \mathbf{v}_N\|\frac{\sqrt{t} + C^{-1}}{2},$$

which yields

$$|u - vN| \approx vN \log\frac{u}{vN} \approx \sqrt{N}\frac{\|\mathbf{v} - \mathbf{v}_N\|}{C}\exp\left(\frac{(\sqrt{t} + C^{-1})\|\mathbf{v} - \mathbf{v}_N\|}{2}\right).$$

In order to get $|u - vN|$ of the order of $p_t^{\alpha + o(1)}$, and assuming $C^{-1} = o(1)$ as $N \to \infty$, we should thus have

$$\|\mathbf{v} - \mathbf{v}_N\| \leqslant 2\log C - \log N + \frac{2\alpha}{\sqrt{t}}\log p_t + o(\log N). \tag{6.10}$$

Schnorr then argues that under reasonable equirepartition assumptions on p_t-smooth numbers, there are sufficiently many vectors in L verifying (6.10). In order to give a feeling of the argument, let us mention that the determinant of our lattice is of the order of $C2^{O(t)}$, so that the number of vectors in a ball of radius R is of the order of $R^t t^{t/2} C^{-1} 2^{O(t)}$; in order to find sufficiently many vectors at distance roughly $\log C^2/N$, we thus need to have $C^{1/t} \approx \log C^2/N$, which means $t \approx \log C/\log\log(C^2/N)$; for $C = N^c$ a power of N, this gives a polynomial dimension t, and thus applying Babai's algorithm to this lattice has a polynomial complexity in N.

However, asking for a fast algorithm finding many vectors verifying (6.10) is far too optimistic in general, which explains why this nice idea has only encountered a very modest success.

Combining Relations

For the sake of completeness, we review shortly how one factors N by combining the relations built in the previous section, in the form

$$\prod_{i=1}^{n} u_i^{\varepsilon_i} = \prod_{i=1}^{n} w_i^{\varepsilon_i} \bmod N,$$

choosing $\varepsilon_i \in \{0, 1\}$ so that both sides are square, say $U^2 = V^2 \bmod N$. With such a recombination, one computes $\gcd(U - V, N)$, and unless one is unlucky or one has built the relations in a unsuitable way, after possibly some failures one ends up with a factor of N.

Recall that we have

$$u_i = \prod_{p_b \in B} p_b^{\alpha_{ib}}, w_i = \prod_{p_b \in B} p_b^{\beta_{ib}}.$$

The recombination above then rewrites, re-arranging terms,

$$\prod_{p_b \in B} p_b^{\sum_{i=1}^{n}(\alpha_{ib} - \beta_{ib})\varepsilon_i} = 1 \bmod N$$

In order that this is of the form square = square, it is thus sufficient that

$$\sum_{i=1}^{n}(\alpha_{ib} - \beta_{ib})\varepsilon_i = 0 \bmod 2$$

for all b, and similarly with β_{ib}. This means that ε_i are found as a vector of the kernel modulo 2 of an integral matrix.

An important remark at that point is the fact that for this method to have a chance of success, one should have *at least as many relations as elements in the factor base*, so that the system one is trying to solve has more unknowns than equations, and has generically a nontrivial kernel.

Analysis

A good feature of this algorithm is the fact that one seemingly only has to reduce a lattice which depends on N in a rather weak way; once this reduction step performed, one just needs to perform "close vector search," by Babai's algorithm (plus some randomized rounding, say). However, note that in order to get an homogeneous problem – and thus a better control in their analyzes – Schnorr and Adleman replace our lattice with the $t + 1$-dimensional lattice $L \oplus \mathbb{Z}\boldsymbol{v}_N$.

However, the fact that both LLL and LLL associated with Babai's nearest plane algorithm loses an exponential factor in the dimension means that the vectors returned by LLL/LLL plus Babai are often too large/far away (by a factor $2^{O(t)}$, whereas the actual expected size is $t \log t \ldots$) to give integers with a significant probability of smoothness.

In order to get very small vectors, one has to reduce the lattice in a stronger sense. In any case, the large dimension of the underlying lattice makes the method somewhat unpractical. Schnorr's estimates show that to factor a 500-bit number, one should reduce a dimension 6300 lattice with 1500-bit integer coefficients.

Remark 10. We must mention two other applications of lattice basis reduction to factoring.

The first is the "2/3 algorithm" by Vallée [38], one of the only subexponential algorithms to actually have proven complexity, while most of the complexities rest on some unproven assumptions on the good repartition of smooth numbers in some sparse sequences.

The second one lies at the core of the number field sieve algorithm, the most efficient factoring algorithm (at least for numbers of the form pq) at the time this survey is written. At the end, one has to compute a huge (exact) square root of $\beta \in K$ in an algebraic number field; Montgomery [39] and then Nguyen [40] showed how to use lattice basis reduction algorithms in order to ease this computation. Indeed, in that setting, LLL can be used to produce many α_i such that $\prod_{i=1}^{n} \alpha_i^2 | \beta$, and $\beta / \prod_{i=1}^{n} \alpha_i^2$ is much smaller than β; hence the final square root is in fact much easier to compute after this "reduction" step.

Remark 11. As is often the case, a slight adaptation of this algorithm gives an algorithm of the same complexity to compute discrete logarithms in $(\mathbb{Z}/p\mathbb{Z})^*$.

Applications to Diophantine Equations

Some of the effective methods of Section "Linear Relations" have been at the core of the applications of LLL to the effective solution of Diophantine equations. The technique involved is rather heavy, and we refer to [41] for an exhaustive account, including detailed bibliography.

We shall thus content ourselves of a rather sketchy account, concentrating on the link with lattice basis reduction algorithms and the machinery developed in the previous sections.

The case of interest is Diophantine equations which can be treated by Baker's method. We first recall Baker's theorem, which is a generic name for a large number of lower bounds obtained by Baker and others from the mid 1960s [42] to nowadays.

Theorem 21 (Baker et al.). *Let $\alpha_1, \dots, \alpha_n$ be algebraic numbers, and λ_i integers, and put $\Lambda := \sum_{i=1}^{n} \lambda_i \operatorname{Log} \alpha_i$. Then, there exists a constant $C(\alpha_1, \dots, \alpha_n)$ such that*

- *Either $\Lambda = 0$;*
- *Or $|\Lambda| \geqslant \exp(-C(\alpha_1, \dots, \alpha_n) \log \max_i |\lambda_i|)$.*

See [43] for the best result currently available.

The key of the success of Baker's method is the dependence on the λ_i, but the price to pay is the huge size of the constant $C(\alpha_1, \dots, \alpha_n)$, which usually yields huge bounds, and makes it almost impossible to completely solve an equation without any further ingredient.

This theorem is at the core of many finiteness results for families of Diophantine equations, by what is called Baker's method. The overall strategy of Baker's method is the following:

- By algebraic number theoretic arguments, reduce the initial equation to the problem of finding very small values of linear forms

$$\Lambda := \sum_{i=1}^{n} \lambda_i \operatorname{Log} \alpha_i + \operatorname{Log} \alpha_{n+1},$$

  where the $\alpha_i \in \mathbb{C}$ are algebraic over $\mathbb{Q}$, and the Log function is the principal determination of the complex logarithm.
  By very small value, we mean that to every solution of the initial equation corresponds an n-uple $\lambda_1, \dots, \lambda_n$ such that

$$|\Lambda| \leqslant \exp(-c(\alpha_1, \dots, \alpha_{n+1}) \max_i |\lambda_i|^\theta),$$

  for some positive real θ (often equals to 1);
- Apply one form or another of Baker's theorem, to get a lower bound on Λ, the case $\Lambda = 0$ being usually discarded as irrelevant or impossible.

Comparing these two bounds yields an explicit upper bound on $\max_i |\lambda_i|$. However, this bound is usually much too large to allow enumeration. The reason why this bound is too large is the fact that Baker's constant $C(\alpha_1, \dots, \alpha_{n+1})$ is often extremely pessimistic.

In some applications – those who are interested in this survey – of Baker's method, are in the favorable situation where *only the* λ_i *are unknown in* Λ. In that case, once the process above is finished and one has an initial bound on the λ_i, the bound obtained by Baker's method can be greatly refined using techniques from Sect. Linear Relations.

Indeed, since the linear form is explicit, once the λ_i are bounded, one can replace Baker's theorem by the estimate of Theorem 19, to get a new lower bound on Λ, and thus a new upper bound on $\max_i |\lambda_i|$. We can start again the process using this new upper bound, etc. In practice, since the upper bound is exponential in λ_i, and since the heuristics developed in Sect. 15 show that the lower bounds we get for Λ with $\lambda_i \leqslant L$ will be of the order of $L^{1-n/r}$, we should have, for some small constant κ,

$$\kappa \times L^{1-n/r} \leqslant \Lambda \leqslant \exp(-c(\alpha_1, \dots, \alpha_{n+1}) \max_i |\lambda_i|^\theta),$$

so that the new bound on $\max_i |\lambda_i|$ is of the order of $(\log L)^{1/\theta}$.

We shall give an example of this reduction process in Section "LLL and ABC"

Baker and Davenport's Approach

The idea of this reduction process goes back to the paper by Baker and Davenport [26] about the solution of a system of simultaneous Pell equations, where they use a special case of Lemma 1. Let us write this system as

$$\begin{cases} x^2 - vy^2 = a \\ x^2 - v'z^2 = a' \end{cases} \tag{6.11}$$

Given a Pell equation $x^2 - vy^2 = a$, one can determine a unit $\eta \in \mathbb{Q}(\sqrt{v})$ and a finite set of elements S of $\mathbb{Q}(\sqrt{v})$ such that the x-value of any solution of the equation is of the form

$$x_d = \tau\eta^d + \frac{a}{\tau\eta^d},$$

where $\tau \in S$ and d is an integer.

Denote by η', S' similar objects for the equation $x^2 - wz^2 = a'$. By equating the two different expressions for x, we thus end up with one equation in unknowns $d, d' \in \mathbb{Z}$ for each pair $(\tau, \tau') \in S \times S'$:

$$\tau\eta^d + \frac{a}{\tau\eta^d} = \tau'\eta'^{d'} + \frac{a'}{\tau'\eta'^{d'}},$$

We can assume $\eta, \eta' > 1$; otherwise, replace η by $\pm 1/\eta$ and similarly for η'. This means that

$$\tau\eta^d - \tau'\eta'^{d'} \ll C^{-\max(d,d')},$$

for some $C > 1$, or,

$$\operatorname{Log}\frac{\tau}{\tau'} + d\operatorname{Log}\eta - d'\operatorname{Log}\eta' \ll C_2^{-\max(d,d')}. \tag{6.12}$$

Baker and Davenport then proceed to use Baker's Theorem 21 to show that the left hand side is $\gg C_3^{-\log\max(d,d')}$, which allows to obtain $\max(d, d') \leqslant \Delta$.

Knowing that $d, d' \leqslant \Delta$, one can apply either Theorem 19 (using CVP on a two-dimensional lattice) or Lemma 1 (using continued fractions) to compute a new, greatly improved, lower bound on

$$\operatorname{Log}\frac{\tau}{\tau'} + d\operatorname{Log}\eta - d'\operatorname{Log}\eta',$$

which in turn yields a new upper bound, etc. until exhaustive enumeration of all $x_d, d \leqslant \Delta$, is possible.

LLL and ABC

The first occurrence of LLL for this reduction task for diophantine equations is probably encountered in the paper by de Weger [44], with, among other applications, the study of the ABC conjecture. This had important consequences on the practical applications of Baker's method for diophantine equations. The idea is the same as in Baker and Davenport, except that instead of having a term with one power $\eta\tau^d$, we now have a term with arbitrarily many powers; this means that we shall end up with a linear form with many variables, and then LLL.

We shall illustrate this approach with de Weger's application to the ABC conjecture, since this is probably the one that uses the smallest amount of algebraic number theory and hence the easier to describe.

The ABC Conjecture

A very important conjecture in diophantine approximation is the so-called ABC conjecture. Its importance stems mostly from the fact that it implies many classical theorems and conjectures and results in diophantine approximation, but also in other areas of number theory. It is concerned with the very simple diophantine equation

$$A + B = C,$$

and the overall meaning is that A, B, and C cannot be simultaneously large and be divisible by large powers of small primes.

More precisely, define the radical of an integer to be its squarefree part:

$$\operatorname{rad} N = \prod_{p|N} p$$

Then the conjecture can be written as:

Conjecture 1 (Masser-Oesterlé's ABC conjecture). [45] For all $\varepsilon > 0$, there exists a $\delta_\varepsilon \in \mathbb{R}^+$ such that if A, B, and C are three integers with $(A, B) = 1$ and $A + B = C$, one has

$$\max(|A|, |B|, |C|) \leqslant \delta_\varepsilon \operatorname{rad}(ABC)^{1+\varepsilon}.$$

The best results known [46] on the subject are quite far from this. See Nitaj's ABC page (http://www.math.unicaen.fr/~nitaj/abc.html) for extensive information on the subject.

Rather than trying to prove the conjecture, another approach is to try to build extremal examples for it. It can be families, or single examples. In the latter case, a reasonable measure of extremality is given by the quantity

$$\log \max(|a|, |b|, |c|) / \log \operatorname{rad}(abc),$$

as large as possible.

A first approach for finding extremal examples is due to de Weger [44], and amounts to find the solutions of $A + B = C$ where A, C have their prime factors in a prescribed set, with B small. This last assumption can be replaced by the fact that B also has its prime factors in the same set (see the same paper), but then since we have no longer any precise control on the sizes of A, B, and C it requires to use linear forms in p-adic logarithms, and we prefer to stick to the simpler case in this survey.

We shall also discuss an elegant and elementary approach due to Dokchitser [47], where one fixes products of powers of small primes a, b, and c and searches for relations of the form $\lambda a + \mu b = \nu c$, for small λ, μ, and ν.

De Weger's Approach

The idea is to search for "large" solutions of $A + B = C$, where the prime factors of A and C are in a small set $S = \{p_i, i \in [1, s]\}$ and B is small. More precisely, we fix $1 > \delta > 0$ and we are looking for solutions of the inequality $0 < C - A \leqslant C^\delta$.

This inequality can be rewritten as

$$\left|1 - \frac{A}{C}\right| \leqslant C^{\delta-1}.$$

Unless $C \leqslant 2^{\frac{1}{1-\delta}}$, for which the solutions can be found by enumeration, the left hand side is $\leqslant 1/2$. Write $C = \prod_{i=1}^s p_i^{c_i}$, $A = \prod_{i=1}^s p_i^{a_i}$.

We can thus use the inequality $|z - 1| \log 4 \geqslant |\operatorname{Log} z|$, valid for $|z - 1| \leqslant 1/2$. We get:

$$\begin{aligned}\left|\sum_{i=1}^s (c_i - a_i) \log p_i\right| &\leqslant C^{\delta-1} \log 4 \leqslant (\min_i p_i)^{(\delta-1)\max_i c_i} \log 4 \\ &=: g(S, \delta)^{\max_i c_i} \log 4, \end{aligned} \tag{6.13}$$

for some $g(S, \delta) < 1$.

Since $(\min_i p_i)^{\max_i a_i} \leqslant A \leqslant C \leqslant (\prod_{i=1}^s p_i)^{\max_i c_i}$, we can replace the upper bound above by

$$\left|\sum_{i=1}^s (c_i - a_i) \log p_i\right| \leqslant g^*(S, \delta)^{\max_i c_i, a_i}, \tag{6.14}$$

Baker's theorem implies, unless $C = A$, that for some constant $f(S)$,

$$\left|\sum_{i=1}^s (c_i - a_i) \log p_i\right| \geqslant \exp(-f(S) \log \max_i (c_i, a_i)).$$

In typical examples, this lower bound is very large. However, notice that once $c_i - a_i \leqslant C$ are bounded, we can substitute to Baker's theorem (6.14) by the bound of Theorem 15 to the real numbers $(\log p_i)_{i\in S}$. The heuristic analysis which follows the theorem shows that the new lower bound is of the order of $\mathcal{C}^{1/s-1}$. Comparing it to (6.13) again, we obtain a new bound which is of the order of $\log C$.

In order to give the reader the "feeling" of what happens in this reduction process, we treat the case where $S = \{2,3,5\}$, and $\delta = 1/2$. In that case, we have either $C = 1$, or rewriting (6.13) in that case,

$$|(c_1 - a_1)\log 2 + (c_2 - a_2)\log 3 + (c_3 - a_3)\log 5| \leqslant 2^{-\max_i c_i/2}\log 4.$$

Now, $2^{\max_i a_i} \leqslant 30^{\max_i c_i}$, which shows that

$$2^{-\max_i c_i/2}\log 4 \leqslant \exp\left(-\frac{\log 2}{\log 30}\max_i(a_i, c_i)\right)\log 4.$$

Matveev's version of Baker's theorem implies that this linear form is either 0 (which implies $A = C$) or

$$\begin{aligned}|(c_1 - a_1)\ &\log 2 + (c_2 - a_2)\log 3 + (c_3 - a_3)\log 5| \\ &\geqslant \exp(-2.2\times 10^7 \log(e\max_i(a_i, c_i))).\end{aligned}$$

Comparing those two bounds, we find that $\max_i c_i \leqslant 2.45\times 10^9 =: \mathcal{B}$. Now, take $\kappa = 200\times(2.45\times 10^9)^3$, and consider the matrix

$$\begin{pmatrix} 1 & 0 & 0 \\ 0 & 1 & 0 \\ \lfloor \kappa\log 2\rceil & \lfloor \kappa\log 3\rceil & \lfloor \kappa\log 5\rceil \end{pmatrix}.$$

The first vector v of an LLL-reduced basis of this lattice has length $\approx 1.03\times 10^{10}$, and hence the lower bound is

$$\frac{1}{\kappa}\sqrt{2^{1-3}\|v\|^2 - 2\mathcal{B}^2} - 3\mathcal{B}/2 \approx 2.04\times 10^{-21},$$

which implies, in view of 6.13,

$$\max_i(a_i, c_i) \leqslant \left\lfloor -\log(2.04\times 10^{-21}/\log 4)\frac{\log 30}{\log 2}\right\rfloor = 235.$$

Repeating the reduction step with $C = 20\times 235^3$, we obtain a lower bound of 9.49×10^{-7}, so that $\max_i(a_i, c_i) \leqslant 69$. The next bounds are 58 and 56, after which the process more or less stops. We now have to enumerate all triples $a_i \leqslant 56$, $c_i \leqslant 56$, with the information that for all i, $a_i c_i = 0$, since we assumed $(A, C) = 1$. We find the solutions

$$\begin{array}{ccc} 1+1=2 & 2+1=3 & 3+1=4 \\ 5+1=6 & 8+1=9 & 9+1=10 \\ 25+2=27 & 27+5=32 & 80+1=81 \\ 243+7=250 & 243+13=256 & 625+23=648 \\ & 15552+73=15625 & \end{array}$$

$$\begin{array}{cc} 3+2=5 & 4+1=5 \\ 15+1=16 & 24+1=25 \\ 125+3=128 & 128+7=135 \\ 2025+23=2048 & 3072+53=3125 \\ 32768+37=32805 & \end{array}$$

Note that if bounds were somewhat larger, one should probably, at least for the large values of c_i, continue to use (6.13) and enumerate the very few vectors in the lattice with sufficiently small norm.

Dokchitser's Approach

To be complete on the subject of the relations between the LLL algorithm and the ABC conjecture, we would like to discuss a recent, good and simple approach due to Dokchitser [47]. The point is to fix integers of small radical and look for a small relation $\lambda a + \mu b = \nu c$, which is a linear relation as studied in Section "Linear Relations".

What can we expect? If we fix C and reduce the lattice generated by the columns of $M_C(a, b, c)$, we expect to find integers

$$\lambda, \mu, \nu \approx (C \max(|a|, |b|, |c|))^{1/3}$$

such that

$$\lambda a + \mu b + \nu c \approx C^{-2/3} \max(|a|, |b|, |c|)^{1/3}.$$

This means that as soon as $C > \max(|a|, |b|, |c|)^{1/2}$, we can expect to find a true relation (not an approximate one) $a\lambda + b\mu + c\nu = 0$.

In practice, rather than picking C larger than this at random, a better strategy is to proceed as in Theorem 14:

- Compute a basis of the free $\mathbb{Z}$-module Ker $((x, y, z) \mapsto ax + by + cz)$;
- LLL-reduce, so that the first vector gives a short relation.

Theorem 14 shows that we shall find $\lambda, \mu, \nu \approx \max(|a|, |b|, |c|)^{1/2}$. If a, b, c are large powers of fixed primes, we thus have

$$\text{rad}\,(\lambda\mu\nu abc) \ll \max(|a|, |b|, |c|)^{3/2}$$

which is the order of $\max(|\lambda a|, |\mu b|, |\nu c|)$ if a, b, c have the same size. We thus construct that way a family of example with the "right order of magnitude" in ABC-conjecture.

Note that the multiplicative structure of λ, μ, ν has a strong influence on the "quality" of the example in view of the measure we chose. This means that one should not always take the smallest possible λ, μ, ν (i.e., the first vector of an LLL-reduced basis), but also try small linear combinations of vectors of an LLL-reduced basis.

This strategy generalizes to the case where a, b, c live in a given number field; we leave it to the reader. The best abc-example (with respect to the classical measure of quality given above) found so far by Dokchitser using this strategy is

$$(A, B, C) = (13^{10}37^2, 3^7 19^5 71^4 223, 2^{26} 5^{12} 1873).$$

Other Diophantine Equations

We discuss shortly the application of Baker's method and LLL to other families of Diophantine equations.

The Thue Equation

The Thue Equation [48] is the natural generalization of Pell's equation, i.e., the equation $P(X, Y) = a$, for P a homogeneous irreducible polynomial of degree $d \geqslant 3$, except that Thue's theorem asserts that this equation has only finitely many solutions. Thue's proof was ineffective, but we shall use Baker's method to sketch an effective proof based on linear forms in logarithms. The first to give such an effective proof was Baker [49], and the seminal paper on algorithms for solution of a general Thue equation is the work by Tzanakis and de Weger [50].

Assume that P is monic (which can be obtained by a suitable change of variables), and let $\alpha = \alpha_1$ be such that $P(1, \alpha) = 0$, with $\alpha_2, \ldots, \alpha_d$ its conjugates.

From an algebraic point of view, Thue equation means that for any solution (x, y), the algebraic integer $x - \alpha y$ has norm a in the number field $\mathbb{Q}(\alpha)$. Hence,

$$x - \alpha y = \eta u_1^{\lambda_1} \ldots u_r^{\lambda_r}, \tag{6.15}$$

for u_i a system of fundamental units of $\mathbb{Q}(\alpha)$, and where η lives in a finite set S. Both S and u_i can be explicitly determined, see [8].

From the fact that the product $\prod_{i=1}^{d}(x - \alpha_i y)$ is bounded, one can deduce that exactly one term, e.g., the one with $i = 1$ in the product, is very small. In fact, one can prove that $\min_{1 \leqslant i \leqslant d} |x - \alpha_1 y| \ll |x|^{1-d}$, and the other terms are then very close to $(\alpha_i - \alpha_1) y \asymp x$.

In particular, if one picks $i, j > 1$, one has

$$(\alpha_i - \alpha_1)(x - \alpha_j y) - (\alpha_j - \alpha_1)(x - \alpha_i y) = (\alpha_j - \alpha_i)(x - \alpha_1 y) \ll |x|^{1-d}. \quad (6.16)$$

Using the decomposition (6.15), we obtain

$$(\alpha_j - \alpha_1)\eta_i u_{1i}^{\lambda_1} \dots u_{ri}^{\lambda_r} - (\alpha_i - \alpha_1)\eta_j u_{1j}^{\lambda_1} \dots u_{rj}^{\lambda_r} \ll |x|^{1-d},$$

which implies

$$\operatorname{Log} \frac{(\alpha_j - \alpha_1)\eta_i}{(\alpha_i - \alpha_1)\eta_j} + \sum_{k=1}^{r} \lambda_k \operatorname{Log} \frac{u_{ki}}{u_{kj}} + \lambda_{r+1} 2i\pi \ll |x|^{1-d}, \quad (6.17)$$

for some integer λ_{r+1}. In this inequality, only the λ_k are unknown. It, thus, has exactly the same form as (6.12) or again as (6.13), once one notices that $x \asymp \log \max_i |\lambda_i|$.

In a similar way as for ABC, Baker's bound will yield a bound on the left hand side of (6.17), which gives in turn a bound on the b_k. In view of (6.17), we are thus left with finding small values of a inhomogeneous linear form with bounded coefficients, or rather, in practice, proving that no small value exists, which is done by means of Theorem 19, and yields a new upper bound on $\max |b_k|$, and so on.

In that setting, notice that we can vary i, j among $r - 1$ pairs giving inhomogeneous independent linear forms, which must be simultaneously small. Doing this, one can use the technique described in Remark 7 to reduce the problem to an inhomogeneous approximation problem in two variables. This is a crucial idea in order to be able to deal with very high degree equations.

There is a wealth of literature on the Thue equation, describing other important algorithmic improvements, as well as the somewhat more difficult relative case. We refer the interested reader to [41] for a more precise description, and for an extensive bibliography.

Remark 12. Notice that (6.16) is, after dividing by the right hand side, a special case of *unit equation*, i.e., $\tau_1 u_1 + \tau_2 u_2 = 1$, where u_1 and u_2 are units in some number field. What we actually do in Thue's equation is reducing to a unit equation, and solving the latter, except that we have a more precise control on the size of the various conjugates in the case of Thue's equation. For instance, the trick of Remark 7 is useless in the case of a general unit equation, since we have control on the size of one single conjugate, and hence one single "small linear form."

Remark 13. Combining the archimedean arguments above with $\mathfrak{p}$-adic ones, one can extend the techniques above to the case of the Thue–Mahler equation [51], i.e., replace the fixed right hand side a by $ap_1^{n_1} \dots p_r^{n_r}$ for fixed a, p_i and unknowns n_i. Or, in the setting of unit equations, dealing with S-unit equations. At that point, one has to use lower bounds for linear forms in p-adic logarithms, and the reduction involves the p-adic techniques that we have developed.

Elliptic Equations

We call elliptic equation an equation $F(x, y) = 0$ such that the curve defined by this equation is an elliptic curve.

In the elliptic logarithm method, we replace the logarithm by the *elliptic logarithm*, namely a morphism $\phi : E_c(\mathbb{R}) \to \mathbb{R}$, associated to each unbounded component $E_c(\mathbb{R})$ of $E(\mathbb{R})$.

An important property of this morphism must be that "large integral solutions" have small logarithms, i.e.,

$$\phi((X, Y)) \ll |X|^{-\alpha}, \tag{6.18}$$

for some α. This depends on the model chosen, since the fact for a point to be integral depends on the model. In practice, when one studies a specific model, one starts by proving a general inequality of the kind (6.18) for that model, where the implied constant depends on the curve.

Afterwards, the strategy is very similar to any application of Baker's method:

- The set of rational solutions of $F(x, y) = 0$ has a group structure, for which one first determines a generating set $P_0, \ldots, P_r$ (this is **not** an easy task, see e.g., [52]);
- One looks for a general solution as a linear combination $\sum n_i P_i$;
- A large integral solution has $\phi((X, Y)) = \sum n_i \phi(P_i) \ll |X|^{-\alpha}$
- One relates X and the n_i using heights, getting $\sum n_i \phi(P_i) \ll \exp(-cn_i^2)$.

We are left in a similar situation as previously, where we are looking for very small values of a linear form in (elliptic) logarithms: apply a Baker type theorem to get a lower bound on the linear form, compare with the upper bound, get a bound on $\max |n_i|$, reduce it using LLL. The technicalities are even tougher than in the case of the Thue equation.

This method was simultaneously introduced by Gebel, Pethő, and Zimmer [53] and by Stroeker and Tzanakis [54]. The former authors used it in their large computation [55] about Mordell's equation $y^2 = x^3 + k$, which they solved for all $|k| \leqslant 10{,}000$. The case of a general plane model is described in [56].

Let us conclude by mentioning that Bilu [57] proposed a technique based on unit equations to treat superelliptic equations $y^p = f(x)$, which were previously solved via reduction to the Thue equations. The algorithmic aspects are very similar to the Thue case, and were studied by Bilu and the author [58].

Again, the methods described and mentioned above can be adapted to find S-integral solutions of elliptic and hyperelliptic equations, see [59, 60].

Approximation by Algebraic Numbers

From the algorithmic results concerning linear relations, we can deduce some applications to approximations of a given real number by an algebraic number.

Finding Polynomial with Small Values at ξ

We give a result about finding a polynomial with a small value in ξ, which is a straightforward application of Theorem 15 about linear relations. We also give a general Lemma which allows one to relate polynomials with small values at x and algebraic approximations of x.

Theorem 22. *Let ξ be real, complex, or integral or in $\mathbb{Z}_p[\theta]$. In the latter case, we assume that θ is integral over $\mathbb{Z}_p$, and write $[\mathbb{Q}_p(\theta) : \mathbb{Q}_p] = \delta$. Given d, and for any integers C and ℓ, one can find in polynomial time a polynomial P with integer coefficients such that* $\deg P \leqslant d$ *and*

$$H(P) \leqslant \begin{cases} 2^{d/4}\left(1+\sum_{i=0}^{d}\lfloor C\xi^i\rceil^2\right)^{\frac{1}{2(d+1)}} & \text{if } \xi \in \mathbb{R} \\ 2^{d/4}\left(1+\frac{1}{2}\sum_{i=0}^{d}(\lfloor C\,\mathrm{Re}\,\xi^i\rceil^2+\lfloor C\,\mathrm{Im}\xi^i\rceil^2)\right)^{\frac{1}{(d+1)}} & \text{if } \xi \in \mathbb{C} \\ 2^{\frac{d+\delta}{4}}\, p^{\frac{\delta\ell}{d+\delta+1}} & \text{if } \xi \in \mathbb{Z}_p[\theta] \end{cases}$$

and

$$|P(\xi)| \leqslant \begin{cases} \frac{2^{(d-4)/4}}{C}\sqrt{d+5}\left(1+\sum_{i=0}^{d}\lfloor C\xi^i\rceil^2\right)^{\frac{1}{2(d+1)}} & \text{if } \xi \in \mathbb{R} \\ \frac{2^{(d-4)/4}}{C}\sqrt{d+5}\left(1+\frac{1}{2}\sum_{i=0}^{d}(\lfloor C\,\mathrm{Re}\,\xi^i\rceil^2+\lfloor C\,\mathrm{Im}\xi^i\rceil^2)\right)^{\frac{1}{d+1}} & \text{if } \xi \in \mathbb{C} \\ p^{-\ell} & \text{if } \xi \in \mathbb{Z}_p[\theta] \end{cases},$$

where $|\cdot|$ is the p-adic valuation in the case of p-adic arguments, the ordinary absolute value otherwise.

Proof. Apply Theorem 15 to the linear form $\sum_{i=0}^{d}\lambda_i\xi^i$ if ξ is real; to the linear forms

$$\sum_{i=0}^{d}\lambda_i\,\mathrm{Re}\,(\xi^i), \quad \sum_{i=0}^{d}\lambda_i\,\mathrm{Im}(\xi^i)$$

if ξ is in $\mathbb{C}-\mathbb{R}$; apply Theorem 17 to the linear forms corresponding to the coefficients of the powers of θ in $\sum_{i=0}^{d}\lambda_i\xi^i$ in the p-adic case. □

In a less technical form, the polynomial obtained has height $H(P) \approx C^{1/(d+1)}$ and $P(\xi) \approx C^{1-1/(d+1)}$, in the real case. In the complex cases, since we have to treat simultaneously the real and imaginary part, one should replace $1/(d+1)$ by $2/(d+1)$.

Remark 14. One can similarly look for an approximation of ξ by an algebraic integer of degree $\leqslant d$ by using the inhomogeneous approach, i.e., by trying to write ξ^d as a linear combination $\sum_{j=0}^{d-1}\lambda_j\xi^j$. As pointed, since we are in the inhomogeneous approach, nothing is guaranteed. For instance, for any monic polynomial P of degree $\leqslant d$, one has $P(1/2) \geqslant 2^{-d}$, whatever the height of P.

Link with Approximation by Algebraic Numbers

A natural question is whether those results on polynomials with small value at ξ can be translated into results on approximation by algebraic numbers of not too large degree and height. In practice, they can, via one of a variety of theorems which relate $\min|\xi-\alpha|$ and $|P(\alpha)|$, where the minimum is taken over the roots of P. Of course, we expect that if ξ is close enough to a root α of P, we have $|\xi-\alpha||P'(\xi)| \approx |P(\xi)|$.

We quote for instance the simplest result of that type:

Theorem 23. *Let $P(X)$ be a nonconstant complex polynomial, and assume $P'(\xi) \neq 0$. Then*

$$\min_{\alpha} |\xi - \alpha| \leqslant n \left| \frac{P(\xi)}{P'(\xi)} \right|,$$

where the minimum is over the roots of P.

Proof. Follows from taking the logarithmic derivative of P,

$$\frac{P'(\xi)}{P(\xi)} = \sum_{\alpha} \frac{1}{\xi - \alpha}.$$ □

In practice, if ξ is chosen at random and P constructed by the techniques of Sect. 15, we expect that P' behaves "randomly" at ξ, i.e., $|P'(\xi)| \approx H(P)|\xi|^d$, and we obtain a lower bound of the order of $|P(\xi)|/(H(P)|\xi|^d)$.

For more general statements suitable for dealing with the case where $|P'(\xi)|$ is also small, we refer to Appendix A.3 of Bugeaud's book [61].

Rational Points Close to Varieties

The results of Section "Linear Relations" can be seen as finding integral points close to a projective or affine linear variety. On the other hand, solving a diophantine equation is finding integral or rational points on an algebraic variety. In this section, we study the problem of finding all, or some points close to a sufficiently smooth variety.

This is an extremely technical subject, and the optimizations are still at quite a preliminary stage of development. We will simply sketch a few ideas on this subject.

Elkies–Lefèvre–Muller Method

This method has been introduced by Lefèvre and Muller [62] in a very specialized setting. Looking closely, one notices that their method in fact involves a continued fraction computation, used to deal with one inhomogeneous linear form in two variables.

The problem they were studying, which is of intrinsic interest, is the *table maker's dilemma*. In a nutshell, one has to evaluate a function f at a point of the form $x/2^p$ (i.e., a floating-point number), $2^{p-1} \leqslant x < 2^p$ a given integer. One wants to find the integers x such that this evaluation is "hard to round," i.e., $f(x/2^p)$ is very close from a number of the same form $y/2^p$, or equivalently, that $|2^p f(x/2^p) - y|$ is very small for integers x, y. Typically one looks for extremal cases, i.e., heuristically

$$|2^p f(x/2^p) - y| \leqslant 2^{-p}.$$

The strategy is to notice that, though the problem seems somewhat untraceable in this form, it becomes an inhomogeneous linear form problem if f is a degree 1 polynomial. Thus, one should split the domain into pieces, on which one replaces the function by a degree 1 polynomial (notice that replacing by a degree 0 polynomial would essentially be the exhaustive approach). The analysis of what can be expected is as follows:

We split the interval $[2^{p-1}, 2^p[$ into N pieces of size $2^p/N$. Over each interval, we shall obtain a lower bound for a linear form $|\alpha x + \beta - y|$, which we can expect to be of the order of $(2^p/N)^{-2}$. Indeed, we have $(2^p/N)^2$ different terms which we expect to be uniformly spaced into $[0, 1]$.

On the other hand, we need this lower bound to be larger than the error made when estimating $2^p f(x/2^p)$ by $\alpha x + \beta$, which is of the order of $1/(2^p N)$. This constraint means that we should choose $N \approx 2^{p/3}$, which gives a total of $2^{2p/3}$ intervals, over which we mainly have to compute a continued fraction (unless we happen to find a solution to our inequality). Thus, up to polynomial factors, the heuristic complexity of the algorithm is $\tilde{O}(2^{2p/3})$.

Elkies' Version

At more or less the same time, Elkies [63] developed a very similar method, though presented in a much more natural way, in the setting of finding rational points close to a curve. This led him to describe what is more or less an homogeneous version of Lefèvre–Muller's method.

Indeed, in that case, one wants to approximate the plane curve $\psi(x, y) = 0$, with ψ regular enough, by the tangent at a smooth point (x_0, y_0)

$$\frac{\partial \psi}{\partial y}(x_0, y_0)(x - x_0) - \frac{\partial \psi}{\partial x}(x_0, y_0)(y - y_0) = 0.$$

Finding a rational point close to the curve in the vicinity of (x_0, y_0) then reduces to finding small values of the *homogeneous* linear form

$$\frac{\partial \psi}{\partial y}(x_0, y_0)x + \frac{\partial \psi}{\partial x}(x_0, y_0)y + \left(\frac{\partial \psi}{\partial x}(x_0, y_0)y_0 - \frac{\partial \psi}{\partial y}(x_0, y_0)x_0\right)z.$$

This problem can be studied from the positive side (finding points close) as well as from the negative side (there is no point very close) by means of the three-dimensional LLL algorithm, and the methods developed in Section "Linear Relations".

Elkies used this method to compute extremal examples for the Hall conjecture

$$|x^3 - y^2| \gg_\varepsilon x^{1/2-\varepsilon},$$

which is related to Mordell's equation already encountered in Section "Elliptic Equations".

Of course, one can trivially extend the Elkies–Lefèvre–Muller to varieties of dimension d embedded in an affine space of dimension n, in a very natural way, simply by cutting the variety in small enough pieces, and replacing each piece by the tangent space at some point $(x_{01}, \ldots, x_{0d})$. From the lattice point of view, one will simply end up with (possibly) more linear forms in more variables.

Coppersmith's Method

Another natural extension of what we have done in the previous paragraph is to try to make the method perform better by using a sharper approximation (i.e., polynomial of larger degree, in practice) of our curve or surface, if the latter is smooth enough. Indeed, that would allow to use a decomposition of our curve or variety in much larger pieces, and hence to have less "small degree subproblems" to solve. This extension is by far more difficult than the previous one, and the limits of it still seem unclear.

The key ingredient to do this is a general method, due to Coppersmith [64], to find small solutions to polynomial equations, which has probably not yet reached its final state of development. We refer to May's paper in this volume for a survey and extensive bibliography on the subject. For the application of Coppersmith's method to the table maker's dilemma, we refer to [65].

Acknowledgements Many thanks to Damien Stehlé for numerous discussions over the last few years about lattices, especially about fine behavior and advanced usage of LLL, and for many corrections, suggestions, and improvements to this survey; thanks also to Nicolas Brisebarre for many useful discussions and for his careful rereading of this survey.

References

1. Lenstra, A.K., Lenstra Jr., H.W., Lovász, L.: Factoring polynomials with rational coefficients. Mathematische Annalen **261**, 513–534 (1982)
2. Martinet, J.: Perfect Lattices in Euclidean Spaces. Springer (2002)
3. Cohn, H., Kumar, A.: The densest lattice in twenty-four dimensions. Electronic Research Announcements. Am. Math. Soc. **10**, 58–67 (2004)

4. Haviv, I., Regev, O.: Tensor-based hardness of the shortest vector problem to within almost polynomial factors. In: ACM Symposium on Theory of Computing, pp. 469–477. ACM (2007)
5. Ajtai, M., Kumar, R., Sivakumar, D.: A Sieve Algorithm for the Shortest Lattice Vector Problem. In: ACM Symposium on Theory of Computing, pp. 601–610. ACM (2001)
6. Kannan, R.: Improved algorithms for integer programming and related lattice problems. In: Proceedings of the 15th Symposium on the Theory of Computing (STOC 1983), pp. 99–108. ACM Press (1983)
7. Schnorr, C.P.: A hierarchy of polynomial lattice basis reduction algorithms. Theor. Comput. Sci. **53**, 201–224 (1987)
8. Cohen, H.: A Course in Computational Algebraic Number Theory, 2nd edition. Springer (1995)
9. Pohst, M.: A modification of the LLL reduction algorithm. J. Symbolic Comput. **4**(1), 123–127 (1987)
10. Babai, L.: On Lovász lattice reduction and the nearest lattice point problem. Combinatorica **6**, 1–13 (1986)
11. Lovász, L.: An Algorithmic Theory of Numbers, Graphs and Convexity. SIAM Publications (1986). CBMS-NSF Regional Conference Series in Applied Mathematics
12. de Weger, B.: Algorithms for Diophantine equations, *CWI-Tract*, vol. 65. CWI (1989)
13. Nguyen, P., Stehlé, D.: LLL on the average. In: Proceedings of the 7th Algorithmic Number Theory Symposium (ANTS VII), *Lecture Notes in Computer Science*, vol. 4076, pp. 238–256. Springer (2006)
14. Hanrot, G., Stehlé, D.: Improved analysis of Kannan enumeration algorithm. In: Proceedings of Crypto'2007, *Lecture Notes in Computer Science*, vol. 4622, pp. 170–186. Springer (2007)
15. Lagarias, J.C.: The computational complexity of simultaneous diophantine approximation problems. In: Proceedings of the 1983 Symposium on the Foundations of Computer Science (FOCS 1983), pp. 32–39. IEEE Computer Society Press (1983)
16. Perron, O.: Kettenbrüche. Chelsea (1950)
17. Khinchin, A.Y.: Continued Fractions. Dover publications (1997)
18. Hardy, G., Wright, H.: An Introduction to the Theory of Numbers. Oxford University Press (1980)
19. Fürer, M.: Faster integer multiplication. In: ACM (ed.) Proceedings of STOC' 2007, pp. 57–66 (2007)
20. von zur Gathen, J., Gerhardt, J.: Modern Computer Algebra, 2nd edition. Cambridge University Press (2003)
21. Gauß, C.: Disquisitiones Arithmeticae. Berlin (1801)
22. Odlyzko, A.M., te Riele, H.: Dispoof of Mertens conjecture. Journal für die reine und angewandte Mathematik **357**, 138–160 (1985)
23. Mertens, F.: Über eine zahlentheorische Funktion. Sitzungberichte Akad. Wien **106**, 761–830 (1897)
24. Kotnik, T., te Riele, H.: The Mertens conjecture revisited. In: F. Heß, S. Pauli, M. Pohst (eds.) Proceedings of ANTS-VII, *LNCS*, vol. 4076, pp. 156–167 (2006)
25. Nguyen, P., Stehlé, D.: Floating-point LLL revisited. In: Proceedings of Eurocrypt 2005, *Lecture Notes in Computer Science*, vol. 3494, pp. 215–233. Springer (2005)
26. Baker, A., Davenport, H.: The equations $3x^2 - 2 = y^2$ and $8x^2 - 7 = z^2$. Quart. J. Math. Oxford (2) **20**, 129–137 (1969)
27. Cassels, J.W.S.: An introduction to diophantine approximation. Cambridge University Press (1957)
28. Ferguson, H., Bailey, D.: A Polynomial Time, Numerically Stable Integer Relation Algorithm (1991). Manuscript
29. Ferguson, H., Bailey, D., Arno, S.: Analysis of PSLQ, An Integer Relation Finding Algorithm. Math. Comp. **68**, 351–369 (1999)
30. Siegel, C.: Über einige Anwendungen diophantischer Approximationen. Abh. der Preuß Akad. der Wissenschaften. Phys-math. kl. **1** (1929) = *Gesammelte Abhandlungen*, I, 209–266
31. Schmidt, W.: Diophantine Approximation, *Lecture Notes in Mathematics*, vol. 785. Springer (1980)

32. Odlyzko, A.M.: The rise and fall of knapsack cryptosystems. In: Proceedings of Cryptology and Computational Number Theory, *Proceedings of Symposia in Applied Mathematics*, vol. 42, pp. 75–88. Am. Math. Soc. (1989)
33. van Hoeij, M.: Factoring polynomials and the knapsack problem. J. Number Th. **95**, 167–189 (2002)
34. Bilu, Y., Hanrot, G.: Solving Thue Equations of Large Degree. J. Number Th. **60**, 373–392 (1996)
35. Schnorr, C.P.: Factoring Integers and computing discrete logarithms via Diophantine approximation. Adv. Comput. Complex. **13**, 171–182 (1993)
36. Adleman, L.: Factoring and lattice reduction (1995). Manuscript
37. Dickman, K.: On the frequency of numbers containing primes of a certain relative magnitude. Ark. Math. Astr. Fys. **22**, 1–14 (1930)
38. Vallée, B.: Provably fast integer factoring with quasi-uniform small quadratic residues. In: Proceedings of the Twenty-First Annual ACM Symposium on Theory of Computing, 15–17 May 1989, Seattle, Washington, USA, pp. 98–106 (1989)
39. Montgomery, P.: Square roots of products of algebraic numbers. In: W. Gautschi (ed.) Mathematics of Computation 1943–1993: a Half-Century of Computational Mathematics, Proceedings of Symposia in Applied Mathematics, pp. 567–571. Am. Math. Soc. (1994)
40. Nguyen, P.: A Montgomery-like square root for the number field sieve. In: J. Buhler (ed.) Algorithmic Number Theory, Third International Symposium, ANTS-III Portland, Oregon, USA, June 21, 1998 Proceedings, *Lecture Notes in Computer Science*, vol. 1423, pp. 151–168. Springer (1998)
41. Smart, N.: The algorithmic solution of Diophantine equations, *London Mathematical Society Students Texts*, vol. 41. Cambridge University Press (1998)
42. Baker, A.: Linear forms in the logarithms of algebraic numbers, I. Mathematika **13**, 204–216 (1966)
43. Matveev, E.: An explicit lower bound for a homogeneous rational linear form in logarithms of algebraic numbers, ii. Izv. Ross. Akad. Nauk, Ser. Math. **64**, 125–180 (2000)
44. de Weger, B.: Solving exponential diophantine equations using lattice basis reduction algorithms. J. Number Th. **26**, 325–367 (1987)
45. Oesterlé, J.: Nouvelles approches du théorème de Fermat. Astérisque **161/162**, 165–186 (1988)
46. Stewart, C., Yu, K.: On the *abc* conjecture. II. Duke Math. J. **108**, 169–181 (2001)
47. Dokchitser, T.: LLL & ABC. J. Number Th. **107**, 161–167 (2004)
48. Thue, A.: Über annäherungswerte algebraischer Zahlen. J. Reine Angew. Math. **135**, 284–305 (1909)
49. Baker, A.: Contributions to the theory of Diophantine equations. I. On the representation of integers by binary forms. Philos. Trans. Roy. Soc. London Ser. A **263**, 173–191 (1968)
50. Tzanakis, N., de Weger, B.: On the Practical Solution of the Thue Equation. J. Number Th. **31**, 99–132 (1989)
51. Tzanakis, N., de Weger, B.: How to explicitly solve a Thue–Mahler Equation. Compositio Math. **84**, 223–288 (1992)
52. Cremona, J.: On the Computation of Mordell-Weil and 2-Selmer Groups of Elliptic Curves. Rocky Mountain J. Math. **32**, 953–966 (2002)
53. Gebel, J., Pethő, A., Zimmer, H.: Computing integral points on elliptic curves. Acta Arith. **68**, 171–192 (1994)
54. Stroeker, R., Tzanakis, N.: Solving elliptic diophantine equations by estimating linear forms in elliptic logarithms. Acta Arith. **67**, 177–196 (1994)
55. Gebel, J., Pethő, A., Zimmer, H.: On Mordell's equation. Compositio Math. pp. 335–367 (1998)
56. Stroeker, R.J., Tzanakis, N.: Computing all integer solutions of a genus 1 equation. Math. Comput. **72**, 1917–1933 (2003)
57. Bilu, Y.: Solving superelliptic Diophantine Equations by Baker's Method (1994). Preprint, Mathématiques stochastiques, université Bordeaux 2
58. Bilu, Y., Hanrot, G.: Solving Superelliptic Diophantine Equations by Baker's Method. Compositio Math. **112**, 273–312 (1998)

59. de Weger, B.: Integral and S-integral solutions of a Weierstrass equation. J. Théor. Nombres Bordx. **9**, 281–301 (1997)
60. Herrmann, E., Pethő, A.: S-integral points on elliptic curves – notes on a paper by B.M.M. de Weger. J. Théor. Nombres Bordx. **13**, 443–451 (2001)
61. Bugeaud, Y.: Approximation by algebraic numbers, *Cambridge Tracts in Mathematics*, vol. 160. Cambridge University Press (2004)
62. Lefèvre, V., Muller, J.M., Tisserand, A.: Towards correctly rounded transcendentals. In: Proceedings of the 13th IEEE Symposium on Computer Arithmetic. IEEE Computer Society Press, Los Alamitos, CA, Asilomar, USA (1997).
URL http://www.acsel-lab.com/arithmetic/arith13/papers/ARITH13_Lefevre.pdf
63. Elkies, N.D.: Rational points near curves and small nonzero $|x^3 - y^2|$ via lattice reduction. In: Proceedings of the 4th Algorithmic Number Theory Symposium (ANTS IV), *Lecture Notes in Computer Science*, vol. 1838, pp. 33–63. Springer (2000)
64. Coppersmith, D.: Small Solutions to Polynomial Equations, and Low Exponent RSA vulnerabilities. Journal of Cryptology **10**(4), 233–260 (1997)
65. Stehlé, D., Lefèvre, V., Zimmermann, P.: Searching worst cases of a one-variable function using lattice reduction. IEEE Trans. Comput. **54**(3), 340–346 (2005)

Chapter 7
Selected Applications of LLL in Number Theory

Denis Simon

Abstract In this survey, I describe some applications of LLL in number theory. I show in particular how it can be used to solve many different linear problems and quadratic equations and to compute efficiently in number fields.

Introduction

The LLL algorithm has really many applications (on MathSciNet, it is cited in the references of at least 118 papers and in at least 50 reviews !!).

Among the most famous ones are of course those in lattice theory (the shortest vector problem: [22] and [17, 23, 32]; the closest vector problem: [1, 3] ...) and also those for factoring polynomials (for example, in [35]), since this was precisely the application Lenstra, Lenstra, and Lovász presented in their original paper [41]. At least as famous is the application to the knapsack problem: [22, 38, 47] or [51, Sect. VI.2].

In this survey, I would like to present other selected applications in number theory, all of which lead to revolutionary results.

1. For linear problems: computing gcd's, kernels, Hermite normal forms, Smith normal forms, integral relations, linear dependence, algebraic dependence ...
2. Solving quadratic equations: Gauss reduction in dimension 3, Shanks' algorithm for the 2-primary part of the class group $Cl(\sqrt{D})$, reduction of indefinite quadratic forms, quadratic equations in dimension 4 and more...
3. Number fields: polynomial reduction, ideal reduction, computing the class group and the unit group, solving the principal ideal problem ...
4. Testing conjectures (Hall, abc, Mertens, ...)

D. Simon
Université de Caen, LMNO, Bd Maréchal Juin BP 5186 – 14032 Caen Cedex, France,
e-mail: simon@math.unicaen.fr

P.Q. Nguyen and B. Vallée (eds.), *The LLL Algorithm*, Information Security and Cryptography, DOI 10.1007/978-3-642-02295-1_7,

Warnings

In most of the applications described in this paper, we only use two properties of the LLL algorithm. The first is that, when it is given a lattice L of dimension n and determinant $d(L)$, then LLL outputs a short vector $\mathbf{b}_1$ bounded by

$$|\mathbf{b}_1| \leqslant 2^{(n-1)/4} \det(L)^{1/n}.$$

The geometry of numbers would give better bounds. The second one is that LLL finds this vector in polynomial time; hence it gives a very efficient algorithm to solve the different problems.

As noticed in [12, Algorithm 2.6.3, Remark 2], LLL only needs to know the Gram matrix of the lattice. This remark implies that it is equivalent to describe the lattice as embedded in $\mathbb{R}^n$ with the euclidean norm or as $\mathbb{Z}^n$ equipped with a positive definite quadratic form q. The result of LLL is then an integral unimodular transformation matrix, or just an n-tuple $(x_1, \dots, x_n) \neq (0, \dots, 0)$ of integers such that

$$q(x_1, \dots, x_n) \leqslant 2^{(n-1)/2} \det(q)^{1/n}.$$

In this paper, LLL will be used mainly in small dimension: almost always $n \leqslant 30$ and very often $n \leqslant 6$. However, the coefficients may have hundreds or thousands of digits.

Reducing Linear Forms

A significant part of the material of this section comes from [12, Sect. 2.7] and [51, Sect. IV.3 and IV.4].

- The best **approximations of a real number α by rational numbers** are usually obtained by the continued fraction algorithm. As suggested in [41], another way to obtain good approximations is to reduce the quadratic form

$$q(x, y) = M(\overline{\alpha}x - y)^2 + \frac{1}{M}x^2,$$

where $\overline{\alpha}$ is usually a decimal approximation of α to precision $\frac{1}{M}$. Indeed, when M is large and (x, y) is a short vector for the quadratic form q, then $q(x, y) = O(1)$, which implies $x = O(\sqrt{M})$ and $\alpha x - y = O\left(\frac{1}{\sqrt{M}}\right)$. Typically, this implies $|x| \approx |y| \approx \sqrt{M}$ and $|\alpha - \frac{y}{x}| \approx \frac{1}{M}$.

More explicitly, it consists of reducing the lattice $\mathbb{Z}^2$ equipped with the quadratic form $q(x, y) = M(\overline{\alpha}x - y)^2 + \frac{1}{M}x^2$, or of applying the LLL algorithm with the 2-dimensional Gram matrix

$$\begin{pmatrix} \overline{\alpha}^2 M + \frac{1}{M} & -\overline{\alpha} M \\ -\overline{\alpha} M & M \end{pmatrix}$$

This Gram matrix has determinant equal to 1, hence corresponds to a lattice of determinant 1. The underlying lattice in the euclidean plane $\mathbb{R}^2$ is given by the matrix

$$\begin{pmatrix} \frac{1}{\sqrt{M}} & 0 \\ \overline{\alpha}\sqrt{M} & -\sqrt{M} \end{pmatrix}.$$

The result of LLL is then a unimodular integral transformation matrix $\begin{pmatrix} a & b \\ c & d \end{pmatrix}$, and the desired short vector (x, y) is just (a, c). Indeed, the bound given by LLL asserts that $q(x, y) \leqslant \sqrt{2}$. This inequality implies two others, namely

$$\begin{cases} |\overline{\alpha}x - y| \leqslant \frac{2^{1/4}}{\sqrt{M}} \\ |x| \leqslant 2^{1/4}\sqrt{M}. \end{cases}$$

Now, using the inequality $|\alpha - \overline{\alpha}| \leqslant \frac{1}{M}$, we find

$$|\alpha x - y| \leqslant \frac{2^{5/4}}{\sqrt{M}}.$$

Example: Assume that we want to find a good rational approximation of $\alpha = \pi$, using the decimal approximation $\overline{\alpha} = 3.1415926536$. We have ten correct decimals, so we choose $M = 10^{10}$. The shortest vector found by LLL is then $(x, y) = (99532, 312689)$. We indeed see that $|x| \approx |y| \approx 10^5 \approx \sqrt{M}$ and $\left|\pi - \frac{y}{x}\right| \approx 0.3 \times 10^{-10} \approx \frac{1}{M}$. This rational approximation is exactly the same as the one given by the continued fraction expansion of π, corresponding to the eight first coefficients $[3, 7, 15, 1, 292, 1, 1, 1]$.

When we use the approximation $\pi \approx 31415926536 \times 10^{-10}$, all the information is contained in the 10 digits of the numerator, and almost nothing in the denominator. In the other approximation $\pi \approx \frac{312689}{99532}$, the information is equally distributed into the numerator and the denominator, but a total of 10 digits is still necessary. There is no gain of storage to use either representation.

If α is close to 1, it might also be interesting to consider the more symmetrical quadratic form

$$q(x, y) = M(\overline{\alpha}x - y)^2 + \frac{1}{M}(x^2 + y^2).$$

- There is an analog of this algorithm for **p-adic approximation** (for example, in [51, Sect. VI.4]). If p is a prime number and α is a p-adic unit, approximated modulo p^m by a rational integer $\overline{\alpha}$, then we can look for a short vector for the quadratic form

$$q(x, y, z) = M^2(\overline{\alpha}x - y - p^m z)^2 + \frac{1}{M}(x^2 + y^2).$$

The Gram matrix of q is

$$\begin{pmatrix} \overline{\alpha}^2 M^2 + \frac{1}{M} & -\overline{\alpha} M^2 & -\overline{\alpha} M^2 p^m \\ -\overline{\alpha} M^2 & M^2 + \frac{1}{M} & M^2 p^m \\ -\overline{\alpha} M^2 p^m & M^2 p^m & M^2 p^{2m} \end{pmatrix}$$

and its determinant is p^{2m}. It corresponds to the lattice in $\mathbb{R}^3$ generated by

$$\begin{pmatrix} \frac{1}{\sqrt{M}} & 0 & 0 \\ 0 & \frac{1}{\sqrt{M}} & 0 \\ \overline{\alpha} M & -M & -p^m M \end{pmatrix}$$

of determinant p^m. Applied to this 3-dimensional Gram matrix, LLL returns a short integral vector (x, y, z) satisfying the inequality

$$q(x, y, z) \leqslant 2^{3/2} p^{2m/3},$$

from which we deduce in particular

$$|\overline{\alpha}x - y - p^m z| \leqslant 2^{3/4} p^{m/3} M^{-1}.$$

If we choose M such that $2^{3/4} p^{m/3} < M < 2p^{m/3}$, we have $|\overline{\alpha}x - y - p^m z| < 1$ which implies that $\overline{\alpha}x - y - p^m z = 0$, since this is an integer. It now remains

$$x^2 + y^2 \leqslant 2^{5/2} p^m,$$

whence

$$\begin{cases} |x| & \leqslant 2^{5/4} p^{m/2} \\ |y| & \leqslant 2^{5/4} p^{m/2}. \end{cases}$$

In summary, we have found rational integers x and y such that $|x|$ and $|y|$ are both $O(p^{m/2})$. Furthermore, when α is a p-adic unit, x is also a p-adic unit, and we have a p-adic approximation $|\alpha - \frac{y}{x}|_p \approx p^{-m}$ with the real bounds $|x| \approx |y| \approx p^{m/2}$.

When $m = 1$, this result proves that any element in $\mathbb{F}_p^*$ can be represented by a fraction $\frac{y}{x}$ with $|x| = O(p^{1/2})$ and $|y| = O(p^{1/2})$. The corresponding algorithm is just an application of LLL in dimension 3.

In this application, we have used a lattice of dimension 3, but a careful reading reveals that because of the relation $\overline{\alpha}x - y - p^m z = 0$, the constructed short vector lies in a dimension 2 sublattice. De Weger [52] describes an algorithm wherein this idea is used and wherein LLL is only applied in dimension 2.

Example: Consider the prime number $p = 10^{10} + 19$ and the p-adic number $\alpha = \frac{16}{17}$. Its p-adic expansion is $\alpha = 7647058839 + 9411764723p + 2352941180p^2 + \dots$. Using the described algorithm, with $m = 3$, $\overline{\alpha} = 7647058839 + 9411764723\,p + 2352941180p^2 = 235294118988235296665882354556$, LLL quickly finds the short vector $(x, y, z) = (17, 16, 4)$, hence recovers the original fraction.

- The previous approach (either real or p-adic) easily generalizes to obtain small values of a linear form $L(x_1, \dots, x_n) = x_n + \sum_{i=1}^{n-1} \alpha_i x_i$ in dimension n, evaluated at small integers $x_1, \dots, x_n$. In the real case, we just have to reduce the quadratic form
$$M^{n-1} L(x_1, \dots, x_n)^2 + \frac{1}{M} \sum_{i=1}^{n-1} x_i^2$$
(this also appears in [41]).

Even more generally, if we are given n linear forms $L_1, \dots L_n$ in $n+m$ variables $x_1, \dots, x_{n+m}$, then we will have a **simultaneous approximation** if we reduce the single quadratic form

$$W_1 L_1(x_1, \dots, x_{n+m})^2 + \dots + W_n L_n(x_1, \dots, x_{n+m})^2 + W_{n+1} x_{n+1}^2 + \dots + W_{n+m} x_{n+m}^2,$$

where the W_i are weights that are to be chosen depending on how much the linear forms and the coefficients have to be reduced compared to each other. This application to Diophantine approximation is described, for example, in [39]. Many other applications of LLL applied to linear forms in logarithms are given in [51, part 2] or in [26].

- For example, if the coefficients of the L_i are integers and if we want to insist on having exactly $L_i(x_1, \dots, x_{n+m}) = 0$ for all $i = 1, \dots, n$, we will choose very large weights W_i, $i = 1, \dots, n$ and the other weights very small. This approach is used by [44] to **compute the kernel of an integral matrix** and to give a modified version of LLL, called MLLL, applicable not only on a basis of a lattice, but also on a generating set of vectors $\mathbf{b}_1, \dots, \mathbf{b}_{n+m}$ of a lattice of dimension n. The result of this algorithm is then a reduced basis of the lattice together with a matrix containing m independent relations among the $\mathbf{b}_i$. See also [12, Sect. 2.7.1].

Example: Consider the matrix

$$M = \begin{pmatrix} 2491 & 5293 & 1032 & 5357 & 9956 \\ 6891 & 4280 & 3637 & 3768 & 4370 \\ 5007 & 4660 & 5712 & 7743 & 4715 \end{pmatrix},$$

which has been chosen randomly with coefficients less than 10^4. Its kernel has dimension 2, and using gaussian elimination, we find that this kernel is generated by

$$\begin{pmatrix} \frac{59229512635}{82248101629} & \frac{33495205035}{82248101629} \\ -\frac{94247651922}{82248101629} & -\frac{180166533113}{82248101629} \\ -\frac{86522262381}{82248101629} & \frac{49731399425}{82248101629} \\ 1 & 0 \\ 0 & 1 \end{pmatrix}.$$

Now, consider the quadratic form $A^4(2491v + 5293w + 1032x + 5357y + 9956z)^2 + A^4(6891v + 4280w + 3637x + 3768y + 4370z)^2 + A^4(5007v + 4660w + 5712x + 7743y + 4715z)^2 + A^{-6}y^2 + A^{-6}z^2$ or equivalently the lattice

$$\begin{pmatrix} 2491A^2 & 5293A^2 & 1032A^2 & 5357A^2 & 9956A^2 \\ 6891A^2 & 4280A^2 & 3637A^2 & 3768A^2 & 4370A^2 \\ 5007A^2 & 4660A^2 & 5712A^2 & 7743A^2 & 4715A^2 \\ 0 & 0 & 0 & A^{-3} & 0 \\ 0 & 0 & 0 & 0 & A^{-3} \end{pmatrix}.$$

This lattice has determinant c, with $c = 82248101629$. We now apply LLL to this quadratic form. If A is large enough (here $A > 13$ works), then the values of the quadratic form are so small that the first two columns of the unimodular transformation matrix

$$U = \begin{pmatrix} 6845 & -300730 & -53974 & 121395 & -138794 \\ 285983 & 782775 & 114176 & -446636 & 387087 \\ -240869 & 202704 & 56844 & 19781 & 73475 \\ 118346 & -306061 & -64578 & 75646 & -131789 \\ -192463 & -197241 & -18341 & 164323 & -107769 \end{pmatrix}$$

are the coordinates of kernel. We can see that the coefficients are much smaller (6 digits) than with the other method (11 digits). Here, we have

$$MU = \begin{pmatrix} 0 & 0 & 0 & -1 & 0 \\ 0 & 0 & 0 & 0 & -1 \\ 0 & 0 & 1 & 0 & 0 \end{pmatrix}.$$

In [28], this idea is used to solve the **extended gcd problem**: given integers $s_1, \ldots, s_m$, find a vector $x = (x_1, \ldots, x_m)$ with integral coefficients and small euclidean norm such that $x_1 s_1 + \cdots + x_m s_m = \gcd(s_1, \ldots, s_m)$. The method is generalized to the problem of producing small unimodular transformation matrices for **computing the Hermite Normal Form** of an integer matrix (in [28]) and for **computing the Smith Normal Form** (in [42] and [30]).

- We can also **find $\mathbb{Z}$-linear relations among real numbers**. In a computer, an integer relation between real numbers can not of course be tested to be exactly 0, but only if it is small. This means that if an integer relation really exists, then

LLL will probably find it. But LLL can also find other $\mathbb{Z}$-linear combinations that are not 0 but just small. In fact, if the relation is not exactly 0, it is usually possible to prove it just by computing it with enough precision. However, it is the responsibility of the mathematician to prove that a relation found by LLL is exactly 0.

Example: We are aware of Machin's formula

$$a \arctan(1) + b \arctan\left(\frac{1}{5}\right) + c \arctan\left(\frac{1}{239}\right) = 0,$$

where a, b, and c are small integers, but we do not know their values. We apply LLL to the quadratic form

$$A^2\left(a \arctan(1) + b \arctan\left(\frac{1}{5}\right) + c \arctan\left(\frac{1}{239}\right)\right)^2 + (b^2 + c^2)$$

or to the lattice

$$\begin{pmatrix} \arctan(1)A & \arctan\left(\frac{1}{5}\right)A & \arctan\left(\frac{1}{239}\right)A \\ 0 & 1 & 0 \\ 0 & 0 & 1 \end{pmatrix}$$

with a large value of A. If A is not large enough, LLL suggests that $\arctan\left(\frac{1}{239}\right) \approx 0$. Is is clearly true, but not exactly 0. If $A > 1{,}000$, LLL suggests the relation

$$\arctan(1) - 4\arctan\left(\frac{1}{5}\right) + \arctan\left(\frac{1}{239}\right) \approx 0.$$

Now, a rigorous proof that this is exactly 0 comes with the observation that through the transformation $\exp(2i \arctan t) = \frac{1+it}{1-it}$, the relation is equivalent to

$$\frac{(1+i)(1+5i)^{-4}(1+239i)}{(1-i)(1-5i)^{-4}(1-239i)} = 1.$$

Other algorithms for detecting integer relations between reals are given in the papers [21] and [20] (without LLL) and [27] (with LLL). See also the `lindep` function in `GP/PARI`, described in [12, Sect. 2.7.2].

Using these algorithms, Borwein and Bradley [5] have for example tried to generalize Apéry's formula

$$\zeta(3) = \frac{5}{2}\sum_{k=1}^{\infty} \frac{(-1)^{k+1}}{k^3\binom{2k}{k}}.$$

After an extensive search, they suggested that there is no formula of the type

$$\zeta(5) = \frac{a}{b} \sum_{k=1}^{\infty} \frac{(-1)^{k+1}}{k^5 \binom{2k}{k}}$$

at least as long as a and b are not astronomically large. However, they found

$$\zeta(7) = \frac{5}{2} \sum_{k=1}^{\infty} \frac{(-1)^{k+1}}{k^7 \binom{2k}{k}} + \frac{25}{2} \sum_{k=1}^{\infty} \frac{(-1)^{k+1}}{k^3 \binom{2k}{k}} \sum_{j=1}^{k-1} \frac{1}{j^4},$$

and similar expressions for $\zeta(9)$, $\zeta(11)$, and proposed the following conjecture

$$\sum_{s=0}^{\infty} \zeta(4s+3) x^{4s} = \frac{1}{2} \sum_{k=1}^{\infty} \frac{(-1)^{k+1}}{\binom{2k}{k} k^3} \frac{5k}{k^4 - x^4} \prod_{n=1}^{k-1} \left(\frac{n^4 + 4x^4}{n^4 - x^4} \right).$$

This was later proved to be true by [2]. Similar methods and results are also obtained about $\zeta(2s+2)$ in [6].

- A special case of integer relations between real (or complex) numbers is the case when the real numbers are powers of the same real number. In this case, a linear relation $\sum_{i=0}^{n} x_i \alpha^i = 0$ proves that α is an algebraic number of degree at most n, and the given polynomial $\sum_{i=0}^{n} x_i X^i \in \mathbb{Z}[X]$ is a multiple of its minimal polynomial. In some sense, LLL is able to **recover the minimal polynomial of an algebraic number**. It is important to remark that in this algorithm, the degree n is fixed, so that it can only find a polynomial if we have a bound on its degree. See for example the `algdep` function in `GP/PARI`, described in [12, Sect. 2.7.2].

In practice, since we only work with real approximations of α, the polynomials found are only candidates. However, it is quite common that checking the vanishing of a polynomial at α is easy. Hence, the algorithm can be used in both ways: either to give evidence that α is a transcendental number, or to build a polynomial taking such a small value at α; there is a high chance that it, in fact, vanishes exactly at α. This method has been used in [34] to give evidence that the numbers $e \pm \pi$ and some other numbers are transcendental. It is described in more details in [33], where a surprising application is given for the **factorization of polynomials in** $\mathbb{Z}[X]$: Start with a polynomial $P \in \mathbb{Z}[X]$ of degree n, and compute sufficiently many digits of the real and imaginary parts of a root α (using your preferred method, for example by Newton's method); then, use the algorithm to look for an integral polynomial Q vanishing at α; if Q divides P, we are done; otherwise, P has no proper factor. See also [40]. Other polynomial factorization algorithms using LLL are given in [45] or [35].

Example: Imagine that we would like to prove that $\alpha = e + \pi$ is an algebraic number of degree 4. This would mean that there are integers a_4, a_3, a_2, a_1, and a_0, not all 0, such that $a_0\alpha^4 + a_1\alpha^3 + a_2\alpha^2 + a_1\alpha + a_0 = 0$. Using an approximation of α to

20 decimals, LLL suggests the relation

$$127\alpha^4 - 399\alpha^3 - 2268\alpha^2 + 1849\alpha - 2417 \approx 0.$$

In order to verify this relation, we increase the precision of α to 40 decimals and observe that the expression is not 0, but close to $-4.6\ 10^{-13}$. Using this new precision, LLL now suggests a new relation

$$3498343\alpha^4 - 940388\alpha^3 - 116624179\alpha^2 + 761230\alpha + 64496487 \approx 0.$$

Repeating this procedure always gives a tentative polynomial with growing coefficients, but never vanishing exactly at α. Of course, this does not prove that α is transcendental, and it does even not prove that is not algebraic of degree 4. However, it proves that there is no quartic polynomial with small coefficients vanishing at α.

Another illustration of this application of LLL is given in [31]. See also [13, Sect. 6.3.2]. For an imaginary quadratic field K of discriminant d, the Hilbert class field of K is of relative degree $h = \#Cl(K)$ over K. If $(\alpha_k)_{1\leq k\leq h}$ are the roots of h inequivalent binary quadratic forms of discriminant d, then $j_k = j(\alpha_k)$ (where the function j is the modular function) are the roots of a degree h polynomial $H_d \in \mathbb{Z}[X]$ defining exactly the **Hilbert class field** of K. Instead of computing each of the h values j_k, it is possible to compute just one such value and to recover the whole polynomial using the previous algorithm. However, since the coefficients of H_d are usually quite large, it is preferable to use the Weber function $f(z)$ instead of the modular function $j(z)$. The corresponding polynomial h_d still defines the Hilbert class field, but the coefficients can be 12 times smaller.

Example: For the discriminant $d = -32$, the class number is 3, and the class group is cyclic of order 3, generated by $P = 2x^2 + x + 3$. Its roots in the upper half plane is

$$\alpha_1 = \frac{-1 + i\sqrt{23}}{4} \approx -0.25 + 1.19895788082817988539935951604i.$$

We can now compute

$$j(\alpha_1) \approx 737.849984966684102752369273665$$
$$+1764.01893861274614164378642717i.$$

This should be an algebraic integer of degree 6 (and in fact of degree 3). Now, LLL suggests that the minimal polynomial of j could be $P = x^3 + 3491750x^2 - 5151296875x + 12771880859375$. It only remains to prove that this polynomial, indeed, defines an unramified extension of $\mathbb{Q}(\sqrt{-23})$, but this is a different story ! A good test is that the norm of j should be a cube. Here, we have $12771880859375 = 23375^3$.

Solving Quadratic Equations Over the Rationals

In this section, we consider the problem of solving quadratic equations over $\mathbb{Q}$. Since testing the solvability of these equations is usually easy, we can always assume that they are indeed solvable.

The first nontrivial family of quadratic equations that are interesting to solve are the ternary quadratic equations $q(x, y, z) = 0$, with rational coefficients and unknowns. Among them is the diagonal equation, also called Legendre's equation: $ax^2 + by^2 + cz^2 = 0$. Usually, the general nondiagonal equation $q(x, y, z) = 0$ is transformed into a diagonal one of Legendre type. For Legendre's equation, a great deal of algorithms exist.

For example, in [46, Chap. IV Sect. 3] or [51, Chap. IV Sect. 3.3], the solution of $ax^2 + by^2 + cz^2 = 0$ is deduced from the solution of $a'x^2 + b'y^2 + c'z^2 = 0$ where the new coefficients are smaller than the old ones. However, this reduction depends on the possibility of extracting square roots modulo a, b, or c, which is only possible if we know the factorization of abc. During the whole algorithm, the total number of factorizations is quite large. The worst drawback is certainly that the numbers that have to be factored may also be quite large. Solving quadratic equations with these algorithms uses only a few lines in theory, but is extremely slow in practice.

Fortunately, a couple of algorithms exist, which do not factor any other integers than a, b, and c. It seems impossible, in general, to avoid these three factorizations. Such an algorithm is given in [16]. In practice, it indeed runs fast.

Other algorithms exist, which use the reduction theory of lattices. They all share the same property of using no other factorization than that of abc. Since the LLL algorithm can reduce quadratic forms, it is not surprising that it can be used to solve quadratic equations over the rationals. However, if a quadratic equation has a solution, it is certainly not positive definite. The problem is that LLL can a priori only handle positive definite quadratic forms. There are two ways to go around this problem :

- Either build a new quadratic form, which is positive definite and the reduction of which can help us in solving the initial quadratic equation
- Or adapt LLL to indefinite quadratic forms.

A positive definite quadratic form attached to the problem is of course $q = |a|x^2 + |b|y^2 + |c|z^2$. However, reducing $\mathbb{Z}^3$ with this quadratic form will not give anything, since it is already orthogonal hence reduced. According to [11], if a, b, and c are coprime squarefree integers, some integral solutions of $ax^2 + by^2 + cz^2 = 0$ lie in a sublattice of $\mathbb{Z}^3$ of index $2|abc|$ defined by the congruences

$$\begin{aligned} by - \lambda_1 z &\equiv 0 \pmod{a} \\ ax - \lambda_2 z &\equiv 0 \pmod{b} \\ ax - \lambda_3 z &\equiv 0 \pmod{c} \end{aligned}$$

plus another condition modulo a power of 2, where λ_1, λ_2, and λ_3 are any choice of square roots of $-bc$, $-ac$, $-ab$ modulo a, b, and c respectively. A smallest vector of this lattice (equipped with the positive definite quadratic form $|a|x^2+|b|y^2+|c|z^2$) will give a solution. Using this algorithm, we see that LLL (in dimension 3) can be used to **solve Legendre's equation**.

The method of Gauss himself in [24, sects. 272, 274, 294] in 1801, was already similar, since he builds the same lattice, using the same congruences. But Gauss reduces directly the corresponding indefinite quadratic form. We can summarize his method in two steps: (1) compute square roots modulo a, b, and c, and build another quadratic form with determinant -1 and (2) Reduce and solve this new quadratic form. The reduction of the indefinite quadratic form suggested by Gauss works simultaneously on the quadratic form and its dual. It is quite different from any version of LLL. This reduction algorithm is analyzed in [37] and proved to run in polynomial time. It is interesting to note that the algorithm is used in [48] and [7] to compute in polynomial time the 2-Sylow subgroup of the class group $Cl(\sqrt{D})$.

The algorithms described up to now are quite specific to solve Legendre equations, that is, diagonal ternary quadratic equations over $\mathbb{Q}$. Some of them can also solve semi-diagonal equations (of the form $ax^2+bxy+cy^2 = dz^2$, but are not able to solve general ternary quadratic equations. Of course, it is always possible to diagonalize any quadratic equation, but if we do so, the integers that are to be factored may be huge (hence, impossible to factor in a reasonable amount of time) compared to the determinant of the original equation. An example is given in [49] of an equation with determinant equal to -1 and coefficients having more than 1, 300 decimal digits: reducing the equation to a diagonal one would require the factorization of integers of that size !

In [49], an algorithm is given to solve general ternary quadratic equations, without requiring other factorizations than that of the determinant. The main strategy follows Gauss:

- Compute first an integral quadratic form, equivalent to the initial one, which has determinant -1 (this is the *minimization* step)
- Then, reduce this indefinite quadratic form (this is the *reduction* step).

The minimization step only uses linear algebra modulo the prime divisors of the determinant. The reduction step could certainly use the reduction of Gauss, but another reduction algorithm is proposed. In fact, when we apply LLL to a quadratic form which is indefinite, without changing LLL, the algorithm may enter into infinite loops. Indeed, the swap condition in the algorithm (also called the Lovász condition) is just

$$|\mathbf{b}_i^* + \mu_{i,i-1}\mathbf{b}_{i-1}^*|^2 < c|\mathbf{b}_{i-1}^*|^2,$$

which tests whether the norm of the vector $\mathbf{b}_{k-1}^*$ decreases when we interchange $\mathbf{b}_{k-1}$ and $\mathbf{b}_k$. In terms of the underlying quadratic form q, this test is equivalent to

$$q(\mathbf{b}_i^* + \mu_{i,i-1}\mathbf{b}_{i-1}^*) < cq(\mathbf{b}_{i-1}^*).$$

In the case where q is not positive definite, these quantities may be negative, and a swap may increase their absolute values. If we just add absolute values in this test

$$| q(\mathbf{b}_i^* + \mu_{i,i-1}\mathbf{b}_{i-1}^*) | < c \,| q(\mathbf{b}_{i-1}^*) |,$$

the new algorithm, which we call the **Indefinite LLL**, has the following properties:

Theorem 1 (IndefiniteLLL). *Let q be a quadratic form over $\mathbb{Z}^n$ defined by $q(\mathbf{x}) = \mathbf{x}^t Q \mathbf{x}$ with a symmetric matrix $Q \in \mathcal{M}_n(\mathbb{Z})$ such that $\det(Q) \neq 0$. The output of the IndefiniteLLL Algorithm applied with a parameter $\frac{1}{4} < c < 1$ to a basis $\mathbf{b}_1, \ldots, \mathbf{b}_n$ of $\mathbb{Z}^n$ is*

- *Either some $\mathbf{x} \in \mathbb{Z}^n$ such that $q(\mathbf{x}) = 0$*
- *Or a reduced basis $\mathbf{b}_1, \ldots, \mathbf{b}_n$ such that*

$$|(\mathbf{b}_{k-1}^*)^2| \leqslant \gamma |(\mathbf{b}_k^*)^2| \quad \text{for } 1 < k \leqslant n,$$

and

$$1 \leqslant |(\mathbf{b}_1)^2|^n \leqslant \gamma^{n(n-1)/2} |\det(Q)|,$$

where $\gamma = \left(c - \frac{1}{4}\right)^{-1} > \frac{4}{3}$.

If furthermore q is indefinite, we have

$$1 \leqslant |(\mathbf{b}_1)^2|^n \leqslant \frac{3}{4} \gamma^{n(n-1)/2} |\det(Q)|.$$

For a better comparison with the standard properties of LLL-reduced bases, we have used the notation $(\mathbf{b})^2$ for $q(\mathbf{b})$, which need not be positive in this situation. In both cases, the algorithm finishes after a polynomial number of steps. Now, we have an efficient way to **solve general ternary quadratic equations**. Combining the algorithm of [48] and [7] with this indefinite LLL, we can now claim that LLL can also **compute the 2-Sylow subgroup of the class group** $Cl\left(\sqrt{D}\right)$.

One of the key points of this algorithm is to reduce indefinite quadratic forms. Up to now, we have seen two algorithms for this : one by Gauss, which only works in dimension 3, and the other in [49], which also works in higher dimensions. In fact, other algorithms exist. For example, in [29] an algorithm very similar to [49] is given. The main difference is in the way it handles isotropic vectors. In one case, the algorithm simply stops, whereas in the other case, it applies a complicated subroutine. Because of this small difference, the quality of the reduction proved in [49] is slightly better than LLL, whereas it is worse than LLL in [29]. I would like to mention a last algorithm used for the reduction of indefinite quadratic forms. Assume that the indefinite quadratic form q is diagonal in some basis (typically obtained by a Gram-Schmidt procedure). Then, we can bound $|q|$ by the quadratic form q' obtained just by taking the absolute values of the coefficients of q in this basis! Since these quadratic forms have the same determinant (up to sign), a reduced basis for q' will also be a reduced basis for q, and the quality of the reduction will be exactly the same as in LLL. This is precisely what is suggested in [11], where the diagonal form $ax^2 + by^2 + cz^2$ is bounded by $|a|x^2 + |b|y^2 + |c|z^2$. The only drawback of this method is that during the diagonalization step, we may introduce large denominators, a fact that can slow down the algorithm. No serious comparison has been done between these different algorithms.

A generalization of the previous algorithms has been recently proposed in [50] to **solve quadratic equations in higher dimensions**. The main strategy of minimization/reduction still works, but in a different way. A quadratic form q in dimension n is minimizable if we can find an integral quadratic form q' with determinant ± 1 and equivalent to q over $\mathbb{Q}$. For an indefinite ternary quadratic form, minimizability is equivalent to solvability. In higher dimension $n \geq 4$, this property is not true any more.

When $n = 4$, we can use a trick of Cassels [10]: there exists a binary quadratic form q_2 such that $q \oplus q_2$ is minimizable and is equivalent to $H \oplus H \oplus H$, where H is an hyperbolic plane (the $\oplus$ notation denotes the direct sum of quadratic modules; in terms of matrices, it just corresponds to taking the larger matrix, which is diagonal by blocks, and the blocks are q and q_2). This binary quadratic form q_2 can be computed explicitly from the local invariants of q and from the 2-Sylow of the class group $Cl(\sqrt{D})$ (D being the determinant of q). Hence, using LLL and [7], we can build this q_2 as soon as we know the factorization of D ! Now, the indefinite LLL algorithm rapidly gives the equivalence between $q \oplus q_2$ and $H \oplus H \oplus H$ (this is the reduction step). The remaining part of the algorithm is just linear algebra : a vector in the intersection of a 3-dimensional isotropic subspace for $H \oplus H \oplus H$ with another 4-dimensional subspace (in the ambient space of dimension 6) will give a solution for $q(\mathbf{x}) = 0$.

When $n \geq 5$, a similar trick applies, except that the minimization works either in dimension $n, n+1, n+2$ or $n+3$.

Number Fields

I see two main reasons why the computation of class groups in imaginary quadratic fields are feasible. The first one is that, thanks to the correspondence between ideals and quadratic forms (see [12, Sect. 5.2]), we can use Gauss reduction. The second one is McCurley's sub-exponential algorithm described in [25]. This algorithm assumes the validity of the Generalized Riemann Hypothesis. It is also described in [9] or [12, Sect. 5.5]. It relies on the notion of *relations*, that is expressions of ideals $\mathfrak{a}$ in the form

$$\mathfrak{a} = \alpha \prod \mathfrak{f}_i^{e_i},$$

where α is an element of K^*, and $\mathfrak{f}_i$ are the ideals of the factor basis.

McCurley's algorithm has been extended to general number fields. In [8] and [15] (see also [12, Sect. 6.5] and [4]), it is explained how one can compute simultaneously the class group and the units from the relations between ideals. It is also explained how one can perform the ideal reduction. The idea is the following.

- Let K be a number field of degree n over $\mathbb{Q}$ with signature (r_1, r_2). Let $\sigma_1, \ldots, \sigma_{r_1}$ be the real embeddings of K and $\sigma_{r_1+1}, \ldots, \sigma_n$ the complex embeddings. For an element $x \in K$, we define the T_2-norm of x to be $\|x\| = \sqrt{T_2(x)}$ where

$$T_2(x) = \sum_{i=1}^{n} |\sigma_i(x)|^2.$$

This T_2-norm can immediately be extended to $\mathbb{R}^n \simeq K \otimes_{\mathbb{Q}} \mathbb{R}$ and defines a positive definite quadratic form. Equipped with this quadratic form, the ring of integers $\mathbb{Z}_K$ of K is a lattice of determinant $\det(\mathbb{Z}_K) = |\operatorname{disc} K|^{1/2}$. Choosing an LLL-reduced basis (w_i) of $\mathbb{Z}_K$ (for the T_2-norm) will give $\mathbb{Z}_K$ the property that elements with small coefficients will also have a small (algebraic-)norm. This is just an application of the arithmetic-geometric mean:

$$n|\mathcal{N}_{K/\mathbb{Q}}(x)|^{2/n} \leqslant T_2\left(\sum_{i=1}^{n} x_i w_i\right) \leqslant \left(\sum_{i=1}^{n} x_i^2\right)\left(\sum_{i=1}^{n} T_2(w_i)\right).$$

Of course the converse is not true since, in general, there are infinitely many units, that is elements of $\mathbb{Z}_K$ such that $\mathcal{N}_{K/\mathbb{Q}}(x) = \pm 1$, but there are only finitely many elements with bounded T_2-norm or bounded coefficients. As indicated in [4], choosing an LLL-reduced basis for the T_2-norm usually gives a **faster arithmetic in the number field**.

- The **polynomial reduction algorithm** described in [14] and [12, Sect. 4.4.2] is a byproduct of this reduction. Indeed, if we have an irreducible polynomial, with splitting field K, one can compute an LLL-reduced integral basis of $\mathbb{Z}_K$, and look for a primitive element with small coefficients. Its minimal polynomial usually has small coefficients. For example, if we look for a defining polynomial for the field $\mathbb{Q}(i, \sqrt{2}, \sqrt{3}, \sqrt{5})$, the function `polcompositum` of `PARI/GP` (which implements the standard idea of the proof of the primitive element theorem) gives

$$x^{16} - 72x^{14} + 1932x^{12} - 22552x^{10} + 154038x^8 - 582456x^6 + 1440748x^4 - 1486824x^2 + 3721041$$

  with discriminant $2^{312}3^{50}5^{8}7^{4}643^{2}$. The function `polredabs` of `PARI/GP` implements this polynomial reduction algorithm and just finds the polynomial

$$x^{16} - 7x^{12} + 48x^8 - 7x^4 + 1$$

  having discriminant $2^{64}3^{32}5^{8}$.
- From my point of view, the most important application of this is the notion of **LLL-reduction of ideals** introduced in [8] and [15], [12, Sect. 6.5.1]. Any integral ideal $\mathfrak{a}$ of $\mathbb{Z}_K$ of norm $\mathcal{N}_{K/\mathbb{Q}}(\mathfrak{a})$ is a sublattice of $\mathbb{Z}_K$ of index $\mathcal{N}_{K/\mathbb{Q}}(\mathfrak{a})$. Hence, we can apply LLL to this sublattice. A short vector for this T_2-norm

is certainly an element of $\mathfrak{a}$, that is, an element $a \in \mathbb{Z}_K$ satisfying $a = \mathfrak{a}\mathfrak{b}$ for some integral ideal $\mathfrak{b}$ such that $\mathcal{N}_{K/\mathbb{Q}}(\mathfrak{b})$ is bounded independently of $\mathfrak{a}$ (notice that this claim gives a proof of the finiteness of the class group). If $\mathfrak{b} = 1$, we have found a generator of the principal ideal $\mathfrak{a}$. If the ideal $\mathfrak{b}$ can be factored in the factor basis, we have a relation. This is exactly the way relations are found in McCurley's algorithm. Hence, we see that combined with McCurley's algorithm, LLL gives a way to **compute class groups** and **solve the principal ideal problem**. Furthermore, relations among principal ideals give units. More precisely, a relation of the type $\alpha\mathbb{Z}_K = \beta\mathbb{Z}_K$ is equivalent to the fact that $\alpha\beta^{-1}$ is a unit. Hence, the algorithm is also able to **compute the unit group**.

Conclusion: Breaking Records

It is a popular belief that when one wants to break records in number theory, one has to use LLL. The applications of LLL given in this section are therefore more isolated, but all serve the same goal: testing conjectures with numerical evidences.

- A revolutionary application was given in 1985 by Odlyzko and te Riele in [43], wherein they **give a disproof of the very old conjecture of Mertens**. This conjecture says that if we consider the Möbius function $\mu(n)$ and its counting function $M(x) = \sum_{n \leqslant x} \mu(n)$, then we should have $|M(x)| < \sqrt{x}$. After having computed the 200 first zeros of the Riemann zeta function and used LLL in dimension 200, they were able to prove that $\limsup M(x)x^{-1/2} > 1.06$ and $\liminf M(x)x^{-1/2} < -1.009$. It has been improved in [36] using the same technics to $\limsup M(x)x^{-1/2} > 1.218$ and $\liminf M(x)x^{-1/2} < -1.229$.
- In a very surprising paper, [19], Elkies applies LLL, not to compute points *on* curves, but to **compute points near curves with small height**. The naive idea to list points of small height on a curve defined by an homogeneous equation $F(x, y, z) = 0$ is to loop on all possible values for x and y with $|x| < N$ and $|y| < N$, compute the corresponding values of z and test for their height. If we also want points near the curve, a loop on z is also needed. If we are looking for points at distance at most δ to the curve, the idea of Elkies is to cut the curve into small pieces, each of length $= O(\delta^{1/2})$. For each piece, the curve is approximated by a segment, and the search region is approximated by a box B of height, length, and width proportional to N, $\delta^{1/2}N$, δN. Hence, as soon as $\delta \gg N^{-2}$, we can expect that the box B of volume $N^3\delta^{3/2}$ will contain $O(N^3\delta^{3/2})$ integral points. Now, **finding all integral points in a box** is a standard application of LLL: find a reduced basis of B and loop over all points with small coefficients in this basis. Using this approach, he is able to find the following surprising relations:

$$386692^7 + 411413^7 \approx 441849^7,$$

$$2063^\pi + 8093^\pi \approx 8128^\pi.$$

This last example raises the question whether there are infinitely many integral solutions to $|x^\pi + y^\pi - z^\pi| < 1$. Using some more tricks leads to the relation

$$5853886516781223^3 - 447884928428402042307918^2 = 1641843$$

which is related to Hall's conjecture, telling that if $k = x^3 - y^2$, then $|k| \gg_\varepsilon x^{1/2-\varepsilon}$. This example satisfies $x^{1/2}/|x^3 - y^2| > 46.6$, improving the previous record by a factor of almost 10.

- See also [18] for examples related the the abc conjecture and the Szpiro conjecture.

References

1. E. Agrell, T. Eriksson, A. Vardy, K. Zeger: *Closest point search in lattices*, IEEE Trans. Inf. Theory **48**, No. 8, 2201–2214 (2002).
2. G. Almkvist, A.J. Granville: *Borwein and Bradley's Apéry-like formulae for* $\zeta(4n+3)$, Exp. Math. **8**, No. 2, 197–203 (1999).
3. L. Babai: *On Lovász lattice reduction and the nearest lattice point problem*, Combinatorica **6**, 1–13 (1986).
4. K. Belabas: *Topics in computational algebraic number theory*, J. Théor. Nombres Bordeaux **16**, No. 1, 19–63 (2004).
5. J. Borwein, D. Bradley: *Empirically determined Apéry-like formulae for* $\zeta(4n+3)$, Exp. Math. **6**, No. 3, 181–194 (1997).
6. J. Borwein, D. Bradley: *Experimental determination of Apéry-like identities for* $\zeta(2n+2)$, Exp. Math. **15**, No. 3, 281–289 (2006).
7. W. Bosma, P. Stevenhagen: *On the computation of quadratic 2-class groups* J. Théor. Nombres Bordeaux. **8**, No. 2, 283–313 (1996); erratum ibid. **9**, No. 1, 249 (1997).
8. J. Buchmann: *A subexponential algorithm for the determination of class groups and regulators of algebraic number fields*, Sémin. Théor. Nombres, Paris/Fr. 1988–89, Prog. Math. **91**, 27–41 (1990).
9. J. Buchmann, S. Düllmann: *A probabilistic class group and regulator algorithm and its implementation*, Computational number theory, Proc. Colloq., Debrecen/Hung. 1989, 53–72 (1991).
10. J.W.S. Cassels: *Note on quadratic forms over the rational field* , Proc. Cambridge Philos. Soc. **55**, 267–270 (1959).
11. T. Cochrane, P. Mitchell: *Small solutions of the Legendre equation*, J. Number Theor. **70**, No. 1, 62–66 (1998).
12. H. Cohen: *A Course in Computational Algebraic Number Theory*, Graduate Texts in Math. **138**, Second corrected printing, Springer, Berlin, (1995).
13. H. Cohen: *Advanced Topics in Computational Algebraic Number Theory*, Graduate Texts in Math. **193**, Springer, Berlin, (2000).
14. H. Cohen, F. Diaz y Diaz: *A polynomial reduction algorithm*, Sémin. Théor. Nombres Bordeaux., Sér. II **3**, No. 2, 351–360 (1991).
15. H. Cohen, F. Diaz y Diaz, M. Olivier: *Subexponential algorithms for class group and unit computations*, J. Symbolic Comput. **24**, No. 3–4, 433–441 (1997), Computational algebra and number theory (London, 1993).
16. J.E. Cremona, D. Rusin: *Efficient solution of rational conics*, Math. Comp. **72**, 1417–1441 (2003).
17. U. Dieter: *Calculating shortest vectors in a lattice* Ber. Math.-Stat. Sekt. Forschungszent. Graz **244**, 14 p. (1985).

18. T. Dokchitser: *LLL & ABC*, J. Number Theor. **107**, No. 1, 161–167 (2004).
19. N.D. Elkies: *Rational points near curves and small nonzero* $|x^3 - y^2|$ *via lattice reduction*, W. Bosma (ed.), Algorithmic number theory. 4th international symposium. ANTS-IV, Leiden, the Netherlands, July 2–7, 2000. Proceedings. Berlin: Springer. Lect. Notes Comput. Sci. **1838**, 33–63 (2000).
20. H.R.P. Ferguson, D. Bailey, S. Arno: *Analysis of PSLQ, an integer relation finding algorithm*, Math. Comput. **68**, No. 225, 351–369 (1999).
21. H.R.P. Ferguson, R.W. Forcade: *Generalization of the Euclidean algorithm for real numbers to all dimensions higher than two*, Bull. Am. Math. Soc., New Ser. **1**, 912–914 (1979).
22. U. Fincke, M. Pohst: *On reduction algorithms in non-linear integer mathematical programming*, Operations research, Proc. 12th Annu. Meet., Mannheim 1983, 289–295 (1984).
23. U. Fincke, M. Pohst: *Improved methods for calculating vectors of short length in a lattice, including a complexity analysis*, Math. Comput. **44**, 463–471 (1985).
24. C.F. Gauss: *Disquisitiones Arithmeticae*, Springer, Berlin, (1986).
25. J. Hafner, K. McCurley: *A rigorous subexponential algorithm for computation of class groups*, J. Amer. Math. Soc. **2**, No. 4, 837–850 (1989).
26. G. Hanrot: *LLL: A tool for effective diophantine approximation*, this volume.
27. J. Håstad, B. Just, J.C. Lagarias, C.P. Schnorr (B. Helfrich): *Polynomial time algorithms for finding integer relations among real numbers*, SIAM J. Comput. **18**, No. 5, 859–881 (1989).
28. G. Havas, B.S. Majewski, K.R. Matthews: *Extended GCD and Hermite normal form algorithms via lattice basis reduction*, Exp. Math. **7**, No. 2, 125–136 (1998); *Addenda and errata: Extended GCD and Hermite normal form algorithms via lattice basis reduction*, Exp. Math. **8**, No. 2, 205 (1999).
29. G. Ivanyos, A. Szánto: *Lattice basis reduction for indefinite forms and an application*, Discrete Math. **153**, No. 1–3, 177–188 (1996).
30. G. Jäger. *Reduction of Smith normal form transformation matrices* Computing **74**, No. 4, 377–388 (2005).
31. E. Kaltofen, N. Yui: *Explicit construction of the Hilbert class fields of imaginary quadratic fields by integer lattice reduction*, Number theory, Proc. Semin., New York/NY (USA) 1989–1990, 149–202 (1991).
32. R. Kannan: *Lattices, basis reduction and the shortest vector problem*, Theory of algorithms, Colloq. Pécs/Hung. 1984, Colloq. Math. Soc. János Bolyai **44**, 283–311 (1986).
33. R. Kannan, A.K. Lenstra, L. Lovász: *Polynomial factorization and nonrandomness of bits of algebraic and some transcendental numbers*, Math. Comput. **50**, No. 181, 235–250 (1988).
34. R. Kannan, L.A. McGeoch: *Basis reduction and evidence for transcendence of certain numbers*, Foundations of software technology and theoretical computer science, Proc. 6th Conf., New Delhi/India 1986, Lect. Notes Comput. Sci. 241, 263–269 (1986).
35. J. Klüners: *The van Hoeij algorithm for factoring polynomials*, this volume.
36. T. Kotnik, H. te Riele: *The Mertens Conjecture Revisited*, F. Hes, S. Pauli, M. Pohst (ed.), ANTS 2006 Berlin, Lect. Notes Comput. Sci. **4076**, 156–167 (2006).
37. J.C. Lagarias: *Worst-case complexity bounds for algorithms in the theory of integral quadratic forms*, J. Algorithm. **1**, 142–186 (1980).
38. J.C. Lagarias: *Knapsack public key cryptosystems and Diophantine approximation*, in *Advances in cryptology* (Santa Barbara, Calif., 1983), 3–23, Plenum, New York, (1984).
39. J.C. Lagarias: *The computational complexity of simultaneous diophantine approximation problems*, SIAM J. Comput. **14**, 196–209 (1985).
40. A.K. Lenstra: *Polynomial factorization by root approximation*, EUROSAM 84, Symbolic and algebraic computation, Proc. int. Symp., Cambridge/Engl. 1984, Lect. Notes Comput. Sci. **174**, 272–276 (1984).
41. A.K. Lenstra, H.W. Lenstra, L. Lovász: *Factoring polynomials with rational coefficients* Math. Ann. **261**, 515–534 (1982).
42. K. Matthews: `www.numbertheory.org/lll.html`
43. A.M. Odlyzko, H. te Riele: *Disproof of the Mertens conjecture* J. Reine Angew. Math. **357**, 138–160 (1985).

44. M. Pohst: *A modification of the LLL reduction algorithm*, J. Symb. Comput. **4**, 123–127 (1987).
45. A. Schönhage: *Factorization of univariate integer polynomials by diophantine approximation and an improved basis reduction algorithm*, in *Automata, languages and programming*, 11th Colloq., Antwerp/Belg. 1984, Lect. Notes Comput. Sci. **172**, 436–447 (1984).
46. J.-P. Serre: *Cours d'arithmétique*, P.U.F. 3rd edition (1988).
47. A. Shamir: *A polynomial time algorithm for breaking the basic Merkle-Hellman cryptosystem*, in *23rd annual symposium on foundations of computer science* (Chicago, Ill., 1982), 145–152, IEEE, New York, (1982)
48. D. Shanks: *Gauss's ternary form reduction and the 2-Sylow subgroup*, Math. Comput. **25**, 837–853 (1971); *Corrigendum: Gauss's ternary form reduction and the 2-Sylow subgroup*, Math. Comput. **32**, 1328–1329 (1978).
49. D. Simon: *Solving quadratic equations using reduced unimodular quadratic forms*, Math. Comput. **74**, No. 251, 1531–1543 (2005).
50. D. Simon: *Quadratic equations in dimensions* 4, 5, *and more*, preprint (2006).
51. N.P. Smart: *The algorithmic resolution of diophantine equations*, London Mathematical Society Student Texts. **41**. Cambridge: Cambridge University Press. (1998).
52. B.M.M. de Weger: *Approcimation lattices of p-adic numbers*, J. Number Theor., **24**(1), 70–88 (1986).

Chapter 8
The van Hoeij Algorithm for Factoring Polynomials

Jürgen Klüners

Abstract In this survey, we report about a new algorithm for factoring polynomials due to Mark van Hoeij. The main idea is that the combinatorial problem that occurs in the Zassenhaus algorithm is reduced to a very special knapsack problem. In case of rational polynomials, this knapsack problem can be very efficiently solved by the LLL algorithm. This gives a polynomial time algorithm, which also works very well in practice.

Introduction

Let $f \in \mathbb{Z}[x]$ be a polynomial of degree n with integral coefficients. One of the classical questions in computer algebra is how to factorize f in an efficient way. About 40 years, ago Hans Zassenhaus [1] developed an algorithm, which was implemented in almost all computer algebra systems until 2002. This algorithm worked very well for many examples, but his worst case complexity was exponential. In the famous LLL–paper [2], it was proved that it is possible to factor polynomials in polynomial time in the degree and the (logarithmic) size of the coefficients. Despite the fact that the new lattice reduction algorithm was very good in theory and in practice, the new polynomial factorization was not used in implementations. For most practical examples, the Zassenhaus algorithm was more efficient than the new algorithm based on LLL.

In 2002, Mark van Hoeij [3] developed a new algorithm, which, for practical examples, was much more efficient than the Zassenhaus algorithm. This new algorithm is also based on the LLL reduction, but it uses a different type of lattices compared to the ones in the original LLL paper [2]. Unfortunately, Mark van Hoeij gave no running time estimates in his original paper. He was only able to show that his algorithm terminates, but he gave very impressive practical examples of factorizations, which have not been possible to compute before. Together with Karim

J. Klüners
Mathematisches Institut, Universität Paderborn, Warburger Str. 100, 30098 Paderborn, Germany.
e-mail: klueners@math.uni-paderborn.de

P.Q. Nguyen and B. Vallée (eds.), *The LLL Algorithm*, Information Security and Cryptography, DOI 10.1007/978-3-642-02295-1_8,

Belabas, Mark van Hoeij, and Allan Steel [4], the author of this survey simplified the presentation of this algorithm and introduced a variant for factoring bivariate polynomials over finite fields. Furthermore, we have been able to show that the new factoring algorithm runs in polynomial time. The worst case estimate is better than the one given in the original LLL paper. Let us remark that we believe that this estimate is still pessimistic.

In this survey, we will study the cases $f \in \mathbb{Z}[x]$ and $f \in \mathbb{F}_p[t][x]$ in parallel, where $\mathbb{F}_p$ denotes the finite field with p elements. We study the second case because the presentation is easier. In the bivariate case, it is not necessary to use LLL reduction. The corresponding step can be solved by computing kernels of systems of linear equations.

The Zassenhaus Algorithm

Before we are able to explain the van Hoeij algorithm, we need to understand the Zassenhaus algorithm. We can assume that our given polynomial is squarefree. Note that multiple factors of f divide the greatest common divisor of f and the derivative f', which can be computed efficiently using the Euclidean algorithm. We remark that we have to be careful in characteristic p since the derivative of f may be 0, e.g., $f(x) = x^p - t$. In this case, all monomials are p-th powers, and we can take the p-th root or we switch the role of t and x.

Let $f \in \mathbb{Z}[x]$ be a squarefree and monic polynomial, i.e., a polynomial with integer coefficients and leading coefficient one. In the Zassenhaus algorithm, we choose a prime number p such that f modulo p has no multiple factors. It is possible to choose every prime that does not divide the discriminant of f. We denote by $\bar{f} \in \mathbb{F}_p[x]$ the polynomial that can be derived out of f by reducing each coefficient modulo p. Using well known algorithms, e.g., see [5, Chap. 14], we can compute the following factorization:

$$\bar{f}(x) = \bar{f}_1(x) \cdots \bar{f}_r(x) \in \mathbb{F}_p[x].$$

Using the so-called Hensel lifting (by solving linear systems of equations), we can efficiently compute for all $k \in \mathbb{N}$ a factorization of the following form: (e.g., see [5, Chap. 15])

$$f(x) \equiv \tilde{f}_1(x) \cdots \tilde{f}_r(x) \bmod p^k,$$

where (using a suitable embedding) $\tilde{f}_i \equiv \bar{f}_i \bmod p$. Let us explain the Zassenhaus algorithm using the example $f(x) = x^4 - 11$. Using $p = 13$, we get that $\bar{f} \in \mathbb{F}_{13}[x]$ is irreducible. Certainly, this implies that $f \in \mathbb{Z}[x]$ is irreducible. When we choose (a little bit unlucky) $p = 5$, then we only get linear factors modulo p, which will be lifted using Hensel lifting:

$$f(x) \equiv (x + 41)(x - 38)(x + 38)(x - 41) \bmod 125.$$

To proceed, we need a bound for the size of the coefficients of a factor of f. The following theorem can be found in [5, p.155ff].

Theorem 1 (Landau-Mignotte). *Let g be a factor of a monic polynomial* $f \in \mathbb{Z}[x]$ *with*

$$f(x) = \sum_{i=0}^{n} a_i x^i \text{ and } g(x) = \sum_{i=0}^{m} b_i x^i.$$

Then: $|b_i| \leq \binom{m}{i} ||f||_2$, *where* $||f||_2 := \sqrt{\sum_{i=0}^{n} a_i^2}$ *denotes the 2–norm.*

In our example, this means that all coefficients of a factor g of f must be less than or equal to 33 in absolute value. Therefore, we see that f has no linear factor, because modulo 125 all linear factors contain a coefficient in the symmetric residue system $\{-62, \ldots, 62\}$, which is bigger than 33 in absolute value. In the next step, we try if the product of two modulo 125 factors corresponds to a true factor of f in $\mathbb{Z}[x]$. We get:

$$\begin{aligned}
(x+41)(x-38) &\equiv x^2 + 3x - 58 \mod 125, \\
(x+41)(x+38) &\equiv x^2 - 46x + 58 \mod 125, \\
(x+41)(x-41) &\equiv x^2 - 56 \mod 125.
\end{aligned}$$

All these quadratic polynomials contain a coefficient that is bigger than 33 in absolute value. This means that the "modular factor" $(x + 41)$ is no divisor of a linear or quadratic factor $g \in \mathbb{Z}[x]$ of f. This implies that the polynomial $f \in \mathbb{Z}[x]$ is irreducible. In case that our given polynomial is reducible, we find modular factors of f such that all coefficients in the symmetric residue system are smaller than the Landau–Mignotte bound. Now, we can use trial division in $\mathbb{Z}[x]$ to check if we have found a factor or not. Since trial divisions are expensive, it is a good idea in actual implementations to choose p^k much bigger than twice the Landau–Mignotte to increase the probability that wrong candidates will be found without a trial division.

The choice of $p = 5$ in the above example was very artificial. We remark that it is easy to construct irreducible polynomials such that for all primes p, we have many modular factors. For example, we can take a polynomial f of degree $n = 2^\ell$ such that the Galois group is isomorphic to the elementary abelian group $(\mathbb{Z}/2Z)^n$. In this case, f is irreducible, but for every prime p, we have at least $2^{\ell-1} = n/2$ modular factors.

If we analyze the Zassenhaus algorithm, we figure out that most parts of the algorithm are very efficient. Since we are able to choose a small prime p, it is not difficult to factor $\bar{f} \in \mathbb{F}_p[x]$. The Hensel lifting can be solved using linear systems of equations, and the Landau-Mignotte bound is sufficiently small. The drawback of the algorithm is the number of tests that have to be performed when the number r of modular factors is big. In this case, we have to perform more or less 2^r tests.

The Knapsack Lattice

This is the place where the new van Hoeij algorithm starts. It reduces the combinatorial problem to a so-called knapsack problem. The resulting knapsack problem can be efficiently solved using lattices and the LLL algorithm. We remark that we use different type of lattices compared to the original LLL paper.

We fix the following notation:

$$f = g_1 \cdots g_s \in \mathbb{Z}[x] \text{ and } f = \tilde{f}_1 \cdots \tilde{f}_r \in \mathbb{Z}_p[x].$$

The factorization over the p-adic numbers $\mathbb{Z}_p$ can only be determined modulo p^k. For the understanding of the following, it is possible to interpret the p-adic numbers as modulo p^k approximations. We write:

$$g_v := \prod_{i=1}^{r} f_i^{v_i} \text{ for } v = (v_1, \ldots, v_r) \in \{0,1\}^r$$

and get a new

Problem 1. For which $v \in \{0,1\}^r$, do we have: $g_v \in \mathbb{Z}[x]$?

To linearize our problem, we consider (more or less) the logarithmic derivative, where $\mathbb{Q}_p(x) := \{\frac{a(x)}{b(x)} \mid a, b \in \mathbb{Q}_p[x]\}$:

$$\Phi : \mathbb{Q}_p(x)^*/\mathbb{Q}_p^* \to \mathbb{Q}_p(x), g \mapsto \frac{fg'}{g}.$$

It is immediately clear that Φ is additive, i.e., $\Phi(g_{v_1}) + \Phi(g_{v_2}) = \Phi(g_{v_1+v_2})$. Furthermore, we have for $v \in \mathbb{Z}^r$ that $\Phi(g_v)$ is a polynomial and therefore an element of $\mathbb{Z}_p[x]$.

The next step is to translate everything into a lattice problem. Let us define vectors $w_1, \ldots, w_s \in \{0,1\}^r$ such that for the true factors $g_1, \ldots, g_s \in \mathbb{Z}[x]$, we have

$$g_i = \prod_{1 \le j \le r} \tilde{f}_j^{w_{i\,j}}.$$

These vectors generate a lattice (the knapsack lattice) $W = \langle w_1, \ldots, w_s \rangle \subseteq \mathbb{Z}^r$. The lattice $\mathbb{Z}^r$ is generated by the standard basis vectors, which correspond to the local factors $\tilde{f}_i$. An important fact for the new method is the property that $v \in \mathbb{Z}^r$ is an element of W only if $\Phi(g_v) \in \mathbb{Z}[x]$ (even in the case that v has negative coefficients). We remark that it is easy to construct the canonical basis vectors $w_1, \ldots, w_s$ of W if we know some generating system of W. As soon as we know the w_i and p^k is at least twice larger than the Landau-Mignotte bound, we are able to reconstruct the corresponding factors g_i like in the Zassenhaus algorithm.

The idea of the algorithm is as follows. We start with the lattice $L = \mathbb{Z}^r$ and know that $W \subseteq L$. Then, we construct a sublattice $L' \subset L$ still containing W. The hope is that after finitely many steps, we will reach $L' = W$.

At this place, we change to the situation $f \in \mathbb{F}_p[t][x]$, because the following part of the algorithm is easier here. The Landau-Mignotte bound simplifies to

$$g \mid f \in \mathbb{F}_p[t][x] \Rightarrow \deg_t(g) \leq \deg_t(f),$$

where $\deg_t(f)$ is the t–degree of the polynomial f. In order to simplify the situation, we assume that $\bar{f}(x) := f(0, x) \in \mathbb{F}_p[x]$ is squarefree. This is a real restriction because it might be the case that $f(a, x)$ has multiple factors for all $a \in \mathbb{F}_p$. For the solution of this problem, we refer the reader to [4]. Using Hensel lifting, we get from the factorization $\bar{f} = \bar{f}_1 \cdots \bar{f}_r \in \mathbb{F}_p[x]$ a factorization $f(t, x) = \tilde{f}_1 \cdots \tilde{f}_r$ in the power series ring $\mathbb{F}_p[[t]][x]$. In practice, we can approximate the power series in t modulo t^k. Now, we define the function Φ in the following way:

$$\Phi : \mathbb{F}_p[[t]](x)^* / \mathbb{F}_p[[t]](x^p)^* \to \mathbb{F}_p[[t]](x), g \mapsto \frac{fg'}{g}.$$

The lattices L and W are defined analogously as in the situation for $\mathbb{Z}[x]$. Assume that we have an element $v \in L \setminus W$. Then, we have:

$$\begin{aligned} \mathrm{Pol}(v) := \Phi(g_v)(x) &= \sum_{i=1}^{r} v_i \Phi(f_i) \\ &= \sum_{i=0}^{n-1} b_i x^i \in \mathbb{F}_p[[t]][x] \setminus \mathbb{F}_p[t][x]. \end{aligned}$$

Additionally, we have for $g_v \in \mathbb{F}_p[t][x]$ the estimate $\deg_t(b_i) \leq \deg_t(f)$. Now, we choose a $k > \deg_t(f)$, and we compute for $v \in L$ the corresponding polynomial

$$g_v \equiv \sum_{i=0}^{n-1} b_i(t) x^i \mod t^k.$$

Here, modulo t^k means that all $b_i(t)$ are reduced modulo t^k, i.e., $\deg_t(b_i) < k$. In case that one of the polynomials b_i has a t-degree that is bigger than $\deg_t(f)$, we know that the corresponding v is not an element of W. In the following, we avoid the combinatorial approach.

Denote by $e_1, \ldots, e_r \in \mathbb{F}_p^r$ the standard basis of $\mathbb{F}_p^r$, and identify the elements of $\mathbb{F}_p$ with $\{0, \ldots, p-1\} \subseteq \mathbb{Z}$. We define $m := \deg_t(f)$ and

$$A_i := \begin{pmatrix} b_{i,m,1} & \cdots & b_{i,m,r} \\ b_{i,m+1,1} & \cdots & b_{i,m+1,r} \\ \vdots & \ddots & \vdots \\ b_{i,k-1,1} & \cdots & b_{i,k-1,r} \end{pmatrix} \in \mathbb{F}_p^{(k-m)\times r},$$

where the $b_{i,j,\ell}$ are given by

$$\mathrm{Pol}(e_\ell) \equiv \sum_{i=0}^{n-1}\sum_{j=0}^{k-1} b_{i,j,\ell} t^j x^i \bmod t^k \quad (1 \le \ell \le r).$$

All $v \in W$ have the property that $A_i v^{\mathrm{tr}} = 0$. Using iterative kernel computation, we are able to determine lattices $L' \supseteq W$, which (hopefully) become smaller.

The Polynomial Complexity Bound

In a slightly improved version of this algorithm, we show in [4] that we finally get $L' = W$:

Theorem 2. *Let $f \in \mathbb{F}_p[t][x]$ be a polynomial of x-degree n and assume $k > (2n-1)\deg_t(f)$. Then, W is the kernel of $A_1, \ldots, A_{n-1}$.*

In the following, we give a sketch of the proof of this theorem. Let $v \in L \setminus W$ be chosen such that g_v is not a p-th power. Then, it is possible to change v using $w_1, \ldots, w_s$ such that the following holds:

1. $f_i \mid \mathrm{Pol}(v)$ for some $1 \le i \le r$.
2. $g_j \nmid \mathrm{Pol}(v)$ for all $1 \le j \le s$.

Take this new v and define $H := \mathrm{Pol}(v) \bmod t^k$ interpreted as a polynomial in $\mathbb{F}_p[t][x]$. Using well known properties of the resultant, we immediately get:

$$\mathrm{Res}(f, \mathrm{Pol}(v)) = 0 \text{ and } \mathrm{Res}(f, H) \neq 0.$$

This implies that $t^k \mid \mathrm{Res}(f, H)$. Choosing k large enough is a contradiction to the definition of the resultant via the Sylvester matrix.

Let us come back to our original problem over $\mathbb{Z}$. We cannot apply the same algorithm because we have overflows when we add, e.g., $(3+1\cdot 5^1)+(3+1\cdot 5^1) = (1+3\cdot 5^1) \neq 1+2\cdot 5^1$ in $\mathbb{Z}_5$. Fortunately, we can show that the errors coming from overflows are small. Instead of solving linear systems of equations, we define a suitable lattice and look for vectors of small length in that lattice. Since finding shortest vectors in lattices is an NP-complete problem, it is important to choose the lattices in a very clever way. For those lattices, we apply LLL reduction and can guarantee that the first basis vectors of the LLL reduced basis will be sufficient to derive a basis of W. We use the analogous definitions as in the bivariate case and get:

$$\mathrm{Pol}(e_\ell) \equiv \sum_{i=0}^{n-1} b_{i,\ell} x^i \bmod p^k \quad (1 \leq \ell \leq r).$$

Now, we define a lattice Λ, which is defined by the columns of the following matrix:

$$A := \begin{pmatrix} I_r & 0 \\ \tilde{A} & p^k I_n \end{pmatrix} \text{ with } \tilde{A} := \begin{pmatrix} b_{0,1} & \cdots & b_{0,r} \\ \vdots & \ddots & \vdots \\ b_{n-1,1} & \cdots & b_{n-1,r} \end{pmatrix}.$$

If we project a vector from Λ to the first r rows, we get a vector in L. Assuming that we choose the precision p^k large enough, we are able to prove that all vectors in Λ, such that the last n entries are smaller than the Landau-Mignotte bound, correspond to a vector in W. We compute an LLL reduced basis of the above lattice and are able to prove that the first s vectors correspond to a basis of W. It is easy to give an upper bound B for the norm for the vectors in Λ, which correspond to $w_1, \ldots, w_s$. For the LLL approach and practical implementations, the following lemma is very useful, since it allows to have some progress, i.e., a new lattice $W \subseteq L' \subseteq L$, if the precision p^k was not large enough to derive $L' = W$.

Lemma 1. *Let Λ be a lattice with basis $b_1, \ldots, b_m$ and Gram–Schmidt–basis $b_1^*, \ldots, b_m^*$. Define $t := \min\{i \mid \forall i < j \leq m : ||b_j^*||_2 > B\}$. Then, all vectors b, such that $||b||_2 \leq B$, are contained in $\mathbb{Z}b_1 + \cdots + \mathbb{Z}b_t$.*

We remark that this lemma is already contained in the original LLL paper [2]. Analogous to the bivariate case, we need an estimate for the precision, which guarantees that the algorithm terminates, i.e., that finally we have $L' = W$. If we use this precision, our algorithm terminates in one (LLL reduction) step and we get the polynomial running time. We remark that, in practice, we do not need this (explicit) estimate, because we start with some precision and increase it until the algorithm terminates.

Theorem 3. *Let $f \in \mathbb{Z}[X]$ of degree n. Then, the above described algorithm terminates if*

$$p^k > c^n \cdot 4^{n^2} ||f||_2^{2n-1} \tag{8.1}$$

holds, where c is an explicit computable constant.

If we determine the running time for this algorithm, we get the same running time as the one in the original LLL paper.

The Original LLL Factoring Method

In order to understand the difference between those two algorithms, we need to understand the original LLL factoring method, at least roughly. As before, let $f \in \mathbb{Z}[x]$ of degree n be the monic polynomial; we would like to factor and assume that we have chosen a prime p such that f mod p has no multiple factors. As before, using Hensel lifting, we get the factorization of the form:

$$f(x) \equiv \tilde{f}_1(x) \cdots \tilde{f}_r(x) \bmod p^k.$$

The idea of the original LLL factorization algorithm is to compute an irreducible factor $g \in \mathbb{Z}[x]$ such that $\tilde{f}_1 \mid g \in \mathbb{Z}_p[x]$. To compute such a g, they write down a tricky lattice that allows to check if there exists such a g of degree $m < n$. After an LLL reduction of this lattice, the first basis vector corresponds to such a g if it exists. If no such g exists, we have to use the theoretical lifting bound (similar to the one we have given in the van Hoeij algorithm) to get a proof that such a g does not exist. This is the practical drawback of the original algorithm. In case that our polynomial is reducible, we might be lucky to find g using a small precision. In case that f is irreducible, we have to use the full theoretical precision in order to get a proof that no nontrivial factor exists. We remark that after some recursive application of this algorithm (we find at most one factor at each LLL-step), we run into this situation. By taking irreducible polynomials, which have many modular factors, we get examples that really need the worst case running time we computed before.

Usually, in the van Hoeij algorithm, such a polynomial will be detected using a very low precision. What we say here is just heuristics and practical experience. We expect that the worst case running time we have given before is just a very bad upper estimate and will never be attained. We remark that we cannot prove such a statement.

One important fact for practical implementations of the van Hoeij algorithm is the fact that it is possible to make partial progress. If we choose a precision that was too small to derive the full factorization, it is very often the case that we are able to compute a smaller lattice L', which means that this "try" was not worthless. If in the original LLL factorization algorithm we use a too small precision and do not succeed, we have nothing.

There is another advantage of the new van Hoeij algorithm compared to the original one. In the original algorithm, we try to compute a factor directly. This means if the coefficients of the given polynomial are big, we need to compute a factor that has big coefficients as well. Therefore, we want to find a short(est) vector in a lattice, which is already huge. In the knapsack approach of van Hoeij, we are looking to find a zero-one combination. The shortest vector we are looking for is really very short (only zeroes and ones and some small errors coming from overflows). In some sense, those lattices are independent on the size of the coefficients of the given polynomial.

In the meantime, the van Hoeij algorithm is implemented in all big computer algebra systems. As already remarked, it is possible to factor polynomials in a few minutes, for which it was impossible to factor those in month before.

References

1. H. Zassenhaus, *On Hensel factorization I*, Journal of Number Theory, **1**, 291–311 (1969)
2. A. K. Lenstra, H. W. Lenstra, Jr. and L. Lovász, *Factoring polynomials with rational coefficients*, Annals of Mathematics, **261**(4), 515–534 (1982)
3. M. van Hoeij, *Factoring polynomials and the knapsack problem*, Journal of Number Theory, **95**, 167–189 (2002)
4. K. Belabas, M. van Hoeij, J. Klüners and A. Steel, *Factoring polynomials over global fields*, Journal de théorie des nombres de Bordeaux, **21**(1), 15–39 (2009)
5. J. von zur Gathen and J. Gerhard, *Modern Computer Algebra*. Cambridge University Press (1999)

Chapter 9
The LLL Algorithm and Integer Programming

Karen Aardal and Friedrich Eisenbrand

Abstract The LLL algorithm has proven to be a powerful theoretical and practical tool in many areas of discrete mathematics. In this chapter, we review some structural and algorithmic results involving basis reduction and integer programming.

Introduction

Let $P = \{\mathbf{x} \in \mathbb{R}^n \mid \mathbf{A}\mathbf{x} \leq \mathbf{d}\}$, where the $m \times n$ matrix $\mathbf{A}$ and the m-vector $\mathbf{d}$ are given by integer input. Assume P is bounded and full-dimensional. The *Integer programming feasibility problem* is defined as:

$$\text{Does there exist a vector } \mathbf{x} \in P \cap \mathbb{Z}^n? \tag{9.1}$$

This problem is NP-complete [1, 2] and is related to the *Integer programming optimization problem*,

$$\max\{\mathbf{c}^T\mathbf{x} \mid \mathbf{x} \in P \cap \mathbb{Z}^n\}, \tag{9.2}$$

where $\mathbf{c}$ is an n-dimensional vector. We call the problem $\max\{\mathbf{c}^T\mathbf{x} \mid \mathbf{x} \in P\}$ the *linear programming relaxation* of (9.2). A *combinatorial optimization problem* is typically an integer optimization problem in which the integer variables take values 0 or 1 only. Well-known examples of combinatorial optimization problems are the subset sum problem, the matching problem and the traveling salesman problem.

In 1981, Lenstra, [3, 4] proved that the integer programming feasibility problem (9.1) can be solved in polynomial time if the dimension n is fixed. The proof was algorithmic, and the main auxiliary algorithm was lattice basis reduction. In the research report [3], a reduction algorithm with polynomial running time for fixed n was used, but in the published version [4], Lenstra used the LLL basis reduction algorithm [5] that had been developed in the meantime.

K. Aardal (✉)
Delft Institute of Applied Mathematics, TU Delft, Mekelweg 4, 2628 CD Delft, The Netherlands and CWI, Science Park 123, 1098 XG Amsterdam, The Netherlands,
e-mail: k.i.aardal@tudelft.nl

P.Q. Nguyen and B. Vallée (eds.), *The LLL Algorithm*, Information Security and Cryptography, DOI 10.1007/978-3-642-02295-1_9,

Not only was Lenstra's result important in that it answered a prominent open complexity question, but it also introduced geometry of numbers to the field of optimization. Many results, inspired by this paper, have since then been obtained.

The purpose of this paper is to provide a glimpse of some of the important theoretical and computational consequences of the LLL algorithm in relation to integer programming, rather than giving a complete overview of all such results. The interested reader can consult the following references for a thorough treatment of the topic. A good undergraduate level introduction to integer programming is given by Wolsey [6]. Graduate textbooks on integer and combinatorial optimization are Grötschel, Lovász, and Schrijver [7], Nemhauser and Wolsey [8], and Schrijver [9,10]. Cassels [11] is a classical book on the geometry of numbers, while the recent books by Barvinok [12] and by Micciancio and Goldwasser [13] focus on algorithmic aspects. Lattices, representations of lattices, and several problems on lattices wherein basis reduction plays a prominent role are presented in the introductory chapter by Lenstra [14]. Lovász [15] treats basis reduction, integer programming, and classical lattice problems such as the shortest vector problem. Kannan [16] provides a nice overview of topics related to lattices and convex bodies, and Aardal and Eisenbrand [17] review results on integer programming in fixed dimension.

Notation

Vectors and matrices are written in boldface. By $\mathbf{x}_j$, we mean the jth vector in a sequence of vectors. The ith element of a vector $\mathbf{x}$ is denoted by x_i. Element (i, j) of the matrix $\mathbf{A}$ is denoted by A_{ij}. The *Euclidean length* of a vector $\mathbf{x} \in \mathbb{R}^n$ is denoted by $\|\mathbf{x}\|$ and is computed as $\|\mathbf{x}\| = \sqrt{\mathbf{x}^T\mathbf{x}}$, where $\mathbf{x}^T$ is the transpose of the vector $\mathbf{x}$.

Let $\mathbf{b}_1, \ldots, \mathbf{b}_l$ be linearly independent vectors in $\mathbb{R}^n$. The set

$$L = \left\{\mathbf{x} \in \mathbb{R}^n \mid \mathbf{x} = \sum_{j=1}^{l} \lambda_j \mathbf{b}_j, \lambda_j \in \mathbb{Z},\ 1 \le j \le l\right\} \tag{9.3}$$

is called a *lattice*. The set of vectors $\{\mathbf{b}_1, \ldots, \mathbf{b}_l\}$ is called a *lattice basis*. If we want to emphasize that we are referring to a lattice L that is generated by the basis $\mathbf{B} = (\mathbf{b}_1, \ldots, \mathbf{b}_l)$, then we use the notation $L(\mathbf{B})$.

The *rank* of L, rk L, is equal to the dimension of the Euclidean vector space generated by a basis of L. The *determinant* of L can be computed as $d(L) = \sqrt{\det(\mathbf{B}^T\mathbf{B})}$, where $\mathbf{B}^T$ is the transpose of the matrix $\mathbf{B}$ that is formed by taking the basis vectors as columns. Notice that if $l = n$, i.e., L is full-dimensional, then $d(L) = |\det(\mathbf{B})|$.

Let $L(\mathbf{B})$ be a full-dimensional lattice in $\mathbb{R}^n$ generated by $\mathbf{B}$. Its *dual lattice* $L^*(\mathbf{B})$ is defined as

$$L^*(\mathbf{B}) = \left\{\mathbf{x} \in \mathbb{R}^n \mid \mathbf{x}^T\mathbf{y} \in \mathbb{R} \text{ for all } \mathbf{y} \in L\right\}.$$

The columns of the matrix $(\mathbf{B}^T)^{-1}$ form a basis for the dual lattice $L^*(\mathbf{B})$. For a lattice L and its dual, we have $d(L) = d(L^*)^{-1}$.

Integer Programming: A Brief Background Sketch

Cutting Planes

The history of integer programming is, compared to many other mathematical subjects, quite brief. The first papers on determining optimal solutions to general integer linear optimization problems were published by Ralph E. Gomory; see, for instance, [18, 19]. It is also interesting to read Gomory's [20] own remarks on how he entered the field and viewed the topic. Gomory, while at Princeton, worked as a consultant for the US Navy, and there he was presented with a problem from the Navy Task Force. It was a linear programming problem with the additional important feature that the answer should be given in integer numbers. After having a thorough look at the problem at hand, Gomory made the following observation: all objective function coefficients are integer, so the optimal value should also be integer. One could solve the linear programming relaxation of the problem first, and if the variable values come out integer, then of course the integer optimum has been found. If not, it is valid to add the restriction that the objective value should be less than or equal to the linear programming objective value rounded down. Gomory describes this as "pushing in" the objective function. After some more thinking, Gomory realized that the same thing can be done with other integer forms as well, and the theory of *cutting planes* was born. Gomory proved the important result that, under certain technical conditions, the integer optimum will be obtained after adding a finite number of the so-called Gomory cutting planes. It is important to notice that an algorithm for solving linear optimization problems had been developed in the 1950s by Dantzig [21], so it was natural to use the linear relaxation as a starting point for solving the integer optimization problem.

In a more problem-specific setting, the idea of cutting planes was introduced by Dantzig, Fulkerson, and Johnson [22, 23], who used this approach to solve a 49-city traveling salesman instance by the combination of linear programming and cutting planes. This approach grew increasingly popular with, for instance, the work on the matching problem by Edmonds [24], the traveling salesman problem by Grötschel [25], and Grötschel and Padberg [26–28], and on the knapsack problem by Balas [29], Hammer et al. [30], and Wolsey [31]. The problem of finding good partial descriptions of the convex hull of feasible solutions to various problems has played a prominent role in the research on integer and combinatorial optimization up to this day.

Branch-and-Bound and Branch-and-Cut

In 1960, Land and Doig [32] introduced branch-and-bound. This is an algorithm for integer optimization that implicitly enumerates solutions. Solving linear programming relaxations is the main engine of the algorithm, and information from these relaxations is used to prune the search, with the aim of avoiding complete enumeration. The algorithm can be illustrated by a search tree as follows. In each node of the tree, the linear relaxation of the problem corresponding to that node is solved. When we start out, we are at the root node, where we solve the linear relaxation of the original problem. Let $\underline{z}$ be the value of the best known integer feasible solution. We can stop investigating further at a certain node k, called *pruning* at node k, if one of the following things happens:

1. The solution of the linear programming relaxation of the subproblem corresponding to node k is integer (prune by optimality). If the solution value is better than $\underline{z}$, then update $\underline{z}$.
2. The linear relaxation at node k is infeasible (prune by infeasibility).
3. The objective function value of the linear relaxation at node k is less than or equal to $\underline{z}$ (prune by bound).

If we cannot prune at node k, we need to *branch*, which simply means that we create two subproblems as follows. Choose a variable that has a fractional value in the optimal linear programming solution. Assume this is variable x_i with current fractional value f. One subproblem is created by adding the constraint $x_i \leq \lfloor f \rfloor$ to the linear relaxation of node k, and the other subproblem is created by adding the constraint $x_i \geq \lceil f \rceil$. In this way, we do not cut off any integer solution. The algorithm continues as long as there are unpruned leaves of the tree. Notice that we can also solve the feasibility problem (9.1) by branch-and-bound by just introducing an arbitrary objective function and terminate the search as soon as a feasible integer solution has been found or when integer infeasibility has been established.

Modern integer programming algorithms use a combination of branch-and-bound and cutting planes, both general and problem-specific, where the cutting planes are used to strengthen the linear relaxation in a selection of the nodes of the search tree. We refer to such algorithms as *branch-and-cut*. Branch-and-cut is not only used in academic research codes, but also in commercial software such as CPLEX [33] and Xpress [34].

Complexity Issues

The early work on integer programming took place before there was a formalization of computational complexity, but it was clear from the very beginning that the number of cutting planes needed in a cutting plane algorithm could grow exponentially, and if we consider branch-and-bound, a 2-dimensional example similar to the one given in Example 1 below, illustrates that a branch-and-bound tree can become arbitrarily deep. With the language of computational complexity at hand, these phenomena could be described more formally. From the cutting plane point-of-view, Karp and Papdimitriou [35] proved that it is not possible to find a concise linear description of the convex hull of feasible solutions for an NP-hard optimization problem unless NP=co-NP. This means that we cannot, a priori, write down such a linear description even if we allow for exponentially sized classes of linear inequalities, such as the subtour elimination constraints for the traveling salesman problem.

Example 1. Consider the integer programming feasibility problem (9.1) with the polytope P, illustrated in Fig. 9.1, as input:

If we solve this feasibility problem by branch-and-bound, we first need to introduce an objective function. Let us choose

$$\max z = x_1 + x_2 .$$

If we solve the linear relaxation of our problem, we obtain the vector $(x_1, x_2)^T = (6\frac{4}{5}, 5)$. We illustrate P and some of the constraints (dashed lines) added during branch-and-bound in Fig. 9.1, and the search tree corresponding to the

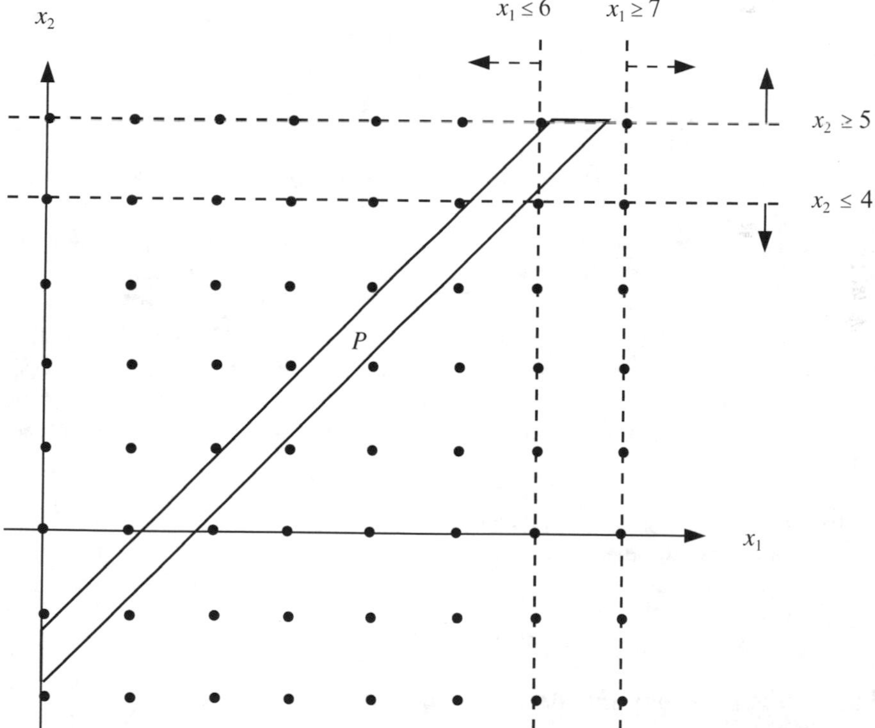

Fig. 9.1 The polytope P of Example 1, and some constraints added in branch-and-bound

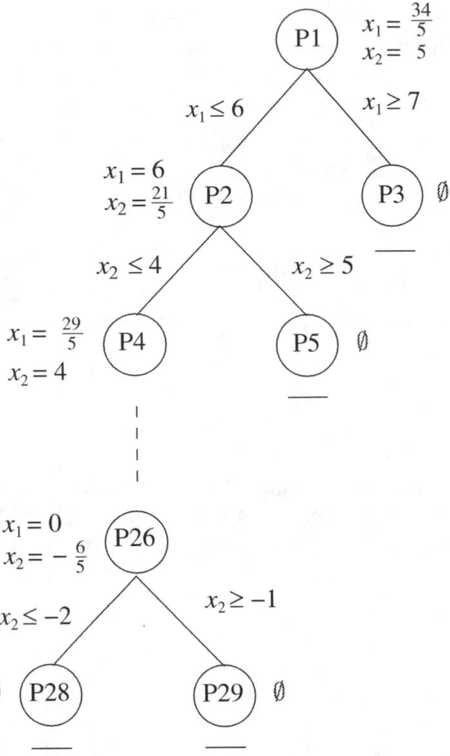

Fig. 9.2 The branch-and-bound search tree

branch-and-bound procedure in Fig. 9.2. Since $(6\frac{4}{5}, 5)$ is not an integer vector, we create two branches at the root node of our search tree: one corresponding to $x_1 \leq 6$ (subproblem P2) and the other corresponding to $x_1 \geq 7$ (subproblem P3). Again, solving the linear relaxation corresponding to subproblem P2 gives the solution $(x_1, x_2)^T = (6, 4\frac{1}{5})$, whereas subproblem P3 is infeasible. Branch-and-bound continues in a similar fashion until subproblems P28 and P29, in which all nodes of the search tree are pruned and it is finally verified that P does not contain any integer vector. □

By "stretching" the polytope given in Example 1 arbitrarily far in both directions, we see that even in dimension $n = 2$, we can obtain a search tree that is arbitrarily deep.

The Integer Linear Feasibility Problem

The above example indicates that branching on variables $x_i \geq \beta$ and $x_i \leq \beta - 1$, for $\beta \in \mathbb{Z}$ can result in an algorithm for integer programming that is exponential

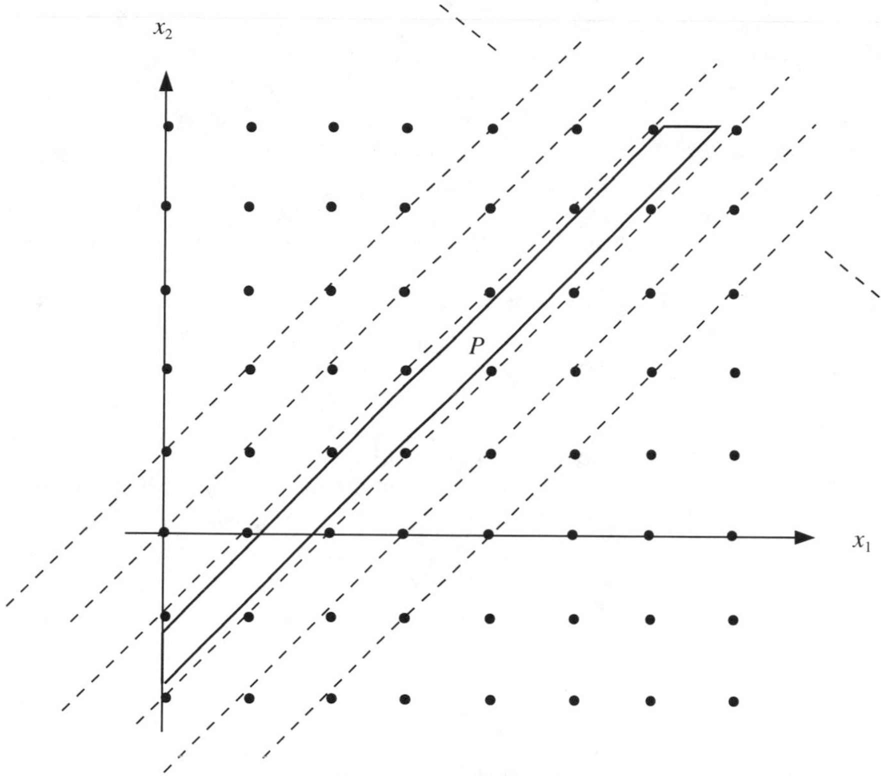

Fig. 9.3 None of the hyperplanes $-x_1 + x_2 = \beta$, $\beta \in \mathbb{Z}$ intersect P

in the binary input encoding of the problem, even in dimension 2. If we allow for hyperplanes that are more general than the single-variable hyperplanes, then we can observe that, for instance, the hyperplanes $-x_1 + x_2 = \beta$, $\beta \in \mathbb{Z}$ do not even intersect with the polytope. Yet, the hyperplanes do contain all points in $\mathbb{Z}^2$. This observation yields a certificate of integer infeasibility of our example; see Fig. 9.3.

The idea of enumerating parallel hyperplanes that cover all lattice points is called *branching on hyperplanes*, and can be described as follows. Let $\mathbf{d} \in \mathbb{Z}^n - \{\mathbf{0}\}$ be a nonzero integer vector. An integer point $\mathbf{x} \in P \cap \mathbb{Z}^n$ satisfies

$$\mathbf{d}^T\mathbf{x} = \beta, \quad \text{where} \quad \beta \in \mathbb{Z} \quad \text{and} \quad \min_{\mathbf{x}\in P} \mathbf{d}^T\mathbf{x} \leq \beta \leq \max_{\mathbf{x}\in P} \mathbf{d}^T\mathbf{x}.$$

This implies that we can continue to search for an integer point in the lower-dimensional polytopes $P \cap (\mathbf{d}^T\mathbf{x} = \beta)$ for each integer $\beta \in \mathbb{Z}$ satisfying

$$\min_{\mathbf{x}\in P} \mathbf{d}^T\mathbf{x} \leq \beta \leq \max_{\mathbf{x}\in P} \mathbf{d}^T\mathbf{x}. \tag{9.4}$$

The question is which direction $\mathbf{d}$ to choose such that the number of integers β satisfying (9.4) is small. Clearly, such an integer direction does not need to exist. Simply consider a ball of sufficiently large radius. The *flatness theorem*, attributed to Khinchin 1948, however, ensures that there exists a nonzero integer vector $\mathbf{d} \in \mathbb{Z}^n$ such that the number of integers in the interval 9.4 is bounded by a *constant* if the polytope does not contain an integer point. A *convex body* is a convex and compact set $K \subseteq \mathbb{R}^n$ with a nonempty interior. If we define the *width* of K *along* $\mathbf{d}$ as $w(K,\mathbf{d}) = \max\{\mathbf{d}^T\mathbf{x} \mid \mathbf{x} \in K\} - \min\{\mathbf{d}^T\mathbf{x} \mid \mathbf{x} \in K\}$, the theorem reads as follows.

Theorem 1 (Khinchin's flatness theorem [36]). *Let $K \subseteq \mathbb{R}^n$ be a closed convex set; then, either K contains an integer point, or there exists a nonzero integer vector $\mathbf{d}$ such that $w(K,\mathbf{d}) \leq f(n)$, where $f(n)$ is a constant depending on the dimension only.*

In the following subsection, we will present Lenstra's algorithm as an algorithmic version of the flatness theorem. This is different from the way the algorithm was presented originally, but our presentation below not only links the algorithm explicitly to the flatness theorem, but also highlights the relationship to other traditional lattice problems such as the closest and the shortest vector problems.

Lenstra's Algorithm

Here, whenever we consider a polytope P, we assume it is *full-dimensional*, and we use the notation $\mathbf{d}$ for a *nonzero integer vector* of appropriate dimension.

Lenstra's algorithm finds either an integer point in the polytope $P \subseteq \mathbb{R}^n$, or an integer direction $\mathbf{d}$ such that P is *flat* in this direction, i.e., a direction $\mathbf{d}$ such that $w(P,\mathbf{d})$ is bounded by a constant in fixed dimension. Thus, Lenstra's algorithm solves the following problem, which we call the *Integer feasibility problem (IP)*

Given a polytope $P \subseteq \mathbb{R}^n$, compute an integer point $\mathbf{x} \in P \cap \mathbb{Z}^n$ or a nonzero

$$\text{integer vector } \mathbf{d} \text{ with } w(P,\mathbf{d}) \leq f(n), \tag{9.5}$$

where $f(n)$ is a constant depending on the dimension only.

If problem IP is solvable in polynomial time in fixed dimension, then the integer programming feasibility problem (9.1) is also solvable in polynomial time in fixed dimension. This follows by induction, since in the case in which the algorithm solving IP returns a direction $\mathbf{d}$, one continues the search for an integer point in P in the *constantly* many *lower-dimensional* polytopes

$$P \cap (\mathbf{d}^T\mathbf{x} = \beta),\;\; \beta \in \mathbb{Z},\;\; \min\left\{\mathbf{d}^T\mathbf{x} \mid \mathbf{x} \in P\right\} \leq \beta \leq \max\left\{\mathbf{d}^T\mathbf{x} \mid \mathbf{x} \in P\right\}.$$

In the remainder of this section, we describe Lenstra's result by a series of reductions that ends up with a problem on a lattice L of finding either a lattice vector close to a given vector $\mathbf{u}$, or a short vector in the dual lattice L^*. We call this problem CSVP. In addition, we highlight the role that the LLL algorithm [5] plays in solving the integer programming feasibility problem.

Problem Reductions

An *ellipsoid* is a set $E(\mathbf{C},\mathbf{c}) = \{\mathbf{x} \in \mathbb{R}^n \mid \|\mathbf{C}(\mathbf{x}-\mathbf{c})\| \leq 1\}$, where $\mathbf{C} \in \mathbb{R}^{n\times n}$ is a nonsingular matrix. In a first step, Lenstra computes an ellipsoid $E(\mathbf{C},\mathbf{c})$ with $\mathbf{C} \in \mathbb{Q}^{n\times n}$ and $\mathbf{c} \in \mathbb{Q}^n$ such that $E(\mathbf{C},\mathbf{c})$ is contained in the polytope $P \subseteq \mathbb{Q}^n$ and such that if $E(\mathbf{C},\mathbf{c})$ is scaled from its center by $2\cdot n^{3/2}$, then it contains P.

Since the width of the scaled ellipsoid is the width of the original ellipsoid scaled by the same factor, we have

$$w(E(\mathbf{C},\mathbf{c}),\mathbf{d}) \leq w(P,\mathbf{d}) \leq 2\cdot n^{3/2}\cdot w(E(\mathbf{C},\mathbf{c}),\mathbf{d}).$$

This shows that we can solve the problem IP in polynomial time in fixed dimension, if we can solve the following analogous problem for ellipsoids, which we call *Integer feasibility of an ellipsoid (EIP)*, in polynomial time in fixed dimension.

Given a nonsingular rational matrix $\mathbf{C} \in \mathbb{Q}^{n\times n}$ and a rational point $\mathbf{c} \in \mathbb{Q}^n$, compute an integer point $\mathbf{x} \in E(\mathbf{C},\mathbf{c}) \cap \mathbb{Z}^n$ or determine an integer nonzero vector $\mathbf{d}$ such that $w(E(\mathbf{C},\mathbf{c}),\mathbf{d}) \leq f_2(n)$,
where $f_2(n)$ is a constant depending on the dimension n only. Following this approach yields $f(n) = 2\cdot n^{3/2}\cdot f_2(n)$ in (9.5).

In problem EIP, we have to compute a lattice point $\mathbf{v} \in L(\mathbf{C})$, such that its Euclidean distance from the point $\mathbf{Cc}$ is at most 1, or find an integer direction $\mathbf{d}$ such that the ellipsoid is flat along this direction. Since the width along $\mathbf{d}$ of an ellipsoid is invariant under translation of the ellipsoid, one has $w(E(\mathbf{C},\mathbf{c}),\mathbf{d}) = w(E(\mathbf{C},\mathbf{0}),\mathbf{d})$.

In other words, if we are not able to find an integer vector $\mathbf{x}$ in $E(\mathbf{C},\mathbf{c})$ we have to compute an integer direction $\mathbf{d}$ such that

$$\max\left\{\mathbf{d}^T\mathbf{x} \mid \mathbf{x} \in \mathbb{R}^n,\ \|\mathbf{Cx}\| \leq 1\right\} - \min\left\{\mathbf{d}^T\mathbf{x} \mid \mathbf{x} \in \mathbb{R}^n,\ \|\mathbf{Cx}\| \leq 1\right\} \leq f_2(n)$$

holds. Now, we have

$$\begin{aligned}\max\left\{\mathbf{d}^T\mathbf{x} \mid \mathbf{x} \in \mathbb{R}^n,\ \|\mathbf{Cx}\| \leq 1\right\} &= \max\left\{\mathbf{d}^T\mathbf{C}^{-1}\mathbf{Cx} \mid \mathbf{x} \in \mathbb{R}^n,\ \|\mathbf{Cx}\| \leq 1\right\} \\ &= \max\left\{\mathbf{d}^T\mathbf{C}^{-1}\mathbf{y} \mid \mathbf{y} \in \mathbb{R}^n,\ \|\mathbf{y}\| \leq 1\right\} \quad (9.6) \\ &= \|(\mathbf{C}^T)^{-1}\mathbf{d}\|. \quad (9.7)\end{aligned}$$

In (9.6) we have used the variable substitution $\mathbf{y} = \mathbf{Cx}$, and in (9.7), we have used the fact that a linear function $\mathbf{f}^T\mathbf{y}$ with $\mathbf{f} \neq \mathbf{0}$ achieves its maximum over the unit ball $B = \{\mathbf{y} \mid \mathbf{y} \in \mathbb{R}^n,\ \|\mathbf{y}\| \leq 1\}$ at the point $\mathbf{y} = \mathbf{f}/\|\mathbf{f}\|$. Similarly, we obtain

$$\min\left\{\mathbf{d}^T\mathbf{x} \mid \mathbf{x} \in \mathbb{R}^n,\ \|\mathbf{Cx}\| \leq 1\right\} = -\|(\mathbf{C}^T)^{-1}\mathbf{d}\|.$$

From this, we can deduce that the width of $E(\mathbf{C}, \mathbf{c})$ along an integer direction $\mathbf{d}$ is twice the length of the vector $(\mathbf{C}^T)^{-1}\mathbf{d}$

$$w(E(\mathbf{C}, \mathbf{c}), \mathbf{d}) = 2 \cdot \|(\mathbf{C}^T)^{-1}\mathbf{d}\|. \tag{9.8}$$

Next, we observe that the vector $\mathbf{v} = (\mathbf{C}^T)^{-1}\mathbf{d}$ is a lattice vector in the dual lattice

$$L^*(\mathbf{C}) = \left\{(\mathbf{C}^T)^{-1}\mathbf{x} \mid \mathbf{x} \in \mathbb{Z}^n\right\}$$

of the lattice $L(\mathbf{C})$. Hence, problem EIP has been reduced to the following problem, which we call *problem CSVP*:

Given a nonsingular rational matrix $\mathbf{B} \in \mathbb{Q}^{n \times n}$ and a rational vector $\mathbf{u} \in \mathbb{Q}^n$, compute a lattice vector $\mathbf{x} \in L(\mathbf{B})$ with $\|\mathbf{x} - \mathbf{u}\| \leq 1$ or determine a nonzero vector $\mathbf{w} \in L^*(\mathbf{B})$ with $\|\mathbf{w}\| \leq f_3(n)$.

In other words, we either have to find a lattice vector *close* to a given vector $\mathbf{u}$, or compute a *short* nonzero vector in the dual lattice. We set $f_2(n) = 2 \cdot f_3(n)$, where the factor of 2 comes from expression (9.8). Tracing back, we have now obtained $f(n) = 2 \cdot n^{3/2} \cdot f_2(n) = 4 \cdot n^{3/2} \cdot f_3(n)$. Notice that finding a short vector in the dual lattice $L^*(\mathbf{B})$ in the Euclidean vector space E is equivalent to finding a hyperplane H in E such that $L(\mathbf{B})$ is contained in widely spaced translates of H; see Lenstra [14].

Using Lenstra's Algorithm to Solve CSVP

Suppose that $\mathbf{B} = (\mathbf{b}_1, \ldots, \mathbf{b}_n) \in \mathbb{Q}^{n \times n}$ is a basis of the full-dimensional rational lattice $L(\mathbf{B})$. The *orthogonality defect* of $\mathbf{B}$ is the number $\gamma \in \mathbb{R}$ such that

$$\|\mathbf{b}_1\| \cdot \cdots \cdot \|\mathbf{b}_n\| = \gamma \cdot d(L(\mathbf{B})) = \gamma \cdot |\det(\mathbf{B})|.$$

Notice that $\gamma = 1$ if and only if the basis vectors are pairwise orthogonal. Hermite showed that every lattice in $\mathbb{R}^n$ has a basis $\mathbf{b}_1, \ldots, \mathbf{b}_n$ such that $\gamma \leq (4/3)^{n(n-1)/4}$, but no polynomial time algorithm is known that can determine a basis with this orthogonality defect guarantee.

Assume further that the longest basis vector is $\mathbf{b}_n$, that is, $\|\mathbf{b}_n\| \geq \|\mathbf{b}_j\|$ for $1 \leq j \leq n-1$. Let $\mathbf{B}^*$ be the matrix such that

$$\mathbf{B} = \mathbf{B}^* \cdot \mathbf{R},$$

where $\mathbf{R} \in \mathbb{Q}^{n \times n}$ is an upper-triangular matrix with $R_{ii} = 1$ for each $1 \leq i \leq n$. The matrix $\mathbf{B}^* = (\mathbf{b}_1^*, \ldots, \mathbf{b}_n^*)$ is the *Gram-Schmidt orthogonalization* of $\mathbf{B}$. Since we have $\|\mathbf{b}_j\| \geq \|\mathbf{b}_j^*\|$ for each $1 \leq j \leq n$ and since

$$d(L) = |\det(\mathbf{B})| = |\det(\mathbf{B}^*)| = \|\mathbf{b}_1^*\| \cdot \cdots \cdot \|\mathbf{b}_n^*\|,$$

it follows that $\gamma \geq 1$, which implies the so-called *Hadamard inequality*

$$\|\mathbf{b}_1\| \cdot \cdots \cdot \|\mathbf{b}_n\| \geq |\det(\mathbf{B})|.$$

Our first goal is to find a vector in the lattice $L(\mathbf{B})$ that is close to the given vector $\mathbf{u} \in \mathbb{Q}^n$. Since $\mathbf{B}$ is a basis of $\mathbb{R}^n$, we can write

$$\mathbf{u} = \sum_{j=1}^{n} \lambda_j \mathbf{b}_j,$$

with $\lambda_j \in \mathbb{R}$. The vector

$$\mathbf{v} = \sum_{j=1}^{n} \lfloor \lambda_j \rceil \mathbf{b}_j$$

belongs to the lattice $L(\mathbf{B})$, where $\lfloor \lambda_j \rceil$ denotes the closest integer to λ_j. We have

$$\begin{aligned}\|\mathbf{v}-\mathbf{u}\| &= \|\sum_{i=1}^{n}(\lfloor \lambda_i \rceil - \lambda_i)\mathbf{b}_i\| \leq \sum_{i=1}^{n}\|(\lfloor \lambda_i \rceil - \lambda_i)\mathbf{b}_i\| \leq \frac{1}{2}\sum_{i=1}^{n}\|\mathbf{b}_i\| \\ &\leq \frac{n}{2}\|\mathbf{b}_n\|, \qquad (9.9)\end{aligned}$$

where inequality (9.9) holds as the last basis vector $\mathbf{b}_n$ is the longest one in the basis.

If $\|\mathbf{v}-\mathbf{u}\| \leq 1$, we have solved problem CSVP as stated at the end of Section "Problem Reductions". Suppose, therefore, that $\|\mathbf{v}-\mathbf{u}\| > 1$. We now need to find a short vector in the dual lattice $L^*(\mathbf{B})$. From inequality (9.9), we obtain

$$\|\mathbf{b}_n\| \geq 2/n\,. \qquad (9.10)$$

If we combine

$$\|\mathbf{b}_1\| \cdot \cdots \cdot \|\mathbf{b}_n\| = \gamma \cdot \|\mathbf{b}_1^*\| \cdot \cdots \cdot \|\mathbf{b}_n^*\|$$

and

$$\|\mathbf{b}_j\| \geq \|\mathbf{b}_j^*\|, \quad 1 \leq j \leq n,$$

we obtain $\|\mathbf{b}_n\| \leq \gamma \cdot \|\mathbf{b}_n^*\|$, which together with 9.10 implies

$$\|\mathbf{b}_n^*\| \geq 2/(n \cdot \gamma).$$

The vector $\mathbf{b}_n^*$ is orthogonal to the vectors $\mathbf{b}_j^*$, $1 \leq j \leq n-1$, and since $\mathbf{R}$ is an upper triangular matrix with only 1's on its diagonal, it follows that $(\mathbf{b}_n^*)^T\mathbf{B}^* \cdot \mathbf{R} = (0, \ldots, 0, \|\mathbf{b}_n^*\|^2)$. Next, let $\mathbf{v} = \mathbf{B}\mathbf{x},\ \mathbf{x} \in \mathbb{Z}$. Notice that $\mathbf{v} \in L(\mathbf{B})$. We now show that the vector $(1/\|\mathbf{b}_n^*\|^2)\mathbf{b}_n^*$ belongs to the dual lattice $L^*(\mathbf{B})$ by showing that $(1/\|\mathbf{b}_n^*\|^2)\mathbf{b}_n^{*T}\mathbf{v} \in \mathbb{Z}$:

$$(1/\|\mathbf{b}_n^*\|^2)\mathbf{b}_n^{*T}\mathbf{v} = (1/\|\mathbf{b}_n^*\|^2)\mathbf{b}_n^{*T}\mathbf{B}^*\mathbf{R}\mathbf{x}$$

$$= (0, \ldots, 0, 1)\,\mathbf{x}$$
$$= x_n \in \mathbb{Z},$$

Hence,

$$\mathbf{w} = (1/\|\mathbf{b}_n^*\|^2)\,\mathbf{b}_n^* \in L^*(\mathbf{B})\,, \tag{9.11}$$

and the norm of $\mathbf{w}$ satisfies

$$\|\mathbf{w}\| \leq (n \cdot \gamma)/2\,. \tag{9.12}$$

The length of $\mathbf{w}$ can be bounded by a constant depending only on n if the orthogonality defect γ can be bounded by such a constant.

The Role of the LLL Algorithm for Solving IP

As described above, Lenstra [4] has shown that *any* basis reduction algorithm that runs in polynomial time in fixed dimension and returns a basis, such that its orthogonality defect is bounded by a constant in fixed dimension suffices to solve CSVP in polynomial time in fixed dimension, and consequently the integer feasibility problem for a rational polytope. If the LLL algorithm is applied to reduce the basis $\mathbf{B}$, then $\gamma \leq 2^{n(n-1)/4}$. Our discussion above shows now that IP can be solved in polynomial time with $f(n) = 4 \cdot n^{3/2} \cdot f_3(n) = 2 \cdot n^{5/2} \cdot 2^{n(n-1)/4}$. In fact, this constant can be slightly improved by a better bound for 9.9. More precisely, for a given $\mathbf{u} \in \mathbb{Q}^n$, one can compute, using the Gram-Schmidt orthogonalization of $\mathbf{B}$, a lattice vector $\mathbf{v} \in L(\mathbf{B})$ with

$$\|\mathbf{v} - \mathbf{u}\| \leq (\sqrt{n}/2) \cdot \|\mathbf{b}_n\|.$$

By propagating this improvement through the constants, we obtain the bound $f(n) \leq 2 \cdot n^2 \cdot 2^{n(n-1)/4}$, yielding Lenstra's main result.

Theorem 2 ([3,4]). *Given a rational polytope $P = \{\mathbf{x} \in \mathbb{Z}^n \mid \mathbf{A}\mathbf{x} \leq \mathbf{b}\}$, one can compute either an integer point $\mathbf{x} \in P \cap \mathbb{Z}^n$ or a nonzero integer vector $\mathbf{d} \in \mathbb{Z}^n$ with $w(P, \mathbf{d}) \leq 2 \cdot n^2 \cdot 2^{n(n-1)/4}$ in polynomial time. The integer linear feasibility problem can be solved in polynomial time, if the dimension is fixed.*

Related Results

Lovász [15] obtained the following result by combining basis reduction and a different way of obtaining an inscribed ellipsoid. His result is more general in the sense that it applies to convex bodies. Let K be a convex body. The unique maximum-volume ellipsoid that is contained in K is called *Löwner-John ellipsoid*. If this ellipsoid is scaled from its center by a factor of n, then it contains the body K. The Löwner-John ellipsoid can be found with the ellipsoid method [7] in polynomial time, provided one can solve the *weak separation problem* [7] for K in polynomial time.

This, together with the LLL algorithm, yields the following result.

Theorem 3 ([15]). *Let $K \subseteq \mathbb{R}^n$ be a convex body for which one can solve the weak separation problem in polynomial time. We can achieve, in polynomial time, one of the following:*

(i) find an integer vector in X, or
(ii) find an integer vector $\mathbf{c} \in \mathbb{Z}^n$ with

$$\max\left\{\mathbf{c}^T\mathbf{x} \mid \mathbf{x} \in X\right\} - \max\left\{\mathbf{c}^T\mathbf{x} \mid \mathbf{x} \in X\right\} \leq 2 \cdot n^2 \cdot 9^n.$$

The polynomial running time in the above theorem depends on the binary encoding length of the radii of a ball, which is inscribed in K and a ball containing K respectively, see [7].

Lovász and Scarf [37] developed a basis reduction algorithm, called *generalized basis reduction*, based on a polyhedral norm, and used it to solve the integer programming feasibility problem. No polynomial algorithm is known to find a generalized reduced basis in the sense of Lovász and Scarf. Such a basis can, however, be derived in polynomial time if the dimension is fixed. Since a reduced basis can be found by solving a sequence of linear programs, this algorithms is still interesting from the implementation point of view. See also the comments at the end of this section.

The *packing radius* $\rho(L)$ of a lattice $L \subseteq \mathbb{R}^n$ is half the length of the shortest nonzero vector of L. It is the largest number α such that the interior of balls centered at lattice points of radius α does not intersect. The *covering radius* $\mu(L)$ is the smallest number β such that the balls of radius β centered at lattice points cover the whole space $\mathbb{R}^n$. The number $\mu(L)$ is the largest distance of a point in $\mathbb{R}^n$ to the lattice L. The flatness theorem implies that $\rho(L) \cdot \mu(L^*) \leq c(n)$, where $c(n)$ is a constant depending on the dimension n only. Lagarias, Lenstra, and Schnorr [38] have shown that $c(n) \leq n^{3/2}/4$. Banaszczyk [39] proved that $c(n) = O(n)$. This shows that a closest vector in L to a vector $\mathbf{u}$ can be computed with $O\left((c \cdot n)!\right)$ shortest vector queries, where c is some constant. The dependence on the dimension n in Lenstra's algorithm is $O(2^{n^3})$. Kannan [40], see also [16], presented an algorithm for integer programming with running time $n^{O(n)}$. Kannan and Lovász [41] have shown that the constant in Kinchines flatness theorem is $O(n^2)$ if K is a rational polytope. The fastest algorithm to compute a shortest vector is by Ajtai [42] is randomized and has an expected running time of $2^{O(n)}$ times a polynomial in the input encoding of the basis. Blömer [43] presented a deterministic algorithm that computes the closest vector in time $n!$ times a polynomial in the input encoding length of the basis.

Barvinok [44] considered the problem of *counting* integer points in a polytope, which is a generalization of the integer feasibility problem. He used an approach based on an identity of Brion for exponential sums over polytopes. Lenstra's algorithm was used as a subroutine, but later Dyer and Kannan [45] showed, in a modification of Barvinok's algorithm, that this subroutine was in fact not needed.

A topic related to the integer feasibility problem (9.1), is the problem of finding the Hermite normal form of a matrix $\mathbf{A}$. The *Hermite normal form* of a matrix $\mathbf{A} \in \mathbb{Z}^{m\times n}$ of full row rank, HNF($\mathbf{A}$), is obtained by multiplying $\mathbf{A}$ by an $n \times n$ unimodular matrix $\mathbf{U}$ to obtain the form $(\mathbf{D},\ \mathbf{0})$, where $\mathbf{D} \in \mathbb{Z}^{m\times m}$ is a nonsingular, nonnegative lower triangular matrix with the unique row maximum along the diagonal. An integer nonsingular matrix $\mathbf{U}$ is *unimodular* if det($\mathbf{U}$) $= \pm 1$. If the matrix $\mathbf{A}$ is rational, then it has a unique Hermite normal form.

Given is a system of rational equations $\mathbf{Ax} = \mathbf{d}$. The question is whether this system has a solution in integers. Frumkin [46,47] and von zur Gathen and Sieveking [48] showed that solving such a system of linear Diophantine equations can be done in polynomial time. Von zur Gathen and Sieveking, and Votyakov and Frumkin [49] showed that it is possible to find a basis for the lattice $L = \{\mathbf{x} \in \mathbb{Z}^n \mid \mathbf{Ax} = \mathbf{0}\}$ is polynomial time. From this result, Frumkin [50] deduced that it is possible to find HNF($\mathbf{A}$) in polynomial time. Kannan and Bachem [51] developed a direct polynomial time algorithm for finding HNF($\mathbf{A}$).

Theorem 4 ([48,49]). *Given a feasible system* $\mathbf{Ax} = \mathbf{d}$ *of rational linear equations, one can find, in polynomial time, integer vectors* $\mathbf{x}_0, \mathbf{x}_1, \ldots, \mathbf{x}_t$ *such that*

$$\left\{\mathbf{x} \in \mathbb{Z}^n \mid \mathbf{Ax} = \mathbf{d}\right\} = \left\{\mathbf{x}_0 + \sum_{j=1}^{t} \lambda_j \mathbf{x}_j \mid \lambda \in \mathbb{Z}^t\right\} \tag{9.13}$$

Let $\mathbf{AU} = (\mathbf{D},\ \mathbf{0})$ be the Hermite normal form of $\mathbf{A}$. Then, we can choose

$$\mathbf{x}_0 = \mathbf{U}\begin{pmatrix} \mathbf{D}^{-1}\mathbf{d} \\ \mathbf{0} \end{pmatrix}, \quad \mathbf{x}_j = \mathbf{U}\begin{pmatrix} \mathbf{0} \\ \mathbf{e}_j \end{pmatrix}, \ 1 \le j \le t\,.$$

Notice that $\mathbf{Ax}_0 = \mathbf{d}$ and that $\mathbf{Ax}_j = \mathbf{0},\ 1 \le j \le t$.

Schrijver [9], p. 74, discusses how one can use the LLL algorithm to find the Hermite normal form of a matrix; see also [52].

Aardal, Hurkens and Lenstra [53] used the representation (9.13) in which the vectors $\mathbf{x}_j$, $1 \le j \le t$ are LLL-reduced basis vectors of the lattice $L_0 = \{\mathbf{y} \in \mathbb{Z}^n \mid \mathbf{Ay} = \mathbf{0}\}$, i.e., they use the reformulation

$$\mathbf{x} = \mathbf{x}_0 + \mathbf{B}_0\lambda,\ \lambda \in \mathbb{Z}^t\,, \tag{9.14}$$

where $\mathbf{x}_0$ satisfies $\mathbf{Ax}_0 = \mathbf{d}$, and $\mathbf{B}_0$ is a reduced basis for the lattice L_0. If $\mathbf{A}$ is an $m \times n$ matrix of full row rank, we have $t = n - m$. Aardal et al. obtain the basis $\mathbf{B}_0$ and the vector $\mathbf{x}_0$ by a single application of the LLL algorithm as follows. Consider the system of linear Diophantine equations: $\mathbf{Ax} = \mathbf{d}$ and let $N_1, N_2 \in \mathbb{N}$. Without loss of generality, we assume that $\gcd(a_{i1}, a_{i2}, \ldots, a_{in}) = 1$ for $1 \le i \le m$, and that $\mathbf{A}$ has full row rank. Furthermore, let

$$\mathbf{B} = \begin{pmatrix} \mathbf{I} & \mathbf{0} \\ \mathbf{0} & N_1 \\ N_2\mathbf{A} & -N_2\mathbf{d} \end{pmatrix}, \tag{9.15}$$

and let $\hat{\mathbf{B}}$ be the basis resulting from applying the LLL algorithm to $\mathbf{B}$ in 9.15. The lattice $L(\mathbf{B}) \in \mathbb{R}^{n+m+1}$ is a lattice of rank $n+1$.

Theorem 5 **([53]).** *Assume that there exists an integer vector* $\mathbf{x}$ *satisfying the rational system* $\mathbf{Ax} = \mathbf{d}$. *There exist numbers* N_{01} *and* N_{02} *such that if* $N_1 > N_{01}$, *and if* $N_2 > 2^{n+m}N_1^2 + N_{02}$, *then the vectors* $\hat{\mathbf{b}}_j \in \mathbb{Z}^{n+m+1}$ *of the reduced basis* $\hat{\mathbf{B}}$ *have the following properties:*

1. $\hat{b}_{n+1,j} = 0$ *for* $1 \le j \le n-m$,
2. $\hat{b}_{ij} = 0$ *for* $n+2 \le i \le n+m+1$ *and* $1 \le j \le n-m+1$,
3. $|\hat{b}_{n+1,n-m+1}| = N_1$.

Moreover, the sizes of N_{01} *and* N_{02} *are polynomially bounded by the sizes of* $\mathbf{A}$ *and* $\mathbf{d}$.

Theorem 5 implies that if N_1 and N_2 are chosen appropriately, then the first $n-m+1$ columns of the reduced basis $\hat{\mathbf{B}}$ are of the following form:

$$\begin{pmatrix} \mathbf{B}_0 & \mathbf{x}_0 \\ \mathbf{0} & \pm N_1 \\ \mathbf{0} & 0 \end{pmatrix},$$

Aardal and Lenstra [54], and Aardal and Wolsey [55] study the lattice reformulation 9.14 for integer equality knapsack problems in more detail.

To conclude this section, we mention some computational results using LLL-inspired techniques to solve integer programming problems. Gao and Zhang [56] have implemented Lenstra's algorithm. Cook et al. [57] implemented the Lovász-Scarf integer programming algorithm based on generalized basis reduction and reported on computational results of solving several, up to then, unsolved telecommunication network design problems. Aardal, Hurkens, Lenstra [53], Aardal et al. [58], and Aardal and Lenstra [54] report on using the LLL-reduced lattice basis formulation 9.14 and a enumerative algorithm inspired by Lenstra [4] to solve several hard integer feasibility problems.

The Integer Linear Optimization Problem

In this section, we want to consider the integer optimization problem in fixed dimension

$$\max\left\{\mathbf{c}^T\mathbf{x} \mid \mathbf{Ax} \le \mathbf{b},\ \mathbf{x} \in \mathbb{Z}^n\right\}. \tag{9.16}$$

In the analysis of the algorithms that follow, we use the parameters m and s, where m is the number of inequalities of the system $\mathbf{A}\mathbf{x} \leq \mathbf{b}$, and s is an upper bound on the binary encoding length of a coefficient of $\mathbf{A}$, $\mathbf{b}$, and $\mathbf{c}$.

The *greatest common divisor* of two integers a and b can be computed with the Euclidean algorithm with $O(s)$ arithmetic operations, where s is an upper bound on the binary encoding length of the integers a and b. On the other hand, we have the following well-known formula

$$\gcd(a,b) = \min\{a\,x_1 + b\,x_2 \mid a\,x_1 + b\,x_2 \geq 1,\ x_1, x_2 \in \mathbb{Z}\}.$$

This implies that the greatest common divisor can be computed with an algorithm for the integer optimization problem in dimension 2 with one constraint.

The integer optimization problem can be reduced to the integer feasibility problem with binary search. The integer feasibility problem in fixed dimension has complexity $O(m+s)$. This follows from an analysis of Lenstra's algorithm in combination with efficient algorithm to compute a Löwner-John ellipsoid; see [59, 60]. With binary search for an optimal point, one obtains a running time of $O(m \cdot s + s^2)$. If, in addition to the dimension, also the number of constraints is fixed, this results in an $O(s^2)$ algorithm for the integer optimization problem, which is in contrast to the linear running time of the Euclidean algorithm.

Clarkson [61] has shown that the integer optimization problem with m constraints can be solved with an expected number of $O(m)$ arithmetic operations and $O(\log m)$ calls to an oracle solving the integer optimization problem on a constant size subset of the input constraints. Therefore, we concentrate now on the integer optimization problem with a fixed number of constraints. In this section, we outline an algorithm that solves the integer optimization problem in fixed dimension with a fixed number of constraints with $O(s)$ arithmetic operations on rational numbers of size $O(s)$. The algorithm relies on the LLL algorithm.

The first step is to reduce the integer optimization problem over a full-dimensional polytope with a fixed number of facets to a disjunction of integer optimization problems over a constant number of *two-layer simplices*. A two layer simplex is a full-dimensional simplex, whose vertices can be partitioned into two sets V and W, such that the objective function values of the elements in each of the sets V and W agree, i.e., for all $\mathbf{v}_1, \mathbf{v}_2 \in V$, one has $\mathbf{c}^T\mathbf{v}_1 = \mathbf{c}^T\mathbf{v}_2$, and for all $\mathbf{w}_1, \mathbf{w}_2 \in W$, one has $\mathbf{c}^T\mathbf{w}_1 = \mathbf{c}^T\mathbf{w}_2$.

How can one reduce the integer optimization problem over a polytope P to a sequence of integer optimization problems over two-layer simplices? Simply consider the hyperplanes $\mathbf{c}^T\mathbf{x} = \mathbf{c}^T\mathbf{v}$ for each vertex $\mathbf{v}$ of P. If the number of constraints defining P is fixed, then these hyperplanes partition P into a constant number of polytopes, whose vertices can be grouped into two groups, according to the value of their first component. Thus, we can assume that the vertices of P itself can be partitioned into two sets V and W, such that the objective function values of the elements in each of the sets V and W agree. Carathéodory's theorem, see Schrijver [9, p. 94], implies that P is covered by the simplices that are spanned by the vertices of P. These simplices are two-layer simplices. Therefore, the integer optimization problem in fixed dimension with a fixed number of constraints can be reduced in constant time to a constant number of integer optimization problems over a two-layer simplex.

The key idea is then to let the objective function slide into the two-layer simplex, until the width of the truncated simplex exceeds the flatness bound. In this way, one can be sure that the optimum of the integer optimization problem lies in the truncation, which is still flat. Thus, one has reduced the *integer optimization problem* in dimension n to a constant number of *integer optimization problems* in dimension $n-1$, and binary search can be avoided.

How do we determine a parameter π such that the truncated two-layer simplex $\Sigma \cap (\mathbf{c}^T\mathbf{x} \geq \pi)$ just exceeds the flatness bound? We explain the idea with the help of the 3-dimensional example in Fig. 9.4.

Here, we have a two-layer simplex Σ in dimension three. The set V consists of the points $\mathbf{0}$ and $\mathbf{v}_1$ and W consists of $\mathbf{w}_1$, and $\mathbf{w}_2$. The objective is to find a highest point in the vertical direction. The picture on the left describes a particular point in time, where the objective function slid into Σ. So we consider the truncation $\Sigma \cap (\mathbf{c}^T\mathbf{x} \geq \pi)$ for some $\pi \geq \mathbf{c}^T\mathbf{w}_1$. This truncation is the convex hull of the points

$$\mathbf{0}, \mathbf{v}_1, \mu\mathbf{w}_1, \mu\mathbf{w}_2, (1-\mu)\mathbf{v}_1 + \mu\mathbf{w}_1, (1-\mu)\mathbf{v}_1 + \mu\mathbf{w}_2,$$

where $\mu = \pi/\mathbf{c}^T\mathbf{w}_1$. Now consider the simplex $\Sigma_{V,\mu W}$, which is spanned by the points $\mathbf{0}, \mathbf{v}_1, \mu\mathbf{w}_1, \mu\mathbf{w}_2$. This simplex is depicted on the right in Fig. 9.4. If this simplex is scaled by 2, then it contains the truncation $\Sigma \cap (\mathbf{c}^T\mathbf{x} \geq \pi)$. This is easy to see, since the scaled simplex contains the points $2(1-\mu)\,\mathbf{v}_1$, $2\,\mu\,\mathbf{w}_1$ and $2\,\mu\,\mathbf{w}_2$. So we have the condition $\Sigma_{V,\mu W} \subseteq \Sigma \cap (\mathbf{c}^T\mathbf{x} \geq \pi) \subseteq 2\,\Sigma_{V,\mu W}$. From this, we can infer the important observation

$$w(\Sigma_{V,\mu W}) \leq w(\Sigma \cap (\mathbf{c}^T\mathbf{x} \geq \pi)) \leq 2\,w(\Sigma_{V,\mu W}).$$

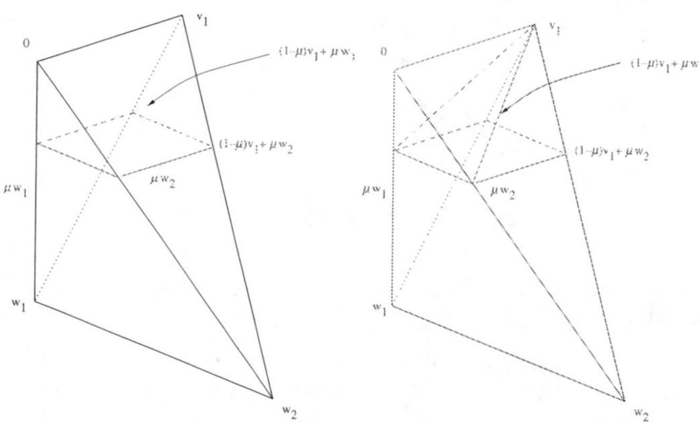

Fig. 9.4 Solving the parametric lattice width problem

This means that we essentially determine the correct π by determining a $\mu \geq 0$, such that the width of the simplex $\Sigma_{V,\mu W}$ just exceeds the flatness bound. The width of $\Sigma_{V,\mu W}$ is roughly (up to a constant factor) the length of the shortest vector of the lattice $L(\mathbf{A}_\mu)$, where $\mathbf{A}_\mu$ is the matrix

$$\mathbf{A}_\mu = \begin{pmatrix} \mu \mathbf{w}_1^T \\ \mu \mathbf{w}_2^T \\ \mathbf{v}_1 \end{pmatrix}.$$

Thus, we have to find a parameter μ, such that the shortest vector of $L(\mathbf{A}_\mu)$ is sandwiched between $f(n)+1$ and $\gamma \cdot (f(n)+1)$ for some constant γ. This problem can be understood as a *parametric shortest vector* problem.

To describe this problem, let us introduce some notation. We define for an $n \times n$-matrix $\mathbf{A} = (a_{ij})_{\forall i,j}$, the matrix $\mathbf{A}^{\mu,k} = (a_{ij})^{\mu,k}_{\forall i,j}$, as

$$a_{ij}^{\mu,k} = \begin{cases} \mu \cdot a_{ij}, & \text{if } i \leq k, \\ a_{ij}, & \text{otherwise.} \end{cases}$$

In other words, the matrix $\mathbf{A}^{\mu,k}$ results from $\mathbf{A}$ by scaling the first k rows with μ. The parametric shortest vector problem is now defined as follows.
Given a nonsingular matrix $\mathbf{A} \in \mathbb{Z}^{n \times n}$ and some $U \in \mathbb{N}$, find a parameter $p \in \mathbb{N}$ such that $U \leq \mathrm{SV}(L(\mathbf{A}^{p,k})) \leq 2^{n+1/2} \cdot U$ or assert that $\mathrm{SV}(L) > U$.
It turns out that the parametric shortest vector problem can be solved in linear time when the dimension is fixed with a cascaded LLL algorithm. From this, it follows that the integer optimization problem in fixed dimension with a fixed number of constraints can be solved in linear time. Together with Clarkson's result, we obtain the following result.

Theorem 6 ([62]). *The integer optimization problem* (9.16) *can be solved with an expected number of* $O(m + s \log m)$ *arithmetic operations on rationals of size* $O(s)$.

Open Problems and Discussion

In the above section, we have sketched a result showing that the integer linear optimization problem can be solved with a linear number of arithmetic operations, if the number of constraints is fixed. The binary encoding length of the numbers in the course of the algorithm remains linear in the input encoding size. Therefore, this result matches the complexity of the Euclidean algorithm if we count arithmetic operations only. When the number m of constraints is arbitrary, Clarkson's algorithm provides a running time of $O(m + s \log m)$, where s is the largest binary encoding length of a coefficient in the input. Clarkson's algorithm is a randomized algorithm. The first question is, whether a deterministic algorithm with running time $O(m + s \log m)$ exists.

In the case of two variables, Eisenbrand and Laue [63] have shown that there exists an algorithm that requires only $O(m + s)$ arithmetic operations. Another question is, whether this result can be extended to any fixed dimension.

The complexity model that reflects the fact that arithmetic operations on large numbers do not come for free is the *bit-complexity* model. Addition and subtraction of s-bit integers take $O(s)$ time. The current state of the art method for multiplication [64] shows that the bit complexity $M(s)$ of multiplication and division is $O(s \log s \log \log s)$.

Recently, Nguyen and Stehlé [65] have presented an LLL-variant that computes an LLL-reduced basis in time $O(n^5(n + \log B) \log B)$ bit-operations, where B is an upper bound on the norm of the vectors in the input. This holds even if the multiplications and divisions are carried out with the straightforward quadratic methods. This means that if the naive algorithms for multiplication and division with remainder are used, the dependence of the running time on the encoding length of the largest binary encoding of a basis-vector component matches exactly the running time of the Euclidean algorithm. In addition, the dependence on the dimension is polynomial. This raises the question, whether these results carry over to the bit-complexity of the integer optimization problem in fixed dimension. In particular, is it possible that this problem can be solved with $O(ms^2)$ bit operations. This would match the complexity of checking whether an integer point is feasible, if the naive methods for multiplication are used.

Acknowledgements This work was partly carried out within the framework of ADONET, a European network in Algorithmic Discrete Optimization, contract no. MRTN-CT-2003-504438. The first author is financed in part by the Dutch BSIK/BRICKS project.

References

1. Borosh, I., Treybig, L.B.: Bounds on positive integral solutions of linear Diophantine equations. Proceedings of the American Mathematical Society **55**, 299–304 (1976)
2. Karp, R.M.: Reducibility among combinatorial problems. In: Complexity of Computer Computations, pp 85–103. Plenum Press, NY (1972)
3. Lenstra, Jr., H.W.: Integer programming with a fixed number of variables. Technical Report 81-03, University of Amsterdam, Amsterdam (1981). Available at `http://staff/science/uva.nl/~peter/mi8103/mi8103c.html`
4. Lenstra, Jr., H.W.: Integer programming with a fixed number of variables. Mathematics of Operations Research **8(4)**, 538–548 (1983)
5. Lenstra, A.K., Lenstra, Jr., H.W., Lovász, L.: Factoring polynomials with rational coefficients. Mahematische Annalen **261**, 515–534 (1982)
6. Wolsey, L.A.: Integer Programming. Wiley, New York (1998)
7. Grötschel, M. Lovász, L., Schrijver, A.: Geometric Algorithms and Combinatorial Optimization. Springer, Berlin (1988)
8. Nemhauser, G.L., Wolsey, L.A.: Integer and Combinatorial Optimization. Wiley, New York (1988)
9. Schrijver, A.: Theory of Linear and Integer Programming. Wiley, Chichester (1986)
10. Schrijver, A.: Combinatorial Optimization: Polyhedra and Efficiency (3 volumes). Algorithms and Combinatorics **24**. Springer, Berlin (2003)

11. Cassels, J.W.S.: An Introduction to the Geometry of Numbers. Classics in Mathematics. Springer, Berlin (1997). Second Printing, Corrected, Reprint of the 1971 ed.
12. Barvinok, A.: A Course in Convexity. Graduate Studies in Mathematics **54**. American Mathematical Society, Providence, RI (2002)
13. Micciancio, D., Goldwasser, S.: Complexity of Lattice Problems: A Cryptographic Perspective. The Kluwer International Series in Engineering and Computer Science **671**. Kluwer Academic Publishers, Boston, Massachusetts (2002)
14. Lenstra, Jr., H.W.: Lattices. Chapter 6 in Algorithmic Number Theory, Mathematical Sciences Research Institute Publications, Vol 44, Cambridge University Press, Cambridge, UK, 127–181, 2008.
15. L. Lovász. An Algorithmic Theory of Numbers, Graphs and Convexity. SIAM, Philadelphia, PA (1986)
16. Kannan, R.: Algorithmic geometry of numbers. Annual Review of Computer Science **2**, 231–267 (1987)
17. Aardal, K., Eisenbrand, F.: Integer programming, lattices and results in fixed dimension. In: Aardal, K., Nemhauser, G.L., Weismantel, R. (eds) Handbook on Discrete Optimization, Chapter 4. North Holland, Amsterdam (2005)
18. Gomory, R.E.: Outline of an algorithm for integer solutions to linear programs. Bulletin of the American Mathematical Society **64**, 275–278 (1958)
19. Gomory, R.E.: An algorithm for integer solutions to linear programs. In: Graves, R.L., Wolfe, P. (eds) Recent Advances in Mathematical Programming, pp 269–302. McGraw-Hill (1963)
20. Gomory, R.E.: Early integer programming. In: Lenstra, J.K., Rinnooy Kan, A.H.G., Schrijver, A. (eds) History of Mathematical Programming: A Collection of Personal Reminiscences, pp 55–61. CWI and North-Holland, Amsterdam (1991)
21. Dantzig, G.B.: Maximization of a linear function of variables subject to linear inequalities. In: Koopmans, T.C. (ed) Activity Analysis of Production and Allocation, pp 339–347. John Wiley & Sons, New York (1951)
22. Dantzig, G.B., Fulkerson, D.R., Johnson, S.M.: Solution of a large-scale traveling-salesman problem. Operations Research **2**, 393–410 (1954)
23. Dantzig, G.B., Fulkerson, D.R., Johnson, S.M.: On a linear-programming, combinatorial approach to the traveling-salesman problem. Operations Research **7**, 58-66 (1959)
24. Edmonds, J.: Paths, trees and flowers. Canadian Journal of Mathematics **17**, 449–467 (1965)
25. Grötschel, M.: On the symmetric traveling salesman problem: Solution of a 120-city problem. Mathematical Programming Study **12**, 61–77 (1980)
26. Grötschel, M., Padberg, M.W.: Partial linear characterizations of the asymmetric traveling salesman problem. Mathematical Programming **8**, 378–381 (1975)
27. Grötschel, M., Padberg, M.W.: On the symmetric traveling salesman problem I: Inequalities. Mathematical Programming **16**, 265–280 (1978)
28. Grötschel, M., Padberg, M.W.: On the symmetric traveling salesman problem I: Lifting theorems and facets. Mathematical Programming **16**, 281–302 (1978)
29. Balas, E.: Facets of the knapsack polytope. Mathematical Programming **8**, 146–164 (1975)
30. Hammer, P.L., Johnson, E., Peled, U.N.: Facets of regular 0-1 polytopes. Mathematical Programming **8**, 179–206 (1975)
31. Wolsey, L.A.: Faces for a linear inequality in 0-1 variables. Mathematical Programming **8**, 165–178 (1975)
32. Land, A., Doig, A.: An automatic method of solving discrete programming problems. Econometrica **28**, 497–520 (1960)
33. ILOG. Cplex. `http://www.ilog.com/products/cplex`
34. Dash Optimization. Xpress-mp optimization software. `http://www.dashoptimization.com/home/index.html`
35. Karp, R.M., Papadimitriou, C.H.: On linear characterizations of combinatorial optimization problems. In: 21st Annual Symposium on Foundations of Computer Science, Syracuse, N.Y., pp 1–9. IEEE, New York (1980)
36. Khinchine, A.: A quantitative formulation of Kronecker's theory of approximation (in russian). Izvestiya Akademii Nauk SSR Seriya Matematika **12**, 113–122 (1948)

37. Lovász, L., Scarf, H.E.: The generalized basis reduction algorithm. Mathematics of Operations Research **17(3)**, 751–764 (1992)
38. Lagarias, J., Lenstra, Jr., H.W., Schnorr, C.: Korkin-zolotarev bases and successive minima of a lattice and its reciprocal lattice. Combinatorica **10(4)**, 333–348 (1990)
39. Banaszczyk, W.: Inequalities for convex bodies and polar reciprocal lattices in $\mathbb{R}^n$. II. Application of K-convexity. Discrete Computational Geometry **16(3)**, 305–311 (1996)
40. Kannan, R.: Minkowski's convex body theorem and integer programming. Mathematics of Operations Research **12(3)**, 415–440 (1987)
41. Kannan, R., Lovász, L.: Covering minima and lattice-point-free convex bodies. Annals of Mathematics **128**, 577–602 (1988)
42. Ajtai, M., Kumar, R., Sivakumar, D.: A sieve algorithm for the shortest lattice vector problem. Proceedings of the 33rd Annual ACM symposium on Theory of Computing, pp 601–610. ACM Press, New York (2001)
43. Blömer, J.: Closest vectors, successive minima, and dual HKZ-bases of lattices. In: Montanari, U., Rolim, J.D.P., Welzl, E. (eds) Automata, Languages and Programming, 27th International Colloquium, ICALP 2000, Geneva, Switzerland, July 9–15, 2000, Proceedings. Lecture Notes in Computer Science **1853**, pp 248–259. Springer, Berlin (2000)
44. Barvinok, A.I.: A polynomial time algorithm for counting integral points in polyhedra when the dimension is fixed. Mathematics of Operations Research **19(4)**, 769–779 (1994)
45. Dyer, M.E., Kannan, R.: On Barvinok's algorithm for counting lattice points in fixed dimension. Mathematics of Operations Research **22(3)**, 545–549 (1997)
46. Frumkin, M.A.: Algorithms for the solution in integers of systems of linear equations. In: Fridman, A.A. (ed) Studies in discrete optimization (Russian), pp 97–127, Izdat. "Nauka", Moscow (1976)
47. Frumkin, M.A.: An application of modular arithmetic to the construction of algorithms for the solution of systems of linear equations. Doklady Akademii Nauk SSSR **229(5)**, 1067–1070 (1976) [English translation: Soviet Mathematics Doklady **17**, 1165–1168 (1976)]
48. Gathen, von zur, J., Sieveking, M.: Weitere zum Erfüllungsproblem polynomial äquivalente kombinatorische Aufgaben. In: Specker, E. Strassen, V. (eds) Komplexität von Entscheidungsproblemen: Ein Seminar, Lecture Notes in Computer Science **43**, pp 49–71. Springer, Berlin (1976)
49. Votjakov, A.A., Frumkin, M.A.: An algorithm for finding the general integer solution of a system of linear equations. In: Studies in discrete optimization (Russian), pp 128–140. Izdat. "Nauka", Moscow (1976)
50. Frumkin, M.A.: An algorithm for the reduction of a matrix of integers to triangular form with power complexity of the computations. Èkonomika i Matematicheskie Metody **12(1)**, 173–178 (1976)
51. Kannan, R., Bachem, A.: Polynomial algorithms for computing the Smith and Hermite normal forms of an integer matrix. SIAM Journal on Computing **8(4)**, 499–507 (1979)
52. Havas, G., Majewski, B.S., Matthews, K.R.: Extended GCD and Hermite normal form algorithms via lattice basis reduction. Experimental Mathematics **7(2)**, 125–136 (1998) (Addenda and errata: Experimental Mathematics **8**, 179–206)
53. Aardal, K., Hurkens, C.A.J., Lenstra, A.K.: Solving a system of linear Diophantine equations with lower and upper bounds on the variables. Mathematics of Operations Research **25(3)**, 427–442 (2000)
54. Aardal, K.A., Lenstra, A.K.: Hard equality constrained integer knapsacks. Mathematics of Operations Research, **29(3)**, 724–738 (2004). Erratum: Mathematics of Operations Research **31(4)**, 846 (2006)
55. Aardal, K., Wolsey, L.A.: Lattice based extended formulations for integer linear equality systems. Mathematical Programming **121**, 337–352 (2010).
56. Gao, L., Zhang, Y.: Computational experience with Lenstra's algorithm. Technical Report TR02-12, Department of Computational and Applied Mathematics, Rice University, Houston, TX (2002)

57. Cook, W., Rutherford, T., Scarf, H.E., Shallcross, D.: An implementation of the generalized basis reduction algorithm for integer programming. ORSA Journal on Computing **5(2)**, 206–212 (1993)
58. Aardal, K., Bixby, R.E., Hurkens, C.A.J., Lenstra, A.K., Smeltink, J.W.: Market split and basis reduction: Towards a solution of the Cornuéjols-Dawande instances. INFORMS Journal on Computing **12(3)**, 192–202 (2000)
59. Matoušek, J., Sharir, M., Welzl, E.: A subexponential bound for linear programming. Algorithmica **16(4–5)**, 498–516 (1996)
60. Welzl, E.: Smallest enclosing disks (balls and ellipsoids). In: New results and new trends in computer science (Graz, 1991), Lecture Notes in Computer Science **555**, pp 359–370. Springer, Berlin (1991)
61. Clarkson, K.L.: Las Vegas algorithms for linear and integer programming when the dimension is small. Journal of the Association for Computing Machinery **42**, 488–499 (1995)
62. Eisenbrand, F.: Fast integer programming in fixed dimension. In: Battista, G.D., Zwick, U. (eds) Algorithms – ESA 2003. Lecture Notes in Computer Science **2832**, 196–207. Springer, Berlin (2003)
63. Eisenbrand, F., Laue, S.: A linear algorithm for integer programming in the plane. Mathematical Programming **102(2)**, 249 – 259 (2005)
64. Schönhage, A., Strassen, V.: Schnelle Multiplikation grosser Zahlen (Fast multiplication of large numbers). Computing **7**, 281–292 (1971)
65. Nguyen, P.Q., Stehlé, D.: Floating-point LLL revisited. In: Cramer, R. (ed) Advances in Cryptology — EUROCRYPT 2005. Lecture Notes in Computer Science **3494**, pp 215–233. Springer, Berlin (2003)

Chapter 10
Using LLL-Reduction for Solving RSA and Factorization Problems

Alexander May

Abstract Twenty five years ago, Lenstra, Lenstra and Lovász presented their celebrated LLL lattice reduction algorithm. Among the various applications of the LLL algorithm is a method due to Coppersmith for finding small roots of polynomial equations. We give a survey of the applications of this root finding method to the problem of inverting the RSA function and the factorization problem. As we will see, most of the results are of a dual nature, they can either be interpreted as cryptanalytic results or as hardness/security results.

Introduction

The RSA cryptosystem invented by Rivest, Shamir, and Adleman in 1977 [1] is today's most important public-key cryptosystem. Let us denote by $N = pq$ an RSA-modulus which is the product of two primes p, q of the same bit-size. Let e be an integer co-prime to Euler's totient function $\phi(N) = (p-1)(q-1)$. The RSA encryption function takes a message m to the eth power in the ring $\mathbb{Z}_N$. The security of RSA relies on the difficulty of inverting the RSA encryption function on the average, i.e., extracting e^{th} roots in the ring $\mathbb{Z}_N$. We call this problem the RSA inversion problem or the RSA problem for short.

Let d be the inverse of e modulo $\phi(N)$. Computing d^{th} powers in $\mathbb{Z}_N$ inverts the RSA encryption function. Since d can be easily computed when the prime factorization of N is known, the RSA cryptosystem is at most as secure as the problem of computing d and the problem of factoring N. Indeed, we will see that the last two problems are polynomial time equivalent. However, it is one of the most challenging problems to prove or disprove the polynomial time equivalence of the RSA problem and the problem of factoring N. There are results that these problems are not equivalent under restricted reductions [2]. On the other hand, one can show that in restricted generic attack models both problems appear to be equivalent [3, 4].

A. May
Horst Görtz Institute for IT-Security, Faculty of Mathematics,
Ruhr-University Bochum, Germany,
e-mail: alex.may@ruhr-uni-bochum.de

P.Q. Nguyen and B. Vallée (eds.), *The LLL Algorithm*, Information Security and Cryptography, DOI 10.1007/978-3-642-02295-1_10,

Despite considerable efforts to attack RSA (see [5, 6] for surveys), currently, the best way is still to factor the RSA modulus. Consequently, researchers focussed for a long time on the construction of factorization algorithms for attacking RSA. In this factorization line of research, the goal is to minimize the computational complexity in the common Turing machine model. The most important milestones in the construction of factorization algorithms in the 80s and 90s are the invention of the Quadratic Sieve [7], the Elliptic Curve Method [8] and the Number Field Sieve (NFS) [9,83]. The NFS is currently the best algorithm for factoring RSA moduli. It factors N in subexponential time and space $L_N[\frac{1}{3}, c] = \mathcal{O}(\exp(c(\log N)^{\frac{1}{3}} (\log\log N)^{\frac{2}{3}}))$ for $c \approx 1.9$.

Of course, ultimately, the cryptanalyst's ultimate goal is the construction of a polynomial time algorithm for either the RSA problem or the factorization problem. Since it is unknown whether there exist algorithms for these problems with Turing complexity $L_N[\alpha, c]$ for $\alpha < \frac{1}{3}$, one might ask for polynomial time algorithms in *other machine models* or for interesting *relaxations* of the RSA and factorization problem.

In 1994, Shor [10] presented an algorithm for solving the factorization problem in time and space polynomial in the bit-length of N, provided that the model of Turing machines is replaced by the model of quantum Turing machines. This ground-breaking theoretical result led to intensive engineering efforts for building quantum computers in practice. However, today, it is still unclear whether quantum computers with a large number of quantum bits can ever be constructed.

In the 90s, another interesting line of research evolved, which uses *polynomial time* algorithms in the *Turing machine model*. However, in order to achieve polynomial complexity, one has to relax the RSA and factorization problem. So instead of changing the model of computation, one relaxes the problems themselves by looking at restricted instances. The most natural restriction is realized by limiting the parameter set of the instances to an interval which is smaller than in the general setting, but still of exponential size.

A variation of this limiting approach addresses full parameter sets but allows additional access to an oracle for parts of the solution, e.g., for some of the bits. Notice that the oracle queries have the effect of cutting down the search space for the solution. The so-called *oracle complexity* measures the number of oracle queries that is required in order to solve the underlying problem in polynomial time. Of course, one is interested in minimizing the number of oracle queries and in restricting the oracle's power, i.e., the type of queries that an oracle replies to. Oracles are motivated by other cryptographical mechanisms, so-called side-channel attacks, that often leak partial information of the secrets and therefore behave in practice like an oracle.

In the following, we will call both approaches, limiting the parameter sets and allowing for an oracle, *relaxations* of the problem instances. In order to solve these relaxed instances, one models them as a polynomial equation and tries to find the integer solutions.

Let us illustrate this approach by a simple example. The RSA factorization problem is the problem of finding p, q on input N. This can be modeled by a polynomial equation $f(x, y) = N - xy$. The positive integer roots of this polynomial equation are $(1, N), (p, q), (q, p), (N, 1)$. Since we assume that p, q are of the same bit-size, finding all integer solutions which are in absolute value smaller than roughly $\sqrt{N}$ suffices to solve the factorization problem. Thus, one only has to find *small solutions*, where *small* means that the size of the root is small compared to the size of the coefficients of the polynomial. Naturally, one can define upper bounds X, Y for the size of the roots in x, y, respectively. The ultimate goal is to find a polynomial time algorithm which succeeds whenever $XY \leq N$. Since we do not know how to achieve this bound, we relax the factorization problem.

A natural relaxation of this problem is to narrow down the search space for the prime factors. Assume that we are given oracle access to the most significant bits of p. This allows us to compute an approximation $\tilde{p}$ of p such that $|p - \tilde{p}|$ is significantly smaller than $\sqrt{N}$. Then, $\tilde{q} = \frac{N}{\tilde{p}}$ defines an approximation of q. Therefore, we obtain the polynomial equation $f(x, y) = N - (\tilde{p} + x)(\tilde{q} + y)$ with a small root $(p - \tilde{p}, q - \tilde{q})$, where the size of the root depends on the quality of the approximation. It was shown by Coppersmith in 1996 [11], that the solution of this problem can be found in polynomial time if $XY \leq N^{\frac{1}{2}}$.

Building on works in the late 80s [12, 13], Coppersmith [11, 14–16] derived a general algorithm for finding small roots of polynomial equations. This root finding algorithm in turn is essentially based on the famous LLL-reduction algorithm by Lenstra, Lenstra and Lovász [17]. The key idea is to encode polynomial equations with small solutions as coefficient vectors that have a small Euclidean norm. These coefficient vectors can efficiently be found by an application of the LLL-reduction algorithm.

We will survey several applications of Coppersmith's algorithm to relaxations of the RSA problem and the factorization problem. Many of these applications naturally allow for a dual interpretation, both as a cryptanalytic result and as a security result. Let us give an example for this duality. In 1996, Coppersmith [14] showed that for RSA with $e = 3$, an attacker who knows $2/3$ of an RSA-encrypted message m can recover the remaining third from the ciphertext in polynomial time. The cryptanalytic interpretation is that knowing only a $2/3$-fraction of the plaintext is already enough to recover the whole. The security interpretation is that recovering a $2/3$-fraction must be hard, provided that solving the RSA problem for $e = 3$ is hard. Thus, this result establishes the security of a $2/3$-fraction of the underlying plaintext under the RSA assumption. This security interpretation was used by Shoup [18] to show the security of RSA-OAEP for $e = 3$ under chosen ciphertext attacks. We will elaborate a bit more on this duality effect in the paper.

This survey is organized as follows. We start in Section "How to Find Small Roots: The Univariate Case" by giving a high-level description of Coppersmith's algorithm for finding small roots of univariate modular polynomials. We state a theorem which provides us with an upper bound for the size of the roots of a univariate polynomial that can efficiently be found.

The details of the theorem's proof are given in Section "Proof of Theorem 1 and Algorithmic Considerations". This section is devoted to people who are interested in the technical details of the method, and those who want to implement a Coppersmith-type univariate root finding algorithm. It is the only section that requires some basic knowledge of lattice theory from the reader. People who are mainly interested in the applications of Coppersmith's method can proceed to the subsequent section.

In Section "Modeling RSA Problems as Univariate Root Finding Problems", we will extensively use our theorem for finding small roots. We will model certain relaxed RSA and factorization problems as univariate polynomial equations. For instance, we present Coppersmith's attack on RSA with stereotyped messages [14] and show its dual use in Shoup's security proof [18] and for the construction of an RSA-based pseudorandom number generator proposed by Steinfeld, Pieprzyk, and Wang [19]. Moreover, we will show a generalization of Håstad's broadcast attack [12] on RSA-encrypted, polynomially related messages that provides a natural link to Coppersmith's attack on stereotyped RSA messages.

We then describe the *factoring with high bits known* results from Coppersmith [11] and Boneh, Durfee, and Howgrave-Graham [20]. Furthermore, we show a deterministic polynomial time reduction of factoring to computing d [21, 22], which establishes the hardness of the so-called RSA secret key recovery problem under the factorization assumption. We conclude this section by stating Boneh's algorithm [23] for finding smooth integers in short intervals. The problem of finding smooth integers is related to classical factorization algorithms such as the Number Field Sieve.

In Section "Applications of Finding Roots of Multivariate Equations", we will turn our focus to multivariate extensions of Coppersmith's LLL-based method. We present Wiener's attack [24] on RSA with $d \leq N^{\frac{1}{4}}$ as a bivariate linear equation, which was originally phrased in terms of the continued fraction algorithm. We then present the bivariate polynomial equation of Boneh and Durfee [25, 26] that led to a heuristic improvement of the bound to $d \leq N^{0.292}$. As an example of an application with more variables, we present a heuristic polynomial time attack of Jochemsz and May [27] for RSA with so-called CRT-exponents $d \bmod p-1, d \bmod q-1$ smaller than $N^{0.073}$. Dually to these attacks, the server-based RSA signature generation proposals of Boneh, Durfee, Frankel [28] and Steinfeld, Zheng [29] are constructive security applications.

Since the number of applications of Coppersmith's LLL-based method for the RSA/factorization problem is already far too large to capture all the different results in this survey, we try to provide a more comprehensive list of references in Section "Survey and References for LLL-Based RSA and Factoring Results". We are aware of the fact that it is impossible to achieve completeness of such a list, but our references will serve the purpose of a good starting point for further reading.

In Section "Open Problems and Speculations", we give some open problems in this area and try to speculate in which direction this line of research will go. Especially, we discuss to which extent we can go from relaxed instances toward general problem instances, and where the limits of the method are. This discussion naturally leads to speculations whether any small root finding algorithm based on LLL-reduction will eventually have the potential to solve general instances of the RSA problem or the factorization problem in polynomial time.

How to Find Small Roots: The Univariate Case

We first introduce the problem of finding solutions of a modular univariate polynomial equation. Then, we argue that this approach extends to polynomials in more variables in a heuristic manner.

Let N be a positive integer of unknown factorization with divisor $b \geq N^{\beta}$, $0 < \beta \leq 1$.[1] Let $f(x)$ be a monic univariate polynomial of degree δ. We are looking for all small roots of the polynomial f modulo b. That is, we want to efficiently find all solutions x_0 satisfying

$$f(x_0) = 0 \bmod b \quad \text{with} \quad |x_0| \leq X,$$

where X is an upper bound on the size of the solutions. Our goal is to maximize the bound X, with the restriction that the running time of our method should be polynomial in the input size, i.e., polynomial in the parameters $(\log N, \delta)$.

We would like to stress that N is an integer of *unknown* factorization, which makes the above root finding problem hard to solve. If the prime factors of N are given, efficient algorithms with finite field arithmetic are known for the problem.

In 1996, Coppersmith [15] proposed an elegant LLL-based method for finding small solutions of univariate polynomial equations. Here, we describe his approach using the notion of Howgrave-Graham's reformulation [30] of the method. Coppersmith's approach is basically a reduction of solving modular polynomial equations to solving univariate polynomials over the integers. That is, one constructs from $f(x)$ another univariate polynomial $g(x)$ that contains all the small modular roots of $f(x)$ over the integers:

$$f(x_0) = 0 \bmod b \quad \Rightarrow \quad g(x_0) = 0 \text{ over } \mathbb{Z} \quad \text{for all } |x_0| \leq X.$$

The algorithmic idea for the construction of $g(x)$ from $f(x)$ can be described via the following two steps:

1. Fix an integer m. Construct a collection C of polynomials $f_1(x), f_2(x), \ldots, f_n(x)$ that all have the small roots x_0 modulo b^m. As an example, take the collection

$$\begin{aligned} f_i(x) &= N^{m-i} f^i(x) \text{ for } i = 1, \ldots, m \\ f_{m+i}(x) &= x^i f^m(x) \quad \text{ for } i = 1, \ldots, m. \end{aligned}$$

2. Construct an integer linear combination $g(x) = \sum_{i=1}^{n} a_i f_i(x)$, $a_i \in \mathbb{Z}$ such that the condition

$$|g(x_0)| < b^m$$

[1] An important special case is $b = N$, i.e., $\beta = 1$.

holds. Notice that b^m divides all $f_i(x_0)$ by construction. Therefore, b^m also divides $g(x_0)$. But then $g(x_0) = 0 \bmod b^m$ and $|g(x_0)| < b^m$, which implies that $g(x_0) = 0$ over the integers.

The construction in step (2) is realized by an LLL-based approach. Namely, one can easily show that every polynomial g whose coefficient vector of $g(xX)$ has sufficiently small norm fulfills the condition $|g(x_0)| < b^m$. The integer linear combinations of the coefficient vectors of $f_i(xX)$, $i = 1 \ldots n$, form a lattice L. Applying a lattice basis reduction algorithm to a basis of L yields a small norm coefficient vector $g(xX)$. One can show that in our case the LLL-reduction algorithm of Lenstra, Lenstra and Lovász [17] outputs a sufficiently small vector. Therefore, $g(x)$ can be computed in polynomial time via LLL-reduction.

Eventually, one has to find the roots of $g(x)$ over the integers. This can be done by standard polynomial factorization methods such as the Berlekamp–Zassenhaus algorithm. Interestingly, the initial application of the LLL algorithm was a deterministic polynomial time algorithm [17] for factoring polynomials in $\mathbb{Q}[X]$. In 2001, van Hoeij [31, 32] proposed an improved, highly efficient LLL-based factorization algorithm (see [33] for an introduction). Thus, we cannot only use LLL to construct g but also to find its integer roots.

The details of the proof of the following result can be found in Section "Proof of Theorem 1 and Algorithmic Considerations".

Theorem 1. *Let N be an integer of unknown factorization, which has a divisor $b \geq N^\beta$, $0 < \beta \leq 1$. Let $f(x)$ be a univariate monic polynomial of degree δ and let $c \geq 1$. Then we can find all solutions x_0 of the equation*

$$f(x) = 0 \bmod b \quad \textit{with} \quad |x_0| \leq cN^{\frac{\beta^2}{\delta}}$$

in time $\mathcal{O}(c\delta^5 \log^9 N)$.

Although LLL reduction only approximates a shortest vector up to some factor that is exponential in the lattice dimension, it is important to point out that lattice reduction techniques which give better approximations do not help improve the bound given in Theorem 1.

Coppersmith proved this result for the special case $\beta = 1$, i.e., $b = N$. The term β^2 first appeared in Howgrave-Graham's work [34] for the special case $\delta = 1$, i.e., for a linear polynomial. A proof of Theorem 1 first appeared in [35].

Coppersmith's method generalizes in a natural way to modular multivariate polynomials $f(x_1, \ldots, x_\ell)$. The idea is to construct ℓ *algebraically independent* polynomials $g^{(1)}, \ldots, g^{(\ell)}$ that all share the desired small roots over the integers. The roots are then computed by resultant computations. For $\ell \geq 2$, this is a heuristic method because although the LLL-algorithm guarantees linear independence of the coefficient vectors, it does not guarantee algebraic independence of the corresponding polynomials.

The case of solving multivariate polynomial equations *over the integers* – not modular – uses similar techniques. In the integer case, the method of finding small roots of bivariate polynomials $f(x, y)$ is rigorous, whereas the extension to more than two variables is again a heuristic. Coron showed in [36, 37], that the case of solving integer polynomials can, in principle, be reduced to the case of solving modular polynomials.

Proof of Theorem 1 and Algorithmic Considerations

In this section, we will give a complete proof of Theorem 1. Readers who are mainly interested in the method's applications can skip this section and proceed to Section "Modeling RSA Problems as Univariate Root Finding Problems".

We provide an algorithm that on input

- An integer N of unknown factorization
- A monic, univariate polynomial $f(x)$ of degree δ
- A bound $\beta \in (0, 1]$, such that $b \geq N^{\beta}$ for some divisor b of N

outputs in time polynomial in $\log N$ and δ all solutions x_0 such that

- $f(x_0) = 0 \bmod b$ and
- $|x_0| \leq N^{\frac{\beta^2}{\delta}}$.

Normally, the property that $f(x)$ is monic is no restriction in practice. Assume that $f(x)$ has a leading coefficient $a_\delta \neq 1$. Then, we can either make $f(x)$ monic by multiplying with the inverse of a_δ modulo N, or we find a non-trivial factorization of N. In the latter case, we can work modulo the factors of N.

The following theorem of Howgrave-Graham [30] gives us two criteria under which we can find a polynomial $g(x)$ that evaluates to zero over the integers at small roots.

Theorem 2 (Howgrave-Graham). *Let $g(x)$ be a univariate polynomial with n monomials. Further, let m be a positive integer. Suppose that*

1. $g(x_0) = 0 \bmod b^m$ *where* $|x_0| \leq X$
2. $\|g(xX)\| < \frac{b^m}{\sqrt{n}}$

Then $g(x_0) = 0$ *holds over the integers.*

Proof. We have

$$|g(x_0)| = \sum_i c_i x_0^i \leq \sum_i |c_i x_0^i|$$
$$\leq \sum_i |c_i| X^i \leq \sqrt{n}\|g(xX)\| < b^m.$$

But $g(x_0)$ is a multiple of b^m, and, therefore, it must be zero.

Using powers of f, we construct a collection $f_1(x), \ldots, f_n(x)$ of polynomials that all have the desired roots x_0 modulo b^m. Thus, for every integer linear combination g, we have

$$g(x_0) = \sum_{i=1}^{n} a_i f_i(x_0) = 0 \bmod b^m, \quad a_i \in \mathbb{Z}.$$

Hence, every integer linear combination satisfies condition (1) of Lemma 2. Among all integer linear combinations, we search for one that also satisfies condition (2). In other words, we have to search among all integer linear combinations of the coefficient vectors $f_i(xX)$ for a vector with Euclidean norm smaller than $\frac{b^m}{\sqrt{n}}$. This can be achieved by finding a short vector in the lattice L spanned by the coefficient vectors of $f_i(xX)$.

Our goal is to ensure that the LLL algorithm finds a vector v with $\|v\| < \frac{b^m}{\sqrt{n}}$ in L. By a theorem of Lenstra, Lenstra and Lovász [17], the norm of a shortest vector v in an LLL-reduced lattice basis can by related to the determinant $\det(L)$ of the corresponding lattice L with dimension n via

$$\|v\| \leq 2^{\frac{n-1}{4}} \det(L)^{\frac{1}{n}}.$$

The determinant $\det(L)$ can be easily computed from the coefficient vectors of $f_i(xX)$. If we could satisfy the condition

$$2^{\frac{n-1}{4}} \det(L)^{\frac{1}{n}} < \frac{N^{\beta m}}{\sqrt{n}}, \tag{10.1}$$

then we obtain the desired inequality $\|v\| < \frac{N^{\beta m}}{\sqrt{n}} \leq \frac{b^m}{\sqrt{n}}$.

Neglecting low-order terms in (10.1), i.e., terms that do not depend on N, we obtain the simplified condition

$$\det(L) < N^{\beta m n}.$$

Let L be a lattice of dimension n with basis B satisfying this condition. Then on average, a basis vector $v \in B$ contributes to the determinant with a factor less than $N^{\beta m}$. We call such a basis vector a *helpful vector*. Helpful vectors will play a central role for the construction of an optimized lattice basis.

The following theorem of Coppersmith states that for a monic polynomial $f(x)$ of degree δ, all roots x_0 with $|x_0| \leq \frac{1}{2} N^{\frac{\beta^2}{\delta} - \epsilon}$ can be found in polynomial time. We will later show that the error term ϵ and the factor $\frac{1}{2}$ can be easily eliminated, which will lead to a proof of Theorem 1.

Theorem 3 (Coppersmith). *Let N be an integer of unknown factorization, which has a divisor $b \geq N^{\beta}$, $0 < \beta \leq 1$. Let $0 < \epsilon \leq \frac{1}{7}\beta$. Furthermore, let $f(x)$ be a univariate monic polynomial of degree δ. Then, we can find all solutions x_0 for the equation*

$$f(x) = 0 \bmod b \quad \textit{with} \quad |x_0| \leq \frac{1}{2} N^{\frac{\beta^2}{\delta} - \epsilon}.$$

The running time is dominated by the time to LLL-reduce a lattice basis of dimension $\mathcal{O}(\epsilon^{-1}\delta)$ with entries of bit-size $\mathcal{O}(\epsilon^{-1}\log N)$. This can be achieved in time $\mathcal{O}(\epsilon^{-7}\delta^5\log^2 N)$.

Proof. Define $X := \frac{1}{2}N^{\frac{\beta^2}{\delta}-\epsilon}$. Let us apply the two steps of Coppersmith's method as described in Section "How to Find Small Roots: The Univariate Case". In the first step, we fix

$$m = \left\lceil \frac{\beta^2}{\delta\varepsilon} \right\rceil. \tag{10.2}$$

Next, we choose a collection C of polynomials, where each polynomial has a root x_0 modulo b^m whenever $f(x)$ has the root x_0 modulo b. In our case, we include in C the polynomials

$$\begin{array}{llll} N^m, & xN^m, & x^2N^m, & \ldots x^{\delta-1}N^m, \\ N^{m-1}f, & xN^{m-1}f, & x^2N^{m-1}f, & \ldots x^{\delta-1}N^{m-1}f, \\ N^{m-2}f^2, & xN^{m-2}f^2, & x^2N^{m-2}f^2, & \ldots x^{\delta-1}N^{m-2}f^2, \\ \vdots & \vdots & \vdots & \vdots \\ Nf^{m-1}, & xNf^{m-1}, & x^2Nf^{m-1}, & \ldots x^{\delta-1}Nf^{m-1}. \end{array}$$

Additionally, we take the polynomials

$$f^m, xf^m, x^2f^m, \ldots, x^{t-1}f^m$$

for some t that has to be optimized as a function of m.

Note that by our ordering the kth polynomial of C is a polynomial of degree k. Thus, it introduces the new monomial x^k. We could also write the choice of our polynomials in C in a more compact form. Namely, we have chosen the polynomials

$$\begin{array}{lll} g_{i,j}(x) = x^jN^if^{m-i}(x) & \text{for } i = 0,\ldots,m-1, \ j = 0,\ldots,\delta-1 \text{ and} \\ h_i(x) \ \ = x^if^m(x) & \text{for } i = 0,\ldots,t-1. \end{array}$$

In Step 2 of Coppersmith's method, we construct the lattice L that is spanned by the coefficient vectors of $g_{i,j}(xX)$ and $h_i(xX)$. As we noticed before, we can order the polynomials $g_{i,j}$ and h_i in strictly increasing order of their degree k. Therefore, the basis B of L, that has as row vectors the coefficient vectors of $g_{i,j}(xX)$ and $h_i(xX)$, can be written as a lower triangular matrix. Let $n := \delta m + t$, then we write B as the $(n \times n)$-matrix given in Table 10.

Since B is in lower triangular form, $\det(L)$ is simply the product of all entries on the diagonal:

$$\det(L) = N^{\frac{1}{2}\delta m(m+1)}X^{\frac{1}{2}n(n-1)}. \tag{10.3}$$

Table 10.1 Basis B of the lattice L. We use the following notation: Every nonspecified entry is zero. The entries marked with "–" may be nonzero, but the determinant of the lattice does not depend on these values

$$
\left(\begin{array}{cccccccccccc}
N^m & & & & & & & & & & & \\
 & N^m X & & & & & & & & & & \\
 & & \ddots & & & & & & & & & \\
 & & & N^m X^{\delta-1} & & & & & & & & \\
\hline
 & \ddots & \ddots & \ddots & \ddots & & & & & & & \\
\hline
- & - & \dots & - & \dots\ N X^{\delta m-\delta} & & & & & & & \\
 & - & \dots & - & \dots & - & N X^{\delta m-\delta+1} & & & & & \\
 & & \ddots & \ddots & \ddots & \ddots & \ddots & \ddots & & & & \\
 & & & - & \dots & - & - & \dots\ N X^{\delta m-1} & & & & \\
\hline\hline
- & - & \dots & - & \dots & - & - & - & - & X^{\delta m} & & \\
 & - & \dots & - & \dots & - & - & - & - & - & X^{\delta m+1} & \\
 & & & & \ddots & \ddots & \ddots & \ddots & \ddots & \ddots & \ddots & \ddots \\
 & & & & & - & - & - & - & - & - & \dots\ X^{\delta m+t-1}
\end{array}\right)
$$

Now, we want to optimize the parameter t, which is equivalent to the optimization of $n = \delta m + t$. Remember that we argued before that every vector which contributes to the determinant by a factor less than $N^{\beta m}$ is *helpful*. In our setting, this means that we have to ensure that the entries of the coefficient vectors $h_i(xX)$ on the diagonal are all less than $N^{\beta m}$, i.e., we have the condition

$$X^{n-1} < N^{\beta m}.$$

Since $X^{n-1} < N^{(\frac{\beta^2}{\delta}-\epsilon)(n-1)} < N^{\frac{\beta^2}{\delta}n}$ this condition is satisfied for the choice

$$n \le \frac{\delta}{\beta} m. \tag{10.4}$$

According to (10.2), we know that $m \le \frac{\beta^2}{\delta\epsilon} + 1$. Then, we immediately have a bound for the lattice dimension

$$n \le \frac{\beta}{\epsilon} + \frac{\delta}{\beta}.$$

Using $7\beta^{-1} \le \epsilon^{-1}$, we obtain $n = \mathcal{O}(\epsilon^{-1}\delta)$. We choose n as the maximal integer that satisfies inequality (10.4). This yields a lower bound of

$$n > \frac{\delta}{\beta} m - 1 \ge \frac{\beta}{\epsilon} - 1 \ge 6.$$

In order to prove the running time, we also need to upper-bound the bit-size of the entries in B. Notice that for every power f^{m-i} in the definition of $g_{i,j}$ and h_i, we can reduce the coefficients modulo N^{m-i}, since x_0 must be a root modulo N^{m-i}. Thus, the largest coefficient in a product $N^i f^{m-i}$ has a bit-size of at most $m\log(N) = \mathcal{O}(\epsilon^{-1}\log N)$. Powers of $X = \frac{1}{2}N^{\frac{\beta^2}{\delta}-\epsilon}$ occur with exponents smaller than n. Thus, the bit-size of powers of X can also be upperbounded by

$$n \cdot \frac{\beta^2}{\delta}\log N = \mathcal{O}\left(\frac{\delta}{\epsilon}\cdot\frac{\beta^2}{\delta}\right)\log N = \mathcal{O}\left(\epsilon^{-1}\log N\right).$$

Nguyen and Stehlé [38, 39] recently proposed a modified version of the LLL-algorithm called L^2-algorithm. The L^2-algorithm achieves the same approximation quality for a shortest vector as the LLL algorithm, but has an improved worst case running time analysis. It takes time $\mathcal{O}(n^5(n+\log b_m)\log b_m)$, where $\log b_m$ is the maximal bit-size of an entry in B. Thus, we obtain for our method a running time of

$$\mathcal{O}\left(\left(\frac{\delta}{\epsilon}\right)^5\left(\frac{\delta}{\epsilon}+\frac{\log N}{\epsilon}\right)\frac{\log N}{\epsilon}\right).$$

Notice that we can assume $\delta \le \log N$, since otherwise our bound $|x_0| \le N^{\frac{\beta^2}{\delta}-\epsilon}$ is vacuous. Therefore, we obtain a running time of $\mathcal{O}(\epsilon^{-7}\delta^5\log^2 N)$.

It remains to show that LLL's approximation quality is sufficient for our purpose. In order to apply the theorem of Howgrave Graham (Theorem 2), we have to ensure that the LLL algorithm finds a vector in L with norm smaller than $\frac{b^m}{\sqrt{n}}$. Since the LLL algorithm finds a vector v in an n-dimensional lattice with $\|v\| \le 2^{\frac{n-1}{4}}\det(L)^{\frac{1}{n}}$, we have to satisfy the condition

$$2^{\frac{n-1}{4}}\det(L)^{\frac{1}{n}} < \frac{b^m}{\sqrt{n}}.$$

Using the term for $\det(L)$ in (10.3) and the fact $b \ge N^\beta$, we obtain the new condition

$$N^{\frac{\delta m(m+1)}{2n}} X^{\frac{n-1}{2}} \le 2^{-\frac{n-1}{4}} n^{-\frac{1}{2}} N^{\beta m}.$$

This gives us a condition on the size of X:

$$X \le 2^{-\frac{1}{2}} n^{-\frac{1}{n-1}} N^{\frac{2\beta m}{n-1}-\frac{\delta m(m+1)}{n(n-1)}}.$$

Notice that $n^{-\frac{1}{n-1}} = 2^{-\frac{\log n}{n-1}} \ge 2^{-\frac{1}{2}}$ for $n > 6$. Therefore, our condition simplifies to

$$X \le \frac{1}{2} N^{\frac{2\beta m}{n-1}-\frac{\delta m(m+1)}{n(n-1)}}.$$

Remember that we made the choice $X = \frac{1}{2}N^{\frac{\beta^2}{\delta}-\epsilon}$. Hence, in order to finish the proof of the theorem, it suffices to show that

$$\frac{2\beta m}{n-1} - \frac{\delta m^2(1+\frac{1}{m})}{n(n-1)} \geq \frac{\beta^2}{\delta} - \epsilon.$$

We obtain a lower bound for the left-hand side by multiplying with $\frac{n-1}{n}$. Then, we use $n \leq \frac{\delta}{\beta}m$ which gives us

$$2\frac{\beta^2}{\delta} - \frac{\beta^2}{\delta}\left(1+\frac{1}{m}\right) \geq \frac{\beta^2}{\delta} - \epsilon.$$

This simplifies to

$$-\frac{\beta^2}{\delta} \cdot \frac{1}{m} \geq -\epsilon.$$

This in turn gives us the condition $m \geq \frac{\beta^2}{\delta\epsilon}$, which holds by the choice of m that we made in (10.2).

Let us briefly summarize the whole algorithm which finds all roots of $f(x)$ modulo b that are in absolute value smaller than X.

Coppersmith's method in the univariate case

INPUT: Polynomial $f(x)$ of degree δ, modulus N of unknown factorization that is a multiple of b, a lower bound $b \geq N^{\beta}$, $\epsilon \leq \frac{1}{7}\beta$

Step 1: Choose $m = \lceil \frac{\beta^2}{\delta\epsilon} \rceil$ and $t = \lfloor \delta m(\frac{1}{\beta} - 1) \rfloor$.
Compute the polynomials

$$\begin{aligned} g_{i,j}(x) &= x^j N^i f^{m-i}(x) \quad \text{for } i = 0,\ldots,m-1,\ j = 0,\ldots,\delta-1 \text{ and} \\ h_i(x) &= x^i f^m(x) \quad\quad\ \ \text{for } i = 0,\ldots,t-1. \end{aligned}$$

Step 2: Compute the bound $X = \frac{1}{2}\lceil N^{\frac{\beta^2}{\delta}-\epsilon} \rceil$. Construct the lattice basis B, where the basis vectors of B are the coefficient vectors of $g_{i,j}(xX)$ and $h_i(xX)$.

Step 3: Apply the LLL algorithm to the lattice basis B. Let v be the shortest vector in the LLL reduced basis. The vector v is the coefficient vector of some polynomial $g(xX)$. Construct $g(x)$ from v.

Step 4: Find the set R of all roots of $g(x)$ over the integers using standard methods. For every root $x_0 \in R$ check whether $\gcd(N, f(x_0)) \geq N^{\beta}$. If this condition is not satisfied then remove x_0 from R.

OUTPUT: Set R, where $x_0 \in R$ whenever $f(x_0) = 0 \bmod b$ for an $|x_0| \leq X$.

As we noticed before, all steps of the algorithm can be done in time $\mathcal{O}(\epsilon^{-7}\delta^5 \log^2 N)$, which concludes the proof of the theorem.

One should remark that the polynomial $g(x)$ that we construct in Coppersmith's method may contain integer roots that are not roots of $f(x)$ modulo b. Therefore, we use in Step 4 of the above algorithm a simple test whether $f(x_0)$ contains a divisor of N of size at least N^β.

It is also worth noticing the following point: The LLL approximation factor of $2^{\frac{n-1}{4}}$ for the shortest vector is exponentially in the lattice dimension n, but this factor essentially translates in the analysis of Theorem 3 to the term $\frac{1}{2}$ for the upper bound of the size of the roots x_0. Thus, computing a shortest vector instead of an LLL approximate version would only improve the bound by a factor of roughly 2 (i.e., only one bit).

Moreover, Theorem 1 is a direct implication of Theorem 3 and shows that we can avoid the terms $\frac{1}{2}$ and ϵ from the upper bound on x_0. The proof uses a simple brute-force search.

Theorem 1. *Let N be an integer of unknown factorization, which has a divisor $b \geq N^\beta$, $0 < \beta \leq 1$. Furthermore, let $f(x)$ be a univariate monic polynomial of degree δ. Then we can find all solutions x_0 for the equation*

$$f(x) = 0 \bmod b \quad \textit{with} \quad |x_0| \leq cN^{\frac{\beta^2}{\delta}}.$$

in time $\mathcal{O}(c\delta^5 \log^9 N)$.

Proof. An application of Theorem 3 with the parameter choice $\epsilon = \frac{1}{\log N}$ shows that we can find all roots x_0 with

$$|x_0| \leq \frac{1}{4} N^{\frac{\beta^2}{\delta}}$$

in time $\mathcal{O}(\delta^5 \log^9 N)$.

In order to find all roots that are of size at most $cN^{\frac{\beta^2}{\delta}}$ in absolute value, we divide the interval $[-cN^{\frac{\beta^2}{\delta}}, cN^{\frac{\beta^2}{\delta}}]$ into $4c$ subintervals of size $\frac{1}{2}N^{\frac{\beta^2}{\delta}}$ centered at some x_i. For each subinterval with center x_i, we apply the algorithm of Theorem 3 to the polynomial $f(x - x_i)$ and output the roots in this subinterval.

For completeness reasons and since it is one of the most interesting cases of Coppersmith's method, we explicitly state the special case $b = N$ and $c = 1$, which is given in the work of Coppersmith [15].

Theorem 4 (Coppersmith). *Let N be an integer of unknown factorization. Furthermore, let $f_N(x)$ be a univariate monic polynomial of degree δ. Then we can find all solutions x_0 for the equation*

$$f_N(x) = 0 \bmod N \quad \textit{with} \quad |x_0| \leq N^{\frac{1}{\delta}}$$

in time $\mathcal{O}(\delta^5 \log^9 N)$.

Modeling RSA Problems as Univariate Root Finding Problems

We address several RSA related problems that can be solved by finding small roots of univariate modular polynomial equations. Throughout this section, we will assume that $N = pq$ is a product of two primes, and that $e \in \mathbb{Z}^*_{\phi(N)}$. Both N and e are publically known.

Relaxed RSA Problem: Stereotyped Messages

The RSA problem is the problem of inverting the RSA function. Given $m^e \bmod N$, one has to find the unique e^{th} root $m \in \mathbb{Z}_N$. The *RSA assumption* states that the RSA problem is difficult to solve for randomly chosen $m \in \mathbb{Z}_N$.

Notice that the RSA problem is trivial to solve for *small m and small e*. Namely, if $m < N^{\frac{1}{e}}$ then $m^e \bmod N = m^e$ over $\mathbb{Z}$. Therefore, computation of the eth roots over the integers yields the desired root.

RSA problem

Given: $m^e \bmod N$
Find : $m \in \mathbb{Z}_N$

Relaxed RSA problem: Small e, High Bits Known

Given: $m^e, \tilde{m}$ with $|m - \tilde{m}| \leq N^{\frac{1}{e}}$
Find : $m \in \mathbb{Z}_N$

Coppersmith extended this result to the case where m is not small, but we know m up to a small part. Namely, we assume the knowledge of an approximation $\tilde{m}$ such that $m = \tilde{m} + x_0$ for some unknown part $|x_0| \leq N^{\frac{1}{e}}$. This can be modeled as the polynomial equation

$$f(x) = (\tilde{m} + x)^e - m^e \bmod N.$$

Let us apply Theorem 1. We set $\beta = 1$, $\delta = e$ and $c = 1$. Therefore, we can recover x_0 as long as $|x_0| \leq N^{\frac{1}{e}}$. This extends the trivial attack where m is small to the inhomogenous case: The most significant bits of m are not zero, but they are known to an attacker.

Clearly, one can interpret this as a cryptanalytic result. For example, if $e = 3$, then an attacker who can guess the first $2/3$-fraction of the message m is able to reconstruct the last $1/3$-fraction of m in polynomial time. This might happen in situations were the plaintext has a stereotype form like "The password for today is: xxxx." Therefore, this is often called an *attack on stereotyped messages*. Loosely speaking, the cryptanalytic meaning is that an attacker gets an $\frac{1}{e}$*-fraction of the RSA message efficiently*. We will see in Section "Related RSA Messages: Extending

Håstad's Attack" that this cryptanalytic interpretation can be generalized to the case where the same message m is sent several times.

On the other hand, one can interpret this result in a dual sense as a security result for a $2/3$-fraction of the plaintext bits in an RSA ciphertext. It is as difficult to compute a $2/3$-fraction of m as inverting the RSA problem for $e = 3$. In general, there is a tight reduction from the RSA problem to the problem of finding an $\frac{e-1}{e}$-fraction of the most significant bits. Under the RSA assumption, this shows that the most significant bits of an RSA plaintext are hard to find. Even stronger results on the security of RSA bits were given by Håstad and Näslund [40].

Constructive Applications of the Relaxed RSA Problem: RSA-OAEP and RSA Pseudorandom Generator

The dual security interpretation of the Relaxed RSA problem was used by Shoup [18] in 2001. He gave a security proof of the padding scheme OAEP [41] when instantiated with the RSA trapdoor function. Here, we only sketch Shoup's proof. More details on the proof and on cryptographic security notations can be found in Gentry's survey [42].

In RSA-OAEP, the plaintext is split into two parts s and t. The first part s depends on the message m, a fixed padding and some randomization parameter r of length k bits. The fixed padding ensures that s fulfills a well-defined format that can be checked. The second part t is simply $h(s) \oplus r$ for some hash function h, which is modeled as a random oracle. One encrypts the padded message $s \cdot 2^k + t$. Let c be the corresponding ciphertext.

Bellare and Rogaway [41] showed that RSA-OAEP is CCA1-secure, i.e., secure against so-called lunch-time attacks. It was widely believed that RSA-OAEP is also CCA2-secure, i.e., that it provides security against adaptive chosen ciphertext attacks. In 2001, Shoup [18] showed that the original proof of Bellare and Rogaway does not offer this level of security. However, using an analogous reasoning as in the *stereotyped message attack*, he could easily derive CCA2-security for RSA-OAEP with exponent 3.

In order to prove CCA2-security, we assume the existence of an adversary that successfully attacks RSA-OAEP under chosen ciphertext attacks. This adversary is then used to invert the RSA function. One defines a simulator in order to answer the adversary's decryption and hash queries. Shoup showed that any adversary that never explicitly queries h on s has a negligible probability to pass the format check for the s-part. Thus, one can assume that the first part s has to appear among the attacker's queries. This in turn is already sufficient to extract t as a root of

$$f(t) = (s \cdot 2^k + t)^e - c \bmod N,$$

provided that $|t| < N^{\frac{1}{e}}$ which is fulfilled whenever $k < \log N/e$. This condition is satisfied for $e = 3$ by the RSA-OAEP parameters. One should notice the correspondence to the Relaxed RSA problem: s plays the role of the known message part $\tilde{m}$, whereas t is the small unknown part.

We have reduced the RSA problem to an algorithm for attacking RSA-OAEP. The reduction is tight up to a factor of q_h, the number of hash queries an adversary is allowed to ask. Namely, the running time is q_h times the time to run the LLL-based algorithm for finding small e^{th} roots. The success probability of the RSA inverter is roughly the same as the success probability of the adversary. This reduction is tighter than the original reduction by Bellare-Rogaway for CCA1-security.

RSA-OAEP was shown to be CCA2-secure for arbitrary e by Fujisaki et al [43] in 2001, using a 2-dimensional lattice technique. However, their reduction is also less tight than Shoup's: If the RSA attacker has success probability ϵ, then the RSA inversion algorithm of [43] has success probability only ϵ^2.

Another constructive application of Coppersmith's attack on stereotyped messages is used for the definition of an efficient RSA-based pseudorandom number generator (PRNG) in a recent paper by Steinfeld, Pieprzyk, and Wang [19], which in turn builds on a work of Fischlin and Schnorr [44]. In the Fischlin-Schnorr RSA-PRNG, one starts with a random seed x_0 and generates a sequence $x_1, x_2, \ldots$ by successively applying the RSA function, i.e., $x_i = x_{i-1}^e \bmod N$. In each iteration, one outputs the r least significant bits of x_i.

In the security proof, Fischlin and Schnorr show that any efficient algorithm that distinguishes the generator's output from the uniform distribution can be used to invert the RSA function, i.e., to solve the RSA problem. However, the reduction is not tight. Namely, if T_D is the running time of the distinguisher, then the inversion algorithm's running time is roughly $2^{2r} T_D$. Therefore, one can only output $r = \mathcal{O}(\log \log N)$ in each iteration in order to preserve a polynomial reduction.

In 2006, Steinfeld, Pieprzyk, and Wang showed that one can securely output $\Theta(\log N)$ bits if one replaces the RSA assumption in the Fischlin-Schnorr proof by a relaxed RSA inversion assumption. Namely, we already know that one can recover an $\frac{1}{e}$-fraction of the message from an RSA ciphertext given the rest of the plaintext. Steinfeld et al. make the assumption that this bound is essentially tight. More precisely, they assume that any algorithm that recovers an $\frac{1}{e} + \epsilon$-fraction for some constant ϵ already requires at least the same running time as the best factoring algorithm for N.

In fact, one replaces the RSA assumption by a stronger assumption which states that the bound $\frac{1}{e}$ for the Coppersmith attack on stereotyped messages cannot be significantly improved. This stronger assumption is sufficient to increase the generator's output rate from $r = \mathcal{O}(\log \log N)$ to the full-size of $r = \Theta(\log N)$ bits. The efficiency of the Steinfeld, Pieprzyk, Wang construction is comparable to the efficiency of the Micali-Schnorr generator [45] from 1988, but uses a weaker assumption than in [45].

Another construction of an efficient PRNG and a MAC based on small root problems was proposed by Boneh, Halevi, and Howgrave-Graham [46]. Its security is proved under the hardness of the so-called modular inversion hidden number problem. The best algorithmic bound for attacking this problem is based on an LLL-approach. The security proofs for the PRNG and the MAC again assume that one cannot go significantly beyond this bound.

Affine Padding: Franklin-Reiter's Attack

The following attack was presented by Franklin and Reiter [47] in 1995. The attack was 1 year later extended by Coppersmith, Franklin, Patarin, and Reiter [48].

Assume that two RSA plaintexts m, m' satisfy an affine relation $m' = m + r$. Let $c = m^3 \bmod N$ and $c' = (m+r)^3 \bmod N$ their RSA ciphertexts, respectively. Franklin and Reiter showed that any attacker with knowledge of c, c', r, and N can efficiently recover m by carrying out the simple computation

$$\frac{c'r + 2cr - r^4}{c' - c + 2r^3} = \frac{3m^3r + 3m^2r^2 + 3mr^3}{3m^2r + 3mr^2 + 3r^3} = m \bmod N.$$

What happens in the case where r is unknown but small?

Affine related messages

Given: $c = m^e \bmod N, c' = (m+r)^e \bmod N$ with $|r| \leq N^{\frac{1}{e^2}}$
Find : m

If one is able to determine r from the ciphertexts, then m can be computed efficiently. The resultant computation

$$\text{Res}_m(c - m^3, c' - (m+r)^3) = r^9 + 3(c - c')r^6 + 3(c^2 + c'^2 + 7cc')r^3 + (c - c')^3 \bmod N$$

yields a monic univariate polynomial $f(r)$ of degree 9. An application of Theorem 1 shows that r can be recovered as long as $|r| \leq N^{\frac{1}{9}}$. For arbitrary e, the bound generalizes to $|r| \leq N^{\frac{1}{e^2}}$.

Related RSA Messages: Extending Håstad's Attack

Assume that we want to broadcast a plain RSA encrypted message to a group of k receivers all having public exponent e and co-prime moduli $N_1, \ldots, N_k$. That is, we send the messages $m^e \bmod N_1, \ldots, m^e \bmod N_k$. From this information, an attacker can compute $m^e \bmod \prod_{i=1}^k N_i$. If m^e is smaller than the product of the moduli, he can compute m by e^{th} root computation over the integers. If all N_i are of the same bit-size, we need $k \geq e$ RSA encrypted messages in order to recover m.

So naturally, an attacker gains more and more information by receiving different encryptions of the same message. Notice that this observation nicely links with the attack on stereotyped RSA messages from Section "Relaxed RSA Problem: Stereotyped Messages". Recall that the cryptanalytic interpretation of the attack in Section "Relaxed RSA Problem: Stereotyped Messages" was that one gets an

$\frac{1}{e}$-fraction of the plaintext efficiently. The above broadcast attack can thus be interpreted as an accumulation of this result. If one gets $k \geq e$ times an $\frac{1}{e}$-fraction of m efficiently, then one eventually obtains the whole m.

The question is whether this is still true when the public exponents are different and when the messages are preprocessed by simple padding techniques, e.g., an affine transformation with a fixed known padding pattern. We show that whenever the messages are polynomially related, then the underlying plaintext can still be discovered given sufficiently many encryptions. This result is an extension of Håstad's original result [12] due to May, Ritzenhofen [49].

Assume that the message m is smaller than $\min_j\{N_j\}$. We preprocess the message by known polynomial relations $g_1, \ldots, g_k$ with degrees $\delta_1, \ldots, \delta_k$, respectively.

Polynomially related RSA messages

Given: $c_i = g_i(m)^{e_i} \bmod N_i$ for $i = 1, \ldots, k$ with $\sum_{i=1}^{k} \frac{1}{\delta_i e_i} \geq 1$.

Find : m

Assume that $g_i(x)$ has leading coefficient $a_i \neq 1$. Compute $a_i^{-1} \bmod N_i$. If this computation fails, we obtain the factorization of N_i, which enables us to compute m. Otherwise, we replace c_i and $g_i(x)$ by $a_i^{-e_i} c_i$ and $a_i^{-1} g_i(x)$, respectively. This makes all $g_i(x)$ monic.

Let $\delta = \mathrm{lcm}_i\{\delta_i e_i\}$ be the least common multiple of all $\delta_i e_i$. Define $N = \prod_{i=1}^{k} N_i^{\frac{\delta}{\delta_i e_i}}$. We know that for all $i = 1, \ldots, k$ we have

$$(g_i(m)^{e_i} - c_i)^{\frac{\delta}{\delta_i e_i}} = 0 \bmod N_i^{\frac{\delta}{\delta_i e_i}}.$$

Let us compute by Chinese Remaindering a polynomial

$$f(x) = \sum_{i=1}^{k} b_i (g_i(x)^{e_i} - c_i)^{\frac{\delta}{\delta_i e_i}} \bmod N,$$

where the b_i are the Chinese remainder coefficients satisfying $b_i \bmod N_j = \begin{cases} 1 & \text{for } i = j \\ 0 & \text{else} \end{cases}$.

Notice that $f(m) = 0 \bmod N$ and that $f(x)$ is by construction a univariate monic polynomial of degree δ. Let us now upper-bound the size of our desired root m. Using the condition $1 \leq \sum_{i=1}^{k} \frac{1}{\delta_i e_i}$, we obtain

$$m < \min_j\{N_j\} \leq \left(\min_j\{N_j\}\right)^{\sum_{i=1}^{k} \frac{1}{\delta_i e_i}} \leq \prod_{i=1}^{k} N_i^{\frac{1}{\delta_i e_i}}.$$

By applying Theorem 1 with the parameters $\beta, c = 1$, we can find all roots m up to the same bound

$$m \leq N^{\frac{1}{\delta}} = \prod_{i=1}^{k} N_i^{\frac{1}{\delta_i e_i}},$$

which completes the description of the attack.

Let us look at our condition $\sum_{i=1}^{k} \frac{1}{\delta_i e_i} \geq 1$ when we encrypt the plain message m without any further transformation. Then $g_i(x) = x$ is the identity with degree $\delta_i = 1$, i.e., we obtain the simplified condition

$$\sum_{i=1}^{k} \frac{1}{e_i} \geq 1.$$

Again this can be interpreted as an accumulation of the results for stereotyped RSA messages in Section "Relaxed RSA Problem: Stereotyped Messages". Recall that for each encryption of m under exponent e_i, we can compute an $\frac{1}{e_i}$-fraction of m efficiently. This information accumulates such that whenever the sum $\sum_i \frac{1}{e_i}$ of all the fractions exceeds 1, we eventually obtain the whole plaintext m.

Factoring with High Bits Known

Let $N = pq$, w.l.o.g. $p > q$. Assume that we are given an oracle for the most significant bits of p. Our task is to find the factorization of N in time polynomial in the bit-size of N with a minimal number of queries to the oracle, i.e., we want to minimize the oracle complexity.

One can view this problem as a natural relaxation of the factorization problem. Without knowing any bits of the prime factor p, i.e., without using the oracle, we have to solve the general factorization problem. For the general problem, it is unclear whether there exists a polynomial time algorithm in the Turing machine model. So, we provide the attacker with an additional sufficiently strong hint given by the oracle answers that allows him to find the factorization in polynomial time.

In 1985, Rivest and Shamir [50] published an algorithm that factors N given a $\frac{2}{3}$-fraction of the bits of p. Coppersmith [51] improved this bound to $\frac{3}{5}$ in 1995. One year later, Coppersmith [11, 15] gave an algorithm using only half of the bits of p.

The *factoring with high bits known* problem can again be reduced to the problem of solving modular univariate polynomial equations with the LLL algorithm. Let us assume that we are given half of the high-order bits of p. Omitting constants, we know an approximation $\tilde{p}$ of p that satisfies $|p - \tilde{p}| \leq N^{\frac{1}{4}}$.

Factorization problem

Given: $N = pq$
Find : p

Relaxed Factorization: High Bits Known

Given: $N = pq$, $\tilde{p}$ with $|p - \tilde{p}| \leq N^{\frac{1}{4}}$
Find : p

Our goal is to recover the least-significant bits of p, i.e., we want to find the root of the univariate, linear modular polynomial

$$f(x) = \tilde{p} + x \bmod p.$$

Observe that $p - \tilde{p}$ is a root of $f(x)$ with absolute value smaller than $N^{\frac{1}{4}}$.

We apply Theorem 1 with $f(x) = \tilde{p} + x$, i.e., we have degree $\delta = 1$, $\beta = \frac{1}{2}$ and $c = 1$. Therefore, we can find all roots x_0 with size

$$|x_0| \leq N^{\frac{\beta^2}{\delta}} = N^{\frac{1}{4}}.$$

This enables us to recover the low-order bits of p in polynomial time with the LLL algorithm, which yields the factorization.

The *factorization with high bits known* approach can be extended to moduli $N = p^r q$, where p and q have the same bit-size. This extension was proposed by Boneh, Durfee, and Howgrave-Graham [20]. For simplicity, we assume that p and q are of the same bit size. For fixed bit-size of N and growing r, these moduli should be – from an information theoretical point of view – easier to factor than usual RSA moduli. Moreover, an attacker should learn from an approximation of p more information than in the standard RSA case. This intuition turns out to be true.

We model this variant of the factorization problem as the univariate polynomial

$$f(x) = (\tilde{p} + x)^r \bmod p^r.$$

Set $\beta = \frac{r}{r+1}$, $\delta = r$ and $c = 1$. An application of Theorem 1 shows that the LLL algorithm recovers all roots x_0 with

$$|x_0| \leq M^{\frac{\beta^2}{\delta}} = N^{\frac{r}{(r+1)^2}}.$$

Since N is roughly of the size p^{r+1}, this means that we need an approximation $\tilde{p}$ with $|p - \tilde{p}| \leq p^{\frac{r}{r+1}}$. Or in other words, we need a $\frac{1}{r+1}$-fraction of the most significant bits in order to factor N in polynomial time. That is, for the RSA case $r = 1$, we need half of the bits, whereas, e.g., for $r = 2$, we only need a third of the most significant bits of p. For $r = \Omega(\sqrt{\frac{\log N}{\log\log N}})$, one only has to guess $\mathcal{O}(\log\log N)$ bits of p, which can be done in polynomial time.

Computing d ≡ Factoring

Our next application of the LLL algorithm addresses the difficulty of computing the RSA secret exponent from the public information (N, e). We show that any algorithm that computes d in *deterministic polynomial time* can be transformed into an algorithm that factors N in *deterministic polynomial time*.

Let $N = pq$ be an RSA-modulus. Let $e, d \in \mathbb{Z}_{\phi(N)}$ be the public/secret exponents, satisfying the equation $ed = 1 \bmod \phi(N)$. If we are given the public information (N, e) and the factorization of N, then d can be computed in polynomial time using the Euclidean algorithm. Rivest, Shamir, and Adleman showed that the converse is also true: Given (N, e, d), one can factor N in *probabilistic* polynomial time by an algorithm due to Miller [52].

In 2004, it was shown in [21, 22] that there is also a *deterministic* reduction of factoring to computing d using Coppersmith's method. This establishes the *deterministic polynomial time equivalence* of both problems.

It is not hard to see that the knowledge of $\phi(N) = N - (p + q - 1)$ yields the factorization of N in polynomial-time. Our goal is to compute $\phi(N)$. Since p, q are of the same bit-size, the term N is an approximation of $\phi(N)$ up to roughly $N^{\frac{1}{2}}$. Therefore, the polynomial

$$f(x) = N - x \mod \phi(N)$$

has a root $x_0 = p + q - 1$ of size $N^{\frac{1}{2}}$. Let $M = ed - 1 = N^{\alpha}$ for some $\alpha \leq 2$. We know that M is a multiple of $\phi(N)$.

Now, we can apply the LLL algorithm via Theorem 1 with the parameter setting $\delta, c = 1$, $b = \phi(N)$, $M = N^{\alpha}$ the integer of unknown factorization and $\beta = \frac{1}{\alpha}$. We conclude that we can find all roots x_0 within the bound

$$|x_0| \leq M^{\frac{\beta^2}{\delta}} = (N^{\alpha})^{\frac{1}{\alpha^2}} = N^{\frac{1}{\alpha}}.$$

Since $\alpha \leq 2$, we can find all roots within the bound $N^{\frac{1}{2}}$, as desired.

Finding Smooth Numbers and Factoring

The following link between finding smooth integers with Coppersmith's LLL-based algorithm and factoring composite integers N was introduced by Boneh [23] in 2001.

Many classical factorization algorithms such as the Quadratic Sieve and the Number Field Sieve have to find values slightly larger than $\sqrt{N}$ such that their square modulo N is *B-smooth*. A number is called B-smooth if it splits into prime factors $p_1, p_2, \ldots, p_n$ smaller than B. We can model this by a univariate polynomial equation

$$f_c(x) = (x + \sqrt{cN})^2 - cN,$$

for small values of c. Given an interval size X, the task is to find all solutions $|x_0| \leq X$ such that $f_c(x_0)$ has a large B-smooth factor. Whenever this factor is as large as $f_c(x_0)$ itself, then $f_c(x_0)$ factors completely over the factor base $p_1, \ldots, p_n$.

Finding Integers with Large Smooth Factor

Given: $f_c(x), B, X$

Find : $|x_0| \leq X$ such that $f_c(x_0)$ has a large B-smooth factor.

Let us define $P = \prod_{i=1}^{n} p_i^{e_i}$. For simplicity reasons, we will assume here $e_i = 1$ for all exponents, although we could handle arbitrary multiplicities as well. We are interested in integers x_0 such that many p_i divide $f_c(x_0)$, i.e., $f_c(x_0) = 0 \bmod b$ for a modulus $b = \prod_{i \in I} p_i$, where $I \subseteq \{1, \ldots, n\}$ is a large index set.

Applying Theorem 1, it is easy to see that $b \geq P^{\sqrt{\frac{2 \log X}{\log P}}}$ is sufficient to find all

$$|x_0| \leq P^{\frac{\beta^2}{\delta}} = P^{\frac{2 \log X}{2 \log P}} = 2^{\log X} = X.$$

Boneh [23] illustrates his result by giving numerical examples where just one application of LLL on a 50-dimensional lattice yields all numbers in an interval of size $X = 2^{500}$ that have a sufficiently large smooth factor.

At the moment, however, the technique does not lead to improvements to classical factorization algorithms, since it is unlikely that randomly chosen intervals of the given size contain sufficiently many smooth numbers. Moreover, classical algorithms usually need fully smooth numbers, whereas with the present method one only finds numbers with a large smooth factor.

Applications of Finding Roots of Multivariate Equations

In this section, we study applications of the LLL algorithm for solving multivariate polynomial equations. We start by presenting the two most famous RSA applications for solving bivariate modular polynomial equations: The attacks of Wiener [24] and Boneh-Durfee [25] on RSA with small secret exponent d.

RSA Key Recovery Problem

Given: N, e
Find : d with $ed = 1 \bmod \phi(N)$

Relaxed RSA Key Recovery Problem: Small key

Given: N, e with $ed = 1 \bmod \phi(N)$ for some $d \leq N^{\delta}$
Find : d

Let us briefly describe Wiener's polynomial time attack on RSA for secret keys $d \leq N^{\frac{1}{4}}$. Although this attack was originally presented using continued fractions, we will describe it within the framework of small solutions to linear bivariate equations.

We can write the RSA key equation $ed = 1 \bmod \phi(N)$ in the form

$$ed + k(p + q - 1) - 1 = kN, \tag{10.5}$$

for some $k \in \mathbb{N}$. This leads to a linear bivariate polynomial $f(x, y) = ex + y$ that has the root $(x_0, y_0) = (d, k(p + q - 1) - 1)$ modulo N. It is not hard to see that $k < d$. In the case of balanced prime factors, we have $p + q \approx \sqrt{N}$. For $d \leq N^{\frac{1}{4}}$, the product $x_0 y_0$ of the desired roots can therefore be upper-bounded by N.

It is well-known that linear modular polynomial equations can be heuristically solved by lattice reduction whenever the product of the unknowns is smaller than the modulus. For the bivariate case, this lattice technique can be made rigorous. In our case, one has to find a shortest vector in the lattice L spanned by the row vectors of the following lattice basis

$$B = \begin{pmatrix} NX & 0 \\ eX & Y \end{pmatrix}, \qquad \text{where } X = N^{\frac{1}{4}} \text{ and } Y = N^{\frac{3}{4}}.$$

Using an argumentation similar to the one in Section "How to Find Small Roots: The Univariate Case", one can see that a shortest vector $v = (c_0, c_1) \cdot B$ yields a polynomial $c_0 Nx + c_1 f(x, y)$ that evaluates to zero over the integers at the point $(x_0, y_0) = (d, k(p + q - 1) - 1)$. Since $f(x_0, y_0) = kN$, we have

$$c_0 Nd = -c_1 Nk.$$

Because v is a shortest vector, the coefficients c_0 and c_1 must be co-prime. Therefore, we conclude that $|c_0| = k$ and $|c_1| = d$. From this information, we can derive via (10.5) the term $p + q$ which in turn yields the factorization of N in polynomial time.

Instead of using a two-dimensional lattice, one could compute the tuple (k, d) by looking at all convergents of the continued fraction expansion of e and N. This approach was taken in Wiener's original work.

In 1999, Boneh and Durfee improved Wiener's bound to $d \leq N^{1-\sqrt{\frac{1}{2}}} \approx N^{0.292}$. This result was achieved by writing the RSA equation as

$$k(N + 1 - (p + q)) + 1 = ed.$$

This in turn yields a bivariate polynomial $f(x, y) = x(N + 1 - y) + 1$ with the root $(x_0, y_0) = (k, p + q)$ modulo e. Notice that f has the monomials x, xy, and 1. As in Wiener's attack, the product $x_0 \cdot x_0 y_0$ can be bounded by N whenever $d \leq N^{\frac{1}{4}}$. Thus, for e of size roughly N, we obtain the same bound as in the Wiener attack if we linearize the polynomial. However, Boneh and Durfee used the polynomial structure of $f(x, y)$ in order to improve the bound to $N^{0.292}$ by a Coppersmith-type approach.

Wiener as well as Boneh and Durfee posed the question whether there is also a polynomial time attack for RSA with small secret CRT-exponent d. We call d a small CRT-exponent if the values $d_p = d \bmod p - 1$ and $d_q = d \bmod q - 1$ are small. This enables a receiver to efficiently decrypt modulo p and q and combine the results using the Chinese remainder theorem (CRT) [53].

RSA Key Recovery Problem

Given: N, e
Find : d with $ed = 1 \bmod \phi(N)$

Relaxed RSA Key Recovery Problem: Small CRT-key

Given: N, e with $ed_p = 1 \bmod p - 1$ and $ed_q = 1 \bmod q - 1$ for $d_p, d_q \leq N^{\delta}$
Find : d with $d = d_p \bmod p - 1$ and $d = d_q \bmod q - 1$

Recently, Jochemsz and May [27] presented a polynomial time attack for RSA with $d_p, d_q \leq N^{0.073}$, building on an attack of Bleichenbacher and May [54]. The basic idea is to write the RSA key equation in the form

$$\left| \begin{array}{c} ed_p + k_p - 1 = k_p p \\ ed_q + k_q - 1 = k_q q \end{array} \right|,$$

with the unknowns d_p, d_q, k_p, k_q, p, and q. We eliminate the unknowns p, q by multiplying both equations. Rearranging terms yields

$$e^2 d_p d_q + e(d_p(k_q - 1) + d_q(k_p - 1)) + k_p k_q (1 - N) + (k_p + k_q + 1) = 0.$$

In [54], the authors linearize this equation and derive attacks for variants of the RSA cryptosystem where e is significantly smaller than N. In [27], the full polynomial structure is exploited using a Coppersmith technique in order to extend the linearization attack to full size e.

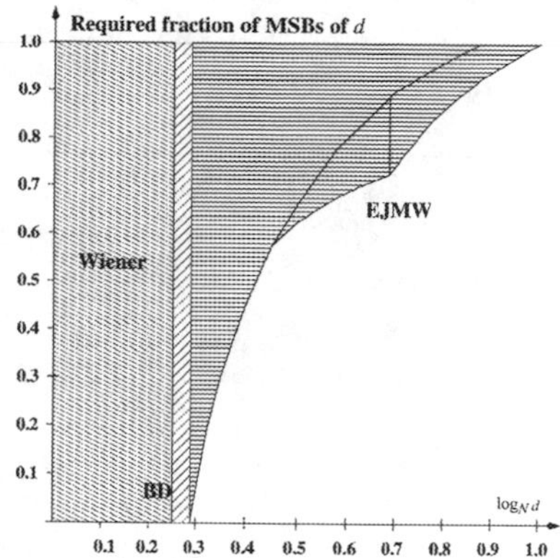

Fig. 10.1 Partial key exposure attack

By assigning the variables x_1, x_2, x_3, x_4 to the unknowns d_p, d_q, k_p, k_q, respectively, one obtains a 4-variate polynomial equation which evaluates to zero over the integers. A Coppersmith-type analysis results in a heuristic polynomial time attack that works for $d_p, d_q \leq N^{0.073}$.

Several results in the literature also address the inhomogenous case of small RSA secret key relaxations, where d is not small but parts of d's bits are known to an attacker. Boneh, Durfee, and Frankel introduced several of these so-called Partial Key Exposure attacks, which were later extended in Blömer, May [55] and EJMW [56]. In the latter work, the authors showed that the Boneh-Durfee attack naturally extends to the inhomogenous case for all d smaller than $\phi(N)$. The larger d is, the more bits of d an attacker has to know (see Fig. 10.1).

Again, the former cryptanalytic results have a dual interpretation as security results. They establish the security of certain parts of the bits of the RSA secret key. More precisely, the results state that recovering these bits is as hard as factoring the RSA modulus given only the public information (N, e). This opens the possibility to publish the remaining bits of the secret key, which can be used, e.g., in server-aided RSA systems, where parts of an RSA signature computation are outsourced to an untrusted server. This dual application was first proposed by Boneh, Durfee, and Frankel [20]. Later, Steinfeld and Zheng [29] proposed another server-based RSA system, which provides provable security against Partial Key Exposure attacks.

Survey and References for LLL-Based RSA and Factoring Results

The following table gives an overview and references of various applications of Coppersmith's LLL-based methods for finding small roots when applied to relaxed RSA or factorization problems. Although not all of these results are originally described in terms of small root problems, they all more or less fit in this framework and might serve as useful pointers for further reading.

Method	
Method/variants	Håstad 88 [12], Girault,Toffin,Vallée 88 [13] Coppersmith 96,97,01 [11, 14–16], Howgrave-Graham 98,01 [30, 57], Jutla 98 [58], May 03 [35], Bernstein 04 [59], Coron 05,07 [36, 37], Bauer, Joux 07 [60]
Optimize bounds	Blömer,May [61] , Jochemsz, May [27]
RSA	
Inverting RSA	Håstad 89 [12], Coppersmith 96 [14, 15], May, Ritzenhofen 08 [49]
Small d	Wiener 90 [24], Boneh, Durfee 98 [25, 26], Durfee, Nguyen 00 [62], Blömer, May 01 [63], de Weger 02 [64], Hinek 02 [65], May 01, 04 [63, 81]
Known bits of d	Boneh, Durfee, Frankel 96 [28, 82], Blömer, May 03 [55], Ernst, Jochemsz, May, de Weger 05 [56]
Key recovery	May 04 [21], Coron, May 07 [22], Kunihiro, Kurosawa 07 [67]
Small CRT-d	May 02 [68], Hinek, Sun, Wu 05[69], Galbraith, Heneghan, McKee 05 [71, 72], Bleichenbacher, May 06 [54], Jochemsz, May 06 [27, 73]
Proving Security	Shoup 01 [18], Boneh 01 [74], Steinfeld, Zheng 04 [29]
PRNG, MAC	Boneh, Halevi, Howgrave-Graham 99 [46], Steinfeld, Pieprzyk, Wang 06 [19]
Factoring	
High Bits known	Rivest, Shamir 86 [50], Coppersmith 95,96 [14, 51], Boneh, Durfee, Howgrave-Graham 99 [20], Crépeau, Slakmon 03 [75], Santoso, Kunihiro, Kanayama, Ohta 06 [76], Herrmann, May 08 [77]
Finding relations	Schnorr 01 [78], Boneh 00 [23]

Open Problems and Speculations

Optimizing Bounds: On Newton Polytopes and Error Terms

In this section, we will explain how to optimize the upper bounds up to which small roots can be found. Here, a polynomial's Newton polytope will play a central role.

We will also see that the upper bounds usually incorporate some error term, which in many cases can be eliminated by splitting the search interval into smaller pieces and treating each subinterval separately (see, e.g., the proof of Theorem 1).

In all applications of Coppersmith's method, one starts with either a polynomial modular equation $f(x_1,\ldots,x_m) = 0 \bmod b$ or a polynomial integer equation $f(x_1,\ldots,x_m) = 0$. Using this equation, one defines algebraic multiples $f_1,\ldots,f_n$ of f which contain the same small roots. For instance, if f is a univariate polynomial equation in x as in Section "How to Find Small Roots: The Univariate Case", this is done by multiplying f with powers of x and by taking powers of f itself. In the univariate case, it is clear which set of algebraic multiples maximizes the size of the roots x_0 that we can efficiently recover. Indeed, we will argue, in Section "What are the Limitations of the Method?", that the bound $|x_0| \le N^{\frac{\beta^2}{\delta}}$ from Section "How to Find Small Roots: The Univariate Case" cannot be improved in general, since beyond this bound f may have too many roots to output them in polynomial time.

For univariate modular polynomial equations $f(x)$, one looks for an integer linear combination $g(x) = \sum_i a_i f_i(x)$, $a_i \in \mathbb{Z}$, such that $g(x_0) = 0$ over the integers for all sufficiently small roots. These roots can then be found by standard root finding methods.

For irreducible bivariate polynomials $f(x, y)$, one similarly defines algebraic multiples $f_1(x, y),\ldots, f_n(x, y)$. The goal is to find a polynomial $g(x, y) = \sum_i a_i f_i(x, y)$ by LLL-reduction such that $g_i(x, y)$ is not a multiple of $f(x, y)$. Then the roots can be found by resultant computations.

Whereas the choice of the algebraic multiples is quite straightforward for univariate polynomials, for multivariate polynomials, the choice of the algebraic multiples appears to be a complex optimization problem. The bounds for the roots that one computes mainly depend on the largest coefficient of the polynomial and the polynomial's Newton polytope – i.e., the convex hull of the monomials' exponents when regarded as points in the Euclidean space.

Let us give an example for this. As explained in Section "Modeling RSA Problems as Univariate Root Finding Problems", we can factor $N = pq$ with known high bits of p by using the univariate polynomial equation $f(x) = \tilde{p} + x \bmod p$, where $\tilde{p}$ is an approximation of p up to $N^{1/4}$. The same result can be achieved by computing $\tilde{q} = \frac{N}{\tilde{p}}$ and solving the bivariate integer polynomial $f(x, y) = (\tilde{p} + x)(\tilde{q} + y) - N$. The largest coefficient in this equation is $\tilde{p}\tilde{q} - N$, which is roughly of the size $W = N^{3/4}$. The monomials of $f(x, y)$ are $1, x, y$, and xy, i.e., the Newton polytope is a square defined by the points $(0,0)$, $(0,1)$, $(1,0)$, and $(1,1)$. Optimizing the upper bounds X, Y for the size of the roots in x, y, respectively, yields the condition $XY \le W^{2/3}$. This is equivalent to $XY \le N^{\frac{1}{2}}$ or $X, Y \le N^{1/4}$. Thus, we achieve the same result as in the univariate modular case.

We could however also look at the bivariate polynomial $f(x, y) = (\tilde{p} + x) y - N$. The largest coefficient is $W = N$ and the Newton polytope defined by $(0,0), (0,1), (1,1)$ is a triangle. Optimizing the bounds for this shape of the Newton polytope yields the condition $(XY)^4 \le W^3$. Setting $Y = N^{1/2}$ and $W = N$ yields $X^4 \le N$ which leads again to $X \le N^{1/4}$.

Interestingly, we do not need the approximation of q for achieving the same result. Since we do not need the bits of q, one should ask whether he or she indeed needs to know the bits of p. Let us look at the polynomial equation $f(x, y) = xy - N$, where $W = N$ and the Newton polytope is a line formed by $(0, 0)$ and $(1, 1)$. Applying a Coppersmith-type analysis to this polynomial yields the bound $XY \leq W^{1-\epsilon} = N^{1-\epsilon}$, for some error term ϵ. Notice that a bound of $XY \leq 2N$ would easily allow to factor $N = pq$ if p, q have equal bit-size, since $p \leq \sqrt{N} = X$ and $q \leq 2\sqrt{N} = Y$.

What does the bound $XY \leq N^{1-\epsilon}$ imply? Can we remove the error term ϵ and derive the desired bound by running the algorithm on $2N^{\epsilon}$ copies, where we search in each copy for the roots in an interval of size $N^{1-\epsilon}$? That is, can we factor in time $\tilde{\mathcal{O}}(N^{\epsilon})$? And provided that the error term ϵ satisfies $\epsilon = \mathcal{O}(\frac{1}{\log N})$, can we factor in polynomial time?

(Un)fortunately, the answer is NO, at least with this approach. The reason is that as opposed to other polynomials, we cannot simply guess a few bits of the desired small root $(x_0, y_0) = (p, q)$, since this would either change the structure of the Newton polytope or the size of W. If we guess bits of x_0, we introduce the monomial y, and symmetrically for y_0, we introduce the x-monomial. But as shown above, this changes our bound to an inferior $XY \leq N^{\frac{3}{4}}$. On the other hand, if we guess bits of $x_0 y_0$, our largest coefficient decreases accordingly.

Notice that, e.g., for the polynomial $(\tilde{p} + x)y - N$ guessing bits of x_0 is doable since the guessing does not introduce new monomials. Thus, in this case a small error term in the bound can be easily eliminated by a brute-force search technique.

Applying the Method to Multivariate Polynomial Equations

Another challenging problem is to obtain provability of the algorithm in the multivariate setting. This problem is not only of theoretical interest. There have been cases reported, where the heuristic method – which computes the roots by resultant computations – for multivariate polynomials systematically fails [63].

Let us see why the method provably works for the bivariate integer case and what causes problems when extending it to a third variable. Coppersmith's original method for bivariate integer polynomials constructs on input $f(x, y)$ a polynomial $g(x, y)$, such that $g(x, y)$ cannot be a polynomial multiple of $f(x, y)$. In other words, $g(x, y)$ does not lie in the ideal $\langle f \rangle$ generated by f, and, therefore, the resultant of f and g cannot be the zero polynomial.

Heuristically, one extends this approach to three variables by constructing two polynomials g_1, g_2 with LLL-reduction. The resultants $r_1 = \mathrm{Res}(f, g_1)$ and $r_2 = \mathrm{Res}(f, g_2)$ are bivariate polynomials. The resultant $\mathrm{Res}(r_1, r_2)$ is then univariate and yields one coordinate of the roots, provided that the resultant does not vanish. The other coordinates of the roots can be found by back-substitution. The resultant is non-vanishing iff g_1 and g_2 are algebraically independent.

Recently, Bauer and Joux [60] proposed a twist in the above construction which in some cases enables to guarantee algebraic independence also for polynomials in three or more variables. Basically, their approach is an iterative application of

Coppersmith's original technique for bivariate polynomials. Given a trivariate polynomial $f(x, y, z)$, one constructs a polynomial $g(x, y, z)$ such that g does not lie in $\langle f \rangle$. Afterward, one uses a Gröbner Basis approach and another iteration of the LLL procedure to construct a third polynomial $h(x, y, z)$, which does not lie in $\langle f, g \rangle$.

Unfortunately, Bauer and Joux's approach still incorporates a heuristic assumption. For trivariate polynomials of a special shape, however, the approach can be made fully rigorous.

What are the Limitations of the Method?

Coppersmith's method outputs *all* sufficiently small solutions of a polynomial equation. Since the method runs in polynomial time, it can only output a polynomial number of solutions. Thus, the method proves in a constructive way a limit for the number of roots within a certain interval. This limit matches for univariate modular polynomials the bounds by Konyagin and Steeger [79]. The number of roots of each polynomial equation thus limits the size of the interval that we are able to search through in polynomial time. Let us demonstrate this effect for univariate modular polynomial equations.

Let $N = p^r$. Assume that we want to solve the equation $f(x) = x^r \mod N$. Clearly, all $x_0 = kp, k \in \mathbb{N}$, are solutions. Hence, solving this equation for solutions $|x_0| \leq p^{1+\epsilon}$ would imply that one has to output p^ϵ solutions, an exponential number.

This argument serves as an explanation why the bound $|x_0| = N^{\frac{1}{\delta}}$ from Section "How to Find Small Roots: The Univariate Case" cannot be improved in general. On the other hand, for the following two reasons, this argument does not fundamentally rule out improvements for any of the applications' current bounds mentioned in this survey.

First, the factorization of $N = p^r$ can be easily determined. Hence, there might be an improved method which exploits this additional information. Indeed, Bleichenbacher and Nguyen [80] describe a lattice-based method for *Chinese Remaindering with errors* that goes beyond the Coppersmith-type bound in cases where the factorization of the modulus is known.

Second, in all the applications we studied so far, an improvement of the bound would not immediately imply an exponential number of solutions. Look for instance at the *factoring with high bits problem* and let us take the polynomial $f(x) = \tilde{p} + x \mod p$. The solution of this polynomial is unique up to the bound $|x_0| \leq p$. So although we have no clue how to solve the factorization problem with the help of lattice reduction techniques, there is also no limiting argument which tells us that it is impossible to extend our bounds to the general case.

As a second example, look at the Boneh-Durfee attack on RSA with $d \leq N^{0.292}$ which introduces the bivariate polynomial equations $f(x, y) = x(N + 1 - y) + 1 \mod e$. Assume that e is roughly of size N. Since y is the variable for $p+q$, its size can be roughly bounded by $\sqrt{N}$. Assume that for a fixed candidate y the mapping

$g : x \mapsto x(N + 1 - y) + 1 \bmod e$ takes on random values in $\mathbb{Z}_e$. If we map $\sqrt{N}$ candidates for x for every of the $\sqrt{N}$ choices of y, we expect to map to zero at most a constant number of times.

This counting argument let Boneh and Durfee conjecture that one can achieve a bound of $d \leq \sqrt{N}$ in polynomial time attacks on RSA with small secret d. Moreover, if one used the fact that y represents $p + q$, which implies that y_0 is already fully determined by N, then the counting argument would not rule out a bound beyond $\sqrt{N}$. If we could make use of this information about y_0, then there would be a unique candidate for x_0 in $\mathbb{Z}_{\phi(N)}$, and recovering this candidate would solve the RSA problem as well as the factorization problem. However, despite considerable research efforts, the bound $d \leq N^{0.292}$ is still the best bound known today. It remains an open problem to further push it.

Summary

The invention of the LLL algorithm in 1982 was the basis for the construction of an efficient algorithm due to Coppersmith for finding small solutions of polynomial equations in 1996. This in turn opened a whole new line of research and enabled new directions for tackling challenging problems such as the RSA problem or the factorization problem from a completely different angle. As opposed to traditional approaches such as the Elliptic Curve Method and the Number Field Sieve, the LLL-based approach is polynomial time but solves only relaxed versions of the RSA and the factorization problem.

Today, the relaxed versions are still pretty far away from the general instances. But, there appears to be a steady progress in finding new interesting applications, and the existing bounds are continuously pushed. From a research point of view, it is likely that the young field of LLL-based root finding still hides many fascinating results that await their discovery.

Acknowledgements The author thanks Mathias Herrmann, Ellen Jochemsz, Phong Nguyen, Maike Ritzenhofen, and Damien Stehlé for comments and discussions.

References

1. R. Rivest, A. Shamir and L. Adleman, A Method for Obtaining Digital Signatures and Public-Key Cryptosystems, Communications of the ACM, Vol. 21(2), pp.120–126, 1978
2. D. Boneh, Venkatesan, Breaking RSA may not be equivalent to factoring, Advances in Cryptology – Eurocrypt '98, Lecture Notes in Computer Science Vol. 1233, Springer, pp. 59–71, 1998
3. D. Brown, Breaking RSA May Be As Difficult As Factoring, Cryptology ePrint Archive Report 2005/380, 2005
4. G. Leander, A. Rupp, On the Equivalence of RSA and Factoring Regarding Generic Ring Algorithms, Advances in Cryptology – Asiacrypt 2006, Lecture Notes in Computer Science Vol. 4284, pp. 241–251, Springer, 2006
5. D. Boneh, Twenty years of attacks on the RSA cryptosystem, Notices of the AMS, 1999

6. S. Katzenbeisser, Recent Advances in RSA Cryptography, Advances in Information Security, Kluwer Academic Publishers, 2001
7. C. Pomerance, The Quadratic Sieve Factoring Algorithm, Advances in Cryptology – Eurocrypt 84, Lecture Notes in Computer Science, pp. 169–182, 1985
8. H. W. Lenstra, Factoring Integers with Elliptic Curves, Mathematische Annalen, Vol. 126, pp. 649–673, 1987
9. A.K. Lenstra, H.W. Lenstra, The Development of the Number Field Sieve, Springer, 1993
10. P.W. Shor, Algorithms for quantum computation: Discrete log and factoring, In Proceedings of the 35th Annual Symposium on Foundations of Computer Science, pages 124–134, 1994
11. D. Coppersmith, Finding a Small Root of a Bivariate Integer Equation; Factoring with High Bits Known, Advances in Cryptology – Eurocrypt '96, Lecture Notes in Computer Science Vol. 1070, Springer, pp. 178–189, 1996
12. J. Håstad, Solving Simultaneous Modular Equations of Low Degree, Siam J. Computing, 1988
13. M. Girault, P. Toffin, B. Vallée, Computation of Approximate L-th Roots Modulo n and Application to Cryptography, Advances in Cryptology – Crypto 1988, Lecture Notes in Computer Science Vol. 403, pp. 100-117, 1988
14. D. Coppersmith, Finding a Small Root of a Univariate Modular Equation, Advances in Cryptology – Eurocrypt '96, Lecture Notes in Computer Science Vol. 1070, Springer, pp. 155–165, 1996
15. D. Coppersmith, Small solutions to polynomial equations and low exponent vulnerabilities, Journal of Cryptology, Vol. 10(4), pp. 223–260, 1997.
16. D. Coppersmith, Finding Small Solutions to Small Degree Polynomials, Cryptography and Lattice Conference (CaLC 2001), Lecture Notes in Computer Science Volume 2146, Springer, pp. 20–31, 2001.
17. A. K. Lenstra, H. W. Lenstra, and L. Lovász, "Factoring polynomials with rational coefficients," Mathematische Annalen, Vol. 261, pp. 513–534, 1982
18. V. Shoup, OAEP Reconsidered, Advances in Cryptology – Crypto 2001, Lecture Notes in Computer Science Vol. 2139, Springer, pp. 239–259, 2001
19. R. Steinfeld, J. Pieprzyk, H. Wang, On the Provable Security of an Efficient RSA-Based Pseudorandom Generator, Advances in Cryptology – Asiacrypt 2006, Lecture Notes in Computer Science Vol. 4284, pp. 194–209, Springer, 2006
20. D. Boneh, G. Durfee, and N. Howgrave-Graham, Factoring $N = p^r q$ for large r, Advances in Cryptology – Crypto '99, Lecture Notes in Computer Science Vol. 1666, Springer, pp. 326–337, 1999
21. A. May, Computing the RSA Secret Key is Deterministic Polynomial Time Equivalent to Factoring, Advances in Cryptology (Crypto 2004), Lecture Notes in Computer Science Volume 3152, pages 213–219, Springer, 2004
22. J.-S. Coron, A. May, Deterministic Polynomial Time Equivalence of Computing the RSA Secret Key and Factoring, Journal of Cryptology, 2007
23. D. Boneh, Finding smooth integers in short intervals using CRT decoding, STOC, pp. 265–272, 2000
24. M. Wiener, Cryptanalysis of short RSA secret exponents, IEEE Transactions on Information Theory, Vol. 36, pp. 553–558, 1990
25. D. Boneh, G. Durfee, Cryptanalysis of RSA with private key d less than $N^{0.292}$, Advances in Cryptology – Eurocrypt'99, Lecture Notes in Computer Science Vol. 1592, Springer, pp. 1–11, 1999.
26. D. Boneh, G. Durfee, Cryptanalysis of RSA with private key d less than $N^{0.292}$, IEEE Trans. on Information Theory, Vol. 46(4), pp. 1339–1349, 2000
27. E. Jochemsz, A. May, A Polynomial Time Attack on Standard RSA with Private CRT-Exponents Smaller than $N^{0.073}$, Advances in Cryptology – Crypto 2007, Lecture Notes in Computer Science Vol. 4622, pp. 395–411, Springer, 2007
28. D. Boneh, G. Durfee, Y. Frankel, An attack on RSA given a small fraction of the private key bits, Advances in Cryptology – Asiacrypt '98, Lecture Notes in Computer Science Vol. 1514, Springer, pp. 25–34, 1998

29. R. Steinfeld, Y. Zheng, On the Security of RSA with Primes Sharing Least-Significant Bits, Applicable Algebra in Engineering, Communication and Computing Vol. 15(3-4), pp. 179–200, 2004
30. N. Howgrave-Graham, Finding small roots of univariate modular equations revisited, Proceedings of Cryptography and Coding, Lecture Notes in Computer Science Vol. 1355, Springer, pp. 131–142, 1997
31. M. van Hoeij, Factoring Polynomials and 0-1 Vectors, Cryptography and Lattice Conference (CaLC 2001), Lecture Notes in Computer Science Vol. 2146, Springer, pp. 45–50, 2001
32. M. van Hoeij, Factoring polynomials and the knapsack problem, J. Number Theory Vol. 95, pp. 167–189, 2002
33. J. Klüners, The van Hoeij algorithm for factoring polynomials, LLL+25 Conference in honour of the 25th birthday of the LLL algorithm, 2007
34. N. Howgrave-Graham, Approximate Integer Common Divisors, Cryptography and Lattice Conference (CaLC 2001), Lecture Notes in Computer Science Vol. 2146, Springer, pp. 51–66, 2001
35. A. May, New RSA Vulnerabilities Using Lattice Reduction Methods, PhD thesis, University of Paderborn, 2003
36. J.-S. Coron, Finding Small Roots of Bivariate Integer Polynomial Equations Revisited, Advances in Cryptology – Eurocrypt 2005, Lecture Notes in Computer Science Vol. 3027, Springer, 2005
37. J.-S. Coron, Finding Small Roots of Bivariate Integer Polynomial Equations: A Direct Approach, Advances in Cryptology – Crypto 2007, Lecture Notes in Computer Science, Springer, 2007
38. P. Nguyen, D. Stehlé, Floating-Point LLL Revisited, Advances in Cryptology – Eurocrypt 2005, Lecture Notes in Computer Science Vol. 3494, pp. 215–233, Springer, 2005
39. D. Stehlé, Floating-Point LLL: Theoretical and Practical Aspects, LLL+25 Conference in honour of the 25th birthday of the LLL algorithm, 2007
40. J. Håstad, M. Näslund, The security of all RSA and discrete log bits, Journal of the ACM Vol. 51(2), pp. 187–230, 2004
41. M. Bellare, P. Rogaway, Optimal Asymmetric Encryption, Advances in Cryptology – Eurocrypt '94, Lecture Notes in Computer Science Vol. 950, Springer, pp. 92–111, 1994
42. G. Gentry, The Geometry of Provable Security: Some Proofs of Security in which Lattices Make a Surprise Appearance, LLL+25 Conference in honour of the 25th birthday of the LLL algorithm, 2007
43. E. Fujisaki, T. Okamoto, D. Pointcheval, J. Stern, RSA-OAEP Is Secure under the RSA Assumption, Advances in Cryptology – Crypto 2001, Lecture Notes in Computer Science Vol. 2139, Springer, pp. 260–274, 2001
44. R. Fischlin, C.-P. Schnorr, Stronger Security Proofs for RSA and Rabin Bits, Journal of Cryptology Vol. 13(2), pp. 221–244, 2000
45. S. Micali, C.P. Schnorr, Efficient, Perfect Random Number Generators, Advances in Cryptology – Crypto 1988, Lecture Notes in Computer Science Vol. 403, pp. 173–198, Springer, 1988
46. D. Boneh, S. Halevi, N. Howgrave-Graham, The Modular Inversion Hidden Number Problem, Advances in Cryptology – Asiacrypt 2001, Lecture Notes in Computer Science Vol. 2248, Springer, pp. 36–51, 2001
47. M.K. Franklin, M. K. Reiter, A linear protocol failure for RSA with exponent three, Rump Session Crypto '95, 1995
48. D. Coppersmith, M. K. Franklin, J. Patarin, and M. K. Reiter, Low-Exponent RSA with Related Messages, Advances in Cryptology – Eurocrypt '96, Lecture Notes in Computer Science Vol. 1070, Springer, pp. 1–9, 1996
49. A. May, M. Ritzenhofen, Solving Systems of Modular Equations in One Variable: How Many RSA-Encrypted Messages Does Eve Need to Know?, Practice and Theory in Public Key Cryptography – PKC 2008, Lecture Notes in Computer Science Vol. 4939, Springer, pp. 37–46, 2008

50. R. Rivest, A. Shamir, Efficient factoring based on partial information, Advances in Cryptology (Eurocrypt '85), Lecture Notes in Computer Science Volume 219, pp. 81–95, Springer, 1985
51. D. Coppersmith, Factoring with a hint, IBM Research Report RC 19905, 1995
52. G.L. Miller, Riemann's hypothesis and test for primality, STOC, pp. 234–239, 1975
53. J.-J. Quisquater, C. Couvreur, Fast decipherment algorithm for RSA public-key cryptosystem, Electronic Letters 18 (21), pp. 905–907, 1982
54. D. Bleichenbacher, A. May, New Attacks on RSA with Small Secret CRT-Exponents, Practice and Theory in Public Key Cryptography – PKC 2006, Lecture Notes in Computer Science Vol. 3958, pp. 1–13, Springer, 2006
55. J. Blömer, A. May, New Partial Key Exposure Attacks on RSA, Advances in Cryptology – Crypto 2003, Lecture Notes in Computer Science Vol. 2729, pp. 27–43, Springer, 2003
56. M. Ernst, E. Jochemsz, A. May, B. de Weger, Partial Key Exposure Attacks on RSA up to Full Size Exponents, Advances in Cryptology – Eurocrypt 2005, Lecture Notes in Computer Science Vol. 3494, pp. 371–386, Springer, 2005
57. N. Howgrave-Graham, Computational Mathematics Inspired by RSA, PhD thesis, University of Bath, 1998
58. C. Jutla, On finding small solutions of modular multivariate polynomial equations, Advances in Cryptology – Eurocrypt '98, Lecture Notes in Computer Science Vol. 1403, Springer, pp. 158–170, 1998
59. D. Bernstein, Reducing lattice bases to find small-height values of univariate polynomials. Algorithmic number theory, 2004
60. A. Bauer, A. Joux, Toward a rigorous variation of Coppersmith's algorithm on three variables, Advances in Cryptology – Eurocrypt 2007, Lecture Notes in Computer Science, Springer, 2007
61. J. Blömer, A. May, A Tool Kit for Finding Small Roots of Bivariate Polynomials over the Integers, Advances in Cryptology – Eurocrypt 2005, Lecture Notes in Computer Science Vol. 3494, pp. 251–267, Springer, 2005
62. G. Durfee, P. Nguyen, Cryptanalysis of the RSA Schemes with Short Secret Exponent from Asiacrypt '99, Advances in Cryptology – Asiacrypt 2000, Lecture Notes in Computer Science Vol. 1976, Springer, pp. 14–29, 2000
63. J. Blömer, A. May, Low Secret Exponent RSA Revisited, Cryptography and Lattice Conference (CaLC 2001), Lecture Notes in Computer Science Volume 2146, Springer, pp. 4–19, 2001.
64. B. de Weger, Cryptanalysis of RSA with small prime difference, Applicable Algebra in Engineering, Communication and Computing ,Vol. 13(1), Springer, pp. 17–28, 2002
65. M.J. Hinek, Another Look at Small RSA Exponents, Topics in Cryptology – CT-RSA 2006, Lecture Notes in Computer Science Vol. 3860, pp. 82–98, 2006
66. A. May, Secret Exponent Attacks on RSA-type Schemes with Moduli $N = p^r q$, Practice and Theory in Public Key Cryptography – PKC 2004, Lecture Notes in Computer Science Vol. 2947, Springer, pp. 218–230, 2004
67. N. Kunihiro, K. Kurosawa, Deterministic Polynomial Time Equivalence between Factoring and Key-Recovery Attack on Takagi's RSA, Practice and Theory in Public Key Cryptography – PKC 2007, Lecture Notes in Computer Science, Springer, 2007
68. A. May, Cryptanalysis of Unbalanced RSA with Small CRT-Exponent, Advances in Cryptology – Crypto 2002, Lecture Notes in Computer Science Vol. 2442, Springer, pp. 242–256, 2002
69. H.-M. Sun, M.J. Hinek, and M.-E. Wu, An Approach Towards Rebalanced RSA-CRT with Short Public Exponent, revised version of [70], online available at http://www.cacr.math.uwaterloo.ca/techreports/2005/cacr2005-35.pdf
70. H.-M. Sun, M.-E. Wu, An Approach Towards Rebalanced RSA-CRT with Short Public Exponent, Cryptology ePrint Archive: Report 2005/053, online available at http://eprint.iacr.org/2005/053
71. S.D. Galbraith, C. Heneghan, and J.F. McKee, Tunable Balancing of RSA, Proceedings of ACISP 2005, Lecture Notes in Computer Science Vol. 3574, pp. 280–292, 2005
72. S.D. Galbraith, C. Heneghan, and J.F. McKee, Tunable Balancing of RSA, full version of [71], online available at http://www.isg.rhul.ac.uk/ sdg/full-tunable-rsa.pdf

73. E. Jochemsz, A. May, A Strategy for Finding Roots of Multivariate Polynomials with New Applications in Attacking RSA Variants, Advances in Cryptology – Asiacrypt 2006, Lecture Notes in Computer Science Vol. 4284, pp. 267–282, Springer, 2006
74. D. Boneh, Simplified OAEP for the RSA and Rabin Functions, Advances in Cryptology – Crypto 2001, Lecture Notes in Computer Science Vol. 2139, pp. 275–291, Springer, 2001
75. C. Crépeau, A. Slakmon, Simple Backdoors for RSA Key Generation, Topics in Cryptology – CT-RSA 2003, Lecture Notes in Computer Science Vol. 2612, pp. 403–416, Springer, 2003
76. B. Santoso, N. Kunihiro, N. Kanayama, K. Ohta, Factorization of Square-Free Integers with High Bits Known, Progress in Cryptology, VIETCRYPT 2006, Lecture Notes in Computer Science Vol. 4341, pp. 115–130, 2006
77. M. Herrmann, A. May, Solving Linear Equations Modulo Divisors: On Factoring Given Any Bits, Advances in Cryptology – Asiacrypt 2008, Lecture Notes in Computer Science, Springer, 2008
78. C.P. Schnorr, Factoring Integers and Computing Discrete Logarithms via Diophantine Approximations, Advances in Cryptology – Eurocrypt 1991, Lecture Notes in Computer Science, pp. 281–293, Springer, 1991
79. S.V. Konyagin, T. Steger, On polynomial congruences, Mathematical Notes Vol. 55(6), pp. 596–600, 1994
80. D. Bleichenbacher, P.G. Nguyen, Noisy Polynomial Interpolation and Noisy Chinese Remaindering, Advances in Cryptology – Eurocrypt 2000, Lecture Notes in Computer Science, pp. 53–69, Springer, 2000
81. J. Blömer, A. May, A Generalized Wiener Attack on RSA, Practice and Theory in Public Key Cryptography – PKC 2004, Lecture Notes in Computer Science Vol. 2947, pp. 1–13, Springer, 2004
82. D. Boneh, G. Durfee, Y. Frankel, Exposing an RSA Private Key Given a Small Fraction of its Bits, Full version of the work from Asiacrypt'98, available at http://crypto.stanford.edu/ dabo/abstracts/bits_of_d.html, 1998
83. A. K. Lenstra, H. W. Lenstra, Jr., M. S. Manasse, and J. M. Pollard, The number field sieve, In Proceedings of the 22nd Annual ACM Symposium on the Theory of Computation, pages 564–572, 1990

Chapter 11
Practical Lattice-Based Cryptography: NTRUEncrypt and NTRUSign

Jeff Hoffstein, Nick Howgrave-Graham, Jill Pipher, and William Whyte

Abstract We provide a brief history and overview of lattice based cryptography and cryptanalysis: shortest vector problems, closest vector problems, subset sum problem and knapsack systems, GGH, Ajtai-Dwork and NTRU. A detailed discussion of the algorithms NTRUEncrypt and NTRUSign follows. These algorithms have attractive operating speed and keysize and are based on hard problems that are seemingly intractable. We discuss the state of current knowledge about the security of both algorithms and identify areas for further research.

Introduction and Overview

In this introduction, we will try to give a brief survey of the uses of lattices in cryptography. Although it is rather a dry way to begin a survey, we should start with some basic definitions related to the subject of lattices. Those with some familiarity with lattices can skip the following section.

Some Lattice Background Material

A lattice L is a discrete additive subgroup of $\mathbb{R}^m$. By discrete, we mean that there exists an $\epsilon > 0$ such that for any $\mathbf{v} \in L$, and all $\mathbf{w} \in \mathbb{R}^m$, if $\|\mathbf{v} - \mathbf{w}\| < \epsilon$, then $\mathbf{w}$ does not belong to the lattice L. This abstract sounding definition transforms into a relatively straightforward reality, and lattices can be described in the following way:

J. Hoffstein (✉)
NTRU Cryptosystems, 35 Nagog Park, Acton, MA 01720, USA,
e-mail: jhoffstein@ntru.com

P.Q. Nguyen and B. Vallée (eds.), *The LLL Algorithm*, Information Security and Cryptography, DOI 10.1007/978-3-642-02295-1_11,

Definition of a lattice

- Let $\mathbf{v}_1, \mathbf{v}_2, \ldots, \mathbf{v}_k$ be a set of vectors in $\mathbb{R}^m$. The set of all linear combinations $a_1\mathbf{v}_1 + a_2\mathbf{v}_2 + \cdots + a_k\mathbf{v}_k$, such that each $a_i \in \mathbb{Z}$, is a lattice. We refer to this as the lattice *generated* by $\mathbf{v}_1, \mathbf{v}_2, \ldots, \mathbf{v}_k$.

Bases and the dimension of a lattice

- If $L = \{a_1\mathbf{v}_1 + a_2\mathbf{v}_2 + \ldots + a_n\mathbf{v}_n | a_i \in \mathbb{Z}, i = 1, \ldots n\}$ and $\mathbf{v}_1, \mathbf{v}_2, \ldots, \mathbf{v}_n$ are n independent vectors, then we say that $\mathbf{v}_1, \mathbf{v}_2, \ldots, \mathbf{v}_n$ is a basis for L and that L has dimension n. For any other basis $\mathbf{w}_1, \mathbf{w}_2, \ldots, \mathbf{w}_k$, we must have $k = n$.

Two different bases for a lattice L are related to each other in almost the same way that two different bases for a vector space V are related to each other. That is, if $\mathbf{v}_1, \mathbf{v}_2, \ldots, \mathbf{v}_n$ is a basis for a lattice L then $\mathbf{w}_1, \mathbf{w}_2, \ldots, \mathbf{w}_n$ is another basis for L if and only if there exist $a_{i,j} \in \mathbb{Z}$ such that

$$\begin{aligned} a_{1,1}\mathbf{v}_1 + a_{1,2}\mathbf{v}_2 + \cdots + \alpha_{1,n}\mathbf{v}_n &= \mathbf{w}_1 \\ a_{2,1}\mathbf{v}_1 + a_{2,2}\mathbf{v}_2 + \cdots + a_{2,n}\mathbf{v}_n &= \mathbf{w}_2 \\ &\vdots \\ a_{n,1}\mathbf{v}_1 + a_{n,2}\mathbf{v}_2 + \cdots + a_{n,n}\mathbf{v}_n &= \mathbf{w}_n \end{aligned}$$

and the determinant of the matrix

$$\begin{pmatrix} a_{1,1} & a_{1,2} & \cdots & a_{1,n} \\ a_{2,1} & a_{2,2} & \cdots & a_{2,n} \\ & & \vdots & \\ a_{n,1} & a_{n,2} & \cdots & a_{n,n} \end{pmatrix}$$

is equal to 1 or -1. The only difference is that the coefficients of the matrix must be integers. The condition that the determinant is nonzero in the vector space case means that the matrix is invertible. This translates in the lattice case to the requirement that the determinant be 1 or -1, the only invertible integers.

A lattice is just like a vector space, except that it is generated by all linear combinations of its basis vectors with integer coefficients, rather than real coefficients. An important object associated to a lattice is the fundamental domain or fundamental parallelepiped. A precise definition is given by:

Let L be a lattice of dimension n with basis $\mathbf{v}_1, \mathbf{v}_2, \ldots, \mathbf{v}_n$. A *fundamental domain* for L corresponding to this basis is

$$\mathcal{F}(\mathbf{v}_1, \ldots, \mathbf{v}_n) = \{t_1\mathbf{v}_1 + t_2\mathbf{v}_2 + \cdots + t_n\mathbf{v}_n \ : \ 0 \le t_i < 1\}.$$

The volume of the fundamental domain is an important invariant associated to a lattice. If L is a lattice of dimension n with basis $\mathbf{v}_1, \mathbf{v}_2, \ldots, \mathbf{v}_n$, the volume of the fundamental domain associated to this basis is called the *determinant* of L and is denoted $\det(L)$.

It is natural to ask if the volume of the fundamental domain for a lattice L depends on the choice of basis. In fact, as was mentioned previously, two different bases for L must be related by an integer matrix W of determinant ± 1. As a result, the integrals measuring the volume of a fundamental domain will be related by a Jacobian of absolute value 1 and will be equal. Thus, the determinant of a lattice is independent of the choice of basis.

Suppose, we are given a lattice L of dimension n. Then, we may formulate the following questions.

1. *Shortest vector problem (SVP)*: Find the shortest non-zero vector in L, i.e., find $0 \neq \mathbf{v} \in L$ such that $\|\mathbf{v}\|$ is minimized.
2. *Closest vector problem (CVP)*: Given a vector $\mathbf{w}$ which is not in L, find the vector $\mathbf{v} \in L$ closest to $\mathbf{w}$, i.e., find $\mathbf{v} \in L$ such that $\|\mathbf{v} - \mathbf{w}\|$ is minimized.

Both of these problems appear to be profound and very difficult as the dimension n becomes large. Solutions, or even partial solutions to these problems also turn out to have surprisingly many applications in a number of different fields. In full generality, the CVP is known to be NP-hard and SVP is NP-hard under a certain "randomized reduction" hypothesis.[1] Also, SVP is NP-hard when the norm or distance used is the l^∞ norm. In practice, a CVP can often be reduced to a SVP and is thought of as being "a little bit harder" than SVP. Reduction of CVP to SVP is used by in [2] to prove that SVP is hard in Ajtai's probabilistic sense. The interested reader can consult Micciancio's book [3] for a more compete treatment of the complexity of lattice problems. In practice it is very hard to achieve "full generality." In a real world scenario, a cryptosystem based on an NP-hard or NP-complete problem may use a particular subclass of that problem to achieve efficiency. It is then possible that this subclass of problems could be easier to solve than the general problem.

Secondary problems, that are also very important, arise from SVP and CVP. For example, one could look for a basis $\mathbf{v}_1, \ldots, \mathbf{v}_n$ of L consisting of all "short" vectors (e.g., minimize $\max \|\mathbf{v}_i\|$). This is known as the Short Basis Problem or SBP. Alternatively, one might search for a nonzero vector $\mathbf{v} \in L$ satisfying

$$\|\mathbf{v}\| \leq \psi(n)\|\mathbf{v}_{\text{shortest}}\|,$$

where ψ is some slowly growing function of n, the dimension of L. For example, for a fixed constant κ, one could try to find $\mathbf{v} \in L$ satisfying

$$\|\mathbf{v}\| \leq \kappa\sqrt{n}\|\mathbf{v}_{\text{shortest}}\|,$$

and similarly for CVP. These generalizations are known as approximate shortest and closest vector problems, or ASVP, ACVP.

[1] Under this hypothesis, the class of polynomial time algorithms is enlarged to include those that are not deterministic but will with high probability terminate in polynomial time. See Ajtai [1]

How big, in fact, is the shortest vector in terms of the determinant and the dimension of L? A theorem of Hermite from the nineteenth century says that for a fixed dimension n there exists a constant γ_n so that in every lattice L of dimension n, the shortest vector satisfies

$$\|\mathbf{v}_{\text{shortest}}\|^2 \le \gamma_n \det(L)^{2/n}.$$

Hermite showed that $\gamma_n \le (4/3)^{(n-1)/2}$. The smallest possible value one can take for γ_n is called *Hermite's constant*. Its exact value is known only for $1 \le n \le 8$ and for $n = 24$ [4]. For example, $\gamma_2 = \sqrt{4/3}$. We now explain why, for large n, Hermite's constant should be no larger than $\mathcal{O}(n)$.

Although exact bounds for the size of the shortest vector of a lattice are unknown for large n, one can make probabilistic arguments using the Gaussian heuristic. One variant of the Gaussian heuristic states that for a fixed lattice L and a sphere of radius r centered at 0, as r tends to infinity, the ratio of the volume of the sphere divided by $\det L$ will approach the number of points of L inside the sphere. In two dimensions, if L is simply $\mathbb{Z}^2$, the question of how precisely the area of a circle approximates the number of integer points inside the circle is a classical problem in number theory. In higher dimensions, the problem becomes far more difficult. This is because as n increases the error created by lattice points near the surface of the sphere can be quite large. This becomes particularly problematic for small values of r. Still, one can ask the question: For what value of r does the ratio

$$\frac{\text{Vol}(S)}{\det L}$$

approach 1. This gives us in some sense an expected value for r, the smallest radius at which the expected number of points of L with length less than r equals 1. Performing this computation and using Stirling's formula to approximate factorials, we find that for large n this value is approximately

$$r = \sqrt{\frac{n}{2\pi e}}\,(\det(L))^{1/n}.$$

For this reason, we make the following definition:

If L is a lattice of dimension n, we define the Gaussian expected shortest length to be

$$\sigma(L) = \sqrt{\frac{n}{2\pi e}}\,(\det(L))^{1/n}.$$

We will find this value $\sigma(L)$ to be useful in quantifying the difficulty of locating short vectors in lattices. It can be thought of as the probable length of the shortest vector of a "random" lattice of given determinant and dimension. It seems to be the case that if the actual shortest vector of a lattice L is significantly shorter than $\sigma(L)$, then LLL and related algorithms have an easier time locating the shortest vector.

A heuristic argument identical to the above can be used to analyze the CVP. Given a vector **w** which is not in L, we again expect a sphere of radius r centered about **w** to contain one point of L after the radius is such that the volume of the sphere equals $\det(L)$. In this case also, the CVP becomes easier to solve as the ratio of actual distance to the closest vector of L over "expected distance" decreases.

Knapsacks

The problems of factoring integers and finding discrete logarithms are believed to be difficult since no one has yet found a polynomial time algorithm for producing a solution. One can formulate the decision form of the factoring problem as follows: does there exist a factor of N less than p? This problem belongs to NP and another complexity class, co-NP. Because it is widely believed that NP is not the same as co-NP, it is also believed that factoring is not an NP-complete problem. Naturally, a cryptosystem whose underlying problem is known to be NP-hard would inspire greater confidence in its security. Therefore, there has been a great deal of interest in building efficient public key cryptosystems based on such problems. Of course, the fact that a certain problem is NP-hard does not mean that every instance of it is NP-hard, and this is one source of difficulty in carrying out such a program.

The first such attempt was made by Merkle and Hellman in the late 70s [5], using a particular NP-complete problem called the subset sum problem. This is stated as follows:

The subset sum problem
Suppose one is given a list of positive integers $\{M_1, M_2, \dots, M_n\}$. An unknown subset of the list is selected and summed to give an integer S. Given S, recover the subset that summed to S, or find another subset with the same property.

Here, there is another way of describing this problem. A list of positive integers $\mathbf{M} = \{M_1, M_2, \dots, M_n\}$ is public knowledge. Choose a secret binary vector $\mathbf{x} = \{x_1, x_2, \dots, x_n\}$, where each x_i can take on the value 1 or 0. If

$$S = \sum_{i=1}^{n} x_i M_i$$

then how can one recover the original vector **x** in an efficient way? (Of course, there might also be another vector $\mathbf{x}'$ which also gives S when dotted with **M**.)

The difficulty in translating the subset sum problem into a cryptosystem is that of building in a trapdoor. Merkle and Hellman's system took advantage of the fact that there are certain subset sum problems that are extremely easy to solve. Suppose that one takes a sequence of positive integers $\mathbf{r} = \{r_1, r_2, \dots, r_n\}$ with the property that $r_{i+1} \geq 2r_i$ for each $1 \leq i \leq n$. Such a sequence is called *super increasing*. Given an integer S, with $S = \mathbf{x} \cdot \mathbf{r}$ for a binary vector **x**, it is easy to recover **x** from S.

The basic idea that Merkle and Hellman proposed was this: begin with a secret super increasing sequence $\mathbf{r}$ and choose two large secret integers A, B, with $B > 2r_n$ and $(A, B) = 1$. Here, r_n is the last and largest element of $\mathbf{r}$, and the lower bound condition ensures that B must be larger than any possible sum of a subset of the r_i. Multiply the entries of $\mathbf{r}$ by A and reduce modulo B to obtain a new sequence $\mathbf{M}$, with each $M_i \equiv Ar_i \pmod{B}$. This new sequence $\mathbf{M}$ is the public key. Encryption then works as follows. The message is a secret binary vector $\mathbf{x}$ which is encrypted to $S = \mathbf{x} \cdot \mathbf{M}$. To decrypt S, multiply by $A^{-1} \pmod{B}$ to obtain $S' \equiv \mathbf{x} \cdot \mathbf{r} \pmod{B}$. If S' is chosen in the range $0 \leq S' \leq B - 1$, one obtains an exact inequality $S' = \mathbf{x} \cdot \mathbf{r}$, as any subset of the integers r_i must sum to an integer smaller than B. The sequence $\mathbf{r}$ is super increasing and $\mathbf{x}$ may be recovered.

A cryptosystem of this type is known as a *knapsack system*. The general idea is to start with a secret super increasing sequence, disguise it by some collection of modular linear operations, then reveal the transformed sequence as the public key. The original Merkle and Hellman system suggested applying a secret permutation to the entries of $A\mathbf{r} \pmod{B}$ as an additional layer of security. Later versions were proposed by a number of people, involving multiple multiplications and reductions with respect to various moduli. For an excellent survey, see the article by Odlyzko [6].

The first question one must ask about a knapsack system is concerns what minimal properties must $\mathbf{r}$, A, and B have to obtain a given level of security? Some very easy attacks are possible if r_1 is too small, so one generally takes $2^n < r_1$. But, what is the minimal value of n that we require? Because of the super increasing nature of the sequence, one has

$$r_n = \mathcal{O}(S) = \mathcal{O}(2^{2n}).$$

The space of all binary vectors $\mathbf{x}$ of dimension n has size 2^n , and thus an exhaustive search for a solution would require effort on the order of 2^n. In fact, a meet in the middle attack is possible, thus the security of a knapsack system with a list of length n is $O(2^{n/2})$.

While the message consists of n bits of information, the public key is a list of n integers, each approximately $2n$ bits long and there requires about $2n^2$ bits. Therefore, taking $n = 160$ leads to a public key size of about 51200 bits. Compare this to RSA or Diffie-Hellman, where, for security on the order of 2^{80}, the public key size is about 1000 bits.

The temptation to use a knapsack system rather than RSA or Diffie-Hellman was very great. There was a mild disadvantage in the size of the public key, but decryption required only one (or several) modular multiplications and none were required to encrypt. This was far more efficient than the modular exponentiations in RSA and Diffie-Hellman.

Unfortunately, although a meet in the middle attack is still the best known attack on the general subset sum problem, there proved to be other, far more effective, attacks on knapsacks with trapdoors. At first, some very specific attacks were announced by Shamir, Odlyzko, Lagarias, and others. Eventually, however, after the publication of the famous LLL paper [7] in 1985, it became clear that a secure knapsack-based system would require the use of an n that was too large to be practical.

A public knapsack can be associated to a certain lattice L as follows. Given a public list $\mathbf{M}$ and encrypted message S, one constructs the matrix

$$\begin{pmatrix} 1 & 0 & 0 & \cdots & 0 & m_1 \\ 0 & 1 & 0 & \cdots & 0 & m_2 \\ 0 & 0 & 1 & \cdots & 0 & m_3 \\ \vdots & \vdots & \vdots & \ddots & \vdots & \vdots \\ 0 & 0 & 0 & \cdots & 1 & m_n \\ 0 & 0 & 0 & \cdots & 0 & S \end{pmatrix}$$

with row vectors $\mathbf{v}_1 = (1,0,0,\ldots,0,m_1), \mathbf{v}_2 = (0,1,0,\ldots,0,m_2),\ldots,\mathbf{v}_n = (0,0,0,\ldots,1,m_n)$ and $\mathbf{v}_{n+1} = (0,0,0,\ldots,0,S)$. The collection of all linear combinations of the $\mathbf{v}_i$ with integer coefficients is the relevant lattice L. The determinant of L equals S. The statement that the sum of some subset of the m_i equals S translates into the statement that there exists a vector $\mathbf{t} \in L$,

$$\mathbf{t} = \sum_{i=1}^{n} x_i \mathbf{v}_i - \mathbf{v}_{n+1} = (x_1, x_2, \ldots, x_n, 0),$$

where each x_i is chosen from the set $\{0, 1\}$. Note that the last entry in $\mathbf{t}$ is 0 because the subset sum problem is solved and the sum of a subset of the m_i is canceled by the S.

The crux of the matter

As the x_i are binary, $\|\mathbf{t}\| \le \sqrt{n}$. In fact, as roughly half of the x_i will be equal to 0, it is very likely that $\|\mathbf{t}\| \approx \sqrt{n/2}$. On the other hand, the size of each $\|\mathbf{v}_i\|$ varies between roughly 2^n and 2^{2n}. The key observation is that it seems rather improbable that a linear combination of vectors that are so large should have a norm that is so small.

The larger the weights m_i were, the harder the subset sum problem was to solve by combinatorial means. Such a knapsack was referred to as a *low density* knapsack. However, for low density knapsacks, S was larger and thus the ratio of the actual smallest vector to the expected smallest vector was smaller. Because of this, the LLL lattice reduction method was more more effective on a low density knapsack than on a generic subset sum problem.

It developed that, using LLL, if n is less than around 300, a secret message $\mathbf{x}$ can be recovered from an encrypted message S in a fairly short time. This meant that in order to have even a hope of being secure, a knapsack would need to have $n > 300$, and a corresponding public key length that was greater than 180000 bits. This was sufficiently impractical that knapsacks were abandoned for some years.

Expanding the Use of LLL in Cryptanalysis

Attacks on the discrete logarithm problem and factorization were carefully analyzed and optimized by many researchers, and their effectiveness was quantified. Curiously, this did not happen with LLL, and improvements in lattice reduction methods such as BKZ that followed it. Although quite a bit of work was done on improving lattice reduction techniques, the precise effectiveness of these techniques on lattices of various characteristics remained obscure. Of particular interest was the question of how the running times of LLL and BKZ required to solve SVP or CVP varied with the dimension of the lattice, the determinant, and the ratio of the actual shortest vector's length to the expected shortest length.

In 1996–1997, several cryptosystems were introduced whose underlying hard problem was SVP or CVP in a lattice L of dimension n. These were, in alphabetical order:

- Ajtai-Dwork, ECCC report 1997 [8]
- GGH, presented at Crypto '97 [9]
- NTRU, presented at the rump session of Crypto '96 [10]

The public key sizes associated to these cryptosystems were $\mathcal{O}(n^4)$ for Ajtai-Dwork, $\mathcal{O}(n^2)$ for GGH, and $\mathcal{O}(n \log n)$ for NTRU.

The system proposed by Ajtai and Dwork was particularly interesting in that they showed that it was provably secure unless a worst case lattice problem could be solved in polynomial time. Offsetting this, however, was the large key size. Subsequently, Nguyen and Stern showed, in fact, that any efficient implementation of the Ajtai-Dwork system was insecure [11].

The GGH system can be explained very simply. The owner of the private key has the knowledge of a special small, reduced basis R for L. A person wishing to encrypt a message has access to the public key B, which is a generic basis for L. The basis B is obtained by multiplying R by several random unimodular matrices, or by putting R into Hermite normal form, as suggested by Micciancio.

We associate to B and R, corresponding matrices whose rows are the n vectors in the respective basis. A plaintext is a row vector of n integers, $\mathbf{x}$, and the encryption of $\mathbf{x}$ is obtained by computing $\mathbf{e} = \mathbf{x}B + \mathbf{r}$, where $\mathbf{r}$ is a random perturbation vector consisting of small integers. Thus, $\mathbf{x}B$ is contained in the lattice L while $\mathbf{e}$ is not. Nevertheless, if $\mathbf{r}$ is short enough, then with high probability, $\mathbf{x}B$ is the unique point in L which is closest to $\mathbf{e}$.

A person with knowledge of the private basis R can compute $\mathbf{x}B$ using Babai's technique [12], from which $\mathbf{x}$ is then obtained. More precisely, using the matrix R, one can compute $\mathbf{e}R^{-1}$ and then round each coefficient of the result to the nearest integer. If $\mathbf{r}$ is sufficiently small, and R is sufficiently short and close to being orthogonal, then the result of this rounding process will most likely recover the point $\mathbf{x}B$.

Without the knowledge of any reduced basis for L, it would appear that breaking GGH was equivalent to solving a general CVP. Goldreich, Goldwasser, and Halevi conjectured that for $n > 300$ this general CVP would be intractable. However, the effectiveness of LLL (and later variants of LLL) on lattices of high dimension had not been closely studied. In [13], Nguyen showed that some information leakage in GGH encryption allowed a reduction to an easier CVP problem, namely one where the ratio of actual distance to the closest vector to expected length of the shortest vector of L was smaller. Thus, he was able to solve GGH challenge problems in dimensions 200, 250, 300, and 350. He did not solve their final problem in dimension 400, but at that point the key size began to be too large for this system to be practical. It also was not clear at this point how to quantify the security of the $n = 400$ case.

The NTRU system was described at the rump session of Crypto '96 as a ring based public key system that could be translated into an SVP problem in a special class of lattices.[2] Specifically, the NTRU lattice L consists of all integer row vectors of the form $(\mathbf{x}, \mathbf{y})$ such that

$$\mathbf{y} \equiv \mathbf{x}H \pmod{q}.$$

Here, q is a public positive integer, on the order of 8 to 16 bits, and H is a public circulant matrix. Congruence of vectors modulo q is interpreted component-wise. Because of its circulant nature, H can be described by a single vector, explaining the shorter public keys.

An NTRU private key is a single short vector $(\mathbf{f}, \mathbf{g})$ in L. This vector is used, rather than Babai's technique, to solve a CVP for decryption. Together with its rotations, $(\mathbf{f}, \mathbf{g})$ yields half of a reduced basis. The vector $(\mathbf{f}, \mathbf{g})$ is likely to be the shortest vector in the public lattice, and thus NTRU is vulnerable to efficient lattice reduction techniques.

At Eurocrypt '97, Coppersmith and Shamir pointed out that any sufficiently short vector in L, not necessarily $(\mathbf{f}, \mathbf{g})$ or one of its rotations, could be used as a decryption key. However, they remarked that this really did not matter as:

"We believe that for recommended parameters of the NTRU cryptosystem, the LLL algorithm will be able to find the original secret key **f**..."

However, no evidence to support this belief was provided, and the very interesting question of quantifying the effectiveness of LLL and its variants against lattices of NTRU type remained.

At the rump session of Crypto '97, Lieman presented a report on some preliminary work by himself and the developers of NTRU on this question. This report, and many other experiments supported the assertion that the time required for LLL-BKZ to find the smallest vector in a lattice of dimension n was at least exponential in n. See [14] for a summary of part of this investigation.

The original algorithm of LLL corresponds to block size 2 of BKZ and provably returns a reasonably short vector of the lattice L. The curious thing is that in low dimensions this vector tends to be the actual shortest vector of L. Experiments have led us to the belief that the BKZ block size required to find the actual shortest vector

[2] NTRU was published in ANTS '98. Its appearance in print was delayed by its rejection by the Crypto '97 program committee.

in a lattice is linear in the dimension of the lattice, with an implied constant depending upon the ratio of the actual shortest vector length over the Gaussian expected shortest length. This constant is sufficiently small that in low dimensions the relevant block size is 2. It seems possible that it is the smallness of this constant that accounts for the early successes of LLL against knapsacks. The exponential nature of the problem overcomes the constant as n passes 300.

Digital Signatures Based on Lattice Problems

In general, it is very straight forward to associate a digital signature process to a lattice where the signer possess a secret highly reduced basis and the verifier has only a public basis for the same lattice. A message to be signed is sent by some public hashing process to a random point $\mathbf{m}$ in $\mathbb{Z}^n$. The signer, using the method of Babai and the private basis, solves the CVP and finds a lattice point $\mathbf{s}$ which is reasonably close to $\mathbf{m}$. This is the signature on the message $\mathbf{m}$. Anyone can verify, using the public basis, that $\mathbf{s} \in L$ and $\mathbf{s}$ is close to $\mathbf{m}$. However, presumably someone without the knowledge of the reduced basis would have a hard time finding a lattice point $\mathbf{s}'$ sufficiently close to $\mathbf{m}$ to count as a valid signature.

However, any such scheme has a fundamental problem to overcome: every valid signature corresponds to a vector difference $\mathbf{s} - \mathbf{m}$. A transcript of many such $\mathbf{s} - \mathbf{m}$ will be randomly and uniformly distributed inside a fundamental parallelepiped of the lattice. This counts as a leakage of information and as Nguyen and Regev recently showed, this vulnerability makes any such scheme subject to effective attacks based on independent component analysis [15].

In GGH, the private key is a full reduced basis for the lattice, and such a digital signature scheme is straightforward to both set up and attack. In NTRU, the private key only reveals half of a reduced basis, making the process of setting up an associated digital signature scheme considerably less straightforward.

The first attempt to base a digital signature scheme upon the same principles as "NTRU encryption" was NSS [16]. Its main advantage, (and also disadvantage) was that it relied *only* on the information immediately available from the private key, namely half of a reduced basis. The incomplete linkage of the NSS signing process to the CVP problem in a full lattice required a variety of ad hoc methods to bind signatures and messages, which were subsequently exploited to break the scheme. An account of the discovery of the fatal weaknesses in NSS can be found in Sect. 7 of the extended version of [17], available at [18].

This paper contains the second attempt to base a signature scheme on the NTRU lattice (NTRUSign) and also addresses two issues. First, it provides an algorithm for generating the full short basis of an NTRU lattice from the knowledge of the private key (half the basis) and the public key (the large basis). Second, it described a method of perturbing messages before signing to reduce the efficiency of transcript leakage (see Section "NTRUSign Signature Schemes: Perturbations"). The learning theory approach of Nguyen and Regev in [15] shows that about 90,000 signatures compromises the security of basic NTRUSign without perturbations. W. Whyte pointed out at the rump session of Crypto '06 that by applying rotations to effectively increase the number of signatures, the number of signatures required to theoretically determine a private key was only about 1000. Nguyen added this approach to his and Regev's technique and was able to, in fact, recover the private key with roughly this number of signatures.

The NTRUEncrypt and NTRUSign Algorithms

The rest of this article is devoted to a description of the NTRUEncrypt and NTRUSign algorithms, which at present seem to be the most efficient embodiments of public key algorithms whose security rests on lattice reduction.

NTRUEncrypt

NTRUEncrypt is typically described as a polynomial based cryptosystem involving convolution products. It can naturally be viewed as a lattice cryptosystem too, for a certain restricted class of lattices.

The cryptosystem has several natural parameters and, as with all practical cryptosystems, the hope is to optimize these parameters for efficiency while at the same time avoiding all known cryptanalytic attacks.

One of the more interesting cryptanalytic techniques to date concerning NTRUEncrypt exploits the property that, under certain parameter choices, the cryptosystem can fail to properly decrypt valid ciphertexts. The *functionality* of the cryptosystem is not adversely affected when these, so-called, "decryption failures" occur with only a very small probability on random messages, but an attacker can choose messages to induce failure, and assuming he knows when messages have failed to decrypt (which is a typical security model in cryptography) there are efficient ways to extract the private key from knowledge of the failed ciphertexts (i.e., the decryption failures are highly key-dependent). This was first noticed in [19, 20] and is an important consideration in choosing parameters for NTRUEncrypt.

Other security considerations for NTRUEncrypt parameters involve assessing the security of the cryptosystem against lattice reduction, meet-in-the-middle attacks based on the structure of the NTRU private key, and hybrid attacks that combine both of these techniques.

NTRUSign

The search for a "zero-knowledge" lattice-based signature scheme is a fascinating open problem in cryptography. It is worth commenting that most cryptographers would assume that anything purporting to be a signature scheme would automatically have the property of "zero-knowledge," i.e., the definition of a signature scheme implies the problems of determining the private key or creating forgeries should become not easier after having seen a polynomial number of valid signatures. However, in the theory of lattices, signature schemes with reduction arguments are just emerging and their computational effectiveness is currently being examined. For most lattice-based signature schemes, there are explicit attacks known which use the knowledge gained from a transcript of signatures.

When considering *practical signature schemes*, the "zero-knowledge" property is not essential for the scheme to be useful. For example, smart cards typically burn out before signing a million times, so if the private key in infeasible to obtain (and a forgery is impossible to create) with a transcript of less than a million signatures, then the signature scheme would be sufficient in this environment. It, therefore, seems that there is value in developing efficient, non-zero-knowledge, lattice-based signature schemes.

The early attempts [16, 21] at creating such practical signature schemes from NTRU-based concepts succumbed to attacks which required transcripts of far too small a size [22, 23]. However, the known attacks on NTRUSign, the currently recommended, signature scheme, require transcript lengths of impractical length, i.e., the signatures scheme does appear to be of practical significance at present.

NTRUSign was invented between 2001 and 2003 by the inventors of NTRUEncrypt together with N. Howgrave-Graham and W. Whyte [17]. Like NTRUEncrypt it is highly parametrizable and, in particular, has a parameter involving the number of perturbations. The most interesting cryptanalytic progress on NTRUSign has been showing that it *must* be used with at least one perturbation, i.e., there is an efficient and elegant attack [15, 24] requiring a small transcript of signatures in the case of zero perturbations.

Contents and Motivation

This paper presents an overview of operations, performance, and security considerations for NTRUEncrypt and NTRUSign. The most up-to-date descriptions of NTRUEncrypt and NTRUSign are included in [25] and [26], respectively. This paper summarizes, and draws heavily on, the material presented in those papers.

This paper is structured as follows. First, we introduce and describe the algorithms NTRUEncrypt and NTRUSign. We then survey known results about the security of these algorithms, and then present performance characteristics of the algorithms.

As mentioned above, the motivation for this work is to produce viable cryptographic primitives based on the theory of lattices. The benefits of this are twofold: the new schemes may have operating characteristics that fit certain environments particularly well. Also, the new schemes are based on different hard problems from the current mainstream choices of RSA and ECC.

The second point is particularly relevant in a post-quantum world. Lattice reduction is a reasonably well-studied hard problem that is currently not known to be solved by any polynomial time, or even subexponential time, quantum algorithms [27, 28]. While the algorithms are definitely of interest even in the classical computing world, they are clearly prime candidates for widespread adoption should quantum computers ever be invented.

NTRUEncrypt: Overview

Parameters and Definitions

An implementation of the NTRUEncrypt encryption primitive is specified by the following parameters:

- N — *Degree Parameter*. A positive integer. The associated NTRU lattice has dimension $2N$.
- q — *Large Modulus*. A positive integer. The associated NTRU lattice is a convolution modular lattice of modulus q.
- p — *Small Modulus*. An integer or a polynomial.
- $\mathcal{D}_f, \mathcal{D}_g$ — *Private Key Spaces*. Sets of small polynomials from which the private keys are selected.
- $\mathcal{D}_m$ — *Plaintext Space*. Set of polynomials that represent encryptable messages. It is the responsibility of the encryption scheme to provide a method for encoding the message that one wishes to encrypt into a polynomial in this space.
- $\mathcal{D}_r$ — *Blinding Value Space*. Set of polynomials from which the temporary blinding value used during encryption is selected.
- `center` — *Centering Method*. A means of performing mod q reduction on decryption.

Definition 1. The *Ring of Convolution Polynomials* is

$$\mathcal{R} = \frac{\mathbb{Z}[X]}{(X^N - 1)}.$$

Multiplication of polynomials in this ring corresponds to the convolution product of their associated vectors, defined by

$$(\mathsf{f} * \mathsf{g})(X) = \sum_{k=0}^{N-1} \Bigg(\sum_{i+j \equiv k \pmod N} \mathsf{f}_i \cdot \mathsf{g}_j \Bigg) X^k .$$

We also use the notation $\mathcal{R}_q = \frac{(\mathbb{Z}/q\mathbb{Z})[X]}{(X^N - 1)}$. Convolution operations in the ring $\mathcal{R}_q$ are referred to as *modular convolutions*.

Definition 2. A polynomial $a(X) = a_0 + a_1 X + \cdots + a_{N-1} X^{N-1}$ is identified with its vector of coefficients $\mathsf{a} = [a_0, a_1, \ldots, a_{N-1}]$. The mean $\bar{\mathsf{a}}$ of a polynomial a is defined by $\bar{a} = \frac{1}{N} \sum_{i=0}^{N-1} \mathsf{a}_i$. The *centered norm* $\|\mathsf{a}\|$ of a is defined by

$$\|\mathsf{a}\|^2 = \sum_{i=0}^{N-1} a_i^2 - \frac{1}{N} \left(\sum_{i=0}^{N-1} a_i \right)^2 . \tag{11.1}$$

Definition 3. The *width* $\mathsf{Width}(\mathsf{a})$ of a polynomial or vector is defined by

$$\mathsf{Width}(\mathsf{a}) = \mathsf{Max}(a_0, \ldots, a_{N-1}) - \mathsf{Min}(a_0, \ldots, a_{N-1}) .$$

Definition 4. A *binary polynomial* is one whose coefficients are all in the set $\{0, 1\}$. A *trinary polynomial* is one whose coefficients are all in the set $\{0, \pm 1\}$. If one of the inputs to a convolution is a binary polynomial, the operation is referred to as a *binary convolution*. If one of the inputs to a convolution is a trinary polynomial, the operation is referred to as a *trinary convolution*.

Definition 5. Define the polynomial spaces $\mathcal{B}_N(d), \mathcal{T}_N(d), \mathcal{T}_N(d_1, d_2)$ as follows. Polynomials in $\mathcal{B}_N(d)$ have d coefficients equal to 1, and the other coefficients are 0. Polynomials in $\mathcal{T}_N(d)$ have $d + 1$ coefficients equal to 1, have d coefficients equal to -1, and the other coefficients are 0. Polynomials in $\mathcal{T}_N(d_1, d_2)$ have d_1 coefficients equal to 1, have d_2 coefficients equal to -1, and the other coefficients are 0.

"Raw" NTRUEncrypt

Key Generation

NTRUEncrypt *key generation* consists of the following operations:

1. Randomly generate polynomials f and g in $\mathcal{D}_f$, $\mathcal{D}_g$, respectively.
2. Invert f in $\mathcal{R}_q$ to obtain f_q, invert f in $\mathcal{R}_p$ to obtain f_p, and check that g is invertible in $\mathcal{R}_q$ [29].
3. The public key $\mathsf{h} = p * \mathsf{g} * \mathsf{f}_q \pmod{q}$. The private key is the pair $(\mathsf{f}, \mathsf{f}_p)$.

Encryption

NTRUEncrypt *encryption* consists of the following operations:

1. Randomly select a "small"polynomial $\mathsf{r} \in \mathcal{D}_r$.
2. Calculate the ciphertext e as $\mathsf{e} \equiv \mathsf{r} * \mathsf{h} + \mathsf{m} \pmod{q}$.

Decryption

NTRUEncrypt *decryption* consists of the following operations:

1. Calculate $\mathsf{a} \equiv \texttt{center}(\mathsf{f} * e)$, where the centering operation center reduces its input into the interval $[A, A+q-1]$.
2. Recover m by calculating $\mathsf{m} \equiv \mathsf{f}_p * \mathsf{a} \pmod{p}$.

To see why decryption works, use $\mathsf{h} \equiv p * \mathsf{g} * \mathsf{f}_q$ and $\mathsf{e} \equiv \mathsf{r} * \mathsf{h} + \mathsf{m}$ to obtain

$$\mathsf{a} \equiv p * \mathsf{r} * \mathsf{g} + \mathsf{f} * \mathsf{m} \pmod{q}\,. \tag{11.2}$$

For appropriate choices of parameters and center, this is an equality over $\mathbb{Z}$, rather than just over $\mathbb{Z}_q$. Therefore, step 2 recovers m: the $p * \mathsf{r} * \mathsf{g}$ term vanishes, and $\mathsf{f}_p * \mathsf{f} * \mathsf{m} = \mathsf{m} \pmod{p}$.

Encryption Schemes: NAEP

To protect against adaptive chosen ciphertext attacks, we must use an appropriately defined *encryption scheme*. The scheme described in [30] gives provable security in the random oracle model [31, 32] with a tight (i.e., linear) reduction. We briefly outline it here.

NAEP uses two hash functions:

$$G : \{0,1\}^{N-l} \times \{0,1\}^l \to \mathcal{D}_r \qquad H : \{0,1\}^N \to \{0,1\}^N$$

To encrypt a message $M \in \{0,1\}^{N-l}$ using NAEP one uses the functions

$$\texttt{compress}(x) = (x \pmod{q}) \pmod{2},$$
$$\texttt{B2P} : \{0,1\}^N \to \mathcal{D}_m \cup \text{“error”}, \quad \texttt{P2B} : \mathcal{D}_m \to \{0,1\}^N$$

The function compress puts the coefficients of the modular quantity $x \pmod{q}$ in to the interval $[0, q)$, and then this quantity is reduced modulo 2. The role of compress is simply to reduce the size of the input to the hash function H for gains in practical efficiency.The function B2P converts a bit string into a binary polynomial, or returns “error” if the bit string does not fulfil the appropriate criteria – for example, if it does not have the appropriate level of combinatorial security. The function P2B converts a binary polynomial to a bit string.

The encryption algorithm is then specified by:

1. Pick $b \overset{R}{\leftarrow} \{0,1\}^l$.
2. Let $\mathsf{r} = G(M, b)$, $\mathsf{m} = \texttt{B2P}(\,(M \| b) \oplus H(\texttt{compress}(\mathsf{r} * \mathsf{h}))\,)$.
3. If B2P returns “error”, go to step 1.
4. Let $\mathsf{e} = \mathsf{r} * \mathsf{h} + \mathsf{m} \in \mathcal{R}_q$.

Step 3 ensures that only messages of the appropriate form will be encrypted.
To decrypt a message $\mathsf{e} \in \mathcal{R}_q$, one does the following:

1. Let $\mathsf{a} = \texttt{center}(\mathsf{f} * \mathsf{e} \pmod{q})$.
2. Let $\mathsf{m} = \mathsf{f}_p^{-1} * \mathsf{a} \pmod{p}$.
3. Let $\mathsf{s} = \mathsf{e} - \mathsf{m}$.
4. Let $M||b = \texttt{P2B}(\mathsf{m}) \oplus H(\texttt{compress}(\texttt{P2B}(\mathsf{s})))$.
5. Let $\mathsf{r} = G(M, b)$.
6. If $\mathsf{r} * \mathsf{h} = \mathsf{s} \pmod{q}$, and $\mathsf{m} \in \mathcal{D}_m$, then return the message M, else return the string "invalid ciphertext."

The use of the scheme NAEP introduces a single additional parameter:

l *Random Padding Length*. The length of the random padding b concatenated with M in step 1.

Instantiating NAEP*:* SVES-3

The EESS#1 v2 standard [21] specifies an instantiation of NAEP known as SVES-3. In SVES-3, the following specific design choices are made:

- To allow variable-length messages, a one-byte encoding of the message length in bytes is prepended to the message. The message is padded with zeroes to fill out the message block.
- The hash function G which is used to produce r takes as input M; b; an OID identifying the encryption scheme and parameter set; and a string h_{trunc} derived by truncating the public key to length l_h bits.

SVES-3 includes h_{trunc} in G so that r depends on the specific public key. Even if an attacker was to find an (M, b) that gave an r with an increased chance of a decryption failure, that (M, b) would apply only to a single public key and could not be used to attack other public keys. However, the current recommended parameter sets do not have decryption failures and so there is no need to input h_{trunc} to G. We will therefore use SVES-3but set $l_h = 0$.

NTRUEncrypt *Coins!*

It is both amusing and informative to view the NTRUEncrypt operations as working with "coins." By coins, we really mean N-sided coins, like the British 50 pence piece.

An element of $\mathcal{R}$ maps naturally to an N-sided coin: one simply write the integer entries of $a \in \mathcal{R}$ on the side-faces of the coin (with "heads" facing up, say). Multiplication by X in $\mathcal{R}$ is analagous to simply rotating the coin, and addition of two elements in $\mathcal{R}$ is analagous to placing the coins on top of each other and summing the faces. A generic multiplication by an element in $\mathcal{R}$ is thus analagous to multiple copies of the same coin being rotated by different amonuts, placed on top of each other, and summed.

The NTRUEncrypt key recovery problem is a binary multiplication problem, i.e., given d_f copies of the h-coin the problem is to pile them on top of eachother (with distinct rotations) so that the faces sum to zero or one modulo q.

The raw NTRUEncrypt encryption function has a similar coin analogy: one piles d_r copies of the h-coin on top of one another with random (but distinct) rotations, then one sums the faces modulo q, and adds a small $\{0, 1\}$ perturbation to faces modulo q (corresponding to the message). The resulting coin, c, is a valid NTRUEncrypt ciphertext.

The NTRUEncrypt decryption function also has a similar coin analogy: one piles d_f copies of a c-coin (corresponding to the ciphertext) on top of each other with rotations corresponding to f. After summing the faces modulo q, centering, and then a reduction modulo p, one should recover the original message m.

These NTRUEncrypt operations are so easy, it seems strong encryption could have been used centuries ago, had public-key encryption been known about. From a number theoretic point of view, the only nontrivial operation is the creation of the h coin (which involves Euclid's algorithm over polynomials).

NTRUSign: Overview

Parameters

An implementation of the NTRUSign primitive uses the following parameters:

N	Polynomials have degree $< N$
q	Coefficients of polynomials are reduced modulo q
$\mathcal{D}_f, \mathcal{D}_g$	Polynomials in $\mathcal{T}(d)$ have $d + 1$ coefficients equal to 1, have d coefficients equal to -1, and the other coefficients are 0.
$\mathcal{N}$	The norm bound used to verify a signature.
β	The balancing factor for the norm $\|\cdot\|_\beta$. Has the property $0 < \beta \le 1$.

"Raw" NTRUSign

Key Generation

NTRUSign *key generation* consists of the following operations:

1. Randomly generate "small" polynomials f and g in $\mathcal{D}_f$, $\mathcal{D}_g$, respectively, such that f and g are invertible modulo q.

2. Find polynomials F and G such that

$$\mathsf{f} * \mathsf{G} - \mathsf{g} * \mathsf{F} = q\,, \tag{11.3}$$

and F and G have size

$$\|\mathsf{F}\| \approx \|\mathsf{G}\| \approx \|\mathsf{f}\| \sqrt{N/12}\,. \tag{11.4}$$

This can be done using the methods of [17]
3. Denote the inverse of f in $\mathcal{R}_q$ by f_q, and the inverse of g in $\mathcal{R}_q$ by g_q. The public key $\mathsf{h} = \mathsf{F} * \mathsf{f}_q \pmod{q} = \mathsf{G} * \mathsf{g}_q \pmod{q}$. The private key is the pair (f, g).

Signing

The signing operation involves *rounding* polynomials. For any $a \in \mathbb{Q}$, let $\lfloor a \rceil$ denote the integer closest to a, and define $\{a\} = a - \lfloor a \rceil$. (For numbers a that are midway between two integers, we specify that $\{a\} = +\frac{1}{2}$, rather than $-\frac{1}{2}$.) If A is a polynomial with rational (or real) coefficients, let $\lfloor A \rceil$ and $\{A\}$ be A with the indicated operation applied to each coefficient.

"Raw" NTRUSign *signing* consists of the following operations:

1. Map the digital document D to be signed to a vector $\mathsf{m} \in [0, q)^N$ using an agreed hash function.
2. Set
$$(\mathsf{x}, \mathsf{y}) = (0, \mathsf{m}) \begin{pmatrix} \mathsf{G} & -\mathsf{F} \\ -\mathsf{g} & \mathsf{f} \end{pmatrix} / q = \left(\frac{-\mathsf{m} * \mathsf{g}}{q}, \frac{\mathsf{m} * \mathsf{f}}{q} \right).$$
3. Set
$$\epsilon = -\{x\} \qquad \text{and} \qquad \epsilon' = -\{y\}\,. \tag{11.5}$$
4. Calculate s, the signature, as
$$\mathsf{s} = \epsilon \mathsf{f} + \epsilon' \mathsf{g}\,. \tag{11.6}$$

Verification

Verification involves the use of a *balancing factor* β and a *norm bound* $\mathcal{N}$. To verify, the recipient does the following:

1. Map the digital document D to be verified to a vector $\mathsf{m} \in [0, q)^N$ using the agreed hash function.
2. Calculate $\mathsf{t} = \mathsf{s} * \mathsf{h} \bmod q$, where s is the signature, and h is the signer's public key.

3. Calculate the norm

$$\nu = \min_{k_1, k_2 \in R} \left(\| \mathsf{s} + k_1 q \|^2 + \beta^2 \| (\mathsf{t} - \mathsf{m}) + k_2 q \|^2 \right)^{1/2}. \tag{11.7}$$

4. If $\nu \leq \mathcal{N}$, the verification succeeds, otherwise, it fails.

Why NTRUSign *Works*

Given any positive integers N and q and any polynomial $h \in R$, we can construct a lattice L_h contained in $R^2 \cong \mathbb{Z}^{2N}$ as follows:

$$L_h = L_h(N, q) = \left\{ (r, r') \in R \times R \;\middle|\; r' \equiv r * h \pmod{q} \right\}.$$

This sublattice of $\mathbb{Z}^{2N}$ is called a *convolution modular lattice*. It has dimension equal to $2N$ and determinant equal to q^N.

Since

$$\det \begin{pmatrix} \mathsf{f} & \mathsf{F} \\ \mathsf{g} & \mathsf{G} \end{pmatrix} = q$$

and we have defined $\mathsf{h} = \mathsf{F}/\mathsf{f} = \mathsf{G}/\mathsf{g} \bmod q$, we know that

$$\begin{pmatrix} \mathsf{f} & \mathsf{F} \\ \mathsf{g} & \mathsf{G} \end{pmatrix} \text{ and } \begin{pmatrix} 1 & \mathsf{h} \\ 0 & q \end{pmatrix}$$

are bases for the same lattice. Here, as in [17], a 2-by-2 matrix of polynomials is converted to a $2N$-by-$2N$ integer matrix matrix by converting each polynomial in the polynomial matrix to its representation as an N-by-N circulant matrix, and the two representations are regarded as equivalent.

Signing consists of finding a close lattice point to the message point $(0, \mathsf{m})$ using Babai's method: express the target point as a real-valued combination of the basis vectors, and find a close lattice point by rounding off the fractional parts of the real coefficients to obtain integer combinations of the basis vectors. The error introduced by this process will be the sum of the rounding errors on each of the basis vectors, and the rounding error by definition will be between $-\frac{1}{2}$ and $\frac{1}{2}$. In NTRUSign, the basis vectors are all of the same length, so the expected error introduced by $2N$ roundings of this type will be $\sqrt{N/6}$ times this length.

In NTRUSign, the private basis is chosen such that $\|\mathsf{f}\| = \|\mathsf{g}\|$ and $\|\mathsf{F}\| \sim \|\mathsf{G}\| \sim \sqrt{N/12}\|\mathsf{f}\|$. The expected error in signing will therefore be

$$\sqrt{N/6}\|\mathsf{f}\| + \beta(N/6\sqrt{2})\|\mathsf{f}\|. \tag{11.8}$$

In contrast, an attacker who uses only the public key will likely produce a signature with N incorrect coefficients, and those coefficients will be distributed randomly mod q. The expected error in generating a signature with a public key is therefore

$$\beta\sqrt{N/12}q\ . \tag{11.9}$$

(We discuss security considerations in more detail in Section "NTRUSign Security Considerations" and onwards; the purpose of this section is to argue that it is plausible that the private key allows the production of smaller signatures than the public key).

It is therefore clear that it is possible to choose $\|\mathsf{f}\|$ and q such that the knowledge of the private basis allows the creation of smaller signing errors than knowledge of the public basis alone. Therefore, by ensuring that the signing error is less than that could be expected to be produced by the public basis, a recipient can verify that the signature was produced by the owner of the private basis and is therefore valid.

NTRUSign *Signature Schemes: Chosen Message Attacks, Hashing, and Message Preprocessing*

To prevent chosen message attacks, the message representative m must be generated in some pseudo-random fashion from the input document D. The currently recommended hash function for NTRUSign is a simple Full Domain Hash. First the message is hashed to a "seed" hash value H_m. H_m is then hashed in counter mode to produce the appropriate number of bits of random output, which are treated as N numbers mod q. Since q is a power of 2, there are no concerns with bias.

The above mechanism is deterministic. If parameter sets were chosen that gave a significant chance of signature failure, the mechanism can be randomized as follows. The additional input to the process is r_{len}, the length of the randomizer in bits.

On signing:

1. Hash the message as before to generate H_m.
2. Select a randomizer r consisting of r_{len} random bits.
3. Hash $H_m\|r$ in counter mode to obtain enough output for the message representative m.
4. On signing, check that the signature will verify correctly.

   a. If the signature does not verify, repeat the process with a different r.
   b. If the signature verifies, send the tuple (r, s) as the signature.

On verification, the verifier uses the received r and the calculated H_m as input to the hash in counter mode to generate the same message representative as the signer used.

The size of r should be related to the probability of signature failure. An attacker who is able to determine through timing information that a given H_m required multiple rs knows that at least one of those rs resulted in a signature that was too big, but does not know which message it was or what the resulting signature was. It is an open research question to quantify the appropriate size of r for a given signature failure probability, but in most cases, $r_{\text{len}} = 8$ or 32 should be sufficient.

NTRUSign *Signature Schemes: Perturbations*

To protect against transcript attacks, the raw NTRUSign signing algorithm defined above is modified as follows.

On key generation, the signer generates a secret *perturbation distribution function*.

On signing, the signer uses the agreed hash function to map the document D to the message representative m. However, before using his or her private key, he or she chooses an error vector e drawn from the perturbation distribution function that was defined as part of key generation. He or she then signs $\mathsf{m} + \mathsf{e}$, rather than m alone.

The verifier calculates m, t, and the norms of s and $\mathsf{t}-\mathsf{m}$ and compares the norms to a specified bound $\mathcal{N}$ as before. Since signatures with perturbations will be larger than unperturbed signatures, $\mathcal{N}$ and, in fact, all of the parameters will in general be different for the perturbed and unpertubed cases.

NTRU currently recommends the following mechanism for generating perturbations.

Key Generation

At key generation time, the signer generates B lattices $L_1 \dots L_B$. These lattices are generated with the same parameters as the private and public key lattice, L_0, but are otherwise independent of L_0 and of each other. For each L_i, the signer stores f_i, g_i, h_i.

Signing

When signing m, for each L_i starting with L_B, the signer does the following:

1. Set $(\mathsf{x}, \mathsf{y}) = \left(\frac{-\mathsf{m}*\mathsf{g}_i}{q}, \frac{\mathsf{m}*\mathsf{f}_i}{q}\right)$.
2. Set $\epsilon = -\{x\}$ and $\epsilon' = -\{y\}$.
3. Set $\mathsf{s}_i = \epsilon \mathsf{f}_i + \epsilon' \mathsf{g}_i$.
4. Set $\mathsf{s} = \mathsf{s} + \mathsf{s}_i$.
5. If $i = 0$ stop and output s; otherwise, continute
6. Set $\mathsf{t}_i = \mathsf{s}_i * \mathsf{h}_i \bmod q$
7. Set $\mathsf{m} = \mathsf{t}_i - (\mathsf{s}_i * \mathsf{h}_{i-1}) \bmod q$.

The final step translates back to a point of the form $(0, \mathsf{m})$ so that all the signing operations can use only the f and g components, allowing for greater efficiency. Note that steps 6 and 7 can be combined into the single step of setting $\mathsf{m} = \mathsf{s}_i * (\mathsf{h}_i - \mathsf{h}_{i-1})$ to improve performance.

The parameter sets defined in [26] take $B = 1$.

NTRUEncrypt **Performance**

NTRUEncrypt *Parameter Sets*

There are many different ways of choosing "small" polynomials. This section reviews NTRU's current recommendations for choosing the form of these polynomials for the best efficiency. We focus here on choices that improve efficiency; security considerations are looked at in Section "NTRUEncrypt Security Considerations".

Form of f

Published NTRUEncrypt parameter sets [25] take f to be of the form $\mathsf{f} = 1 + p\mathsf{F}$. This guarantees that $\mathsf{f}_p = 1$, eliminating one convolution on decryption.

Form of F, g, r

NTRU currently recommends several different forms for F and r. If F and r take *binary* and *trinary* form, respectively, they are drawn from $\mathcal{B}_N(d)$, the set of binary polynomials with d 1s and $N - d$ 0s or $\mathcal{T}_N(d)$, the set of trinary polynomials with $d+1$ 1s, d -1s and $N-2d-1$ 0s. If F and r take *product* form, then $\mathsf{F} = \mathsf{f}_1 * \mathsf{f}_2 + \mathsf{f}_3$, with $\mathsf{f}_1, \mathsf{f}_2, \mathsf{f}_3 \stackrel{R}{\leftarrow} \mathcal{B}_N(d), \mathcal{T}_N(d)$, and similarly for r. (The value d is considerably lower in the product-form case than in the binary or trinary case).

A binary or trinary convolution requires on the order of dN adds mod q. The best efficiency is therefore obtained when d is as low as possible consistent with the security requirements.

Plaintext Size

For k-bit security, we want to transport $2k$ bits of message and we require $l \geq k$, l the random padding length. SVES-3 uses 8 bits to encode the length of the transported message. N must therefore be at least $3k + 8$. Smaller N will in general lead to lower bandwidth and faster operations.

Form of p, q

The parameters p and q must be relatively prime. This admits of various combinations, such as ($p = 2, q =$ prime), ($p = 3, q = 2^m$), and ($p = 2 + X, q = 2^m$).

The B2P Function

The polynomial m produced by the B2P function will be a random trinary polynomial. As the number of 1s, (in the binary case), or 1s and -1s (in the trinary case), decreases, the strength of the ciphertext against both lattice and combinatorial attacks will decrease. The B2P function therefore contains a check that the number of 1s in m is not less than a value d_{m_0}. This value is chosen to be equal to df. If, during encryption, the encrypter generates m that does not satisfy this criterion, they must generate a different value of b and re-encrypt.

NTRUEncrypt *Performance*

Table 11.1 and Table 11.2 give parameter sets and running times (in terms of operations per second) for size optimized and speed optimized performance, respectively, at different security levels corresponding to k bits of security. "Size" is the size of the public key in bits. In the case of NTRUEncrypt and RSA, this is also the size of the ciphertext; in the case of some ECC encryption schemes, such as ECIES, the ciphertext may be a multiple of this size. Times given are for unoptimized C implementations on a 1.7 GHz Pentium and include time for all encryption scheme operations, including hashing, random number generation, as well as the primitive operation. d_{m_0} is the same in both the binary and product-form case and is omitted from the product-form table.

For comparison, we provide the times given in [33] for raw elliptic curve point multiplication (not including hashing or random number generation times) over the

Table 11.1 Size-optimized NTRUEncrypt parameter sets with trinary polynomials

k	N	d	d_{m_0}	q	size	RSA size	ECC size	enc/s	dec/s	ECC mult/s	Enc ECC ratio	Dec ECC ratio
112	401	113	113	2,048	4,411	2,048	224	2,640	1,466	1,075	4.91	1.36
128	449	134	134	2,048	4,939	3,072	256	2,001	1,154	661	6.05	1.75
160	547	175	175	2,048	6,017	4,096	320	1,268	718	n/a	n/a	n/a
192	677	157	157	2,048	7,447	7,680	384	1,188	674	196	12.12	3.44
256	1,087	120	120	2,048	11,957	15, 360	512	1,087	598	115	18.9	5.2

Table 11.2 Speed-optimized NTRUEncrypt parameter sets with trinary polynomials

k	N	d	d_{m_0}	q	$size$	RSA size	ECC size	enc/s	dec/s	ECC mult/s	Enc ECC ratio	Dec ECC ratio
112	659	38	38	2,048	7, 249	2,048	224	4,778	2,654	1,075	8.89	2.47
128	761	42	42	2,048	8, 371	3,072	256	3,767	2,173	661	11.4	3.29
160	991	49	49	2048	10, 901	4,096	320	2,501	1,416	n/a	n/a	n/a
192	1, 087	63	63	2,048	11, 957	7,680	384	1,844	1,047	196	18.82	5.34
256	1, 499	79	79	2,048	16, 489	15,360	512	1,197	658	115	20.82	5.72

NIST prime curves. These times were obtained on a 400 MHz SPARC and have been converted to operations per second by simply scaling by 400/1700. Times given are for point multiplication without precomputation, as this corresponds to common usage in encryption and decryption. Precomputation improves the point multiplication times by a factor of 3.5–4. We also give the speedup for NTRUEncrypt decryption vs. a single ECC point multiplication.

NTRUSign Performance

NTRUSign *Parameter Sets*

Form of f, g

The current recommended parameter sets take f and g to be trinary, i.e., drawn from $\mathcal{T}_N(d)$. Trinary polynomials allow for higher combinatorial security than binary polynomials at a given value of N and admit efficient implementations. A trinary convolution requires $(2d + 1)N$ adds and one subtract mod q. The best efficiency is therefore obtained when d is as low as possible consistent with the security requirements.

Form of p, q

The parameters q and N must be relatively prime. For efficiency, we take q to be a power of 2.

Signing Failures

A low value of $\mathcal{N}$, the norm bound, gives the possibility that a validly generated signature will fail. This affects efficiency, as if the chance of failure is non-negligible, the signer must randomize the message before signing and check for failure on signature generation. For efficiency, we want to set $\mathcal{N}$ sufficiently high to make the chance of failure negligible. To do this, we denote the expected size of a signature by $\mathcal{E}$ and define the *signing tolerance* ρ by the formula

$$\mathcal{N} = \rho\mathcal{E} .$$

As ρ increases beyond 1, the chance of a signing failure appears to drop off exponentially. In particular, experimental evidence indicates that the probability that a validly generated signature will fail the normbound test with parameter ρ is smaller than $e^{-C(N)(\rho-1)}$, where $C(N) > 0$ increases with N. In fact, under the assumption that each coefficient of a signature can be treated as a sum of independent identically distributed random variables, a theoretical analysis indicates that $C(N)$ grows quadratically in N. The parameter sets below were generated with $\rho = 1.1$, which appears to give a vanishingly small probability of valid signature failure for N in the ranges that we consider. It is an open research question to determine precise signature failure probabilities for specific parameter sets, i.e., to determine the constants in $C(N)$.

NTRUSign *Performance*

With one perturbation, signing takes time equivalent to two "raw" signing operations (as defined in Section "Signing") and one verification. Research is ongoing into alternative forms for the perturbations that could reduce this time.

Table 11.3 gives the parameter sets for a range of security levels, corresponding to k-bit security, and the performance (in terms of signatures and verifications per second) for each of the recommended parameter sets. We compare signature times to a single ECC point multiplication with precomputation from [33]; without precomputation, the number of ECC signatures/second goes down by a factor of 3.5–4. We compare verification times to ECDSA verification times without memory constraints from [33]. As in Tables 11.1 and 11.2, NTRUSign times given are for the entire scheme (including hashing, etc.), not just the primitive operation, while ECDSA times are for the primitive operation alone.

Above the 80-bit security level, NTRUSign signatures are smaller than the corresponding RSA signatures. They are larger than the corresponding ECDSA signatures by a factor of about 4. An NTRUSign private key consists of sufficient space to store f and g for the private key, plus sufficient space to store f_i, g_i, and h_i for each of the B perturbation bases. Each f and g can be stored in $2N$ bits, and each h can be stored in $N \log_2(q)$ bits, so the total storage required for the one-perturbation

Table 11.3 Performance measures for different NTRUSign parameter sets. (Note: parameter sets have not been assessed against the hybrid attack of Section "The Hybrid Attack" and may give less than k bits of security)

Parameters				Public key and				sign/s			vfy/s		
k	N	d	q	NTRU	ECDSA key	ECDSA sig	RSA	NTRU	ECDSA	Ratio	NTRU	ECDSA	Ratio
80	157	29	256	1,256	192	384	1,024	4,560	5,140	0.89	15,955	1,349	11.83
112	197	28	256	1,576	224	448	~2,048	3,466	3,327	1.04	10,133	883	11.48
128	223	32	256	1,784	256	512	3,072	2,691	2,093	1.28	7,908	547	14.46
160	263	45	512	2,367	320	640	4,096	1,722	–	–	5,686	–	–
192	313	50	512	2,817	384	768	7,680	1,276	752	1.69	4,014	170	23.61
256	349	75	512	3,141	512	1024	15,360	833	436	1.91	3,229	100	32.29

case is $16N$ bits for the 80- to 128-bit parameter sets below and $17N$ bits for the 160- to 256-bit parameter sets, or approximately twice the size of the public key.

Security: Overview

We quantify security in terms of bit strength k, evaluating how much effort an attacker has to put in to break a scheme. All the attacks we consider here have variable running times, so we describe the strength of a parameter set using the notion of *cost*. For an algorithm $\mathcal{A}$ with running time t and probability of success ε, the cost is defined as

$$C_{\mathcal{A}} = t/\varepsilon \, .$$

This definition of cost is not the only one that could be used. For example, in the case of indistinguishability against adaptive chosen-ciphertext attack, the attacker outputs a single bit $i \in \{0, 1\}$, and obviously has a chance of success of at least $\frac{1}{2}$. Here, the probability of success is less important than the attacker's *advantage*, defined as

$$\mathrm{adv}(\mathcal{A}(\mathsf{ind})) = 2.(\mathbb{P}[\mathsf{Succ}[\mathcal{A}]] - 1/2) \, .$$

However, in this paper, the cost-based measure of security is appropriate.

Our notion of cost is derived from [34] and related work. An alternate notion of cost, which is the definition above multiplied by the amount of memory used, is proposed in [35]. The use of this measure would allow significantly more efficient parameter sets, as the meet-in-the-middle attack described in Section "Combinatorial Security" is essentially a time-memory tradeoff that keeps the product of time and memory constant. However, current practice is to use the measure of cost above.

We also acknowledge that the notion of comparing public-key security levels with symmetric security levels, or of reducing security to a single headline measure, is inherently problematic – see an attempt to do so in [36], and useful comments on this in [37]. In particular, extrapolation of breaking times is an inexact science, the behavior of breaking algorithms at high security levels is by definition untested, and one can never disprove the existence of an algorithm that attacks NTRUEncrypt (or any other system) more efficiently than the best currently known method.

Common Security Considerations

This section deals with security considerations that are common to NTRUEncrypt and NTRUSign.

Most public key cryptosystems, such as RSA [38] or ECC [39, 40], are based on a one-way function for which there is one best-known method of attack: factoring in the case of RSA, Pollard-rho in the case of ECC. In the case of NTRU, there are *two* primary methods of approaching the one-way function, both of which must be considered when selecting a parameter set.

Combinatorial Security

Polynomials are drawn from a known space $\mathcal{S}$. This space can best be searched by using a combinatorial technique originally due to Odlyzko [41], which can be used to recover f or g from h or r and m from e. We denote the combinatorial security of polynomials drawn from $\mathcal{S}$ by Comb[$\mathcal{S}$]

$$\text{Comb}[\mathcal{B}_N(d)] \geq \frac{\binom{N/2}{d/2}}{\sqrt{N}}. \tag{11.10}$$

For trinary polynomials in $\mathcal{T}_N(d)$, we find

$$\text{Comb}[\mathcal{T}(d)] > \binom{N}{d+1}/\sqrt{N}. \tag{11.11}$$

For product-form polynomials in $\mathcal{P}_N(d)$, defined as polynomials of the form $\mathsf{a} = \mathsf{a}_1 * \mathsf{a}_2 + \mathsf{a}_3$, where $\mathsf{a}_1, \mathsf{a}_2, \mathsf{a}_3$ are all binary with $d_{a_1}, d_{a_2}, d_{a_3}$ 1s respectively, $d_{a1} = d_{a2} = d_{a3} = d_a$, and there are no further constraints on a, we find [25]:

$$\begin{aligned}\text{Comb}[\mathcal{P}_N(d)] \geq \min\Bigg(&\binom{N - \lceil N/d \rceil}{d-1}^2, \\ &\max\left(\binom{N - \lceil \frac{N}{d} \rceil}{d-1}\binom{N - \lceil \frac{N}{d-)} \rceil}{d-2}, \binom{N}{2d}\right), \\ &\max\left(\binom{N}{d}\binom{N}{d-1}, \binom{N - \lceil \frac{N}{2d} \rceil}{2d-1}\right)\Bigg)\end{aligned}$$

Lattice Security

An NTRU public key h describes a $2N$-dimensional NTRU lattice containing the private key (f, g) or (f, F). When f is of the form $\mathsf{f} = 1 + p\mathsf{F}$, the best lattice attack on the private key involves solving a Close Vector Problem (CVP).[3] When f is not of the

[3] Coppersmith and Shamir [42] propose related approaches which turn out not to materially affect security.

form $\mathsf{f} = 1 + \mathsf{pF}$, the best lattice attack involves solving an Approximate Shortest Vector Problem (apprSVP). Experimentally, it has been found that an NTRU lattice of this form can usefully be characterized by two quantities

$$\begin{aligned} a &= N/q, \\ c &= \sqrt{4\pi e \|\mathsf{F}\| \|\mathsf{g}\|/q} \quad \text{(NTRUEncrypt)}, \\ &= \sqrt{4\pi e \|\mathsf{f}\| \|\mathsf{F}\|/q} \quad \text{(NTRUSign)}. \end{aligned}$$

(For product-form keys the norm $\|\mathsf{F}\|$ is variable but always obeys $|\mathsf{F}| \geq \sqrt{D(N-D)/N}$, $D = d^2 + d$. We use this value in calculating the lattice security of product-form keys, knowing that in practice the value of c will typically be higher.)

This is to say that for constant (a, c), the experimentally observed running times for lattice reduction behave roughly as

$$\log(T) = AN + B\ ,$$

for some experimentally-determined constants A and B.

Table 11.4 summarizes experimental results for breaking times for NTRU lattices with different (a, c) values. We represent the security by the constants A and B. The breaking time in terms of bit security is $AN + B$. It may be converted to time in MIPS-years using the equality 80 bits$\sim 10^{12}$ MIPS-years.

For constant (a, c), increasing N increases the breaking time exponentially. For constant (a, N), increasing c increases the breaking time. For constant (c, N), increasing a decreases the breaking time, although the effect is slight. More details on this table are given in [14].

Note that the effect of moving from the "standard" NTRUEncrypt lattice to the "transpose" NTRUSign lattice is to increase c by a factor of $(N/12)^{1/4}$. This allows for a given level of lattice security at lower dimensions for the transpose lattice than for the standard lattice. Since NTRUEncrypt uses the standard lattice, NTRUEncrypt key sizes given in [25] are greater than the equivalent NTRUSign key sizes at the same level of security.

The technique known as *zero-forcing* [14,43] can be used to reduce the dimension of an NTRU lattice problem. The precise amount of the expected performance gain is heavily dependent on the details of the parameter set; we refer the reader to [14, 43] for more details. In practice, this reduces security by about 6–10 bits.

Table 11.4 Extrapolated bit security constants depending on (c, a)

c	a	A	B
1.73	0.53	0.3563	−2.263
2.6	0.8	0.4245	−3.440
3.7	2.7	0.4512	+0.218
5.3	1.4	0.6492	−5.436

The Hybrid Attack

In this section, we will review the method of [44]. The structure of the argument is simpler for the less efficient version of NTRU where the public key has the form $h \equiv f^{-1} * g \pmod{q}$. The rough idea is as follows. Suppose one is given N, q, d, e, h and hence implicitly an NTRUEncrypt public lattice L of dimension $2N$. The problem is to locate the short vector corresponding to the secret key (f, g). One first chooses $N_1 < N$ and removes a $2N_1$ by $2N_1$ lattice L_1 from the center of L. Thus, the original matrix corresponding to L has the form

$$\left(\begin{array}{c|c} qI_N & 0 \\ \hline H & I_N \end{array}\right) = \left(\begin{array}{c|c|c} qI_{N-N_1} & 0 & 0 \\ \hline * & L_1 & 0 \\ \hline * & * & I_{N-N_1} \end{array}\right) \tag{11.12}$$

and L_1 has the form

$$\left(\begin{array}{c|c} qI_{N_1} & 0 \\ \hline H_1 & I_{N_1} \end{array}\right). \tag{11.13}$$

Here, H_1 is a truncated piece of the circulant matrix H corresponding to h appearing in (11.12). For increased flexibility, the upper left and lower right blocks of L_1 can be of different sizes, but for ease of exposition, we will consider only the case where they are equal.

Let us suppose that an attacker must use a minimum of k_1 bits of effort to reduce L_1 until all N_1 of the q-vectors are removed. When this is done and L_1 is put in lower triangular form, the entries on the diagonal will have values $\{q^{\alpha_1}, q^{\alpha_2}, \ldots, q^{\alpha_{2N_1}}\}$, where $\alpha_1 + \cdots + \alpha_{2N_1} = N_1$, and the α_i will come very close to decreasing linearly, with

$$1 \approx \alpha_1 > \cdots > \alpha_{2N_1} \approx 0.$$

That is to say, L_1 will roughly obey the geometric series assumption or GSA. This reduction will translate back to a corresponding reduction of L, which when reduced to lower triangular form will have a diagonal of the form

$$\{q, q, \ldots, q, q^{\alpha_1}, q^{\alpha_2}, \ldots, q^{\alpha_{2N_1}}, 1, 1, \ldots, 1\}.$$

The key point here is that it requires k_1 bits of effort to achieve this reduction, with $\alpha_{2N_1} \approx 0$. If $k_2 > k_1$ bits are used, then the situation can be improved to achieve $\alpha_{2N_1} = \alpha > 0$. As k_2 increases the value of α is increased.

In the previous work, the following method was used to launch the meet in the middle attack. It was assumed that the coefficients of f are partitioned into two blocks. These are of size N_1 and $K = N - N_1$. The attacker guesses the coefficients of f that fall into the K block and then uses the reduced basis for L to check if his or her guess is correct. The main observation of [44] is that a list of guesses can be made about half the coefficients in the K block and can be compared to a list of guesses about the other half of the coefficients in the K block. With a probability $p_s(\alpha)$, a correct matching of two half guesses can be confirmed, where $p_s(0) = 0$ and $p_s(\alpha)$ increases monotonically with α. In [44], a value of $\alpha = 0.182$ was used with a corresponding probability $p_s(0.182) = 2^{-13}$. The probability $p_s(0.182)$ was computed by sampling and the bit requirement, k_2 was less than 60.3. In general, if one used k_2 bits of lattice reduction work to obtain a given $p_s(\alpha)$ (as large as possible), then the number of bits required for a meet in the middle search through the K block decreases as K decreases and as $p_s(\alpha)$ increases.

A very subtle point in [44] was the question of how to optimally choose N_1 and k_2. The objective of an attacker was to choose these parameters so that k_2 equalled the bit strength of a meet in the middle attack on K, given the $p_s(\alpha)$ corresponding to N_1. It is quite hard to make an optimal choice, and for details we refer the reader to [44] and [45].

One Further Remark

For both NTRUEncrypt and NTRUSign the degree parameter N must be prime. This is because, as Gentry observed in [46], if N is the composite, the related lattice problem can be reduced to a similar problem in a far smaller dimension. This reduced problem is then comparatively easy to solve.

NTRUEncrypt Security Considerations

Parameter sets for NTRUEncrypt at a k-bit security level are selected subject to the following constraints:

- The work to recover the private key or the message through lattice reduction must be at least k bits, where bits are converted to MIPS-years using the equality 80 bits $\sim 10^{12}$ MIPS-years.
- The work to recover the private key or the message through combinatorial search must be at least 2^k binary convolutions.
- The chance of a decryption failure must be less than 2^{-k}.

Decryption Failure Security

NTRU decryption can fail on validly encrypted messages if the `center` method returns the wrong value of A, or if the coefficients of $\mathsf{prg} + \mathsf{fm}$ do not lie in an interval of width q. Decryption failures leak information about the decrypter's private key [19, 20]. The recommended parameter sets ensure that decryption failures will not happen by setting q to be greater than the maximum possible width of $\mathsf{prg} + \mathsf{m} + \mathsf{pFm}$. q should be as small as possible while respecting this bound, as lowering q increases the lattice constant c and hence the lattice security. Centering then becomes simply a matter of reducing into the interval $[0, q-1]$.

It would be possible to improve performance by relaxing the final condition to require only that the probability of a decryption failure was less than 2^{-K}. However, this would require improved techniques for estimating decryption failure probabilities.

N, q, and p

The small and large moduli p and q must be relatively prime in the ring $\mathcal{R}$. Equivalently, the three quantities

$$p, \quad q, \quad X^N - 1$$

must generate the unit ideal in the ring $\mathbb{Z}[X]$. (As an example of why this is necessary, in the extreme case that p divides q, the plaintext is equal to the ciphertext reduced modulo p.)

Factorization of $X^N - 1$ (mod q)

If $\mathsf{F}(X)$ is a factor of $X^N - 1 \pmod{q}$, and if $\mathsf{h}(X)$ is a multiple of $\mathsf{F}(X)$, i.e., if $\mathsf{h}(X)$ is zero in the field $K = (\mathbb{Z}/q\mathbb{Z})[X]/\mathsf{F}(X)$, then an attacker can recover the value of $\mathsf{m}(X)$ in the field K.

If q is prime and has order $t \pmod{N}$, then

$$X^N - 1 \equiv (X-1)\mathsf{F}_1(X)\mathsf{F}_2(X)\cdots\mathsf{F}_{(N-1)/t}(X) \quad \text{in } (\mathbb{Z}/q\mathbb{Z})[X],$$

where each $\mathsf{F}_i(X)$ has degree t and is irreducible mod q. (If q is composite, there are corresponding factorizations.) If $\mathsf{F}_i(X)$ has degree t, the probability that $\mathsf{h}(X)$ or $\mathsf{r}(X)$ is divisible by $\mathsf{F}_i(X)$ is presumably $1/q^t$. To avoid attacks based on the factorization of h or r, we will require that for each prime divisor P of q, the order of $P \pmod{N}$ must be $N-1$ or $(N-1)/2$. This requirement has the useful side-effect of increasing the probability that randomly chosen f will be invertible in $\mathcal{R}_q$ [47].

Information Leakage from Encrypted Messages

The transformation $\mathsf{a} \to \mathsf{a}(1)$ is a ring homomorphism, and so the ciphertext e has the property that

$$\mathsf{e}(1) = \mathsf{r}(1)\mathsf{h}(1) + \mathsf{m}(1)\,.$$

An attacker will know $\mathsf{h}(1)$, and for many choices of parameter set $\mathsf{r}(1)$ will also be known. Therefore, the attacker can calculate $\mathsf{m}(1)$. The larger $|\mathsf{m}(1) - N/2|$ is, the easier it is to mount a combinatorial or lattice attack to recover the msssage, so the sender should always ensure that $\|\mathsf{m}\|$ is sufficiently large. In these parameter sets, we set a value d_{m_0} such that there is a probability of less than 2^{-40} that the number of 1s or 0s in a randomly generated m is less than d_{m_0}. We then calculate the security of the ciphertext against lattice and combinatorial attacks in the case where m has exactly this many 1s and require this to be greater than 2^k for k bits of security.

NTRUEncrypt *Security: Summary*

In this section, we present a summary of the security measures for the parameter sets under consideration. Table 11.5 gives security measures optimized for size. Table 11.6 gives security measures optimized for speed. The parameter sets for NTRUEncrypt have been calculated based on particular conservative assumptions about the effectiveness of certain attacks. In particular, these assumptions assume the attacks will be improved in certain ways over the current best known attacks, although we do not know yet exactly how these improvements will be implemented. The tables below show the strength of the current recommended parameter sets against the best attacks that are currently known. As attacks improve, it will be instructive to watch the "known hybrid strength" reduce to the recommended security level. The "basic lattice strength" column measures the strength against a pure lattice-based (nonhybrid) attack.

NTRUSign **Security Considerations**

This section considers security considerations that are specific to NTRUSign.

Table 11.5 NTRUEncrypt security measures for size-optimized parameters using trinary polynomials

Recommended security level	N	q	d_f	Known hybrid strength	c	Basic lattice strength
112	401	2,048	113	154.88	2.02	139.5
128	449	2,048	134	179.899	2.17	156.6
160	547	2,048	175	222.41	2.44	192.6
192	677	2,048	157	269.93	2.5	239
256	1,087	2,048	120	334.85	2.64	459.2

Table 11.6 NTRUEncrypt security measures for speed-optimized parameters using trinary polynomials

Recommended security level	N	q	d_f	Known hybrid strength	c	Basic lattice strength
112	659	2,048	38	137.861	1.74	231.5
128	761	2,048	42	157.191	1.85	267.8
160	991	2,048	49	167.31	2.06	350.8
192	1,087	2,048	63	236.586	2.24	384
256	1,499	2,048	79	312.949	2.57	530.8

Security Against Forgery

We quantify the probability that an adversary, without the knowledge of f, g, can compute a signature s on a given document D. The constants $N, q, \delta, \beta, \mathcal{N}$ must be chosen to ensure that this probability is less than 2^{-k}, where k is the desired bit level of security. To investigate this, some additional notation will be useful:

1. Expected length of s: $\mathcal{E}_s$
2. Expected length of $t - m$: $\mathcal{E}_t$

By $\mathcal{E}_s$, $\mathcal{E}_t$, we mean, respectively, the expected values of $\|s\|$ and $\|t - m\|$ (appropriately reduced mod$q$) when generated by the signing procedure described in Section "Signing". These will be independent of m but dependent on N, q, δ. A genuine signature will then have expected length

$$\mathcal{E} = \sqrt{\mathcal{E}_s^2 + \beta^2 \mathcal{E}_t^2}$$

and we will set

$$\mathcal{N} = \rho \sqrt{\mathcal{E}_s^2 + \beta^2 \mathcal{E}_t^2}. \tag{11.14}$$

As in the case of recovering the private key, an attack can be made by combinatorial means, by lattice reduction methods or by some mixing of the two. By balancing these approaches, we will determine the optimal choice of β, the public scaling factor for the second coordinate.

Combinatorial Forgery

Let us suppose that $N, q, \delta, \beta, \mathcal{N}, h$ are fixed. An adversary is given m, the image of a digital document D under the hash function H. His or her problem is to locate an s such that

$$\|(s \bmod q, \beta(h * s - m) \bmod q)\| < \mathcal{N}.$$

In particular, this means that for an appropriate choice of $k_1, k_2 \in R$

$$(\|(s + k_1 q\|^2 + \beta^2 \|h * s - m + k_2 q)\|^2)^{1/2} < \mathcal{N}.$$

A purely combinatorial attack that the adversary can take is to choose s at random to be quite small, and then to hope that the point $h * s - m$ lies inside of a sphere of radius $\mathcal{N}/\beta$ about the origin after its coordinates are reduced mod$q$. The attacker can also attempt to combine guesses. Here, the attacker would calculate a series of random s_i and the corresponding t_i and $t_i - m$, and file the t_i and the $t_i - m$ for future reference. If a future s_j produces a t_j that is sufficiently close to $t_i - m$, then $(s_i + s_j)$ will be a valid signature on m. As with the previous meet-in-the-middle attack, the core insight is that filing the t_i and looking for collisions allow us to check l^2 t-values while generating only l s-values.

An important element in the running time of attacks of this type is the time that it takes to file a t value. We are interested not in exact collisions, but in two t_i that lie close enough to allow forgery. In a sense, we are looking for a way to file the t_i in a spherical box, rather than in a cube as is the case for the similar attacks on private keys. It is not clear that this can be done efficiently. However, for safety, we will assume that the process of filing and looking up can be done in constant time, and that the running time of the algorithm is dominated by the process of searching the s-space. Under this assumption, the attacker's expected work before being able to forge a signature is:

$$p(N, q, \beta, \mathcal{N}) < \sqrt{\frac{\pi^{N/2}}{\Gamma(1 + N/2)} \cdot \left(\frac{\mathcal{N}}{q\beta}\right)^N}. \tag{11.15}$$

If k is the desired bit security level it will suffice to choose parameters so that the right hand side of (11.15) is less than 2^{-k}.

Signature Forgery Through Lattice Attacks

On the other hand, the adversary can also launch a lattice attack by attempting to solve a closest vector problem. In particular, he can attempt to use lattice reduction methods to locate a point $(s, \beta t) \in L_h(\beta)$ sufficiently close to $(0, \beta m)$ that $\|(s, \beta(t - m))\| < \mathcal{N}$. We will refer to $\|(s, \beta(t - m))\|$ as the norm of the intended forgery.

The difficulty of using lattice reduction methods to accomplish this can be tied to another important lattice constant:

$$\gamma(N, q, \beta) = \frac{\mathcal{N}}{\sigma(N, q, \delta, \beta)\sqrt{2N}}. \tag{11.16}$$

Table 11.7 Bit security against lattice forgery attacks, ω_{lf}, based on experimental evidence for different values of $(\gamma, N/q)$

Bound for γ and N/q	$\omega_{\text{lf}}(N)$
$\gamma < 0.1774$ and $N/q < 1.305$	$0.995113N - 82.6612$
$\gamma < 0.1413$ and $N/q < 0.707$	$1.16536N - 78.4659$
$\gamma < 0.1400$ and $N/q < 0.824$	$1.14133N - 76.9158$

This is the ratio of the required norm of the intended forgery over the norm of the expected smallest vector of $L_h(\beta)$, scaled by $\sqrt{2N}$. For usual NTRUSign parameters, the ratio, $\gamma(N,q,\beta)\sqrt{2N}$, will be larger than 1. Thus, with high probability, there will exist many points of $L_h(\beta)$ that will work as forgeries. The task of an adversary is to find one of these without the advantage that knowledge of the private key gives. As $\gamma(N,q,\beta)$ decreases and the ratio approaches 1, this becomes measurably harder.

Experiments have shown that for fixed $\gamma(N,q,\beta)$ and fixed N/q the running times for lattice reduction to find a point $(s,t) \in L_h(\beta)$ satisfying

$$\|(s, t - m)\| < \gamma(N,q,\beta)\sqrt{2N}\sigma(N,q,\delta,\beta)$$

behave roughly as

$$\log(T) = AN + B$$

as N increases. Here, A is fixed when $\gamma(N,q,\beta), N/q$ are fixed, increases as $\gamma(N,q,\beta)$ decreases and increases as N/q decreases. Experimental results are summarized in Table 11.7.

Our analysis shows that lattice strength against forgery is maximized, for a fixed N/q, when $\gamma(N,q,\beta)$ is as small as possible. We have

$$\gamma(N,q,\beta) = \rho\sqrt{\frac{\pi e}{2N^2 q} \cdot (\mathcal{E}_s^2/\beta + \beta\mathcal{E}_t^2)} \tag{11.17}$$

and so clearly the value for β which minimizes γ is $\beta = \mathcal{E}_s/\mathcal{E}_t$. This optimal choice yields

$$\gamma(N,q,\beta) = \rho\sqrt{\frac{\pi e \mathcal{E}_s \mathcal{E}_t}{N^2 q}}. \tag{11.18}$$

Referring to (11.15), we see that increasing β has the effect of improving combinatorial forgery security. Thus, the optimal choice will be the minimal $\beta \geq \mathcal{E}_s/\mathcal{E}_t$ such that $p(N,q,\beta,\mathcal{N})$ defined by (11.15) is sufficiently small.

An adversary could attempt a mixture of combinatorial and lattice techniques, fixing some coefficients and locating the others via lattice reduction. However, as explained in [17], the lattice dimension can only be reduced a small amount before a solution becomes very unlikely. Also, as the dimension is reduced, γ decreases, which sharply increases the lattice strength at a given dimension.

Transcript Security

NTRUSign is not zero-knowledge. This means that, while NTRUEncrypt can have provable security (in the sense of a reduction from an online attack method to a purely offline attack method), there is no known method for establishing such a reduction with NTRUSign. NTRUSign is different in this respect from established signature schemes such as ECDSA and RSA-PSS, which have reductions from online to offline attacks. Research is ongoing into quantifying what information is leaked from a transcript of signatures and how many signatures an attacker needs to observe to recover the private key or other information that would allow the creation of forgeries. This section summarizes existing knowledge about this information leakage.

Transcript Security for Raw NTRUSign

First, consider raw NTRUSign. In this case, an attacker studying a long transcript of valid signatures will have a list of pairs of polynomials of the form

$$\mathsf{s} = \epsilon \mathsf{f} + \epsilon' \mathsf{g}, \quad \mathsf{t} - \mathsf{m} = \epsilon \mathsf{F} + \epsilon' \mathsf{G}$$

where the coefficients of ϵ, ϵ' lie in the range $[-1/2, 1/2]$. In other words, the signatures lie inside a parallopiped whose sides are the good basis vectors. The attacker's challenge is to discover one edge of this parallelopiped.

Since the ϵs are random, they will average to 0. To base an attack on averaging s and $t-m$, the attacker must find something that does not average to zero. To do this, he uses the *reversal* of s and $t-m$. The reversal of a polynomial a is the polynomial

$$\bar{\mathsf{a}}(X) = \mathsf{a}(X^{-1}) = \mathsf{a}_0 + \sum_{i=1}^{N-1} \mathsf{a}_{N-i} X^i .$$

We then set

$$\hat{\mathsf{a}} = \mathsf{a} * \bar{\mathsf{a}}.$$

Notice that $\hat{\mathsf{a}}$ has the form

$$\hat{\mathsf{a}} = \sum_{k=0}^{N-1} \left(\sum_{i=0}^{N-1} \mathsf{a}_i \mathsf{a}_{i+k} \right) X^k .$$

In particular, $\hat{\mathsf{a}}_0 = \sum_i \mathsf{a}^2$. This means that as the attacker averages over a transcript of $\hat{\mathsf{s}}, \widehat{\mathsf{t} - \mathsf{m}}$, the cross-terms will essentially vanish and the attacker will recover

$$\langle \hat{\epsilon}_0 \rangle (\hat{\mathsf{f}} + \hat{\mathsf{g}}) = \frac{N}{12} (\hat{\mathsf{f}} + \hat{\mathsf{g}})$$

for s and similarly for $\mathsf{t} - \mathsf{m}$, where $\langle . \rangle$ denotes the average of . over the transcript.

We refer to the product of a measurable with its reverse as its *second moment*. In the case of raw NTRUSign, recovering the second moment of a transcript reveals the Gram Matrix of the private basis. Experimentally, it appears that significant information about the Gram Matrix is leaked after 10,000 signatures for all of the parameter sets in this paper. Nguyen and Regev [15] demonstrated an attack on parameter sets without perturbations that combines Gram matrix recovery with creative use of averaging moments over the signature transcript to recover the private key after seeing a transcript of approximately 70,000 signatures. This result has been improved to just 400 signatures in [24], and so the use of unperturbed NTRUSign is strongly discouraged.

Obviously, something must be done to reduce information leakage from transcripts, and this is the role played by perturbations.

Transcript Security for NTRUSign *with Perturbations*

In the case with B perturbations, the expectation of $\hat{\mathsf{s}}$ and $\hat{\mathsf{t}}-\hat{\mathsf{m}}$ is (up to lower order terms)

$$E(\hat{\mathsf{s}}) = (N/12)(\hat{\mathsf{f}}_0 + \hat{\mathsf{g}}_0 + \cdots + \hat{\mathsf{f}}_B + \hat{\mathsf{g}}_B)$$

and

$$E(\hat{\mathsf{t}} - \hat{\mathsf{m}}) = (N/12)(\hat{\mathsf{f}}_0 + \hat{\mathsf{g}}_0 + \cdots + \hat{\mathsf{f}}_B + \hat{\mathsf{g}}_B).$$

Note that this second moment is no longer a Gram matrix but the sum of $(B + 1)$ Gram matrices. Likewise, the signatures in a transcript do not lie within a parallelopiped but within the sum of $(B + 1)$ parallelopipeds.

This complicates matters for an attacker. The best currently known technique for $B = 1$ is to calculate

$$\begin{aligned} &\text{the second moment } \langle \hat{\mathsf{s}} \rangle \\ &\text{the fourth moment } \langle \hat{\mathsf{s}}^2 \rangle \\ &\text{the sixth moment } \langle \hat{\mathsf{s}}^3 \rangle \;. \end{aligned}$$

Since, for example, $\langle \hat{\mathsf{s}} \rangle^2 \neq \langle \hat{\mathsf{s}}^2 \rangle$, the attacker can use linear algebra to eliminate f_1 and g_1 and recover the Gram matrix, whereupon the attack of [15] can be used to recover the private key. It is an interesting open research question to determine whether there is any method open to the attacker that enables them to eliminate the perturbation bases without recovering the sixth moment (or, in the case of B perturbation bases, the $(4B + 2)$-th moment). For now, the best known attack is this algebraic attack, which requires the recovery of the sixth moment. It is an open research problem to discover analytic attacks based on signature transcripts that improve on this algebraic attack.

We now turn to estimate τ, the length of transcript necessary to recover the sixth moment. Consider an attacker who attempts to recover the sixth moment by averaging over τ signatures and rounding to the nearest integer. This will give a reasonably correct answer when the error in many coefficients (say at least half) is less than $1/2$. To compute the probability that an individual coefficient has an error less than $1/2$, write $(12/N)\hat{s}$ as a main term plus an error, where the main term converges to $\hat{f}_0 + \hat{g}_0 + \hat{f}_1 + \hat{g}_1$. The error will converge to 0 at about the same rate as the main term converges to its expected value. If the probability that a given coefficient is further than $1/2$ from its expected value is less than $1/(2N)$, then we can expect at least half of the coefficients to round to their correct values (Note that this convergence cannot be speeded up using lattice reduction in, for example, the lattice $\hat{h}$, because the terms $\hat{f}$, $\hat{g}$ are unknown and are larger than the expected shortest vector in that lattice).

The rate of convergence of the error and its dependence on τ can be estimated by an application of Chernoff-Hoeffding techniques [48], using an assumption of a reasonable amount of independence and uniform distribution of random variables within the signature transcript. This assumption appears to be justified by experimental evidence and, in fact, benefits the attacker by ensuring that the cross-terms converge to zero.

Using this technique, we estimate that, to have a single coefficient in the $2k$-th moment with error less than $\frac{1}{2}$, the attacker must analyze a signature transcript of length $\tau > 2^{2k+4} d^{2k}/N$. Here, d is the number of 1s in the trinary key. Experimental evidence for the second moment indicates that the required transcript length will in fact be much longer than this. For one perturbation, the attacker needs to recover the sixth moment accurately, leading to required transcript lengths $\tau > 2^{30}$ for all the recommended parameter sets in this paper.

NTRUSign Security: Summary

The parameter sets in Table 11.8 were generated with $\rho = 1.1$ and selected to give the shortest possible signing time σ_S. These security estimates do *not* take the hybrid attack of [44] into account and are presented only to give a rough idea of the parameters required to obtain a given level of security.

The security measures have the following meanings:

ω_{lk}	The security against key recovery by lattice reduction
c	The lattice characteristic c that governs key recovery times
ω_{cmb}	The security against key recovery by combinatorial means
ω_{frg}	The security against forgery by combinatorial means
γ	The lattice characteristic γ that governs forgery times
ω_{lf}	The security against forgery by lattice reduction

Table 11.8 Parameters and relevant security measures for trinary keys, one perturbation, $\rho = 1.1$, q = power of 2

Parameters						Security measures						
k	N	d	q	β	$\mathcal{N}$	ω_{cmb}	c	ω_{lk}	ω_{frg}	γ	ω_{lf}	$\log_2(\tau)$
80	157	29	256	0.38407	150.02	104.43	5.34	93.319	80	0.139	102.27	31.9
112	197	28	256	0.51492	206.91	112.71	5.55	117.71	112	0.142	113.38	31.2
128	223	32	256	0.65515	277.52	128.63	6.11	134.5	128	0.164	139.25	32.2
160	263	45	512	0.31583	276.53	169.2	5.33	161.31	160	0.108	228.02	34.9
192	313	50	512	0.40600	384.41	193.87	5.86	193.22	192	0.119	280.32	35.6
256	349	75	512	0.18543	368.62	256.48	7.37	426.19	744	0.125	328.24	38.9

Quantum Computers

All cryptographic systems based on the problems of integer factorization, discrete log, and elliptic curve discrete log are potentially vulnerable to the development of an appropriately sized quantum computer, as algorithms for such a computer are known that can solve these problems in time polynomial in the size of the inputs. At the moment, it is unclear what effect quantum computers may have on the security of the NTRU algorithms.

The paper [28] describes a quantum algorithm that square-roots asymptotic lattice reduction running times for a specific lattice reduction algorithm. However, since, in practice, lattice reduction algorithms perform much better than they are theoretically predicted to, it is not clear what effect this improvement in asymptotic running times has on practical security. On the combinatorial side, Grover's algorithm [49] provides a means for square-rooting the time for a brute-force search. However, the combinatorial security of NTRU keys depends on a meet-in-the-middle attack, and we are not currently aware of any quantum algorithms to speed this up. The papers [50–54] consider potential sub-exponential algorithms for certain lattice problems. However, these algorithms depend on a subexponential number of coset samples to obtain a polynomial approximation to the shortest vector, and no method is currently known to produce a subexponential number of samples in subexponential time.

At the moment, it seems reasonable to speculate that quantum algorithms will be discovered that will square-root times for both lattice reduction and meet-in-the-middle searches. If this is the case, NTRU key sizes will have to approximately double, and running times will increase by a factor of approximately 4 to give the same security levels. As demonstrated in the performance tables in this paper, this still results in performance that is competitive with public key algorithms that are in use today. As quantum computers are seen to become more and more feasible, NTRUEncrypt and NTRUSign should be seriously studied with a view to wide deployment.

References

1. M. Ajtai, *The shortest vector problem in L2 is NP-hard for randomized reductions (extended abstract)*, in Proc. 30th ACM symp on Theory of Computing, pp. 10–19, 1998
2. O. Goldreich, D. Micciancio, S. Safra, J.-P. Seifert, *Approximating shortest lattice vectors is not harder than approximating closest lattice vectors*, in Inform. Process. Lett. 71(2), 55–61, 1999
3. D. Micciancio, *Complexity of Lattice Problems*, Kluwer International Series in Engineering and Computer Science, vol. 671 Kluwer, Dordrecht, March 2002
4. H. Cohn, A. Kumar, *The densest lattice in twenty-four dimensions* in Electron. Res. Announc. Amer. Math. Soc. 10, 58–67, 2004
5. R.C. Merkle, M.E. Hellman, *Hiding information and signatures in trapdoor knapsacks*, in Secure communications and asymmetric cryptosystems, AAAS Sel. Sympos. Ser, 69, 197–215, 1982
6. A.M. Odlyzko, *The rise and fall of knapsack cryptosystems*, in Cryptology and computational number theory (Boulder, CO, 1989), Proc. Sympos. Appl. Math. 42, 75–88, 1990
7. A.K. Lenstra, A.K., H.W. Lenstra, L. Lovász, *Factoring polynomials with rational coefficients*, Math. Ann. 261, 515–534, 1982
8. M. Ajtai, C. Dwork, *A public-key cryptosystem with worst- case/average-case equivalence*, in Proc. 29th Annual ACM Symposium on Theory of Computing (STOC), pp. 284–293, ACM Press, New York, 1997
9. O. Goldreich, S. Goldwasser, S. Halevi, *Public-key cryptosystems from lattice reduction problems, advances in cryptology*, in Proc. Crypto 97, Lecture Notes in Computer Science, vol. 1294, pp. 112–131, Springer, Berlin, 1997
10. J. Hoffstein, J. Pipher, J.H. Silverman, *NTRU: A new high speed public key cryptosystem*, in J.P. Buhler (Ed.), Algorithmic Number Theory (ANTS III), Portland, OR, June 1998, Lecture Notes in Computer Science 1423, pp. 267–288, Springer, Berlin, 1998
11. P. Nguyen, J. Stern, *Cryptanalysis of the Ajtai-Dwork cryptosystem*, in Proc. of Crypto '98, vol. 1462 of LNCS, pp. 223–242, Springer, Berlin, 1998
12. L. Babai, *On Lovasz Lattice Reduction and the Nearest Lattice Point Prob- lem*, Combinatorica, vol. 6, pp. 113, 1986
13. P. Nguyen, *Cryptanalysis of the Goldreich-Goldwasser-Halevi Cryptosystem from Crypto '97*, in Crypto'99, LNCS 1666, pp. 288–304, Springer, Berlin, 1999
14. J. Hoffstein, J.H. Silverman, W. Whyte, *Estimated Breaking Times for NTRU Lattices*, Technical report, NTRU Cryptosystems, June 2003 Report #012, version 2, Available at http://www.ntru.com
15. P. Nguyen, O. Regev, *Learning a Parallelepiped: Cryptanalysis of GGH and NTRU Signatures*, Eurocrypt, pp. 271–288, 2006
16. J. Hoffstein, J. Pipher, J.H. Silverman, *NSS: The NTRU signature scheme*, in B. Pfitzmann (Ed.), Eurocrypt '01, Lecture Notes in Computer Science 2045, pp. 211–228, Springer, Berlin, 2001
17. J. Hoffstein, N. Howgrave-Graham, J. Pipher, J. Silverman, W. Whyte, *NTRUSign: Digital Signatures Using the NTRU Lattice*, CT-RSA, 2003
18. J. Hoffstein, N. Howgrave-Graham, J. Pipher, J. Silverman, W. Whyte, *NTRUSign: Digital Signatures Using the NTRU Lattice, extended version*, Available from http://ntru.com/cryptolab/pdf/NTRUSign-preV2.pdf
19. N. Howgrave-Graham, P. Nguyen, D. Pointcheval, J. Proos, J.H. Silverman, A. Singer, W. Whyte, *The Impact of Decryption Failures on the Security of NTRU Encryption*, Advances in Cryptology – Crypto 2003, Lecture Notes in Computer Science 2729, pp. 226–246, Springer, Berlin, 2003
20. J. Proos, *Imperfect Decryption and an Attack on the NTRU Encryption Scheme*, IACR ePrint Archive, report 02/2003, Available at http://eprint.iacr.org/2003/002/
21. Consortium for Efficient Embedded Security, *Efficient Embedded Security Standard #1 version 2*, Available from http://www.ceesstandards.org

22. C. Gentry, J. Jonsson, J. Stern, M. Szydlo, *Cryptanalysis of the NTRU signature scheme, (NSS), from Eurocrypt 2001*, in Proc. of Asiacrypt 2001, Lecture Notes in Computer Science, pp. 1–20, Springer, Berlin, 2001
23. C. Gentry, M Szydlo, *Cryptanalysis of the Revised NTRU SignatureScheme*, Advances in Cryptology – Eurocrypt '02, Lecture Notes in Computer Science, Springer, Berlin, 2002
24. P.Q. Nguyen, *A Note on the Security of NTRUSign*, Cryptology ePrint Archive: Report 2006/387
25. N. Howgrave-Graham, J.H. Silverman, W. Whyte, *Choosing Parameter Sets for* NTRUEncrypt *with* NAEP *and* SVES-3, CT-RSA, 2005
26. J. Hoffstein, N. Howgrave-Graham, J. Pipher, J. Silverman, W. Whyte, *Performance Improvements and a Baseline Parameter Generation Algorithm for* NTRUSign, Workshop on Mathematical Problems and Techniques in Cryptology, Barcelona, Spain, June 2005
27. P. Shor, *Polynomial time algorithms for prime factorization and discrete logarithms on a quantum computer*, Preliminary version appeared in Proc. of 35th Annual Symp. on Foundations of Computer Science, Santa Fe, NM, Nov 20–22, 1994. Final version published in SIAM J. Computing 26 (1997) 1484, Published in SIAM J. Sci. Statist. Comput. 26, 1484, 1997, e-Print Archive: quant-ph/9508027
28. C. Ludwig, *A Faster Lattice Reduction Method Using Quantum Search*, TU-Darmstadt Cryptography and Computeralgebra Technical Report No. TI-3/03, revised version published in Proc. of ISAAC 2003
29. J. Hoffstein, J.H. Silverman, *Invertibility in truncated polynomial rings*, Technical report, NTRU Cryptosystems, October 1998, Report #009, version 1, Available at http://www.ntru.com
30. N. Howgrave-Graham, J.H. Silverman, A. Singer, W. Whyte, *NAEP: Provable Security in the Presence of Decryption Failures*, IACR ePrint Archive, Report 2003-172, http://eprint.iacr.org/2003/172/
31. M. Bellare, P. Rogaway, *Optimal asymmetric encryption*, in Proc. of Eurocrypt '94, vol. 950 of LNCS, IACR, pp. 92–111, Springer, Berlin, 1995
32. D. Boneh, *Simplified OAEP for the RSA and Rabin functions*, in Proc. of Crypto '2001, Lecture Notes in Computer Science, vol. 2139, pp. 275–291, Springer, Berlin, 2001
33. M. Brown, D. Hankerson, J. López, A. Menezes, *Software Implementation of the NIST Elliptic Curves Over Prime Fields* in D. Naccache (Ed.), CT-RSA 2001, LNCS 2020, pp. 250–265, Springer, Berlin, 2001
34. A.K. Lenstra, E.R. Verheul, *Selecting cryptographic key sizes*, J. Cryptol. 14(4), 255–293, 2001, Available from http://www.cryptosavvy.com
35. R.D. Silverman, *A Cost-Based Security Analysis of Symmetric and Asymmetric Key Lengths*, RSA Labs Bulletin 13, April 2000, Available from http://www.rsasecurity.com/rsalabs
36. NIST Special Publication 800-57, *Recommendation for Key Management, Part 1: General Guideline*, January 2003, Available from http://csrc.nist.gov/CryptoToolkit/kms/guideline-1-Jan03.pdf
37. B. Kaliski, *Comments on SP 800-57, Recommendation for Key Management, Part 1: General Guidelines*, Available from http://csrc.nist.gov/CryptoToolkit/kms/CommentsSP800-57Part1.pdf
38. R. Rivest, A. Shamir, L.M. Adleman, *A method for obtaining digital signatures and public-key cryptosystems*, Commun. ACM 21, 120–126, 1978
39. N. Koblitz, *Elliptic curve cryptosystems*, Mathematics of Computation, 48, pp. 203–209, 1987
40. V. Miller, *Uses of elliptic curves in cryptography*, in Advances in Cryptology: Crypto '85, pp. 417–426, 1985
41. N. Howgrave-Graham, J.H. Silverman, W. Whyte, *A Meet-in-the-Middle Attack on an NTRU Private key*, Technical report, NTRU Cryptosystems, June 2003, Report #004, version 2, Available at `http://www.ntru.com`
42. D. Coppersmith, A. Shamir, *Lattice Attack on NTRU*, Advances in Cryptology – Eurocrypt 97, Springer, Berlin

43. A. May, J.H. Silverman, *Dimension reduction methods for convolution modular lattices*, in J.H. Silverman (Ed.), Cryptography and Lattices Conference (CaLC 2001), Lecture Notes in Computer Science 2146, Springer, Berlin, 2001
44. N. Howgrave-Graham, *A Hybrid Lattice-Reduction and Meet-in-the-Middle Attack Against NTRU*, Lecture Notes in Computer Science, Springer, Berlin, in Advances in Cryptology – CRYPTO 2007, vol. 4622/2007, pp. 150–169, 2007
45. P. Hirschhorn, J. Hoffstein, N. Howgrave-Graham, W. Whyte, *Choosing NTRU Parameters in Light of Combined Lattice Reduction and MITM Approaches*
46. C. Gentry, Key Recovery and Message Attacks on NTRU-Composite, *Advances in Cryptology – Eurocrypt '01*, LNCS 2045, Springer, Berlin, 2001
47. J.H. Silverman, *Invertibility in Truncated Polynomial Rings*, Technical report, NTRU Cryptosystems, October 1998, Report #009, version 1, Available at http://www.ntru.com
48. Kirill Levchenko, *Chernoff Bound*, Available at http://www.cs.ucsd.edu/˜klevchen/techniques/chernoff.pdf
49. L. Grover, *A fast quantum mechanical algorithm for database search*, in Proc. 28th Annual ACM Symposium on the Theory of Computing, 1996
50. O. Regev, *Quantum computation and lattice problems*, in Proc. 43rd Annual Symposium on the Foundations of Computer Science, pp. 520–530, IEEE Computer Society Press, Los Alamitos, California, USA, 2002, http://citeseer.ist.psu.edu/regev03quantum.html
51. T. Tatsuie, K. Hiroaki, *Efficient algorithm for the unique shortest lattice vector problem using quantum oracle*, IEIC Technical Report, Institute of Electronics, Information and Communication Engineers, vol. 101, No. 44(COMP2001 5–12), pp. 9–16, 2001
52. Greg Kuperberg, *A Sub-Exponential-Time Quantum Algorithm For The Dihedral Hidden Subgroup Problem*, 2003, http://arxiv.org/abs/quant-ph/0302112
53. O. Regev, *A Sub-Exponential Time Algorithm for the Dihedral Hidden Subgroup Problem with Polynomial Space*, June 2004, http://arxiv.org/abs/quant-ph/0406151
54. R. Hughes, G. Doolen, D. Awschalom, C. Caves, M. Chapman, R. Clark, D. Cory, D. DiVincenzo, A. Ekert, P. Chris Hammel, P. Kwiat, S. Lloyd, G. Milburn, T. Orlando, D. Steel, U. Vazirani, B. Whaley, D. Wineland, *A Quantum Information Science and Technology Roadmap, Part 1: Quantum Computation*, Report of the Quantum Information Science and Technology Experts Panel, Version 2.0, April 2, 2004, Advanced Research and Development Activity, http://qist.lanl.gov/pdfs/qc_roadmap.pdf
55. ANSI X9.62, Public Key Cryptography for the Financial Services Industry: The Elliptic Curve Digital Signature Algorithm (ECDSA), 1999
56. D. Hankerson, J. Hernandez, A. Menezes, *Software implementation of elliptic curve cryptography over binary fields*, in Proc. CHES 2000, Lecture Notes in Computer Science, 1965, pp. 1–24, 2000
57. J. Hoffstein, J.H. Silverman, *Optimizations for NTRU*, In Publickey Cryptography and Computational Number Theory. DeGruyter, 2000, Available from http://www.ntru.com
58. J. Hoffstein, J.H. Silverman, *Random Small Hamming Weight Products with Applications to Cryptography*, Discrete Applied Mathematics, Available from http://www.ntru.com
59. E. Kiltz, J. Malone-Lee, *A General Construction of IND-CCA2 Secure Public Key Encryption*, in Cryptography and Coding, pp. 152–166, Springer, Berlin, December 2003
60. T. Meskanen, A. Renvall, *Wrap Error Attack Against NTRUEncrypt*, in Proc. of WCC '03, 2003
61. NIST, *Digital Signature Standard*, FIPS Publication 186-2, February 2000

Chapter 12
The Geometry of Provable Security: Some Proofs of Security in Which Lattices Make a Surprise Appearance

Craig Gentry

Abstract We highlight some uses of lattice reduction in security proofs of nonlattice-based cryptosystems. In particular, we focus on RSA-OAEP, the Rabin partial-domain hash signature scheme, techniques to compress Rabin signatures and ciphertexts, the relationship between the RSA and Paillier problems and Hensel lifting, and the hardness of the most significant bits of a Diffie–Hellman secret.

Introduction

In modern cryptography, we try to design and use cryptosystems that are "provably secure." That is, we try to *prove* (via *reductio ad absurdum*) that if an *adversary* can efficiently attack a cryptosystem within the framework of a specified *security model*, then the attacker can be used to help efficiently solve a specified *hard problem* (or, more properly, a problem that is *assumed* to be hard). In short, a "proof of security" is a *reduction* of putative hard problem to a cryptosystem. A provably secure cryptosystem might not actually be secure in the real world. Even if the security proof is correct, the hardness assumption might turn out to be false, the security model might not account for all feasible real-world attacks, or the cryptosystem might not be used in the real world as envisioned in the security model. Still, an approach based on provable security has tangible value. Typically (or at least preferably), a security proof uses hardness assumptions and security models that have been honed and analyzed by researchers for years. In this case, one can be reasonably confident that the system is secure in the real world, if the security proof is correct and the system is not misused.

This survey focuses on some "surprising" uses of lattices in proofs of security. Of course, there are a number of "lattice-based cryptosystems" that directly use a problem over lattices, such as the unique shortest vector problem, as the assumed hard problem; this is not what we will be discussing here. Rather, we will focus primarily

C. Gentry
Stanford University, USA,
e-mail: cgentry@cs.stanford.edu

P.Q. Nguyen and B. Vallée (eds.), *The LLL Algorithm*, Information Security and Cryptography, DOI 10.1007/978-3-642-02295-1_12,

on cryptosystems based on more "conventional" assumptions, such as the hardness of factoring, the RSA problem, and the Diffie–Hellman problem. For most of the cryptosystems we consider, the actual implementation of the cryptosystem does not involve lattices either. Instead, lattices only appear in the security reduction, which is somewhat surprising, at least to someone less familiar with the role of lattices in attacking the Diophantine problems that often underlie "classical" cryptosystems.

In Section "Preliminaries: Basic Notions in Provable Security", we review some basic cryptographic concepts, such as standard hardness assumptions (like factoring and RSA), security models for signature and encryption schemes, and the random oracle model. To illustrate the concepts, we give the reduction, in the random oracle model, of factoring large numbers to the problem of forging a Rabin signature.

We describe the flawed security proof of RSA-OAEP in Section "The Security of RSA-OAEP and Rabin-OAEP", as well as several lattice-based approaches for patching the proof. A couple of these approaches use Coppersmith's method to get efficient security reductions for Rabin and low-exponent RSA encryption. For general-exponent RSA, the reduction is also lattice-based, but far less efficient, since it solves RSA only after running the RSA-OAEP-breaking algorithm twice.

In Section "Compressing Rabin Signatures and Ciphertexts", we describe how to compress Rabin signatures or ciphertexts down to $c \log N$ bits, $c \in \{1/2, 2/3\}$, while retaining provable security based on the hardness of factoring N. Interestingly, one can define an efficiently computable trapdoor one-way "quasi-permutation" over a subinterval $[0, c'N^{\alpha}]$ of $[0, N]$ for $\alpha = 2/3$ based on modular squaring. Evaluating this quasi-permutation involves finding lattice points that lie in the region between two parabolas.

Section "The Relationship Among the RSA, Paillier, and RSA-Paillier Problems" describes how the hardness of the RSA, Paillier and RSA-Paillier problems can all be solved by Hensel lifting (to an appropriate power of the modulus N), and then applying lattice reduction to the result. Curiously, the Hensel power needed to break Paillier encryption is smaller than that needed to break RSA with encryption exponent N, suggesting a separation.

Finally, in Section "The Bit Security of Diffie–Hellman", we review a result that suggests that the most significant bits of a Diffie–Hellman secret are hard to compute, since otherwise lattice reductions could be used to recover the entire secret.

Preliminaries: Basic Notions in Provable Security

Here, we review some basic notions in provable security – namely, the security models for signature and encryption schemes; standard complexity assumptions such as factoring, RSA, and Diffie–Hellman; and an idealization of hash functions called the "random oracle model." To illustrate how these notions are used in security proofs and to illustrate the importance of "concrete security" (as opposed to asymptotic security), we review the rather simple reduction of factoring to the security of full-domain-hash Rabin signatures in the random oracle model. This reduction will also help understand the lattice-based security proof for partial-domain Rabin signatures in the random oracle model, discussed in Section "The Security of Rabin-PDH".

Security Models

Clearly, no cryptographic algorithm can be secure against an adversary with unbounded resources, that can access all cryptographic secrets in the system. Consequently, in modelling the adversary, one must limit its capabilities before one can hope to prove the security of the cryptographic algorithm. However, the adversarial model should include only those limitations that one can justifiably claim that a real-world adversary will have in practice.

First, let us consider how to model security for digital signature schemes. Recall the definition of a signature scheme.

Definition 1 (Signature Scheme). A signature scheme has three algorithms (Gen, Sign, Verify), defined as follows:

- Gen is a probabilistic key generation algorithm, which given 1^κ for security parameter κ, outputs secret signing key SK and public verification key PK.
- Sign takes (SK, m) as input, where m is the message to be signed, and outputs a signature $\sigma = \mathsf{Sign}_{SK}(m)$.
- Verify takes (m, σ', PK) as input. It outputs $\mathsf{Verify}_{PK}(m, \sigma') \in \{0, 1\}$, where "1" means the signature is accepted, and "0" rejected. It is required that if $\sigma \leftarrow \mathsf{Sign}_{SK}(m)$, then $\mathsf{Verify}_{PK}(m, \sigma) = 1$.

For concreteness, one may think of κ as related to the bit-length of PK.

A protocol's security is typically defined in terms of an interactive "game" played between a challenger and the adversary. Here is the commonly accepted notion of security for a digital signature scheme.

Definition 2 (Security against existential forgery under an adaptive chosen-message attack). A signature scheme (Gen,Sign,Verify) is (t, q, ε)-secure against existential forgery under an adaptive chosen-message attack if an adversary limited to computation time t has probability, at most ε, of winning the following game.

1. The challenger runs Gen and gives PK to the adversary.
2. For $1 \leq i \leq q$:
   The adversary adaptively picks m_i and requests a signature on m_i.
   The challenger returns $\sigma_i = \mathsf{Sign}_{SK}(m_i)$.
3. Finally, the adversary outputs a signature σ for (PK, m).

The adversary wins if σ is a valid signature for (PK, m) – i.e., $\mathsf{Verify}_{PK}(m, \sigma) = 1$ – and the adversary did not ask the challenger to sign m during the interactive phase.

If the scheme is not (t, q, ε)-secure, we say that there is an adversary that (t, q, ε)-breaks the scheme – i.e., there is an adversary limited to t computation and q queries that wins the above game with probability greater than ε.

Of course, a (t, q, ε)-secure scheme is not actually "secure" unless ε is very small for reasonable values of t and q. Since one would at least to defend against polynomial-time adversaries, one would like ε to be *negligible* (i.e., asymptotically smaller than the inverse of any polynomial) in the security parameter κ whenever t and q are polynomial in κ.

This "game" may seem rather abstract, but it actually models reality quite well. To use a weaker security model, one would have to justify why a real-world adversary would be unable – for all settings in which the signature scheme is used – to collect a long transcript of its target's signatures on messages of its choice. On the other hand, using a stronger security model would be overkill for most settings, and would make it difficult or impossible to prove the security of a given signature scheme. Nonetheless, there are stronger security models that account for even more powerful adversaries. For example, an adversary may be able to learn significant information – even an entire secret key – simply by measuring the amount of time the device takes to perform a cryptographic operation or by measuring the amount of power that the device consumes [1, 2]. (See [3] and [4] for a description of how such "side-channel" attacks may be included in the adversarial model.)

Now, let us consider encryption schemes. Recall the definition of an encryption scheme.

Definition 3 (Public-Key Encryption Scheme). A public-key encryption scheme has three algorithms (Gen, $\mathsf{Encrypt}$, $\mathsf{Decrypt}$), defined as follows:

- Gen is a probabilistic key generation algorithm, which given 1^κ for security parameter κ, outputs secret decryption key SK and public encryption key PK.
- $\mathsf{Encrypt}$ takes (PK, m) as input, where $m \in \mathcal{M}$ is the plaintext message message space $\mathcal{M}$, and outputs a ciphertext $C = \mathsf{Encrypt}_{PK}(m)$.
- $\mathsf{Decrypt}$ takes (C', SK) as input. It outputs $\mathsf{Decrypt}_{SK}(C')$, which may be either a candidate plaintext or an error symbol $\perp$. It is required that if $C \leftarrow \mathsf{Encrypt}_{PK}(m)$, then $\mathsf{Decrypt}_{SK}(C) = m$.

Defining the "right" security model for encryption schemes is more difficult. Here is a weak notion of security.

Definition 4 (One-wayness). A public-key encryption scheme (Gen, $\mathsf{Encrypt}$, $\mathsf{Decrypt}$) is (t, ε)-one-way, if an adversary limited to computation time t has probability, at most ε, of winning the following game.

1. The challenger runs Gen and gives PK to the adversary.
2. The challenger picks random message $m \in \mathcal{M}$ and sends $C = \mathsf{Encrypt}_{PK}(m)$ to the adversary.
3. The adversary outputs a candidate plaintext m'.

The adversary wins if $m' = m$.

One-wayness is too weak for most applications for several reasons. It does not account for an adversary that can obtain decryptions from the challenger of ciphertexts other than C. It also does not account for the fact that the adversary might obtain significant information about m from C, even if it cannot recover m completely. Finally, one-wayness leaves open the possibility that ciphertexts are "malleable" – i.e., an adversary can, without knowing m, modify C to construct a ciphertext C' that encrypts $f(m)$ for some nontrivial function f.

Here is the preferred notion of security for a public-key encryption scheme, defined in [5].

Definition 5 (Semantic security against adaptive chosen-ciphertext attack). A public-key encryption scheme (Gen, Encrypt, Decrypt) is (t, q, ε) semantically secure against adaptive chosen-ciphertext attack if, when limited to computation time t, the adversary wins the following game with probability at least $1/2 - \varepsilon$ and at most $1/2 + \varepsilon$.

1. The challenger runs Gen and gives PK to the adversary.
2. Phase 1: For $i = 1$ to $q' \leq q$:
   The adversary chooses a ciphertext C_i.
   The challenger returns $\mathsf{Decrypt}_{SK}(C_i)$ to the adversary.
3. Challenge: The adversary chooses two messages $m_0, m_1 \in \mathcal{M}$ and sends them to the challenger. The challenger randomly sets $b \in \{0, 1\}$. It sends $C^* = \mathsf{Encrypt}_{PK}(m_b)$ to the adversary.
4. Phase 2: Similar to Phase 1 for $i \in [q' + 1, q]$, subject to the constraint that $C_i \neq C^*$.
5. Guess: The adversary outputs $b' \in \{0, 1\}$.

The adversary wins if $b' = b$.

Chosen-ciphertext security turns out to be equivalent to "non-malleability," as defined in [6, 7]. Sometimes, the following intermediate notion of security is useful.

Definition 6 (Semantic security against chosen-plaintext attack). A public-key encryption scheme (Gen, Encrypt, Decrypt) is (t, ε)-semantically secure against chosen-plaintext attack if it is $(t, 0, \varepsilon)$-semantically secure against adaptive chosen-ciphertext attack.

Note that (Gen, Encrypt, Decrypt) cannot be semantically-secure, in either sense, for a reasonable choice t and ε, unless Encrypt is probabilistic.

Complexity Assumptions

Here, we review some "hard problems" frequently used in cryptography, beginning with the "Diffie–Hellman" problem that initiated public-key cryptography [8].

Definition 7 (Diffie–Hellman Problem). Let G be a cyclic group of prime order q, and let $g \in G$ be a generator. Given g, g^a, g^b for random $a, b \in \mathbb{Z}_q^2$, compute g^{ab}.

The Diffie–Hellman assumption (for G) is that the Diffie–Hellman problem is intractible – i.e., there is no efficient (polynomial in $\log q$) algorithm for solving it. This assumption underlies the security of the Diffie–Hellman key agreement scheme, in which one party (typically called "Alice") picks random a and transmits g^a, the other party (typically called "Bob") picks random b and transmits g^b, and their shared secret is $(g^b)^a = (g^a)^b = g^{ab}$. This assumption also implies the one-wayness of the ElGamal encryption scheme [9], in which the recipient's secret/public key pair is (a, g^a) and the sender encrypts $m \in G$ by picking b and sending $C = (g^b, m \cdot (g^a)^b) \in G^2$. If the Diffie–Hellman problem is hard, then so is the following problem.

Definition 8 (Discrete Logarithm Problem). Let G be a cyclic group of prime order q and let $g \in G$ be a generator. Given g, g^a for random $a \in \mathbb{Z}_q$, compute a.

The first public-key signature and encryption schemes used the following problem [10].

Definition 9 (RSA Problem). Let N be a composite integer, and let e be an integer coprime to $\phi(N)$. Given $s \in \mathbb{Z}_N^*$, compute $r \in \mathbb{Z}_N^*$ such that $r^e = s \bmod N$.

(Note: $\mathbb{Z}_N$ is often used in cryptography as shorthand for $\mathbb{Z}/N\mathbb{Z}$.) The RSA function $f(x) = r^e \bmod N$ is the classic example of a one-way (i.e., hard to invert) trapdoor permutation. With the trapdoor information – i.e., the factorization of N – the RSA function becomes easy to invert. In the RSA cryptosystem, the recipient picks two random large primes p and q and sets $N = pq$; its secret/public key pair is $((p, q), N)$. The sender encrypts $m \in \mathbb{Z}_N^*$ as $C = m^e \bmod N$. The RSA assumption implies the one-wayness of the RSA cryptosystem.

When N is chosen as the product of two primes (which is normally the case in cryptography), the RSA problem obviously can be reduced efficiently the following problem.

Definition 10 (Factoring). Given a composite integer N, output a nontrivial factor of N.

By the abovementioned reduction, the factoring assumption is *weaker* than the RSA assumption. The security of Rabin signing and encryption [11], discussed later, is based on factoring.

Random Oracle Model

Many natural cryptographic constructions, including some discussed below, use cryptographic hash functions. Roughly speaking, a cryptographic hash function is a deterministic and efficiently computable function whose output distribution is as "random-looking" as possible. A cryptographic hash function is typically used, for example, to create a short "digest" of a message before applying the signing operation in a signature scheme, rather than (inefficiently) applying the signing operation directly to a long message. For this to be secure, it is obviously necessary that the hash function H be "collision-resistant"; for, if it is easy to find distinct messages m_1 and m_2 such that $H(m_1) = H(m_2)$, then clearly the adversary can break the signature scheme by requesting a signature σ on m_1 and offering σ as a forgery on m_2. However, mere collision resistance is often not sufficient to achieve security.

The random oracle model, introduced by Bellare and Rogaway [12], is a heuristic that often simplifies the analysis of constructions that use cryptographic hash functions. In the random oracle model, one proves the security of a scheme in an idealized world in which one pretends that hash functions behave like truly random functions. This entails modifications to the definition of security and the attack game. Instead of each hash function $H_i : \{0,1\}^{m_i} \rightarrow \{0,1\}^{n_i}$ that would normally be given to the adversary as part of the public key, the adversary is given *oracle* access to a function $f_i : \{0,1\}^{m_i} \rightarrow \{0,1\}^{n_i}$, controlled by the challenger, that is chosen randomly from the set of all functions from $\{0,1\}^{m_i} \rightarrow \{0,1\}^{n_i}$. (Since the description of such a function requires space exponential in m_i, the challenger assigns the function incrementally as the adversary queries the oracle.) The intuition motivating this heuristic is that, if the hash function's output "looks random" to the adversary, why should the adversary be better able to break a scheme that uses the hash function than one that uses a truly random function?

It turns out that this intuition is not entirely correct; security in the random oracle model does not necessarily imply security in the real world [13]. However, known counterexamples to the heuristic are somewhat unnatural. So, security proofs in the random oracle model still seem useful for validating natural cryptographic constructions.

Reduction and Concrete Security

We illustrate how the notions that we have discussed so far – the security model, complexity assumptions, and the random oracle model – come together by reviewing the rather simple reduction of factoring to the security of the Rabin full-domain-hash signature (Rabin-FDH) scheme in the random oracle model. This reduction is also directly relevant to the lattice-based reduction of factoring to the security of the Rabin partial-domain-hash (Rabin-PDH) signature scheme in the random oracle model, described in Section "The Security of Rabin-PDH".

A Rabin Full-Domain-Hash (Rabin-FDH) Signature Scheme:

Gen: Generate suitably large random primes p, q with $p = 3 \bmod 8$ and $q = 7 \bmod 8$. Set $N = pq$. Let $H : \{0,1\}^* \rightarrow \mathbb{Z}_N^*$ be a hash function. The secret signing key is (p, q). The public key is (N, H).

Sign: Uniformly (but deterministically) pick σ s.t. $\sigma^2 = c \cdot H(m) \bmod N$ for $c \in \{\pm 1, \pm 2\}$.

Verify: Confirm that $\sigma^2 = c \cdot H(m) \bmod N$ for $c \in \{\pm 1, \pm 2\}$.

Remark 0.1. Since $p = 3 \bmod 8$ and $q = 7 \bmod 8$, $c \cdot H(m)$ is a quadratic residue for exactly one $c \in \{\pm 1, \pm 2\}$ for any $H(m) \in \mathbb{Z}_N^*$. For any quadratic residue in $\mathbb{Z}_N^*$, there are four square roots. For reasons that will become clear shortly, the scheme is secure only if, in the signing algorithm, the signer picks the same square root of $c \cdot H(m)$ each time it signs m. (Actually, it is fine if the signer sends $-\sigma$ instead, but it cannot send one of the other two square roots.) The scheme is "full-domain-hash," since H's output range is *all* of $\mathbb{Z}_N^*$.

Now, suppose that there is an adversary $\mathcal{A}$ that (t, q, ε)-breaks the Rabin-FDH signature scheme in the random oracle model. Then, we can construct an algorithm $\mathcal{B}$ that (t', ε')-solves the factoring problem, where $\varepsilon' > (\varepsilon - 4/\phi(N))/2$ and $t' = t + q \cdot t_{\text{mult}} + t_{\text{gcd}}$ (where t_{mult} is essentially the time needed to multiply two numbers modulo N and t_{gcd} the time needed to perform a certain gcd computation), by interacting with $\mathcal{A}$ as follows.

The adversary $\mathcal{A}$ can make signature queries and queries to the random oracle H. When $\mathcal{A}$ makes either a signature query or an H-query on m_i and if $\mathcal{A}$ has not queried m_i before, $\mathcal{B}$ picks uniformly random values $r_i \in \mathbb{Z}_N^*$ and $c_i \in \{\pm 1, \pm 2\}$ and sets $H(m_i) = r_i^2/c_i \bmod N$. It returns $H(m_i)$ as a response to $\mathcal{A}$'s H-query on m_i. If $\mathcal{A}$ requests a signature on m_i, $\mathcal{B}$ returns r_i. Finally, at the end of this game, $\mathcal{A}$ gives $\mathcal{B}$ a pair (m^*, σ).

The interaction between $\mathcal{B}$ and $\mathcal{A}$ is called a *simulation*. Note that $\mathcal{B}$ is not a "real signer," since it does not know the factorization of N, like a real signer would. Despite this handicap, $\mathcal{B}$ tries to provide a perfect simulation for $\mathcal{A}$ of what would happen if $\mathcal{A}$ interacted with a real signer. For this simulation to be perfect, the distribution of H-outputs and signatures in $\mathcal{B}$'s simulation should be *indistinguishable* from the real-world distribution; it is easy to verify that this is in fact the case in the above simulation for Rabin-FDH, in the random oracle model. Since the simulation is indistinguishable from the real world, $\mathcal{A}$ must win the simulated game in time t with probability greater than ε.

So, with probability greater than ε, we have that $\sigma^2 = c^* \cdot H(m^*)$ for some $c^* \in \{\pm 1, \pm 2\}$, where $\mathcal{A}$ did not query a signature on m^*. If $\mathcal{A}$ made no H-query on m^*, then $\mathcal{A}$ knows nothing about the value of $H(m^*)$; in this case, σ can be a valid signature on m^* only with probability $4/\phi(N)$. If $\mathcal{A}$ did make an H-query on m^*, then $\mathcal{B}$ knows a value r^* such that $r^{*2} = c^* \cdot H(m^*) = \sigma^2 \bmod N$. Since $\mathcal{A}$ did not make a signature query on m^*, $\mathcal{A}$ does not know which modular square root of $c^* \cdot H(m^*)$ is known by $\mathcal{B}$. Thus, with probability $1/2$, $\gcd(N, \sigma - r^*)$ is a nontrivial factor of N. So, we obtain our desired result: $\mathcal{B}$ factors N with probability $\varepsilon' > (\varepsilon - 4/\phi(N))/2$ in time $t' = t + q \cdot t_{\text{mult}} + t_{\text{gcd}}$.

Notice that this reduction is quite *tight*. That is, $\mathcal{B}$ can (t', ε')-factor for values of (t', ε') that are quite close to the values (t, ε) for which $\mathcal{A}$ can (t, q, ε)-break Rabin-FDH. ($\mathcal{A}$'s q queries are implicitly included in its time t.) Often in security proofs, either t' or ε' degrades by a multiplicative factor of q or q^2 from t or ε. In a security proof that we will see later, $\mathcal{B}$ solves its hard problem only with probability $\varepsilon' \approx \varepsilon^2$. Obviously, all other things being equal, a tight reduction is preferable to a *loose* one, because the former gives a better security guarantee. If a reduction is loose, one should adjust the security parameter of the system upward to ensure that the system has sufficient "concrete security"; this makes the system less efficient. For example, see [14] for a description of how, in the context of Schnorr signature schemes (like DSA), one must use security parameters much larger than those used in practice before the security proof becomes meaningful, since the security reduction is so loose. See [15] for a survey on the concrete security approach.

We note that there are other ways of instantiating Rabin signatures. For example, one can turn the signing operation into a permutation over $\mathbb{Z}_N^*$, for N generated as above, as follows. Let u be a modular square root of 1 with Jacobi symbol -1. Then, $f(x) = abr^2$ is a one-way permutation over $\mathbb{Z}_N^*$, where (a, b, r) are the unique values satisfying $x = au^b r \in \mathbb{Z}_N^*$ with $a \in \{\pm 1\}$, $b \in \{1, 2\}$, and $r \in [1, N/2]$ having Jacobi symbol 1. The signature for m is simply $f^{-1}(H(m))$. The disadvantage of this approach is that, for technical reasons, the reduction to factoring is loose; if $\mathcal{A}$ succeeds with probability ε and is permitted to make q signature queries, $\mathcal{B}$'s success probability is only about ε/q.

Preliminaries: Coppersmith's Algorithm

Several proofs of security discussed here are based on the following important result due to Coppersmith [16].

Theorem 1 (Coppersmith). *Let N be an integer and let $f(x) \in \mathbb{Z}_N[x]$ be a monic polynomial of degree d. Then, there is an efficient algorithm to find all $x_0 \in \mathbb{Z}$ such that $f(x_0) = 0 \bmod N$ and $|x_0| < N^{1/d}$.*

We will not discuss how Coppersmith's algorithm works in this survey; instead, we will use it as a black box. We denote the running time of Coppersmith's algorithm by $T_C(N, d)$ when finding roots of polynomial $f \in \mathbb{Z}[x]$ of degree d.

Caveat Emptor

In this survey, we will provide informal "explanations of security." Please do not mistake these for genuine proofs of security, or infer that these explanations reveal anything about the style or rigor of cryptographic proofs. Our explanations are merely intended to be mostly convincing and to illustrate how some security proofs use lattices.

The Security of RSA-OAEP and Rabin-OAEP

RSA is the most widely used public-key encryption scheme, and OAEP (optimal asymmetric encryption padding) [17] is the most widely used method of "padding" the plaintext message before applying the RSA permutation. Padding is necessary to prevent malleability attacks on RSA. If m were encrypted simply as $c = m^e \bmod N$, an adversary could, for any $a \in \mathbb{Z}_N^*$ and without knowing m, easily change the ciphertext to $c' = a^e \cdot m^e \bmod N$, which encrypts $a \cdot m \bmod N$. This is an attack we would like to prevent.

Bellare and Rogaway introduced OAEP and provided a proof that if f is a one-way permutation, then f-OAEP is a chosen-ciphertext-secure encryption scheme (in the random oracle model). Shoup [18] found an irreparable gap in their proof. He proposed a new padding scheme, OAEP+, with a valid proof. However, it turns out that one can prove (in the random oracle model) that f-OAEP *is* chosen-ciphertext secure assuming the *set partial-domain one-wayness* of f – i.e., the stronger assumption that, given $f(s\|t)$, it is hard to output a list (of reasonable length) that contains s. For the RSA permutation, one can show that one-wayness implies partial-domain one-wayness in two different ways, each of which uses lattices. For low-exponent RSA (e.g., $e = 3$) and Rabin, one can use Coppersmith's method to recover $s\|t$ from s very efficiently [18, 19]. For general-exponent RSA, the reduction is much looser, since one must recover *two* partial pre-images (namely, s_1 and s_2 of $(s_1\|t_1)^e \bmod N$ and $(s_2\|t_2)^e \bmod N$, where $s_2\|t_2 = a(s_1\|t_1) \bmod N$ for randomly chosen a) to recover one full pre-image by using lattices [20]. (Hence, the probability of recovering a full pre-image is only about ε^2, if the probability of recovering a partial pre-image is ε.)

We review these results in more detail below.

Shoup Reconsiders OAEP

Let f be a one-way permutation on k-bit strings and f^{-1} its inverse. From any such permutation f, the OAEP padding scheme, introduced in [17], induces an encryption scheme as follows.

f-OAEP Encryption Scheme: The message space is $\{0,1\}^n$ for $n = k - k_0 - k_1$, where k_0 and k_1 satisfy $k_0 + k_1 < k$ and where 2^{-k_0} and 2^{-k_1} are very small. The scheme uses two cryptographic hash functions $G : \{0,1\}^{k_0} \to \{0,1\}^{n+k_1}$ and $H : \{0,1\}^{n+k_1} \to \{0,1\}^{k_0}$, modeled as random oracles in the security analysis.

Gen: Run the key generation algorithm to obtain f and f^{-1}, where (f, G, H) is the public key and f^{-1} is the private key.

Encrypt: Given the plaintext $m \in \{0,1\}^n$, pick a random $r \in \{0,1\}^{k_0}$ and compute:

$$s \in \{0,1\}^{n+k_1}, \quad t \in \{0,1\}^{k_0}, \quad w \in \{0,1\}^k, \quad y \in \{0,1\}^k$$

as follows

$$\begin{aligned} s &= G(r) \oplus (m\|0^{k_1}), \\ t &= H(s) \oplus r, \\ w &= s\|t, \\ y &= f(w) . \end{aligned}$$

The ciphertext is y.

Decrypt: Given a ciphertext y, the decryption algorithm sets $w = f^{-1}(y)$, splits w appropriately into s and t, sets $r = H(s) \oplus t$, and sets $m\|c = G(r) \oplus s$. If $c = 0^{k_1}$, then the algorithm outputs m as the plaintext; otherwise, the algorithm rejects the ciphertext, outputting only the error symbol $\perp$.

In [17], Bellare and Rogaway claim that if f is a one-way permutation, then f-OAEP is secure against adaptive chosen-ciphertext attack in the random oracle. Shoup [18] demonstrates that the proof is flawed. Let us try to prove Bellare and Rogaway's claim ourselves to see where the flaw arises.

Let $\mathcal{A}$ be an algorithm that breaks the chosen-ciphertext security of f-OAEP; from $\mathcal{A}$, we would like to construct an algorithm $\mathcal{B}$ that inverts f – i.e., an algorithm that, given random $y^* \in \{0,1\}^k$, returns $w^* = f^{-1}(y^*)$ with non-negligible probability. In the simulation, $\mathcal{B}$'s responses to $\mathcal{A}$'s queries to the G-oracle or H-oracle are trivial. Specifically, $\mathcal{B}$ checks its log to see whether $\mathcal{A}$ has made the given query before. If so, $\mathcal{B}$ responds as it did before; if not, it simply generates a random string of the appropriate length, returns it to $\mathcal{A}$ as the result, and adds an entry in its log recording $\mathcal{A}$'s query and the response.

The difficult part of the proof is to show that $\mathcal{B}$ can respond appropriately to $\mathcal{A}$'s decryption queries and can gain information from $\mathcal{A}$'s final output. Borrowing Shoup's notation, we observe that each ciphertext y_i queried by $\mathcal{A}$ corresponds implicitly to a unique value $w_i = s_i\|t_i = f^{-1}(y_i)$, which in turn induces values for $r_i = H(s_i) \oplus t_i$ and $m_i\|c_i = G(r_i) \oplus s_i$. Similarly, if y^* is the ciphertext that $\mathcal{B}$ gives to $\mathcal{A}$ in the "Challenge" phase, it also induces values for $w^* = s^*\|t^*$, r^*, m^*, and c^*. Now, we can complete the proof if the following two related claims were true.

- (Claim 1): The adversary $\mathcal{A}$ has an extremely small chance (specifically, a 2^{-k_1} probability) of constructing a valid ciphertext y_i – i.e., one for which $c_i = 0^{k_1}$ – unless it queries s_i to the H-oracle and r_i to the G-oracle.
- (Claim 2): The adversary $\mathcal{A}$ cannot have any advantage in distinguishing which message m_b is encrypted by the challenge ciphertext y^* unless it queries s^* to the H-oracle and r^* to the G-oracle.

If these claims were true, then $\mathcal{B}$ could give a valid response to $\mathcal{A}$'s decryption query on y_i with high probability as follows. By Claim 1, if the ciphertext is valid, $\mathcal{A}$ must have (with high probability) queried s_i to H and r_i to G. $\mathcal{B}$ searches in its log for (r_i, s_i) by computing, for each pair (r_j, s_ℓ) queried to the G and H oracles respectively, the values $t_{j\ell} = H(s_\ell) \oplus r_j$, $w_{j\ell} = s_\ell\|t_{j\ell}$, and $y_{j\ell} = f(w_{j\ell})$. If $y_{j\ell} = y_i$, then $\mathcal{B}$ concludes that $(r_j, s_\ell) = (r_i, s_i)$, and therefore the correct decryption of y_i is $m_i\|c_i = G(r_j) \oplus s_\ell$; if $c_i = 0^{k_1}$, it returns m_i to $\mathcal{A}$. Otherwise, or if no $y_{j\ell} = y_i$, $\mathcal{B}$ outputs $\perp$, because it is overwhelmingly likely that $\mathcal{A}$'s ciphertext y_i is invalid. $\mathcal{B}$ uses a similar strategy to compute $f^{-1}(y^*)$ from its interaction with $\mathcal{A}$. By Claim 2, if $\mathcal{A}$ wins with non-negligible advantage, $\mathcal{A}$ must have queried s^* to H and r^* to G with non-negligible probability; $\mathcal{B}$ finds these values in its log and outputs $w^* = f^{-1}(y^*) = s^*\|t^*$, where $t^* = H(s^*) \oplus r^*$.

These claims seem to make sense intuitively: even if $\mathcal{A}$ knows the f-inverse $w = s\|t$ of its query y_i or the challenge y^*, how can it know r unless it queries s to H (to compute $r = H(s) \oplus t$), and how can it know anything about the value of $G(r) \oplus s$ – i.e., either that its last k_1 bits are 0 (as needed for a valid ciphertext) or whether its first n bits correspond to m_0 or m_1 (as needed to "win" the game) – unless it knows r and queries r to G?

Unfortunately, though, the claims are false. While it is true, for example, that $\mathcal{A}$ *cannot* (except with extremely small probability) generate a valid ciphertext y_i *from scratch* without querying the implicit values s_i to H and r_i to G, it *might be able* to do so without querying r_i to G by *modifying* the ciphertext y^* that $\mathcal{B}$ gives to $\mathcal{A}$ in the Challenge phase. Specifically, suppose that $\mathcal{A}$ can compute $w^* = f^{-1}(y^*) = s^*\|t^*$, and that it sets:

$$\begin{aligned} s_i &= s^* \oplus (\Delta\|0^{k_1}), \\ t_i &= t^* \oplus H(s^*) \oplus H(s_i), \\ w_i &= s_i\|t_i, \\ y_i &= f(w_i), \end{aligned}$$

where Δ is any n-bit string. $\mathcal{A}$ can make these computations without querying G, and yet y_i is a valid ciphertext if y^* is. In particular, if y^* encrypts m^*, then y_i encrypts $m^* \oplus \Delta$, since $t_i = t^* \oplus H(s^*) \oplus H(s_i) = H(s_i) \oplus r^*$ and $s_i = s^* \oplus (\Delta\|0^{k_1}) = G(r^*) \oplus ((m^* \oplus \Delta)\|0^{k_1})$. Basically, $\mathcal{A}$ borrows the r^*-value from $\mathcal{B}$'s ciphertext. The reason that $\mathcal{A}$ can do this is that, in contrast to the "intuition" behind the OAEP proof, $\mathcal{A}$ *does* know something about the value r_i implicit in its chosen ciphertext y_i – namely, that $r_i = r^*$. It also knows something about $s_i \oplus G(r_i)$ despite not knowing r_i – namely, that the last k_1 bits of $s_i \oplus G(r_i)$ are 0^{k_1}, assuming y^* is a valid ciphertext. Since we cannot upper-bound the probability that $\mathcal{A}$ can construct a valid ciphertext y_i without querying r_i to G, $\mathcal{B}$'s strategy for answering decryption queries – i.e., searching its log for (r_i, s_i) – no longer works. Similarly, one can no longer claim that $\mathcal{A}$ has negligible advantage in the game unless it queries r^* to G with non-negligible probability, since $\mathcal{A}$ *can* gain information about m_b by querying the ciphertext described above, which ostensibly encrypts $m_b \oplus \Delta$. Shoup [18] proves that this gap in the proof cannot be filled; there exists an oracle relative to which the OAEP scheme is actually insecure.

OAEP+: A Way to Fix OAEP

Shoup [18] proposed a way to fix OAEP, called OAEP+. The f-OAEP+ encryption scheme is the same as the f-OAEP encryption scheme, except that in f-OAEP+ one sets

$$s = (G(r) \oplus m)\|H'(r\|m)\,,$$

where $H' : \{0,1\}^{n+k_0} \to \{0,1\}^{k_1}$ is an additional hash function, and the output range of G is changed to $\{0,1\}^n$. In the decryption algorithm, one checks that the last k_1 bits of s equal $H'(r\|m)$; if not, the output is $\perp$. Shoup provides a rigorous proof that f-OAEP+ is secure (in the random oracle model) against chosen-ciphertext attack if f is a one-way permutation. The proof is lengthy and does not use lattices, but for completeness, we provide an informal argument for its security here.

Roughly speaking, the reason f-OAEP+ is secure is that Claims 1 and 2 are true for f-OAEP+ (except that r is queried to either G or H' and the definition of a valid ciphertext is different), which allows $\mathcal{B}$ to compute w by searching its logs for (r, s), as in the OAEP proof. We have already shown that the claims are true, even for f-OAEP, for the s term; it remains to show that the claims are true for the r term. Consider how $\mathcal{B}$ responds to $\mathcal{A}$'s decryption query y_i in the f-OAEP+ simulation. Notice that there is a bijection between pairs (r_i, m_i) and valid ciphertexts y_i. When $\mathcal{A}$ queries $r_i\|m_i$ to H', $\mathcal{B}$ logs the corresponding ciphertext y_i. When $\mathcal{A}$ queries y_i, $\mathcal{B}$ responds with m_i, if y_i is in the log; otherwise, it outputs $\perp$. The reason that this response is valid with overwhelming probability is that, to ensure that its query y_i has better than a 2^{-k_1} probability of being a valid ciphertext, $\mathcal{A}$ must query H' at $(r_i\|m_i)$, since the values of $H'(r_j\|m_j)$ for $(r_j, m_j) \neq (r_i, m_i)$ (including the value of $H'(r^*\|m^*)$) give $\mathcal{A}$ no information about the value of $H'(r_i\|m_i)$.

Now, consider how $\mathcal{B}$ computes w^* from $\mathcal{A}$'s queries. If $\mathcal{A}$ queries r^* to G, then $\mathcal{B}$ can find (r^*, s^*), and we are done. Suppose $\mathcal{A}$ does not query r^* to G. Querying a ciphertext y_i, where $r_i \neq r^*$, gives $\mathcal{A}$ no information about m^*, since $G(r^*)$ remains completely random from $\mathcal{A}$'s perspective; thus, $\mathcal{A}$ must query $y_i \neq y^*$ such that $r_i = r^*$ to have non-negligible advantage. But, as we saw above, $\mathcal{A}$ must query H' at $r_i\|m_i = r^*\|m_i$ for y_i to have a non-negligible chance of being valid.

OAEP is Still Secure When Applied to Low-Exponent RSA

Shoup uses a lattice-based approach to show that RSA-OAEP is nonetheless secure when the encryption exponent e is very small (e.g., 3) and $k_0 \leq (\log N)/3$. The proof basically relies on the following weaker claims.

- (Claim 1b): The adversary $\mathcal{A}$ has an extremely small chance (specifically, a 2^{-k_1} probability) of constructing a valid ciphertext y_i – i.e., one for which $c_i = 0^{k_1}$ – unless it queries s_i to the H-oracle.
- (Claim 2b): The adversary $\mathcal{A}$ cannot have any advantage in distinguishing which message m_b is encrypted by the challenge ciphertext y^*, unless it queries s^* to the H-oracle.

To see that they are true, note that if $\mathcal{A}$ does not query s_i to H prior to offering y_i as a ciphertext, then the value of r_i is random and independent from $\mathcal{A}$'s perspective, even if $\mathcal{A}$ knows w_i. Thus, $G(r_i)$ is random and independent from $\mathcal{A}$'s perspective, and $\mathcal{A}$ can expect that the last k_1 bits of $H(s_i) \oplus t_i$ are 0 only with probability 2^{-k_1}. Similarly, if $\mathcal{A}$ does not query s^*, $G(r^*)$ appears random and independent to $\mathcal{A}$, and $\mathcal{A}$ has no information about m_b.

Here is the simulation. $\mathcal{B}$ responds to $\mathcal{A}$'s decryption query y_i as follows. By Claim 1b, if y_i is a valid ciphertext, $\mathcal{A}$ must have (with high probability) queried s_i to H. The pair (s_i, y_i) satisfies the equation $(2^{k_0} s_i + t_i)^e = y_i \bmod N$ for some k_0-bit integer t_i. To find s_i in its log, and to compute t_i, $\mathcal{B}$ applies Coppersmith's algorithm to the equation $(2^{k_0} s_j + x)^e = y_i \bmod N$ for each s_j in its H-log. When $s_j = s_i$, Coppersmith's algorithm returns $x = t_i$, since $(2^{k_0})^e \leq N$. If no invocation of Coppersmith's algorithm returns a suitable t_i, $\mathcal{B}$ returns $\perp$, since s_i is not in its H-log, and therefore y_i is an invalid ciphertext with overwhelming probability. Otherwise, having recovered the pair (s_i, t_i) corresponding to y_i, $\mathcal{B}$ computes $r_i = H(s_i) \oplus t_i$ and $m_i \| c_i = G(r_i) \oplus s_i$ and returns m_i or $\perp$, depending on whether or not $c_i = 0^{k_1}$. (If r_i was not previously queried to G, $\mathcal{B}$ assigns a value to $G(r_i)$ and makes an entry in its log.) By Claim 2b, if $\mathcal{A}$ has non-negligible advantage in the game, s^* is in $\mathcal{B}$'s H-log with non-negligible probability, and $\mathcal{B}$ can similarly recover $w^* = s^* \| t^*$ using Coppersmith's algorithm. The simulation might be imperfect if $\mathcal{B}$ assigns a value to $G(r_i)$ for $r_i = r^*$ that is inconsistent with the implicit value of $G(r^*)$ in y^* (this is precisely the "attack" on OAEP described above); then, $\mathcal{A}$ might abort. But, this can happen only with negligible probability, if $\mathcal{A}$ does not query s^* to H, since in this case the value r^* is independent of $\mathcal{A}$'s view.

Notice that the reduction here is tighter than for general-exponent OAEP+, since $\mathcal{B}$ can respond to decryption queries by stepping through just its H-log, rather than testing pairs of queries from two different logs. The reduction is similarly tight for low-exponent RSA-OAEP+.

RSA-OAEP is Secure for General Exponent

Fujisaki et al. provide two significant results in [20]. First, they prove that f-OAEP is chosen-ciphertext secure in the random oracle model, if f has a "set partial-domain one-wayness" property. Next, using lattices, they show that one-wayness implies set partial-domain one-wayness in the case of RSA. Combining these results, we have that RSA-OAEP is chosen-ciphertext secure in the random oracle assuming the RSA problem is hard.

Basically, f is partial-domain one-way if, given $f(s,t)$, it is hard to recover s. More formally, the (ℓ, t, ε)-set partial-domain one-wayness of f, means that for any adversary $\mathcal{A}$ that outputs a set S of ℓ elements in time t, the success probability $Succ^{s-pd-ow}(\mathcal{A})$ is upper-bounded by ε, where $Succ^{s-pd-ow}(\mathcal{A}) = \Pr_{s,t}[s \in \mathcal{A}(f(s,t))]$.

Their first result is as follows:

Theorem 2. *Let $\mathcal{A}$ be an adaptive chosen-ciphertext adversary against f-OAEP that runs in time t, has advantage ε, and makes q_D, q_G and q_H queries to the decryption oracle and the hash functions G and H respectively. Then one can can construct an algorithm $\mathcal{B}$ such that $Succ^{s-pd-pw}(\mathcal{B})$ is greater than*

$$\frac{\varepsilon}{2} - \frac{2q_Dq_G + q_D + q_G}{2^{k_0}} - \frac{2q_D}{2^{k_1}},$$

that runs in time $t' \leq t + q_Gq_H(T_f + O(1))$, *where* T_f *denotes the time complexity of* f.

The simulation requires a stronger first claim than that for RSA-OAEP with low exponent.

- (Claim 1c): Until the adversary $\mathcal{A}$ queries H at s^*, $\mathcal{A}$ has an extremely small chance (specifically, a 2^{-k_1} probability) of constructing a valid ciphertext y_i – i.e., one for which $c_i = 0^{k_1}$ – unless it queries s_i to the H-oracle and r_i to the G-oracle.
- (Claim 2c): The adversary $\mathcal{A}$ cannot have any advantage in distinguishing which message m_b is encrypted by the challenge ciphertext y^*, unless it queries s^* to the H-oracle.

Using these stronger claims, the simulation works as follows. To respond to $\mathcal{A}$'s decryption query y_i before $\mathcal{A}$ has queried H at s^*, $\mathcal{B}$ responds as in the proof of f-OAEP+; it searches its logs for (r_i, s_i) and responds appropriately. By Claim 1c, (r_i, s_i) will be in $\mathcal{B}$'s logs with high probability, if y_i is valid. $\mathcal{B}$ need not respond appropriately to $\mathcal{A}$'s decryption queries after $\mathcal{A}$ has queried s^* to H, for by that point $\mathcal{B}$ will already know s^*. At the send of the simulation, $\mathcal{B}$ outputs its list of $\mathcal{A}$'s H-queries; if $\mathcal{A}$'s advantage is non-negligible, then by Claim 2c, s^* will be in this list with non-negligible probability.

Here is an informal description of why Claim 1c is true. Unless $\mathcal{A}$ queries G at r_i or unless $r_i = r^*$, the value of $G(r_i)$ is completely random and independent from $\mathcal{A}$'s perspective. If $\mathcal{A}$ does not query G at r_i and if $r_i \neq r^*$, there is therefore only a 2^{-k_1} chance that the last k_1 bits of $G(r_i) \oplus s_i$ are 0^{k_1}, as required for a valid ciphertext. To query r_i, $\mathcal{A}$ must compute $r_i = H(s_i) \oplus t_i$ – i.e., query H at s_i; otherwise, it has only a negligible chance of guessing r_i, even if it knows $w_i = s_i \| t_i$. Thus, either $\mathcal{A}$ queries s_i to H and r_i to G, as claimed, or $r_i = r^*$. However, r^* is random and independent of $\mathcal{A}$'s view until $\mathcal{A}$ queries H at s^*; thus, there is only a negligible probability that r_i happens to equal r^*.

So far, we have shown that f-OAEP is secure against adaptive chosen ciphertext attack in the random oracle model assuming that f is set-partial-domain one-way. Now, suppose $\mathcal{A}$ is an algorithm that (t, ε)-breaks the set-partial-domain one-wayness of f; how do we construct an algorithm $\mathcal{B}$ from $\mathcal{A}$ that (t', ε')-inverts f, where (t', ε') are polynomially related to (t, ε)?

We already know how to construct $\mathcal{B}$ when f is the low-exponent RSA permutation: $\mathcal{B}$ simply runs $\mathcal{A}$ once, obtains from $\mathcal{A}$ a list that contains s^* with probability ε, and applies Coppersmith's algorithm sequentially to the items in the list until it computes $s^* \| t^*$. But, when the encryption exponent is large, Coppersmith's algorithm no longer works. How do we compute $s^* \| t^*$ from a list containing s^*, when the encryption exponent is large?

We don't. As described in [20], $\mathcal{B}$ runs $\mathcal{A}$ *twice* (or perhaps more times if needed). Specifically, to compute w such that $w^e = y$, it runs $\mathcal{A}$ first on y and then on $a^e y$

for random $a \in \mathbb{Z}_N$. $\mathcal{A}$ returns two lists – the first containing s_1 such that $w = 2^{k_0}s_1 + t_1 \bmod N$ and the second containing s_2 such that $aw = 2^{k_0}s_2 + t_2 \bmod N$, where $t_1, t_2 \in [0, 2^{k_0} - 1]$. This gives us the equations:

$$(2^{k_0}s_2 + t_2) = a(2^{k_0}s_1 + t_1) \bmod N$$
$$\Rightarrow at_1 - t_2 = 2^{k_0}(s_2 - as_1) \bmod N \ ,$$

which is a linear modular equation in the unknowns t_1 and t_2, which are known to have solutions smaller than 2^{k_0}. If 2^{k_0} is small enough in comparison to N and if there are not "too many" candidate solutions, we can find (t_1, t_2) using lattices. Unfortunately, the reduction is quite loose; since $\mathcal{A}$ is run twice, $\mathcal{B}$'s success probability is quadratic in $\mathcal{A}$'s success probability, and $\mathcal{B}$'s computation is potentially much higher than $\mathcal{A}$'s, since it must reduce lattices for pairs of items on the two lists. Nonetheless, it completes the first proof of security for RSA-OAEP. We give the details below.

For the above equation, let us call a solution (t_1', t_2') "small" if $(t_1, t_2) \in [0, 2^{k_0} - 1]^2$. The following lemma [20] bounds the probability that (t_1, t_2) is not a unique small solution to the above equation.

Lemma 1. *Let t_1 and t_2 be smaller than 2^{k_0}. Let a be randomly chosen from $\mathbb{Z}_N^*$, and let $c = at_1 - t_2$. Then, the probability (over the choice of a) that (t_1, t_2) is not a unique small solution to the equation $ax - y = c \bmod N$ is at most $2^{2k_0+6}/N$. If unique, the solution can be found in time $O(\log^2 N)$.*

To see that this lemma is true, consider the lattice $L_a = \{(x, y) \in \mathbb{Z}^2 : ax - y = 0 \bmod N\}$. Let $P = \{(0, 0) \neq (x, y) \in \mathbb{Z}^2 : x^2 + y^2 \leq 2^{2k_0+4}\}$. Clearly, if L_a has no vector in P, then if a small solution to $at_1 - t_2 = c \bmod N$ exists, it is unique. Moreover, roughly speaking, each P-vector (x_0, y_0) is associated to a lattice L_a for only one value $a \in \mathbb{Z}_N$ – namely $a = y_0/x_0 \bmod N$. (We have oversimplified a bit here; this fails in the rare case that $\gcd(x_0, N) > 1$.) Thus, the number of lattices L_a with a P-vector is at most the number of P-vectors, which is approximately $\pi 2^{2k_0+4} < 2^{2k_0+6}$, the desired result.

If L_a has no P-vector, finding $T = (t_1, t_2)$ amounts to a closest vector problem. Specifically, let T' be *some* solution to $ax - y = c \bmod N$, and let v be the L_a-vector closest to T'. Then, $T = T' - v$, since T is the only vector in $T + L_a$ shorter than $2^{k_0+1/2}$ (since L_a has no P-vector). Solving the closest lattice vector problem for a two dimensional lattice is not difficult, but Fujisaki et al. show that it is particularly easy for the lattice L_a and point T'. First, apply the Gaussian algorithm to L_a to obtain a reduced basis (b_1, b_2). Then, compute v simply by computing the coefficients (c_1', c_2') such that $T' = c_1' b_1 + c_2' b_2$ and then setting $v = \lfloor c_1' \rceil b_1 + \lfloor c_2' \rceil b_2$, where $\lfloor c_i' \rceil$ is the integer closest to c_i'. This works because if one expresses T as $c_1 b_1 + c_2 b_2$, then $-1/2 < c_1, c_2 < 1/2$. To see this, note that since $\langle b_1, b_2 \rangle \leq \|b_1\|^2/2$, we get:

$$\|T\|^2 = c_1^2\|b_1\|^2 + c_2^2\|b_2\|^2 + 2c_1c_2\langle b_1, b_2\rangle \geq (c_1^2 + c_2^2 - c_1c_2)\|b_1\|^2$$
$$= ((c_1 - c_2/2)^2 + 3c_2^2/4)\|b_1\|^2 \geq (3c_2^2/4)\|b_1\|^2 \ .$$

Since $\|T\|^2 < 2^{2k_0+1}$ and $\|b_1\|^2 \geq 2^{2k_0+4}$, we get $(3c_2^2/4) < 1/8 \Rightarrow |c_2| < \sqrt{1/6}$. By switching c_1 and c_2 in the inequalities above, we also conclude that $|c_1| < \sqrt{1/6}$.

Lemma 1, combined with the abovementioned approach for reducing the one-wayness to the set partial-domain one-wayness of RSA, leads to their main theorem:

Theorem 3. *Let $\mathcal{A}$ be an adaptive chosen-ciphertext adversary for RSA-OAEP, with a k-bit modulus with $k > 2k_0$, that runs in time t, has advantage ε, and makes q_D, q_G and q_H queries to the decryption oracle and the hash functions G and H respectively. Then one can can construct an algorithm $\mathcal{B}$ such that solves the RSA problem with probability at least*

$$\frac{\varepsilon^2}{4} - \varepsilon \cdot \left(\frac{2q_D q_G + q_D + q_G}{2^{k_0}} + \frac{2q_D}{2^{k_1}} + \frac{32}{2^{k-2k_0}} \right),$$

that runs in time $t' \leq 2t + q_H \cdot (q_H + 2q_G) \times O(k^3)$.

SAEP: A Simplification of OAEP

Boneh [19] noticed that OAEP, which can be viewed as a two-round Feistel cipher, can actually be simplified to a one-round Feistel cipher. He proposed simplified padding schemes, SAEP and SAEP+, as follows. Let $G : \{0,1\}^{k_0} \to \{0,1\}^{n+k_1}$ and $H : \{0,1\}^{n+k_0} \to \{0,1\}^{k_1}$ be hash functions. Let the message space be $\{0,1\}^n$. Then:

$$\text{SAEP}(m,r) = s\|r, \quad \text{where } s = G(r) \oplus (m\|0^{k_1}),$$
$$\text{SAEP+}(m,r) = s\|r, \quad \text{where } s = G(r) \oplus (m\|H(m,r)).$$

Here, we will focus on SAEP, and particularly on Rabin-SAEP.

In Rabin-SAEP, the sender encrypts a message m by choosing random $r \in \{0,1\}^{k_0}$, computing $w = \text{SAEP}(m,r)$, and sending the ciphertext $y = w^2 \bmod N$. To decrypt, the recipient uses its knowledge of N's factorization to compute all of the modular square roots w of y. (If N is the product of two primes, there are four of them.) For each $w = s\|r$, it sets $m\|c = G(r) \oplus s$ and tests whether $c = 0^{k_1}$. If all tests fail, it outputs $\perp$. If a test succeeds, it outputs m. (If k_1 is reasonably large, it is unlikely that more than one test will succeed.)

Boneh provides a very efficient reduction of factoring to the security of Rabin-SAEP. His main theorem is as follows:

Theorem 4. *Let $\mathcal{A}$ be an adaptive chosen-ciphertext adversary for Rabin-SAEP a k-bit modulus N with $k > 2k_0$, $n < k/4$, and $n + k_1 < k/2$ that runs in time t, has advantage ε, and makes q_D and q_G queries to the decryption oracle and the hash function G respectively. Then one can can construct an algorithm $\mathcal{B}$ that factors N with probability at least*

$$\frac{1}{6} \cdot \varepsilon \left(1 - \frac{2q_D}{2^{k_1}} - \frac{2q_D}{2^{k_0}}\right),$$

that runs in time $t' = t + O(q_D q_H T_C + q_D T'_C)$. *Here* $T_C = T_C(N,2)$ *and* $T'_C = T_C(N,4)$.

Recall that $T_C(N,d)$ denotes the time complexity of Coppersmith's algorithm for finding small roots of a d-degree polynomial modulo N. Notice that the requirement $n < k/4$ severely constrains the message space. However, this is fine for the typical setting of using 1024-bit modulus to encrypt a 128-bit session key.

From here onward, assume N is the product of two primes. The proof of this theorem relies on the following (slightly numerically inaccurate) claims:

- (Claim 1d): $\mathcal{A}$ has an extremely small chance (specifically, a 2^{-k_1+2} probability) of constructing a valid ciphertext y_i – i.e., one for which $c_i = 0^{k_1}$ – unless it queries some r_i to the G-oracle or some r_i equals some r^*.
- (Claim 2d): The adversary $\mathcal{A}$ cannot have any advantage in distinguishing which message m_b is encrypted by the challenge ciphertext y^*, unless it queries some r^* to the G-oracle or makes a decryption query y_i such that some r_i equals some r^*.

Since Rabin encryption – i.e., modular squaring – is not a permutation, we use the phrase "some w_i" to refer to any one of the several modular square roots of y_i. (There are four possible values of w_i when N is the product of 2 primes.) Once we have fixed a value for w_i, it induces values for s_i and r_i. By "some r^*," we mean one of the (four) r^*-values induced by the challenge ciphertext y^*.

Informally, the first claim is true, since unless $\mathcal{A}$ queries G at some r_i or unless some r_i equals some r^*, the values of $G(r_i)$ for all r_i are completely random and independent of what $\mathcal{A}$ knows. So, if $\mathcal{A}$ does not query G at some r_i and if no r_i equals no r^*, there is therefore only a 2^{-k_1+2} chance that the last k_1 bits of $G(r_i) \oplus s_i$ are 0^{k_1} for any r_i, as required for a valid ciphertext. The second claim is true basically for the same reason: unless $\mathcal{A}$ queries G at some r^* or gains indirect information about some $G(r^*)$ by querying a ciphertext y_i, where some r_i equals some r^*, the value of $G(r^*)$ for all r^* is completely random and independent of what $\mathcal{A}$ knows, and therefore m_b as well.

Assuming the claims are true, here is how the simulation works. $\mathcal{B}$ picks a random $w \in \mathbb{Z}_N^*$ and sets $y^* = w^2 \bmod N$. When $\mathcal{A}$ queries the ciphertext y_i, $\mathcal{B}$ searches its logs for an r_i by applying Coppersmith's algorithm to the equation $(2^{k_0}x + r_j)^2 = y_i \bmod N$ for every value r_j that $\mathcal{A}$ has queried to the G-oracle. $\mathcal{B}$ concludes that $r_j = r_i$ when Coppersmith's algorithm outputs a solution x satisfying $0 \leq x < 2^{k-k_0} = 2^{n+k_1} < \sqrt{N}$; it sets $w_i = r_j \| x$ and then finishes decryption in the usual way. If $\mathcal{B}$ fails to find r_i, it concludes (by Claim 1d) that $r_i = r^*$, or y_i is almost certainly invalid. Dealing with the first possibility is the tricky, and interesting, part of this simulation. If $r_i = r^*$ (we are supressing the term "some"), we have that $w_i = w^* + 2^{k_0+k_1}\Delta$ for $|\Delta| < 2^n < N^{1/4}$. In other words, if we define the polynomials

$$f(x) = x^2 - y^* \quad \text{and} \quad g(x,z) = (x + 2^{k_0+k_1}z)^2 - y_i \ ,$$

then we have that $f(w^*) = g(w^*, \Delta) = 0 \bmod N$. Unfortunately, w^* is not necessarily small, which seems to block Coppersmith's algorithm. We would like to find a univariate polynomial $h(z)$ such that Δ is a (small) root of $h(z) \bmod N$.

Fortunately, we can do exactly that by setting $h(z)$ to be $\text{Res}_x(f, g_z)$, the *resultant* of $f(x)$ and $g_z(x) = g(x,z)$ (eliminating the variable x). Recall that the resultant h of polynomials $f(x) = f_{d_f} x^{d_f} + \cdots + f_0$ and $g_z(x) = g_{z,d_g} x^{d_g} + \cdots + g_{z,0}$ is defined as:

$$\text{Res}_x(f, g_z) = f_{d_f}^{d_g} g_{d_g}^{d_f} \prod_{i=1}^{d_f} \prod_{j=1}^{d_g} (\alpha_i - \beta_j) \ ,$$

where $\{\alpha_i\}$ and $\{\beta_j\}$ are the roots of f and g_z, respectively. Of course, $\mathcal{B}$ does not know the roots of f and g_z. Fortunately, the resultant can be computed without knowing the roots. In particular, the resultant is the determinant of the following *Sylvester matrix* (for $d_f = d_g = 2$), which uses only the coefficients of f and g_z:

$$S_{f,g_z} = \begin{bmatrix} f_2 & f_1 & f_0 & 0 \\ 0 & f_2 & f_1 & f_0 \\ g_{z,2} & g_{z,1} & g_{z,0} & 0 \\ 0 & g_{z,2} & g_{z,1} & g_{z,0} \end{bmatrix} .$$

Since each $g_{z,i}$ is a polynomial of degree at most 2 in z, the resultant $h(z)$ has degree at most 4. We know that $\text{Res}_x(f, g_\Delta) = 0$, since f and g_Δ have the common root w^*; thus, $h(\Delta) = 0$. Since $\Delta < N^{1/4}$, $\mathcal{B}$ can use Coppersmith's algorithm on $h(z)$ to recover Δ, after which it is easy to recover w^*. If Coppersmith's algorithm fails to return a satisfactory Δ, $\mathcal{B}$ concludes that y_i is an invalid ciphertext and returns $\perp$. At last, we have finished describing how $\mathcal{B}$ responds to decryption queries.

By Claim 2d, if $\mathcal{A}$ has non-negligible advantage in the game, $\mathcal{A}$ must query some r^* to G, in which case, $\mathcal{B}$ can recover some w^* using Coppersmith's algorithm for degree 2, as described above, or $\mathcal{A}$ must query a ciphertext y_i for which some r_i equals some r^*, in which case, $\mathcal{B}$ recovers some w^* using Coppersmith's algorithm for degree 4. But, notice that $\mathcal{B}$ reveals nothing (even information-theoretically) in the simulation about *which* square root w of y^* that it used to generate y^*, and it does not use its knowledge of w to compute w^*. Thus, it is likely that $w^* \neq w$, and in fact (ignoring some messiness caused by the fact that $[0, N-1] \neq [0, 2^k - 1]$) there is basically a 1/2 probability that $\mathcal{B}$ obtains a nontrivial factor of N by computing $\gcd(w - w^*, N)$.

Compressing Rabin Signatures and Ciphertexts

Simple Techniques to Get Short Rabin Signatures

Recall that the signature σ of a message m in a Rabin full-domain-hash (Rabin-FDH) signature scheme is essentially a modular square root of $H(m)$, up to some "fudge factor." For example, in the scheme described in Section "Reduction and Concrete Security", σ must satisfy $\sigma^2 = c \cdot H(m) \bmod N$ for fudge factor $c \in \{\pm 1, \pm 2\}$. Here, the hash function $H : \{0,1\}^* \rightarrow \mathbb{Z}_N^*$ maps messages to essentially the "full domain" modulo N.

Bernstein [21] mentions that one can simply remove the $\frac{1}{e} \log N$, least significant bits of a Rabin (i.e., $e = 2$) or RSA signature, and the verifier can use Coppersmith's algorithm to recover those bits. The verifier then completes signature verification in the usual way. This technique cuts down the number of bits in a Rabin-FDH signature down to $(\log N)/2$.

Bleichenbacher [22] describes a different technique that achieves the same space-efficiency. In his scheme, the signer uses continued fractions to express the signature σ as $a/b (\text{mod} N)$, where a is about $\frac{e-1}{e} \log N$ bits and b is at most $\frac{1}{e} \log N$ bits; the signer sends a as the signature. The advantage of Bleichenbacher's approach over Bernstein's is that it preserves an advantage of Rabin signatures: fast signature verification. The verifier simply checks that, for some $c \in \{\pm 1, \pm 2\}$, $B = a^e / cH(m) (\text{mod} N)$ is an eth power (namely b^e) in $\mathbb{Z}$.

The security of both schemes follows immediately from the security of the underlying Rabin-FDH scheme, since anyone, without knowing any secrets, can express a Rabin-FDH signature compactly (whether by truncation or continued fractions) and obtain the original Rabin-FDH signature from its compact expression (as part of the verification process).

Disadvantages of the Simple Techniques

Bernstein's and Bleichenbacher's simple techniques, while allowing short Rabin signatures, are not really lossless "compression" algorithms in the usual sense. The verifier cannot recover the Rabin-FDH signature solely from the compact Rabin signature; it also needs the value of $H(m)$. This requirement has some disadvantages.

First, it is incompatible with a different technique for constructing space-efficient signature schemes, called *message recovery*. In a signature scheme with message recovery, the signer does not need to send the message m along with its signature σ, because the verifier can recover m from σ during verification. The motivation for message recovery is that it reduces the total bit-length of the *information needed to verify* the signature, which includes not only the signature itself, but also the message being signed. For the simple techniques above, the total verification information is at least $(\log N)/2$ bits longer than the message. On the other hand, suppose that $m \in \{0,1\}^n$ is signed by setting σ to be a modular square root (up to the fudge factor) of $G(m)\|(m \oplus H(G(m)))$, where G and H are hash functions modeled as random oracles, and G's output is 160 bits. In this case, the total verification information can be as little as 160 bits longer than m. (However, we remark that Bernstein's and Bleichenbacher's simple techniques certainly perform better than basic Rabin-FDH, as outlined in Section "Reduction and Concrete Security", which does not allow message recovery.)

Second, and perhaps more importantly, the simple techniques above are not very versatile; e.g., they do not lead to "compressed versions" of encryption, signcryption, ring signature, or aggregate signature schemes (the last being a scheme that allows multiple signatures to be aggregated into a single short signature). The reason, essentially, is that the techniques above ruin an important property of the Rabin and RSA operations – that they are permutations (well, *almost*, in the case of some Rabin instantiations). Trapdoor one-way permutations are a very versatile cryptographic tool that can be used to construct many different types of cryptosystems.

The rest of this section will be directed toward describing a more versatile compression technique. In particular, we will describe a trapdoor "quasi-bijection" that remains one-way, assuming factoring N is hard, even though the domain and range are small subsets of $\mathbb{Z}/N\mathbb{Z}$ (e.g., $[0, cN^{2/3}]$ for constant c). This enables compressed versions of a variety of cryptographic schemes that use modular squaring as a one-way function. For example, one can reduce the length of Rabin ciphertexts by about 33% without reducing security. Along the way, we review Vallée's elegant analysis of the distribution of numbers with small modular squares and the proof of security for the Rabin partial-domain-hash (Rabin-PDH) signature scheme (in which H hashes to a small subset of $\mathbb{Z}/N\mathbb{Z}$).

The Distribution of Numbers with Small Modular Squares

In the process of constructing a factoring algorithm with low provable runtime, Vallée [23, 24] constructed a polynomial-time lattice-based algorithm for sampling elements "quasi-uniformly" from $B_{N,h,h'} = \{x \in [0, N/2) : h \le x^2 (\text{mod} N) < h'\}$ for $h' - h = 8N^{2/3}$. By sampling "quasi-uniformly," we mean that, for each $x \in B_{N,h,h'}$, the probability that x is sampled is between $\ell_1/|B_{N,h,h'}|$ and $\ell_2/|B_{N,h,h'}|$ for constants ℓ_1 and ℓ_2 independent of N. In the rest of this section, we will use "quasi-" to mean "up to a multiplicative constant."

The first step in Vallée's quasi-uniform algorithm for sampling elements of $B_{N,h,h'}$ is to reduce it to a set of local problems by using Farey sequences.

Definition 11 (Farey Sequence). The Farey sequence $\mathcal{F}_k$ of order k is the ascending sequence $(\frac{0}{1}, \frac{1}{k}, \ldots, \frac{1}{1})$ of fractions $\frac{a_i}{b_i}$ with $1 \le a_i \le b_i \le k$ and gcd $(a_i, b_i) = 1$.

The characteristic property of Farey sequences is the following [25]:

Fact: *If* $\frac{a_i}{b_i}$ *and* $\frac{a_{i+1}}{b_{i+1}}$ *are consecutive in* $\mathcal{F}_k$*, then* $b_i a_{i+1} - a_i b_{i+1} = 1$.

Farey sequences lead naturally to the notion of a Farey partition, in which the set of mediants partition the interval $[0, N/2)$ into subintervals.

Definition 12 (Farey Partition). The Farey partition of order k of the interval $[0, N/2)$ is the set of Farey partition intervals

$$J(a_i, b_i) = \left[\frac{(a_{i-1} + a_i)N}{2(b_{i-1} + b_i)}, \frac{(a_i + a_{i+1})N}{2(b_i + b_{i+1})}\right]$$

where $\frac{a_i}{b_i}$ is the i-th term in $\mathcal{F}_k$, together with an interval at each end covering the uncovered portion of $[0, N/2)$.

Vallée found it convenient to use another set of intervals $I(a_i, b_i)$, called "Farey intervals," that are related to Farey partition intervals.

Definition 13 (Farey Interval). The Farey interval $I(a_i, b_i)$ of order k is the open interval with center $\frac{a_i N}{2b_i}$ and radius $\frac{N}{2kb_i}$, where $\frac{a_i}{b_i}$ is the i-th term in $\mathcal{F}_k$.

One can easily prove that $I(a_i, b_i)$ contains $J(a_i, b_i)$, and that the interval $I(a_i, b_i)$ is no more than twice as wide as the interval $J(a_i, b_i)$. One can also prove that every number in $[0, N/2)$ is covered by at least one and at most two Farey intervals. Vallée probably favored using the Farey intervals rather than the $J(a_i, b_i)$ in her analysis, because (roughly speaking) $I(a_i, b_i)$'s symmetry about $a_i N/2b_i$ permits cleaner computations. A "Farey covering" is then defined in the expected way.

Definition 14 (Farey Covering). The Farey covering of order k of the interval $[0, N/2)$ is the set Farey intervals $I(a_i, b_i)$ of order k.

Although Vallée's analysis and her sampling algorithm focus on Farey intervals, we will state her algorithm here with respect to Farey partition intervals, since it is slightly simpler to state this way (and it works equally well).

Vallée's Sampling Algorithm (High Level):

1. Set $k = N^{1/3}/4$ and set h and h', so that $h' - h \geq 8N^{2/3}$.
2. Sample an interval $J(a_i, b_i)$ from the Farey partition of order k with probability quasi-proportional to $|J(a_i, b_i) \cap B_{N,h,h'}|$.
3. Sample an element from $J(a_i, b_i) \cap B_{N,h,h'}$, quasi-uniformly.

Vallée demonstrates two facts regarding Farey intervals (for suitable k, h and h' as above), which also apply to Farey partition intervals:

- Fact 1: $|J(a_i, b_i) \cap B_{N,h,h'}|$ is quasi-proportional to $|J(a_i, b_i)|$.
- Fact 2: There is an efficient algorithm to sample from $J(a_i, b_i) \cap B_{N,h,h'}$ quasi-uniformly.

By Fact 1, the second step of Vallée's algorithm is easy: select x uniformly from $[0, N/2)$ and use continued fractions to compute $0 \leq a_i \leq b_i \leq k$, such that $x \in J(a_i, b_i)$. Fact 2 is nontrivial. Together, these facts imply that Vallée's algorithm samples quasi-uniformly from $B_{N,h,h'}$. Both facts stem from Vallée's analysis of how $B_{N,h,h'}$-elements are distributed in $I(a_i, b_i)$, which we now describe.

Consider the problem of sampling a $B_{N,h,h'}$-element quasi-uniformly from $I(a_i, b_i)$. Let x_0 be the integer closest to $\frac{a_i}{2b_i}$ and let $u_0 = x_0 - \frac{a_i}{2b_i}$. Let $L(x_0)$ be the lattice generated by the vectors $(1, 2x_0)$ and $(0, N)$. Let $x \in I(a_i, b_i)$ and let $x = x_0 + u$. Then, x is in $B_{N,h,h'} \cap I(a_i, b_i)$ precisely when $h \leq x_0^2 + 2x_0u + u^2 (\text{mod} N) < h'$. But, this is true precisely when there is a w such that $(u, w) \in L(x_0)$ and $h \leq x_0^2 + w + u^2 < h'$. The latter requirement implies that (u, w) is in between the two parabolas defined, in variables u' and w', by the formulas $x_0^2 + w' + u'^2 = h$ and $x_0^2 + w' + u'^2 = h'$. Thus, we have reduced the problem of sampling $B_{N,h,h'}$-elements from $I(a_i, b_i)$ to the problem of sampling $L(x_0)$-points that lie the region bounded by two parabolas and the vertical lines marking the beginning and end of $I(a_i, b_i)$. Denote this region by $R(a_i, b_i)$, and denote the set of $L(x_0)$-points in $R(a_i, b_i)$ by $P(a_i, b_i)$.

How do we sample quasi-uniformly from $P(a_i, b_i)$? Indeed, this seems like a difficult task, since finding all of the $L(x_0)$-points on a *single* parabola is equivalent to finding all of a number's modular square roots, which is equivalent to factoring. However, since $h' - h > 0$, the region $R(a_i, b_i)$ has some "thickness." How thick does $R(a_i, b_i)$ need to be before we can efficiently sample $P(a_i, b_i)$-points quasi-uniformly? It depends in part on how short our basis $(\mathbf{r}, \mathbf{s})$ of $L(x_0)$ is. Vallée gives the following short basis of $L(x_0)$, with one basis vector being "quasi-horizontal" (since $|2b_i u_0| \leq |b_i|$) and the other being "quasi-vertical":

$$\mathbf{r} = b_i(1, 2x_0) - a_i(0, N) = (b_i, 2b_i u_0),$$
$$\mathbf{s} = b_{i-1}(1, 2x_0) - a_{i-1}(0, N) = (b_{i-1}, \frac{N}{b_i} + 2b_{i-1}u_0).$$

Now, the intuition is that if we set $h' - h$ to be large enough, $R(a_i, b_i)$ eventually becomes so thick that most quasi-horizontal lines – by which we mean lines parallel to $\mathbf{r}$ that contain $P(a_i, b_i)$-points – will intersect $R(a_i, b_i)$ in (at least) one line segment of length $\geq \|\mathbf{r}\|$. Clearly, any line intersects $R(a_i, b_i)$ in at most two segments. If we know that a quasi-horizontal line ℓ intersects $R(a_i, b_i)$ in segments of lengths s_1, s_2 with some $s_i \geq \|\mathbf{r}\|$, we obtain fairly tight lower and upper bounds on the number of $P(a_i, b_i)$-points on ℓ without needing to count those points directly – namely, the bounds $\lfloor s_1/\|\mathbf{r}\| \rfloor + \lfloor s_2/\|\mathbf{r}\| \rfloor$ and $\lceil s_1/\|\mathbf{r}\| \rceil + \lceil s_2/\|\mathbf{r}\| \rceil$, respectively, which differ by at most a factor of 3. These quasi-tight bounds on the number of $P(a_i, b_i)$-points on most of the quasi-horizontal lines will help us sample from $P(a_i, b_i)$ quasi-uniformly.

But, is there a single small value of $h' - h$ that works for all (a_i, b_i)? To convince ourselves there is, let us consider the shape of $R(a_i, b_i)$ in more detail. Like Vallée, we can view $R(a_i, b_i)$ as being composed of three subregions: the "chest" (the top part), the "legs" (the two subregions extending downward and outward), and the "feet" (the left foot basically begins where the vertical line marking the beginning of $I(a_i, b_i)$ intersects the left leg). Notice that, for a fixed value of $h' - h$, a smaller value of b_i means that $I(a_i, b_i)$ has a larger radius and thus the legs extend further and are narrower (in the quasi-horizontal direction) toward the feet. At first, it may seem that a large value of $h' - h$ is necessary for the quasi-tight bounds above to hold for $R(a_i, b_i)$'s long tapered legs when b_i is small. But, fortunately, $\mathbf{r}$ is shorter when b_i is smaller, in exactly the right proportion so that the same value of $h' - h$ will work for all (a_i, b_i). So, in retrospect, it becomes clear why using Farey intervals was a good choice.

For suitable k, h, and h' as above, Vallée shows that the quasi-tight bounds hold for quasi-horizontal lines intersecting the legs. She then shows that only a constant number of quasi-horizontal lines intersect the chest or feet. Since there are so few quasi-horizontal lines in the chest and feet, Vallée obtains quasi-tight bounds on the total number of $P(a_i, b_i)$-points (via integration), despite having no lower bounds on the number of $P(a_i, b_i)$-points in the chest and feet. These quasi-tight bounds show that the number of $P(a_i, b_i)$-points is quasi-proportional to the width of $I(a_i, b_i)$. (See [23,24] for the computations.) Moreover, they allow us to complete Step 3 of her algorithm, as follows. (A similar algorithm pertains to $J(a_i, b_i)$.)

Algorithm to Sample Quasi-Uniformly from $I(a_i, b_i) \cap B_{N,h,h'}$:

1. *Approximate the number of points in $P(a_i, b_i)$:* Compute $x_0 = \lfloor \frac{a_i N}{b_i} \rceil$, count exactly the number n_{c+f} of points in the chest and feet, and obtain a lower bound n_l on the number of points in the legs using Vallée's lower bounds.
2. *Pick a point from $P(a_i, b_i)$:* Randomly select an integer in $t \in [1, n_{c+f} + n_l]$ with uniform distribution. If $t \leq n_{c+f}$, output the appropriate point from the chest or feet. Else, determine which quasi-horizontal line would contain the $(t - n_{c+f})^{th}$ point in the legs if each line met Vallée's lower bounds and randomly choose a point in $P(a_i, b_i)$ on that line with uniform distribution.
3. *Output element associated to chosen $P(a_i, b_i)$-point:* Let (u, w) be the lattice point output by the previous step. Set $x = x_0 + u$.

The Security of Rabin-PDH

Building on Vallée's work, Coron proved the security of the Rabin partial-domain hash (Rabin-PDH) signature scheme in the random oracle model, which, despite the absence of a security proof, had already been included in the IS0 9796-2 standard. In ISO 9796-2, a signature on m is a modular square root (up to a fudge factor) of $\mu(m) = 4A_{16} \| m \| H(m) \| BC_{16}$, where the right side represents an element of $\mathbb{Z}/N\mathbb{Z}$ in hexadecimal. This is a "partial-domain-hash" scheme, since H does not hash onto all of $\mathbb{Z}/N\mathbb{Z}$. An advantage of Rabin-PDH over Rabin-FDH is that it allows some message recovery.

To see how Vallée's algorithm is relevant to Rabin-PDH, reconsider the security proof for Rabin-FDH given in Section "Reduction and Concrete Security". Let

$S = \mathbb{Z}/N\mathbb{Z}$ and $T_S = \{(r,c) : (r^2/c \bmod N) \in S, r \in \mathbb{Z}/N\mathbb{Z}, c \in \{\pm 1, \pm 2\}\}$. In the simulation for Rabin-FDH, $\mathcal{B}$ responds to $\mathcal{A}$'s H-query on message m_i by picking an element $(r_i, c_i) \in T_S$ uniformly at random and setting $H(m_i) = r_i^2/c_i \bmod N \in S$. The distribution of $\mathcal{B}$'s H-query responses is identical to the distribution of H outputs in the "real world," when H is modeled as a full-domain hash function. Eventually, $\mathcal{A}$ presents a forgery σ on some m_i. Since, from $\mathcal{A}$'s perspective, the modular square root r_i of $c_i \cdot H(m_i)$ known by $\mathcal{B}$ is uniformly random, there is a 1/2 probability that $\gcd(\sigma - r_i, N)$ is a nontrivial factor of N, allowing $\mathcal{B}$ to factor. Notice that the above security proof works even if S is a proper subset of $\mathbb{Z}/N\mathbb{Z}$, *if* $\mathcal{B}$ *can sample uniformly from* T_S and H is modeled as a random oracle hashing onto S.

It is not hard to see that Vallée's algorithm can be used to sample quasi-uniformly from T_S for $S = [h, h']$ when $h' - h = 8N^{2/3}$, and similarly to sample quasi-uniformly from T_S for $S = \{4\text{A}_{16}\|m\|x\|\text{BC}_{16} : x \in [0, 8N^{2/3}]\}$. However, this is not sufficient to prove the security of Rabin-PDH, since the sampling does not necessarily give a *perfectly* uniform distribution of H-outputs or even a distribution that is statistically indistinguishable from uniform. The distribution must be very close to uniform, so that an adversary cannot distinguish a real attack from a simulated attack.

Coron [26] addresses this problem and thereby proves the security of Rabin-PDH, essentially by showing that Vallée's quasi-uniform algorithm for sampling elements from $B_{N,h,h'}$ for $h' - h = 8N^{2/3}$ can be transformed into an algorithm that samples elements from $B_{N,h,h'}$ for $h' - h = N^{2/3+\varepsilon}$ with a distribution whose distance from uniform is at most $16N^{\frac{-3\varepsilon}{13}}$. Intuitively, a larger value of $h' - h$ makes $R(a_i, b_i)$ thicker, making Vallée's lower and upper bounds for the legs extremely tight and permitting an almost uniform sampling algorithm. For the statistical distance to be at most 2^{-k}, we must have that $4 - \frac{3\varepsilon}{13} \log N \leq -k$, which implies that $\varepsilon \geq \frac{13(k+4)}{3 \log N}$. When $k = 80$, for example, $h' - h$ must be at least $\frac{2}{3} \log N + 364$ bits, which becomes less than $\log N$ bits when $\log N > 1092$.

Gentry [27] addresses the problem in a different way, showing that Vallée's algorithm can be made perfectly uniform (for $h' - h = 8N^{2/3}$) with a simple rejection sampling technique. The technique relies on the fact that anyone – e.g., the simulator in the security proof – can efficiently compute the *exact* probability P_x that Vallée's quasi-uniform sampling algorithm will output x. If x is in $J(a_i, b_i)$ on quasi-horizontal line ℓ_j (the j-th line passing through $R(a_i, b_i)$), the probability that x is chosen is simply $(2 \cdot |J(a_i, b_i)|/N) \cdot \P[\ell_j | J(a_i, b_i)] \cdot (1/n(\ell_j))$, where $n(\ell_j)$ is the number of $P(a_i, b_i)$-points on ℓ_j, and the probability term is dictated by Vallée's lower bounds. Then, letting $P_{\min}$ be a lower bound on such probabilities over all $x \in B_{N,h,h'}$, the perfectly uniform algorithm is as follows:

1. Use Vallée's method to pick an $x \in B_{N,h,h'}$ quasi-uniformly.
2. Compute P_x.
3. Goto Step 1 with probability $(P_x - P_{\min})/P_x$.
4. Otherwise, output x.

Since Vallée's sampling algorithm is quasi-uniform, the expected number of "Goto" loops per sample is a (small) constant; thus, the simulator's estimated time complexity increases only by a constant factor. The probability that x is chosen in Step 1 and that it "survives" Step 3 is the same for all x – namely, $P_x \cdot (1 - \frac{P_x - P_{\min}}{P_x}) = P_{\min}$; for this reason, and since each run of Vallée's algorithm is independent, the algorithm is perfectly uniform. This shows that Rabin-PDH is secure when H's output is $3 + (2/3)\log_2 N$ bits.

A Trapdoor Quasi-Bijection Over a Small Interval Modulo N

We showed that H's output can be "compressed," but the Rabin-PDH signature itself – i.e., the modular square root of $H(m_i)$ (possibly concatenated with other material) – is still $\log N$ bits. Since the "entropy" of the hash output is just over $(2/3)\log N$ bits, however, it is theoretically possible that the signature could be similarly short.

For example, suppose we have injective function $\theta_{N,h} : B_{N,-h,h} \to [-\alpha h, \alpha h]$ for some constant $\alpha \geq 1$ and $h \geq 4N^{2/3}$. Furthermore, suppose that both $\theta_{N,h}(x)$ and $\theta^{-1}_{N,h}(x')$ (when defined) are both efficiently computable without knowledge of N's factorization. Then, the following scheme allows short Rabin signatures:

- Signing: To sign m, uniformly (but deterministically) pick σ' s.t. $\sigma'^2 = c \cdot H(m) \bmod N$ for $c \in \{\pm 1, \pm 2\}$, where $H : \{0,1\}^* \to [-h, h]$. Output $\sigma = \theta_{N,2h}(\sigma')$.
- Verification: Set $\sigma' = \theta^{-1}_{N,2h}(\sigma)$. Confirm that $\sigma'^2 = c \cdot H(m) \bmod N$ for $c \in \{\pm 1, \pm 2\}$.

In this scheme, we let H's range be $[-h, h]$ and used $\theta_{N,2h}$ because this is a simple way of handling the fudge factor c; this is not meant to be limiting. The security of this scheme follows easily from the security of the underlying Rabin-PDH scheme (that does not use θ).

Signatures in the above scheme are $(2/3)(\log N) + 2 + \log \alpha$ bits when $h = 4N^{2/3}$. This is longer than the signatures obtained using Bernstein's or Bleichenbacher's techniques. However, the above approach is compatible with message recovery. If m is encoded with a reversible padding scheme allowing message recovery, the total number of bits needed to verify is about $\max\{|m| + 160, (2/3)\log N\}$. This is less than $|m| + (1/2)\log N$ when $\log N > 320$ and $|m| > (1/6)\log N$.

More interestingly, the compression function θ allows compression of more than just basic signatures. For example, consider the following "compressed" Rabin variant of Lysyanskaya et al. [28] sequential aggregate signature scheme for trapdoor homomorphic permutations. Below, we let N_i be the ith signer's public key, f_i^{-1} be the Rabin permutation over $\mathbb{Z}^*_{N_i}$ mentioned at the end of Section "Reduction and Concrete Security", $H_i : \{0,1\}^* \to [-h_i, h_i]$ be the partial domain hash used by the ith signer, $\theta_{N_i,h_i} : B_{N_i,-h_i,h_i} \to [-\alpha h_i, \alpha h_i]$ be the ith compression function, and m_i be the message signed by the ith signer; σ_0 is initialized to 0. Assume that $h_i = 2^{k_i} - 1$ for integer k_i.

- Signing: To sign m_i, the ith signer sets $\sigma_i' = f_i^{-1}(\sigma_{i-1} \oplus H_i(m_i, \ldots, m_1))$. It is required that $\sigma_{i-1} \in [-h_i, h_i]$. It outputs $\sigma_i = \theta_{N_i,2h_i}(\sigma_i')$.
- Verification: To verify that σ_n shows that the signer with key N_i signed message m_i for all $i \in [1, n]$, confirm that $\sigma_0 = 0$, where σ_0 is computed via the recursion $\sigma_{i-1} = f_i(\theta_{N_i,2h_i}^{-1}(\sigma_i)) \oplus H_i(m_i, \ldots, m_1)$.

The compression functions $\theta_{N_i,2h_i}$ allow the aggregate signature σ_n to be about 33% shorter than the RSA-based aggregate signature proposed in [28]. Note: we omitted some technical details here – e.g., the ith signer must provide a proof, as part of its public key, that N_i is a product of two primes to ensure that f_i is a permutation.

Now, we describe Gentry's approach for constructing the compression function $\theta_{N,h,h'} : B_{N,h,h'} \to [0, \alpha(h' - h)]$. (For other domains and ranges, the construction is similar.) Roughly speaking, the intuition is to express $x \in B_{N,h,h'}$ according to its Farey partition interval and its "address" (using Vallée's lattice) within the interval. The naive way of doing this – expressing x as (a_i, b_i, j, k), where x is in $J(a_i, b_i)$ in the kth position on the quasi-horizontal line with index j – does not work well, since the number of $B_{N,h,h'}$-elements associated to an interval or a line varies widely. Gentry provides the following more efficient alternative (h'' is a parameter whose value will be calibrated later).

Computing $\theta(x)$:

1. Determine (a_i, b_i) for which x is in $J(a_i, b_i)$.
2. Compute x_{left}, the smallest integer in $[0, h'']$ with $(x_{\text{left}} + 1) \cdot \frac{N}{h''}$ in $J(a_i, b_i)$, and x_{right}, the largest integer in $[0, h'']$ with $x_{\text{right}} \cdot \frac{N}{h''}$ in $J(a_i, b_i)$.
3. Compute n_{c+f}, the number of lattice points in the chest and feet, and n_l, an upper bound for the number of points in the legs.
4. Using Vallée's upper bounds, select one integer in $x_{\text{right}} - x_{\text{left}}$ (there may be several) that corresponds to the lattice point (u, w) that is associated to x. More specifically:

   (a) If (u, w) is the lth point in the chest or feet, set $c = l$.
   (b) Otherwise, let s_v be Vallée's upper bound for the number of leg lattice points on quasi-horizontal lines with index at most v. Compute the index v of the line containing (u, w). Let n_v be the actual number of lattice points on the line with index v and let $n_v' = s_v - s_{v-1}$ be Vallée's upper-bound estimate. Suppose that x is the kth lattice point on the line. Pick an integer $c \in (n_{c+f} + s_{v-1} + n_v' \frac{k-1}{n_v}, n_{c+f} + s_{v-1} + n_v' \frac{k}{n_v}]$.
   (c) Pick an integer $c' \in ((x_{right} - x_{left}) \frac{c-1}{n_{c+f}+n_l}, (x_{right} - x_{left}) \frac{c}{n_{c+f}+n_l}]$. Set $x = x_{left} + c'$.

Although not mentioned explicitly above, the algorithm depends on the values of h and h' (which we assume to be publicly available). Given $x' = \theta(x)$, one can recover the value of x as follows:

Computing $\theta^{-1}(x')$:

1. Determine (a_i, b_i) for which $x' \cdot \frac{N}{h''}$ is in $J(a_i, b_i)$.
2. Compute x_{left}, the smallest integer in $[0, h'']$ with $(x_{\text{left}} + 1) \cdot \frac{N}{h''}$ in $J(a_i, b_i)$, and x_{right}, the largest integer in $[0, h'']$ with $x_{\text{right}} \cdot \frac{N}{h''}$ in $J(a_i, b_i)$.
3. Compute n_{c+f}, the number of lattice points in the chest and feet of $P(a_i, b_i)$, and n_l, an upper bound for the number of points in the legs.
4. Compute $c' = x' - x_{\text{left}}$. From c' and $n_{c+f} + n_l$, compute the value of c. If $c \leq n_{c+f}$, let (u, w) be the cth point in the chest or feet. Otherwise, compute the index v such that $c \in (n_{c+f} + s_{v-1}, n_{c+f} + s_v]$, as well as the value of k (defined as above), and let (u, w) be the kth point on the quasi-horizontal line with index v.
5. Set $x = \theta^{-1}(x') = \lfloor \frac{a_i N}{b_i} \rceil + u$.

The value h'' is set to be as small as possible, subject to the constraint that $x_{\text{right}} - x_{\text{left}}$ is larger than $n_{c+f} + n_l$, which is necessary for injectivity. Using Vallée's bounds, one finds that $h'' = 8(h' - h)$ suffices. For this value of h'', the θ mapping compresses $B_{N,h,h'}$-elements to within 3 bits of the theoretical minimum.

Gentry uses similar techniques to construct a Rabin encryption scheme with short ciphertexts. In particular, the scheme uses an efficiently computable function $\pi_{N,h} : [-h, h] \times \mathcal{D} \to B_{N,-\alpha h,\alpha h}$ for constant $\alpha \geq 1$ and $h \geq 4N^{2/3}$ and some space $\mathcal{D}$, such that $\pi_{N,h}(x, d)$ is quasi-uniform in $B_{N,-\alpha h,\alpha h}$, if x and d are sampled uniformly, and such that x can be recovered efficiently and uniquely from $\pi_{N,h}(x, d)$. Encryption and decryption then proceed as follows:

- Encryption: To encrypt a message m, first compute a reversible encoding $x' \in [-h, h]$ of m – e.g., OAEP+. Pick random $d \in \mathcal{D}$, and set $x = \pi_{N,h}(x', d)$. Output the ciphertext $c = x^2 \bmod N$.
- Decryption: To decrypt c, compute each of the four values of x satisfying $c = x^2 \bmod N$. For each x, set $x' = \pi_{N,h}^{-1}(x)$, undo the encoding, and confirm that the resulting message m is encoded correctly. If it is, output m.

We refer the reader to [27] for additional details. We note that it remains an open problem to prove the security of partial-domain hash for low-exponent RSA signatures or to construct a compression algorithm for low-exponent RSA analogous to Gentry's, for Rabin.

The Relationship Among the RSA, Paillier, and RSA-Paillier Problems

Hensel Lifting

In the RSA problem, one is asked to compute $r \in [0, N-1]$ from $r^e \bmod N$, where N is an integer that is hard to factor and e is an integer coprime to $\phi(N)$.

One way to approach this problem is to attempt to *Hensel lift* – i.e., to somehow compute $r^e \bmod N^\ell$. Obviously, if $\ell \geq e$, the Hensel lift completely solves the RSA problem. But, what can we say about the relationship between Hensel lifting and the RSA problem when $\ell < e$?

Catalano et al. [29] define the Hensel-RSA(N, e, ℓ) problem as follows (similarly defined in [30]).

Definition 15 (Hensel-RSA(N, e, ℓ)). Given $r^e \bmod N$, $r \in [0, N-1]$, compute $r^e \bmod N^\ell$.

They have the following result.

Theorem 5. *Let N and e be integers, where $e = fN^\ell$ with f is co-prime to $\phi(N^2)$ and $\ell \geq 0$. Then, Hensel-RSA$(N, e, \ell+2)$ is hard iff the RSA(N, e) problem is hard.*

In Section "Security Implications of Catalano et al. Result on Hensel-RSA", we will describe how they apply this result to characterize the relationship among the RSA, Paillier, and RSA-Paillier problems. First, let us get to the proof of Theorem 5.

It is obvious that if Hensel-RSA$(N, e, \ell+2)$ is hard, then RSA(N, e) is hard. Now, given an algorithm $\mathcal{A}$ that solves Hensel-RSA$(N, e, \ell+2)$, we construct an algorithm $\mathcal{B}$ that solves RSA(N, e), as follows. As in Fujisaki et al.'s security proof for RSA-OAEP [20] given in Section "RSA-OAEP is Secure for General Exponent", $\mathcal{B}$ runs $\mathcal{A}$ *twice* – once on $r^e \bmod N$ and once on $u^e \bmod N$, where $u = (ra \bmod N) \in [0, N-1]$ for random $a \in [0, N-1]$. If $\mathcal{A}$ (t, ε)-breaks Hensel-RSA$(N, e, \ell+2)$, then $\mathcal{A}$ returns $r^e \bmod N^{\ell+2}$ and $u^e \bmod N^{\ell+2}$ in time $2t$ with probability ε^2.

From $\mathcal{A}$'s output, $\mathcal{B}$ recovers r as follows. Let z be such that

$$ra = u(1 + zN) \bmod N^{\ell+2}$$

Then, it is the case that

$$r^e a^e = u^e(1 + zN)^{fN^\ell} = u^e(1 + fzN^{\ell+1}) \bmod N^{\ell+2}$$

Knowing f, $\mathcal{B}$ can thus recover $z_0 = (z \bmod N) \in [0, N-1]$. Next, $\mathcal{B}$ reduces the lattice $L = \{(x, y) : ra = u(1 + z_0 N) \bmod N^2\}$ to obtain a vector $(r', u') \in [0, N-1]^2$ in time polynomial in $\log N$. Since r, u, r', and u' are all in $[0, N-1]$, and since $ru' = r'u \bmod N^2$, we get that $ru' = r'u$ (over $\mathbb{Z}$), and thus $r = r' \cdot \frac{c}{\gcd(r', u')}$ and $u = u' \cdot \frac{c}{\gcd(r', u')}$ for some integer c. Catalano et al. show that, with high probability, $\mathcal{B}$ can find (r, u) after a short exhaustive search.

Note that, like the security proof for RSA-OAEP, the reduction is quite *loose*. Although the reduction implies that Hensel-RSA$(N, e, \ell+2)$ and RSA(N, e) are polynomially equivalent, it gives an RSA-breaking algorithm whose success probability may be much lower than the success probability of the Hensel lifting algorithm.

Security Implications of Catalano et al. Result on Hensel-RSA

While Theorem 5 has direct implications for RSA, it also has interesting implications for other cryptosystems – specifically, the Paillier and RSA-Paillier (RSAP) encryption schemes, which we now review.

In 1999, Paillier [31] proposed an encryption scheme based on the (new and unstudied) "composite residuosity class problem," defined as follows:

Definition 16 (Composite Residuosity Class Problem). Let $N = pq$ be an RSA modulus and let g be an element whose order is a multiple of N in $\mathbb{Z}^*_{N^2}$. Given $c \in \mathbb{Z}^*_{N^2}$, find m such that cg^{-m} mod N^2 is an N-residue. We say that $h \in \mathbb{Z}^*_{N^2}$ is an N-residue, if there exists $r \in \mathbb{Z}^*_{N^2}$, such that $r^N = h$ mod N^2, and we say that $m = Class_{g,N}(c)$.

There is a decisional variant of this problem: given c and m, decide whether $m = Class_{g,N}(c)$.

In Paillier's encryption scheme, one encrypts $m \in \mathbb{Z}_N$ by generating random $r \in \mathbb{Z}^*_N$ and setting $c = r^N g^m$ mod N^2 to be the ciphertext. To decrypt, one sets $s = c^{\phi(N)} = r^{N\phi(N)} g^{m\phi(N)} = g^{m\phi(N)}$ mod N^2. Then, one computes $m = \frac{(s-1)/N}{(g^{\phi(N)}-1)/N}$ mod N. The final step works, since $g^{\phi(N)} = kN + 1$ mod N^2 for some k, and thus $g^{m\phi(N)} = (kN + 1)^m = 1 + kmN$ mod N^2.

Paillier's encryption scheme is one-way, if the composite residuousity class problem is hard. It is semantically secure against chosen plaintext attack, if the decisional variant is hard. An advantage of Paillier's cryptosystem is that it is additively homomorphic – i.e., if c_1 and c_2 encrypt m_1 and m_2 respectively, then $c_1 c_2$ mod N^2 encrypts $m_1 m_2$ mod N; this is useful, for example, in e-voting applications. A disadvantage is that the scheme is computationally slow, since encryption and decryption both require exponentiation with respect to a large modulus.

Catalano et al. [32] proposed a mix of Paillier's scheme and the RSA scheme, called RSA-Paillier (RSAP), that is nearly as efficient as plain RSA, but remains semantically secure assuming the hardness of a certain decision problem. Unfortunately, the scheme loses Paillier's homomorphic property.

In the RSAP encryption scheme, one encrypts $m \in \mathbb{Z}_N$ by generating random $r \in \mathbb{Z}^*_N$ and setting $c = r^e(1 + mN)$ mod N^2 to be the ciphertext, where the encryption exponent e satisfies $\gcd(e, \phi(N^2)) = 1$. To decrypt, one first solves an RSA problem: find $r \in \mathbb{Z}^*_N$ such that $r^e = c$ mod N. Then, one computes $m = \frac{(c/r^e)-1 \bmod N^2}{N}$. Note that Paillier's scheme with $g = 1 + N$ looks like RSAP with $e = N$.

Since it is easy to show RSAP is one-way iff Hensel-RSA$(N, e, 2)$ is hard (since decryption in RSAP is basically just Hensel lifting), Theorem 5 implies that RSAP(N, e) is one-way iff RSA(N, e) is, assuming $\gcd(e, \phi(N^2)) = 1$.

The connection between Paillier's scheme and RSA is less straightforward. Theorem 5 implies that Hensel-RSA$(N, N, 3)$ is equivalent to RSA(N, N). On the other hand, Catalano et al. show that Paillier's scheme is equivalent to Hensel

$(N, N, 2)$, which might be an easier problem. This separation leads Catalano et al. to conjecture that the one-wayness of the Paillier scheme is not equivalent to the RSA assumption with exponent N.

Theorem 6. *Let N be an integer. Hensel-RSA$(N, N, 2)$ is hard iff $Class_{g,N}$ is hard.*

The proof of Theorem 6 is straightforward. Assume $g = 1 + N$, since all instances of $Class_g$ are computationally equivalent. Given a Hensel-RSA$(N, N, 2)$ oracle and $c = r^N g^m = r^N(1 + mN) \bmod N^2$, one first recovers $r^N \bmod N^2$ using the oracle, and then recovers $1 + mN$ and then m. Given a Class$_g$ oracle and $d = (r^N \bmod N) \in [0, N - 1]$, one uses the oracle to recover a value $m \in \mathbb{Z}_N$, such that $d = r^N(1 + mN) \bmod N^2$, after which one recovers $r^N \bmod N^2$ easily.

Hensel Lifting Results for the Discrete Logarithm Problem

Although it does not say much about the security of any particular cryptosystem, Catalano et al. also describe an interesting relationship between the discrete logarithm problem and the following Hensel lifting problem.

Definition 17 (Hensel-Dlog(p, g, ℓ)). Let g have prime order ω modulo p. Let ℓ be such that $g^\omega = 1 \bmod p^{\ell-1}$, but $g^\omega \neq 1 \bmod p^\ell$. Given $g^x \bmod p$, $x \in [0, \omega - 1]$, compute $g^x \bmod p^\ell$.

In particular, they prove:

Theorem 7. *Hensel-Dlog(p, g, ℓ) is hard iff the discrete logarithm problem in $\mathbb{Z}_p^*$ is hard for generator g, where ℓ is defined as above.*

The "only if" implication is obvious.

Now, suppose we are given $g^x \bmod p$ and access to a Hensel-Dlog(p, g, ℓ) oracle that outputs $g^x \bmod p^\ell$ with probability ε. To recover x, we generate random $a \in [0, w - 1]$, set $u = xa \bmod w \in [0, w - 1]$, and send $g^x \bmod p$ and $g^u \bmod p$ to the oracle. With probability ε^2, we get back $g^x \bmod p^\ell$ and $g^u \bmod p^\ell$.

Since $u = xa - zw$ for some integer z, we can compute $g^{zw} \bmod p^\ell$. From $g^{zw} = (g^w)^z = (1 + mp^{\ell-1})^z = 1 + mzp^{\ell-1} \bmod p^\ell$, we can recover $z \bmod p$. It turns out that $z \bmod p$ reveals z completely, since the fact that u, x, and a are all in $[0, w - 1]$ implies that z is also in $[0, w - 1]$ and since $w < p$. Finally, we use lattices on the equation $u = xa + zw$ to recover u and x. (The lattice reduction may return multiple possible pairs $(u, x) \in [0, w - 1]^2$, but they show that, with non-negligible probability, the possibilities can be searched exhaustively in polynomial time.)

The Bit Security of Diffie–Hellman

Many cryptographic schemes are based on the Diffie–Hellman problem over $\mathbb{F}_p^*$ for prime p. Apart from the Diffie–Hellman protocol itself [8], we mention ElGamal's encryption scheme [9], Shamir's message passing scheme [33], and Okamoto's conference key sharing scheme [34]. In practice, these schemes are used to establish a short (e.g., 128-bit) session key for a symmetric cryptosystem. Since p, and hence the shared secret g^{ab} mod p, must be quite long (e.g., 1024 bits) to defend against subexponential attacks, it is natural to ask whether it is secure to use a subsequence of bits from the shared secret as the session key.

For prime p and $x \neq 0 \bmod p$, let $[x]_p$ denote the value x', such that $x - x' = 0 \bmod p$ and $|x'|$ is minimal, and let $\mathsf{MSB}_{k,p}(x)$ denote any integer u, such that $|[x]_p - u| \leq (p-1)/2^{k+1}$ – i.e., informally, the k most significant bits (MSBs) of $[x]_p$. In this section, we review the following result by Boneh and Venkatesan [35]:

Theorem 8. *Let $c > 0$ be a constant. Set $k = \lceil c\sqrt{(\log p)/(\log\log p)} \log\log\log p \rceil$. Given an efficient algorithm to compute $\mathsf{MSB}_{k,p}(g^{ab})$ from input (g, g^a, g^b), there is an efficient algorithm to compute $[g^{ab}]_p$ itself with probability at least $1/2$.*

We have stated their result with a slightly smaller value of k than they provided in the original paper. There are analogous results for extension fields [36] and bilinear maps (e.g., the Weil pairing) [37], elliptic curve bits [38], and the XTR and LUC cryptosystems [39].

Theorem 8 follows from the following more abstract statement.

Theorem 9. *Let $c > 0$ be a constant. Set $k = \lceil c\sqrt{(\log p)/(\log\log p)} \log\log\log p \rceil$ and $d = 2\lceil (\log p)/k \rceil$. Let α be an integer not divisible by p. Let $\mathcal{O}$ be an oracle defined by $\mathcal{O}_\alpha(t) = \mathsf{MSB}_{k,p}(\alpha t)$. Then, there exists a deterministic polynomial time algorithm A (the polynomial's degree depends on c) such that*

$$\mathbb{P}\left[(t_1, \ldots, t_d) \stackrel{R}{\leftarrow} [1, p-1]^d, A(t_1, \ldots, t_d, \mathcal{O}_\alpha(t_1), \ldots, \mathcal{O}_\alpha(t_d)) = [\alpha]_p\right] \geq \frac{1}{2}.$$

The problem of determining $[\alpha]_p$ from the MSBs of $[\alpha t_i]_p$ for multiple random t_i is called the "hidden number problem." Theorem 8 follows from Theorem 9, since an algorithm that outputs $\mathsf{MSB}_{k,p}(g^{a(b+x_i)})$ given (g, g^a, g^{b+x_i}) can serve as the oracle $\mathcal{O}_\alpha$ in Theorem 9, where $\alpha = g^{ab}$ and $t_i = g^{ax_i}$ and the x_i's are selected randomly. (Actually, this statement is true only if $\gcd(a, p-1) = 1$, since otherwise the t_i's all come from a subgroup of $\mathbb{F}_p^*$. González Vasco and Shparlinski [40] extended the result to general case.)

The reduction in Theorem 9 involves a closest vector problem over a $(d+1)$-dimensional lattice. To show that the reduction is efficient, we use Theorem 10 below, which follows from Schnorr's modification [41] to LLL [42] and Kannan's [43] reduction of approximate CVP to approximate SVP. (This is also Theorem 1 in [44].)

Theorem 10. *There is a polynomial time algorithm which, given a s-dimensional full rank lattice L and a vector $\mathbf{u} \in \mathbb{R}^s$, finds a lattice vector $\mathbf{v}$ satisfying the inequality*

$$\|\mathbf{v} - \mathbf{u}\| \leq 2^{O(s \log^2 \log s / \log s)} \min\{\|\mathbf{z} - \mathbf{u}\|, \mathbf{z} \in L\}.$$

Given t_i and $u_i = \mathsf{MSB}_{k,p}(\alpha t_i)$ for $i = 1, \ldots, d$, set $\mathbf{u} = (u_1, \ldots, u_d, 0)$ and let L be the lattice generated by the rows of the following matrix

$$\begin{bmatrix} p & 0 & \ldots & 0 & 0 \\ 0 & p & \ldots & 0 & 0 \\ \vdots & \vdots & \vdots & \vdots & \vdots \\ 0 & 0 & \ldots & p & 0 \\ t_1 & t_2 & \ldots & t_d & 1/2^k \end{bmatrix} \tag{12.1}$$

Now, we use Theorem 10's algorithm to recover $\mathbf{w} = ([\alpha t_1]_p, \ldots, [\alpha t_d]_p, [\alpha]_p/2^k)$, which is a vector in L whose distance from $\mathbf{u}$ is at most $(\sqrt{d+1})(p-1)/2^{k+1}$ and from which we can recover the hidden number $[\alpha]_p$.

To show $\mathbf{w}$ is in fact recoverable, Boneh and Venkatesan's approach is essentially to show that if L-vector $\mathbf{v}$ satisfies

$$\|\mathbf{v} - \mathbf{u}\| \leq 2^{O(d \log^2 \log d / \log d)} \|\mathbf{w} - \mathbf{u}\| = 2^{O(d \log^2 \log d / \log d)} (p-1)/2^{k+1} ,$$

which must be satisfied by the output of the CVP algorithm indicated in Theorem 10, then it is overwhelmingly likely that $\mathbf{v}_i = \mathbf{w}_i \bmod p$ for $i = 1, \ldots, d$, in which case we can recover $\mathbf{w}$ from $\mathbf{v}$. If we assume the contrary – i.e., that there is a close L-vector $\mathbf{v}$ that does not satisfy the congruence – then there must exist $\gamma \neq 0 \bmod p$ such that

$$\mathbf{y} = ([\gamma t_1]_p, \ldots, [\gamma t_d]_p, [\gamma]_p/2^k)$$

is a nonzero L-vector whose length is at most $h = 2^{O(d \log^2 \log d / \log d)}(p-1)/2^k$. This implies that each coefficient $\mathbf{y}_i \in [-h, h]$. The probability (over the choices of $(t_1, \ldots, t_d)$) that there exists $\gamma \neq 0 \bmod p$ for which $[\gamma t_i]_p \in [-h, h]$ for all $i = 1, \ldots, d$ is at most

$$(p-1)(2h)^d/(p-1)^d < p 2^{O(d^2 \log^2 \log d / \log d)}/2^{dk} .$$

For some constant $c > 0$, this probability is negligible when

$$k = \left\lceil c \sqrt{\frac{\log p}{\log \log p}} \log \log \log p \right\rceil \quad \text{and} \quad d = 2 \left\lceil \frac{\log p}{k} \right\rceil ,$$

from which Theorem 9 follows.

Boneh and Venkatesan observe that the reduction also holds when $\mathcal{O}$'s responses are correct with probability only $1 - 1/(\log p)$. In this case, all of $\mathcal{O}$'s d responses are correct with high probability.

What does the result say about the security of using the MSBs of a Diffie-Hellman value as a session key? Certainly, an adversary that can compute a session key with non-negligible probability $\varepsilon < 1 - 1/(\log p)$ "breaks" a system's security, though it may not be clear that such an adversary can be used to recover an entire Diffie–Hellman value. However, if we run such an adversary an appropriate polynomial number of times, then at least one of the adversary's outputs is correct with probability $1 - 1/(\log p)$. (The trials can be made independent, thanks to the random self-reducibility property of Diffie–Hellman.) But, how do we remove the adversary's incorrect outputs? One can often verify that a session key is correct simply by trying to use that session key in the protocol with the other party; if the session key is incorrect, the other party will typically indicate an error and abort. In such a setting, an adversary that guesses the MSBs with non-negligible probability can be used to break the entire Diffie–Hellman secret. However, this still does not imply that it is secure to use the MSBs as a key. This remains an open problem.

Another way of viewing Boneh and Venkatesan's result is as (nondispositive) evidence of the hardness of the *decision* Diffie–Hellman (DDH) problem – i.e., given (g, g^a, g^b, h), decide whether h equals g^{ab} or is a random element of the group. DDH is a strong, but very useful, assumption. Notably, DDH was used by Cramer and Shoup [45] to prove, *without random oracles*, the security of their encryption scheme against adaptive chosen ciphertext attack.

Acknowledgements We thank Phong Nguyen and the reviewers for their helpful suggestions and comments.

References

1. P. Kocher. Timing Attacks on Implementations of Diffie-Hellman, RSA, DSS and Other Systems. In *Proc. of Crypto 1996*, LNCS 1109, pages 104–113. Springer, 1996
2. P. Kocher, J. Jaffe, and B. Jun. Differential Power Analysis. In *Proc. of Crypto 1999*, LNCS 1666, pages 388-397. Springer, 1999
3. Y. Ishai, A. Sahai and D. Wagner. Private Circuits: Securing Hardware Against Probing Attacks. In *Proc. of Crypto 2003*, LNCS 2729, pages 463–481. Springer, 2003
4. S. Micali and L. Reyzin. A Model for Physically Observable Cryptography. In *Proc. of TCC*, LNCS 2951, pages 278–296. Springer, 2004
5. C. Rackoff and D. Simon. Noninteractive Zero-Knowledge Proof of Knowledge and Chosen Ciphertext Attack. In *Proc. of Crypto 1991*, pages 433–444, 1991
6. D. Dolev, C. Dwork, and M. Naor. Non-malleable Cryptography. In *Proc. of STOC*, 542–552, 1991
7. D. Dolev, C. Dwork, and M. Naor. Non-malleable Cryptography. *SIAM J. Computing*, 30(2):391–437, 2000
8. W. Diffie and M. Hellman. New Directions in Cryptography. *IEEE Transactions on Information Theory*, 22(6):644–654, 1976
9. T. ElGamal. A Public Key Cryptosystem and a Signature Scheme Based on the Discrete Logarithm. *IEEE Transactions on Information Theory*, 31(4):469–472, 1985
10. R.L. Rivest, A. Shamir, and L.M. Adleman. A Method for Obtaining Digital Signatures and Public-Key Cryptosystems. In *Comm. of the ACM*, pages 120–126, 1978
11. M.O. Rabin. Digitalized Signatures and Public-Key Functions as Intractable as Factorization. MIT/LCS/TR-212, MIT Laboratory for Computer Science, 1979
12. M. Bellare and P. Rogaway. Random Oracles are Practical: A Paradigm for Designing Efficient Protocols. In *Proc. of ACM CCS*, pages 62–73, 1993

13. R. Canetti, O. Goldreich, and S. Halevi. The Random Oracle Model, Revisited. J. ACM 51(4): 557–594, 2004
14. E.-J. Goh and S. Jarecki. A Signature Scheme as Secure as the Diffie Hellman Problem. In *Proc. of Eurocrypt 2003*, LNCS 2656, pages 401–415. Springer, 2003
15. M. Bellare. Practice-Oriented Provable Security. In *Proc. of International Workshop on Information Security (ISW) 1997*, LNCS 1396, pages 221–231. Springer, 1998
16. D. Coppersmith. Finding a Small Root of a Univariate Modular Equation. In *Proc. of Eurocrypt 1996*, pages 155–165, 1996
17. M. Bellare and P. Rogaway. Optimal Asymmetric Encryption. In *Proc. of Eurocrypt 1994*, pages 92–111. Springer, 1994
18. V. Shoup. OAEP Reconsidered. In *Proc. of Crypto 2001*, pages 239–259. Springer, 2003
19. D. Boneh. Simplified OAEP for the RSA and Rabin Functions. In *Proc. of Crypto 2001*, LNCS 2139, pages 275–291. Springer, 2001
20. E. Fujisaki, T. Okamoto, D. Pointcheval, and J. Stern. RSA-OAEP Is Secure under the RSA Assumption. In *J. Cryptology*, 17(2): 81–104 (2004)
21. D.J. Bernstein. Reducing Lattice Bases to Find Small-Height Values of Univariate Polynomials. 2003. Available at http://cr.yp.to/djb.html
22. D. Bleichenbacher. Compressing Rabin Signatures. In *Proc. of CT-RSA 2004*, LNCS 2964, pages 126–128. Springer, 2004
23. B. Vallée. Provably Fast Integer Factoring with Quasi-Uniform Small Quadratic Residues. In *Proc. of STOC 1989*, pages 98–106
24. B. Vallée. Generation of Elements with Small Modular Squares and Provably Fast Integer Factoring Algorithms. In *Mathematics of Computation*, 56(194): 823–849, 1991
25. G.H. Hardy and E.M. Wright, *An Introduction to the Theory of Numbers*, Oxford Science Publications (5th edition)
26. J.-S. Coron. Security Proof for Partial-Domain Hash Signature Schemes. In *Proc. of Crypto 2002*, LNCS 2442, pages 613–626. Springer, 2002
27. C. Gentry. How to Compress Rabin Ciphertexts and Signatures (and More). In *Proc. of Crypto 2004*, LNCS 3152, pages 179–200. Springer, 2004
28. A. Lysyanskaya, S. Micali, L. Reyzin, and H. Shacham. Sequential Aggregate Signatures from Trapdoor Homomorphic Permutations. In *Proc. of Eurocrypt 2004*, LNCS 3027, pages 74–90. Springer, 2004
29. D. Catalano, P.Q. Nguyen, and J. Stern. The Hardness of Hensel Lifting: The Case of RSA and Discrete Logarithm. In *Proc. of Asiacrypt 2002*, pages 299–310. Springer, 2002
30. K. Sakurai and T. Takagi. New Semantically Secure Public Key Cryptosystems from the RSA Primitive. In *Proc. of Public Key Cryptography*, pages 1–16, 2002
31. P. Paillier. Public-Key Cryptosystems Based on Composite Degree Residuosity Classes. In *Proc. of Eurocrypt 1999*, pages 223–238, 1999
32. D. Catalano, R. Gennaro, N. Howgrave-Graham, and P.Q. Nguyen. Paillier's Cryptosystem Revisited. In *Proc. of ACM CCS 2001*, pages 206–214, 2001
33. N. Koblitz. *A Course in Number Theory and Cryptography*. Springer, 1987
34. T. Okamoto. *Encryption and Authentication Schemes Based on Public Key Systems*. Ph.D. Thesis, University of Tokyo, 1988
35. D. Boneh and R. Venkatesan. Hardness of Computing the Most Significant Bits of Secret Keys in Diffie-Hellman and Related Schemes. In *Proc. of Crypto 1996*, LNCS 1109, pages 129–142. Springer, 1996
36. M.I. Gonzáles Vasco, M. Naslund, and I.E. Shparlinski. The hidden number problem in extension fields and its applications. In *Proc. of LATIN 2002*, LNCS 2286, pages 105–117. Springer, 2002
37. S.D. Galbraith, H.J. Hopkins, and I.E. Shparlinski. Secure Bilinear Diffie-Hellman Bits. In *Proc. of ACISP 2004*, LNCS 3108, pages 370–378. Springer, 2004
38. D. Boneh and I. Shparlinski. On the Unpredictability of Bits of the Elliptic Curve Diffie–Hellman Scheme. In *Proc. of Crypto 2001*, LNCS 2139, pages 201–212. Springer, 2001

39. W.-C.W. Li, M. Naslund, and I.E. Shparlinski. The Hidden Number Problem with the Trace and Bit Security of XTR and LUC. In *Proc. of Crypto 2002*, LNCS 2442, pages 433–448. Springer, 2002
40. M.I. Gonzáles Vasco and I.E. Shparlinski. On the security of Diffie-Hellman bits. In *Proc. of Workshop on Cryptography and Computational Number Theory*, 1999
41. C.-P. Schnorr. A Hierarchy of Polynomial Time Lattice Basis Reduction Algorithms. *Theoretical Computer Science*, 53:201–224, 1987
42. A. Lenstra, H. Lenstra, and L. Lovasz. Factoring Polynomials with Rational Coefficients. Math. Ann. 261, pages 515–534, 1982
43. R. Kannan. Algorithmic geometry of numbers. In *Annual Review of Computer Science*, vol. 2, pages 231–267, 1987
44. I.E. Shparlinski. Exponential Sums and Lattice Reduction: Applications to Cryptography. In *Finite Fields with Applications to Coding Theory, Cryptography and Related Areas*, pages 286–298. Springer, 2002
45. R. Cramer and V. Shoup. A Practical Public Key Cryptosystem Provably Secure against Adaptive Chosen Ciphertext Attack. In *Proc. of Crypto 1998*, LNCS 1462, pages 13–25. Springer, 1998

Chapter 13
Cryptographic Functions from Worst-Case Complexity Assumptions

Daniele Micciancio

Abstract Lattice problems have been suggested as a potential source of computational hardness to be used in the construction of cryptographic functions that are provably hard to break. A remarkable feature of lattice-based cryptographic functions is that they can be proved secure (that is, hard to break on the average) based on the assumption that the underlying lattice problems are computationally hard in the *worst-case*. In this paper we give a survey of the constructions and proof techniques used in this area, explain the importance of basing cryptographic functions on the *worst-case* complexity of lattice problems, and discuss how this affects the traditional approach to cryptanalysis based on random challenges.

Introduction

A lattice is the set of integer linear combinations of n linearly independent vectors $\mathbf{b}_1, \ldots, \mathbf{b}_n$, and can be pictorially described as the set of intersection points of an infinite regular (but not necessarily orthogonal) grid (see Fig. 13.1). A typical algorithmic problem on lattices is the following: given a lattice (represented by a basis $\{\mathbf{b}_1, \ldots, \mathbf{b}_n\}$), find a nonzero lattice vector that is as short as possible, or (in case approximate solutions are allowed) not much longer than the shortest.

Traditionally, in cryptography, lattices have been used mostly as an algorithmic tool for cryptanalysis. Since the development of the LLL basis reduction algorithm of Lenstra, Lenstra and Lovász [1] in the early 80s, lattices have been used to attack a wide range of public-key cryptosystems (see survey papers [2–4] and references therein). Moreover, much work on improving lattice basis reduction algorithms and heuristics (e.g., [5–11]) has been directly motivated by cryptanalysis applications. Quoting [4], the success of basis reduction algorithms at breaking various

D. Micciancio
Department of Computer Science and Engineering, University of California at San Diego, La Jolla CA 92093, USA,
e-mail: daniele@cs.ucsd.edu

P.Q. Nguyen and B. Vallée (eds.), *The LLL Algorithm*, Information Security and Cryptography, DOI 10.1007/978-3-642-02295-1_13,

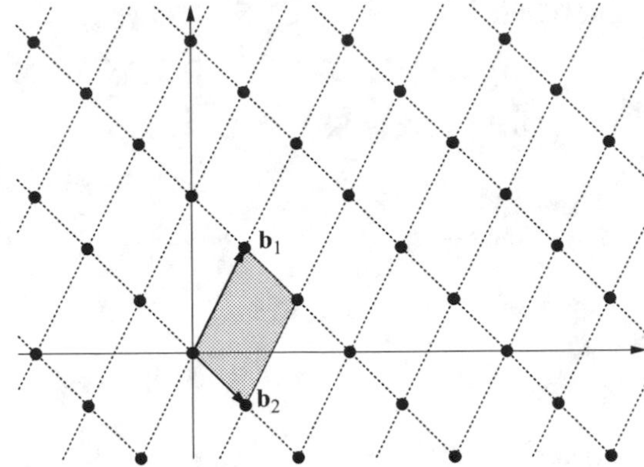

Fig. 13.1 The 2-dimensional lattice generated by the basis $[\mathbf{b}_1, \mathbf{b}_2]$. The lattice vectors are the intersection points of the grid generated by $\mathbf{b}_1$ and $\mathbf{b}_2$

cryptographic schemes over the past 20 years has arguably established lattice reduction techniques as the most popular tool in public-key cryptanalysis.

In this survey we are not concerned with the many applications of the LLL algorithm in cryptanalysis. Rather, based on the fact that after 25 years from its discovery, the LLL algorithm is still essentially unsurpassed, we consider the use of lattices as a source of computational hardness to be used in cryptography. More specifically, we consider the design of cryptographic schemes that are provably hard to break based on the conjecture that no efficient algorithm solves various lattice problems substantially better than LLL. We remark that while the LLL algorithm has been substantially improved in terms of running time, progress on improving the approximation factor achieved by LLL (while maintaining polynomial running time) has been fairly limited so far: the approximation factor achieved by the currently best polynomial time (randomized) lattice approximation algorithms [6,9,12] is $2^{O(n \log \log n / \log n)}$, only a modest improvement over the exponential approximation factor $2^{O(n)}$ achieved by LLL. So, it is reasonable to conjecture that no polynomial time algorithm can approximate lattice problems within factors $n^{O(1)}$ that are polynomial in the rank of the lattice.

In the late 90s, the computational complexity of lattice problems attracted renewed attention, largely stimulated by Ajtai's surprising discovery [13] of a connection between the worst-case and average-case complexity of certain lattice approximation problems. The importance of such connection relies on the potential applicability of lattices to the *design* of secure cryptosystems. It is a well established fact that cryptography requires problems that are hard to solve *on the average*, so that when a cryptographic key is chosen at random, the corresponding function is hard to break with high probability. So, it is not surprising that, till recently, all known cryptographic functions relied on average-case complexity assumptions. Ajtai's connection showed, for the first time, that hard-on-average problems (and therefore, secure cryptographic functions), could be obtained from the qualitatively weaker assumption that lattice problems are intractable in the *worst case*. This discovery attracted a lot of interest both within the theoretical cryptography and computational complexity communities, and stimulated substantial research efforts in the area. Following Ajtai's initial discovery, research has progressed on several fronts:

- Determining weaker and weaker worst-case assumptions on the complexity of lattice problems that still allow to distill average-case hardness [14–16].
- Improving the efficiency of lattice based functions both in terms of key size and computation time [17–21].
- Building more complex cryptographic primitives than simple one-way functions, like pubic key encryption schemes [22–25], identification protocols [26, 27], digital signatures [28, 29] and more [30, 31].

These various lines of research, besides being individually interesting from both a theoretical and practical point of view, have a lot in common at a technical level. Ideas and techniques originally introduced in one setting have often found applications in the other contexts (e.g., the Gaussian perturbation techniques originally introduced by Regev [23] to improve the analysis of public-key encryption schemes have been further developed and used by Micciancio and Regev [16, 17] to study other cryptographic functions). In this paper, we give a self contained introduction to this general area, review the current state of the art and describe the main open problems related to the construction of cryptographic functions based on the worst-case hardness of lattice problems. In particular, we highlight two important issues that have not received much attention so far in the traditional algorithms/complexity and cryptanalysis literature: the study of lattices with special properties, and the development of an appropriate methodology for the cryptanalysis of functions based on worst-case assumptions. For simplicity, at the technical level, we focus on collision resistant hash functions, which are described and analyzed in some depth. We also illustrate the main ideas behind the construction of public key encryption schemes and efficient cryptographic functions based on special classes of lattices, but at a more informal level. For an overview of the zero knowledge proof systems underlying the lattice based identification schemes of [26] the reader is referred to Regev's survey [32] in this volume.

The rest of the paper is organized as follows. In Section "Background" we give some general background about lattices. In Section "Collision Resistant Hashing" we describe and analyze (a simplified version of) the collision resistant hash function of Micciancio and Regev [16]. In particular, in Section "Simple Cryptographic Hash Function with Worst-Case/Average-Case Connection", we give a detailed and essentially self-contained description of a cryptographic function (and relative security proof) based on a worst-case complexity assumption. In Section "Using Special Lattices to Improve Efficiency" we also informally discuss the work of Micciancio [17] and subsequent developments [19, 20] on efficient hash functions based on special lattices. Next, in Section "Public-Key Encryption Scheme", we describe the main ideas behind the public-key cryptosystems of Ajtai and Dwork [22] and Regev [23]. Section "Concrete Security of Lattice Based Cryptography" concludes the paper with a discussion of the aforementioned issues related to the security evaluation of cryptographic functions with average-case/worst-case security guarantees.

Background

In this section we briefly define the concepts and notation used in the algorithmic study of lattices. For additional background, the reader is referred to [33].

We use $\tilde{O}(f(n))$ to denote the set of all functions $g(n)$ that are asymptotically bounded by $f(n)$ up to poly-logarithmic terms, i.e., $g(n) \leq f(n) \log^c f(n)$ for some constant c and all sufficiently large n. A function $\varepsilon(n)$ is *negligible* if $\varepsilon(n) < 1/n^c$ for any $c > 0$ and all sufficiently large n.

A lattice (see Fig. 13.1) is the set of all integer linear combinations $\mathcal{L}(\mathbf{b}_1, \ldots, \mathbf{b}_n) = \sum_i \mathbf{b}_i \cdot x_i$ (where $x_i \in \mathbb{Z}$) of a set of linearly independent vectors $\mathbf{b}_1, \ldots, \mathbf{b}_n$, called the basis of the lattice. The integer n is called the *rank* of the lattice. Using matrix notation, if $\mathbf{B} = [\mathbf{b}_1, \ldots, \mathbf{b}_n]$ is the matrix with the basis vectors as columns, lattice vectors can be written as $\mathbf{Bx}$, and $\mathcal{L}(\mathbf{B}) = \{\mathbf{Bx} \colon \mathbf{x} \in \mathbb{Z}^n\}$. In this paper, we will be mostly concerned with the Euclidean (also known as the ℓ_2) norm $\|\mathbf{x}\| = \sqrt{\sum_i x_i^2}$. Another norm that will be occasionally used is the ℓ_∞ norm $\|\mathbf{x}\|_\infty = \max_i |x_i|$. Define the half-open parallelepiped $\mathcal{P}(\mathbf{B}) = \{\sum_i x_i \mathbf{b}_i \ : 0 \leq x_i < 1 \text{ for } 1 \leq i \leq n\}$. For any lattice basis $\mathbf{B}$ and point $\mathbf{x}$ in the linear span of $\mathbf{B}$, there exists a unique vector $\mathbf{y} \in \mathcal{P}(\mathbf{B})$ such that $\mathbf{y} - \mathbf{x} \in \mathcal{L}(\mathbf{B})$. This vector is denoted $\mathbf{y} = \mathbf{x} \bmod \mathbf{B}$, and it can be computed in polynomial time given $\mathbf{B}$ and $\mathbf{x}$. The dual of a lattice Λ is the set

$$\Lambda^* = \{\mathbf{x} \in \operatorname{linspan}(\Lambda) \colon \forall \mathbf{y} \in \Lambda \ \langle \mathbf{x}, \mathbf{y} \rangle \in \mathbb{Z}\}$$

of all vectors that have integer scalar product ($\langle \mathbf{x}, \mathbf{y} \rangle = \sum_i x_i y_i$) with all lattice vectors. It is easy to see that for any lattice Λ and $c \in \mathbb{R}$, the dual of the lattice obtained scaling all vectors in Λ by a factor c is $(c\Lambda)^* = (1/c)\Lambda^*$. The *determinant* of a lattice $\det(\mathcal{L}(\mathbf{B}))$ is the (n-dimensional) volume of the fundamental parallelepiped $\mathcal{P}(\mathbf{B})$, and does not depend on the choice of the basis. The determinant of the dual lattice is $\det(\Lambda^*) = 1/\det(\Lambda)$.

The *minimum distance* of a lattice Λ, denoted $\lambda_1(\Lambda)$, is the minimum distance between any two distinct lattice points, and equals the length of the shortest nonzero lattice vector:

$$\begin{aligned} \lambda_1(\Lambda) &= \min\{\|\mathbf{x} - \mathbf{y}\| : \mathbf{x} \neq \mathbf{y} \in \Lambda\} \\ &= \min\{\|\mathbf{x}\| : \mathbf{x} \in \Lambda \setminus \{\mathbf{0}\}\} . \end{aligned}$$

This definition can be generalized to define the ith successive minimum as the smallest λ_i such that the ball $\lambda_i \mathcal{B} = \{\mathbf{x} \colon \|\mathbf{x}\| \leq \lambda_i\}$ contains at least i linearly independent lattice points. With some abuse of notation, we will often write $\lambda_i(\mathbf{B})$ to denote the successive minima of the lattice generated by basis $\mathbf{B}$.

A central problem in the computational study of lattices is the *Shortest Vector Problem* (SVP): given a lattice basis $\mathbf{B}$, find a nonzero lattice vector $\mathbf{Bx} \neq \mathbf{0}$ achieving the minimum distance $\|\mathbf{Bx}\| = \lambda_1$. The problem can be defined with respect to any norm, but the Euclidean norm ℓ_2 is the most common. A γ-approximate solution to the shortest vector problem (SVP_γ) is a nonzero lattice vector of length at most $\gamma\lambda_1(\mathbf{B})$. In complexity theory it is common to consider the decision version of SVP_γ (typically denoted $GapSVP_\gamma$), defined below.

Definition 1 (Shortest Vector Decision Problem). An input to $GapSVP_1$ is a pair $(\mathbf{B}, d)$ where $\mathbf{B}$ is an n-dimensional lattice basis and d is a rational number. The problem is to decide whether $\lambda_1(\mathbf{B}) \leq d$ or $\lambda_1(\mathbf{B}) > d$. More generally, for any $\gamma \geq 1$, an input to $GapSVP_\gamma$ is a pair $(\mathbf{B}, d)$ where $\mathbf{B}$ is an n-dimensional lattice basis and d is a rational number. The problem is to decide whether $\lambda_1(\mathbf{B}) \leq d$ or $\lambda_1(\mathbf{B}) > \gamma(n) \cdot d$. (If neither condition is satisfied, any answer is acceptable.)

Clearly, the promise problem $GapSVP_\gamma$ reduces to the problem of finding nonzero vectors of length at most $\gamma \cdot \lambda_1$. In the opposite direction, showing that solving $GapSVP_\gamma$ is at least as hard as computing vectors of length at most $\gamma \cdot \lambda_1$ is a classic open problem. In Section "Public-Key Encryption Scheme" we will also consider a restricted version of $GapSVP_\gamma$ (denoted $uGapSVP_\gamma$) where the second minimum of the lattice satisfies $\lambda_2(\mathbf{B}) > \gamma d$. Interestingly, $uGapSVP_\gamma$ is polynomial time equivalent (in both directions) to the corresponding search problem $uSVP_\gamma$: given a lattice basis $\mathbf{B}$ such that $\lambda_2(\mathbf{B}) > \gamma\lambda_1(\mathbf{B})$, find the shortest nonzero lattice vector.

Another classic problem in the study of lattices is the *Closest Vector Problem* (CVP_γ): Given a lattice basis $\mathbf{B}$ and a target vector $\mathbf{t}$, find the lattice point closest to the target. In the approximate version of the problem CVP_γ, the goal is to find a lattice vector within distance from the target no more than γ times the distance of the optimal solution. For any approximation factor γ, CVP_γ is at least as hard as SVP_γ [34].

The last successive minimum λ_n (where n is the rank of the lattice) and corresponding search problem defined below, play a central role in the construction of cryptographic functions with average-case/worst-case connection.

Definition 2 (Shortest Independent Vectors Problem). An input to $SIVP_\gamma$ is a lattice basis $\mathbf{B}$. The goal is to output a set of n linearly independent lattice vectors $\mathbf{S} \subset \mathcal{L}(\mathbf{B})$ such that $\|\mathbf{S}\| \leq \gamma(n) \cdot \lambda_n(\mathbf{B})$ where $\|\mathbf{S}\|$ is the maximum length of a vector in $\mathbf{S}$, and n is the rank of the lattice.

The shortest independent vectors problem is closely related to the following variant of the closest vector problem.

Definition 3 (Guaranteed Distance Decoding). An input to GDD_γ is a lattice basis $\mathbf{B}$ and a target point $\mathbf{t}$. The goal is to output a lattice point $\mathbf{x} \in \mathcal{L}(\mathbf{B})$ such that $\|\mathbf{t} - \mathbf{x}\| \leq \gamma(n) \cdot \nu(\mathbf{B})$, where $\nu = \max_{\mathbf{t} \in \mathcal{P}(\mathbf{B})} \min_{\mathbf{x} \in \mathcal{L}(\mathbf{B})} \|\mathbf{t} - \mathbf{x}\|$ is the covering radius of the lattice.

The difference between GDD_γ and CVP_γ is that the quality of the solution is measured with respect to the worst possible distance ν, rather than the distance of the given target. (A related and more common variant of CVP_γ that arises in coding theory applications, but not used in this survey, is the "Bounded Distance Decoding," where the promise is that the target is very close to the lattice, typically within distance $\lambda_1/2$.) The relation between SIVP_γ and GDD_γ is easily explained. First, we recall (e.g., see [33, Theorem 7.9]) that for any n-dimensional lattice, ν and λ_n satisfy

$$\lambda_n \leq 2\nu \leq \sqrt{n}\lambda_n. \tag{13.1}$$

Next, we observe that the proof of (13.1) is constructive in the sense that it gives efficient reductions from $\text{GDD}_{\sqrt{n}\gamma}$ to SIVP_γ and from $\text{SIVP}_{\sqrt{n}\gamma}$ to GDD_γ. Specifically, given a solution to SIVP_γ instance $\mathbf{B}$, and a target vector $\mathbf{t}$, one can efficiently find (e.g., using Babai's nearest plane algorithm [35]) a lattice point within distance $(\sqrt{n}/2)\gamma \cdot \lambda_n(\mathbf{B})$ from the target. (When $\gamma = 1$, this proves that $2\nu \leq \sqrt{n}\lambda_n$.) Conversely, given a lattice $\mathbf{B}$, one can efficiently find a set of linearly independent lattice vectors $\mathbf{v}_1, \ldots, \mathbf{v}_n$ of length $2\gamma\nu$ making n calls to a GDD_γ oracle on input the same lattice $\mathbf{B}$ and n distinct target vectors, where each $\mathbf{t}_i$ is chosen adaptively as a vector orthogonal to $\mathbf{v}_1, \ldots, \mathbf{v}_{i-1}$ of length (slightly larger than) $\gamma\nu$. (Again, when $\gamma = 1$, this proves that $\lambda_n \leq 2\nu$.)

Collision Resistant Hashing

In this section, we give a general introduction to the problem of designing cryptographic functions that are provably secure, based on the worst-case intractability of lattice problems. For simplicity, in this section, we concentrate on collision resistant hashing (defined below), a fairly simple, but still useful, cryptographic primitive. Public-key encryption schemes are discussed separately in Section "Public-Key Encryption Scheme".

A collision resistant hash function family is a collection of keyed functions $\{h_k \colon D \to R\}_{k \in K}$ with common domain D and range R. The hash functions are efficiently computable in the sense that there is an efficient algorithm that on input a key $k \in K$ and a value $x \in D$, computes the corresponding value $f_k(x) \in R$ in the range of the functions. Usually, the domain D is larger than the range R, so that by the pigeon-hole principle, each function f_k admits collisions, i.e., pairs $x \neq y$ of distinct inputs that are mapped to the same output $f_k(x) = f_k(y)$. A hash function family is called *collision resistant*, if such collisions are hard to find, i.e., no efficient algorithm on input a randomly chosen key $k \in K$ will find a collision for function f_k. In order to study hash function families in the asymptotic computational setting, one considers sequences $\{h_k \colon D_n \to R_n\}_{k \in K_n}$ of hash function families (indexed by a security parameter n,) with larger and larger domain, range and key space, e.g., $K_n = \{0,1\}^n$, $D_n = \{0,1\}^{2n}$ and $R_n = \{0,1\}^n$.

Most lattice based cryptographic functions have a subset-sum structure, and the constructions are fairly easy to describe. Let $(G, +, 0)$ be a commutative group with binary operation $+$ and identity element 0. Any sequence $\mathbf{a} = (a_1, \ldots, a_m) \in G^m$ of group elements defines a subset-sum function $f_{\mathbf{a}}(\mathbf{x})$ mapping binary string $\mathbf{x} \in \{0,1\}^m$ to group element

$$f_{\mathbf{a}}(\mathbf{x}) = \sum_{i=1}^{m} a_i \cdot x_i. \tag{13.2}$$

A collision for function $f_{\mathbf{a}}$ is a pair of binary vectors $\mathbf{x} \neq \mathbf{y}$ such that $f_{\mathbf{a}}(\mathbf{x}) = f_{\mathbf{a}}(\mathbf{y})$. Equivalently, collisions can be represented by a single vector $\mathbf{z} = \mathbf{x} - \mathbf{y}$ such that $\|\mathbf{z}\|_\infty = 1$ and $f_{\mathbf{a}}(\mathbf{z}) = 0$ is the identity element in G.

Notice that if $m > \log_2 |G|$, then $f_{\mathbf{a}}$ is a hash function, i.e., it compresses the input, and collisions are guaranteed to exist. We want to prove that for appropriate choice of the group G and parameter m, finding collisions to function $f_{\mathbf{a}}$ with non-negligible probability (when the key $\mathbf{a} \in G^m$ is chosen uniformly at random) is at least as hard as solving various lattice approximation problems in the worst case.

We remark that, for any subset-sum function (13.2), the set of integer vectors $\mathbf{z} \in \mathbb{Z}^m$ satisfying $f_{\mathbf{a}}(\mathbf{z}) = 0$ forms a lattice. So, the problem of breaking (i.e., finding collisions to) hash function $f_{\mathbf{a}}$ can be equivalently formulated as the problem of finding shortest nonzero vectors (in the ℓ_∞ norm) in a random lattice. In this sense, the construction of lattice based hash functions establishes a connection between the worst-case and average-case complexity of (different) lattice problems.

In the rest of this section we first review some mathematical material that will be used in the analysis of the hash functions. Next, we present an oversimplified instantiation and analysis of a lattice based function, which, although technically inaccurate, conveys most of the intuition behind the actual constructions. In Section "Simple Cryptographic Hash Function with Worst-Case/Average-Case Connection" we give a self contained analysis (using the mathematical facts stated in Section "Background") of a simple lattice based hash function. In this section we strive for simplicity, rather than achieving the best quantitative results. The current state of the art in the construction of lattice based hash functions is discussed in Section "Improving the Construction and Analysis", followed in Section "Using Special Lattices to Improve Efficiency" by a description of recent work on the construction of very efficient cryptographic functions based on special classes of lattices.

Background

The mathematical techniques used in the analysis of the currently best results in the area of lattice based cryptography involve Gaussian distributions and the n-dimensional Fourier transform. Fortunately, just a few basic facts in the area are enough to analyze a simplified version of the hash function of [16]. In this section we briefly review all necessary definitions and properties used in our analysis in Section "Simple Cryptographic Hash Function with Worst-Case/Average-Case Connection".

For any vector $\mathbf{c} \in \mathbb{R}^n$ and parameter $s > 0$, let $D_{s,\mathbf{c}}(\mathbf{x}) = e^{-\pi\|(\mathbf{x}-\mathbf{c})/s\|^2}/s^n$ be (the probability density function of) the Gaussian distribution with center $\mathbf{c}$ and variance $ns^2/(2\pi)$. For any n-dimensional lattice Λ, let $D_{\Lambda,s,\mathbf{c}}$ be the discrete probability distribution obtained conditioning $D_{s,\mathbf{c}}(\mathbf{x})$ on the event that $\mathbf{x} \in \Lambda$. More precisely, for any $\mathbf{x} \in \Lambda$ let

$$D_{\Lambda,s,\mathbf{c}}(\mathbf{x}) = \frac{D_{s,\mathbf{c}}(\mathbf{x})}{\sum_{\mathbf{y}\in\Lambda} D_{s,\mathbf{c}}(\mathbf{y})}.$$

In [16], Micciancio and Regev introduce a lattice parameter $\eta_\varepsilon(\Lambda)$, called the *smoothing parameter* of the lattice. (To be precise, η_ε is a *family* of parameters, indexed by a real number ε. Typically, one considers families of lattices Λ_n in increasing dimension, and smoothing parameter η_ε where $\varepsilon(n)$ is some fixed negligible function of n.) Informally, the smoothing parameter satisfies the following fundamental property [16, Lemma 4.1]: if $s \geq \eta_\varepsilon(\Lambda)$, then adding Gaussian noise with distribution $D_{s,\mathbf{c}}$ to a lattice Λ results in an almost uniform distribution over $\mathbb{R}^n$. More precisely, the property satisfied by the smoothing parameter is that for any n-dimensional lattice $\mathcal{L}(\mathbf{B})$, vector $\mathbf{c} \in \mathbb{R}^n$, and parameter $s \geq \eta_\varepsilon(\Lambda)$, the statistical distance between $D_{s,\mathbf{c}}$ mod $\mathcal{P}(\mathbf{B})$ and the uniform distribution over $\mathcal{P}(\mathbf{B})$ is at most

$$\Delta(D_{s,\mathbf{c}} \bmod \mathcal{P}(\mathbf{B}), U(\mathcal{P}(\mathbf{B}))) \leq \varepsilon/2. \tag{13.3}$$

Here, the operation of adding noise D to a lattice $\mathcal{L}(\mathbf{B})$ is expressed as reducing D (which is a distribution over $\mathbb{R}^n$) modulo the lattice, yielding a distribution over $\mathcal{P}(\mathbf{B})$. Intuitively, this represents a distribution over the entire space obtained by tiling $\mathbb{R}^n$ with copies of $\mathcal{P}(\mathbf{B})$. We refer the reader to [16] for the exact definition of the smoothing parameter, which involves the dual lattice, and is somehow technical. The reader can think of (13.3) essentially as the definition of the smoothing parameter. Below we state the two only other important properties of the smoothing parameter that will be used in this paper. The first property [16, Lemma 4.2] is that the smoothing parameter is not much larger than λ_n: for any n-dimensional lattice Λ and positive real $\varepsilon > 0$

$$\eta_\varepsilon(\Lambda) \leq \sqrt{\frac{\ln(2n(1+1/\varepsilon))}{\pi}} \cdot \lambda_n(\Lambda). \tag{13.4}$$

As a special case, if $\varepsilon = n^{-\log n}$, then $\eta_\varepsilon(\Lambda) \leq \log(n) \cdot \lambda_n(\Lambda)$. The second property is that if $s \geq \eta_\varepsilon(\Lambda)$, then the discrete distribution $D_{\Lambda,s,\mathbf{c}}$ behaves in many respect like the continuous distribution $D_{s,\mathbf{c}}$. In particular, (and this is all we need in this paper) [16, Lemma 4.4] shows that if $s \geq \eta_\varepsilon(\Lambda)$, then

$$\mathbb{P}_{\mathbf{x} \sim D_{\Lambda,s,\mathbf{c}}}\{\|\mathbf{x} - \mathbf{c}\| > s\sqrt{n}\} \leq \frac{1+\varepsilon}{1-\varepsilon} \cdot 2^{-n}, \tag{13.5}$$

i.e., the discrete distribution $D_{\Lambda,s,\mathbf{c}}$ is highly concentrated in a sphere of radius $s\sqrt{n}$ around $\mathbf{c}$.

An Oversimplified Construction

In the limit, the hash function family that we are going to describe and analyze, corresponds to the subset-sum problem over the additive group $\mathbb{R}^n$ of n-dimensional real vectors. Using matrix notation, the subset-sum functions are indexed by a real matrix $\mathbf{A} \in \mathbb{R}^{n \times m}$ and map input $\mathbf{x} \in \{0,1\}^m$ to $f_{\mathbf{A}}(\mathbf{x}) = \mathbf{A}\mathbf{x} \in \mathbb{R}^n$. Clearly, using $\mathbb{R}^n$ as underlying group is not a valid choice, because it results in a function with finite domain $\{0,1\}^m$ and infinite range $\mathbb{R}^n$ that cannot possibly compress the input and be a hash function. Still, the intuition behind the actual construction can be easily illustrated using real numbers. So, let us ignore this finiteness issue for now, and observe that the range of the function does not depend on the parameter m. So, if the range were finite, we could easily make the domain $\{0,1\}^m$ bigger than the range by choosing a sufficiently large m.

We want to prove that collisions are hard to find, in the sense that any procedure that finds collisions to $f_{\mathbf{A}}$ for random $\mathbf{A}$ (with arbitrarily small, but non-negligible probability) can be converted into a worst-case approximation algorithm for lattice problems, e.g., finding short vectors in an arbitrary input lattice $\mathbf{B}$. Technically, we want to give a reduction that on input a lattice basis $\mathbf{B}$ selects a key $\mathbf{A}$ such that the ability to find a collision for $f_{\mathbf{A}}$ helps to find short vectors in $\mathcal{L}(\mathbf{B})$. The challenge is that in order for this to be a valid worst-case to average-case reduction, the key $\mathbf{A}$ selected by the reduction should be (almost) independent from the input lattice $\mathbf{B}$. In other words, no matter what lattice $\mathcal{L}(\mathbf{B})$ we start from, we should end up generating keys $\mathbf{A}$ with essentially the same (uniformly random) probability distribution.

This can be achieved as follows. Consider the result of selecting a random lattice point $\mathbf{y} \in \mathcal{L}(\mathbf{B})$ and adding a small amount of noise $\mathbf{r}$ to it. Again, here we are being informal. Since $\mathcal{L}(\mathbf{B})$ is a countably infinite set of points, we cannot choose $\mathbf{y} \in \mathcal{L}(\mathbf{B})$ uniformly at random. We will solve this and other technical problems in the next section. Going back to the proof, if the amount of noise is large enough (say, $\mathbf{r}$ is chosen uniformly at random from a sphere of radius $n\lambda_n$ sufficiently larger than the smoothing parameter of the lattice) the resulting point $\mathbf{y} + \mathbf{r}$ will be very close to being uniformly distributed over the entire space $\mathbb{R}^n$. So, we can select a random key $\mathbf{A}$ by choosing, for $i = 1, \ldots, m$, a random lattice point $\mathbf{y}_i \in \mathcal{L}(\mathbf{B})$ and random small error vector $\mathbf{r}_i$, and setting $\mathbf{a}_i = \mathbf{y}_i + \mathbf{r}_i$. The resulting key $\mathbf{A} = [\mathbf{a}_1, \ldots, \mathbf{a}_n]$ won't be distributed exactly at random, but will be sufficiently close to uniform that the hypothetical collision finder algorithm, on input $\mathbf{A}$, will find a collision $f_{\mathbf{A}}(\mathbf{z}) = \mathbf{0}$ (where $\|\mathbf{z}\|_\infty = 1$) with non-negligible probability. (If the collision finder does not produce a valid collision, we simply repeat the procedure again by choosing a new $\mathbf{A}$ independently at random. Since the success probability of the collision finder is non-negligible, a collision will be found with very high probability after at most a polynomial number of iterations.) Let $\mathbf{z}$ be the collision found by the algorithm. We have $\sum_i \mathbf{a}_i z_i = \mathbf{0}$. Substituting $\mathbf{a}_i = \mathbf{y}_i + \mathbf{r}_i$ and rearranging we get

$$\sum_{i=1}^{m} z_i \cdot \mathbf{y}_i = \sum_{i=1}^{m} (-z_i) \cdot \mathbf{r}_i . \tag{13.6}$$

Notice that the vector on the left hand side of (13.6) is a lattice vector because it is an integer linear combination (with coefficients $z_i \in \mathbb{Z}$) of lattice vectors $\mathbf{y}_i \in \mathcal{L}(\mathbf{B})$. At the same time, the vector on the right hand side of (13.6) is a short vector because it is a small linear combination (with coefficients $-z_i \in \{0, 1, -1\}$) of short vectors $\mathbf{r}_i$. In particular, if the error vectors $\mathbf{r}_i$ have length at most $n\lambda_n$, then the vector in

(13.6) has length at most $mn\lambda_n$. So, we have found a lattice vector of length not much bigger than λ_n.

The construction we just described and informally analyzed has several shortcomings. The main problem is that the range of the function is infinite, so the assumption that no algorithm can find collisions is vacuous. (Specifically, when $\mathbf{A} \in \mathbb{R}^{n\times m}$ is chosen at random, the corresponding function is injective with probability 1.) The conclusion that we can find a short vector in a lattice is also trivial, as our proof sketch does not rule out the possibility that the vector $\sum_i \mathbf{y}_i \cdot z_i$ equals $\mathbf{0}$. All these problems are solved in the next section by replacing $\mathbb{R}^n$ with the finite group $\mathbb{Z}_q^n$ of n-dimensional vectors modulo q, and using a variant of the closest vector problem (GDD_γ) as the worst-case problem to be solved.

Simple Cryptographic Hash Function with Worst-Case/Average-Case Connection

Consider the subset-sum function family over the group $\mathbb{Z}_q^n$ of integer vectors modulo q. Using matrix notation, this function family is indexed by $\mathbf{A} \in \mathbb{Z}_q^{n\times m}$, and each function maps binary string $\mathbf{x} \in \{0,1\}^m$ to vector

$$f_{\mathbf{A}}(\mathbf{x}) = \mathbf{A}\mathbf{x} \bmod q \in \mathbb{Z}_q^n. \tag{13.7}$$

For concreteness, let us fix $q = 2^n$ and $m = 2n^2$. Notice that the corresponding subset-sum function has a domain of size $|\{0,1\}^m| = 2^{2n^2}$ and range of size $|\mathbb{Z}_q^n| = 2^{n^2}$. So, the function compresses the length of the input by a factor 2, and collisions $f_{\mathbf{A}}(\mathbf{z}) = \mathbf{0}$ (with $\|\mathbf{z}\|_\infty = 1$) are guaranteed to exist. We prove that finding collisions is computationally hard, even on the average when $\mathbf{A}$ is chosen uniformly at random.

Theorem 1. *Let $\{f_{\mathbf{A}}\colon \{0,1\}^m \to \mathbb{Z}_q^n\}_{\mathbf{A}}$ be the function family defined in (13.7), with $q(n) = 2^n$, $m(n) = 2n^2$, and $\mathbf{A} \in \mathbb{Z}^{n\times m}$. If no polynomial time algorithm can solve GDD_γ in the worst case within a factor $\gamma = n^3$ (where n is the rank of the input lattice), then $\{f_{\mathbf{A}}\}_{\mathbf{A}}$ is collision resistant.*

Proof. Assume for contradiction that there exists an algorithm $\mathcal{F}$ that on input a uniformly random matrix $\mathbf{A} \in \mathbb{Z}_q^{n\times m}$, finds a collision $\mathbf{z}$ with non-negligible probability p. We can assume, without loss of generality, that $\mathcal{F}$ always outputs a vector $\mathbf{z} = \mathcal{F}(\mathbf{A})$ of norm $\|\mathbf{z}\|_\infty = 1$, which may or may not satisfy $\mathbf{A}\mathbf{z} = \mathbf{0}$ (mod q). We know that $\mathbb{P}\{\mathbf{A}\mathbf{z} \bmod q = \mathbf{0}\} = p$ is non-negligible. Notice that for any possible output $\mathbf{z} = \mathcal{F}(\mathbf{A})$, there exists an index $j \in \{1,\ldots,m\}$ such that $z_j \neq 0$. So, by union bound, there is an index $j_0 \in \{1,\ldots,m\}$ such that $\mathbb{P}\{\mathbf{A}\mathbf{z} \bmod q = \mathbf{0} \wedge z_{j_0} \neq 0\} \geq p/m = p/2n^2$ is also non-negligible, where the probability is computed over the random selection of $\mathbf{A} \in \mathbb{Z}_q^{n\times m}$ and $\mathbf{z} = \mathcal{F}(\mathbf{A})$.

We use $\mathcal{F}$ to approximate the GDD_γ variant of the closest vector problem within a factor n^3. Specifically, given a lattice basis $\mathbf{B} \in \mathbb{R}^{n\times n}$ and target $\mathbf{t} \in \mathbb{R}^n$, we find a lattice vector in $\mathcal{L}(\mathbf{B})$ within distance $n^3\nu(\mathbf{B})$ from $\mathbf{t}$, where $\nu(\mathbf{B})$ is the covering radius of the lattice. As remarked in Section "Background", it is easy to see that this problem is at least as hard as approximating SIVP_γ within (slightly larger) polynomial factors $\gamma' = n^{3.5}$. Assume that a value $r \in [n^3\nu(\mathbf{B})/2, n^3\nu(\mathbf{B})]$ is known. (In general, one can try all possible values $r = 2^k$, and stop as soon as the reduction is successful.) Oracle $\mathcal{F}$ is used as follows:

1. Apply the LLL basis reduction algorithm to $\mathbf{B}$, so that $\|\mathbf{B}\| \le (2/\sqrt{3})^n\lambda_n(\mathbf{B})$.
2. Choose a random index $j \in \{1, \ldots, m\}$.
3. Choose error vectors $\mathbf{r}_i$ with distribution $D_{s,\delta_{ij}\mathbf{t}}$ where $s = r/(2m\sqrt{n})$, and $\delta_{ij} \in \{0, 1\}$ equals 1 if and only if $i = j$. In other words, $\mathbf{r}_i$ is chosen according to a Gaussian distribution of parameter s centered around either $\mathbf{0}$ or $\mathbf{t}$.
4. Let $\mathbf{c}_i = \mathbf{r}_i \bmod \mathbf{B}$ and $\mathbf{y}_i = \mathbf{c}_i - \mathbf{r}_i \in \mathcal{L}(\mathbf{B})$.
5. Let $\mathbf{a}_i = \lfloor q\mathbf{B}^{-1}\mathbf{c}_i \rceil$, and call the oracle $\mathcal{F}(\mathbf{A})$ on input $\mathbf{A} = [\mathbf{a}_1, \ldots, \mathbf{a}_m]$.
6. Output the vector

$$\mathbf{x} = z_j(\mathbf{BA}/q - \mathbf{Y})\mathbf{z}$$

where $\mathbf{Y} = [\mathbf{y}_1, \ldots, \mathbf{y}_m]$ and $\mathbf{z} = \mathcal{F}(\mathbf{A})$ is the collision returned by the oracle.

We need to prove that the reduction is correct, i.e., it outputs a lattice vector $\mathbf{x} \in \mathcal{L}(\mathbf{B})$ within distance r from the target $\mathbf{t}$ with non-negligible probability. (Since the success probability depends only on the internal randomness of the reduction algorithm, the failure probability can be made exponentially small using standard repetition techniques.)

Since j is chosen uniformly at random, $j = j_0$ with non-negligible probability $1/m = 1/(2n^2)$. In the rest of the proof, we consider the conditional success probability of the reduction given that $j = j_0$, and show that this conditional probability is still non-negligible.

Notice that, by our assumption on r, and using (13.4), the Gaussian parameter satisfies $s = r/(2m\sqrt{n}) \ge n^3\nu/4m\sqrt{n} \ge (\sqrt{n}/16)\cdot\lambda_n(\mathbf{B}) \ge \eta_\varepsilon(\mathbf{B})$ for any $\varepsilon = 2^{-o(n)}$. So, by (13.3), the distribution of the vectors $\mathbf{c}_i = \mathbf{r}_i \bmod \mathbf{B}$ is within statistical distance $\varepsilon/2$ from the uniform distribution over $\mathcal{P}(\mathbf{B})$. Since the function $\mathbf{c} \mapsto \lfloor q\mathbf{B}^{-1}\mathbf{c}\rfloor$ maps the uniform distribution over $\mathcal{P}(\mathbf{B})$ to the uniform distribution over $\mathbb{Z}_q^n$, we get that each vector $\mathbf{a}_i = \lfloor q\mathbf{B}^{-1}\mathbf{c}_i \rceil$ is also within statistical distance $\varepsilon/2$ from the uniform distribution over $\mathbb{Z}_q^n$. Overall, the key $\mathbf{A} = [\mathbf{a}_1, \ldots, \mathbf{a}_m]$ is within negligible distance $\varepsilon m/2$ from uniform and

$$\mathbb{P}\{\mathbf{Az} = \mathbf{0} \pmod{q} \wedge z_{j_0} \neq 0\} \ge \frac{p}{m} - \frac{\varepsilon m}{2} \ge \frac{p}{2m}.$$

Notice that if $\mathbf{Az} = \mathbf{0} \pmod{q}$, then the output of the reduction

$$\mathbf{x} = z_j\mathbf{B}\frac{\mathbf{Az}}{q} - z_j\mathbf{Yz}$$

is certainly a lattice vector because it is a linear combination of lattice vectors $\mathbf{B}$ and $\mathbf{Y}$ with integer coefficients $z_j(\mathbf{Az}/q)$ and $-z_j\mathbf{z}$. It remains to prove that the conditional probability that $\|\mathbf{t}-\mathbf{x}\| \le r$ (given that $j = j_0$, $\mathbf{Az} = \mathbf{0} \pmod q$ and $z_{j_0} \neq 0$) is non-negligible. We will prove something even stronger: the conditional probability, given $j = j_0$, $\mathbf{C}$, $\mathbf{A}$ and $\mathbf{z}$ satisfying $\mathbf{Az} = \mathbf{0} \pmod q$ and $z_{j_0} \neq 0$, is exponentially close to 1. We bound the distance $\|\mathbf{t}-\mathbf{x}\|$ as follows:

$$\begin{aligned}
\|\mathbf{t}-\mathbf{x}\| &\le \|\mathbf{t} - z_j(\mathbf{BA}/q - \mathbf{Y})\mathbf{z}\| \\
&= \|z_j \cdot \mathbf{t} + (\mathbf{C} - \mathbf{BA}/q)\mathbf{z} + (\mathbf{Y}-\mathbf{C})\mathbf{z}\| \\
&\le \frac{n}{q}\|\mathbf{B}\| \cdot \|(q\mathbf{B}^{-1}\mathbf{C} - \lfloor q\mathbf{B}^{-1}\mathbf{C}\rceil)\mathbf{z}\|_\infty + \sum_i |z_i| \cdot \|\mathbf{y}_i - (\mathbf{c}_i - \delta_{ij}\mathbf{t})\| \\
&\le \frac{nm}{q}\|\mathbf{B}\| + m \cdot \max_i \cdot \|\mathbf{y}_i - (\mathbf{c}_i - \delta_{ij}\mathbf{t})\|.
\end{aligned}$$

Finally, observe that the distribution of $\mathbf{y}_i$, given $\mathbf{c}_i$ and j, is $D_{\mathcal{L}(\mathbf{B}),s,\mathbf{c}_i-\delta_{ij}\mathbf{t}}$. So, by (13.5), $\|\mathbf{y}_i - (\mathbf{c}_i - \delta_{ij}\mathbf{t})\| \le \sqrt{n}s$ with probability exponentially close to 1, and

$$\|\mathbf{t}-\mathbf{x}\| \le \frac{2n^3(2/\sqrt{3})^n}{2^n}\lambda_n(\mathbf{B}) + m\sqrt{n}\frac{r}{2m\sqrt{n}} = \frac{2n^3}{3^{n/2}}\lambda_n(\mathbf{B}) + \frac{r}{2} < r.$$

□

Improving the Construction and Analysis

The main difference between the construction described in Section "Simple Cryptographic Hash Function with Worst-Case/Average-Case Connection" and those studied and analyzed in [14–16, 36] is that we used a very large value of q (exponential in n), while [14–16, 36] use $q = n^{O(1)}$. Using a large value of q seems necessary if one starts from an LLL reduced basis and wants to find a short vector with just one application of the reduction. The approach taken in [14–16, 36] is to use the reduction to find lattice vectors that are shorter than the vectors in the input basis $\mathbf{B}$ by just a small (constant or polynomial) factor. A set of very short lattice vectors can then be found by successive improvements, starting from a basis $\mathbf{B}$ that contains vectors potentially as large as $2^n\lambda_n$, and progressively finding shorter and shorter bases for the lattice.

Using a large value of q has a negative impact on both the efficiency and provable security of the hash function. On the efficiency side, since $2^m > q^n$, using $q = 2^n$ yields hash functions with key size and computation time at least $n \cdot m \cdot \log(q) > n^4$. By contrast, using $q = n^{O(1)}$ yields key size and computation time $\tilde{O}(n^2)$. On the security side, our proof shows that breaking the hash function is at least as hard as approximating SIVP_γ within a factor $\gamma = n^{3.5}$. In fact, with little more effort, one can reduce the factor to $\gamma = \tilde{O}(\sqrt{mn})$. Still, using $q = 2^n$ results in factors larger than $n^{1.5}$. Using smaller $q = n^{O(1)}$ yields collision resistant hash functions that are at least as hard to break as approximating SIVP_γ within factors $\gamma = \tilde{O}(n)$ almost linear in the dimension of the lattice. This is the best currently known result, proved by Micciancio and Regev in [16]. (See also [29] for a method to reduce the parameter q to as low as $q(n) = \tilde{O}(n)$.)

Theorem 2. *For any sufficiently large polynomial q (e.g., $q(n) = n^3$) the subset-sum function $f_{\mathbf{A}}\colon \{0,1\}^m \to \mathbb{Z}_q^n$ defined in (13.2) is collision resistant, assuming that no polynomial time algorithm can solve $SIVP_\gamma$ (as well as various other lattice problems, e.g., $GapSVP_\gamma$) in the worst case within a factor $\gamma = \tilde{O}(n)$ almost linear in the rank of the lattice.*

We remark that the collision resistant hash function of Micciancio and Regev [16] is essentially the same as the one-way function originally proposed by Ajtai [13]. The difference across [13–16] is mostly in the techniques used to analyze the function, choice of the parameters, and corresponding worst-case factor achieved. Further reducing the factor in the worst-case inapproximability assumption to $\sqrt{n}$ (or even to $n^{1-\varepsilon}$ for some constant $\varepsilon > 0$) is currently one of the main open problems in the area.

Using Special Lattices to Improve Efficiency

From a practical point of view, a drawback of the cryptographic functions of [13–16] is that they require a key size $\tilde{O}(n^2)$, approximately quadratic in the natural security parameter n. Unfortunately, it seems hard to build lattice based hash functions with key size sub-quadratic in the dimension of the underlying lattice. Intuitively, the reason is that an n-dimensional lattice is described by an $n \times n$ integer matrix, and this representation is essentially optimal. (For example, it can be easily shown that there are $2^{\Omega(n^2)}$ distinct integer lattices with determinant 2^n.) More concretely, the keys to the hash functions studied in Section "Simple Cryptographic Hash Function with Worst-Case/Average-Case Connection" and "Improving the Construction and Analysis" are matrices

$$\mathbf{A} = \begin{bmatrix} a_{1,1} & \cdots & a_{1,n} & a_{1,n+1} & \cdots & a_{1,2n} & a_{1,2n+1} & \cdots & a_{1,m} \\ \vdots & \ddots & \vdots & \vdots & \ddots & \vdots & \vdots & \cdots & \vdots \\ a_{n,1} & \cdots & a_{n,n} & a_{n,n+1} & \cdots & a_{n,2n} & a_{n,2n+1} & \cdots & a_{n,m} \end{bmatrix}$$

where the number of rows is the rank n of the underlying lattice, and the number of columns m is strictly larger than n in order to achieve compression. So, even if the entries $a_{i,j}$ are small integers, the key size is still at least $O(n^2)$. A possible approach to overcome this difficulty and get more efficient and *provably secure* lattice based cryptographic functions was suggested by Micciancio in [17], inspired in part by NTRU [37], a very fast commercial cryptosystem. NTRU is a ring based cryptosystem that can equivalently be described using certain lattices with special structure, but for which no proof of security is known. The idea common to [17] and [37] is to use lattices with special structure that can be described with only $\tilde{O}(n)$ bits. For example, Micciancio [17] considers cyclic lattices, i.e., lattices that are invariant under cyclic rotations of the coordinates. Many such lattices can be described by giving a single lattice vector $\mathbf{v}$, and a basis for the corresponding lattice

$$\mathbf{C_v} = \begin{bmatrix} v_1 & v_n & v_{n-1} & \cdots & v_2 \\ v_2 & v_1 & v_n & \cdots & v_3 \\ v_3 & v_2 & v_1 & \cdots & v_4 \\ \vdots & \vdots & \vdots & \ddots & \vdots \\ v_n & v_{n-1} & v_{n-2} & \cdots & v_1 \end{bmatrix}$$

can be obtained by taking $\mathbf{v}$ together with its $n-1$ cyclic rotations.

The novelty in Micciancio's work [17] (which sets it apart from [37]), is that lattices with special structure are used to build very efficient cryptographic functions that also have strong *provable security* guarantees, similar to those of Ajtai's original proposal [13] and subsequent work [14–16]. Specifically, Micciancio [17] gives a much more efficient one-way function than the one of [13, 16], and still provably secure based on the worst-case intractability of SIVP_γ (and other lattice problems) within almost linear factors over the class of *cyclic lattices*. In particular, the one-way function of [17] has key size and computation time $\tilde{O}(n)$ almost linear in the security parameter n. The adaptation of the proof of [16] to the cyclic setting is non-trivial, and several subtle issues come up during the proof. For example, Micciancio [17] only shows that the proposed cryptographic function is one-way, but not necessarily collision resistant. In fact, it was later shown (independently, by Peikert and Rosen [20] and Lyubashevsky and Micciancio [19]) that the function of [17] is *not* collision resistant, but it can be easily modified to become a collision resistant hash function provably secure essentially under the same assumptions as in [17]. Lyubashevsky and Micciancio [19] also extend the construction to a wider class of lattices, named "ideal" lattices in that paper.

We only describe the construction of efficient cryptographic functions (called "generalized compact knapsacks"), and refer the reader to the original papers for the proofs of security, which combine the geometric techniques described in Section "Simple Cryptographic Hash Function with Worst-Case/Average-Case Connection" with new algebraic methods. The fundamental idea of [17] is that the key $\mathbf{A}$ to function $f_{\mathbf{A}}(\mathbf{x}) = \mathbf{Ax} \pmod q$ does not have to be chosen at random from the set $\mathbb{Z}_q^{n\times m}$ of *all* possible matrices. One can restrict, for example, matrix $\mathbf{A}$ to be the concatenation

$$\mathbf{A} = [\mathbf{C}_{\mathbf{a}^1}\mathbf{C}_{\mathbf{a}^2}\cdots\mathbf{C}_{\mathbf{a}^{m/n}}] = \begin{bmatrix} a_1^1 & \cdots & a_2^1 & a_1^2 & \cdots & a_2^2 & a_1^3 & \cdots & a_2^{m/n} \\ \vdots & \ddots & \vdots & \vdots & \ddots & \vdots & \vdots & \cdots & \vdots \\ a_n^1 & \cdots & a_1^1 & a_n^2 & \cdots & a_1^2 & a_n^3 & \cdots & a_1^{m/n} \end{bmatrix}$$

of a small number of circulant matrices, i.e., matrices $\mathbf{C}_{\mathbf{a}^i}$ where each column is the cyclic rotation of the previous column. Perhaps unexpectedly, an algorithm to invert random instances of the corresponding function $f_{\mathbf{A}}$ can be used to approximate SIVP_γ and other lattice problems in the worst case, when restricted to the special case of cyclic lattices, i.e., lattices that are invariant under cyclic rotation of the coordinates. As already remarked, adapting the reduction of [16] from general lattices to the cyclic case is non-trivial, and, in particular achieving collision resistance requires some new algebraic ideas. The nice feature of the new function is that the key $\mathbf{A}$ can now be represented using only $m \log q$ bits (e.g., just the vectors $\mathbf{a}^1, \ldots, \mathbf{a}^{m/n}$, rather than the entire matrix $\mathbf{A}$), and the corresponding function $f_{\mathbf{A}}$ can also be computed in $\tilde{O}(m \log q)$ time using the fast Fourier transform.

The idea of using structured matrices is further developed in [19, 20] where it is shown how to turn $f_{\mathbf{A}}$ into a collision resistant hash function. The approaches followed in the two (independent) papers are closely related, but different. In [20], collision resistance is achieved by restricting the domain of the function to a subset of all binary strings. In [19], it is shown how to achieve collision resistance by appropriately changing the constraint on matrix $\mathbf{A}$. Here we follow the approach of [19], which better illustrates the algebraic ideas common to both papers. The fundamental idea (already present in [17, 37]) is that the ring of circulant matrices is isomorphic to the ring of polynomials $\mathbb{Z}[X]$ modulo $(X^n - 1)$. It turns out that the collision resistance properties of $f_{\mathbf{A}}$ are closely related to the factorization of $X^n - 1$: the linear factor $(X - 1)$ allows to efficiently find collisions, while if we replace $(X^n - 1)$ with an irreducible polynomial $p(X)$, we get a collision resistant function. The proof of security is based on the worst-case intractability of lattice problems over the corresponding class of "ideal" lattices: lattices that can be expressed as ideals of the ring $\mathbb{Z}[X]/p(X)$. (Notice that when $p(X) = X^n - 1$, the ideals of $\mathbb{Z}[X]/p(X)$ correspond exactly to cyclic lattices.) A specific choice of $p(X)$ that results in very efficient implementation [21] is $p(X) = X^n + 1$, which is irreducible when n is a power of 2. In terms of matrices, $X^n + 1$ corresponds to using a variant of circulant matrices

$$\mathbf{C}_{\mathbf{v}}^{-} = \begin{bmatrix} v_1 & -v_n & -v_{n-1} & \cdots & -v_2 \\ v_2 & v_1 & -v_n & \cdots & -v_3 \\ v_3 & v_2 & v_1 & \cdots & -v_4 \\ \vdots & \vdots & \vdots & \ddots & \vdots \\ v_n & v_{n-1} & v_{n-2} & \cdots & v_1 \end{bmatrix}$$

where each column is a cyclic shift of the previous one, with the element wrapping around negated.

We emphasize that the lattice intractability assumption underlying the constructions of [17, 19, 20] is a worst-case one, but over a restricted class of lattices. The main open question is whether lattice problems over cyclic (or, more generally, ideal) lattices are indeed hard. Very little is known about them, but state of the art lattice reduction algorithms do not seem to perform any better over cyclic lattices than arbitrary ones, supporting the conjecture that lattice problems over cyclic (or similarly restricted) lattices are as hard as the general case.

Public-Key Encryption Scheme

One-way functions and collision resistant hash functions (treated in the previous section) are useful cryptographic primitives, and can be used (at least in theory, via polynomial time but not necessarily practical constructions) to realize many other cryptographic operations, like pseudo-random generation, private key encryption, message authentication, commitments schemes, and digital signatures. Unfortunately, this is not the case for *public-key encryption*, one of the most fundamental operations of modern cryptography, defined below.

Definition 4. A public-key encryption scheme is a triple of probabilistic polynomial time algorithms (G, E, D) where

- G, the key generation algorithm, on input a security parameter n, outputs (in time polynomial in n) a pair of keys (pk, sk), called the public and secret key.
- E, the encryption algorithm, on input the public key pk and a message string m (called plaintext), outputs a string $E(pk, m)$, called ciphertext.
- D, the decryption algorithm, on input the secret key sk and a ciphertext $c = E(pk, m)$, recovers the original message $D(sk, E(pk, m)) = m$.

The typical application of public-key encryption schemes is the transmission of confidential information over a public network. Here, the intended recipient generates a pair of keys (pk, sk) using the key generation algorithm, and makes his public key pk widely available, e.g., by publishing it next to his name in a directory. Anybody wishing to send a message to this recipient can use the public key pk to encode the message m into a corresponding ciphertext $E(pk, m)$, which is transmitted over the network. The intended recipient can recover the underlying message m using the decryption algorithm and his knowledge of the secret key sk, but it is assumed that nobody else can efficiently perform the same task. This security property is formulated as follows: when the public key pk is generated at random using G, for any two messages m_0, m_1 no efficient (probabilistic polynomial time) adversary, given pk and the encryption $E(pk, m_b)$ of a randomly selected message, can guess the bit b with probability substantially better than $1/2$. This is essentially the classic notion of security introduced by Goldwasser and Micali in [38], and typically referred to as security against *chosen plaintext attack* (or, CPA-security).

No public-key encryption scheme based on an arbitrary one-way or collision resistant hash function family is known, and any such construction must necessarily be non black-box [39]. Still, public-key encryption schemes can be built from many specific (average-case) computational hardness assumptions, e.g., the hardness of factoring random numbers, or computing discrete logarithms in finite fields, etc. Can a public-key encryption scheme be constructed and proved secure based on the assumption that SVP_γ or SIVP_γ is hard to approximate in the worst-case?

Inspired by Ajtai's work on lattice-based one-way functions [13], Ajtai and Dwork [22] proposed a public-key encryption scheme (subsequently improved by Regev [23], whose proof techniques are followed in this survey) that is provably secure based on the worst-case hardness of a lattice problem, although a seemingly

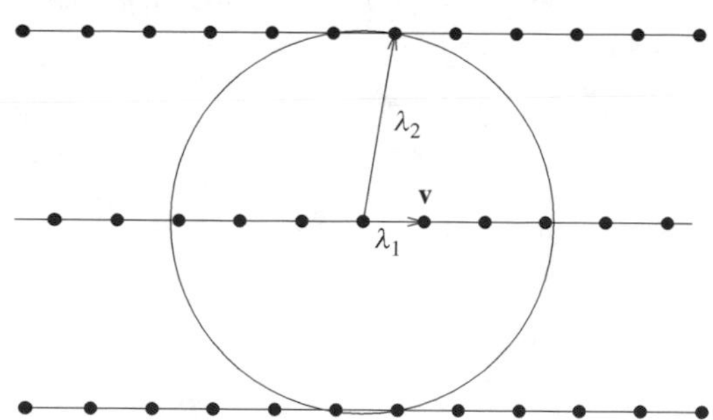

Fig. 13.2 A lattice with "unique" shortest vector. The length λ_1 of the shortest nonzero vector $\mathbf{v}$ in the lattice is much smaller than the length λ_2 of all lattice vectors that are not parallel to $\mathbf{v}$

easier one that those underlying the construction of one-way hash functions. The problem underlying the Ajtai-Dwork cryptosystem can be described as a restriction of SVP_γ to a special class of lattices, namely lattices such that $\lambda_2 > \gamma\lambda_1$ for some factor $\gamma = n^{O(1)}$ polynomial in the rank of the lattice (see fig. 13.2). The restriction of SVP_γ to such lattices is usually referred to as the "unique shortest vector problem" (uSVP_γ).

Definition 5 (Unique Shortest Vector Problem, uSVP_γ). On input a lattice basis $\mathbf{B}$ such that $\lambda_2(\mathbf{B}) > \gamma\lambda_1(\mathbf{B})$, find a nonzero lattice vector $\mathbf{v} \in \mathcal{L}(\mathbf{B})$ of length $\lambda_1(\mathbf{B})$.

The name of this problem is motivated by the fact that in such lattices the shortest nonzero vector $\mathbf{v}$ is unique, in the sense that any vector of length less than $\gamma\lambda_1$ is parallel to $\mathbf{v}$. It is also easy to see that for such lattices, finding a γ-approximate solution to SVP_γ is equivalent to finding the shortest nonzero lattice vector exactly: given a γ-approximate solution $\mathbf{v}$, the shortest vector in the lattice is necessarily of the form $\mathbf{v}/c$ for some $c \in \{1, \ldots, \gamma\}$, and can be found in polynomial time by checking all possible candidates for membership in the lattice. (Here we are using the fact that γ is polynomially bounded. Better ways to find the shortest vector exist, which work for any factor γ.) Interestingly, it can be shown [23] that solving uSVP_γ is also equivalent to the decision problem GapSVP_γ under the additional promise that the input instance $(\mathbf{B}, d)$ satisfies $\lambda_2(\mathbf{B}) > \gamma d$. We refer to this problem as $\mathrm{uGapSVP}_\gamma$.

Definition 6 (Unique Shortest Vector Decision Problem, uSVP_γ). Given a lattice basis $\mathbf{B}$ and a real d, distinguish between these two cases

- $\lambda_1(\mathbf{B}) \le d$ and $\lambda_2(\mathbf{B}) > \gamma d$,
- $\lambda_1(\mathbf{B}) > \gamma d$ (and $\lambda_2(\mathbf{B}) > \gamma d$).

If the input satisfies neither condition, any answer is acceptable.

We remark that this is different from the variant of GapSVP_γ considered in [40], and proved not to be NP-hard for factor $\gamma = n^{1/4}$ under standard complexity assumptions. The problem studied in [40] corresponds to GapSVP_γ with the stronger additional promise that $\lambda_2(\mathbf{B}) > \gamma\lambda_1(\mathbf{B})$, and is not known to be equivalent to uSVP_γ.

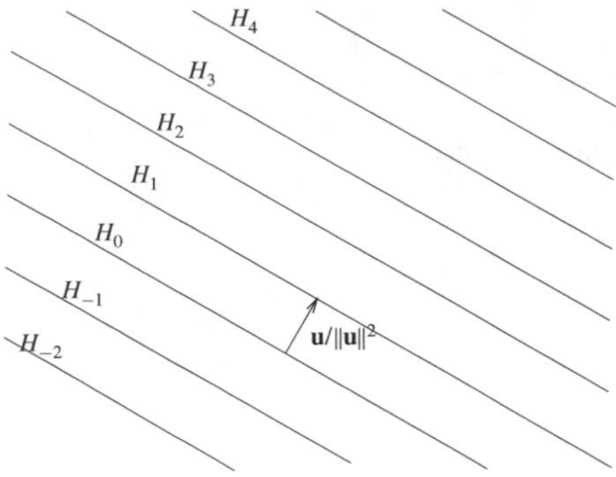

Fig. 13.3 The vector $\mathbf{u}$ defines a collection of equally spaced hyperplanes $H_i = \{\mathbf{x} : \langle \mathbf{x}, \mathbf{u} \rangle = i\}$ at distance $1/\|\mathbf{u}\|$ one from the other

Theorem 3 **([23]).** *For any polynomially bounded factor γ, uSVP$_\gamma$ and uGapSVP$_\gamma$ are equivalent under polynomial time reductions.*

Another equivalent formulation of uSVP$_\gamma$ is the "hidden hyperplane problem," described below, which is the problem directly underlying the Ajtai-Dwork cryptosystem. Informally, the "hidden hyperplane problem" is the problem of distinguishing the uniform distribution over $\mathbb{R}^n$ from a distribution concentrated nearby a set of equally spaced parallel hyperplanes $H_i = \{\mathbf{v}\colon \langle \mathbf{v}, \mathbf{u} \rangle = i\}$ (where $\mathbf{u}$ is a vector orthogonal to the hyperplanes of length inversely proportional to the distance between consecutive hyperplanes, see Fig. 13.3). The relation between the hidden hyperplane problem and uSVP$_\gamma$ is easily explained, but it requires the use of dual lattices. Recall that the dual lattice $\mathcal{L}(\mathbf{B})^*$ is the set of all vectors $\mathbf{u}$ that have integer scalar product with all lattice vectors $\mathbf{v} \in \mathcal{L}(\mathbf{B})$. So, any dual vector $\mathbf{u}$ partitions the lattice $\mathcal{L}(\mathbf{B})$ into consecutive hyperplanes

$$H_i = \{\mathbf{v}\colon \langle \mathbf{u}, \mathbf{v} \rangle = i\}$$

(for $i \in \mathbb{Z}$) at distance $1/\|\mathbf{u}\|$ from each other. If the dual lattice contains a very short vector $\mathbf{u}$, then the distance between the hyperplanes defined by $\mathbf{u}$ will be very large. Moreover, if the successive minimum $\lambda_2(\mathcal{L}(\mathbf{B})^*)$ is much larger than $\|\mathbf{u}\|$ (i.e., $\mathcal{L}(\mathbf{B})^*$ is a uSVP$_\gamma$ lattice), then the distance between the hyperplanes will be much larger than the characteristic distance between lattice points within each hyperplane. So, a uGapSVP$_\gamma$ instance $(\mathbf{B}, d)$ can be reduced to the hidden hyperplane problem as follows:

1. Pick a random lattice point in the dual lattice $\mathbf{u} \in \mathcal{L}(\mathbf{B})^*$. (Here we are being rather informal. Technical problems like the impossibility of choosing lattice points uniformly at random can be easily solved by methods similar to those described in Section "Collision Resistant Hashing".)
2. Perturb the dual lattice point by a random error vector $\mathbf{r}$.
3. Run the hidden hyperplane distinguisher on the perturbed dual lattice point $\mathbf{u}+\mathbf{r}$.

We observe that if the length of the error vector is appropriately chosen, the reduction is correct. We only describe the main idea here, and refer the reader to [22, 23] for the technical details. If $(\mathbf{B}, d)$ is a "yes" instance of uGapSVP$_\gamma$, then $\mathcal{L}(\mathbf{B})$ contains a vector $\mathbf{u}$ of length at most d, and the dual lattice $\mathcal{L}(\mathbf{B}^*)$ can be partitioned into hyperplanes at distance $1/d$ from each other. Moreover, the additional uGapSVP$_\gamma$ promise that $\lambda_2 > \gamma d$ guarantees that the distance between dual lattice points within each hyperplane is much smaller than $1/d$ (essentially proportional to $1/\gamma d$). So, by adding some noise (sufficiently bigger than $1/\gamma d$, but smaller than $1/d$) to the dual lattice points, we can erase the fine structure of the dual lattice within each hyperplane, and obtain a distribution concentrated nearby (and almost uniform over) these hyperplanes. On the other hand, if $(\mathbf{B}, d)$ is a "no" instance of uGapSVP$_\gamma$, then $\lambda_1(\mathbf{B}) > \gamma d$, and the characteristic distance between *all* dual lattice points is proportional to $1/\gamma d$. So, by adding an amount of noise much larger than $1/\gamma d$ we get a distribution that is very close to uniform over the entire space. This shows that the perturbed dual lattice point $\mathbf{u} + \mathbf{r}$ is chosen according to one of the two hidden hyperplane distributions, almost uniform either over the entire space or nearby the hidden hyperplanes. So, the reduction is correct.

The nice feature of the hidden hyperplane problem is that it is, in a sense, random self reducible: given an arbitrary instance of the problem, we can obtain a random instance by applying a random rotation and scaling. (Technically, the hidden hyperplane problem is not a computational problem in the standard sense, as its instances are not strings, but probability distributions. What we mean by reducing a given instance to a random instance of the hidden hyperplane problem is that there is an efficient oracle algorithm that, given black-box access to a hidden hyperplane distribution corresponding to a fixed vector $\mathbf{u}$, produces samples according to the distribution corresponding to a random rotation $\mathbf{Qu}$.) So, if we can solve the hidden hyperplane problem for a non-negligible fraction of the "hidden" vectors $\mathbf{u}$, then we can efficiently solve it for all vectors $\mathbf{u}$ with very high probability. Equivalently, solving the hidden hyperplane problem on the average is at least as hard as solving uSVP$_\gamma$ in the worst case. From the average-case hardness of the hidden hyperplane problem it is easy to derive a secure public-key cryptosystem for single bit messages (longer messages can be encrypted bit by bit):

- The secret key is a random vector $\mathbf{u}$.
- The public key is a set of polynomially many points $\mathbf{p}_1, \ldots, \mathbf{p}_n$ chosen at random nearby the hyperplanes defined by $\mathbf{u}$.
- The bit 0 is encrypted by adding up a small subset of the public vectors, and perturbing it by a small amount of noise. Notice that this results in a vector distributed at random nearby the hyperplanes.
- The bit 1 is encrypted by picking a random point in space, which, with high probability, will be far from the hyperplanes.

Notice that on the one hand, the encrypted bit can be efficiently recovered using the secret key $\mathbf{u}$ by computing the scalar product $\langle \mathbf{u}, \mathbf{c} \rangle$, where $\mathbf{c}$ is the received cipher-text. If $\mathbf{c}$ is the encryption of 0, then the product will be close to an integer, while if $\mathbf{c}$ is the encryption of 1, the product will be close to $\mathbb{Z} + \frac{1}{2}$ with high probability. On the other hand, distinguishing encryptions of 0 from encryptions of 1 is essentially a random instance of the hidden hyperplane problem, which is hard to solve on the average. So, the encryption scheme is secure based on the worst-case hardness of uSVP$_\gamma$. We remark that many applications require a stronger notion of security than the CPA-security considered here. Informally, CPA-security corresponds to passive adversary that can eavesdrop, but not alter, the transmitted ciphertexts. Security with respect to active adversaries that are able to alter transmitted messages, and trick the legitimate receiver into decrypting adversarially chosen ciphertexts (called security against *chosen ciphertext attacks*, or CCA-security), is often desirable.

The Ajtai-Dwork cryptosystem introduces the possibility of decryption errors: it is possible that the bit 1 is encrypted (by chance) as a vector close to the hyperplanes, which would subsequently be decrypted as 0. This problem can be easily solved (as suggested by Goldreich, Goldwasser and Halevi, [41]) by encrypting 0's and 1's as points close to alternating sequences of hyperplanes. Another problem with Ajtai-Dwork cryptosystem is that it relies on the hardness of uSVP$_\gamma$ for a fairly large value of γ. This has also been substantially improved by Regev [23] to $\gamma = n^{1.5}$ using Fourier analysis techniques similar to those described in Section "Collision Resistant Hashing". Regev [23] also shows that the hidden hyperplane problem can be reduced to a one-dimensional problem, yielding a subset-sum style cryptosystem where the public vectors $\mathbf{p}_1, \ldots, \mathbf{p}_n$ are replaced by single numbers $p_1, \ldots, p_n$. The use of numbers p_i rather than vectors $\mathbf{p}_i$ seems to be essentially a matter of style, without much of an impact on performance, because the numbers p_i require very high precision.

A very interesting result is another public-key cryptosystem of Regev [24], which can be proved secure under the assumption that no *quantum* algorithm can efficiently approximate SVP$_\gamma$ (in the worst case over arbitrary lattices) within polynomial factors. Another recent interesting result is the work of Peikert and Waters [25] who, building on the cryptosystem of [24], were able to design a lattice-based cryptosystem achieving CCA security. We remark that the cryptosystems of [24, 25] are entirely classical: encryption and decryption can be efficiently implemented on standard computers. Only the reduction from SVP$_\gamma$ to the problem of breaking the cryptosystems involves quantum computation.

Concrete Security of Lattice Based Cryptography

The importance of basing lattice cryptography on worst-case complexity assumptions cannot be overemphasized. The worst-case approximation factor achieved by lattice reduction algorithms [1,6,12] is known to be exponential in the dimension of the lattice [1,36]. However, lattice reduction algorithms have often been reported to perform much better in practice than the worst-case theoretical upper bound, when run on random problem instances like those arising in cryptanalysis applications. (See also [42,43] for a recent empirical study showing that the approximation factor achieved by the LLL algorithm (and improvements) on the average, when the input lattice is random, is still exponential in the rank, but with a much smaller constant in the exponent than the worst-case factor.) So, while it seems reasonable to conjecture, based on our current knowledge, that many computational problems on lattices may be hard to approximate in the worst-case even within moderately large factors, extreme caution should be exercised when making conjectures on the average-case complexity of the same problems. In summary, if worst-case intractability assumptions are preferable to average-case ones in general, this is especially true when dealing with point lattices.

The use of worst-case intractability assumptions also frees us from a major burden associated with making average-case complexity conjectures: finding appropriate distributions on problem instances for which no heuristics perform satisfactorily. In the case of lattice problems, heuristics (e.g., [5, 44]) seem to perform reasonably well in practice, and the lattices demonstrating the worst-case performance of known algorithms [1, 36] seem to be more of a pathological example than a source of real practical concern. Worst-case intractability assumptions do not require the selection of an appropriate distribution over input lattices, and the performance of heuristic approaches does not say much about the validity of the assumption. Since the conjecture is a worst-case intractability one, if the algorithm fails to achieve good approximation guarantees even on just a few examples, the conjecture still stands.

All this seems good news for the cryptographic designer, but it also raises new issues when it comes to instantiating the function with specific values of the security parameter. Traditionally, the concrete hardness of average-case problems has been evaluated through a challenge/cryptanalysis process: the cryptographic designer proposing a new intractability assumption produces a list of random problem instances (for increasing values of the security parameter) as a challenge to the rest of the cryptography community. If cryptanalysts are successful in breaking a challenge, the corresponding value of the security parameter is considered too small to achieve security in practice. Typically, there is a cost associated to breaking challenges in varying dimension, which can be experimentally evaluated for moderately small values of the security parameter, and extrapolated to larger values as an estimate of the time required to break bigger instances (see for example [45]).

Now consider worst-case intractability assumptions. If heuristic approaches to lattice cryptanalysis cannot be used to disprove worst-case intractability assumptions, how can we possibly gain confidence in the validity of the assumption? And how can we select appropriate values of the security parameter to be used in practice? Several approaches come to mind:

1. *The worst-case challenge approach*: We ask the cryptographer to come up with a list of (not necessarily random) challenge problems. Since the cryptographer has complete freedom in the choice of the challenges, he can choose the *worst* problem instances that are hardest to solve by any algorithm. The problem with this approach is that the cryptographer may not know how to select hard instances. In fact, worst-case intractability assumptions do not even require

such hard instances to be easy to find. One of the nicest features of worst-case intractability assumptions is that they do not require to find such hard instances. By asking the cryptographer to come up with hard challenges, the advantage of basing cryptography on worst-case problems would be largely lost.

2. *The reverse challenge approach*: After the cryptographer formulates a worst-case intractability conjecture, the cryptanalyst produces a challenge to the conjecture in the form of a heuristic algorithm and a claim on its worst-case performance. The challenge is for the cryptographer to come up with an input instance for which the heuristic algorithm does not meet the promised performance bound.
3. *Direct cryptanalysis of the cryptographic function*: Instead of evaluating the strength of the underlying worst-case complexity assumption, directly attack the cryptographic application.

We remark that the second approach, although unconventional, seems at least more appropriate than the first one. Since the intractability assumption is that no efficient algorithm solves every single problem instance, in an experimental setting it seems more appropriate to first fix the algorithm and then search for the input that triggers the worst-case performance, rather then the other way around. Still, producing a counterexample each time somebody comes up with a new heuristic algorithm with worst-case performance claims may seem too much of a burden for the cryptographers.

A disadvantage of both the first and second approach is that it does not necessarily give an indication of the true security of the cryptographic scheme. It may well be the case that breaking the cryptographic function is even harder than solving the worst case problem underlying the security proof. This issue is addressed by the last approach, which focuses directly on the cryptographic function rather than the underlying complexity assumption. This last approach has also the advantage that the proof of security already gives a precise probability distribution according to which challenges should be chosen. For example, in the case of lattice based hash functions, for any set of parameters n, m, q, the challenge should be a matrix $\mathbf{M} \in \mathbb{Z}_q^{n \times m}$ chosen uniformly at random. There is no need to have the cryptographer to come up with a list of challenges. The cryptanalyst can select the challenges on her own, as long as the challenge is selected at random according to the prescribed probability distribution.

A possible objection to the last approach is that it bypasses the proof of security. If the concrete security of the cryptographic function is evaluated directly by cryptanalysis, what is the value of providing a security proof? We believe that even in this scenario, security proofs are very valuable. They may not help in assessing the concrete security of the function for specific values of the security parameter, but they ensure that the construction has no structural flaw, and (in the case of worst-case to average-case reductions) they provide invaluable guidance when selecting the appropriate distribution over random keys.

A more serious disadvantage of the last approach is that any new cryptographic primitive requires a separate cryptanalytic effort to assess its concrete security. This way, one of the main advantages of the provable security approach is lost: the ability to concentrate cryptanalytic efforts on a small number of selected computational problems, and derive the (concrete) security of old and new cryptographic constructions by means of mathematical proofs.

One possible way to address some of the issues brought up in the previous discussion is to study the tightness of worst-case to average-case security proofs. For example, we know that breaking the cryptographic hash functions described in Section "Collision Resistant Hashing" with non-negligible probability is at least as hard as approximating SIVP_γ within a factor $\tilde{O}(n)$. Is it possible to prove also that given an oracle that solves SIVP_γ within a factor $\tilde{O}(n)$ (in the worst-case) one can find collisions to the hash function for randomly chosen keys? Some results of this kind are proved by Nguyen and Stern [46], who show that solving SVP_γ within a factor $n^{0.5-\varepsilon}$ allows to break the Ajtai-Dwork cryptosystem. However, there are several gaps between the two reductions: in [46] the worst-case lattice problems are different (SVP_γ or CVP_γ, rather than SIVP_γ or GapSVP_γ), the required polynomial approximation factors are smaller, and the lattice dimension is larger than in the proof of security. Giving a tight worst-case to average-case reduction between lattice problems is an important open problem as it would allow to evaluate the concrete worst-case complexity of lattice problems by cryptanalyzing random challenges of the average-case problem. Well understood worst-case assumptions could then be used to prove the security of other average-case problems, without the need of new challenges and cryptanalysis.

We conclude the paper with a discussion of two other issues that have received little attention so far, but are certainly very important in the context of cryptanalysis of lattice based functions. The first issue is that most lattice cryptanalysis problems are more naturally formulated as lattice approximation problems in the ℓ_∞ norm. For example, we observe that finding collisions to the hash function described in Section "Collision Resistant Hashing" is exactly the same as solving SVP_γ in the ℓ_∞ norm in the lattice naturally associated to the function. Still, most algorithmic work has focused on the Euclidean norm ℓ_2. While the ℓ_2 norm may be more convenient when designing algorithms, there are several reasons to prefer the ℓ_∞ norm when working with applications of lattices: the ℓ_∞ norm often leads to faster and easier to implement functions, and there is theoretical evidence [47, 48] that lattice problems in the ℓ_∞ norm are harder (or at least not easier) than the same problems in the Euclidean norm. For example, Regev and Rosen [48] gives reductions from many lattice problems in the ℓ_2 norm to the same problems in the ℓ_∞ norm. Moreover, hardness results for SVP_γ in the ℓ_2 norm lag behind similar results in the ℓ_∞ norm. For example, SVP_γ in the ℓ_∞ norm can be shown to be hard to approximate within almost polynomial factors $n^{1/O(\log\log n)}$ under the assumption that NP $\neq$ P [47]. A similar result can be proved for the ℓ_2 norm [49], but only under the much stronger assumption that NP does not have randomized sub-exponential time algorithms.

Despite the importance of the ℓ_∞ norm in cryptanalysis, the study of lattice reduction algorithms in norms different from ℓ_2 has seen little progress so far.

For example, the lattice reduction algorithm of [50] for general norms is not even known to terminate in polynomial time for arbitrary dimension.

A second very interesting issue that requires further investigation is the complexity of lattice problems when restricted to special classes of lattices, as those arising in the design of lattice based public-key cryptosystems, and efficient hash functions. Most work on lattice algorithms has been focused on the solution of SVP_γ or other lattice problems in the ℓ_2 norm on arbitrary lattices. (See [51] for an interesting exception.) We remark that the classes of lattices underlying the construction of efficient one-way hash functions and public-key encryption schemes are restricted in very different ways:

- In the case of one-way functions, the restriction is, in a sense, *algebraic*, as clearly illustrated by the most recent work on collision resistance and ideal lattices [19, 20, 52].
- In the case of public-key encryption, the restriction is more *geometric* (e.g., there is a gap between λ_1 and λ_2).

Here we may call *geometric* those properties that are (approximately) preserved under small perturbations of the lattice. It is easy to see that even a small perturbation to a cyclic or ideal lattice would immediately destroy the algebraic structure. Is there any reason to prefer one kind of restriction over the other one? Ajtai's work [53] (which essentially conjectures that any non-trivial geometric property of a lattice is hard to decide) seems to express more confidence in the hardness of lattice problems under geometric restrictions. On the other hand, cryptanalysis experiments [4] suggest that lattice reduction algorithms perform better on the average when there is a big gap between λ_1 and λ_2. Improving our understanding of the computational complexity of lattice approximation problems when restricted to special classes of lattices, as well as classifying the restrictions into meaningful categories (like, algebraic or geometric restrictions) is a very interesting open problem, both from a structural point of view (identifying properties that make computational problems on lattices easier or harder to solve) and for the cryptographic applications of lattices.

Acknowledgements This work was supported in part by NSF grant CCF-0634909. Any opinions, findings, and conclusions or recommendations expressed in this material are those of the author(s) and do not necessarily reflect the views of the National Science Foundation.

References

1. Lenstra, A.K., Lenstra, Jr., H.W., Lovász, L.: Factoring polynomials with rational coefficients. Mathematische Annalen **261**, 513–534 (1982)
2. Brickell, E.F., Odlyzko, A.M.: Cryptanalysis: A survey of recent results. In: G.J. Simmons (ed.) Contemporary Cryptology, chap. 10, pp. 501–540. IEEE Press (1991)
3. Joux, A., Stern, J.: Lattice reduction: A toolbox for the cryptanalyst. Journal of Cryptology **11**(3), 161–185 (1998)

4. Nguyen, P., Stern, J.: The two faces of lattices in cryptology. In: Proceedings of CaLC '01, *LNCS*, vol. 2146, pp. 146–180. Springer (2001)
5. Schnorr, C.P., Euchner, M.: Lattice basis reduction: Improved practical algorithms and solving subset sum problems. Mathematical programming **66**(1–3), 181–199 (1994). Preliminary version in FCT 1991
6. Schnorr, C.P.: A hierarchy of polynomial time lattice basis reduction algorithms. Theoretical Computer Science **53**(2–3), 201–224 (1987)
7. Schnorr, C.P.: A more efficient algorithm for lattice basis reduction. Journal of Algorithms **9**(1), 47–62 (1988)
8. Schnorr, C.P.: Fast LLL-type lattice reduction. Information and Computation **204**(1), 1–25 (2006)
9. Ajtai, M., Kumar, R., Sivakumar, D.: A sieve algorithm for the shortest lattice vector problem. In: Proceedings of STOC '01, pp. 266–275. ACM (2001)
10. Kumar, R., Sivakumar, D.: On polynomial-factor approximations to the shortest lattice vector length. SIAM Journal on Discrete Mathematics **16**(3), 422–425 (2003)
11. Nguyen, P., Stehlé, D.: Floating-point LLL revisited. In: Proceedings of EUROCRYPT '05, *LNCS*, vol. 3494, pp. 215–233. Springer (2005)
12. Gama, N., Nguyen, P.Q.: Finding short lattice vectors within mordell's inequality. In: Proceedings of STOC '08, pp. 207–216. ACM (2008)
13. Ajtai, M.: Generating hard instances of lattice problems. Complexity of Computations and Proofs, Quaderni di Matematica **13**, 1–32 (2004). Preliminary version in STOC 1996
14. Cai, J.Y., Nerurkar, A.P.: An improved worst-case to average-case connection for lattice problems (extended abstract). In: Proceedings of FOCS '97, pp. 468–477. IEEE (1997)
15. Micciancio, D.: Almost perfect lattices, the covering radius problem, and applications to Ajtai's connection factor. SIAM Journal on Computing **34**(1), 118–169 (2004). Preliminary version in STOC 2002.
16. Micciancio, D., Regev, O.: Worst-case to average-case reductions based on Gaussian measure. SIAM Journal on Computing **37**(1), 267–302 (2007). Preliminary version in FOCS 2004
17. Micciancio, D.: Generalized compact knapsacks, cyclic lattices, and efficient one-way functions. Computational Complexity **16**(4), 365–411 (2007). Preliminary version in FOCS 2002
18. Ajtai, M.: Representing hard lattices with O(n log n) bits. In: Proceedings of STOC '05, pp. 94–103. ACM (2005)
19. Lyubashevsky, V., Micciancio, D.: Generalized compact knapsacks are collision resistant. In: Proceedings of ICALP '06, *LNCS*, vol. 4052, pp. 144–155. Springer (2006)
20. Peikert, C., Rosen, A.: Efficient collision-resistant hashing from worst-case assumptions on cyclic lattices. In: Proceedings of TCC '06, *LNCS*, vol. 3876, pp. 145–166. Springer (2006)
21. Lyubashevsky, V., Micciancio, D., Peikert, C., Rosen, A.: Swifft: a modest proposal for fft hashing. In: Proceedings of FSE '08, *LNCS*, vol. 5086, pp. 54–72. Springer (2008)
22. Ajtai, M., Dwork, C.: A public-key cryptosystem with worst-case/average-case equivalence. In: Proceedings of STOC '97, pp. 284–293. ACM (1997)
23. Regev, O.: New lattice based cryptographic constructions. Journal of the ACM **51**(6), 899–942 (2004). Preliminary version in STOC 2003
24. Regev, O.: On lattices, learning with errors, random linear codes, and cryptography. In: Proceedings of STOC '05, pp. 84–93. ACM (2005)
25. Peikert, C., Waters, B.: Lossy trapdoor functions and their applications. In: Proceedings of STOC '08, pp. 187–196. ACM (2008)
26. Micciancio, D., Vadhan, S.: Statistical zero-knowledge proofs with efficient provers: lattice problems and more. In: Proceedings of CRYPTO '03, *LNCS*, vol. 2729, pp. 282–298. Springer (2003)
27. Lyubashevsky, V.: Lattice-based identification schemes secure under active attacks. In: Proceedings of PKC '08, no. 4939 in LNCS, pp. 162–179. Springer (2008)
28. Lyubashevsky, V., Micciancio, D.: Asymptotically efficient lattice-based digital signatures. In: Proceedings of TCC '08, *LNCS*, vol. 4948, pp. 37–54. Springer (2008)

29. Gentry, C., Peikert, C., Vaikuntanathan, V.: Trapdoors for hard lattices and new cryptographic constructions. In: Proceedings of STOC '08, pp. 197–206. ACM (2008)
30. Peikert, C., Vaikuntanathan, V.: Noninteractive statistical zero-knowledge proofs for lattice problems. In: Proceedings of CRYPTO '08, *LNCS*, vol. 5157, pp. 536–553. Springer (2008)
31. Peikert, C., Vaikuntanathan, V., Waters, B.: A framework for efficient and composable oblivious transfer. In: Proceedings of CRYPTO '08, *LNCS*, vol. 5157, pp. 554–571. Springer (2008)
32. Regev, O.: On the complexity of lattice problems with polynomial approximation factors. In: this volume (2008)
33. Micciancio, D., Goldwasser, S.: Complexity of Lattice Problems: a cryptographic perspective, The Kluwer International Series in Engineering and Computer Science, vol. 671. Kluwer Academic Publishers (2002)
34. Goldreich, O., Micciancio, D., Safra, S., Seifert, J.P.: Approximating shortest lattice vectors is not harder than approximating closest lattice vectors. Information Processing Letters **71**(2), 55–61 (1999)
35. Babai, L.: On Lovasz' lattice reduction and the nearest lattice point problem. Combinatorica **6**(1), 1–13 (1986)
36. Ajtai, M.: The worst-case behavior of Schnorr's algorithm approximating the shortest nonzero vector in a lattice. In: Proceedings of STOC '03, pp. 396–406. ACM (2003)
37. Hoffstein, J., Pipher, J., Silverman, J.H.: NTRU: a ring based public key cryptosystem. In: Proceedings of ANTS-III, *LNCS*, vol. 1423, pp. 267–288. Springer (1998)
38. Goldwasser, S., Micali, S.: Probabilistic encryption. Journal of Computer and System Sciences **28**(2), 270–299 (1984). Preliminary version in Proc. of STOC 1982
39. Impagliazzo, R., Rudich, S.: Limits on the provable consequences of one-way permutations. In: Proceedings of STOC '89, pp. 44–61. ACM (1989)
40. Cai, J.Y.: A relation of primal-dual lattices and the complexity of the shortest lattice vector problem. Theoretical Computer Science **207**(1), 105–116 (1998)
41. Goldreich, O., Goldwasser, S., Halevi, S.: Eliminating decryption errors in the Ajtai-Dwork cryptosystem. In: Proceedings of CRYPTO '97, *LNCS*, vol. 1294, pp. 105–111. Springer (1997)
42. Nguyen, P., Stehlé, D.: LLL on the average. In: Proceedings of ANTS-VII, *LNCS*, vol. 4076, pp. 238–256. Springer (2006)
43. Gama, N., Nguyen, P.Q.: Predicting lattice reduction. In: Proceedings of EUROCRYPT '08, *LNCS*, vol. 4965, pp. 31–51. Springer (2008)
44. Schnorr, C.P., Hörner, H.H.: Attacking the Chor-Rivest cryptosystem by improved lattice reduction. In: Proceedings of EUROCRYPT '95, *LNCS*, vol. 921, pp. 1–12. Springer (1995)
45. Lenstra, A.K., Verheul, E.R.: Selecting cryptographic key sizes. Journal of Cryptology **14**(4), 255–293 (2001)
46. Nguyen, P., Stern, J.: Cryptanalysis of the Ajtai-Dwork cryptosystem. In: Proceedings of CRYPTO '98, *LNCS*, vol. 1462, pp. 223–242. Springer (1998)
47. Dinur, I.: Approximating SVP_∞ to within almost-polynomial factors is NP-hard. Theoretical Computer Science **285**(1), 55–71 (2002)
48. Regev, O., Rosen, R.: Lattice problems and norm embeddings. In: Proceedings of STOC '06, pp. 447–456. ACM (2006)
49. Haviv, I., Regev, O.: Tensor-based hardness of the shortest vector problem to within almost polynomial factors. In: Proceedings of STOC '07, pp. 469–477. ACM (2007)
50. Lovász, L., Scarf, H.: The generalized basis reduction algorithm. Mathematics of Operations Research **17**(3), 754–764 (1992)
51. Gama, N., Howgrave-Graham, N., Nguyen, P.: Symplectic lattice reduction and NTRU. In: Proceedings of EUROCRYPT '06, *LNCS*, vol. 4004, pp. 233–253. Springer (2006)
52. Lyubashevsy, V., Micciancio, D., Peikert, C., Rosen, A.: Provably secure FFT hashing. In: 2nd NIST cryptographic hash workshop. Santa Barbara, CA, USA (2006)
53. Ajtai, M.: Random lattices and a conjectured 0-1 law about their polynomial time computable properties. In: Proceedings of FOCS '02, pp. 733–742. IEEE (2002)

Chapter 14
Inapproximability Results for Computational Problems on Lattices

Subhash Khot

Abstract In this article, we present a survey of known inapproximability results for computational problems on lattices, viz. the Shortest Vector Problem (SVP), the Closest Vector Problem (CVP), the Closest Vector Problem with Preprocessing (CVPP), the Covering Radius Problem (CRP), the Shortest Independent Vectors Problem (SIVP), and the Shortest Basis Problem (SBP).

Introduction

An n-dimensional lattice $\mathcal{L}$ is a set of vectors $\{\sum_{i=1}^{n} x_i \mathbf{b}_i \mid x_i \in \mathbb{Z}\}$ where $\mathbf{b}_1, \mathbf{b}_2, \ldots, \mathbf{b}_n \in \mathbb{R}^m$ is a set of linearly independent vectors called the basis for the lattice (the same lattice could have many bases). In this article, we survey known results regarding the complexity of several computational problems on lattices. Most of these problems turn out to be intractable, and even computing approximate solutions remains intractable. Excellent references on the subject include Micciancio and Goldwasser's book [1], an expository article by Kumar and Sivakumar [2], and a survey of Regev [3] in the current proceedings.

The Shortest Vector Problem (SVP)

The most studied computational problem on lattices is the Shortest Vector Problem (SVP),[1] where given a basis for an n-dimensional lattice, we seek the shortest non-zero vector in the lattice.[2]

S. Khot
New York University, NY 10012, USA,
e-mail: khot@cs.nyu.edu.

[1] Formal definitions for all problems appear in Section "Notation and Problem Definitions" where we also clarify the issue of how the input is represented.
[2] In this article, we use ℓ_2 norm unless stated otherwise.

P.Q. Nguyen and B. Vallée (eds.), *The LLL Algorithm*, Information Security and Cryptography, DOI 10.1007/978-3-642-02295-1_14,

The problem has been studied since the time of Gauss ([4], 1801) who gave an algorithm that works for 2-dimensional lattices. The general problem for arbitrary dimensions was formulated by Dirichlet in 1842. A well-known theorem of Minkowski [5] deals with the existence of short non-zero vectors in lattices. In a celebrated result, Lenstra, Lenstra, and Lovász [6] gave a polynomial time algorithm for approximating SVP within factor $2^{n/2}$. This algorithm has numerous applications, e.g., factoring rational polynomials [6], breaking knapsack-based codes [7], checking solvability by radicals [8] and integer programming in a fixed number of variables [6, 9, 10]. Schnorr [11] improved the approximation factor to $2^{O(n(\log\log n)^2/\log n)}$. It is a major open problem whether SVP has an efficient polynomial factor approximation. Exact computation of SVP in exponential time is also investigated, see for instance Kannan [12] and Ajtai, Kumar, and Sivakumar [13]. The latter paper also gave a polynomial time $2^{O(n\log\log n/\log n)}$ factor approximation, an improvement over Schnorr's algorithm.

In 1981, van Emde Boas [14] proved that SVP in ℓ_∞ norm is NP-hard and conjectured that the same is true in any ℓ_p norm. However, proving NP-hardness in ℓ_2 norm (or in any finite ℓ_p norm for that matter) was an open problem for a long time. A breakthrough result by Ajtai [15] in 1998 finally showed that SVP is NP-hard under randomized reductions. Cai and Nerurkar [16] improved Ajtai's result to a hardness of approximation result showing a hardness factor of $\left(1 + \frac{1}{n^\varepsilon}\right)$. Micciancio [17] showed that SVP is NP-hard to approximate within some constant factor, specifically any factor less than $\sqrt{2}$. Recently, Khot [18] proved that SVP is NP-hard to approximate within any constant factor and hard to approximate within factor $2^{(\log n)^{1/2-\varepsilon}}$ for any $\varepsilon > 0$, unless NP has randomized quasipolynomial time algorithms[1]. This hardness result was further improved to an almost polynomial factor, i.e., $2^{(\log n)^{1-\varepsilon}}$, by Haviv and Regev [19].

Showing hardness of approximation results for SVP was greatly motivated by Ajtai's discovery [20] of worst-case to average-case reduction for SVP and subsequent construction of a lattice-based public key cryptosystem by Ajtai and Dwork [21]. Ajtai showed that if there is a randomized polynomial time algorithm for solving (exact) SVP on a non-negligible fraction of lattices from a certain natural class of lattices, then there is a randomized polynomial time algorithm for approximating SVP on *every* instance within some polynomial factor n^c (he also presented a candidate one-way function). In other words, if approximating SVP within factor n^c is hard in the worst case, then solving SVP exactly is hard on average. Based on this reduction, Ajtai and Dwork [21] constructed a public-key cryptosystem whose security depends on (conjectured) worst-case hardness of approximating SVP (cryptography in general relies on average-case hardness of problems, but for SVP, it is same as worst-case hardness via Ajtai's reduction).

Cai and Nerurkar [22] and Cai [23] brought down the constant c to $9 + \varepsilon$ and $4 + \varepsilon$ respectively.

[1] Quasipolynmial (randomized) Time is the class $\cup_{C>0}\mathrm{BPTIME}(2^{(\log n)^C})$.

Recently, Regev [24] gave an alternate construction of a public key cryptosystem based on $n^{1.5}$-hardness of SVP.[2] Thus, in principle, one could show that approximating SVP within factor $n^{1.5}$ is NP-hard, and it would imply cryptographic primitives whose security relies on the widely believed conjecture that $P \neq NP$, attaining the holy grail of cryptography! Unfortunately, there are barriers to showing such strong hardness results. We summarize the so-called *limits to inapproximability* results in Section "Limits to Inapproximability" and refer to Regev's article [3] in the current proceedings for a more detailed exposition.

The Closest Vector Problem (CVP)

Given a lattice and a point $\mathbf{z}$, the Closest Vector Problem (CVP) is to find the lattice point that is closest to $\mathbf{z}$. Goldreich, Micciancio, Safra, and Seifert [25] gave a Turing reduction from SVP to CVP, showing that any hardness for SVP implies the same hardness for CVP (but not vice versa). CVP was shown to be NP-hard by van Emde Boas [14]. Arora, Babai, Sweedyk, and Stern [26] used the PCP machinery to show that approximating CVP within factor $2^{\log^{1-\varepsilon} n}$ is hard unless NP has quasipolynomial time algorithms. This was improved to a NP-hardness result by Dinur, Kindler, and Safra [27]; their result gives even a subconstant value of ε, i.e., $\varepsilon = (\log \log n)^{-t}$ for any $t < \frac{1}{2}$.

The Closest Vector Problem with Preprocessing (CVPP)

The Closest Vector Problem with Preprocessing (CVPP) is the following variant of CVP: Given a lattice, one is allowed to do arbitrary preprocessing on it and store polynomial amount of information. The computational problem is to compute the closest lattice point to a given point $\mathbf{z}$. The motivation for studying this problem comes from cryptoraphic applications. In a common scenario, the encryption key is a lattice, the received message is viewed as a point $\mathbf{z}$ and decryption consists of computing the closest lattice point to $\mathbf{z}$. Thus, the lattice is fixed and only the received message changes as an input. A natural question to ask is whether the hardness of CVP arises because one needs to solve the problem on *every* lattice, or whether the problem remains hard even for some fixed lattice when arbitrary preprocessing is allowed.

CVPP was shown to be NP-hard by Micciancio [28] and NP-hard to approximate within any factor less than $\sqrt{5/3}$ by Feige and Micciancio [29]. This was improved to any factor less than $\sqrt{3}$ by Regev [30]. Alekhnovich, Khot, Kindler,

[2] Actually all these results assume hardness of a variant called unique-SVP, see [24] for its definition.

and Vishnoi [31] showed that for every $\varepsilon > 0$, CVPP cannot be approximated in polynomial time within factor $(\log n)^{1/2-\varepsilon}$ unless NP has quasipolynomial time algorithms.[3] Their reduction is from the problem of finding vertex cover on k-uniform hypergraphs. On the other hand, Aharonov and Regev [32] gave a polynomial time $\sqrt{n/\log n}$-approximation.

The Covering Radius Problem (CRP)

The Covering Radius Problem (CRP) asks for a minimum radius r such that balls of radius r around all lattice points cover the whole space. CRP is (clearly) in Π_2, but not even known to be NP-hard. Recently, Haviv and Regev [33] showed that for every large enough p, there is a constant $c_p > 1$ such that CRP under ℓ_p norm is Π_2-hard to approximate within factor c_p. For $p = \infty$, they achieve inapproximability factor of $c_\infty = 1.5$. Their reduction is from a Π_2-hard problem called GroupColoing.

The Shortest Independent Vectors Problem (SIVP) *and the Shortest Basis Problem* (SBP)

The Shortest Independent Vectors Problem (SIVP) asks for the minimum length r such that the given n-dimensional lattice has n linearly independent vectors each of length at most r. The Shortest Basis Problem (SBP) asks for the minimum length r such that the given lattice has a basis with each vector of length at most r. Blömer and Seifert [34] showed that both SIVP and SBP are NP-hard and inapproximable within almost polynomial factor unless NP has quasipolynomial time algorithms. Their reduction is from CVP, and they use specific properties of hard CVP instances produced by Arora et al. [26] reduction.

Results in ℓ_p Norms

Regev and Rosen [35] showed a reduction from lattice problems in ℓ_2 norm to corresponding problems in ℓ_p norm for any $1 \le p \le \infty$. The reduction preserves the inapproximability gap upto $1 + \varepsilon$ for any $\varepsilon > 0$. Thus, all hardness results for CVP, SVP, CVPP, SIVP, SBP mentioned above apply to the respective problems in ℓ_p norm for every $1 \le p \le \infty$. The idea behind Regev and Rosen's reduction

[3] Because of the peculiar definition of CVPP, the hardness results actually rely on the assumption that NP does not have (quasi)polynomial size circuits.

is the well-known fact that ℓ_2^n *embeds* into $\ell_p^{\text{poly}(n)}$ with distortion $1+\varepsilon$ for every $1 \leq p < \infty$, and moreover the embedding is linear. Thus, a lattice in ℓ_2^n space can be mapped to a lattice in $\ell_p^{\text{poly}(n)}$ space, essentially preserving all distances.

In ℓ_∞ norm, stronger inapproximability results are known for SVP and CVP; both are NP-hard to approximate within factor $n^{c/\log\log n}$ for some constant $c > 0$, as proved by Dinur [36].

Limits to Inapproximability

For all the lattice problems, there is a limit to how strong an inapproximately result can be proved. For example, Banaszczyk [37] showed that GapSVP_n is in coNP.[4] Thus, if GapSVP_n is NP-hard then NP = coNP. We state the best known results along this line (see Aharonov and Regev [32], Goldreich and Goldwasser [38], Guruswami, Micciancio, and Regev [39]). We note that AM is the class of languages that have a constant round interactive proof system. A well-known complexity theoretic result is that if NP $\subseteq$ coAM, then polynomial hierarchy collapses.

- $\mathsf{GapCVP}_{\sqrt{n}} \in \mathsf{coNP}$ [32], $\mathsf{GapCVP}_{\sqrt{n/\log n}} \in \mathsf{coAM}$ [38].
- $\mathsf{GapSVP}_{\sqrt{n}} \in \mathsf{coNP}$ [32], $\mathsf{GapSVP}_{\sqrt{n/\log n}} \in \mathsf{coAM}$ [38].
- $\mathsf{GapCVPP}_{\sqrt{n/\log n}} \in \mathsf{P}$ [32].
- $\mathsf{GapCRP}_2 \in \mathsf{AM}$, $\mathsf{GapCRP}_{\sqrt{n/\log n}} \in \mathsf{coAM}$, $\mathsf{GapCRP}_{\sqrt{n}} \in \mathsf{NP} \cap \mathsf{coNP}$ [39].
- $\mathsf{GapSIVP}_{\sqrt{n/\log n}} \in \mathsf{coAM}$, $\mathsf{GapSIVP}_{\sqrt{n}} \in \mathsf{coNP}$ [39].

In short, CVP, SVP, CRP, SIVP cannot be NP-hard to approximate within $\sqrt{n/\log n}$ unless NP $\subseteq$ coAM (and polynomial hierarchy collapses). CRP cannot be Π_2-hard to approximate within factor 2 unless $\Pi_2 = \mathsf{AM}$. CVPP has a polynomial time $\sqrt{n/\log n}$-approximation.

Overview of the Article

After introducing the necessary notation and definitions, in the rest of the article, we present inapproximability results for CVP and SVP. For CVP, we include essentially complete proofs and for SVP, only a sketch of the proofs. We refrain from presenting inapproximability results for the remaining problems. A more comprehensive treatment of the subject is beyond the scope of this article.

In Section "Inapproximability of CVP", we present inapproximability results for CVP. We present two results: one gives an arbitrarily large constant factor hardness

[4] See Section "Notation and Problem Definitions" for the definitions of gap-versions of problems.

via a polynomial time reduction from Set Cover and and the other gives almost polynomial factor hardness (i.e., $2^{(\log n)^{1-\varepsilon}}$ for every $\varepsilon > 0$) via a quasipolynomial time reduction from the Label Cover Problem. Both results are due to Arora, Babai, Stern, and Sweedyk [26], though our presentation is somewhat different.

In Section "Inapproximability of SVP", we sketch inapproximability results for SVP. We note that computing SVP exactly was proved NP-hard only in 1998, a breakthrough result of Ajtai [15]. We skip Ajtai's proof from this article (see [2] for a nice sketch) and jump directly to inapproximability results. First we present a reduction of Micciancio [17] showing that GapSVP_γ is NP-hard for any constant $1 < \gamma < \sqrt{2}$.

Next, we present a result of Khot [18] and Haviv and Regev [19] showing that $\mathsf{GapSVP}_{2^{(\log n)^{1-\varepsilon}}}$ is hard via a quasipolynomial time reduction.

Notation and Problem Definitions

In this section, we formally define all the lattice problems considered in this article. We also define their gap-versions which are useful towards proving inapproximability results.

All vectors are column vectors and denoted by bold face letters. A lattice $\mathcal{L}$ generated by a basis $\mathbf{B}$ is denoted as $\mathcal{L}(\mathbf{B})$. $\mathbf{B}$ is a $m \times n$ real matrix whose columns are the basis vectors. The columns are linearly independent (and hence $m \geq n$). The n-dimensional lattice $\mathcal{L}$ in $\mathbb{R}^m$ is given by

$$\mathcal{L} = \mathcal{L}(\mathbf{B}) := \{\mathbf{Bx} \mid \mathbf{x} \in \mathbb{Z}^n\}.$$

We call $\mathbf{x}$ as the coefficient vector (with respect to the specific basis) and any $\mathbf{z} = \mathbf{Bx}$ as the lattice vector. The norm $\|\mathbf{z}\|$ denotes ℓ_2 norm. We restrict to the ℓ_2-norm for much of the article, but Section "Results in ℓ_p Norms" does mention known results for other ℓ_p norms.

Let $\lambda_1(\mathcal{L})$ denote the length of the shortest vector in a lattice, i.e.,

$$\lambda_1(\mathcal{L}(\mathbf{B})) := \min_{\mathbf{x} \in \mathbb{Z}^n,\ \mathbf{x} \neq \mathbf{0}} \|\mathbf{Bx}\|.$$

Definition 1. The Shortest Vector Problem (SVP) asks for the value of $\lambda_1(\mathcal{L}(\mathbf{B}))$ when a lattice basis $\mathbf{B}$ is given as input.

Remark 1. In this article, the dimension m of the ambient space will always be polynomial in the dimension n of the lattice. All real numbers involved are either integers with poly(n) bits or represented by an approximation with poly(n) bits, but we hide this issue for the ease of presentation. Thus, the input size for all the problems is parameterized by the dimension n of the lattice.

Let $\mathrm{dist}(\mathbf{z}, \mathcal{L}(\mathbf{B}))$ denote the minimum distance between a vector $\mathbf{z} \in \mathbb{R}^m$ and any vector in lattice $\mathcal{L}(\mathbf{B})$, i.e.,

$$\mathrm{dist}(\mathbf{z}, \mathcal{L}(\mathbf{B})) := \min_{\mathbf{x} \in \mathbb{Z}^n} \|\mathbf{z} - \mathbf{B}\mathbf{x}\|.$$

Definition 2. The Closest Vector Problem (CVP) asks for the value of $\mathrm{dist}(\mathbf{z}, \mathcal{L}(\mathbf{B}))$ when a lattice basis $\mathbf{B}$, and a vector $\mathbf{z}$ are given.

Definition 3. The Closest Vector Problem with Preprocessing (CVPP) is the following variant: Given a lattice $\mathcal{L}(\mathbf{B})$, one is allowed to do arbitrary preprocessing on it and store polynomial (in the dimension of the lattice) amount of information. The computational problem is to compute $\mathrm{dist}(\mathbf{z}, \mathcal{L}(\mathbf{B}))$ for a given point $\mathbf{z} \in \mathbb{R}^m$.

Let $\mathrm{span}(\mathbf{B})$ denote the linear span of the columns of $\mathbf{B}$. This is a n-dimensional linear subspace of $\mathbb{R}^m$. Let $\rho(\mathcal{L}(\mathbf{B}))$ denote the covering radius of a lattice, i.e., the least radius r such that balls of radius r around lattice points cover $\mathrm{span}(\mathbf{B})$. Equivalently, it is the maximum distance of any point in $\mathrm{span}(\mathbf{B})$ from the lattice:

$$\rho(\mathcal{L}(\mathbf{B})) := \max_{\mathbf{z} \in \mathrm{span}(\mathbf{B})} \mathrm{dist}(\mathbf{z}, \mathcal{L}(\mathbf{B})).$$

Definition 4. The Covering Radius Problem (CRP) asks for the value of $\rho(\mathcal{L}(\mathbf{B}))$ when a lattice basis $\mathbf{B}$ is given.

Let $\lambda_n(\mathcal{L})$ denote the minimum length r such that ball of radius r around the origin contains n linearly independent vectors from the (n-dimensional) lattice $\mathcal{L}$.

Definition 5. The Shortest Independent Vectors Problem (SIVP) asks for the value of $\lambda_n(\mathcal{L}(\mathbf{B}))$ when a lattice basis $\mathbf{B}$ is given.

Definition 6. The Shortest Basis Problem (SBP) asks for the minimum length r such that given lattice $\mathcal{L}(\mathbf{B})$ has a basis whose every vector has length at most r.

We note that CVP, SIVP, SBP are NP-complete and SVP is NP-complete under randomized reductions.[5] CVPP is NP-complete in the following sense: there is a polynomial time reduction from a SAT instance ϕ to CVPP instance $(\mathcal{L}(\mathbf{B}), \mathbf{z})$ such that the lattice $\mathcal{L}(\mathbf{B})$ depends only on $|\phi|$ and not on ϕ itself. This implies that if there is a polynomial time algorithm for CVPP, then SAT has polynomial size circuits (and polynomial hierarchy collapses). Finally, CRP is in Π_2, but not known even to be NP-hard (but it is known to be Π_2-hard for ℓ_p norms with large p).

In this article, we focus on inapproximability results for lattice problems. Such results are proved by a reduction from a *hard* problem (such as SAT) to the gap-version of the lattice problem. Towards this end, we define the gap-versions of all the problems under consideration. In the following $g(n) > 1$ is a function of the dimension of the lattice that corresponds to the *gap-function*. In general, a gap-version $\mathsf{GapX}_{g(n)}$ of an optimization problem X is a promise problem where the

[5] We defined all problems as search problems, so to be precise, one considers their natural decision versions while talking about NP-completeness.

instance is guaranteed to either have a good optimum (the YES instances) or is far from it (the NO instances). The ratio between the optimum value in the YES and the NO cases is at least $g(n)$. An inapproximability result for problem X is typically proved by exhibiting a polynomial time reduction from SAT to $\mathsf{GapX}_{g(n)}$ that preserves the YES and NO instances. Such a reduction clearly implies that it is NP-hard to approximate X within a factor of $g(n)$.

Definition 7. $\mathsf{GapSVP}_{g(n)}$ is a promise problem $(\mathcal{L}(\mathbf{B}), r)$ whose YES instances satisfy $\lambda_1(\mathcal{L}(\mathbf{B})) \leq r$, and NO instances satisfy $\lambda_1(\mathcal{L}(\mathbf{B})) \geq g(n)r$.

Definition 8. $\mathsf{GapCVP}_{g(n)}$ is a promise problem $(\mathcal{L}(\mathbf{B}), \mathbf{t}, r)$ whose YES instances satisfy $\mathrm{dist}(\mathbf{t}, \mathcal{L}(\mathbf{B})) \leq r$, and NO instances satisfy $\mathrm{dist}(\mathbf{t}, \mathcal{L}(\mathbf{B})) \geq g(n)r$.

Definition 9. $\mathsf{GapCVPP}_{g(n)}$ is a promise problem $(\mathcal{L}(\mathbf{B}), \mathbf{t}, r)$ whose YES instances satisfy $\mathrm{dist}(\mathbf{t}, \mathcal{L}(\mathbf{B})) \leq r$, and NO instances satisfy $\mathrm{dist}(\mathbf{t}, \mathcal{L}(\mathbf{B})) \geq g(n)r$. The lattice $\mathcal{L}(\mathbf{B})$ is fixed once the dimension n is fixed.

Definition 10. $\mathsf{GapCRP}_{g(n)}$ is a promise problem $(\mathcal{L}(\mathbf{B}), r)$ whose YES instances satisfy $\rho(\mathcal{L}(\mathbf{B})) \leq r$, and NO instances satisfy $\rho(\mathcal{L}(\mathbf{B})) \geq g(n)r$.

Definition 11. $\mathsf{GapSIVP}_{g(n)}$ is a promise problem $(\mathcal{L}(\mathbf{B}), r)$ whose YES instances satisfy $\lambda_n(\mathcal{L}(\mathbf{B})) \leq r$, and NO instances satisfy $\lambda_n(\mathcal{L}(\mathbf{B})) \geq g(n)r$.

Definition 12. $\mathsf{GapSBP}_{g(n)}$ is a promise problem $(\mathcal{L}(\mathbf{B}), r)$ whose YES instances have a basis with each basis vector of length at most r, and for NO instances, there is no basis with each basis vector of length at most $g(n)r$.

Inapproximability of CVP

In this section, we present two results:

Theorem 1. *For any constant $\eta > 0$, $\mathsf{GapCVP}_{1/\sqrt{\eta}}$ is NP-hard. Thus, CVP is NP-hard to approximate within any constant factor.*

Theorem 2. *For any constant $\varepsilon > 0$, there is a reduction from SAT instance ϕ to $\mathsf{GapCVP}_{2^{(\log n)^{1-\varepsilon}}}$ that runs in time $2^{(\log|\phi|)^{O(1/\varepsilon)}}$. Thus CVP is hard to approximate within almost polynomial factor unless $\mathsf{NP} \subseteq \mathsf{DTIME}(2^{(\log n)^{O(1)}})$.*

Both results are due to Arora, Babai, Stern, and Sweedyk [26], though our presentation is different, especially for the second result.

Proof of Theorem 1

We prove the following theorem which implies Theorem 1 along with some additional properties of GapCVP instance that we need later.

Theorem 3. *For any constant $\eta > 0$, there are constants C, C', C'', and a reduction from* SAT *instance of size n to a* CVP *instance $(\mathcal{L}(\mathbf{B}_{\text{cvp}}), \mathbf{t})$ with the following properties:*

1. *$\mathbf{B}_{\text{cvp}}$ is an integer matrix with size $C'd \times Cd$. The vector $\mathbf{t}$ also has integer co-ordinates and it is linearly independent of the columns of matrix $\mathbf{B}_{\text{cvp}}$.*
2. *The reduction runs in time $n^{C''}$ and therefore $d \leq n^{C''}$.*
3. *If the* SAT *instance is a* YES *instance, then there is a coefficient vector $\mathbf{y} \in \{0,1\}^{Cd}$ such that the vector $\mathbf{B}_{\text{cvp}}\mathbf{y} - \mathbf{t}$ is also a $\{0,1\}$-vector and has exactly ηd co-ordinates equal to 1. In particular,* $\text{dist}(\mathbf{t}, \mathcal{L}(\mathbf{B}_{\text{cvp}})) \leq \|\mathbf{B}_{\text{cvp}}\mathbf{y} - \mathbf{t}\| = \sqrt{\eta d}$.
4. *If the SAT instance is a NO instance, then for any coefficient vector $\mathbf{y} \in \mathbb{Z}^{Cd}$, and any non-zero integer j_0, the vector $\mathbf{B}_{\text{cvp}}\mathbf{y} - j_0\mathbf{t}$ either has a co-ordinate equal to d^{4d}, or has at least d non-zero co-ordinates. In particular,* $\text{dist}(\mathbf{t}, \mathcal{L}(\mathbf{B}_{\text{cvp}})) \geq \sqrt{d}$.

Proof. The reduction is from Exact Set Cover. It is known that for any constant $\eta > 0$, there is a polynomial time reduction from SAT to the Set Cover problem such that : If the SAT instance is a YES instance, then there are ηd sets that cover each element of the universe exactly once. If the SAT instance is a NO instance then there is no set-cover of size d. Let the universe for the set cover instance be $[n']$ and the sets be $S_1, S_2, \ldots, S_{n''}$. It holds that $n' = C_1 d$ and $n'' = Cd$ for some constants C_1, C.

Let the matrix $\mathbf{B}_{\text{cvp}}$ and vector $\mathbf{t}$ be as shown in Fig. 14.1. Here Q is a large integer, say $Q = d^{4d}$. The matrix $\mathbf{B}_{\text{cvp}}$ has $n' + n'' = C'd$ rows and n'' columns. $\mathbf{B}_{\text{cvp}}$ is Q-multiple of the element-set inclusion matrix appended by an identity matrix. The vector $\mathbf{t}$ has first n' co-ordinates equal to Q and the rest are 0.

Let $\mathbf{y} = (y_1, y_2, \ldots, y_{n''}) \in \mathbb{Z}^{n''}$ be the coefficient vector. If the Set Cover instance has an exact cover consisting of ηd sets, then define $y_j = 1$ if the set S_j is

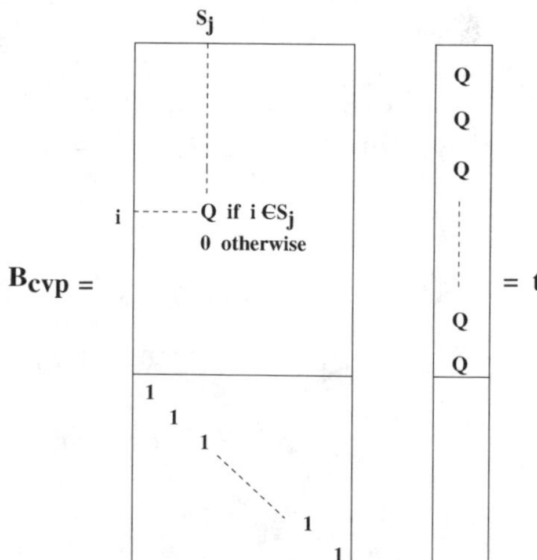

Fig. 14.1 The CVP instance

included in the set cover and $y_j = 0$ otherwise. Clearly, $\mathbf{B}_{\text{cvp}}\mathbf{y} - \mathbf{t}$ has exactly ηd co-ordinates equal to 1 and the rest are zero.

Now assume there is no set cover of size d. Let $\mathbf{y}$ be an arbitrary coefficient vector and $j_0 \in \mathbb{Z}$, $j_0 \neq 0$. If at least d of the co-ordinates y_j are non-zero, we are done. Otherwise the family of sets S_j such that $y_j \neq 0$ has fewer than d sets. This family cannot cover the universe and therefore there is a coordinate in $\mathbf{B}_{\text{cvp}}\mathbf{y} - j_0\mathbf{t}$ that is a non-zero multiple of Q. This coordinate corresponds to an element that is not covered.

Proof of Theorem 2

We prove Theorem 2 using a reduction from the Label Cover problem (see Def. 13). The reduction is essentially from Arora et al. paper [26] which also defined the Label Cover problem.

We first give a reformulation of CVP as the following problem: Let $\mathbf{y} = (y_1, y_2, \ldots, y_n)$ be a vector of integer valued variables. For $1 \leq i \leq m$, $\phi_i(\mathbf{y}) = \sum_j a_{ij} y_j$ be (homogeneous) linear forms and for $1 \leq k \leq t$, $\psi_k(\mathbf{y}) = c_k + \sum_j b_{kj} y_j$ be (non-homogeneous) linear forms. Then CVP (in ℓ_2 norm) is same as the following optimization problem:

$$\min_{\mathbf{y} \in \mathbb{Z}^n} \left(\sum_{i=1}^{m} |\phi_i(\mathbf{y})|^2 \right)^{1/2}$$
$$\text{subject to} \quad \psi_k(\mathbf{y}) = 0 \;\; \forall\, 1 \leq k \leq t.$$

To see that this is just a reformulation of CVP, one can think of the constraints $\psi_k(\mathbf{y}) = 0$ as defining an affine subspace of $\mathbb{R}^n$. The set of all points on this affine subspace corresponding to $\mathbf{y} \in \mathbb{Z}^n$ is of the form $\{\mathbf{z}_0 - \mathbf{v} \mid \mathbf{v} \in \mathcal{L}_0\}$ for a suitable point $\mathbf{z}_0$ and a suitable lattice $\mathcal{L}_0$. If Φ denotes the matrix whose rows are the linear forms ϕ_i, then we are minimizing $\|\Phi\mathbf{y}\|$ over $\mathbf{y} \in \mathbb{Z}^n$. This is same as minimizing $\|\Phi(\mathbf{z}_0 - \mathbf{v})\|$ over $\mathbf{v} \in \mathcal{L}_0$. This, in turn, is same as minimizing the distance of point $\Phi\mathbf{z}_0$ from the lattice $\Phi\mathcal{L}_0$ (whose basis is obtained by linearly transforming the basis for $\mathcal{L}_0$ via matrix Φ).

The Label Cover Problem

Definition 13. (Label Cover Problem): An instance of label cover is specified as:

$$LC(G(V, W, E), n, m, \{\pi_{v,w}\}_{(v,w)\in E}, [R], [S]).$$

$G = (V, W, E)$ is a bipartite graph with left side vertices V, right side vertices W and a set of edges E. The graph is left regular, i.e., all vertices in V have the same degree. $n = |V|$ and $m = |W|$.

The goal is to assign one "label" to every vertex, where the vertices in V are required to receive a label from set $[R]$ and the vertices in W are required to receive a label from set $[S]$. Thus, a labeling A is just a map $A : V \mapsto [R], A : W \mapsto [S]$. The labeling is supposed to satisfy certain constraints given by maps $\pi_{v,w} : [R] \mapsto [S]$. There is one such map for every edge $(v, w) \in E$. A labeling A "satisfies" an edge (v, w), if

$$\pi_{v,w}(A(v)) = A(w).$$

The optimum $OPT(LC)$ of the instance is defined to be the maximum fraction of edges satisfied by any labeling. We assume that $n \geq m$, and $R \geq S$ (thus the left side is viewed as *larger*).

The following theorem can be obtained by combining the PCP Theorem (Arora and Safra [40], Arora et al. [41]) with Raz's Parallel Repetition Theorem [42]. This theorem is the starting point for most of the recent PCP constructions and hardness results.

Theorem 4. *There exists an absolute constant $\beta > 0$ such that for every integer $R \geq 7$, there is a reduction from* SAT *instance ϕ to Label Cover instance $LC(G(V, W, E), n, m, \{\pi_{v,w}\}, [R], [S])$ with the following property. The* YES *instances of* SAT *map to label cover instances with* OPT(LC)$= 1$ *and the* NO *instances map to label cover instances with* OPT(LC) $\leq R^{-\beta}$. *The running time of the reduction and size of the label cover instance are bounded by $|\phi|^{O(\log R)}$.*

Reduction from Label Cover to GapCVP

Let $LC(G(V, W, E), n, m, \{\pi_{v,w}\}, [R], [S])$ be the instance of Label Cover given by Theorem 4. We describe a reduction from this instance to GapCVP$_g$ where the gap $g = \frac{1}{20} R^{\beta/2}$. We construct the CVP instance according to the new CVP-formulation described in the beginning of this section. The set of integer valued variables is:

$$Y := \{y_{v,j} \mid v \in V, j \in [R]\} \bigcup \{z_{w,i} \mid w \in W, i \in [S]\}.$$

The function to be minimized is

$$OBJ := \left(m \cdot \sum_{v \in V, j \in [R]} y_{v,j}^2 + n \cdot \sum_{w \in W, i \in [S]} z_{w,i}^2 \right)^{1/2}.$$

The affine *constraints* are:

$$\forall\, v \in V, \qquad \sum_{j \in [R]} y_{v,j} = 1. \tag{14.1}$$

$$\forall\, w \in W, \qquad \sum_{i \in [S]} z_{w,i} = 1. \tag{14.2}$$

$$\forall\, (v,w) \in E, \forall\, i \in [S], \qquad z_{w,i} = \sum_{j \in [R]:\pi_{v,w}(j)=i} y_{v,j}. \tag{14.3}$$

The YES Case:

We prove that if the Label Cover instance has a labeling that satisfies all edges (i.e., $OPT(LC) = 1$), then there is an integer assignment to variables in Y such that $OBJ \le \sqrt{2mn}$. Indeed, let $A : V \mapsto [R]$, $A : W \mapsto [S]$ be such a labeling. Define

$$y_{v,j} := \begin{cases} 1 & \text{if } j = A(v) \\ 0 & \text{otherwise.} \end{cases}$$

Similarly, define $z_{w,i} = 1$ if $i = A(w)$ and $z_{w,i} = 0$ otherwise. Clearly, for every $v \in V$ ($w \in W$ resp.), there is exactly one $j \in [R]$ ($i \in [S]$ resp.) such that $y_{v,j}$ ($z_{w,i}$ resp.) is non-zero, and its value equals 1. Therefore

$$OBJ = \sqrt{m \cdot |V| + n \cdot |W|} = \sqrt{2mn}.$$

The above reasoning also shows that all constraints in (14.1) and (14.2) are satisfied. Now we show that all constraints in (14.3) are satisfied. Fix any such constraint, i.e., fix $(v,w) \in E$ and $i \in [S]$. We will show that

$$z_{w,i} = \sum_{j \in [R]:\pi_{v,w}(j)=i} y_{v,j}. \tag{14.4}$$

Let $i_0 = A(w)$ and $j_0 = A(v)$. Since the labeling satisfies the edge (v,w), we have $\pi_{v,w}(j_0) = i_0$. Clearly, if $i \neq i_0$, then both sides of (14.4) evaluate to zero, and if $i = i_0$, both sides evaluate to 1.

The NO Case:

We prove that if the Label Cover instance has no labeling that satisfies even α fraction of its edges for $\alpha < 0.1$, then for any integer assignment to variables in Y that satisfies constraints (14.1)–(14.3), one must have $OBJ \ge 0.1\sqrt{mn/\alpha}$. Note that, once proven, it implies that if $OPT(LC) \le R^{-\beta}$, then $OBJ \ge 0.1 \cdot R^{\beta/2}\sqrt{mn}$. Thus the gap between the YES and NO cases is

$$\frac{0.1 \cdot R^{\beta/2}\sqrt{mn}}{\sqrt{2mn}} \geq \frac{1}{20} R^{\beta/2} \qquad \text{as claimed.}$$

Consider any integer assignment to variables in Y that satisfies constraints (14.1)–(14.3). Define sets $T_v \subseteq [R]$, $T_w \subseteq [S]$ as:

$$T_v := \{j \in [R] \mid y_{v,j} \neq 0\}, \qquad T_w := \{i \in [S] \mid z_{w,i} \neq 0\}.$$

Due to constraints (14.1) and (14.2), the sets T_v, T_w are non-empty for all $v \in V$, $w \in W$.

Lemma 1. *For any $(v,w) \in E$ and $i \in T_w$, there exists $j^* \in T_v$ such that $\pi_{v,w}(j^*) = i$.*

Proof. Consider the constraint $z_{w,i} = \sum_{j \in [R]: \pi_{v,w}(j)=i} y_{v,j}$. Since $i \in T_w$, $z_{w,i} \neq 0$. Hence, one of the variables on the right side must be non-zero, say the variable y_{v,j^*}. Thus $j^* \in T_v$ and $\pi_{v,w}(j^*) = i$.

We consider two scenarios depending on whether the typical size of sets T_v is *small* or *large*. Towards this end, let

$$V^* := \{v \in V \mid |T_v| \geq 0.1/\alpha\}.$$

Case (i): $|V^*| \geq 0.1|V| = 0.1 \cdot n$. In this case,

$$OBJ \geq \left(m \cdot \sum_{v \in V, j \in [R]} y_{v,j}^2\right)^{1/2} \geq \left(m \cdot \sum_{v \in V^*, j: y_{v,j} \neq 0} 1\right)^{1/2} = \left(m \cdot \sum_{v \in |V^*|} |T_v|\right)^{1/2}$$
$$\geq \sqrt{m \cdot |V^*| \cdot 0.1/\alpha} \geq 0.1\sqrt{mn/\alpha}.$$

Case (ii): $|V^*| \leq 0.1|V|$. In this case, ignore the edges (v,w) that are incident on V^*. Since the graph of label cover instance is regular, we ignore only 0.1 fraction of its edges. Define the following labeling to V and W. The label of $w \in W$ is an arbitrary label from set T_w. The label of $v \in V$ is a random label from set T_v. We show that this labeling satisfies, in expectation, at least $0.9\alpha/0.1$ fraction of edges of label cover (arriving at a contradiction, since we know that there is no labeling that satisfies even α fraction of edges). Indeed, if (v,w) is any edge such that $v \notin V^*$, then $|T_v| \leq 0.1/\alpha$. Let label of w be some $i^* \in T_w$. By Lemma 1, we know that there exists a label $j^* \in T_v$ such that $\pi_{v,w}(j^*) = i^*$. With probability $1/|T_v| \geq \alpha/0.1$, we select j^* as the label of v and the edge (v,w) is satisfied. This completes the proof.

Finishing the Proof of Theorem 2

Let n be the size of SAT instance. Combining reduction from SAT to Label Cover in Theorem 4 with our reduction from Label Cover to GapCVP, we get a reduction from SAT to $\mathsf{GapCVP}_{1/20\cdot R^{\beta/2}}$ that runs in time $n^{C\log R}$ for some constant C.

Choose $R = 2^{(\log n)^k}$ for some large integer k. The size of CVP instance is $N \leq n^{C\log R}$. Thus $\log N \leq C \log R \log n \leq (\log n)^{2+k}$. The inapproximability factor for CVP is

$$\frac{1}{20}R^{\beta/2} = \frac{1}{20}2^{\beta/2\log R} = \frac{1}{20}2^{\beta/2(\log n)^k} \geq 2^{(\log n)^{k-1}} \geq 2^{(\log N)^{(k-1)/(k+2)}}.$$

When $k \approx 1/\varepsilon$, the hardness factor is $\approx 2^{(\log N)^{1-\varepsilon}}$ which proves Theorem 2.

Inapproximability of SVP

In this section, we present two results:

Theorem 5. *For any constant* $1 < \gamma' < \sqrt{2}$, $\mathsf{GapSVP}_{\gamma'}$ *is* NP*-hard. Thus,* SVP *is* NP*-hard to approximate within any constant factor less than* $\sqrt{2}$.

Theorem 6. *For any constant* $\varepsilon > 0$*, there is a reduction from* SAT *instance* ϕ *to* $\mathsf{GapSVP}_{2^{(\log n)^{1-\varepsilon}}}$ *that runs in time* $2^{(\log|\phi|)^{O(1/\varepsilon)}}$. *Thus* SVP *is hard to approximate within almost polynomial factor unless* $\mathsf{NP} \subseteq \mathsf{DTIME}(2^{(\log n)^{O(1)}})$.

The first result is due to Micciancio [17] and the second is a combination of results of Khot [18] and Haviv and Regev [19]. We present only a sketch of both proofs.

Proof of Theorem 5

The reduction is from GapCVP. Let $(\mathbf{B}_{\text{cvp}}, \mathbf{t})$ be an instance of $\mathsf{GapCVP}_{1/\sqrt{\eta}}$ given by Theorem 1. Micciancio constructs GapSVP instance $\mathcal{L}(\mathbf{B}')$ as follows:

$$\mathbf{B}' = \left[\begin{array}{c|c} \alpha\mathbf{B}_{\text{cvp}}\mathbf{T} & \alpha\mathbf{t} \\ \hline \mathbf{B}_{\text{gad}} & \mathbf{s} \end{array}\right] \tag{14.5}$$

Here, α is a suitable constant, $\mathbf{B}_{\text{gad}}, \mathbf{T}$ are matrices and $\mathbf{s}$ is a vector (of appropriate dimensions). The crucial ingredient of Micciancio's reduction is construction of the *gadget* $(\mathbf{B}_{\text{gad}}, \mathbf{T}, \mathbf{s})$. Here, $\mathcal{L}(\mathbf{B}_{\text{gad}})$ is a lattice and $\mathbf{s}$ is a (non-lattice) point such that: (1) $\lambda_1(\mathcal{L}(\mathbf{B}_{\text{gad}})) \geq \gamma r$ for a parameter r, $1 < \gamma < \sqrt{2}$ and (2) the ball of radius r

around $\mathbf{s}$ contains exponentially many lattice points of $\mathcal{L}(\mathbf{B}_{\text{gad}})$. The set of all lattice points in this ball also satisfy an extra property, as made precise in the statement of the lemma below (and this is where the matrix $\mathbf{T}$ enters into the picture). The construction is quite involved, based on sphere packings, Schnorr-Adleman prime number lattice, and a probabilistic version of Sauer's Lemma.

Micciancio's Gadget Construction

Lemma 2. *For every $1 < \gamma < \sqrt{2}$ and integer m, one can construct in probabilistic polynomial time, matrices $\mathbf{B}_{\text{gad}}, \mathbf{T}$, a vector $\mathbf{s}$, and parameters k, ℓ, r such that:*

1. *$\mathbf{T}$ has size $m \times k$, $\mathbf{B}_{\text{gad}}$ has size $\ell \times k$, and $\mathbf{s}$ is a column vector of size ℓ. Here, $k, \ell \leq \text{poly}(m)$.*
2. *The lattice $\mathcal{L}(\mathbf{B}_{\text{gad}})$ has no non-zero vector of length less than γr, i.e.,*

$$\forall\, \mathbf{x} \in \mathbb{Z}^k, \mathbf{x} \neq 0, \quad \|\mathbf{B}_{\text{gad}}\mathbf{x}\| \geq \gamma r.$$

3. *For every $\mathbf{y} \in \{0,1\}^m$, there exists $\mathbf{x} \in \mathbb{Z}^k$ such that $\mathbf{T}\mathbf{x} = \mathbf{y}$ and $\|\mathbf{B}_{\text{gad}}\mathbf{x} - \mathbf{s}\| \leq r$. In particular, the ball of radius r around $\mathbf{s}$ contains at least 2^m points from the lattice $\mathcal{L}(\mathbf{B}_{\text{gad}})$.*

Micciancio's Reduction

We now present a reduction from GapCVP to GapSVP. Let $(\mathbf{B}_{\text{cvp}}, \mathbf{t})$ be the $\mathsf{GapCVP}_{1/\sqrt{\eta}}$ instance as in Theorem 1. We will choose η to be a small enough constant later. Let $m' \times m$ denote the size of matrix $\mathbf{B}_{\text{cvp}}$ (and hence $\mathbf{t}$ is a column vector of size m). Let $(\mathbf{B}_{\text{gad}}, \mathbf{T}, \mathbf{s})$ be the gadget given by Lemma 2 with parameters m and $1 < \gamma < \sqrt{2}$. Parameters k, ℓ, r are as in that lemma.

Construct matrix $\mathbf{B}'$ as in Equation (14.5) where $\alpha = \gamma r / \sqrt{d}$. Let us denote the coefficient vector for lattice $\mathcal{L}(\mathbf{B}')$ by $\mathbf{x}'$ and write $\mathbf{x}' = (\mathbf{x}, j)$ with $j \in \mathbb{Z}$. Note that

$$\mathbf{B}'\mathbf{x}' = \big(\alpha(\mathbf{B}_{\text{cvp}}\mathbf{T}\,\mathbf{x} + j\mathbf{t}),\ \ \mathbf{B}_{\text{gad}}\,\mathbf{x} + j\mathbf{s}\big)\,. \tag{14.6}$$

Note that the GapCVP instance satisfies Property 3 (the YES case) or Property 4 (the NO case) in Theorem 1. We show that in the YES case, the lattice $\mathcal{L}(\mathbf{B}')$ has a *short* non-zero vector, whereas in the NO case, every non-zero vector is *long*.

The YES Case:

In the YES case, we know that that there exists $\mathbf{y} \in \{0,1\}^m$ such that $\|\mathbf{B}_{\text{cvp}}\mathbf{y} - \mathbf{t}\| \leq \sqrt{\eta d}$. We prove that $\mathcal{L}(\mathbf{B}')$ has a non-zero vector of length at most $\sqrt{1 + \gamma^2\eta} \cdot r$. Indeed, Lemma 2 guarantees existence of $\mathbf{x} \in \{0,1\}^k$ such that $\mathbf{T}\mathbf{x} = \mathbf{y}$ and $\|\mathbf{B}_{\text{gad}}\mathbf{x} - \mathbf{s}\| \leq r$. We let $\mathbf{x}' = (\mathbf{x}, -1)$. Clearly,

$$\begin{aligned}\|\mathbf{B}'\mathbf{x}'\|^2 &= \alpha^2\|\mathbf{B}_{\text{cvp}}\mathbf{T}\,\mathbf{x} - \mathbf{t}\|^2 + \|\mathbf{B}_{\text{gad}}\,\mathbf{x} - \mathbf{s}\|^2 \\ &\leq \alpha^2\|\mathbf{B}_{\text{cvp}}\mathbf{y} - \mathbf{t}\|^2 + r^2 \leq \alpha^2\eta d + r^2 = (1+\gamma^2\eta)r^2,\end{aligned}$$

by the choice of $\alpha = \gamma r/\sqrt{d}$. Note that $\|\mathbf{B}'\mathbf{x}'\| \approx r$ by choosing η sufficiently small.

The NO Case:

In the NO case, for every $\mathbf{y} \in \mathbb{Z}^m$ and $j_0 \neq 0$, $\|\mathbf{B}_{\text{cvp}}\mathbf{y} + j_0\mathbf{t}\| \geq \sqrt{d}$. We prove that every non-zero vector in $\mathcal{L}(\mathbf{B}')$ has length at least γr.

Let $\mathbf{B}'\mathbf{x}'$ be an arbitrary non-zero lattice vector with $\mathbf{x}' = (\mathbf{x}, j_0)$. First consider the case when $j_0 \neq 0$. In this case

$$\|\mathbf{B}'\mathbf{x}'\| \geq \alpha\|\mathbf{B}_{\text{cvp}}(\mathbf{T}\,\mathbf{x}) + j_0\mathbf{t}\| \geq \alpha\sqrt{d} = \gamma r.$$

Now consider the case when $j_0 = 0$. In this case $\mathbf{x} \neq 0$ and from Lemma 2, Property (2),

$$\|\mathbf{B}'\mathbf{x}'\| \geq \|\mathbf{B}_{\text{gad}}\,\mathbf{x}\| \geq \gamma r.$$

Thus, the instance of GapSVP has a gap of $\gamma' = \frac{\gamma}{\sqrt{1+\gamma^2\eta}}$ which can be made arbitrarily close to $\sqrt{2}$ by choosing γ to be close enough to $\sqrt{2}$ and then choosing η small enough. This proves Theorem 5.

Proof of Theorem 6

Proof of Theorem 6 proceeds by first giving a basic reduction from GapCVP to $\mathsf{GapSVP}_{1/\zeta}$ for some constant $\zeta < 1$ and then boosting the SVP-hardness by tensoring operation on the lattice. Let $\mathcal{L}_0$ be the instance of $\mathsf{GapSVP}_{1/\zeta}$ produced by the basic reduction, i.e., for some parameter d, either $\lambda_1(\mathcal{L}_0(\mathbf{B})) \leq \zeta\sqrt{d}$ (YES case) or $\lambda_1(\mathcal{L}_0(\mathbf{B})) \geq \sqrt{d}$ (NO case). By taking the k-wise tensored lattice $\mathcal{L}_0^{\otimes k}$, it is easy to see that in the YES case,

$$\lambda_1(\mathcal{L}_0(\mathbf{B})) \leq \zeta\sqrt{d} \implies \lambda_1(\mathcal{L}_0^{\otimes k}(\mathbf{B})) \leq \zeta^k\sqrt{d}^k.$$

On the other hand, in the NO case, suppose it were true that

$$\lambda_1(\mathcal{L}_0(\mathbf{B})) \geq \sqrt{d} \implies \lambda_1(\mathcal{L}_0^{\otimes k}(\mathbf{B})) \geq \sqrt{d}^k. \tag{14.7}$$

The resulting gap would be boosted to $(1/\zeta)^k$ and the size of instance $\mathcal{L}_0^{\otimes k}$ would be $(size(\mathcal{L}_0))^k$. By choosing k appropriately, it would prove $2^{(\log n)^{1-\varepsilon}}$ hardness for SVP, i.e., Theorem 6. But, as we shall see, the implication in (14.7) is false for a general lattice. However, the implication does hold for the *specific* lattice $\mathcal{L}_0(\mathbf{B})$ produced in the NO Case in Khot's [18] reduction. Though he did not prove that (14.7) holds for his lattice, by using a slight variant of the tensor product, he was able to boost hardness to $2^{(\log n)^{1/2-\varepsilon}}$. In a subsequent paper, Haviv and Regev [19] proved that (14.7) holds for Khot's lattice. This boosts hardness to $2^{(\log n)^{1-\varepsilon}}$. Let us first define the tensor product operation.

Tensor Product of Lattices

For two column vectors $\mathbf{u}$ and $\mathbf{v}$ of dimensions m_1 and m_2 respectively, we define their tensor product $\mathbf{u} \otimes \mathbf{v}$ as the m_1m_2-dimensional column vector

$$\begin{pmatrix} u_1\mathbf{v} \\ \vdots \\ u_{m_1}\mathbf{v} \end{pmatrix}.$$

If we think of the coordinates of $\mathbf{u} \otimes \mathbf{v}$ as arranged in an $m_1 \times m_2$ matrix, we obtain the equivalent description of $\mathbf{u} \otimes \mathbf{v}$ as the matrix $\mathbf{u} \cdot \mathbf{v}^T$. Finally, for an $m_1 \times n_1$ matrix $\mathbf{A}$ and an $m_2 \times n_2$ matrix $\mathbf{B}$, one defines their tensor product $\mathbf{A} \otimes \mathbf{B}$ as the $m_1m_2 \times n_1n_2$ matrix

$$\begin{pmatrix} A_{11}\mathbf{B} & \cdots & A_{1n_1}\mathbf{B} \\ \vdots & & \vdots \\ A_{m_11}\mathbf{B} & \cdots & A_{m_1n_1}\mathbf{B} \end{pmatrix}.$$

Let $\mathcal{L}_1$ be a lattice generated by $m_1 \times n_1$ matrix $\mathbf{B}_1$ and $\mathcal{L}_2$ be a lattice generated by $m_2 \times n_2$ matrix $\mathbf{B}_2$. Then the tensor product of $\mathcal{L}_1$ and $\mathcal{L}_2$ is defined as the n_1n_2-dimensional lattice generated by the $m_1m_2 \times n_1n_2$ matrix $\mathbf{B}_1 \otimes \mathbf{B}_2$ and is denoted by $\mathcal{L} = \mathcal{L}_1 \otimes \mathcal{L}_2$. Equivalently, $\mathcal{L}$ is generated by the n_1n_2 vectors obtained by taking the tensor of two column vectors, one from $\mathbf{B}_1$ and one from $\mathbf{B}_2$.

We are interested in the behavior of the shortest vector in a tensor product of lattices. It is easy to see that for any two lattices $\mathcal{L}_1$ and $\mathcal{L}_2$, we have

$$\lambda_1(\mathcal{L}_1 \otimes \mathcal{L}_2) \leq \lambda_1(\mathcal{L}_1) \cdot \lambda_1(\mathcal{L}_2). \tag{14.8}$$

Indeed, any two vectors $\mathbf{v}_1$ and $\mathbf{v}_2$ satisfy $\|\mathbf{v}_1 \otimes \mathbf{v}_2\| = \|\mathbf{v}_1\| \cdot \|\mathbf{v}_2\|$. Applying this to shortest nonzero vectors of $\mathcal{L}_1$ and $\mathcal{L}_2$ implies Inequality (14.8).

Inequality (14.8) has an analogue for linear codes, with λ_1 replaced by the minimum distance of the code under the Hamming metric. There, it is not too hard to show that the inequality is in fact an equality: the minimal distance of the tensor product of two linear codes always equals the product of their minimal distances. However, contrary to what one might expect, there exist lattices for which Inequality (14.8) is *strict*. The following lemma due to Steinberg shows this fact (his lattice is actually self-dual).

Lemma 3 ([43, Page 48]). *For any large enough n there exists an n-dimensional lattice $\mathcal{L}$ satisfying*

$$\lambda_1(\mathcal{L} \otimes \mathcal{L}) \leq \sqrt{n} \text{ and } \lambda_1(\mathcal{L}) = \Omega(\sqrt{n}).$$

Khot's Reduction

Let us imagine a hypothetical reduction from CVP to an instance $\mathcal{L}_0(\mathbf{B})$ of SVP that has the following properties (we assume w.l.o.g. that all lattice vectors have integer co-ordinates):

1. If the CVP instance is a YES instance, then there is a non-zero lattice vector with norm at most $\zeta\sqrt{d}$ where $\zeta < 1$ is a constant.
2. If the CVP instance is a NO instance, then any non-zero lattice vector has at least d non-zero co-ordinates.

In particular, this gives a gap-instance of SVP with gap $1/\zeta$. It is not hard to see that if we had such a *magic reduction*, then the k-wise tensor product of the lattice $\mathcal{L}_0$ in NO case would satisfy implication (14.7) and lead to a gap-instance with gap $(1/\zeta)^k$. Thus the tensor product would work provided that in the NO case, every non-zero lattice vector is not only *long*, but also has *many* non-zero co-ordinates. However, we do not know whether such a reduction exists. Nevertheless, Khot [18] gives a reduction that achieves somewhat weaker properties, but still good enough for boosting purposes. The following theorem summarizes his reduction (with a minor modification by Haviv and Regev [19]).

Theorem 7. *There is a constant $\zeta < 1$ and a polynomial-time randomized reduction from* SAT *to* SVP *that outputs a lattice basis* $\mathbf{B}$ *and integers n, d such that, $\mathcal{L}(\mathbf{B}) \subseteq \mathbb{Z}^n$, and w.h.p. the following holds:*

1. *If the* SAT *instance is a* YES *instance, then $\lambda_1(\mathcal{L}(\mathbf{B})) \leq \zeta \cdot \sqrt{d}$.*
2. *If the* SAT *instance is a* NO *instance, then every nonzero vector $\mathbf{v} \in \mathcal{L}(\mathbf{B})$*
   - *Either has at least d nonzero coordinates*
   - *Or has all coordinates even and at least $d/4$ of them are nonzero*
   - *Or has all coordinates even and $\|\mathbf{v}\|_2 \geq d$*
   - *Or has a coordinate with absolute value at least $Q := d^{4d}$*

In particular, $\lambda_1(\mathcal{L}(\mathbf{B})) \geq \sqrt{d}$.

Boosting the SVP Hardness Factor

We boost the hardness factor using the standard tensor product of lattices. If $(\mathcal{L}_0(\mathbf{B}), d)$ is a YES instance of the SVP instance in Theorem 7, then clearly

$$\lambda_1(\mathcal{L}_0^{\otimes k}) \leq \zeta^k d^{k/2}. \tag{14.9}$$

When $(\mathcal{L}_0(\mathbf{B}), d)$ is a NO instance, Haviv and Regev [19] show that any nonzero vector of $\mathcal{L}_0^{\otimes k}$ has norm at least $d^{k/2}$, i.e.,

$$\lambda_1(\mathcal{L}_0^{\otimes k}) \geq d^{k/2}. \tag{14.10}$$

This yields a gap of ζ^k between the two cases. Inequality (14.10) easily follows by induction from the central lemma of Haviv and Regev stated below, which shows that NO instances "tensor nicely." We skip the proof of this lemma.

Lemma 4. *Let $(\mathcal{L}_0(\mathbf{B}), d)$ be a* NO *instance of* SVP *given in Theorem 7. Then for any lattice $\mathcal{L}$,*

$$\lambda_1(\mathcal{L}_0 \otimes \mathcal{L}) \geq \sqrt{d} \cdot \lambda_1(\mathcal{L}).$$

References

1. D. Micciancio, S. Goldwasser. Complexity of lattice problems, A cryptographic perspective. Kluwer Academic Publishers, 2002
2. R. Kumar, D. Sivakumar. Complexity of SVP – A reader's digest. SIGACT News, 32(3), Complexity Theory Column (ed. L. Hemaspaandra), 2001, pp 40–52
3. O. Regev. On the Complexity of Lattice Problems with polynomial Approximation Factors. In Proc. of the LLL+25 Conference, Caen, France, June 29-July 1, 2007
4. C.F. Gauss. Disquisitiones arithmeticae. (leipzig 1801), art. 171. Yale University. Press, 1966. English translation by A.A. Clarke
5. H. Minkowski. Geometrie der zahlen. Leizpig, Tuebner, 1910
6. A.K. Lenstra, H.W. Lenstra, L. Lovász. Factoring polynomials with rational coefficients. Mathematische Ann., 261, 1982, pp 513–534
7. J.C. Lagarias, A.M. Odlyzko. Solving low-density subset sum problems. Journal of the ACM, 32(1), 1985, pp 229–246
8. S. Landau, G.L. Miller. Solvability of radicals is in polynomial time. Journal of Computer and Systems Sciences, 30(2), 1985, pp 179–208
9. H.W. Lenstra. Integer programming with a fixed number of variables. Tech. Report 81–03, Univ. of Amsterdam, Amstredam, 1981
10. R. Kannan. Improved algorithms for integer programming and related lattice problems. In Proc. of the 15th Annual ACM Symposium on Theory of Computing, 1983, pp 193–206
11. C.P. Schnorr. A hierarchy of polynomial-time basis reduction algorithms. In Proc. of Conference on Algorithms, Péecs (Hungary), 1985, pp 375–386
12. R. Kannan. Minkowski's convex body theorem and integer programming. Mathematics of Operations Research, 12:415–440, 1987
13. M. Ajtai, R. Kumar, D. Sivakumar. A sieve algorithm for the shortest lattice vector problem. In Proc. of the 33*rd* Annual ACM Symposium on the Theory of Computing, 2001, pp 601–610

14. P. van Emde Boas. Another NP-complete problem and the complexity of computing short vectors in a lattice. Tech. Report 81-04, Mathematische Instiut, University of Amsterdam, 1981
15. M. Ajtai. The shortest vector problem in L_2 is NP-hard for randomized reductions. In Proc. of the 30th Annual ACM Symposium on the Theory of Computing, 1998, pp 10–19
16. J.Y. Cai, A. Nerurkar. Approximating the SVP to within a factor $(1 + 1/\dim^{\epsilon})$ is NP-hard under randomized reductions. In Proc. of the 13th Annual IEEE Conference on Computational Complexity, 1998, pp 151–158
17. D. Micciancio. The shortest vector problem is NP-hard to approximate to within some constant. In Proc. of the 39th IEEE Symposium on Foundations of Computer Science, 1998
18. S. Khot. Hardness of approximating the shortest vector problem in lattices. Journal of the ACM, 52(5), 2005, pp 789–808
19. I. Haviv, O. Regev. Tensor-based hardness of the Shortest Vector Problem to within almost polynomial factors. To appear in Proc. of the 39th Annual ACM Symposium on the Theory of Computing, 2007
20. M. Ajtai. Generating hard instances of lattice problems. In Proc. of the 28th Annual ACM Symposium on the Theory of Computing, 1996, pp 99–108
21. M. Ajtai, C. Dwork. A public-key cryptosystem with worst-case/average-case equivalence. In Proc. of the 29th Annual ACM Symposium on the Theory of Computing, 1997, pp 284–293
22. J.Y. Cai, A. Nerurkar. An improved worst-case to average-case connection for lattice problems. In 38th IEEE Symposium on Foundations of Computer Science, 1997
23. J.Y. Cai. Applications of a new transference theorem to Ajtai's connection factor. In Proc. of the 14th Annual IEEE Conference on Computational Complexity, 1999
24. O. Regev. New lattice based cryptographic constructions. To appear in Proc. of the 35th Annual ACM Symposium on the Theory of Computing, 2003
25. O. Goldreich, D. Micciancio, S. Safra, J.P. Seifert. Approximating shortest lattice vectors is not harder than approximating closest lattice vectors. Information Processing Letters, 1999
26. S. Arora, L. Babai, J. Stern, E.Z. Sweedyk. The hardness of approximate optima in lattices, codes and systems of linear equations. Journal of Computer and Systems Sciences (54), 1997, pp 317–331
27. I. Dinur, G. Kindler, S. Safra. Approximating CVP to within almost-polynomial factors is NP-hard. In Proc. of the 39th IEEE Symposium on Foundations of Computer Science, 1998
28. D. Micciancio. The hardness of the closest vector problem with preprocessing. IEEE Transactions on Information Theory, vol 47(3), 2001, pp 1212–1215
29. U. Feige and D. Micciancio. The inapproximability of lattice and coding problems with preprocessing. Computational Complexity, 2002, pp 44–52
30. O. Regev. Improved inapproximability of lattice and coding problems with preprocessing. IEEE Transactions on Information Theory, 50(9), 2004, pp 2031–2037
31. M. Alekhnovich, S. Khot, G. Kindler, N. Vishnoi. Hardness of approximating the closest vector problem with pre-processing. In Proc. of the 46th IEEE Symposium on Foundations of Computer Science, 2005
32. D. Aharonov, O. Regev. Lattice problems in NP $\cap$ coNP. Journal of the ACM, 52(5), 2005, pp 749–765
33. I. Haviv, O. Regev. Hardness of the covering radius problem on lattices. In Proc. of the $21st$ Annual IEEE Computational Complexity Conference, 2006
34. J. Blömer, J.P. Seifert. On the complexity of compuing short linearly independent vectors and short bases in a lattice. In Proc. of the $31st$ Annual ACM Symposium on the Theory of Computing, 1999, pp 711–720
35. O. Regev, R. Rosen. Lattice problems and norm embeddings. In Proc. of the 38th Annual ACM Symposium on the Theory of Computing, 2006
36. I. Dinur. Approximating SVP_{∞} to within almost polynomial factors is NP-hard. Proc. of the 4th Italian Conference on Algorithms and Complexity, LNCS, vol 1767, Springer, 2000
37. W. Banaszczyk. New bounds in some transference theorems in the geometry of numbers. Mathematische Annalen, vol. 296, 1993, pp 625–635
38. O. Goldreich, S. Goldwasser. On the limits of non-approximability of lattice problems. In Proc. of the 30th Annual ACM Symposium on the Theory of Computing, 1998, pp 1–9

39. V. Guruswami, D. Micciancio, O. Regev. The complexity of the covering radius problem on lattices. Computational Complexity 14(2), 2005, pp 90–121
40. S. Arora and S. Safra. Probabilistic checking of proofs : A new characterization of NP. Journal of the ACM, 45(1), 1998, pp 70–122
41. S. Arora, C. Lund, R. Motwani, M. Sudan, M. Szegedy. Proof verification and the hardness of approximation problems. Journal of the ACM, 45(3), 1998, pp 501–555
42. R. Raz. A parallel repetition theorem. SIAM Journal of Computing, 27(3), 1998, pp 763–803
43. J. Milnor, D. Husemoller. Symmetric bilinear forms. Springer, Berlin, 1973
44. J.C. Lagarias, H.W. Lenstra, C.P. Schnorr. Korkine-Zolotarev bases and successive minima of a lattice and its reciprocal lattice. Combinatorica, vol 10, 1990, pp 333–348

Chapter 15
On the Complexity of Lattice Problems with Polynomial Approximation Factors

Oded Regev

Abstract Lattice problems are known to be hard to approximate to within sub-polynomial factors. For larger approximation factors, such as $\sqrt{n}$, lattice problems are known to be in complexity classes, such as $\mathsf{NP} \cap \mathsf{coNP}$, and are hence unlikely to be NP-hard. Here, we survey known results in this area. We also discuss some related zero-knowledge protocols for lattice problems.

Introduction

A *lattice* is the set of all integer combinations of n linearly independent vectors $v_1, \ldots, v_n$ in $\mathbb{R}^n$. These vectors are known as a *basis* of the lattice. Lattices have been investigated by mathematicians for decades and have recently also attracted considerable attention in the computer science community following the discovery of the LLL algorithm by Lenstra, Lenstra, and Lovász [1]. Many different problems can be phrased as questions about lattices, such as integer programming [2], factoring polynomials with rational coefficients [1], integer relation finding [3], integer factoring, and Diophantine approximation [4]. More recently, the study of lattices attracted renewed attention due to the fact that lattice problems were shown, by Ajtai [5], to possess a particularly desirable property for cryptography: worst-case to average-case reducibility.

Lattice problems, such as the shortest vector problem (SVP) and the closest vector problem (CVP), are fascinating from a computational complexity point of view (see Fig. 15.1). On one hand, by the LLL algorithm [1] and subsequent improvements [6], we are able to efficiently approximate lattice problems to within essentially exponential factors, namely $2^{n(\log\log n)^2/\log n}$, where n is the dimension of the lattice. In fact, if we allow randomization, the approximation factor improves slightly to $2^{n\log\log n/\log n}$ [7]. On the other hand, we know that for some $c > 0$, no efficient algorithm can approximate lattice problems to within $n^{c/\log\log n}$, unless

O. Regev
School of Computer Science, Tel-Aviv University, Tel-Aviv 69978, Israel

P.Q. Nguyen and B. Vallée (eds.), *The LLL Algorithm*, Information Security and Cryptography, DOI 10.1007/978-3-642-02295-1_15,

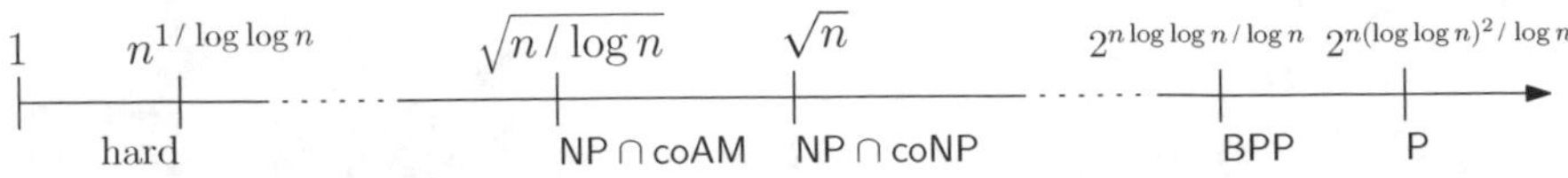

Fig. 15.1 The complexity of lattice problems (some constants omitted)

P = NP or another unlikely event occurs. This was established in a long sequence of works, including [8–14]. See also Khot's chapter [15] in these proceedings.

Considering the above results, one immediate question arises: what can we say about approximation factors in between these two extremes? There is a very wide gap between the approximation factor achieved by the best known algorithm ($2^{n\log\log n/\log n}$) and the best known hardness result ($n^{c/\log\log n}$). Of particular importance is the range of polynomial approximation factors. The reason for this is that the security of lattice-based cryptographic constructions following Ajtai's seminal work [5] is based on the worst-case hardness of approximating lattice problems in this region (see also [16–18] and Micciancio's chapter [19] in these proceedings). If, for instance, we could prove that approximating lattice problems to within $O(n^2)$ is NP-hard, then this would have the tremendous implication of a public key cryptosystem whose security is based solely on the P $\neq$ NP conjecture.

This scenario, however, is unlikely to happen. There are several results indicating that approximating lattice problems to within polynomial factors is unlikely to be NP-hard. These results are sometimes known as "limits on inapproximability." They are established by showing containment in complexity classes such as NP ∩ coNP. As is well known, if a problem in NP ∩ coNP is NP-hard, then NP = coNP and the polynomial hierarchy collapses. For lattice problems, this is true even under Cook-reductions, as we show in Appendix 15.

To state these results precisely, let us first recall the promise problems associated with the shortest vector problem and the closest vector problem. Below, we use $\mathcal{L}(B)$ to denote the lattice generated by the basis B. Moreover, all distances and lengths in this survey are with respect to the ℓ_2 norm (but see [20] for an interesting extension of the results described here to other ℓ_p norms).

Definition 1. GapCVP$_\gamma$

YES instances: triples (B, v, d), such that $\text{dist}(v, \mathcal{L}(B)) \leq d$

NO instances: triples (B, v, d), such that $\text{dist}(v, \mathcal{L}(B)) > \gamma d$,

where B is a basis for a lattice in $\mathbb{Q}^n$, $v \in \mathbb{Q}^n$ is a vector, and $d \in \mathbb{Q}$ is some number.

Definition 2. GapSVP$_\gamma$

YES instances: pairs (B, d), such that $\lambda_1(\mathcal{L}(B)) \leq d$

NO instances: pairs (B, d), such that $\lambda_1(\mathcal{L}(B)) > \gamma d$,

where B is a basis for a lattice in $\mathbb{Q}^n$, $d \in \mathbb{Q}$ is some number, and λ_1 denotes the length of the shortest nonzero vector in a lattice.

Note that in both cases, setting d to some fixed value (say 1) leads to an essentially equivalent definition (as one can easily rescale the input).

The oldest result showing a limit on the inapproximability of lattice problems is by Lagarias, Lenstra, and Schnorr [21], who showed that $\mathsf{GapCVP}_{n^{1.5}}$ is in $\mathsf{NP} \cap \mathsf{coNP}$. As we mentioned above, this shows that $\mathsf{GapCVP}_{n^{1.5}}$ is highly unlikely to be NP-hard. Let us remark at the outset that showing containment in NP is trivial: a witness for $\mathrm{dist}(v, \mathcal{L}(B)) \leq d$ is simply a vector $u \in \mathcal{L}(B)$, such that $\|v-u\| \leq d$. The more interesting part is providing a witness for the fact that a point is *far* from the lattice. Some thought reveals that this is no longer a trivial task: there is a huge number of lattice vectors that can potentially be very close to v. The way containment in coNP is usually shown is by utilizing properties of the *dual lattice*. Let us also mention that although we state this result and the results below only for GapCVP, they all hold also for GapSVP. This follows from a simple approximation preserving reduction from GapSVP to GapCVP [22], which we include for completeness in Appendix 15.

An improvement of the Lagarias et al. result was obtained by Banaszczyk [23] who showed that GapCVP_n is in $\mathsf{NP} \cap \mathsf{coNP}$. This was recently further improved by Aharonov and Regev [24] to $\mathsf{GapCVP}_{\sqrt{n}}$.

Theorem 1 ([24]). *There exists $c > 0$ such that $\mathsf{GapCVP}_{c\sqrt{n}}$ is in $\mathsf{NP} \cap \mathsf{coNP}$.*

In their coNP proof, the witness simply consists of a list of short vectors in the dual lattice. The verifier then uses these vectors to determine the distance of the target vector v from the lattice. A sketch of this proof appears in Section "Containment in coNP".

Another "limits on inapproximability" result is by Goldreich and Goldwasser [25], who showed that $\mathsf{GapCVP}_{\sqrt{n/\log n}}$ is in $\mathsf{NP} \cap \mathsf{coAM}$ (where containment in coAM means that the complement of the problem is in the class AM defined in Definition 3).

Theorem 2 ([25]). *For any $c > 0$, $\mathsf{GapCVP}_{c\sqrt{n/\log n}}$ is in $\mathsf{NP} \cap \mathsf{coAM}$.*

We present a proof of this theorem in Section "The Goldreich–Goldwasser Protocol". The proof uses an elegant protocol in which an all-powerful prover convinces a computationally limited verifier that a point v is far from the lattice. We note that their result is incomparable with that of [24] since it involves a slightly harder problem ($\mathsf{GapCVP}_{\sqrt{n/\log n}}$), but shows containment in a somewhat wider class (coAM). It is an interesting open question whether containment in $\mathsf{NP} \cap \mathsf{coNP}$ holds also for gaps between $\sqrt{n/\log n}$ and $\sqrt{n}$.

In Section "Zero-Knowledge Proof Systems", we will discuss the topic of *zero-knowledge protocols*. We will observe that the Goldreich–Goldwasser protocol is zero-knowledge (against honest verifiers). We will then describe two zero-knowledge protocols with efficient provers, one for coGapCVP and one for GapCVP.

We can summarize our current state of knowledge by saying that for approximation factors beyond $\sqrt{n/\log n}$, lattice problems are unlikely to be NP-hard. This naturally brings us to one of the most important questions regarding the complexity of lattice problems: is there an efficient algorithm for approximating lattice problem to within polynomial factors? Given how difficult it is to come up with algorithms that perform even slightly better than the exponential factor achieved by the LLL algorithm, many people conjecture that the answer is negative. This conjecture lies at the heart of latticed-based cryptographic constructions, such as Ajtai's [5], and is therefore of central importance. How can we hope to show such hardness, if we do not believe the problem is NP-hard? One promising direction is by relating lattice problems to other problems that are believed to be hard. For instance, a reduction from factoring to, say, GapSVP_{n^2} would give a strong evidence to the conjecture, and would also establish the remarkable fact that lattice-based cryptosystems are at least as secure as factoring-based cryptosystems.

Outline:

In Section"The Goldreich–Goldwasser Protocol", we present a proof of Theorem 2, including some of the technical details that go into making the proof completely rigorous. These technical details, especially how to work with periodic distributions, appear in many other lattice-related results and are therefore discussed in detail. Then, in Section "Containment in coNP", we present a sketch of the proof of Theorem 1. This sketch contains all the important ideas of the proof, but proofs of technical claims are omitted. The two sections are independent. Then, in Section "Zero-Knowledge Proof Systems", we discuss zero-knowledge proof systems for lattice problems, and in particular, sketch the prover-efficient zero-knowledge protocol of Micciancio and Vadhan [26]. This section requires a basic understanding of Section"The Goldreich–Goldwasser Protocol". Finally, in Appendix 15, we show in what sense the two theorems above imply "limits on inapproximability," and in Appendix 15, we show how to extend our results to GapSVP.

The Goldreich–Goldwasser Protocol

In this section, we prove Theorem 2. For simplicity, we will show that $\mathsf{GapCVP}_{\sqrt{n}} \in \mathsf{coAM}$. A slightly more careful analysis of the same protocol yields a gap of $c\sqrt{n/\log n}$ for any constant $c > 0$. First, let us define the class AM.

Definition 3. A promise problem is in AM, if there exists a protocol with a constant number of rounds between a BPP machine Arthur and a computationally unbounded machine Merlin, and two constants $0 \le a < b \le 1$ such that

- *Completeness*: For any YES input, there exists a strategy for Merlin such that Arthur accepts with probability at least b, and
- *Soundness*: For any NO input, and any strategy for Merlin, Arthur accepts with probability at most a.

To prove Theorem 2, we present a protocol that allows Arthur to verify that a point is far from the lattice. Specifically, given (B, v, d), Arthur accepts with probability 1, if $\mathrm{dist}(v, \mathcal{L}(B)) > \sqrt{n}d$, and rejects with some positive probability, if $\mathrm{dist}(v, \mathcal{L}(B)) \le d$.

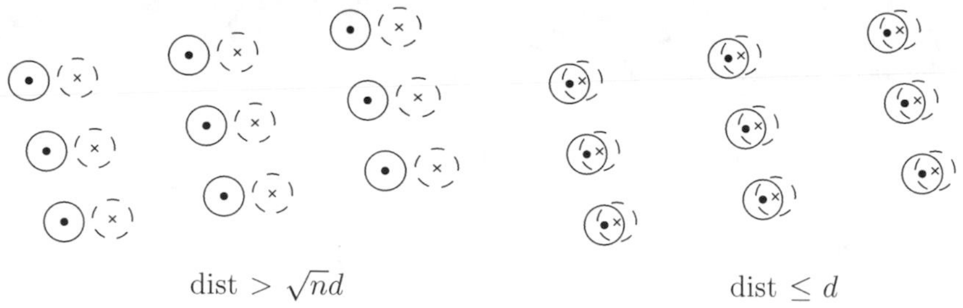

Fig. 15.2 The two distributions

Informally, the protocol is as follows. Arthur first flips a fair coin. If it comes up heads, he randomly chooses a "uniform" point in the lattice $\mathcal{L}(B)$; if it comes up tails, he randomly chooses a "uniform" point in the shifted lattice $v + \mathcal{L}(B)$. Let w denote the resulting point. Arthur randomly chooses a uniform point x from the ball of radius $\frac{1}{2}\sqrt{n}d$ around w and then sends x to Merlin. Merlin is supposed to tell Arthur if the coin came up heads or not.

The correctness of this protocol follows from the following two observations (see Fig. 15.2). If $\text{dist}(v, \mathcal{L}(B)) > \sqrt{n}d$, then the two distributions are disjoint and Merlin can answer correctly with probability 1. On the other hand, if $\text{dist}(v, \mathcal{L}(B)) \leq d$, then the overlap between the two distributions is large and Merlin must make a mistake with some positive probability.

This informal description hides two technical problems. First, we cannot really work with the point x, since it is chosen from a continuous distribution (and hence cannot be represented precisely in any finite number of bits). This is easy to take care of by working with an approximation of x with some polynomial number of bits. Another technical issue is the choice of a "random" point from $\mathcal{L}(B)$. This is an infinite set and there is no uniform distribution on it. One possible solution is to take the uniform distribution on points in the intersection of $\mathcal{L}(B)$ with, say, some very large hypercube. This indeed solves the problem, but introduces some unnecessary complications to the proof, since one needs to argue that the probability to fall close to the boundary of the hypercube is low. The solution we choose here is different and avoids this problem altogether by working with distributions on the basic parallelepiped of the lattice. We describe this solution in Section "Working with Periodic Distributions".

In the next few subsections, we present the necessary preliminaries for the proof.

Statistical Distance

Definition 4. The *statistical distance* between two distributions X, Y on some set Ω is defined as

$$\Delta(X, Y) = \max_{A \subseteq \Omega} |\mathbb{P}(X \in A) - \mathbb{P}(Y \in A)|.$$

One useful special case of this definition is the case where X and Y are discrete distributions over some countable set Ω. In this case, we have

$$\Delta(X,Y) = \frac{1}{2}\sum_{\omega\in\Omega} |\mathbb{P}(X=\omega) - \mathbb{P}(Y=\omega)|.$$

Another useful special case is when X and Y are distributions on $\mathbb{R}^n$ with density functions f, g. In this case, we have

$$\Delta(X,Y) = \frac{1}{2}\int_{\mathbb{R}^n} |f(x) - g(x)|\,\mathrm{d}x.$$

For any distributions X, Y, $\Delta(X,Y)$ obtains values between 0 and 1. It is 0 if and only if X and Y are identical and 1 if and only if they are disjoint. It is helpful to consider the following interpretation of statistical distance. Assume we are given a sample that is taken from X with probability $\frac{1}{2}$ or from Y with probability $\frac{1}{2}$. Our goal is to decide which distribution the sample comes from. Then, it can be seen that our best strategy succeeds with probability $\frac{1}{2} + \frac{1}{2}\Delta(X,Y)$.

One important fact concerning the statistical distance is that it cannot increase by the application of a possibly randomized function. In symbols, $\Delta(f(X), f(Y)) \leq \Delta(X,Y)$ for any (possibly randomized) function f. This fact follows easily from the above interpretation of Δ.

Balls in n-Dimensional Space

Let $\mathbf{B}(v,r)$ denote a ball of radius r around v. It is known that the volume of the unit ball $\mathbf{B}(0,1)$ in n dimensions is

$$V_n \stackrel{def}{=} \frac{\pi^{n/2}}{(n/2)!},$$

where we define $n! = n(n-1)!$ for $n \geq 1$ and $\frac{1}{2}! = \frac{1}{2}\sqrt{\pi}$. It can be shown that

$$\frac{(n+\frac{1}{2})!}{n!} \approx \frac{n!}{(n-\frac{1}{2})!} \approx \sqrt{n}.$$

Lemma 1. *For any $\varepsilon > 0$ and any vector v of length $\|v\| \leq \varepsilon$, the relative volume of the intersection of two unit balls whose centers are separated by v satisfies*

$$\frac{\mathrm{vol}(\mathbf{B}(0,1)\cap\mathbf{B}(v,1))}{\mathrm{vol}(\mathbf{B}(0,1))} \geq \varepsilon\frac{(1-\varepsilon^2)^{\frac{n-1}{2}}}{3}\sqrt{n}$$

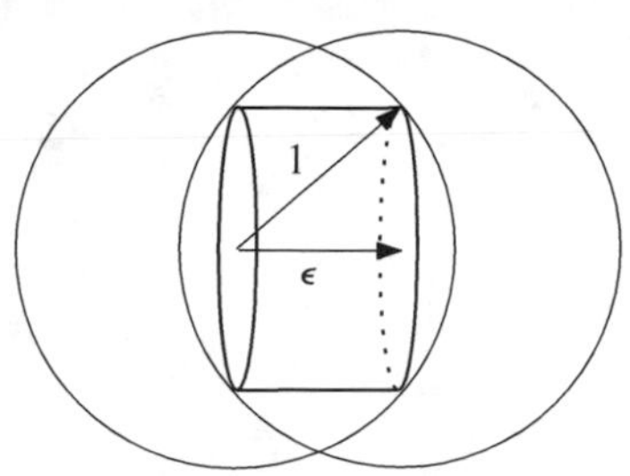

Fig. 15.3 A cylinder in the intersection of two balls

Proof. It suffices to consider the case $\|v\| = \varepsilon$. As shown in Fig. 15.3, the intersection contains a cylinder of height ε and radius $\sqrt{1-\varepsilon^2}$ centered around $v/2$. Hence, the volume of the intersection satisfies:

$$\frac{\mathrm{vol}(\mathbf{B}(0,1)\cap\mathbf{B}(v,1))}{\mathrm{vol}(\mathbf{B}(0,1))} > \frac{\varepsilon V_{n-1}(\sqrt{1-\varepsilon^2})^{n-1}}{V_n}$$
$$= \varepsilon(1-\varepsilon^2)^{\frac{n-1}{2}}\frac{\pi^{\frac{n-1}{2}}/(\frac{n-1}{2})!}{\pi^{\frac{n}{2}}/(\frac{n}{2})!} \approx \varepsilon(1-\varepsilon^2)^{\frac{n-1}{2}}\frac{\sqrt{n/2}}{\sqrt{\pi}}.$$

□

Notice that for $\varepsilon = \frac{2}{\sqrt{n}}$, the right hand side of the expression in Lemma 1 is bounded from below by some positive constant independent of n. This yields the following corollary.

Corollary 1. *There exists a constant $\delta > 0$ such that for any $d > 0$ and any $y \in \mathbb{R}^n$ such that $\|y\| \le d$,*

$$\Delta\left(U(\mathbf{B}(0,\tfrac{1}{2}\sqrt{n}d)),\ U(\mathbf{B}(y,\tfrac{1}{2}\sqrt{n}d))\right) < 1-\delta,$$

where $U(\cdot)$ denotes the uniform distribution on a set.

Proof. This statistical distance is exactly the volume of the symmetric difference of two balls divided by the sum of their volumes. According to the above lemma, this is bounded away from 1. □

Remark. When $\varepsilon = c\sqrt{\log n/n}$ for some $c > 0$, the right hand side of the expression in Lemma 1 is still greater than some $1/\mathrm{poly}(n)$. Using this, one can obtain the improved result $\mathsf{GapCVP}_{c\sqrt{n/\log n}} \in \mathsf{coAM}$.

Working with Periodic Distributions

In the informal description above, we talked about the "uniform distribution" on the lattice. This is clearly not defined. One possible solution is to restrict our attention

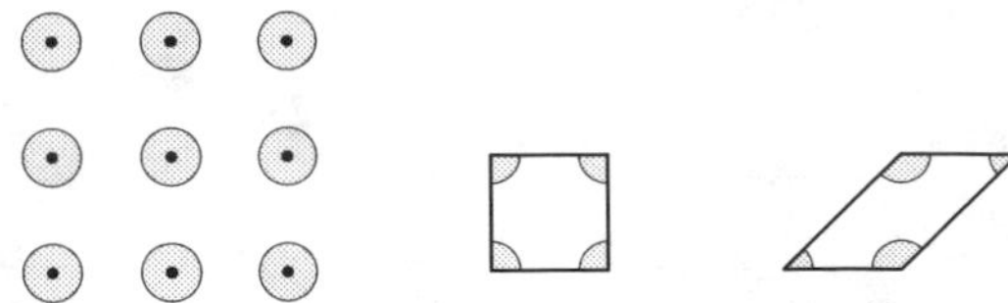

Fig. 15.4 A periodic distribution on $\mathbb{Z}^2$ (*left*), restricted $\mathcal{P}((0,1),(1,0))$ (*center*), and to $\mathcal{P}((0,1),(1,1))$ (*right*)

to some large enough cube $[-K, K]^n$. While possible, this solution introduces some technical annoyances as one has to argue that the probability to fall too close to the boundary of the cube (where the protocol might behave badly) is small.

Instead, our solution will be to work with only one period of the distribution. To demonstrate this approach, let us first consider the one-dimensional case. Assume we want to represent the distribution intuitively described as follows: choose a random point from the lattice $3\mathbb{Z}$ and add to it a number chosen uniformly from $[-0.1, 0.1]$. The first solution above would require us to take some large segment, say, $[-1000, 1000]$, and to restrict our distribution to it. Instead, we take one period of the distribution, say the segment $[0, 3]$, and consider the distribution on it. Hence, we obtain the uniform distribution on $[0, 0.1] \cup [2.9, 3]$. Notice that we could take another period, say the segment $[-3, 0]$, and work with it instead. Crucially, the transformation from one representation to another can be performed efficiently (by subtracting or adding 3 as needed).

A similar idea works for higher dimensions (see Fig. 15.4). If we want to represent a periodic distribution on a lattice, we consider it as a distribution on some period of the lattice. A common choice is to take a *basic parallelepiped* of the lattice, defined as

$$\mathcal{P}(B) = \mathcal{P}(v_1, \ldots, v_n) = \left\{ \sum_{i=1}^{n} x_i v_i \;\middle|\; x_i \in [0, 1) \right\},$$

where $B = (v_1, \ldots, v_n)$ is some basis of the lattice. As before, we have several possible representations, depending on the choice of basis B. The transformation from a representation using B_1 to one using B_2 can be done efficiently by reducing points modulo $\mathcal{P}(B_2)$ (see Definition 5 below). Mathematically speaking, the objects we work with are distributions on the quotient $\mathbb{R}^n/\mathcal{L}(B)$, and $\mathcal{P}(B)$ is its set of representatives.

We emphasize that it is much easier to imagine "periodic distributions" on $\mathbb{R}^n$. However, technically, it is much easier to work with distributions on $\mathcal{P}(B)$.

The Protocol

We will now show using Protocol 1 that $\mathsf{GapCVP}_{\sqrt{n}} \in \mathsf{coAM}$. The protocol uses the following definition.

Definition 5. For $x \in \mathbb{R}^n$, $x \bmod \mathcal{P}(B)$ is the unique $y \in \mathcal{P}(B)$ satisfying $x - y \in \mathcal{L}(B)$.

Protocol 1 The Goldreich–Goldwasser AM protocol

1. Arthur selects $\sigma \in \{0, 1\}$ uniformly and a random point t in the ball $\mathbf{B}(0, \frac{1}{2}\sqrt{n}d)$. He then sends $x = (\sigma v + t) \bmod \mathcal{P}(B)$ to Merlin.
2. Merlin checks if $\text{dist}(x, \mathcal{L}(B)) < \text{dist}(x, v + \mathcal{L}(B))$. If so, he responds with $\tau = 0$; otherwise, he responds with $\tau = 1$.
3. Arthur accepts if and only if $\tau = \sigma$.

Remark. For simplicity, we ignore issues of finite precision; these can be dealt with by standard techniques. One issue that we do want to address is how to choose a point from the ball $\mathbf{B}(0, R)$ uniformly at random. One option is to use known algorithms for sampling (almost) uniformly from arbitrary convex bodies and apply them to the case of a ball. A simpler solution is the following. Take n independent samples $u_1, \ldots, u_n \in \mathbb{R}$ from the standard normal distribution and let u be the vector $(u_1, \ldots, u_n) \in \mathbb{R}^n$. Then, u is distributed according to the standard n-dimensional Gaussian distribution, which is rotationally invariant. Now, choose r from the distribution on $[0, R]$ whose probability density function is proportional to r^{n-1} (this corresponds to the $(n-1)$-dimensional surface area of a sphere of radius r). The vector $\frac{r}{\|u\|}u$ is distributed uniformly in $\mathbf{B}(0, R)$.

Claim (Completeness). If $\text{dist}(v, \mathcal{L}(B)) > \sqrt{n}d$, then Arthur accepts with probability 1.

Proof. Assume $\sigma = 0$. Then,

$$\text{dist}(x, \mathcal{L}(B)) = \text{dist}(t, \mathcal{L}(B)) \leq \|t\| \leq \frac{1}{2}\sqrt{n}d.$$

On the other hand,

$$\begin{aligned}\text{dist}(x, v + \mathcal{L}(B)) &= \text{dist}(t, v + \mathcal{L}(B)) = \text{dist}(t - v, \mathcal{L}(B)) \\ &\geq \text{dist}(v, \mathcal{L}(B)) - \|t\| > \frac{1}{2}\sqrt{n}d.\end{aligned}$$

Hence, Merlin answers correctly and Arthur accepts. The case $\sigma = 1$ is similar. □

Claim (Soundness). If $\text{dist}(v, \mathcal{L}(B)) \leq d$, then Arthur rejects with some constant probability.

Proof. Let y be the difference between v and its closest lattice point. So, y is such that $v - y \in \mathcal{L}(B)$ and $\|y\| \leq d$. Let η_0 be the uniform distribution on $\mathbf{B}(0, \frac{1}{2}\sqrt{n}d)$ and let η_1 be the uniform distribution on $\mathbf{B}(y, \frac{1}{2}\sqrt{n}d)$. Notice that the point Arthur sends can be equivalently seen as a point chosen from η_σ reduced modulo $\mathcal{P}(B)$.

According to Corollary 1, $\Delta(\eta_0, \eta_1)$ is smaller than $1 - \delta$. Since statistical distance cannot increase by the application of a function,

$$\Delta(\eta_0 \bmod \mathcal{P}(B), \eta_1 \bmod \mathcal{P}(B)) \leq \Delta(\eta_0, \eta_1) < 1 - \delta$$

and Arthur rejects with probability at least δ. □

Containment in coNP

In this section, we sketch the proof of Theorem 1. For more details, see [24]. As mentioned in the introduction, containment in NP is trivial and it suffices to prove, e.g., that $\mathsf{GapCVP}_{100\sqrt{n}}$ is in coNP (we make no attempt to optimize the constant 100 here). To show this, we construct an NP verifier that, given a witness of polynomial size, verifies that the given point v is *far* from the lattice. There are three steps to the proof.

1. **Define** f
   In this part, we define a function $f : \mathbb{R}^n \to \mathbb{R}^+$ that is periodic over the lattice $\mathcal{L}$, i.e., for all $x \in \mathbb{R}^n$ and $y \in \mathcal{L}$, we have $f(x) = f(x + y)$ (see Fig. 15.5). For any lattice $\mathcal{L}$, the function f satisfies the following two properties: it is non-negligible (i.e., larger than some $1/\mathrm{poly}(n)$) for any point that lies within distance $\sqrt{\log n}$ from a lattice point and is exponentially small at distance $\geq \sqrt{n}$ from the lattice. Hence, given the value $f(v)$, one can tell whether v is far or close to the lattice.
2. **Encode** f
   We show that there exists a succinct description (which we denote by W) of a function f_W that approximates f at *any* point in $\mathbb{R}^n$ to within polynomially small additive error (see Fig. 15.5). We use W as the witness in the NP proof.

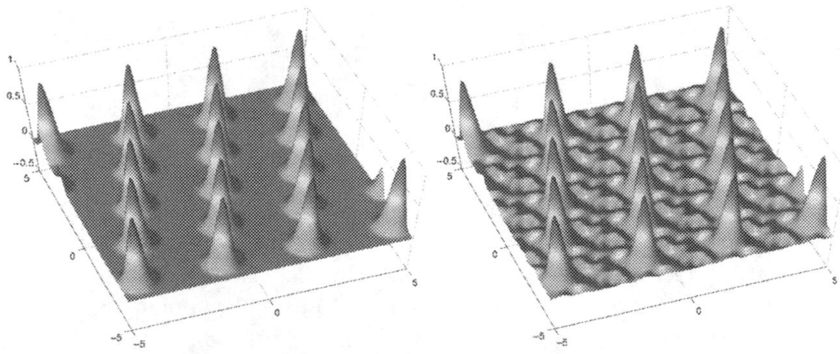

Fig. 15.5 The function f (*left*) and its approximation f_W (*right*) for a two-dimensional lattice

3. **Verify** f
   We construct an efficient NP verifier that, given a witness W, verifies that v is *far* from the lattice. The verifier verifies first that $f_W(v)$ is small and also that $f_W(x) \geq 1/2$, for any x that is close to the lattice.

We now explain each of these steps in more detail. For all missing proofs and more details, see [24].

Step 1: Define f

Define the function $g : \mathbb{R}^n \to \mathbb{R}$ as

$$g(x) = \sum_{y \in \mathcal{L}} e^{-\pi \|x-y\|^2},$$

and let

$$f(x) = \frac{g(x)}{g(0)}.$$

Hence, f is a sum of Gaussians centered around each lattice point and is normalized to be 1 at lattice points. See Fig. 15.5 for a plot of f. The function f was originally used by Banaszczyk [23] to prove "transference theorems," i.e., theorems relating parameters of a lattice to those of its dual.

The two properties mentioned above can be stated formally as follows.

Lemma 2. *Let $c > \frac{1}{\sqrt{2\pi}}$ be a constant. Then for any $x \in \mathbb{R}^n$, if $d(x, \mathcal{L}) \geq c\sqrt{n}$ then $f(x) = 2^{-\Omega(n)}$.*

Lemma 3. *Let $c > 0$ be a constant. Then for any $x \in \mathbb{R}^n$, if $d(x, \mathcal{L}) \leq c\sqrt{\log n}$ then $f(x) > n^{-10c^2}$.*

Step 2: Encode f

This step is the core of the proof. Here, we show that the function f can be approximated pointwise by a polynomial size circuit with only an inverse polynomial additive error. A naive attempt would be to store f's values on some finite subset of its domain and use these points for approximation on the rest of the domain. However, it seems that for this to be meaningful, we would have to store an exponential number of points.

Instead, we consider the *Fourier series* of f, which is a function $\hat{f}$ whose domain is the dual lattice $\mathcal{L}^*$ (defined as the set of all points in $\mathbb{R}^n$ with integer inner product with all lattice points). For any $w \in \mathcal{L}^*$, it is given by

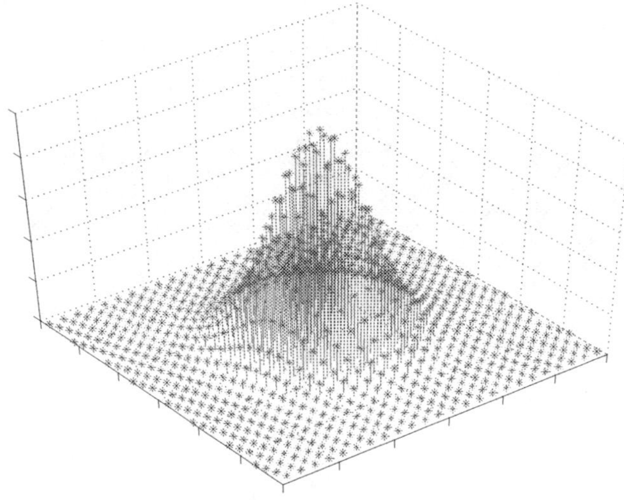

Fig. 15.6 The Fourier series $\hat{f}$ of f

$$\hat{f}(w) = \frac{1}{\det(B)} \int_{z \in \mathcal{P}(B)} f(z) e^{-2\pi i \langle w,z \rangle} dz,$$

where B is some basis of $\mathcal{L}$. (It can be shown that this definition is independent of the basis we choose for $\mathcal{L}$.) A short calculation, which we omit here, shows that $\hat{f}$ has a nice form, namely

$$\hat{f}(w) = \frac{e^{-\pi \|w\|^2}}{\sum_{z \in \mathcal{L}^*} e^{-\pi \|z\|^2}}.$$

See Fig. 15.6 for a plot of $\hat{f}$. One very useful and crucial property of $\hat{f}$ is that it is a probability distribution over the dual lattice $\mathcal{L}^*$. In other words, it is a non-negative function and the sum of all its values is 1.

A basic result in Fourier analysis is the Fourier inversion formula. It says that a function f can be recovered from its Fourier series $\hat{f}$ by using the formula

$$f(x) = \sum_{w \in \mathcal{L}^*} \hat{f}(w) e^{2\pi i \langle w,x \rangle}.$$

Since in our case, both f and $\hat{f}$ are real, we can simplify it to

$$f(x) = \sum_{w \in \mathcal{L}^*} \hat{f}(w) \cos(2\pi \langle w, x \rangle)$$

by taking the real part of both sides. By thinking of $\hat{f}$ as a probability distribution, we can rewrite this as

$$f(x) = \mathrm{E}_{w \sim \hat{f}}\left[\cos(2\pi \langle w, x \rangle)\right].$$

Hence, $f(x)$ can be seen as the expectation of $\cos(2\pi \langle w, x \rangle)$ (whose values range between -1 and 1), where w is chosen according to the probability distribution $\hat{f}$.

This brings us to the main idea of this step: we can approximate f by replacing the expectation with an average over a large enough sample from $\hat{f}$. More formally, for some large enough $N = \mathrm{poly}(n)$, let $W = (w_1, \ldots, w_N)$ be N vectors in the dual lattice chosen randomly and independently from the distribution $\hat{f}$ and define

$$f_W(x) \stackrel{def}{=} \frac{1}{N} \sum_{i=1}^{N} \cos(2\pi \langle x, w_i \rangle). \tag{15.1}$$

See Fig. 15.5 for a plot of f_W. Then, one can show that with high probability, $|f_W(x) - f(x)| \leq n^{-10}$ for all $x \in \mathbb{R}^n$. The proof of this statement is based on the Chernoff–Hoeffding bound.

Given the above, it is natural to choose our NP witness to be the list $W = (w_1, \ldots, w_N)$ of vectors in the dual lattice. We note that these vectors are typically short and hence computing them directly seems difficult.

Step 3: Verify f

Here, we construct an efficient NP verifier that, given the witness W, verifies that a point is *far* from the lattice. More precisely, given a lattice $\mathcal{L}$ and a vector v, it accepts if the distance of v from $\mathcal{L}$ is greater than $\sqrt{n}$ and rejects if this distance is less than $1/100$. This shows that $\mathrm{GapCVP}_{100\sqrt{n}}$ is in coNP (after appropriate rescaling).

The verifier starts by performing the following test: compute $f_W(v)$, as defined in (15.1), and reject if it is at least, say, $1/2$. We can do this because when the distance of v from $\mathcal{L}$ is greater than $\sqrt{n}$, $f(v)$ is exponentially small by Lemma 2 and hence $f_W(v)$ must be at most $1/\mathrm{poly}(n) < 1/2$ (assuming the witness W is chosen from $\hat{f}$, as it should be).

This verifier, however, is clearly not strong enough: the prover can "cheat" by sending w_i's that have nothing to do with $\hat{f}$ or with the lattice, and for which $f_W(v)$ is small even though v is within distance $1/100$ of the lattice. One might try to avoid such cheating strategies by verifying that f_W is close to f everywhere, or, alternatively, that the w_i's were indeed chosen from the correct distribution $\hat{f}$. It is not known how to construct such a verifier. Instead, we will now show a somewhat weaker verifier. (This weaker verifier is what limits the proof to a gap of $\sqrt{n}$ and not $\sqrt{n/\log n}$ as one could expect, given the properties of f stated in Lemmas 2 and 3.)

To test the witness W, we verify that the w_i's "look like" vectors chosen from $\hat{f}$, according to some simple statistical tests. We will later see that these tests suffice to provide soundness. But, what do vectors chosen from $\hat{f}$ look like? We identify two important properties. First, by definition, we see that all the w_i's are in $\mathcal{L}^*$. Second, it turns out that with high probability, for any unit vector $u \in \mathbb{R}^n$, it holds that $\frac{1}{N}\sum_{i=1}^{N}\langle u, w_i\rangle^2$ is bounded from above by some constant, say 3. Intuitively, this follows from the fact that the length of the w_i's is roughly $\sqrt{n}$ and that they are not concentrated in any particular direction (the proof of this fact is not trivial and is based on a lemma by Banaszczyk [23]).

Fortunately, the verifier can check these two properties efficiently. The first property is easy to check by, say, solving linear equations. But, how can we check the second property efficiently? It seems that we have to check it for all unit vectors u. The main observation here is that we can equivalently check that the largest eigenvalue of the $n \times n$ matrix $W \cdot W^T$, where W is the $n \times N$ matrix whose columns are the vectors $w_1, \dots, w_N$, is at most $3N$. This can be done in polynomial time by known algorithms for computing the eigenvalues of a matrix.

To summarize, the verifier performs the following three tests and accepts if and only if all of them are satisfied:

1. Checks that $f_W(v) < 1/2$;
2. Checks that W consists of vectors in the dual lattice $\mathcal{L}^*$;
3. Checks that the maximal eigenvalue of the $n \times n$ positive semidefinite matrix WW^T is at most $3N$.

As mentioned above, if v is a YES instance, i.e., its distance from $\mathcal{L}$ is at least $\sqrt{n}$, then a witness W chosen according to $\hat{f}$ satisfies all the tests with high probability. Hence, completeness holds. To complete the proof, we need to prove soundness. We will show that any witness W that passes tests (2) and (3) must satisfy $f_W(x) \geq 1/2$, for all x within distance $1/100$ from the lattice. In particular, if v is a NO instance, i.e., its distance from $\mathcal{L}$ is at most $1/100$, then test (1) must reject.

To see this, we note that by the definition of f_W, the fact that W consists of vectors in $\mathcal{L}^*$ guarantees that the function f_W is periodic on $\mathcal{L}$. Indeed, for any $v \in \mathcal{L}$,

$$\langle v + x, w_i\rangle = \langle v, w_i\rangle + \langle x, w_i\rangle$$

with the first term being integer by the definition of a dual lattice. Hence, it suffices to show that $f_W(x) \geq 1/2$ for any x satisfying $\|x\| \leq 1/100$. For such x, the eigenvalue test implies that for most i's, $|\langle x, w_i\rangle|$ is small. Therefore, for such x, most of the cosines in the definition of $f_W(x)$ are close to 1. This implies that $f_W(x)$ is greater than $1/2$ and soundness follows. In more detail, let x be such that $\|x\| \leq 1/100$. Since test (c) accepts, we have that

$$\frac{1}{N}\sum_{j=1}^{N}\langle x, w_j\rangle^2 = \frac{1}{N}x^T W W^T x \leq \frac{1}{N}\frac{3N}{10000} = \frac{3}{10000}$$

where the inequality follows by expressing x in the eigenvector basis of WW^T.

Using the inequality $\cos x \geq 1 - x^2/2$ (valid for any $x \in \mathbb{R}$), we get

$$f_W(x) = \frac{1}{N}\sum_{j=1}^{N}\cos(2\pi\langle x, w_j\rangle) \geq 1 - \frac{4\pi^2}{2N}\sum_{j=1}^{N}\langle x, w_j\rangle^2 \geq 1 - \frac{6\pi^2}{10000} > \frac{1}{2}.$$

Zero-Knowledge Proof Systems

The containments in NP, coNP, and coAM discussed in the previous sections can be stated equivalently in terms of proof systems between a computationally unbounded prover and a polynomial time verifier. For instance, Theorem 1 gives a proof system for $\text{coGapCVP}_{\sqrt{n}}$, in which the prover simply sends one message to the verifier who then decides whether to accept or reject. Similarly, Theorem 2 gives a proof system for $\text{coGapCVP}_{\sqrt{n/\log n}}$, in which the prover and verifier exchange a small number of messages. Finally, for any γ, GapCVP_γ clearly has a proof system in which the prover simply sends the nearby lattice point.

In addition to the usual requirements of completeness and soundness, one can ask for proof systems that satisfy the *zero-knowledge* property. Intuitively, we say that a proof system is zero-knowledge, if in the case of a true statement, the verifier learns nothing beyond the validity of the statement. There are in fact two natural notions of zero-knowledge: the first is zero-knowledge against *honest verifiers*, which are verifiers that obey the protocol but still try to extract some information from the interaction: the second and stronger notion is zero-knowledge against all verifiers, which says that even if the verifier deviates from the protocol he can still learn nothing from the interaction with the prover.

Although for our purposes the above intuitive description suffices, let us mention that the formal definition of zero-knowledge uses the notion of a *simulator*. Specifically, one says that a proof system is (statistical) zero-knowledge against honest verifiers, if there exists an efficient algorithm, known as a simulator, that produces communication transcripts whose distribution is statistically close to that of the actual transcripts of communication between the verifier and the prover. The existence of such a simulator captures the intuitive idea that the verifier learns nothing from the interaction. A similar definition exists for zero-knowledge against all verifiers. The concept of zero-knowledge has led to many important developments in cryptography and complexity over the past two decades. For the formal definition and further discussion, see [27].

Among the three proof systems mentioned above, the only one that is zero-knowledge is the one by Goldreich and Goldwasser. (The other two are clearly not zero-knowledge, since the verifier receives the witness.) Indeed, consider the protocol described in Section "The Protocol" in the case of a true statement, i.e., $\mathrm{dist}(v, \mathcal{L}(B)) > \sqrt{n}d$. Notice that the answer τ received by the verifier is always identical to his bit σ. Hence, the verifier *already knows* the answer the prover is about to send him, and therefore can learn nothing from the protocol (beyond the fact that $\mathrm{dist}(v, \mathcal{L}(B)) > \sqrt{n}d$). This argument (once written formally) establishes that the Goldreich–Goldwasser protocol is a statistical (and in fact perfect) zero-knowledge protocol against honest verifiers, or in complexity-theoretic terms, that the class $\mathsf{coGapCVP}_{\sqrt{n/\log n}}$ is contained in a complexity class known as Honest Verifier Statistical Zero Knowledge, or HVSZK. This protocol is not zero-knowledge against dishonest verifiers, since by deviating from the protocol, a dishonest verifier can find out if certain points are close to the lattice or not (which seems to be something he cannot do without the help of the prover). Still, using the remarkable fact that $\mathsf{HVSZK} = \mathsf{SZK}$ [27], we obtain that $\mathsf{coGapCVP}_{\sqrt{n/\log n}} \in \mathsf{SZK}$, i.e., that $\mathsf{coGapCVP}_{\sqrt{n/\log n}}$ has a zero-knowledge proof system that is secure also against dishonest verifiers. Another truly remarkable fact regarding zero-knowledge proof systems is that SZK is closed under complement [27, 28]. This implies that we also have that $\mathsf{GapCVP}_{\sqrt{n/\log n}} \in \mathsf{SZK}$, i.e., there exists a zero-knowledge proof system that allows a prover to convince a verifier that a point is close to the lattice.

Proof Systems with Efficient Provers

In the traditional complexity-theoretic definition of zero-knowledge protocols, the complexity of the prover does not play any role. However, from a cryptographic standpoint, in order for these proof systems to be useful, the prover must be efficiently implementable. This gives rise to the following question: do all problems in $\mathsf{NP} \cap \mathsf{SZK}$ have a statistical zero-knowledge proof system in which the prover can be implemented efficiently when given an NP witness? Note that without providing the prover with an NP witness, this task is clearly impossible. This is also the reason the question makes sense only for problems in $\mathsf{NP} \cap \mathsf{SZK}$.

In the context of lattice problems, this question was raised by Micciancio and Vadhan [26], who also made some progress toward answering the question for general problems in $\mathsf{NP} \cap \mathsf{SZK}$. Building on their work, Nguyen and Vadhan [29] very recently gave a positive answer to the question: any problem in $\mathsf{NP} \cap \mathsf{SZK}$ has a statistical zero-knowledge proof system with an efficient prover. Their protocol is secure even against dishonest verifiers.

From a theoretical point of view, Nguyen and Vadhan's exciting result gives a complete answer to our question. Yet, their construction is very complicated and does not seem to yield protocols that are efficient in practice. For this reason, we will now describe two examples of "practical" proof systems for lattice problems. Such direct constructions of proof systems with efficient provers have applications in cryptography, as described in [26].

The first problem we consider is coGapCVP. As we have seen, $\text{coGapCVP}_{\sqrt{n}}$ is in NP ∩ SZK. However, in the Goldreich–Goldwasser proof system, the prover is required to solve a nontrivial problem, namely to tell whether a point x is within distance $\frac{1}{2}\sqrt{n}d$ from $\mathcal{L}(B)$ or within distance $\frac{1}{2}\sqrt{n}d$ from $v + \mathcal{L}(B)$, under the assumption that $\text{dist}(v, \mathcal{L}(B)) > \sqrt{n}d$. This seems like a hard problem, even when given the NP witness described in Section "Containment in coNP". However, the Goldreich–Goldwasser protocol as described in Section "The Protocol" *does* have an efficient prover, if we consider it as a protocol for the (easier) problem coGapCVP_n. Indeed, the prover's task in this protocol is to tell whether a point x is within distance $\frac{1}{2}\sqrt{n}d$ from $\mathcal{L}(B)$ or within distance $\frac{1}{2}\sqrt{n}d$ from $v + \mathcal{L}(B)$, under the assumption that $\text{dist}(v, \mathcal{L}(B)) > nd$. Notice that in the latter case, the distance of x from $\mathcal{L}(B)$ is at least $nd - \frac{1}{2}\sqrt{n}d \geq nd/2$. Hence, the gap between the two cases is at least $\sqrt{n}$ and therefore the prover can distinguish between them by using the witness described in Section "Containment in coNP". This proof system, just like the original Goldreich–Goldwasser protocol, is secure only against honest verifiers.

The second problem we consider is $\text{GapCVP}_{\sqrt{n}}$. Here, the prover's task is to convince the verifier through a zero-knowledge protocol that a point v is close to the lattice. An elegant protocol for this task was presented by Micciancio and Vadhan in [26]. Their protocol is secure even against dishonest verifiers, and in addition, the prover's strategy can be efficiently implemented, given a lattice point close to v. The main component in their protocol is given as Protocol 2. We use D_0 to denote the set of points that are within distance $\frac{1}{2}\sqrt{n}d$ of the lattice $\mathcal{L}(B)$ and D_1 to denote the set of points that are within distance $\frac{1}{2}\sqrt{n}d$ of the shifted lattice $v + \mathcal{L}(B)$ (see Fig. 15.2).

Protocol 2 Part of the Micciancio–Vadhan zero-knowledge protocol for $\text{GapCVP}_{\sqrt{n}}$

1. The prover chooses uniformly a bit $\sigma \in \{0, 1\}$ and sends to the verifier a point x chosen "uniformly" from D_σ.
2. The verifier then challenges the prover by sending him a uniformly chosen bit τ.
3. The prover is supposed to reply with a point y.
4. The verifier accepts if and only if $\text{dist}(x, y) \leq \frac{1}{2}\sqrt{n}d$ and $y \in \tau v + \mathcal{L}(B)$ (i.e., y is a lattice point if $\tau = 0$, and a point in the shifted lattice, if $\tau = 1$).

The soundness of this protocol is easy to establish: if $\text{dist}(v, \mathcal{L}(B)) > \sqrt{n}d$ then the verifier accepts with probability at most $\frac{1}{2}$, no matter what strategy is played by the prover, since no point x can be within distance $\frac{1}{2}\sqrt{n}d$ both from $\mathcal{L}(B)$ and from $v + \mathcal{L}(B)$. To prove completeness, consider the case $\text{dist}(v, \mathcal{L}(B)) \leq d/10$. Using a proof similar to the one of Lemma 1, one can show that the relative volume of the intersection of two balls of radius $\frac{1}{2}\sqrt{n}d$, whose centers differ by at most $d/10$ is at least 0.9. This means that with probability at least 0.9, the point x chosen by the prover from D_σ is also in $D_{1-\sigma}$. In such a case, the prover is able to reply to both possible challenges τ and the verifier accepts. Notice, moreover, that the prover can be efficiently implemented, if given a lattice point w within distance $d/10$ of v: by adding or subtracting $w - v$ as necessary, the prover can respond to both challenges in case x falls in $D_0 \cap D_1$.

Unfortunately, Protocol 2 is *not* zero-knowledge. Intuitively, the reason for that is when the prover is unable to answer the verifier's challenge, the verifier learns that x is outside $D_0 \cap D_1$, a fact which he most likely could not have established alone. We can try to mend this by modifying the prover to only send points x that are in $D_0 \cap D_1$. This still does not help, since now the verifier obtains a uniform point x in $D_0 \cap D_1$, and it seems that he could not sample from this distribution alone. (This modification does, however, allow us to obtain perfect completeness.)

Instead, the solution taken by [26] is to "amplify" Protocol 2, so as to make the information leakage negligible. Instead of just sending one point x, the prover now sends a list of $2k$ points $x_1, \ldots, x_{2k}$, each chosen independently as in the original protocol, where k is some parameter. The verifier again challenges the prover with a random bit τ. The prover is then supposed to reply with a list of points $y_1, \ldots, y_{2k}$. The verifier accepts if and only if for all i, $\text{dist}(x_i, y_i) \leq \frac{1}{2}\sqrt{n}d$ and y_i is either in $\mathcal{L}(B)$ or in $v + \mathcal{L}(B)$, and moreover, the number of y_i's contained in $\mathcal{L}(B)$ is even, if $\tau = 0$, and odd, otherwise. The idea in this modified protocol is to allow the prover to respond to the challenge whenever there is at least one point x_i that falls in $D_0 \cap D_1$. This reduces the probability of failure from a constant to an exponentially small amount in k. The soundness, completeness, prover efficiency, and zero-knowledge property of the modified protocol are established similarly to those of the original protocol. For further details, see [26].

NP-Hardness

In this section we show that Theorem 1 implies that $\mathsf{GapCVP}_{\sqrt{n}}$ is unlikely to be NP-hard, even under Cook reductions. One can also show that Theorem 2 implies that $\mathsf{GapCVP}_{\sqrt{n/\log n}}$ is unlikely to be NP-hard. However, for simplicity, we show this only for a $\sqrt{n}$ gap. Our proof is based on [17, 30, 31].

First, let us consider the simpler case of Karp reductions. If a problem in coNP is NP-hard under a Karp reduction (i.e., there is a many-to-one reduction from SAT to our problem) then the following easy claim shows that $\mathsf{NP} \subseteq \mathsf{coNP}$ (and hence the polynomial hierarchy collapses).

Claim. If a promise problem $\Pi = (\Pi_{\text{YES}}, \Pi_{\text{NO}})$ is in coNP and is NP-hard under Karp reductions, then $\mathsf{NP} \subseteq \mathsf{coNP}$.

Proof. Take any language L in NP. By assumption, there exists an efficient procedure R that maps any $x \in L$ to $R(x) \in \Pi_{\text{YES}}$ and any $x \notin L$ to $R(x) \in \Pi_{\text{NO}}$. Since $\Pi \in \mathsf{coNP}$, we have an NP verifier V such that for any $y \in \Pi_{\text{NO}}$ there exists a w such that $V(y, w)$ accepts, and for any $y \in \Pi_{\text{YES}}$ and any w, $V(y, w)$ rejects. Consider the verifier $U(x, w)$ given by $V(R(x), w)$. Notice that for all $x \notin L$ there exists a w such that $U(x, w)$ accepts and moreover, for all $x \in L$ and all w $U(x, w)$ rejects. Hence, $L \in \mathsf{coNP}$. □

The case of Cook reductions requires some more care. For starters, there is nothing special about a problem in coNP that is NP-hard under Cook reductions (for example, coSAT is such a problem). Instead, we would like to show that if a problem in NP $\cap$ coNP is NP-hard under Cook reductions, the polynomial hierarchy collapses. This implication is not too difficult to show for *total* problems (i.e., languages). However, we are dealing with *promise* problems and for such problems this implication is not known to hold (although still quite believable). In a nutshell, the difficulty arises because a Cook reduction might perform queries that are neither a YES instance nor a NO instance and for such queries we have no witness.

This issue can be resolved by using the fact that not only $\mathsf{GapCVP}_{\sqrt{n}} \in \mathsf{NP}$ but also $\mathsf{GapCVP}_1 \in \mathsf{NP}$. In other words, no promise is needed to show that a point is close to the lattice. In the following, we show that any problem with the above properties is unlikely to be NP-hard.

Lemma 4. *Let $\Pi = (\Pi_{\mathrm{YES}}, \Pi_{\mathrm{NO}})$ be a promise problem and let Π_{MAYBE} denote all instances outside $\Pi_{\mathrm{YES}} \cup \Pi_{\mathrm{NO}}$. Assume that Π is in* coNP *and that the (non-promise) problem $\Pi' = (\Pi_{\mathrm{YES}} \cup \Pi_{\mathrm{MAYBE}}, \Pi_{\mathrm{NO}})$ is in* NP. *Then, if Π is NP-hard under Cook reductions then* NP $\subseteq$ coNP *and the polynomial hierarchy collapses.*

Proof. Take any language L in NP. By assumption, there exists a Cook reduction from L to Π. That is, there exists a polynomial time procedure T that solves L given access to an oracle for Π. The oracle answers YES on queries in Π_{YES} and NO on queries in Π_{NO}. Notice, however, that its answers on queries from Π_{MAYBE} are arbitrary and should not affect the output of T.

Since $\Pi \in$ coNP, there exists a verifier V_1 and a witness $w_1(x)$ for every $x \in \Pi_{\mathrm{NO}}$ such that V_1 accepts $(x, w_1(x))$. Moreover, V_1 rejects (x, w) for any $x \in \Pi_{\mathrm{YES}}$ and any w. Similarly, since $\Pi' \in$ NP, there exists a verifier V_2 and a witness $w_2(x)$ for every $x \in \Pi_{\mathrm{YES}} \cup \Pi_{\mathrm{MAYBE}}$ such that V_2 accepts $(x, w_2(x))$. Moreover, V_2 rejects (x, w) for any $x \in \Pi_{\mathrm{NO}}$ and any w.

We now show that L is in coNP by constructing an NP verifier. Let Φ be an input to L and let $x_1, \ldots, x_k$ be the set of oracle queries which T performs on input Φ. Our witness consists of k pairs, one for each x_i. For $x_i \in \Pi_{\mathrm{NO}}$ we include the pair $(\mathrm{NO}, w_1(x_i))$ and for $x_i \in \Pi_{\mathrm{YES}} \cup \Pi_{\mathrm{MAYBE}}$ we include the pair $(\mathrm{YES}, w_2(x_i))$. The verifier simulates T; for each query x_i that T performs, the verifier reads the pair corresponding to x_i in the witness. If the pair is of the form (YES, w) then the verifier checks that $V_2(x_i, w)$ accepts and then returns YES to T. Similarly, if the pair is of the form (NO, w) then the verifier checks that $V_1(x_i, w)$ accepts and then returns NO to T. If any of the calls to V_1 or V_2 rejects, then the verifier rejects. Finally, if T decides that $\Phi \in L$, the verifier rejects and otherwise it accepts.

The completeness follows easily. More specifically, if $\Phi \notin L$ then the witness described above will cause the verifier to accept. To prove soundness, assume that $\Phi \in L$ and let us show that the verifier rejects. Notice that for each query $x_i \in \Pi_{\mathrm{NO}}$ the witness must include a pair of the form (NO, w) because otherwise V_2 would reject. Similarly, for each query $x_i \in \Pi_{\mathrm{YES}}$ the witness must include a pair of the form (YES, w) because otherwise V_1 would reject. This implies that T receives the correct answers for all of its queries inside $\Pi_{\text{NO}} \cup \Pi_{\text{YES}}$ and must therefore output the correct answer, i.e., that $\Phi \in L$ and then the verifier rejects. □

We just saw that the promise problem $\mathsf{GapCVP}_{\sqrt{n}}$ is unlikely to be NP-hard, even under Cook reductions. Consider now the *search problem* $\mathsf{CVP}_{\sqrt{n}}$ where given a lattice basis B and a vector v, the goal is to find a lattice vector $w \in \mathcal{L}(B)$ such that $\text{dist}(v, w) \leq \sqrt{n}\,\text{dist}(v, \mathcal{L}(B))$. This problem is clearly at least as hard as $\mathsf{GapCVP}_{\sqrt{n}}$. Can it possibly be NP-hard (under Cook reductions)? A similar argument to the one used above shows that this is still unlikely, as it would imply NP ⊆ coNP. Let us sketch this argument. Assume we have a Cook reduction from any NP language L to the search problem $\mathsf{CVP}_{\sqrt{n}}$. Then we claim that $L \in$ coNP. The witness used to show this is a list of valid answers by the $\mathsf{CVP}_{\sqrt{n}}$ oracle to the questions asked by the reduction, together with a witness that each answer is correct. More precisely, for each question (B, v), the witness is supposed to contain the vector $w \in \mathcal{L}(B)$ closest to v together with an NP proof that the instance $(B, v, \text{dist}(v, w)/\sqrt{n})$ is a NO instance of $\mathsf{GapCVP}_{\sqrt{n}}$. Having the NP proof for each answer w assures us that $\text{dist}(v, w) \leq \sqrt{n}\,\text{dist}(v, \mathcal{L}(B))$ and hence w is a valid answer of the $\mathsf{CVP}_{\sqrt{n}}$ oracle.

Reducing GapSVP to GapCVP

Both Theorem 1 and Theorem 2 hold also for GapSVP. The following lemma shows this for Theorem 1. A similar argument shows this for Theorem 2.

Lemma 5. *If for some* $\beta = \beta(n)$, GapCVP_β *is in* coNP *then so is* GapSVP_β.

Proof. Consider an instance of GapSVP_β given by the lattice $\mathcal{L}$ whose basis is $(b_1, \ldots, b_n)$ (in this proof we use Definitions 1 and 2 with d fixed to 1). We map it to n instances of GapCVP_β where the ith instance, $i = 1, \ldots, n$, is given by the lattice $\mathcal{L}_i$ spanned by $(b_1, \ldots, b_{i-1}, 2b_i, b_{i+1}, \ldots, b_n)$ and the target vector b_i. In the following we show that this mapping has the property that if $\mathcal{L}$ is a YES instance of GapSVP_β then at least one of $(\mathcal{L}_i, b_i)$ is a YES instance of GapCVP_β and if $\mathcal{L}$ is a NO instance then all n instances $(\mathcal{L}_i, b_i)$ are NO instances. This will complete the proof of the lemma since a NO witness for $\mathcal{L}$ can be given by n NO witnesses for $(\mathcal{L}_i, b_i)$.

Consider the case where $\mathcal{L}$ is a YES instance. In other words, if

$$u = a_1 b_1 + a_2 b_2 + \cdots + a_n b_n$$

denotes the shortest vector, then its length is at most 1. Notice that not all the a_i's are even for otherwise the vector $u/2$ is a shorter lattice vector. Let j be such that a_j is odd. Then the distance of b_j from the lattice $\mathcal{L}_j$ is at most $\|u\| \leq 1$ since $b_j + u \in \mathcal{L}_j$. Hence, $(\mathcal{L}_j, b_j)$ is a YES instance of GapCVP_β. Now consider the case where $\mathcal{L}$ is a NO instance of GapSVP_β, i.e., the length of the shortest vector in $\mathcal{L}$ is more than β. Fix any $i \in [n]$. By definition, $b_i \notin \mathcal{L}_i$ and therefore for any $w \in \mathcal{L}_i$ the vector $b_i - w \neq 0$. On the other hand, $b_i - w \in \mathcal{L}$ and hence $\|b_i - w\| > \beta$. This shows that $d(b_i, \mathcal{L}_i) > \beta$ and hence $(\mathcal{L}_i, b_i)$ is a NO instance of GapCVP_β. □

Acknowledgements This chapter is partly based on lecture notes scribed by Michael Khanevsky as well as on the paper [24] coauthored with Dorit Aharonov. I thank Ishay Haviv and the anonymous reviewers for their comments on an earlier draft. I also thank Daniele Micciancio for pointing out that the argument in Section "NP-Hardness" extends to the search version. Supported by the Binational Science Foundation, by the Israel Science Foundation, by the European Commission under the Integrated Project QAP funded by the IST directorate as Contract Number 015848, and by a European Research Council (ERC) Starting Grant.

References

1. Lenstra, A.K., Lenstra, H.W., and Lovász, L.: Factoring polynomials with rational coefficients. *Math. Ann.*, 261:515–534 (1982)
2. Kannan, R.: Improved algorithms for integer programming and related lattice problems. In *Proc. 15th ACM Symp. on Theory of Computing (STOC)*, pages 193–206. ACM (1983)
3. Haastad, J., Just, B., Lagarias, J.C., and Schnorr, C.P.: Polynomial time algorithms for finding integer relations among real numbers. *SIAM J. Comput.*, 18(5):859–881 (1989)
4. Schnorr, C.P.: Factoring integers and computing discrete logarithms via diophantine approximation. In *Proc. of Eurocrypt '91*, volume 547, pages 171–181. Springer (1991)
5. Ajtai, M.: Generating hard instances of lattice problems. In *Complexity of computations and proofs*, volume 13 of *Quad. Mat.*, pages 1–32. Dept. Math., Seconda Univ. Napoli, Caserta (2004)
6. Schnorr, C.P.: A hierarchy of polynomial time lattice basis reduction algorithms. *Theoretical Computer Science*, 53(2–3):201–224 (1987)
7. Ajtai, M., Kumar, R., and Sivakumar, D.: A sieve algorithm for the shortest lattice vector problem. In *Proc. 33rd ACM Symp. on Theory of Computing*, pages 601–610. ACM (2001)
8. van Emde Boas, P.: Another NP-complete problem and the complexity of computing short vectors in a lattice. Technical report, University of Amsterdam, Department of Mathematics, Netherlands (1981). Technical Report 8104
9. Ajtai, M.: The shortest vector problem in l_2 is NP-hard for randomized reductions (extended abstract) 10–19. In *Proc. 30th ACM Symp. on Theory of Computing (STOC)*, pages 10–19. ACM (1998)
10. Cai, J.Y. and Nerurkar, A.: Approximating the SVP to within a factor $(1+1/\dim^{\varepsilon})$ is NP-hard under randomized reductions. *J. Comput. System Sci.*, 59(2):221–239 (1999). ISSN 0022-0000
11. Dinur, I., Kindler, G., Raz, R., and Safra, S.: Approximating CVP to within almost-polynomial factors is NP-hard. *Combinatorica*, 23(2):205–243 (2003)
12. Micciancio, D.: The shortest vector problem is NP-hard to approximate to within some constant. *SIAM Journal on Computing*, 30(6):2008–2035 (2001). Preliminary version in FOCS 1998
13. Khot, S.: Hardness of approximating the shortest vector problem in lattices. In *Proc. 45th Annual IEEE Symp. on Foundations of Computer Science (FOCS)*, pages 126–135. IEEE (2004)
14. Haviv, I. and Regev, O.: Tensor-based hardness of the shortest vector problem to within almost polynomial factors. In *Proc. 39th ACM Symp. on Theory of Computing (STOC)* (2007)
15. Khot, S.: Inapproximability results for computational problems on lattices (2007). These proceedings
16. Ajtai, M. and Dwork, C.: A public-key cryptosystem with worst-case/average-case equivalence. In *Proc. 29th ACM Symp. on Theory of Computing (STOC)*, pages 284–293. ACM (1997)

17. Micciancio, D. and Goldwasser, S.: *Complexity of Lattice Problems: a cryptographic perspective*, volume 671 of The Kluwer International Series in Engineering and Computer Science. Kluwer Academic Publishers, Boston, MA (2002)
18. Regev, O.: Lattice-based cryptography. In *Advances in cryptology (CRYPTO)*, pages 131–141 (2006)
19. Micciancio, D.: Cryptographic functions from worst-case complexity assumptions (2007). These proceedings
20. Peikert, C.J.: Limits on the hardness of lattice problems in ℓ_p norms. In *Proc. of 22nd IEEE Annual Conference on Computational Complexity (CCC)* (2007)
21. Lagarias, J.C., Lenstra, Jr., H.W., and Schnorr, C.P.: Korkin-Zolotarev bases and successive minima of a lattice and its reciprocal lattice. *Combinatorica*, 10(4):333–348 (1990)
22. Goldreich, O., Micciancio, D., Safra, S., and Seifert, J.P.: Approximating shortest lattice vectors is not harder than approximating closest lattice vectors. *Inform. Process. Lett.*, 71(2):55–61 (1999). ISSN 0020-0190
23. Banaszczyk, W.: New bounds in some transference theorems in the geometry of numbers. *Mathematische Annalen*, 296(4):625–635 (1993)
24. Aharonov, D. and Regev, O.: Lattice problems in NP intersect coNP. In *Proc. 45th Annual IEEE Symp. on Foundations of Computer Science (FOCS)*, pages 362–371 (2004)
25. Goldreich, O. and Goldwasser, S.: On the limits of nonapproximability of lattice problems. *J. Comput. System Sci.*, 60(3):540–563 (2000)
26. Micciancio, D. and Vadhan, S.: Statistical zero-knowledge proofs with efficient provers: lattice problems and more. In D. Boneh, editor, *Advances in cryptology - CRYPTO 2003, Proc. of the 23rd annual international cryptology conference*, volume 2729 of *Lecture Notes in Computer Science*, pages 282–298. Springer, Santa Barbara, CA, USA (2003)
27. Vadhan, S.P.: *A Study of Statistical Zero-Knowledge Proofs*. Ph.D. thesis, MIT (1999)
28. Okamoto, T.: On relationships between statistical zero-knowledge proofs. In *Proc. 28th ACM Symp. on Theory of Computing (STOC)*, pages 649–658. ACM (1996)
29. Nguyen, M.H. and Vadhan, S.: Zero knowledge with efficient provers. In *Proc. 38th ACM Symp. on Theory of Computing (STOC)*, pages 287–295. ACM (2006)
30. Cai, J.Y. and Nerurkar, A.: A note on the non-NP-hardness of approximate lattice problems under general Cook reductions. *Inform. Process. Lett.*, 76(1–2):61–66 (2000)
31. Goldreich, O.: (2003). A comment available online at http://www.wisdom. weizmann.ac.il/~oded/p_lp.html